Computational Neuroscience of Vision

Computational Neuroscience of Vision

EDMUND T. ROLLS

University of Oxford, Department of Experimental Psychology, Oxford, UK

and

GUSTAVO DECO

Siemens, Munich; and University of Munich, Germany

OXFORD

UNIVERSITY PRESS

OXFORD

UNIVERSITY PRESS

Great Clarendon Street, Oxford OX2 6DP

Oxford University Press is a department of the University of Oxford.
It furthers the University's objective of excellence in research, scholarship,
and education by publishing worldwide in

Oxford New York

Athens Auckland Bangkok Bogotá Buenos Aires Cape Town
Chennai Dar es Salaam Delhi Florence Hong Kong Istanbul Karachi
Kolkata Kuala Lumpur Madrid Melbourne Mexico City Mumbai Nairobi
Paris São Paulo Shanghai Singapore Taipei Tokyo Toronto Warsaw
with associated companies in Berlin Ibadan

Oxford is a registered trade mark of Oxford University Press
in the UK and in certain other countries

Published in the United States
by Oxford University Press Inc., New York

A catalogue record for this book is available from the British Library

British Library Cataloguing in Publication Data
Data available
Library of Congress Cataloging in Publication Data

ISBN 0 19 852489 7 (hbk.)
ISBN 0 19 852488 9 (pbk.)

1 3 5 7 9 10 8 6 4 2

Typeset in LaTex by the author
Printed in Great Britain
on acid-free paper by
Biddles Ltd, Guildford and King's Lynn

Preface

To understand how the brain works, including how it functions in vision, it is necessary to combine different approaches, including neural computation. Neurophysiology at the single neuron level is needed because this is the level at which information is exchanged between the computing elements of the brain. Evidence from the effects of brain damage, including that available from neuropsychology, is needed to help understand what different parts of the system do, and indeed what each part is necessary for. Neuroimaging is useful to indicate where in the human brain different processes take place, and to show which functions can be dissociated from each other. Knowledge of the biophysical and synaptic properties of neurons is essential to understand how the computing elements of the brain work, and therefore what the building blocks of biologically realistic computational models should be. Knowledge of the anatomical and functional architecture of the cortex is needed to show what types of neuronal network actually perform the computation. And finally the approach of neural computation is needed, as this is required to link together all the empirical evidence to produce an understanding of how the system actually works. This book utilizes evidence from all these disciplines to develop an understanding of how vision is implemented by processing in the brain.

A test of whether one's understanding is correct is to simulate the processing on a computer, and to show whether the simulation can perform the tasks of vision, and whether the simulation has similar properties to the real brain. The approach of neural computation leads to a precise definition of how the computation is performed, and to precise and quantitative tests of the theories produced. How vision works is a paradigm example of this approach, because it is a sufficiently complex problem that it requires a computational approach, and indeed has not been solved in artificial vision systems operating in natural scenes. At the same time, vision raises fundamental issues in cognitive neuroscience, such as how attention works.

One of the distinctive properties of this book is that it links the neural computation approach not only firmly to neuronal neurophysiology, which provides much of the primary data about how the brain operates, but also to psychophysical studies (for example of attention); to neuropsychological studies of patients with brain damage; and to functional magnetic resonance imaging (fMRI) (and other neuroimaging) approaches. The empirical evidence that is brought to bear is largely from non-human primates and from humans, because of the considerable similarity of their visual systems, and the overall aims to understand how visual perception is implemented in the human brain, and the disorders that arise after brain damage. The focus of the book is on the processes that underlie high-level vision, and in particular visual object recognition and attention, and on how the object-processing visual system interfaces to other brain areas. However, the intermediate level of visual processing that takes place in earlier cortical areas is described in early Chapters of the book, to provide a foundation for what follows. We also describe some of the processes that take place in the dorsal visual system and that are involved in the perception of the 3D world and the visual control of action, and how the object and spatial processing streams may interface to each other.

The overall plan of the book is as follows. Chapter 1 provides an introduction to information processing in neural systems in the brain, especially with respect to visual information processing. Chapters 2–4 provide an overview of processing in early cortical areas (V1 in Chapter 2, intermediate areas such as V2, V4, MT and MST in Chapter 3), and in the parietal cortex (Chapter 4). We regard these areas as providing the preprocessing for many of the higher visual functions which are the most important issues at the heart of this book. Chapter 5 describes the higher visual information processing related to invariant object and face representation and recognition in the inferior temporal cortical visual areas. Chapter 6 describes approaches to another aspect of higher visual function, namely visual attentional processes. Chapter 7 then lays the foundation for the neurocomputational approaches taken in later Chapters, by providing a summary of many of the main classes of biologically plausible networks, which are incorporated into many of the neurocomputational models described in later Chapters. Chapter 7 provides quite a self-contained overview of neural networks, in the sense that those not already familiar with neural computation can find many of the fundamentals and foundations here. It is complemented by Appendix A, which summarizes some of the straightforward aspects of linear algebra which are useful when developing a quantitative understanding of neural computation. Chapters 7 and 5 are also complemented by Appendix B, which describes how information theory can be applied to understanding the representation and processing of information in the brain. This Appendix is especially directed towards understanding and measuring the information made available by the spikes of single neurons, and by populations of neurons, and, more generally, to understanding neural encoding.

Chapter 8 builds on earlier Chapters to describe approaches to invariant object and face recognition, and to describe one particular approach which seems to be close to how the brain solves invariant object recognition. Chapters 5 and 8 thus cover one of the key aspects of this book, how invariant face and object representation and recognition are performed in the brain.

Another key aspect of this book is how visual attention operates in the brain. After the introduction to attentional processes in Chapter 6, Chapter 9 describes a model of how attention could be implemented in the brain using a number of key features of the architecture of the visual system, including biased competition within a brain area, feedback ('top-down') as well as feedforward ('bottom-up') processing, and at least partly separate 'what' and 'where' visual processing streams connected by early visual cortical areas. The approach taken is able to account for spatial attention, object attention, and both serial search and parallel search for visual stimuli, within a unified system of interacting networks with the same general architecture as that of the visual system. This attentional architecture is applied to understanding more difficult processes in attention such as feature binding in Chapter 10, and to the neuropsychology of attention in Chapter 11.

We show in Chapter 12 how the representations formed in the ventral, object-based, visual processing stream are appropriate for, and interface to, the brain systems that receive the outputs of this visual information processing about objects.

The computational approach described uses largely feedforward processing for Chapters 2–5 and 8. For the model of invariant object recognition described in Chapter 8, dynamics and backprojections are not incorporated in order to focus on the computational issues of how invariant representations could be computed, and indeed how object recognition could operate when there is only time for feedforward processing. We introduce inter alia cortical backprojections and interacting networks in Chapter 7. We introduce top-down processing and attentional effects in Chapter 6, and develop dynamic computational models of these in

Chapter 9, and of their effects in Chapters 10 and 11. Dynamical models, and top-down as well as bottom up effects, are needed to understand the operation of attentional processes, including issues such as the speed with which visual search takes place. These approaches, the feedforward analysis of invariant object recognition adopted in Chapter 8, and the dynamical feedback model of attention described in Chapters 9–11, are complementary, and enable different issues to be investigated in quite tractable models. However, we have shown that the two approaches can be combined in a single integrated model, which displays the combined properties of each model, as summarized in Chapter 13.

The book thus focuses on the principles of processing in neural networks in cortical areas, and on providing a foundation for how these can to applied to understanding especially the formation of representations of objects in the ventral visual system, and how attentional processes that link the ventral visual object processing stream to the dorsal spatial processing visual stream operate in the brain. The book does not stop though with purely visual processing, but considers also how visual inputs reach and are involved in the computations underlying a wide range of behaviours, including short term memory, long term memory, emotion and motivation, and the initiation of action. The book is relatively unique in combining evidence from the neurophysiology of the high-level visual processing systems in the brain and their connected output systems with biologically plausible computational models. For an introduction to visual science, we recommend Palmer (1999), and for an account of the computational processes involved in intermediate-level vision Mallot (2000).

The material in this text is the copyright of E.T. Rolls (who was mainly responsible for Chapters 1, 5, 7, 8, 12 and 13) and G. Deco (who was mainly responsible for Chapters 4, 6 and 9–11). Part of the material described in the book reflects work done over many years in collaboration with many colleagues, whose tremendous contributions are warmly appreciated. The contributions of many will be evident from references cited in the text. We are particularly grateful to our collaborators on the most recent research work, including S. M. Stringer, T. P. Trappenberg, C. Hölscher, and N. Aggelopoulos, who have agreed to include material in this book at a time when part of it has not yet been published in research journals. Much of the research presented in Chapters 6, 9, 10 and 11 is the result of years of very fruitful, strong and enthusiastic collaboration between the computational neuroscience group of Siemens (G. Deco), the neuropsychology group of J. Zihl (Max-Planck-Institute for Psychiatry and Ludwig-Maximilians-University Munich), D. Heinke and G. Humphreys (Birmingham University), and T. S. Lee (Carnegie Mellon University). G. Deco would like particularly to acknowledge the radical enthusiasm of J. Zihl, who introduced him to, and motivated and taught him how to do research in, this new field of computational neuroscience-based modelling of visual perception and neuropsychology. In addition, we have benefited enormously from the discussions we have had with a large number of colleagues and friends, many of whom we hope will see areas of the text that they have been able to illuminate. E. T. Rolls wishes to record especial thanks to Simon M. Stringer for the contributions he made to some of the work described in Chapter 8, to Alessandro Treves for an always interesting and fruitful research collaboration, and to Morten L. Kringelbach for contributing to the design of the cover and of the book. We also wish to thank W. P. C. Mills for reading parts of the book and for helpful comments. Much of the work described would not have been possible without financial support from a number of sources, particularly the Medical Research Council of the UK, the Human Frontier Science Program, the Wellcome Trust, the McDonnell-Pew Foundation, and the Commission of the European Communities.

The cover shows part of the picture Pandora painted in 1896 by J.W.Waterhouse.

Updates to the publications cited in this book are available at http://www.cns.ox.ac.uk.

Gustavo Deco dedicates this work to his wife, María Eugenia, and sons, Nikolas, Sebastián and Martín.

Edmund T. Rolls dedicates this work to the overlapping group: his family, friends, and colleagues – in salutem praesentium, in memoriam absentium.

Contents

1	**Introduction**	**1**
	1.1 Introduction and overview	1
	1.2 Neurons	2
	1.3 Neurons in a network	2
	1.4 Synaptic modification	4
	1.5 Long-Term Potentiation and Long-Term Depression	7
	1.6 Distributed representations	11
	1.6.1 Definitions	11
	1.6.2 Advantages of different types of coding	12
	1.7 Neuronal network approaches versus connectionism	13
	1.8 Introduction to three neuronal network architectures	14
	1.9 Systems-level analysis of brain function	16
	1.10 The fine structure of the cerebral neocortex	21
	1.10.1 The fine structure and connectivity of the neocortex	21
	1.10.2 Excitatory cells and connections	21
	1.10.3 Inhibitory cells and connections	23
	1.10.4 Quantitative aspects of cortical architecture	25
	1.10.5 Functional pathways through the cortical layers	27
	1.10.6 The scale of lateral excitatory and inhibitory effects, and the concept of modules	29
	1.11 Backprojections in the cortex	30
	1.11.1 Architecture	30
	1.11.2 Learning	31
	1.11.3 Recall	33
	1.11.4 Semantic priming	34
	1.11.5 Attention	34
	1.11.6 Autoassociative storage, and constraint satisfaction	34
2	**The primary visual cortex**	**36**
	2.1 Introduction and overview	36
	2.2 Retina and lateral geniculate nuclei	37
	2.3 Striate cortex: Area V1	43
	2.3.1 Classification of V1 neurons	43
	2.3.2 Organization of the striate cortex	45
	2.3.3 Visual streams within the striate cortex	48

2.4 Computational processes that give rise to V1 simple cells 49

    2.4.1 Linsker's method: Information maximization 50

    2.4.2 Olshausen and Field's method: Sparseness maximization 53

2.5 The computational role of V1 for form processing 55

2.6 Backprojections to the lateral geniculate nucleus 55

3 Extrastriate visual areas **57**

3.1 Introduction 57

3.2 Visual pathways in extrastriate cortical areas 57

3.3 Colour processing 61

    3.3.1 Trichromacy theory 61

    3.3.2 Colour opponency, and colour contrast: Opponent cells 61

3.4 Motion and depth processing 65

    3.4.1 The motion pathway 65

    3.4.2 Depth perception 67

4 The parietal cortex **70**

4.1 Introduction 70

4.2 Spatial processing in the parietal cortex 70

    4.2.1 Area LIP 71

    4.2.2 Area VIP 73

    4.2.3 Area MST 74

    4.2.4 Area 7a 74

4.3 The neuropsychology of the parietal lobe 75

    4.3.1 Unilateral neglect 75

    4.3.2 Balint's syndrome 77

    4.3.3 Gerstmann's syndrome 79

5 Inferior temporal cortical visual areas **81**

5.1 Introduction 81

5.2 Neuronal responses in different areas 81

5.3 The selectivity of one population of neurons for faces 83

5.4 Combinations of face features 84

5.5 Distributed encoding of object and face identity 84

    5.5.1 Distributed representations evident in the firing rate distributions 85

    5.5.2 The representation of information in the responses of single neurons to a set of stimuli 90

    5.5.3 The representation of information in the responses

		of a population of inferior temporal visual cortex neurons	94
	5.5.4	Advantages for brain processing of the distributed representation of objects and faces	98
	5.5.5	Should one neuron be as discriminative as the whole organism, in object encoding systems?	103
	5.5.6	Temporal encoding in the spike train of a single neuron	105
	5.5.7	Temporal synchronization of the responses of different cortical neurons	108
	5.5.8	Conclusions on cortical encoding	111
5.6	Invariance in the neuronal representation of stimuli		112
	5.6.1	Size and spatial frequency invariance	112
	5.6.2	Translation (shift) invariance	113
	5.6.3	Reduced translation invariance in natural scenes	113
	5.6.4	A view–independent representation of objects and faces	115
5.7	Face identification and face expression systems		118
5.8	Learning in the inferior temporal cortex		120
5.9	Cortical processing speed		122
5.10	Conclusions		125
6	**Visual attentional mechanisms**		**126**
6.1	Introduction		126
6.2	The classical view		126
	6.2.1	The spotlight metaphor and feature integration theory	126
	6.2.2	Computational models of visual attention	129
6.3	Biased competition – single cell studies		132
	6.3.1	Neurophysiology of attention	133
	6.3.2	The role of competition	135
	6.3.3	Evidence of attentional bias	136
	6.3.4	Non-spatial attention	136
	6.3.5	High-resolution buffer hypothesis	139
6.4	Biased competition – fMRI		140
	6.4.1	Neuroimaging of attention	140
	6.4.2	Attentional effects in the absence of visual stimulation	141
6.5	The computational role of top-down feedback connections		142
7	**Neural network models**		**145**

7.1	Introduction	145
7.2	Pattern association memory	145
	7.2.1 Architecture and operation	146
	7.2.2 The vector interpretation	149
	7.2.3 Properties	150
	7.2.4 Prototype extraction, extraction of central tendency, and noise reduction	151
	7.2.5 Speed	151
	7.2.6 Local learning rule	152
	7.2.7 Implications of different types of coding for storage in pattern associators	158
7.3	Autoassociation memory	159
	7.3.1 Architecture and operation	160
	7.3.2 Introduction to the analysis of the operation of autoassociation networks	161
	7.3.3 Properties	163
	7.3.4 Use of autoassociation networks in the brain	170
7.4	Competitive networks, including self-organizing maps	171
	7.4.1 Function	171
	7.4.2 Architecture and algorithm	171
	7.4.3 Properties	173
	7.4.4 Utility of competitive networks in information processing by the brain	178
	7.4.5 Guidance of competitive learning	180
	7.4.6 Topographic map formation	182
	7.4.7 Radial Basis Function networks	187
	7.4.8 Further details of the algorithms used in competitive networks	188
7.5	Continuous attractor networks	192
	7.5.1 Introduction	192
	7.5.2 The generic model of a continuous attractor network	195
	7.5.3 Learning the synaptic strengths between the neurons that implement a continuous attractor network	196
	7.5.4 The capacity of a continuous attractor network	198
	7.5.5 Continuous attractor models: moving the activity packet of neuronal activity	198
	7.5.6 Stabilization of the activity packet within the continuous attractor network when the agent is stationary	202
	7.5.7 Continuous attractor networks in two or more dimensions	203

		7.5.8 Mixed continuous and discrete attractor networks	203
7.6	Network dynamics: the integrate-and-fire approach		204
	7.6.1	From discrete to continuous time	204
	7.6.2	Continuous dynamics with discontinuities	205
	7.6.3	Conductance dynamics for the input current	207
	7.6.4	The speed of processing of one-layer attractor networks with integrate-and-fire neurons	209
	7.6.5	The speed of processing of a four-layer hierarchical network with integrate-and-fire attractor dynamics in each layer	212
	7.6.6	Spike response model	215
7.7	Network dynamics: introduction to the mean field approach		216
7.8	Mean-field based neurodynamics		218
	7.8.1	Population activity	218
	7.8.2	A basic computational module based on biased competition	220
	7.8.3	Multimodular neurodynamical architectures	221
7.9	Interacting attractor networks		224
7.10	Error correction networks		228
	7.10.1	Architecture and general description	229
	7.10.2	Generic algorithm (for a one-layer network taught by error correction)	229
	7.10.3	Capability and limitations of single-layer error-correcting networks	230
	7.10.4	Properties	234
7.11	Error backpropagation multilayer networks		236
	7.11.1	Introduction	236
	7.11.2	Architecture and algorithm	237
	7.11.3	Properties of multilayer networks trained by error backpropagation	238
7.12	Biologically plausible networks		239
7.13	Reinforcement learning		240
7.14	Contrastive Hebbian learning: the Boltzmann machine		241
8	**Models of invariant object recognition**		**243**
8.1	Introduction		243
8.2	Approaches to invariant object recognition		244
	8.2.1	Feature spaces	244
	8.2.2	Structural descriptions and syntactic pattern recognition	245

	8.2.3	Template matching and the alignment approach	247
	8.2.4	Invertible networks that can reconstruct their inputs	248
	8.2.5	Feature hierarchies	249
8.3	Hypotheses about object recognition mechanisms		253
8.4	Computational issues in feature hierarchies		257
	8.4.1	The architecture of VisNet	258
	8.4.2	Initial experiments with VisNet	266
	8.4.3	The optimal parameters for the temporal trace used in the learning rule	274
	8.4.4	Different forms of the trace learning rule, and their relation to error correction and temporal difference learning	275
	8.4.5	The issue of feature binding, and a solution	284
	8.4.6	Operation in a cluttered environment	295
	8.4.7	Learning 3D transforms	301
	8.4.8	Capacity of the architecture, and incorporation of a trace rule into a recurrent architecture with object attractors	307
	8.4.9	Vision in natural scenes — effects of background versus attention	313
8.5	Synchronization and syntactic binding		319
8.6	Further approaches to invariant object recognition		320
8.7	Processes involved in object identification		321

9 The cortical neurodynamics of visual attention – a model **323**

9.1	Introduction		323
9.2	Physiological constraints		324
	9.2.1	The dorsal and ventral paths of the visual cortex	324
	9.2.2	The biased competition hypothesis	326
	9.2.3	Neuronal receptive fields	327
9.3	Architecture of the model		328
	9.3.1	Overall architecture of the model	328
	9.3.2	Formal description of the model	331
	9.3.3	Performance measures	336
9.4	Simulations of basic experimental findings		336
	9.4.1	Simulations of single-cell experiments	337
	9.4.2	Simulations of fMRI experiments	339
9.5	Object recognition and spatial search		341
	9.5.1	Dynamics of spatial attention and object recognition	343
	9.5.2	Dynamics of object attention and visual search	345

9.6   Evaluation of the model   348
    9.6.1 Spatial attention and object attention   348
    9.6.2 Translation-invariant object recognition   350
    9.6.3 Contributions and limitations   351

10 Visual search: Attentional neurodynamics at work   **353**
10.1 Introduction   353
10.2 Simple visual search   354
10.3 Visual search of hierarchical patterns   358
    10.3.1 The spatial resolution hypothesis   358
    10.3.2 Neurodynamics of the resolution hypothesis   361
    10.3.3 Visual search in the framework of the resolution hy-
           pothesis   363
10.4 Visual conjunction search   369
    10.4.1 The binding problem   369
    10.4.2 The time course of conjunction search: experimental
           evidence   371
    10.4.3 Extension of the computational cortical architecture   373
    10.4.4 Computational results   376
10.5 Conclusion   381

11 A computational approach to the neuropsychology of visual
attention   **383**
11.1 Introduction   383
11.2 The neglect syndrome   383
    11.2.1 A model of visual spatial neglect   384
    11.2.2 Spatial cueing effect on neglect   388
    11.2.3 Extinction and visual search   390
    11.2.4 Effect on neglect of top-down knowledge about ob-
           jects   392
11.3 Hierarchical patterns – neuropsychology   398
11.4 Conjunction search – neuropsychology   400
    11.4.1 Simulations and predictions   400
    11.4.2 Experimental test of the predictions in human sub-
           jects   401
11.5 Conclusion   403

12 Outputs of visual processing   **404**
12.1 Visual outputs to Short Term Memory systems   406
    12.1.1 Prefrontal cortex short term memory networks, and
           their relation to temporal and parietal perceptual net-

works   406

12.1.2 Computational details of the model of short term memory   409

12.1.3 Computational necessity for a separate, prefrontal cortex, short term memory system   412

12.1.4 Role of prefrontal cortex short term memory systems in visual search and attention   412

12.1.5 Synaptic modification is needed to set up but not to reuse short term memory systems   413

12.2 Visual outputs to Long Term Memory systems in the brain   413

12.2.1 Effects of damage to the hippocampus and connected structures on object-place and episodic memory   414

12.2.2 Neurophysiology of the hippocampus and connected areas   415

12.2.3 Hippocampal models   418

12.2.4 The perirhinal cortex, recognition memory, and familiarity   421

12.3 Visual stimulus–reward association, emotion, and motivation   424

12.3.1 Emotion   425

12.3.2 Reward is not processed in the temporal cortical visual areas   429

12.3.3 Why the reward and punishment associations of stimuli are not represented early in information processing in the primate brain   430

12.3.4 Amygdala   434

12.3.5 Orbitofrontal cortex   439

12.3.6 Effects of mood on memory and visual processing   447

12.4 Output to object selection and action systems   448

12.5 Visual search   452

12.6 Visual outputs to behavioral response systems   453

12.7 Multimodal representations in different brain areas   453

12.8 Visuo–spatial scratchpad, and change blindness   454

12.9 Conscious visual perception   454

13 Principles and Conclusions   **456**

13.1 Transform invariance in the inferior temporal visual cortex   456

13.2 Representation of information in IT   456

13.3 IT information processing is fast   457

13.4 Continuous neuronal dynamics allows fast network processing | 457

13.5 Hierarchical feature analysis | 457

13.6 Trace learning rule for invariant representations | 459

13.7 Spatial feature binding by feature combination neurons | 460

13.8 IT provides a representation for later memory networks | 461

13.9 Face expression and object motion | 462

13.10 Attentional mechanisms | 462

13.11 Visual search | 464

13.12 Egocentric vs allocentric representations | 464

13.13 Short term memory as the controller of attention | 465

13.14 Output to object selection and action systems | 466

13.15 'What' versus 'where' processing streams | 466

13.16 Short term memory must be separated from perception | 467

13.17 Backprojections must be weak | 468

13.18 Long-term potentiation and short-term memory | 469

13.19 "Executive control" by the prefrontal cortex | 469

13.20 Reward processing occurs after object identification | 470

13.21 Effects of mood on memory and visual processing | 471

13.22 Visual outputs to Long Term Memory systems | 471

13.23 Episodic memory and the operation of mixed discrete and continuous attractor networks | 472

13.24 Visual outputs to behavioural response systems | 472

13.25 Multimodal representations in different brain areas | 472

13.26 Visuo–spatial scratchpad and change blindness | 472

13.27 Invariant object recognition and attention | 473

13.28 Conscious visual perception | 473

13.29 Attention – future directions | 473

13.30 Integrated approaches to understanding vision | 475

13.31 Apostasis | 475

A **Introduction to linear algebra for neural networks** | **477**

A.1 Vectors | 477

A.1.1 The inner or dot product of two vectors | 477

A.1.2 The length of a vector | 478

A.1.3 Normalizing the length of a vector | 479

A.1.4 The angle between two vectors: the normalized dot product | 479

A.1.5 The outer product of two vectors | 480

A.1.6 Linear and non-linear systems | 481

A.1.7 Linear combinations of vectors, linear independence, and linear separability 482

A.2 Application to understanding simple neural networks 484

A.2.1 Capability and limitations of single-layer networks: linear separability and capacity 484

A.2.2 Non-linear networks: neurons with non-linear activation functions 487

A.2.3 Non-linear networks: neurons with non-linear activations 488

B Information theory **490**

B.1 Basic notions 490

B.1.1 The information conveyed by definite statements 491

B.1.2 Information conveyed by probabilistic statements 491

B.1.3 Information sources, information channels, and information measures 492

B.1.4 The information carried by a neuronal response and its averages 494

B.1.5 The information conveyed by continuous variables 496

B.2 The information carried by neuronal responses 498

B.2.1 The limited sampling problem 498

B.2.2 Correction procedures for limited sampling 500

B.2.3 The information from multiple cells: decoding procedures 501

B.2.4 Information in the correlations between the spikes of different cells 504

B.3 Information theory results 507

B.3.1 Temporal codes versus rate codes within the spike train of a single neuron 507

B.3.2 The speed of information transfer from single neurons 509

B.3.3 The information from multiple cells: independent information versus redundancy across cells 512

B.3.4 The information from multiple cells: the effects of cross-correlations between cells 514

B.3.5 Conclusions 517

B.4 Information theory terms — a short glossary 518

References **520**

Index **565**

1 Introduction

1.1 Introduction and overview

To understand how the brain works, including how it functions in vision, it is necessary to combine different approaches, including neural computation. Neurophysiology at the single neuron level is needed because this is the level at which information is exchanged between the computing elements of the brain. Evidence from the effects of brain damage, including that available from neuropsychology, is needed to help understand what different parts of the system do, and indeed what each part is necessary for. Neuroimaging is useful to indicate where in the human brain different processes take place, and to show which functions can be dissociated from each other. Knowledge of the biophysical and synaptic properties of neurons is essential to understand how the computing elements of the brain work, and therefore what the building blocks of biologically realistic computational models should be. Knowledge of the anatomical and functional architecture of the cortex is needed to show what types of neuronal network actually perform the computation. And finally the approach of neural computation is needed, as this is required to link together all the empirical evidence to produce an understanding of how the system actually works. This book utilizes evidence from all these disciplines to develop an understanding of how vision is implemented by processing in the brain.

A test of whether one's understanding is correct is to simulate the processing on a computer, and to show whether the simulation can perform the tasks of vision, and whether the simulation has similar properties to the real brain. The approach of neural computation leads to a precise definition of how the computation is performed, and to precise and quantitative tests of the theories produced. How vision works is a paradigm example of this approach, because it is a sufficiently complex problem that it requires a computational approach, and indeed has not been solved in artificial vision systems operating in natural scenes. At the same time, vision raises fundamental issues in cognitive neuroscience, such as how attention works.

One of the distinctive properties of this book is that it links the neural computation approach not only firmly to neuronal neurophysiology, which provides much of the primary data about how the brain operates, but also to psychophysical studies (for example of attention); to neuropsychological studies of patients with brain damage; and to functional magnetic resonance imaging (fMRI) (and other neuroimaging) approaches. The empirical evidence that is brought to bear is largely from non-human primates and from humans, because of the considerable similarity of their visual systems, and the overall aims to understand how visual perception is implemented in the human brain, and the disorders that arise after brain damage. The focus of the book is on the processes that underlie high-level vision, and in particular visual object recognition and attention, and on how the object processing visual system interfaces to other brain areas. However, the intermediate level of visual processing that takes place in earlier cortical areas is described in early Chapters of the book, to provide a foundation for what follows. We also describe some of the processes that take place in the dorsal visual system and that are involved in the perception of the 3D world and the visual control of action, and how the object and spatial processing streams may interface to each other.

The computational approach described uses largely feedforward processing for Chapters 2–5 and 8. We introduce inter alia cortical backprojections and interacting networks in Chapter 7. We introduce top-down processing and attentional effects in Chapter 6, and develop models of these in Chapter 9, and of their effects in Chapters 10 and 11.

In the rest of this Chapter, we introduce some of the background for understanding brain computation, such as how single neurons operate, how some of the essential features of this can be captured by simple formalisms, and some of the biological background to what it can be taken happens in the nervous system, such as synaptic modification based on information available locally at each synapse.

1.2 Neurons in the brain, and their representation in neuronal networks

Neurons in the vertebrate brain typically have, extending from the cell body, large dendrites which receive inputs from other neurons through connections called synapses. The synapses operate by chemical transmission. When a synaptic terminal receives an all-or-nothing action potential from the neuron of which it is a terminal, it releases a transmitter that crosses the synaptic cleft and produces either depolarization or hyperpolarization in the postsynaptic neuron, by opening particular ionic channels. (A textbook such as Kandel, Schwartz and Jessel (2000) gives further information on this process.) Summation of a number of such depolarizations or excitatory inputs within the time constant of the receiving neuron, which is typically 15–25 ms, produces sufficient depolarization that the neuron fires an action potential. There are often 5,000–20,000 inputs per neuron. Examples of cortical neurons are shown in Fig. 1.1, and further examples are shown in Rolls and Treves (1998) and Shepherd (1998). Once firing is initiated in the cell body (or axon initial segment of the cell body), the action potential is conducted in an all-or-nothing way to reach the synaptic terminals of the neuron, whence it may affect other neurons. Any inputs the neuron receives that cause it to become hyperpolarized make it less likely to fire (because the membrane potential is moved away from the critical threshold at which an action potential is initiated), and are described as inhibitory. The neuron can thus be thought of in a simple way as a computational element that sums its inputs within its time constant and, whenever this sum, minus any inhibitory effects, exceeds a threshold, produces an action potential that propagates to all of its outputs. This simple idea is incorporated in many neuronal network models using a formalism of a type described in the next Section.

1.3 A formalism for approaching the operation of single neurons in a network

Let us consider a neuron i as shown in Fig. 1.2, which receives inputs from axons that we label j through synapses of strength w_{ij}. The first subscript (i) refers to the receiving neuron, and the second subscript (j) to the particular input. j counts from 1 to C, where C is the

Fig. 1.1 Examples of neurons found in the brain. Cell types in the cerebral neocortex are shown. The different laminae of the cortex are designated I–VI, with I at the surface. Cells A–E are pyramidal cells in the different layers. Cell E is a spiny stellate cell, and F is a double bouquet cell. (After Jones, 1981; see Jones and Peters, 1984, p. 7.)

number of synapses or connections received. The firing rate of the ith neuron is denoted as y_i, and that of the jth input to the neuron as x_j. To express the idea that the neuron makes a simple linear summation of the inputs it receives, we can write the activation of neuron i, denoted h_i, as

$$h_i = \sum_j x_j w_{ij} \tag{1.1}$$

where $\sum_j$ indicates that the sum is over the C input axons (or connections) indexed by j to each neuron. The multiplicative form here indicates that activation should be produced by an axon only if it is firing, and depending on the strength of the synapse w_{ij} from input axon j onto the dendrite of the receiving neuron i. Equation 1.1 indicates that the strength of the activation reflects how fast the axon j is firing (that is x_j), and how strong the synapse w_{ij} is. The sum of all such activations expresses the idea that summation (of synaptic currents in real neurons) occurs along the length of the dendrite, to produce activation at the cell body, where the activation h_i is converted into firing y_i. This conversion can be expressed as

$$y_i = f(h_i) \tag{1.2}$$

which indicates that the firing rate is a function (f) of the postsynaptic activation. The function is called the activation function in this case. The function at its simplest could be linear, so that the firing rate would be proportional to the activation (see Fig. 1.3). Real neurons have thresholds, with firing occurring only if the activation is above the threshold. A threshold linear activation function is shown in Fig. 1.3. This has been useful in formal analysis of the properties of neural networks. Neurons also have firing rates that become saturated at a

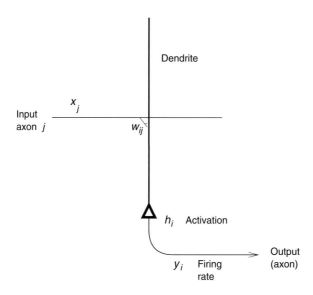

Fig. 1.2 Notation used to describe an individual neuron in a network model. By convention, we generally represent the dendrite as thick, and vertically oriented (as this is the normal way that neuroscientists view cortical pyramidal cells under the microscope); and the axon as thin. The cell body or soma is indicated between them. The firing rate we also call the activity of the neuron.

maximum rate, and we could express this as the sigmoid activation function shown in Fig. 1.3c. Another simple activation function, used in some models of neural networks, is the binary threshold function (Fig. 1.3d), which indicates that if the activation is below threshold, there is no firing, and that if the activation is above threshold, the neuron fires maximally. Some non-linearity in the activation function is an advantage, for it enables many useful computations to be performed in neuronal networks, including removing interfering effects of similar memories, and enabling neurons to perform logical operations, such as firing only if several inputs are present simultaneously.

A property implied by equation 1.1 is that the postsynaptic membrane is electrically short, and so summates its inputs irrespective of where on the dendrite the input is received. In real neurons, the transduction of current into firing frequency (the analogue of the transfer function of equation 1.2) is generally studied not with synaptic inputs but by applying a steady current through an electrode into the soma. An example of the resulting curves, which illustrate the additional phenomenon of firing rate adaptation, is reproduced in Fig. 1.4.

1.4 Synaptic modification

For a neuronal network to perform useful computation, that is to produce a given output when it receives a particular input, the synaptic weights must be set up appropriately. This is often performed by synaptic modification occurring during learning.

A simple learning rule that was originally presaged by Donald Hebb (1949) proposes that synapses increase in strength when there is conjunctive presynaptic and postsynaptic activity. The Hebb rule can be expressed more formally as follows

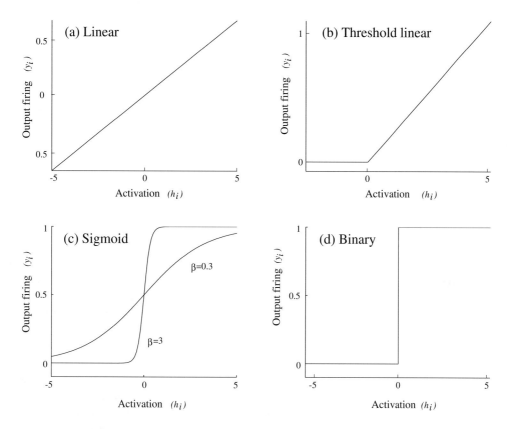

Fig. 1.3 Different types of activation function. The activation function relates the output activity (or firing rate), y_i, of the neuron (i) to its activation, h_i. (a) Linear. (b) Threshold linear. (c) Sigmoid. (One mathematical exemplar of this class of activation function is $y_i = 1/(1 + exp(-2\beta h_i))$. The output of this function, also sometimes known as the logistic function, is 0 for an input of $-\infty$, 0.5 for 0, and 1 for $+\infty$. The function incorporates a threshold at the lower end, followed by a linear portion, and then an asymptotic approach to the maximum value at the top end of the function. The parameter β controls the steepness of the almost linear part of the function round $h_i = 0$. If β is small, the output goes smoothly and slowly from 0 to 1 as h_i goes from $-\infty$ to $+\infty$. If β is large, the curve is very steep, and approximates a binary threshold activation function.) d) Binary threshold.

$$\delta w_{ij} = \alpha y_i x_j. \tag{1.3}$$

where δw_{ij} is the change of the synaptic weight w_{ij} which results from the simultaneous (or conjunctive) presence of presynaptic firing x_j and postsynaptic firing y_i (or strong depolarization), and α is a learning rate constant that specifies how much the synapses alter on any one pairing. The presynaptic and postsynaptic activity must be present approximately simultaneously (to within perhaps 100–500 ms in the real brain).

The Hebb rule is expressed in this multiplicative form to reflect the idea that both presynaptic and postsynaptic activity must be present for the synapses to increase in strength. The multiplicative form also reflects the idea that strong pre- and postsynaptic firing will produce a larger change of synaptic weight than smaller firing rates. The Hebb rule thus captures what is

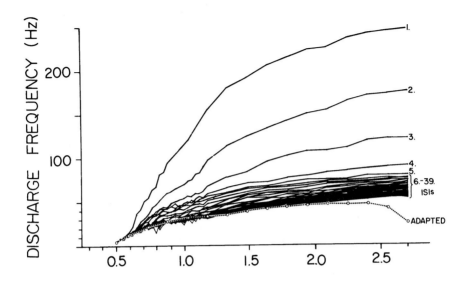

Fig. 1.4 Frequency current plot (the closest experimental analogue of the activation function) for a CA1 pyramidal cell. The firing frequency (in Hz) in response to the injection of 1.5 s long, rectangular depolarizing current pulses has been plotted against the strength of the current pulses (in nA) (abscissa). The first 39 interspike intervals (ISIs) are plotted as instantaneous frequency (1/ISI, where ISI is the inter-stimulus interval), together with the average frequency of the adapted firing during the last part of the current injection (circles and broken line). The plot indicates a current threshold at approximately 0.5 nA, a linear range with a tendency to saturate, for the initial instantaneous rate, above approximately 200 Hz, and the phenomenon of adaptation, which is not reproduced in simple non-dynamical models (see further Appendix A5 of Rolls and Treves, 1998). (Reprinted with permission from Lanthorn, Storm and Andersen, 1984.)

typically found in studies of associative Long-Term Potentiation (LTP) in the brain, described in Section 1.5.

One useful property of large neurons in the brain, such as cortical pyramidal cells, is that with their short electrical length, the postsynaptic term, y_i, is available on much of the dendrite of a cell. The implication of this is that once sufficient postsynaptic activation has been produced, any active presynaptic terminal on the neuron will show synaptic strengthening. This enables associations between coactive inputs, or correlated activity in input axons, to be learned by neurons using this simple associative learning rule.

If, in contrast, a group of coactive axons made synapses close together on a small dendrite, then the local depolarization might be intense, and only these synapses would modify onto the dendrite. (A single distant active synapse might not modify in this type of neuron, because of the long electrotonic length of the dendrite.) The computation in this case is described as Sigma-Pi ($\Sigma\Pi$), to indicate that there is a local product computed during learning; this allows a particular set of locally active synapses to modify together, and then the output of the neuron can reflect the sum of such local multiplications (see Rumelhart and McClelland (1986), Koch (1999)).

1.5 Long-Term Potentiation and Long-Term Depression as models of synaptic modification

Long-Term Potentiation (LTP) and Long-Term Depression (LTD) provide useful models of some of the synaptic modifications that occur in the brain. The synaptic changes found appear to be synapse–specific, and to depend on information available locally at the synapse. LTP and LTD may thus provide a good model of the biological synaptic modifications involved in real neuronal network operations in the brain. We next therefore describe some of the properties of LTP and LTD, and evidence that implicates them in learning in at least some brain systems. Even if they turn out not to be the basis for the synaptic modifications that occur during learning, they have many of the properties that would be needed by some of the synaptic modification systems used by the brain.

Long-term potentiation (LTP) is a use-dependent and sustained increase in synaptic strength that can be induced by brief periods of synaptic stimulation. It is usually measured as a sustained increase in the amplitude of electrically evoked responses in specific neural pathways following brief trains of high-frequency stimulation (see Fig. 1.5b). For example, high frequency stimulation of the Schaffer collateral inputs to the hippocampal CA1 cells results in a larger response recorded from the CA1 cells to single test pulse stimulation of the pathway. LTP is long-lasting, in that its effect can be measured for hours in hippocampal slices, and in chronic in vivo experiments in some cases it may last for months. LTP becomes evident rapidly, typically in less than 1 minute. LTP is in some brain systems associative. This is illustrated in Fig. 1.5c, in which a weak input to a group of cells (e.g. the commissural input to CA1) does not show LTP unless it is given at the same time as (i.e. associatively with) another input (which could be weak or strong) to the cells. The associativity arises because it is only when sufficient activation of the postsynaptic neuron to exceed the threshold of NMDA receptors (see below) is produced that any learning can occur. The two weak inputs summate to produce sufficient depolarization to exceed the threshold. This associative property is shown very clearly in experiments in which LTP of an input to a single cell only occurs if the cell membrane is depolarized by passing current through it at the same time as the input arrives at the cell. The depolarization alone or the input alone is not sufficient to produce the LTP, and the LTP is thus associative. Moreover, in that the presynaptic input and the postsynaptic depolarization must occur at about the same time (within approximately 500 ms), the LTP requires temporal contiguity. LTP is also synapse–specific, in that for example an inactive input to a cell does not show LTP even if the cell is strongly activated by other inputs (Fig. 1.5b, input B).

These spatiotemporal properties of LTP can be understood in terms of actions of the inputs on the postsynaptic cell, which in the hippocampus has two classes of receptor, NMDA (N-methyl-D-aspartate) and K-Q (kainate-quisqualate), activated by the glutamate released by the presynaptic terminals. The NMDA receptor channels are normally blocked by Mg^{2+}, but when the cell is strongly depolarized by strong tetanic stimulation of the type necessary to induce LTP, the Mg^{2+} block is removed, and Ca^{2+} entering via the NMDA receptor channels triggers events that lead to the potentiated synaptic transmission (see Fig. 1.6). Part of the evidence for this is that NMDA antagonists such as AP5 (D-2-amino-5-phosphonopentanoate) block LTP. Further, if the postsynaptic membrane is voltage clamped to prevent depolarization by a strong input, then LTP does not occur. The voltage–dependence of the NMDA receptor channels introduces a threshold and thus a non-linearity that contributes to a number of the phenomena of some types of LTP, such as cooperativity (many small inputs together produce

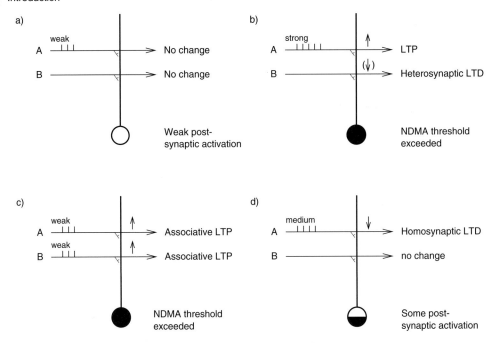

Fig. 1.5 Schematic illustration of synaptic modification rules as revealed by Long-Term Potentiation (LTP) and Long-Term Depression (LTD). The activation of the postsynaptic neuron is indicated by the extent to which its soma is black. There are two sets of inputs to the neuron: A and B. (a) A weak input (indicated by 3 spikes) on the set A of input axons produces little postsynaptic activation, and there is no change in synaptic strength. (b) A strong input (indicated by 5 spikes) on the set A of input axons produces strong postsynaptic activation, and the active synapses increase in strength. This is LTP. It is homosynaptic in that the synapses that increase in strength are the same as those through which the neuron is activated. LTP is synapse–specific, in that the inactive axons, B, do not show LTP. They either do not change in strength, or they may weaken. The weakening is called heterosynaptic LTD, because the synapses that weaken are other than those through which the neuron is activated (hetero- is Greek for other). (c) Two weak inputs present simultaneously on A and B summate to produce strong postsynaptic activation, and both sets of active synapses show LTP. (d) Intermediate strength firing on A produces some activation, but not strong activation, of the postsynaptic neuron. The active synapses become weaker. This is homosynaptic LTD, in that the synapses that weaken are the same as those through which the neuron is activated (homo- is Greek for same).

sufficient depolarization to allow the NMDA receptors to operate), associativity (a weak input alone will not produce sufficient depolarization of the postsynaptic cell to enable the NMDA receptors to be activated, but the depolarization will be sufficient if there is also a strong input), and temporal contiguity between the different inputs that show LTP (in that if inputs occur non-conjunctively, the depolarization shows insufficient summation to reach the required level, or some of the inputs may arrive when the depolarization has decayed). Once the LTP has become established (which can be within one minute of the strong input to the cell), the LTP is expressed through the K–Q receptors, in that AP5 blocks only the establishment of LTP, and not its subsequent expression (Bliss and Collingridge 1993, Nicoll and Malenka 1995, Fazeli and Collingridge 1996).

There are a number of possibilities about what change is triggered by the entry of Ca^{2+}

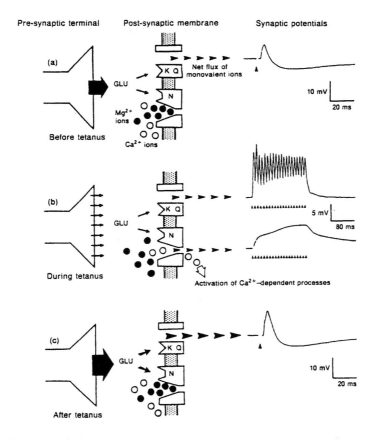

Fig. 1.6 The mechanism of induction of LTP in the CA1 region of the hippocampus. (a) Neurotransmitter (e.g. L-glutamate) is released and acts upon both K–Q (kainate–quisqualate) and NMDA (N) receptors. The NMDA receptors are blocked by magnesium and the excitatory synaptic response (EPSP) is therefore mediated primarily by ion flow through the channels associated with K–Q receptors. (b) During high-frequency activation, the magnesium block of the ion channels associated with NMDA receptors is released by depolarization. Activation of the NMDA receptor by transmitter now results in ions moving through the channel. In this way, calcium enters the postsynaptic region to trigger various intracellular mechanisms which eventually result in an alteration of synaptic efficacy. (c) Subsequent low-frequency stimulation results in a greater EPSP. See text for further details. (Reprinted with permission from Collingridge and Bliss, 1987.)

to the postsynaptic cell to mediate LTP. One possibility is that somehow a messenger reaches the presynaptic terminals from the postsynaptic membrane and, if the terminals are active, causes them to release more transmitter in future whenever they are activated by an action potential. Consistent with this possibility is the observation that, after LTP has been induced, more transmitter appears to be released from the presynaptic endings. Another possibility is that the postsynaptic membrane changes just where Ca^{2+} has entered, so that K–Q receptors become more responsive to glutamate released in future. Consistent with this possibility is the observation that after LTP, the postsynaptic cell may respond more to locally applied glutamate (using a microiontophoretic technique).

The rule that underlies associative LTP is thus that synapses connecting two neurons become stronger if there is conjunctive presynaptic and (strong) postsynaptic activity. This learning rule for synaptic modification is sometimes called the Hebb rule, after Donald Hebb of McGill University who drew attention to this possibility, and its potential importance in learning, in 1949.

In that LTP is long–lasting, develops rapidly, is synapse–specific, and is in some cases associative, it is of interest as a potential synaptic mechanism underlying some forms of memory. Evidence linking it directly to some forms of learning comes from experiments in which it has been shown that the drug AP5 infused so that it reaches the hippocampus to block NMDA receptors blocks spatial learning mediated by the hippocampus (see Morris (1989), Martin, Grimwood and Morris (2000)). The task learned by the rats was to find the location relative to cues in a room of a platform submerged in an opaque liquid (milk). Interestingly, if the rats had already learned where the platform was, then the NMDA infusion did not block performance of the task. This is a close parallel to LTP, in that the learning, but not the subsequent expression of what had been learned, was blocked by the NMDA antagonist AP5. Although there is still some uncertainty about the experimental evidence that links LTP to learning (see for example Martin, Grimwood and Morris (2000)), there is a need for a synapse-specific modifiability of synaptic strengths on neurons if neuronal networks are to learn (see Chapter 7). If LTP is not always an exact model of the synaptic modification that occurs during learning, then something with many of the properties of LTP is nevertheless needed, and is likely to be present in the brain given the functions known to be implemented in many brain regions (see Rolls and Treves (1998)).

In another model of the role of LTP in memory, Davis (2000) has studied the role of the amygdala in learning associations to fear-inducing stimuli. He has shown that blockade of NMDA synapses in the amygdala interferes with this type of learning, consistent with the idea that LTP also provides a useful model of this type of learning (see further Chapter 12).

Long-Term Depression (LTD) can also occur. It can in principle be associative or non-associative. In associative LTD, the alteration of synaptic strength depends on the pre- and post-synaptic activities. There are two types. Heterosynaptic LTD occurs when the postsynaptic neuron is strongly activated, and there is low presynaptic activity (see Fig. 1.5b input B, and Table 7.1). Heterosynaptic LTD is so-called because the synapse that weakens is other than (hetero-) the one through which the postsynaptic neuron is activated. Heterosynaptic LTD is important in associative neuronal networks, and in competitive neuronal networks (see Chapter 7). In competitive neural networks it would be helpful if the degree of heterosynaptic LTD depended on the existing strength of the synapse, and there is some evidence that this may be the case (see Chapter 7). Homosynaptic LTD occurs when the presynaptic neuron is strongly active, and the postsynaptic neuron has some, but low, activity (see Fig. 1.5d and Table 7.1). Homosynaptic LTD is so-called because the synapse that weakens is the same as (homo-) the one that is active. Heterosynaptic and homosynaptic LTD are found in the neocortex (Artola and Singer 1993, Singer 1995, Frégnac 1996) and hippocampus (Christie 1996), and in many cases are dependent on activation of NMDA receptors (see also Fazeli and Collingridge (1996)). LTD in the cerebellum is evident as weakening of active parallel fibre to Purkinje cell synapses when the climbing fibre connecting to a Purkinje cell is active (Ito 1984, Ito 1989, Ito 1993b, Ito 1993a).

An interesting time–dependence of LTP and LTD has been observed, with LTP occurring especially when the presynaptic spikes precede by a few ms the post–synaptic activation, and LTD occurring when the pre–synaptic spikes follow the post–synaptic activation by a few ms

(Markram, Lübke, Frotscher and Sakmann 1997, Bi and Poo 1998). This type of temporally asymmetric Hebbian learning rule, demonstrated in the neocortex and the hippocampus, can induce associations over time, and not just between simultaneous events. Networks of neurons with such synapses can learn sequences (Minai and Levy 1993), enabling them to predict the future state of the postsynaptic neuron based on past experience (Abbott and Blum 1996) (see further Koch (1999), Markram, Pikus, Gupta and Tsodyks (1998) and Abbott and Nelson (2000)). This mechanism, because of its apparent time–specificity for periods in the range of tens of ms, could also encourage neurons to learn to respond to temporally synchronous pre–synaptic firing (Gerstner, Kreiter, Markram and Herz 1997), and indeed to decrease the synaptic strengths from neurons that fire at random times with respect to the synchronized group. This mechanism might also play a role in the normalization of the strength of synaptic connection strengths onto a neuron. Under the somewhat steady state conditions of the firing of neurons in the higher parts of the ventral visual system on the 10 ms timescale that are observed not only when single stimuli are presented for 500 ms (see Fig. 5.5), but also when macaques have found a search target and are looking at it (in the experiments described in Section 5.6.3), the average of the pre-synaptic and postsynaptic rates are likely to be the important determinants of synaptic modification. Part of the reason for this is that correlations between the firing of simultaneously recorded inferior temporal cortex neurons are not common, and if present are not very strong or typically restricted to a short time window in the order of 10 ms (see Section B.3.4). This point is also made in the context that each neuron has thousands of inputs, several tens of which are normally likely to be active when a cell is firing above its spontaneous firing rate and is strongly depolarized. This may make it unlikely statistically that there will be a strong correlation between a particular pre-synaptic spike and postsynaptic firing, and thus that this is likely to be a main determinant of synaptic strength under these natural conditions.

1.6 Distributed representations

When considering the operation of many neuronal networks in the brain, it is found that many useful properties arise if each input to the network (arriving on the axons as **x**) is encoded in the activity of an ensemble or population of the axons or input lines (distributed encoding), and is not signalled by the activity of a single input, which is called local encoding. We start off with some definitions, and then highlight some of the differences, and summarize some evidence that shows the type of encoding used in some brain regions. Then in Chapter 7 (e.g. Table 7.2), we show how many of the useful properties of the neuronal networks described depend on distributed encoding. In Chapter 5, we review evidence on the encoding actually found in visual cortical areas.

1.6.1 Definitions

A *local representation* is one in which all the information that a particular stimulus or event occurred is provided by the activity of one of the neurons. In a famous example, a single neuron might be active only if one's grandmother was being seen. An implication is that most neurons in the brain regions where objects or events are represented would fire only very rarely. A problem with this type of encoding is that a new neuron would be needed for every object or event that has to be represented. There are many other disadvantages of this type of encoding, many of which will become apparent in this book. Moreover, there is evidence that objects are represented in the brain by a different type of encoding.

A *fully distributed representation* is one in which all the information that a particular stimulus or event occurred is provided by the activity of the full set of neurons. If the neurons are binary (e.g. either active or not), the most distributed encoding is when half the neurons are active for any one stimulus or event.

A *sparse distributed representation* is a distributed representation in which a small proportion of the neurons is active at any one time. In a sparse representation with binary neurons, less than half of the neurons are active for any one stimulus or event. For binary neurons, we can use as a measure of the sparseness the proportion of neurons in the active state. For neurons with real, continuously variable, values of firing rates, the sparseness a of the representation can be measured, by extending the binary notion of the proportion of neurons that are firing, as

$$a = \frac{(\sum\limits_{i=1}^{N} y_i/N)^2}{\sum\limits_{i=1}^{N} y_i^2/N} \tag{1.4}$$

where y_i is the firing rate of the ith neuron in the set of N neurons (Treves and Rolls 1991).

Coarse coding utilizes overlaps of receptive fields, and can compute positions in the input space using differences between the firing levels of coactive cells (e.g. colour-tuned cones in the retina). The representation implied is very distributed. Fine coding (in which for example a neuron may be 'tuned' to the exact orientation and position of a stimulus) implies more local coding.

1.6.2   Advantages of different types of coding

One advantage of distributed encoding is that the similarity between two representations can be reflected by the correlation between the two patterns of activity that represent the different stimuli. We have already introduced the idea that the input to a neuron is represented by the activity of its set of input axons x_j, where j indexes the axons, numbered from $j = 1, C$ (see Fig. 1.2 and equation 1.1). Now the set of activities of the input axons is a vector (a vector is an ordered set of numbers; Appendix 1 of Rolls and Treves (1998) provides a summary of some of the concepts involved). We can denote as $\mathbf{x}^1$ the vector of axonal activity that represents stimulus 1, and $\mathbf{x}^2$ the vector that represents stimulus 2. Then the similarity between the two vectors, and thus the two stimuli, is reflected by the correlation between the two vectors. The correlation will be high if the activity of each axon in the two representations is similar; and will become more and more different as the activity of more and more of the axons differs in the two representations. Thus the similarity of two inputs can be represented in a graded or continuous way if (this type of) distributed encoding is used. This enables generalization to similar stimuli, or to incomplete versions of a stimulus (if it is for example partly seen or partly remembered), to occur. With a local representation, either one stimulus or another is represented, and similarities between different stimuli are not encoded.

Another advantage of distributed encoding is that the number of different stimuli that can be represented by a set of C components (e.g. the activity of C axons) can be very large. A simple example is provided by the binary encoding of an 8-element vector. One component can code for which of two stimuli has been seen, 2 components (or bits in a computer byte) for 4 stimuli, 3 components for 8 stimuli, 8 components for 256 stimuli, etc. That is, the number of stimuli increases exponentially with the number of components (or in this case, axons) in the representation. (In this simple binary illustrative case, the number of stimuli that can be encoded is 2^C.) Put the other way round, even if a neuron has only a limited number of inputs (e.g. a few thousand), it can nevertheless receive a great deal of information about which stimulus was present. This ability of a neuron with a limited number of inputs to receive information about which of potentially very many input events is present is probably one factor that makes computation by the brain possible. With local encoding, the number of stimuli that can be encoded increases only linearly with the number C of axons or components (because a different component is needed to represent each new stimulus). (In our example, only 8 stimuli could be represented by 8 axons.)

In the real brain, there is now good evidence that in a number of brain systems, including the high-order visual and olfactory cortices, and the hippocampus, distributed encoding with the above two properties, of representing similarity, and of exponentially increasing encoding capacity as the number of neurons in the representation increases, is found (Rolls and Tovee 1995b, Abbott, Rolls and Tovee 1996, Rolls, Treves and Tovee 1997b, Rolls, Treves, Robertson, Georges-François and Panzeri 1998). For example, in the primate inferior temporal visual cortex, the number of faces or objects that can be represented increases approximately exponentially with the number of neurons in the population (see Chapter 5). If we consider instead the information about which stimulus is seen, we see that this rises approximately linearly with the number of neurons in the representation (see Chapter 5). This corresponds to an exponential rise in the number of stimuli encoded, because information is a log measure (see Appendix B). A similar result has been found for the encoding of position in space by the primate hippocampus (Rolls, Treves, Robertson, Georges-François and Panzeri 1998). It is particularly important that the information can be read from the ensemble of neurons using a simple measure of the similarity of vectors, the correlation (or dot product, see Appendix A) between two vectors. The importance of this is that it is essentially vector similarity operations that characterize the operation of many neuronal networks (see Chapter 7). The neurophysiological results show that both the ability to reflect similarity by vector correlation, and the utilization of exponential coding capacity, are a property of real neuronal networks found in the brain.

To emphasize one of the points being made here, although the binary encoding used in the 8-bit vector described above has optimal capacity for binary encoding, it is not optimal for vector similarity operations. For example, the two very similar numbers 127 and 128 are represented by 01111111 and 10000000 with binary encoding, yet the correlation or bit overlap of these vectors is 0. The brain in contrast uses a code that has the attractive property of exponentially increasing capacity with the number of neurons in the representation, though it is different from the simple binary encoding of numbers used in computers; and at the same time the brain codes stimuli in such a way that the code can be read off with simple dot product or correlation-related decoding, which is what is specified for the elementary neuronal network operation shown in equation 1.2 (see Section 5.5).

1.7   Neuronal network approaches versus connectionism

The approach taken in this book is to introduce how real neuronal networks in the brain may compute, and thus to achieve a fundamental and realistic basis for understanding brain function. This may be contrasted with connectionism, which aims to understand cognitive function by analysing processing in neuron-like computing systems. Connectionist systems are neuron–like in that they analyze computation in systems with large numbers of computing elements in which the information which governs how the network computes is stored in the connection strengths between the nodes (or "neurons") in the network. However, in many connectionist models the individual units or nodes are not intended to model individual neurons, and the variables that are used in the simulations are not intended to correspond to quantities that can be measured in the real brain. Moreover, connectionist approaches use learning rules in which the synaptic modification (the strength of the connections between the nodes) is determined by algorithms that require information that is not local to the synapse, that is, evident in the pre- and post-synaptic firing rates (see further Chapter 7). Instead, in many connectionist systems, information about how to modify synaptic strengths is propagated backwards from the output of the network to affect neurons hidden deep within the network (see Section 7.11). Because it is not clear that this is biologically plausible, we have instead in this text concentrated on introducing neuronal network architectures which are more biologically plausible, and which use a local learning rule. Connectionist approaches (see for example McClelland and Rumelhart (1986), McLeod, Plunkett and Rolls (1998)) are very valuable, for they show what can be achieved computationally with networks in which the connection strength determines the computation that the network achieves with quite simple computing elements. However, as models of brain function, many connectionist networks achieve almost too much, by solving problems with a carefully limited number of "neurons" or nodes, which contributes to the ability of such networks to generalize successfully over the problem space. Connectionist schemes thus make an important start on understanding how complex computations (such as language) could be implemented in brain-like systems. In doing this, connectionist models often use simplified representations of the inputs and outputs, which are often crucial to the way in which the problem is solved. In addition, they may use learning algorithms that are really too powerful for the brain to perform, and therefore they can be taken only as a guide to how cognitive functions might be implemented by neuronal networks in the brain. In this book, we focus on more biologically plausible neuronal networks.

1.8   Introduction to three neuronal network architectures

With neurons of the type outlined in Section 1.3, and an associative learning rule of the type described in Section 1.4, three neuronal network architectures arise that appear to be used in many different brain regions. The three architectures will be described in Chapter 7, and a brief introduction is provided here.

In the first architecture (see Fig. 1.7a and b), pattern associations can be learned. The output neurons are driven by an unconditioned stimulus. A conditioned stimulus reaches the output neurons by associatively modifiable synapses w_{ij}. If the conditioned stimulus is paired during learning with activation of the output neurons produced by the unconditioned stimulus, then later, after learning, due to the associative synaptic modification, the conditioned stimulus alone will produce the same output as the conditioned stimulus. Pattern associators are described in Chapter 7.

In the second architecture, the output neurons have recurrent associatively modifiable synaptic connections w_{ij} to other neurons in the network (see Fig. 1.7c). When an external

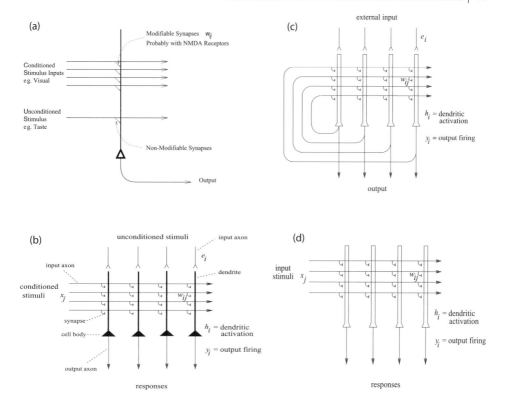

Fig. 1.7 Three network architectures that use local learning rules: (a) Pattern association introduced with a single output neuron; (b) Pattern association network; (c) Autoassociation network; (d) Competitive network.

input causes the output neurons to fire, then associative links are formed through the modifiable synapses that connect the set of neurons that is active. Later, if only a fraction of the original input pattern is presented, then the associative synaptic connections or weights allow the whole of the memory to be retrieved. This is called completion. Because the components of the pattern are associated with each other as a result of the associatively modifiable recurrent connections, this is called an autoassociative memory. It is believed to be used in the brain for many purposes, including episodic memory, in which the parts of a memory of an episode are associated together; and helping to define the response properties of cortical neurons, which have collaterals between themselves within a limited region.

In the third architecture, the main input to the output neurons is received through associatively modifiable synapses w_{ij} (see Fig. 1.7d). Because of the initial values of the synaptic strengths, or because every axon does not contact every output neuron, different patterns tend to activate different output neurons. When one pattern is being presented, the most strongly activated neurons tend via lateral inhibition to inhibit the other neurons. For this reason the network is called competitive. During the presentation of that pattern, associative modification of the active axons onto the active postsynaptic neuron takes place. Later, that or similar patterns will have a greater chance of activating that neuron or set of neurons. Other neurons learn to respond to other input patterns. In this way, a network is built that can categorize

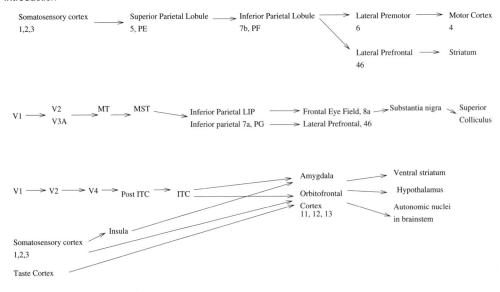

Fig. 1.8 Schematic diagram showing the major connection of three separate processing systems in the brain (see text). The top pathway shows the connections from the primary somatosensory cortex, areas 1, 2 and 3, via area 5 in the parietal cortex, to area 7b. The middle pathway, also shown in Fig. 1.11, shows the connections in the 'dorsal or where visual pathway' from V1 to V2, MT, MST, 7a etc, with some connections reaching the dorsolateral prefrontal cortex and frontal eye fields. The lower pathway, also shown in Fig. 1.9, shows the connections in the 'ventral or what visual pathway' from V1 to V2, V4, the inferior temporal visual cortex, etc., with some connections reaching the amygdala and orbitofrontal cortex.

patterns, placing similar patterns into the same category. This is useful as a preprocessor for sensory information, self-organizes to produce feature analyzers, and finds uses in many other parts of the brain too (see Chapter 12).

These are three fundamental building blocks for neural architectures in the brain. They are often used in combination with each other. Because they are some of the building blocks of some of the architectures found in the brain, they are described in Chapter 7.

1.9   Systems-level analysis of brain function

To understand the neuronal network operations of any one brain region, it is useful to have an idea of the systems-level organization of the brain, in order to understand how the networks in each region provide a particular computational function as part of an overall computational scheme. In the context of vision, it is very useful to appreciate the different processing streams, and some of the outputs that each has. Some of the processing streams are shown in Fig. 1.8. Some of these regions are shown in the drawings of the primate brain in the next few Figures. Each of these routes is described in turn. The description is based primarily on studies in non-human primates, for they have well-developed cortical areas that in many cases correspond to those found in humans, and it has been possible to analyze their connectivity and their functions by recording the activity of neurons in them.

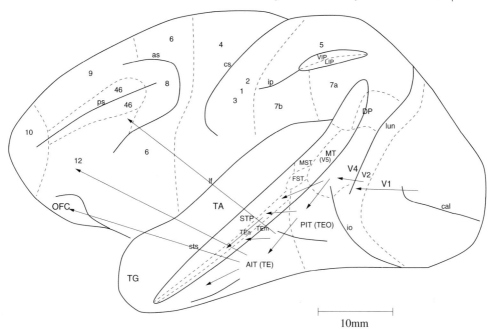

Fig. 1.9 Lateral view of the macaque brain showing the connections in the 'ventral or what visual pathway' from V1 to V2, V4, the inferior temporal visual cortex, etc., with some connections reaching the amygdala and orbitofrontal cortex. as, arcuate sulcus; cal, calcarine sulcus; cs, central sulcus; lf, lateral (or Sylvian) fissure; lun, lunate sulcus; ps, principal sulcus; io, inferior occipital sulcus; ip, intraparietal sulcus (which has been opened to reveal some of the areas it contains); sts, superior temporal sulcus (which has been opened to reveal some of the areas it contains). AIT, anterior inferior temporal cortex; FST, visual motion processing area; LIP, lateral intraparietal area; MST, visual motion processing area; MT, visual motion processing area (also called V5); OFC, orbitofrontal cortex; PIT, posterior inferior temporal cortex; STP, superior temporal plane; TA, architectonic area including auditory association cortex; TE, architectonic area including high order visual association cortex, and some of its subareas TEa and TEm; TG, architectonic area in the temporal pole; V1 – V4, visual areas 1 – 4; VIP, ventral intraparietal area; TEO, architectonic area including posterior visual association cortex. The numerals refer to architectonic areas, and have the following approximate functional equivalence: 1,2,3, somatosensory cortex (posterior to the central sulcus); 4, motor cortex; 5, superior parietal lobule; 7a, inferior parietal lobule, visual part; 7b, inferior parietal lobule, somatosensory part; 6, lateral premotor cortex; 8, frontal eye field; 12, inferior convexity prefrontal cortex; 46, dorsolateral prefrontal cortex.

Information in the *'ventral or what'* visual cortical processing stream projects after the primary visual cortex, area V1, to the secondary visual cortex (V2), and then via area V4 to the posterior and then to the anterior inferior temporal visual cortex (lower stream in Fig. 1.8; Fig. 1.9; Fig. 1.10).

Information processing along this stream is primarily unimodal, as shown by the fact that inputs from other modalities (such as taste or smell) do not anatomically have significant inputs to these regions, and by the fact that neurons in these areas respond primarily to visual stimuli, and not to taste or olfactory stimuli, etc (Rolls 2000a, Baylis, Rolls and Leonard 1987, Ungerleider 1995). The representation built along this pathway is mainly about what object is being viewed, independently of exactly where it is on the retina, of its size, and even of the angle with which it is viewed (see Chapter 5), and for this reason it is frequently referred to as the 'what' visual pathway. The representation is also independent of whether the object is associated with reward or punishment, that is the representation is about objects per se (Rolls, Judge and Sanghera 1977). The computation that must be performed along this stream is thus primarily to build a representation of objects that shows invariance. After this processing, the visual representation is interfaced to other sensory systems in areas in which simple associations must be learned between stimuli in different modalities (see Chapter 12). The representation must thus be in a form in which the simple generalization properties of associative networks can be useful. Given that the association is about what object is present (and not where it is on the retina), the representation computed in sensory systems must be in a form that allows the simple correlations computed by associative networks to reflect similarities between objects, and not between their positions on the retina. The way in which such invariant sensory representations could be built in the brain is the subject of Chapter 8.

The ventral visual stream converges with other mainly unimodal information processing streams for taste, olfaction, touch, and hearing in a number of areas, particularly the amygdala and orbitofrontal cortex (see Figs. 1.8, 1.9, and 1.10). These areas appear to be necessary for learning to associate sensory stimuli with other reinforcing (rewarding or punishing) stimuli. For example, the amygdala is involved in learning associations between the sight of food and its taste. (The taste is a primary or innate reinforcer.) The orbitofrontal cortex is especially involved in rapidly relearning these associations, when environmental contingencies change (see Rolls (1999a), Rolls (2000e)). They thus are brain regions in which the computation at least includes simple pattern association (e.g. between the sight of an object and its taste). In the orbitofrontal cortex, this association learning is also used to produce a representation of flavour, in that neurons are found in the orbitofrontal cortex that are activated by both olfactory and taste stimuli (Rolls and Baylis 1994), and in that the neuronal responses in this region reflect in some cases olfactory to taste association learning (Rolls, Critchley, Mason and Wakeman 1996b, Critchley and Rolls 1996). In these regions too, the representation is concerned not only with what sensory stimulus is present, but for some neurons, with its hedonic or reward-related properties, which are often computed by association with stimuli in other modalities. For example, many of the visual neurons in the orbitofrontal cortex respond to the sight of food only when hunger is present. This probably occurs because the visual inputs here have been associated with a taste input, which itself in this region only occurs to a food if hunger is present, that is when the taste is rewarding (see Chapter 12 and Rolls (1999a), Rolls (2000e)). The outputs from these associative memory systems, the amygdala and orbitofrontal cortex, project onwards to structures such as the hypothalamus, through which they control autonomic and endocrine responses such as salivation and insulin release to the sight of food; and to the striatum, including the ventral striatum, through which behaviour to learned reinforcing stimuli is produced.

The *'dorsal or where' visual processing stream* shown in Figs. 1.8, 1.11, and 1.10 is that from V1 to MT, MST and thus to the parietal cortex (see Ungerleider (1995); Ungerleider and Haxby (1994); Section 10.5). This 'where' pathway for primate vision is involved in representing where stimuli are relative to the animal (i.e. in egocentric space), and the motion of these stimuli. Neurons here respond for example to stimuli in visual space around the animal, including the distance from the observer, and also respond to optic flow or to moving stimuli. Outputs of this system control eye movements to visual stimuli (both slow pursuit and

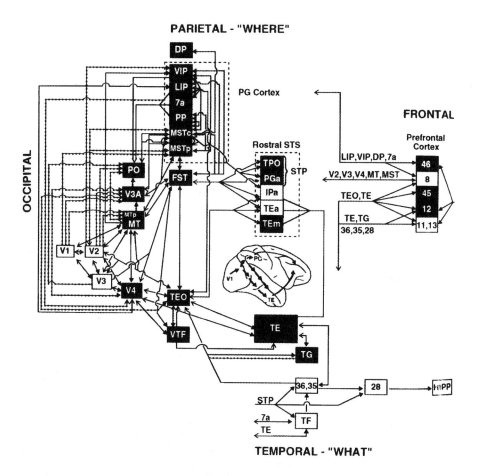

Fig. 1.10 Visual processing pathways in monkeys. Solid lines indicate connections arising from both central and peripheral visual field representations; dotted lines indicate connections restricted to peripheral visual field representations. Shaded boxes in the 'ventral (lower) or what' stream indicate visual areas related primarily to object vision; shaded boxes in the 'dorsal or where' stream indicate areas related primarily to spatial vision; and white boxes indicate areas not clearly allied with only one stream. Abbreviations: DP, dorsal prelunate area; FST, fundus of the superior temporal area; HIPP, hippocampus; LIP, lateral intraparietal area; MSTc, medial superior temporal area, central visual field representation; MSTp, medial superior temporal area, peripheral visual field representation; MT, middle temporal area; MTp, middle temporal area, peripheral visual field representation; PO, parieto-occipital area; PP, posterior parietal sulcal zone; STP, superior temporal polysensory area; V1, primary visual cortex; V2, visual area 2; V3, visual area 3; V3A, visual area 3, part A; V4, visual area 4; and VIP, ventral intraparietal area. Inferior parietal area 7a; prefrontal areas 8, 11 to 13, 45 and 46 are from Brodmann (1925). Inferior temporal areas TE and TEO, parahippocampal area TF, temporal pole area TG, and inferior parietal area PG are from Von Bonin and Bailey (1947). Rostral superior temporal sulcal (STS) areas are from Seltzer and Pandya (1978) and VTF is the visually responsive portion of area TF (Boussaoud, Desimone and Ungerleider, 1991). (Reprinted with permission from Ungerleider, 1995.)

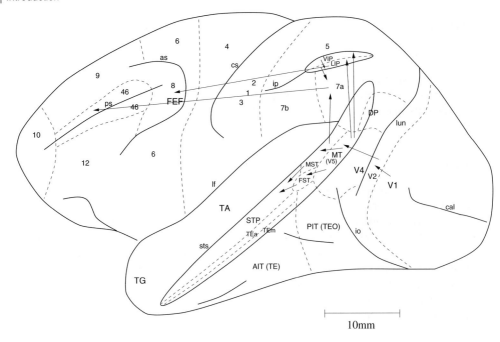

Fig. 1.11 Lateral view of the macaque brain showing the connections in the 'dorsal or where visual pathway' from V1 to V2, MST, LIP, VIP, and parietal cortex area 7a, with some connections then reaching the dorsolateral prefrontal cortex. Abbreviations as in Fig. 1.9. FEF - frontal eye field.

saccadic eye movements). These outputs proceed partly via the frontal eye fields, which then project to the striatum, and then via the substantia nigra reach the superior colliculus (see Fig. 1.8) (Goldberg 2000). Other outputs of these regions are to the dorsolateral prefrontal cortex, area 46, which is important as a short term memory for where fixation should occur next, as shown by the effects of lesions to the prefrontal cortex on saccades to remembered targets, and by neuronal activity in this region (Goldman-Rakic 1996). The dorsolateral prefrontal cortex short term memory systems in area 46 with spatial information received from the parietal cortex play an important role in attention, by holding on-line the target being attended to, as described in Chapters 9–11 and 13.

The hippocampus receives inputs from both the 'what' and the 'where' visual systems (see Section 12.2 and Rolls and Treves (1998)). By rapidly learning associations between conjunctive inputs in these systems, it is able to form memories of particular events occurring in particular places at particular times. To do this, it needs to store whatever is being represented in each of many cortical areas at a given time, and to later recall the whole memory from a part of it. The types of network it contains that are involved in this simple memory function are described in Chapter 6 of Rolls and Treves (1998) and summarized in Section 12.2.

1.10 Introduction to the fine structure of the cerebral neocortex

An important part of the approach to understanding how the cerebral cortex could implement the computational processes that underlie visual perception is to take into account as much as possible its fine structure and connectivity, as these provide important indicators of and constraints on how it computes.

1.10.1 The fine structure and connectivity of the neocortex

The neocortex consists of many areas that can be distinguished by the appearance of the cells (cytoarchitecture) and fibres or axons (myeloarchitecture), but nevertheless, the basic organization of the different neocortical areas has many similarities, and it is this basic organization that is considered here. Useful sources for more detailed descriptions of neocortical structure and function are the book 'Cerebral Cortex' edited by Jones and Peters (Jones and Peters (1984) and Peters and Jones (1984)); and Douglas and Martin (1990). Approaches to quantitative aspects of the connectivity are provided by Braitenberg and Schuz (1991) and by Abeles (1991). Some of the connections described in Sections 1.10.2 and 1.10.3 are shown schematically in Fig. 1.13.

1.10.2 Excitatory cells and connections

Some of the cell types found in the neocortex are shown in Fig. 1.1. Cells A–D are pyramidal cells. The dendrites (shown thick in Fig. 1.1) are covered in spines, which receive the excitatory synaptic inputs to the cell. Pyramidal cells with cell bodies in different laminae of the cortex (shown in Fig. 1.1 as I–VI) not only have different distributions of their dendrites, but also different distributions of their axons (shown thin in Fig. 1.13), which connect both within that cortical area and to other brain regions outside that cortical area (see labelling at the bottom of Fig. 1.13).

The main information-bearing afferents to a cortical area have many terminals in layer 4. (By these afferents, we mean primarily those from the thalamus or from the preceding cortical area. We do not mean the cortico-cortical backprojections, nor the subcortical cholinergic, noradrenergic, dopaminergic, and serotonergic inputs, which are numerically minor, although they are important in setting cortical cell thresholds, excitability, and adaptation, see for example Douglas and Martin (1990).) In primary sensory cortical areas only there are spiny stellate cells in a rather expanded layer 4, and the thalamic terminals synapse onto these cells (Lund 1984, Martin 1984, Douglas and Martin 1990, Levitt, Lund and Yoshioka 1996). (Primary sensory cortical areas receive their inputs from the primary sensory thalamic nucleus for a sensory modality. An example is the primate striate cortex which receives inputs from the lateral geniculate nucleus, which in turn receives from the retinal ganglion cells. Spiny stellate cells are so-called because they have radially arranged, star-like, dendrites. Their axons usually terminate within the cortical area in which they are located.) Each thalamic axon makes 1,000–10,000 synapses, not more than several (or at most 10) of which are onto any one spiny stellate cell. In addition to these afferent terminals, there are some terminals of the thalamic afferents onto pyramidal cells with cell bodies in layers 6 and 3 (Martin 1984) (and terminals onto inhibitory interneurons such as basket cells, which thus provide for a feedforward inhibition) (see Fig. 1.12). Even in layer 4, the thalamic axons provide less than

20% of the synapses. The spiny stellate neurons in layer 4 have axons which terminate in layers 3 and 2, at least partly on dendrites of pyramidal cells with cell bodies in layers 3 and 2. (These synapses are of Type I, that is are asymmetrical and are on spines, so that they are probably excitatory. Their transmitter is probably glutamate.) These layer 3 and 2 pyramidal cells provide the onward cortico-cortical projection, with axons which project into layer 4 of the next cortical area. For example, layer 3 and 2 pyramidal cells in the primary visual (striate) cortex of the macaque monkey project into the second visual area (V2), layer 4.

In non-primary sensory areas, important information-bearing afferents from a preceding cortical area terminate in layer 4, but there are no or few spiny stellate cells in this layer (Lund 1984, Levitt, Lund and Yoshioka 1996). Layer 4 still looks 'granular' (due to the presence of many small cells), but these cells are typically small pyramidal cells (Lund 1984). (It may be noted here that spiny stellate cells and small pyramidal cells are similar in many ways, with a few main differences including the absence of a major apical dendrite in a spiny stellate which accounts for its non-pyramidal, star-shaped, appearance; and for many spiny stellate cells, the absence of an axon that projects outside its cortical area.) The terminals presumably make synapses with these small pyramidal cells, and also presumably with the dendrites of cells from other layers, including the basal dendrites of deep layer 3 pyramidal cells (see Fig. 1.13).

The axons of the *superficial (layer 2 and 3) pyramidal cells* have collaterals and terminals in layer 5 (see Fig. 1.13), and synapses are made with the dendrites of the layer 5 pyramidal cells (Martin 1984). The axons also typically project out of that cortical area, and on to the next cortical area in sequence, where they terminate in layer 4, forming the forward cortico-cortical projection. It is also from these pyramidal cells that projections to the amygdala arise in some sensory areas that are high in the hierarchy (Amaral, Price, Pitkanen and Carmichael 1992).

The axons of the *layer 5 pyramidal cells* have many collaterals in layer 6 (see Fig. 1.1), where synapses could be made with the layer 6 pyramidal cells (based on indirect evidence, see Fig. 13 of Martin (1984)), and axons of these cells typically leave the cortex to project to subcortical sites (such as the striatum), or back to the preceding cortical area to terminate in layer 1. It is remarkable that there are as many of these backprojections as there are forward connections between two sequential cortical areas. The possible computational significance of this connectivity is considered below in Section 1.11.

The *layer 6 pyramidal cells* have prolific dendritic arborizations in layer 4 (see Fig. 1.1), and receive synapses from thalamic afferents (Martin 1984), and also presumably from pyramidal cells in other cortical layers. The axons of these cells form backprojections to the thalamic nucleus which projects into that cortical area, and also axons of cells in layer 6 contribute to the backprojections to layer 1 of the preceding cortical area (see Jones and Peters (1984) and Peters and Jones (1984); see Figs. 1.1 and 1.13).

Although the pyramidal and spiny stellate cells form the great majority of neocortical neurons with excitatory outputs, there are in addition several further cell types (see Peters and Jones (1984), Chapter 4). Bipolar cells are found in layers 3 and 5, and are characterized by having two dendritic systems, one ascending and the other descending, which, together with the axon distribution, are confined to a narrow vertical column often less than 50 μm in diameter (Peters 1984a). Bipolar cells form asymmetrical (presumed excitatory) synapses with pyramidal cells, and may serve to emphasize activity within a narrow vertical column.

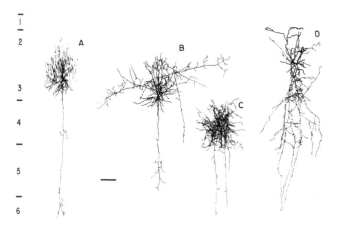

Fig. 1.12 Smooth cells from cat visual cortex. (A) Chandelier or axoaxonic cell. (B) Large basket cell of layer 3. Basket cells, present in layers 3–6, have few spines on their dendrites so that they are described as smooth, and have an axon which participates in the formation of weaves of preterminal axons which surround the cell bodies of pyramidal cells and form synapses directly onto the cell body. (C) Small basket or clutch cell of layer 3. The major portion of the axonal arbor is confined to layer 4. (D) Double bouquet cell. The axon collaterals run vertically. The cortical layers are as indicated. Bar = 100 μm. (Reproduced with permission from Douglas and Martin, 1990, Fig. 12.4.)

1.10.3 Inhibitory cells and connections

There are a number of types of neocortical inhibitory neurons. All are described as smooth in that they have no spines, and use GABA (gamma-amino-butyric acid) as a transmitter. (In older terminology they were called Type II.) A number of types of inhibitory neuron can be distinguished, best by their axonal distributions (Szentagothai 1978, Peters and Regidor 1981, Douglas and Martin 1990). One type is the *basket cell*, present in layers 3–6, which has few spines on its dendrites so that it is described as smooth, and has an axon that participates in the formation of weaves of preterminal axons which surround the cell bodies of pyramidal cells and form synapses directly onto the cell body, but also onto the dendritic spines (Somogyi, Kisvarday, Martin and Whitteridge 1983) (Fig. 1.12). Basket cells comprise 5–7% of the total cortical cell population, compared with approximately 72% for pyramidal cells (Sloper and Powell 1979b, Sloper and Powell 1979a). Basket cells receive synapses from the main extrinsic afferents to the neocortex, including thalamic afferents (Fig. 1.12), so that they must contribute to a feedforward type of inhibition of pyramidal cells. The inhibition is feedforward in that the input signal activates the basket cells and the pyramidal cells by independent routes, so that the basket cells can produce inhibition of pyramidal cells that does not depend on whether the pyramidal cells have already fired. Feedforward inhibition of this type not only enhances stability of the system by damping the responsiveness of the pyramidal cell simultaneously with a large new input, but can also be conceived of as a mechanism which normalizes the magnitude of the input vector received by each small region of neocortex (see further Chapters 7 and 8). In fact, the feedforward mechanism allows the pyramidal cells to be set at the appropriate sensitivity for the input they are about to receive. Basket cells can also be polysynaptically activated by an afferent volley in the thalamo-cortical projection (Martin 1984), so that they may receive inputs from pyramidal cells, and thus participate in feedback inhibition of pyramidal cells.

The transmitter used by the basket cells is gamma-amino-butyric acid (GABA), which opens chloride channels in the postsynaptic membrane. Because the reversal potential for Cl^- is approximately -10 mV relative to rest, opening the Cl^- channels does produce an inhibitory postsynaptic potential (IPSP), which results in some hyperpolarization, especially in the dendrites. This is a subtractive effect, hence it is a linear type of inhibition (Douglas and Martin 1990). However, a major effect of the opening of the Cl^- channels in the cell body is that this decreases the membrane resistance, thus producing a shunting effect. The importance of shunting is that it decreases the magnitude of excitatory postsynaptic potentials (EPSPs) (cf. Andersen, Dingledine, Gjerstad, Langmoen and Laursen (1980) for hippocampal pyramidal cells), so that the effect of shunting is to produce division (i.e. a multiplicative reduction) of the excitatory inputs received by the cell, and not just to act by subtraction (see further Bloomfield (1974), Martin (1984), Douglas and Martin (1990)). Thus, when modelling the normalization of the activity of cortical pyramidal cells, it is common to include division in the normalization function (cf. Chapter 7). It is notable that the dendrites of basket cells can extend laterally 0.5 mm or more (primarily within the layer in which the cell body is located), and that the axons can also extend laterally from the cell body 0.5–1.5 mm. Thus the basket cells produce a form of lateral inhibition which is quite spatially extensive. There is some evidence that each basket cell may make 4–5 synapses with a given pyramidal cell, that each pyramidal cell may receive from 10–30 basket cells, and that each basket cell may inhibit approximately 300 pyramidal cells (Martin 1984, Douglas and Martin 1990). The basket cells are sometimes called clutch cells.

A second type of GABA-containing inhibitory interneuron is the *axoaxonic (or 'chandelier') cell*, named because it synapses onto the initial segment of the axon of pyramidal cells. The pyramidal cells receiving this type of inhibition are almost all in layers 2 and 3, and much less in the deep cortical layers. One effect that axoaxonic cells probably produce is thus prevention of outputs from layer 2 and 3 pyramidal cells reaching the pyramidal cells in the deep layers, or from reaching the next cortical area. Up to five axoaxonic cells converge onto a pyramidal cell, and each axoaxonic cell may project to several hundred pyramidal cells scattered in a region that may be several hundred microns in length (see (Martin 1984, Peters 1984b)). This implies that axoaxonic cells provide a rather simple device for preventing runaway overactivity of pyramidal cells, but little is known yet about the afferents to axoaxonic cells, so that the functions of these neurons are very incompletely understood.

A third type of (usually smooth and inhibitory) cell is the *double bouquet cell*, which has primarily vertically organized axons. These cells have their cell bodies in layer 2 or 3, and have an axon traversing layers 2–5, usually in a tight bundle consisting of varicose, radially oriented collaterals often confined to a narrow vertical column 50 μm in diameter (Somogyi and Cowey 1984). Double bouquet cells receive symmetrical, type II (presumed inhibitory) synapses, and also make type II synapses, perhaps onto the apical dendrites of pyramidal cells, so that these neurons may serve, by this double inhibitory effect, to emphasize activity within a narrow vertical column.

Another type of GABA-containing inhibitory interneuron is the smooth and sparsely spinous non-pyramidal (multipolar) neuron with local axonal plexuses (Peters and Saint Marie 1984). In addition to extrinsic afferents, these neurons receive many type I (presumed excitatory) terminals from pyramidal cells, and have inhibitory terminals on pyramidal cells, so that they may provide for the very important function of feedback or recurrent lateral inhibition

Table 1.1 Typical quantitative estimates for neocortex (partly after Abeles (1991) and reflecting estimates in macaques)

Neuronal density	20,000–40,000/mm^3
Neuronal composition:	
Pyramidal	75%
Spiny stellate	10%
Inhibitory neurons, for example smooth stellate, chandelier	15%
Synaptic density	8 x 10^8/mm^3
Numbers of synapses on pyramidal cells:	
Excitatory synapses from remote sources onto each neuron	9,000
Excitatory synapses from local sources onto each neuron	9,000
Inhibitory synapses onto each neuron	2,000
Pyramidal cell dendritic length	10 mm
Number of synapses made by axons of pyramidal cells	18,000
Number of synapses on inhibitory neurons	2,000
Number of synapses made by inhibitory neurons	300
Dendritic length density	400 m/mm^3
Axonal length density	3,200 m/mm^3
Typical cortical thickness	2 mm
Cortical area	
human (assuming 3 mm for cortical thickness)	300,000 mm^2
macaque (assuming 2 mm for cortical thickness)	30,000 mm^2
rat (assuming 2 mm for cortical thickness)	300 mm^2

(see Chapter 7).

1.10.4 Quantitative aspects of cortical architecture

Some quantitative aspects of cortical architecture are described, because, although only pre-liminary data are available, they are crucial for developing an understanding of how the neocortex could work. Further evidence is provided by Braitenberg and Schuz (1991), and by Abeles (1991). Typical values, many of them after Abeles (1991), are shown in Table 1.1. The figures given are for a rather generalized case, and indicate the order of magnitude. The number of synapses per neuron (20,000) is an estimate for monkeys; those for humans may be closer to 40,000, and for the mouse, closer to 8,000. The number of 18,000 excitatory synapses made by a pyramidal cell is set to match the number of excitatory synapses received by pyramidal cells, for the great majority of cortical excitatory synapses are made from axons of cortical, principally pyramidal, cells.

Microanatomical studies show that pyramidal cells rarely make more than one connection with any other pyramidal cell, even when they are adjacent in the same area of the cerebral cortex. An interesting calculation takes the number of local connections made by a pyramidal cell within the approximately 1 mm of its local axonal arborization (say 9,000), and the number of pyramidal cells with dendrites in the same region, and suggests that the probability that a pyramidal cell makes a synapse with its neighbour is low, approximately 0.1 (Braitenberg and Schuz 1991, Abeles 1991). This fits with the estimate from simultaneous recording of nearby pyramidal cells using spike-triggered averaging to monitor time-locked EPSPs (Abeles 1991, Thomson and Deuchars 1994).

Now the implication of the pyramidal cell to pyramidal cell connectivity just described is that within a cortical area of perhaps 1 mm^2, the region within which typical pyramidal cells have dendritic trees and their local axonal arborization, there is a probability of excitatory-to-excitatory cell connection of 0.1. Moreover, this population of mutually interconnected neurons is served by 'its own' population of inhibitory interneurons (which have a spatial receiving and sending zone in the order of 1 mm^2), enabling local threshold setting and optimization of the set of neurons with 'high' (0.1) connection probability in that region. Such an architecture is effectively recurrent or re-entrant. It may be expected to show some of the properties of recurrent networks, including the fast dynamics described in Chapter 7. Such fast dynamics may be facilitated by the fact that cortical neurons in the awake behaving monkey generally have a low spontaneous rate of firing (personal observations; see for example Rolls and Tovee (1995b), and Rolls, Treves, Tovee and Panzeri (1997d)), which means that even any small additional input may produce some spikes sooner than would otherwise have occurred, because some of the neurons may be very close to a threshold for firing. It might also show some of the autoassociative retrieval of information typical of autoassociation networks, if the synapses between the nearby pyramidal cells have the appropriate (Hebbian) modifiability. In this context, the value of 0.1 for the probability of a connection between nearby neocortical pyramidal cells is of interest, for the connection probability between hippocampal CA3 pyramidal is approximately 0.02–0.04, and this is thought to be sufficient to sustain associative retrieval (see Chapter 7 and Rolls and Treves (1998)).

In the neocortex, each 1 mm^2 region within which there is a relatively high density of recurrent collateral connections between pyramidal cells, probably overlaps somewhat continuously with the next. This raises the issue of modules in the cortex, described by many authors as regions of the order of 1 mm^2 (with different authors giving different sizes), in which there are vertically oriented columns of neurons that may share some property (for example, responding to the same orientation of a visual stimulus), and that may be anatomically marked (for example Powell (1981), Mountcastle (1984), see Douglas, Mahowald and Martin (1996)). The anatomy just described, with the local connections between nearby (1 mm) pyramidal cells, and the local inhibitory neurons, may provide a network basis for starting to understand the columnar architecture of the neocortex, for it implies that local recurrent connectivity on this scale implementing local re-entrancy is a feature of cortical computation. We can note that the neocortex could not be a single, global, autoassociation network, because the number of memories that could be stored in an autoassociation network, rather than increasing with the number of neurons in the network, is limited by the number of recurrent connections per neuron, which is in the order of 10,000 (see Table 1.1), or less, depending on the species, as pointed out by O'Kane and Treves (1992). This would be an impossibly small capacity for the whole cortex. It is suggested that instead a principle of cortical design is that it does have in part local connectivity, so that each part can have its own processing and storage, which may be triggered by other modules, but is a distinct operation from that which occurs simultaneously in other modules.

An interesting parallel between the hippocampus and any small patch of neocortex is the allocation of a set of many small excitatory (usually non-pyramidal, spiny stellate or granular)

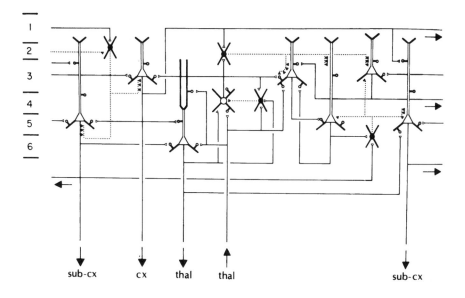

Fig. 1.13 Basic circuit for visual cortex. Excitatory neurons, which are spiny and use glutamate as a transmitter, and include the pyramidal and spiny stellate cells, are indicated by open somata; their axons are indicated by solid lines, and their synaptic boutons by open symbols. Inhibitory (smooth, GABAergic) neurons are indicated by black (filled) somata; their axons are indicated by dotted lines, and their synaptic boutons by solid symbols. thal, thalamus; cx, cortex; sub-cx, subcortical. Cortical layers 1–6 are as indicated. (Reproduced with permission from Douglas and Martin, 1990, Fig. 12.7.)

cells at the input side. In the neocortex this is layer 4, in the hippocampus the dentate gyrus. In both cases, these cells receive the feedforward inputs and relay them to a population of pyramidal cells (in layers 2–3 of the neocortex and in the CA3 field of the hippocampus) which have extensive recurrent collateral connections. In both cases, the pyramidal cells receive inputs both as relayed by the preprocessing array and directly. Such analogies might indicate that the functional roles of neocortical layer 4 cells and of dentate granule cells could be partially the same (see Rolls and Treves (1998)).

The short-range high density of connectivity may also contribute to the formation of cortical topographic maps, as described in Chapter 7, which may help to ensure that different parameters of the input space are represented in a nearly continuous fashion across the cortex, to the extent that the reduction in dimensionality allows it, or by the clustering of cells with similar response properties, when preserving strict continuity is not possible, as illustrated for example by colour 'blobs' in striate cortex.

1.10.5 Functional pathways through the cortical layers

Because of the complexity of the circuitry of the cerebral cortex, some of which is summarized in Fig. 1.13, there are only preliminary indications available now of how information is processed by the cortex. In primary sensory cortical areas, the main extrinsic 'forward' input is from the thalamus, and ends in layer 4, where synapses are formed onto spiny stellate cells. These in turn project heavily onto pyramidal cells in layers 3 and 2, which in turn send projections forward to the next cortical area. The situation is made more complex than this by the fact that the thalamic afferents synapse also onto the basal dendrites in or close to the layer 2 pyramidal cells, as well as onto layer 6 pyramidal cells and inhibitory interneurons. Given that the functional implications of this particular architecture are not fully clear, it would be of interest to examine the strength of the functional links between thalamic afferents and different classes of cortical cell using cross-correlation techniques, to determine which neurons are strongly activated by thalamic afferents with monosynaptic or polysynaptic delays. Given that this is technically difficult, an alternative approach has been to use electrical stimulation of the thalamic afferents to classify cortical neurons as mono- or poly-synaptically driven, then to examine the response properties of the neuron to physiological (visual) inputs, and finally to fill the cell with horseradish peroxidase so that its full structure can be studied (see for example Martin (1984)). Using these techniques, it has been shown in the cat visual cortex that spiny stellate cells can indeed be driven monosynaptically by thalamic afferents to the cortex. Further, many of these neurons have S-type receptive fields, that is they have distinct on and off regions of the receptive field, and respond with orientation tuning to elongated visual stimuli (Martin 1984) (see Chapter 2). Further, consistent with the anatomy just described, pyramidal cells in the deep part of layer 3, and in layer 6, could also be monosynaptically activated by thalamic afferents, and had S-type receptive fields (Martin 1984). Also consistent with the anatomy just described, pyramidal cells in layer 2 were di- (or poly-) synaptically activated by stimulation of the afferents from the thalamus, but also had S-type receptive fields.

Inputs could reach the layer 5 pyramidal cells from the pyramidal cells in layers 2 and 3, the axons of which ramify extensively in layer 5, in which the layer 5 pyramidal cells have widespread basal dendrites (see Fig. 1.1), and also perhaps from thalamic afferents. Many layer 5 pyramidal cells are di- or trisynaptically activated by stimulation of the thalamic afferents, consistent with them receiving inputs from monosynaptically activated deep layer 3 pyramidal cells, or from disynaptically activated pyramidal cells in layer 2 and upper layer 3 (Martin 1984). Interestingly, many of the layer 5 pyramidal cells had C-type receptive fields, that is they did not have distinct on and off regions, but did respond with orientation tuning to elongated visual stimuli (Martin 1984) (see Chapter 2).

Studies on the function of inhibitory pathways in the cortex are also beginning. The fact that basket cells often receive strong thalamic inputs, and that they terminate on pyramidal cell bodies where part of their action is to shunt the membrane, suggests that they act in part as a feedforward inhibitory system that normalizes the thalamic influence on pyramidal cells by dividing their response in proportion to the average of the thalamic input received (see Chapter 7). The smaller and numerous smooth (or sparsely spiny) non-pyramidal cells that are inhibitory may receive inputs from pyramidal cells as well as inhibit them, so that these neurons could perform the very important function of recurrent or feedback inhibition (see Chapter 7). It is only feedback inhibition that can take into account not only the inputs received by an area of cortex, but also the effects that these inputs have, once multiplied by the synaptic weight vector on each neuron, so that recurrent inhibition is necessary for competition and contrast enhancement (see Chapter 7).

Another way in which the role of inhibition in the cortex can be analyzed is by applying a drug such as bicuculline using iontophoresis (which blocks GABA receptors to a single neuron), while examining the response properties of the neuron (see Sillito (1984)). With this technique, it has been shown that in the visual cortex of the cat, layer 4 simple cells lose their orientation and directional selectivity. Similar effects are observed in some complex cells, but the selectivity of other complex cells may be less affected by blocking the effect of endogenously released GABA in this way (Sillito 1984). One possible reason for this is that the inputs to complex cells must often synapse onto the dendrites far from the cell body, and distant synapses will probably be unaffected by the GABA receptor blocker released near the cell body. The experiments reveal that inhibition is very important for the normal selectivity of many visual cortex neurons for orientation and the direction of movement. Many of the cells displayed almost no orientation selectivity without inhibition. This implies that not only is the inhibition important for maintaining the neuron on an appropriate part of its activation function, but also that lateral inhibition between neurons is important because it allows the responses of a single neuron (which need not be markedly biased by its excitatory input) to have its responsiveness set by the activity of neighbouring neurons (see Chapter 7).

1.10.6 The scale of lateral excitatory and inhibitory effects, and the concept of modules

The forward cortico-cortical afferents to a cortical area sometimes have a columnar pattern to their distribution, with the column width 200–300 μm in diameter (see Eccles (1984)). Similarly, individual thalamo-cortical axons often end in patches in layer 4 which are 200–300 μm in diameter (Martin 1984). The dendrites of spiny stellate cells are in the region of 500 μm in diameter, and their axons can distribute in patches 200–300 μm across separated by distances of up to 1 mm (Martin 1984). The dendrites of layer 2 and 3 pyramidal cells can be approximately 300 μm in diameter, but after this the relatively narrow column appears to become less important, for the axons of the superficial pyramidal cells can distribute over 1 mm or more, both in layers 2 and 3, and in layer 5 (Martin 1984). Other neurons that may contribute to the maintenance of processing in relatively narrow columns are the double bouquet cells, which because they receive inhibitory inputs, and themselves produce inhibition, all within a column perhaps 50 μm across (see above), would tend to enhance local excitation. The bipolar cells, which form excitatory synapses with pyramidal cells, may also serve to emphasize activity within a narrow vertical column approximately 50 μm across. These two mechanisms for enhancing local excitation operate against a much broader-ranging set of lateral inhibitory processes, and could it is suggested have the effect of increasing contrast between the firing rates of pyramidal cells 50 μm apart, and thus be very important in competitive interactions between pyramidal cells. Indeed, the lateral inhibitory effects are broader than the excitatory effects described so far, in that for example the axons of basket cells spread laterally 500 μm or more (see above) (although those of the small, smooth non-pyramidal cells are closer to 300 μm – see Peters and Saint Marie (1984)). Such short-range local excitatory interactions with longer range inhibition not only provide for contrast enhancement and for competitive interactions, but also can result in the formation of maps in which neurons with similar responses are grouped together and neurons with dissimilar response are more widely separated (see Chapter 7). Thus these local interactions are consistent with the possibilities that cortical pyramidal cells form a competitive network (see Chapter 7 and below), and that cortical maps are formed at least partly as a result of local interactions of this kind in a competitive network (see Section 7.4.6).

In contrast to the relatively localized terminal distributions of forward cortico-cortical and thalamo-cortical afferents, the cortico-cortical backward projections that end in layer 1 have a much wider horizontal distribution, of up to several mm. It is suggested below that this enables the backward-projecting neurons to search over a larger number of pyramidal cells in the preceding cortical area for activity that is conjunctive with their own (see below).

1.11 Theoretical significance of backprojections in the neocortex

In Chapter 8 a possible way in which processing could operate through a hierarchy of cortical stages is described. Convergence and competition are key aspects of the processing. This processing could act in a feedforward manner, and indeed, in the experiments on backward masking described in Chapter 5, it is shown that there is insufficient time for top-down processing to occur when objects can just be recognized (see Sections 5.5.6 and 5.9). (Neurons in each cortical stage respond for 20–30 ms when an object can just be seen, and given that the time for processing to travel from V1 to inferior temporal visual cortex (IT) is approximately 50 ms, there is insufficient time for a return projection from IT to reach V1, influence processing there, and in turn for V1 to project up to IT to alter processing there.) Nevertheless, backprojections are a major feature of cortical connectivity, and we next consider hypotheses about their possible function.

1.11.1 Architecture

The forward and backward projections in the neocortex that will be considered are shown in Fig. 1.14 (for further anatomical information see Jones and Peters (1984) and Peters and Jones (1984)). As described above, in primary sensory cortical areas, the main extrinsic 'forward' input is from the thalamus and ends in layer 4, where synapses are formed onto spiny stellate cells. These in turn project heavily onto pyramidal cells in layers 3 and 2, which in turn send projections forwards to terminate strongly in layer 4 of the next cortical layer (on small pyramidal cells in layer 4 or on the basal dendrites of the layer 2 and 3 (superficial) pyramidal cells, and also onto layer-6 pyramidal cells and inhibitory interneurons). Inputs reach the layer 5 (deep) pyramidal cells from the pyramidal cells in layers 2 and 3 (Martin 1984), and it is the deep pyramidal cells that send backprojections to end in layer 1 of the preceding cortical area (see Fig. 1.14), where there are apical dendrites of pyramidal cells. It is important to note that in addition to the axons and their terminals in layer 1 from the succeeding cortical stage, there are also axons and terminals in layer 1 in many stages of the cortical hierarchy from the amygdala and (via the subiculum, entorhinal cortex, and parahippocampal cortex) from the hippocampal formation (see Figs. 1.14, and 12.5) (Van Hoesen 1981, Turner 1981, Amaral and Price 1984, Amaral 1986, Amaral 1987, Amaral, Price, Pitkanen and Carmichael 1992).

A feature of the cortico-cortical forward and backprojection connectivity shown schematically in Figs. 1.14, and 12.5 is that it is 'keyed', in that the origin and termination of the connections between cortical areas provide evidence about which one is forwards or higher in the hierarchy. The reasons for the asymmetry (including the need for backprojections not to dominate activity in preceding cortical areas, see Rolls and Stringer (2001b)) are described below, but the nature of the asymmetry between two cortical areas provides additional evidence about the hierarchical nature of processing in visual cortical areas especially in the

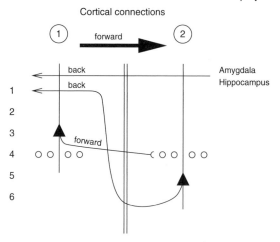

Fig. 1.14 Schematic diagram of forward and backward connections between adjacent neocortical areas. (Area 1 projects forwards to area 2 in the diagram. Area 1 would in sensory pathways be closest to the sense organs.) The superficial pyramidal cells (triangles) in layers 2 and 3 project forwards (in the direction of the arrow) to the next cortical area. The deep pyramidal cells in layer 5 project backwards to end in layer 1 of the preceding cortical area (on the apical dendrites of pyramidal cells). The hippocampus and amygdala are also sources of backprojections that end mainly in layer 1 of the higher association cortical areas. Spiny stellate cells are represented by small circles in layer 4. See text for further details.

ventral stream, for which there is already much other evidence (see Chapters 5 and 8, and Panzeri, Rolls, Battaglia and Lavis (2001)).

One point made in Figs. 12.12 and 12.5 is that the amygdala and hippocampus are stages of information processing at which the different sensory modalities (such as vision, hearing, touch, taste, and smell for the amygdala) are brought together, so that correlations between inputs in different modalities can be detected in these regions, but not at prior cortical processing stages in each modality, as these cortical processing stages are mainly unimodal. As a result of bringing together any two sensory modalities, significant correspondences between the two modalities can be detected. One example might be that a particular visual stimulus is associated with the taste of food. Another example might be that another visual stimulus is associated with painful touch. Thus at these limbic (and orbitofrontal cortex, see Chapter 12) stages of processing, but not before, the significance of, for example, visual and auditory stimuli can be detected and signalled. Sending this information back to the neocortex could thus provide a signal that indicates to the cortex that information should be stored. Even more than this, the backprojection pathways could provide patterns of firing that could help the neocortex to store the information efficiently, one of the possible functions of backprojections within and to the neocortex considered next.

1.11.2 Learning

The way in which the backprojections could assist learning in the cortex can be considered using the architecture shown in Figs. 1.14 and 7.11 (see also Section 7.4.5). The input stimulus occurs as a vector applied to (layer 3) cortical pyramidal cells through modifiable synapses in the standard way for a competitive net (input A in the schematic Fig. 7.11). If it is a primary cortical area, the input stimulus is at least partly relayed through spiny stellate cells, which may help to normalize and orthogonalize the input patterns in a preliminary way before the patterns are applied to the layer 3 pyramidal cells. (The details of these computational operations will be made clear in Chapter 7). If it is a non-primary cortical area, the cortico-cortical forward axons may end more strongly on the basal dendrites of neurons in the superficial cortical layers (3 and possibly 2). The lower set of synapses on the pyramidal cells would then start by competitive learning to set up representations on these neurons that would represent correlations in the input information space, and could be said to correspond to features in the input information space, where a feature is defined simply as the representation of a correlation in the input information space (see Chapter 7).

Consider now the conjunctive application of a pattern vector via the backprojection axons with terminals in layer 1 (B in Fig. 7.11) with the application of one of the input stimulus vectors. If, to start with, all synapses can be taken to have random weights, some of the pyramidal cells will by chance be strongly influenced both by the input stimulus and by the backprojecting vector. These strongly activated neurons will then compete with each other as in a standard competitive net, to produce contrast enhancement of their firing patterns. The relatively short-range (50 μm) excitatory operations produced by the bipolar and double bouquet cells, together with more widespread (300–500 μm) recurrent lateral inhibition produced by the smooth non-pyramidal cells and perhaps the basket cells, may be part of the mechanism of this competitive interaction. Next, Hebbian learning takes place as in a competitive net (see Chapter 7), with the addition that not only the synapses between forward projecting axons and active neurons are modified, but also the synapses between the backward projecting axons and the active neurons, which are in layer 1, are associatively modified.

This functional architecture has the following properties (see also Section 7.4.5). First, orthogonal backprojecting inputs can help the neurons to separate input stimuli (on the forward projection lines, A in Fig. 7.11) even when the input stimuli are very similar. This is achieved by pairing two somewhat different input stimuli A with very different (for example orthogonal) backprojection stimuli B. This is easily demonstrated in simulations (for example Rolls (1989b) and Rolls (1989f)). Conversely, if two somewhat different input stimuli A are paired with the same backprojection stimulus B, then the outputs of the network to the two input stimuli are more similar than they would otherwise be (see Section 7.11). This is also easily demonstrated in simulations.

In the neocortex, the backprojecting 'tutors' (see Rolls (1989b) and Rolls (1989f)) can be of two types. One originates from the amygdala and hippocampus, and by benefiting from cross-modal comparison, can for example provide orthogonal backprojected vectors. This backprojection, moreover, may only be activated if the multimodal areas detect that the visual stimulus is significant, because for example it is associated with a pleasant taste. This provides one way in which guidance can be provided for a competitive learning system as to what it should learn, so that it does not attempt to lay down representations of all incoming sensory information. The type of guidance is to influence which categories are formed by the competitive network.

The second type of backprojection 'tutor' is that from the next cortical area in the hierarchy. The next cortical area could operate in the same manner, and because it is a competitive system, is able to further categorize or orthogonalize the stimuli it receives, benefiting also from additional convergence (see for example Chapter 8). It then projects back these more orthogonal representations as tutors to the preceding stage, to effectively build better filters for the categories it is finding (cf. Fig. 7.12). These categories might be higher order (such as two parallel lines on the retina), and even though the receptive fields of neurons in the preceding area might never receive inputs about both lines because the receptive fields are too small, the backprojections could still help to build feature analyzers at the earlier stage that would be tuned to the components of what can be detected as higher order features at the next stage (see Chapter 8 and Fig. 7.12).

Another way for what is laid down in neocortical networks to be influenced is by neurons which 'strobe' the cortex when new or significant stimuli are shown. The cholinergic system originating in the basal forebrain, and the noradrenergic input to layer 1 of the cortex from the locus coeruleus, may contribute to this function (see Rolls and Treves (1998) Section 7.1.5). By modulating whether storage occurs according to the arousal or activation being produced by the environment, the storage of new information can be promoted at important times only, thus making good use of inevitably limited storage capacity. This influence on neocortical storage is not explicit guidance about the categories formed, as could be produced by backprojections, but instead consists of influencing which patterns of neuronal activity in the cortex are stored.

1.11.3 Recall

Evidence that during recall neural activity does occur in cortical areas involved in the original processing comes, for example, from investigations that show that when humans are asked to recall visual scenes in the dark, blood flow is increased in visual cortical areas, stretching back from association cortical areas as far as early (possibly even primary) visual cortical areas (Roland and Friberg 1985, Kosslyn 1994). Recall is a function that could be produced by cortical backprojections.

If in Figs. 1.14 or 7.11 only the backprojection input (B in Fig. 7.11) is presented after the type of learning just described, then the neurons originally activated by the forward-projecting input stimuli (A in Fig. 7.11) are activated. This occurs because the synapses from the backprojecting axons onto the pyramidal cells have been associatively modified only where there was conjunctive forward and backprojected activity during learning. This thus provides a mechanism for recall. The crucial requirement for recall to operate in this way is that, in the backprojection pathways, the backprojection synapses would need to be associatively modifiable, so that the backprojection input when presented alone could operate effectively as a pattern associator to produce recall. Some aspects of neocortical architecture consistent with this hypothesis (Rolls 1989b, Rolls 1989f, Rolls and Treves 1998) are as follows. First, there are many NMDA receptors on the apical dendrites of cortical pyramidal cells, where the backprojection synapses terminate. These receptors are implicated in associative modifiability of synapses, and indeed plasticity is very evident in the superficial layers of the cerebral cortex (Diamond, Huang and Ebner 1994). Second, the backprojection synapses in ending on the apical dendrite, quite far from the cell body, might be expected to be sufficient to dominate the cell firing when there is no forward input close to the cell body. In contrast, when there is forward input to the neuron, activating synapses closer to the cell body than the backprojecting inputs, this would tend to electrically shunt the effects received on the apical dendrites. This could be beneficial during the original learning, in that during the original learning the forward input would have the stronger effect on the activation of the cell, with mild guidance then being provided by the backprojections.

An example of how this recall could operate is provided next. Consider the situation when in the visual system the sight of food is forward projected onto pyramidal cells in higher order visual cortex, and conjunctively there is a backprojected representation of the taste of the food from, for example, the amygdala or orbitofrontal cortex. Neurons which have conjunctive inputs from these two stimuli set up representations of both, so that later if only the taste representation is backprojected, then the visual neurons originally activated by the sight of that food will be activated. In this way many of the low-level details of the original visual stimulus might be recalled. Evidence that during recall relatively early cortical processing stages are activated comes from cortical blood flow studies in humans, in which it has been found, for example, that quite posterior visual areas are activated during recall of visual (but not auditory) scenes (Kosslyn 1994). The backprojections are probably in this situation acting as pattern associators.

The quantitative analysis of the recall that could be implemented through the hippocampal backprojection synapses to the neocortex, and then via multiple stages of cortico-cortical backprojections, makes it clear that the most important quantitative factor influencing the number of memories that can be recalled is the number of backprojecting synapses onto each cortical neuron in the backprojecting pathways (see Section 12.2, Fig. 12.5, Treves and Rolls (1994) and Treves and Rolls (1991)). This provides an interpretation of why there are in general as many backprojecting synapses between two adjacent cortical areas as forward connections. The number of synapses on each neuron devoted to the backprojections needs to be large to recall as many memories as possible, but need not be larger than the number of forward inputs to each neuron, which influences the number of possible classifications that the neuron can perform with its forward inputs (see Chapter 7 and Section 12.2).

An implication of these ideas is that if the backprojections are used for recall, as seems likely as just discussed, then this would place severe constraints on their use for functions such as error backpropagation (see Chapter 7). It would be difficult to use the backprojections in cortical architecture to convey an appropriate error signal from the output layer back to the earlier, hidden, layers if the backprojection synapses are also to be set up associatively to implement recall.

1.11.4   Semantic priming

A third property of this backprojection architecture is that it could implement semantic priming, by using the backprojecting neurons to provide a small activation of just those neurons that are appropriate for responding to that semantic category of input stimulus.

1.11.5   Attention

In the same way, attention could operate from higher to lower levels, to selectively facilitate only certain pyramidal cells by using the backprojections. Indeed, the backprojections described could produce many of the 'top-down' influences that are common in perception (cf. Fig. 7.12 and Chapters 9–11).

1.11.6   Autoassociative storage, and constraint satisfaction

If the forward connections from one cortical area to the next, and the return backprojections, are both associatively modifiable, then the coupled networks could be regarded as, effectively, an autoassociative network. (Autoassociation networks are described in Section 7.3). A pattern of activity in one cortical area would be associated with a pattern in the next that occurred regularly with it. This could enable higher cortical areas to influence the state of earlier cortical areas, and could be especially influential in the type of situation shown in Fig. 7.12 in which some convergence occurs at the higher area. For example, if one of the earlier stages (for example the olfactory stage in Fig. 7.12) had a noisy input on a particular occasion, its representation could be cleaned up if a taste input normally associated with it was present. The higher cortical area would be forced into the correct pattern of firing by the taste input, and this would feed back as a constraint to affect the state into which the olfactory area settled. This could be a useful general effect in the cerebral cortex, in that constraints arising only after information has converged from different sources at a higher level could feed back to influence the representations that earlier parts of the network settle into. This is a way in which top-down processing could be implemented, and is analyzed in Section 7.9.

The autoassociative effect between two forward and backward connected cortical areas could also be used in short term memory functions (see Section 12.1), to implement the types of short term memory effect described in Chapter 7 and Section 12.1. Such connections could also be used to implement a trace learning rule as described in Chapter 8.

With this overview, it is now time to consider in depth the information processing in the visual pathways in the next few Chapters.

2 The primary visual cortex

2.1 Introduction and overview

Anatomical and neurophysiological evidence shows that there is partially segregated processing of information in the visual pathways starting at the retina. The visual pathways from
the cones in the retina to the primary visual cortex and onwards are shown schematically
in Fig. 2.1. Consistent with this evidence, patients with localized damage to parts of these
visual pathways often show selective visual impairments, such as loss of colour discrimination
without impairment of form perception, or loss of motion perception without impairment of
form and colour perception.

The responses of the neural elements to visual stimuli alter through these pathways. The
cone receptors respond optimally to spots of light. The retinal ganglion cells have centre-
surround receptive fields. They respond, for example, to a spot of light in the 'on' centre,
are switched off if there is also light in the surround, and respond best if light covers the
centre but not (much of) the surround (illustrated below in Fig. 2.4). They thus respond best
if there is spatial discontinuity in the image on the retina within their receptive field. The
lateral geniculate cells also have centre-surround organization, and use lateral inhibition to
produce contrast enhancement. There is a division of processing streams as early as the retinal
geniculate cells into a parvocellular (small cell, P) pathway and a magnocellular (large cell,
M) pathway (see Fig. 2.1).

Within the P pathways, many cells in the primary visual cortex (V1), for example pyramidal
cells in layers 2 and 3, and 5 and 6, have receptive fields that are elongated, responding best,
for example, to a bar or edge in the receptive field. Cells with responses of this type that
have definite 'on' and 'off' regions of the elongated receptive field, and that do not respond
to uniform illumination that covers the whole receptive field, are called simple cells (Hubel
and Wiesel 1962). They are orientation-sensitive, with approximately four orientations being
sufficient to account for their orientation sensitivity. Complex cells are also found in V1. They
do not have distinct 'on' and 'off' regions that can be mapped with spots of light, but do respond
to oriented edges or bars in their receptive fields. The 'goals' of this early visual processing
include making information explicit about the location of characteristic features of the world,
such as elongated edges and their orientation and colour; and removing redundancy from the
visual input by not responding to areas of uniform brightness (Marr 1982, Barlow 1989).

Cells in the layers of V1 that receive primarily M inputs (see Fig. 2.1) may not have
orientation tuning, but do have properties such as retinal disparity used in stereoscopic vision,
and responsiveness to rapidly moving or changing stimuli. They give rise to the paths that
reach area MT and eventually parietal cortical areas.

Evidence on the nature of the neuronal responses in V1 and the processing streams that
lead to V1, and on some of their computational properties and the computational processes
by which they arise, are described in the remainder of this Chapter.

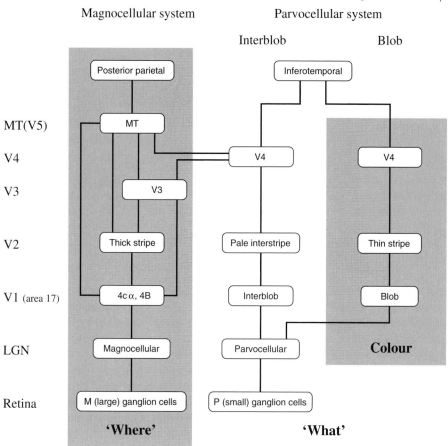

Magnocellular system   Parvocellular system

Interblob   Blob

Fig. 2.1 A schematic diagram of the visual pathways from the retina to visual cortical areas. LGN, lateral geniculate nucleus; V1, primary visual cortex; V2, V3, V4, and MT – other visual cortical areas; M, magnocellular; P, parvocellular.

2.2 Retina and lateral geniculate nuclei

The first step in the information processing chain occurs in the eye, namely in the photoreceptors and retinal cells that transform the visual information contained in the reflected light from objects into neural signal. The photoreceptors are in the retina, which is located on the inner surface of the eye. The flow of electrical current in the photoreceptors is activated by light-sensitive molecules called photopigments. There are two types of photoreceptors: rods and cones. They differ in their spectral (see Fig. 2.2) and brightness sensitivity. Rods are widely distributed throughout the retina and are colour insensitive. They show high sensitivity to a low level of brightness and therefore are mainly in use in dark environments, for example at night. On the other hand, cones have their highest density in the fovea, the central region of the retina, and require high levels of brightness, so that they are used mainly in bright conditions of illumination (i.e. during daytime vision situations). The latency and dynamics of the activation of the photopigments in rods are slow and in cones are much faster. Cones are

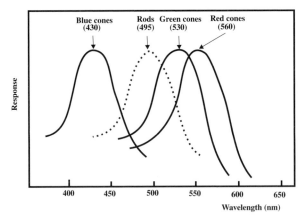

Fig. 2.2 Spectral sensitivity functions for the three different types of cones and for rods. 'Blue' cones are sensitive to a wavelength band peaked around the blue colour. The corresponding observation holds for the 'green' and 'red' cones. White light contains all wavelengths and therefore activates all three cone types and the rods. The numbers in parentheses indicate the wavelength at which there is maximum sensitivity.

sensitive to different wavelength bands and consequently are essential for colour vision. As is shown in Fig. 2.2, one can distinguish three types of cone: 'blue', 'green', and 'red'. Each type of cone has broad tuning, so that together they span the visible spectrum, and provide a distributed representation about which wavelengths are present.

The output from the photoreceptors is connected to horizontal and bipolar cells. Intracellular experimental measurements (Werblin and Dowling 1969) have been required for studying the neural responses of bipolar cells. The bipolar cells respond by generating continuous graded activations of the membrane electrical potential. This atypical form of neural response for mammalian neurons can be only transmitted with accuracy and speed over short distances, and is therefore only suitable for neurons with short axons and therefore very local connectivity. There are two pathways for connections to a bipolar cell: direct and indirect. The direct path projects directly from photoreceptors to the bipolar cells, whereas the indirect path projects from the photoreceptors first into the horizontal cells and from there to the bipolar cells (see Fig. 2.3). The receptive field of a bipolar cell is formed by the connection with a small central region of photoreceptors via the direct pathway and a broader and weaker region of photoreceptors via the indirect pathways. The direct pathway can be either excitatory or inhibitory, while the associated indirect path is always the opposite. These two regions are integrated by the bipolar cell to generate a centre/surround or 'Mexican hat' receptive field, as illustrated in Fig. 2.3.

The continuous graded output of bipolar cells is further processed by the ganglion cells that form the last layer of the retina. The axons of the retinal ganglion cells constitute the optic nerve that transmits the retinal visual information into the central nervous system. The responses of retinal ganglion cells are characterized by a train of discrete action potentials or spikes. Consequently, the activity of a ganglion cell can be characterized by its firing rate. Kuffler (1953) and Barlow (1953) performed experimental extracellular recordings of the firing activity of ganglion cells. They observed that the firing rate was highest for a spot of light of a particular size at a specific location on the retina. Even more, they observed that the firing rate decreased if the size of the spot was either increased or decreased. This behaviour

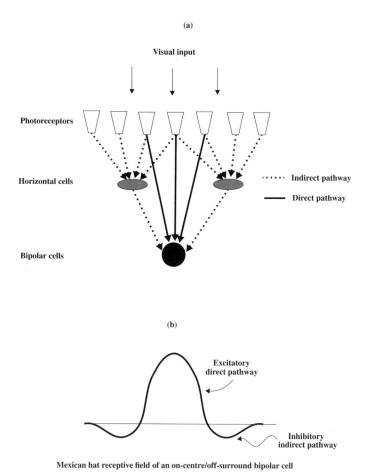

Fig. 2.3 Anatomy of the retina. (a) Light coming through the cornea activates the photoreceptors located at the rear surface of the retina. The receptive fields of bipolar cells have a centre/surround or 'Mexican hat' shape. The direct pathway from the photoreceptors to the bipolar cells defines the sensitivity of the centre of the receptive field. The indirect pathway of opposite polarity comes from the photoreceptors in the surround of the receptive field via the horizontal cells. (b) An example of the receptive field of an on-centre/off-surround bipolar cell.

can be interpreted by postulating an antagonism between an inner circular centre and the surrounding ring. Fig. 2.4 shows the two different types of responses and the corresponding receptive field. On-centre/off surround ganglion cells increase their firing to a light spot at the centre and decrease their firing if the surround is stimulated. Conversely, off-centre/on-surround ganglion cells decrease their firing rate to a light spot at the centre and increase their firing rate to light in the surround. In addition to this distinction, further studies (Shapley and Perry 1986) have shown that there are two different kinds of primate retinal ganglion cells that project selectively to different processing systems in the cortex. They are called P (Parvo, small) ganglion cells and M (Magno, large) ganglion cells. P ganglion cells are sensitive to colour, and M ganglion cells are not. This can be explained by their different wiring with the

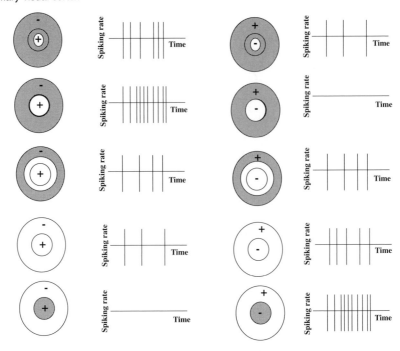

Fig. 2.4 Left: on-centre/off surround retinal ganglion cells increase their firing rate to a light spot at the centre and decrease their firing rate to light in the surround. Right: off-center/on-surround ganglion cells have the opposite type of response. A bright spot is shown as white; a dark area is shown shaded.

input photoreceptors. P cells receive input connections just from the cones, whereas M cells receive input connections from both rods and cones.

The optic nerve leaving the eyes passes through the optic chiasm, where some axons cross to the opposite side, to provide synaptic input both to the lateral geniculate nuclei (LGN) of the thalamus, and to the superior colliculi. For conscious perception, the geniculo-cortical path is needed, and pathways to the colliculus (and other brainstem regions) seem to be primarily involved in eye movement control, and support only residual visual function (called blindsight, Weiskrantz (1998)) in which patients report seeing nothing. We consider here the perception-related geniculo-cortical visual path. As shown in Fig. 2.5, visual information coming from the right hemifield is projected into the left hemisphere (i.e. left LGN and left cortex), and the left visual hemifield is projected into the right hemisphere. In primates the lateral geniculate nuclei have a laminar structure consisting of a six-layered three-dimensional structure, which provides a clear segregation of the visual information into two paths. The four dorsal layers contain small cell bodies (and are therefore called parvocellular layers), and receive from the P system in the retina. The two ventral layers contain large cell bodies (and are called magnocellular)[1], and receive from the M retinal ganglion cells. Each individual LGN neuron is monocular, responding to stimulation from just one eye. Each eye projects to three of the six layers in an alternating fashion as shown in Fig. 2.5 (i.e. each half retina projects three times onto one geniculate body, twice to the parvocellular layers and once to the

---

[1] In Latin, magnus means large and parvus small

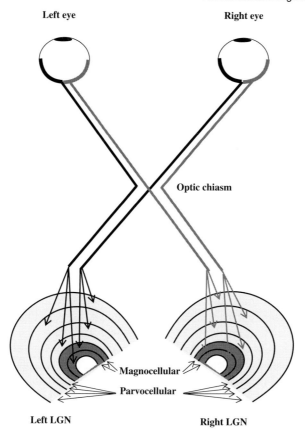

Fig. 2.5 Diagram of the functional segregation of the pre-cortical visual system. Visual information coming from the right hemifield is projected into the left lateral geniculate nucleus (LGN) of the thalamus, and the left visual hemifield is projected into the right lateral geniculate nucleus of the thalamus. The four dorsal layers of the lateral geniculate nuclei contain small cell bodies and are part of the P (parvocellular) system. The two ventral layers contain large cell bodies and are part of the M (magnocellular) system.

magnocellular layers). The spatial organization of each LGN layer is retinotopic, preserving the topographical relative location of cells from retina to LGN, so that neighboring regions on the retina project to nearby regions of the LGN.

Both magno- and parvo- cells of the LGN have circularly symmetric receptive fields, and most of them (around 90%) present centre/surround opponency. But they differ physiologically in four major aspects: colour, contrast sensitivity, spatial resolution, and latency. Let us consider the first difference, colour sensitivity. Livingstone and Hubel (1988) report that 90% of the neurons in the parvocellular layers of the LGN are very sensitive to differences in wavelength, whereas neurons in the magnocellular layers are not. In fact, parvo neurons combine different cone inputs. Receptive fields for typical colour-opponent parvocellular geniculate neurons are for example red on-centre/green off-surround, showing an excitatory response when a red spot stimulates the centre, and inhibitory when green light is on the surround. They thus respond best to colour contrast within their receptive field. On the other hand, typical broadband magnocellular neurons integrate the input of the different colour-sensitive cones, so that the response to stimulation at the centre or surround, either on ('excitatory') or off ('inhibitory') is independent of the wavelength. This is also the case for the remaining 10% of colour-insensitive parvo neurons.

The second difference is contrast sensitivity (Shapley and Perry 1986, Shapley 1995). Magno neurons are much more sensitive than parvo neurons to low-contrast stimulation. Both parvo and magno cells start to fire when the centre and surround brightness differ by 1 or 2%. In the case of magno neurons, the firing activity saturates rapidly and levels off at about 10 to 15% contrast. On the other hand, parvo neurons' responses increase more slowly, and saturate at much higher contrast levels.

The third difference is spatial resolution. The size of the receptive field of both magno and parvo LGN cells increases with the distance from the fovea. This fact accounts for the acuity differences between foveal and peripheral vision (cf. Rolls and Cowey (1970)). Nevertheless, at a particular eccentricity, the size of the receptive fields of magno cells is a factor of 2 or 3 times larger than that of parvo cells.

The fourth difference is the latency, in that magno cells respond faster than parvo cells. Consequently, the magnocellular system is more suitable for analyzing temporal aspects of the visual stimuli, like motion, due to its rapid response dynamics. Related to this function, the M system tends to respond transiently, and is thus sensitive to changes in light on the receptive field, whereas the P system has more sustained responses, which are suitable for shape perception which may involve prolonged viewing of objects.

A fifth difference is linearity, in that the parvo system is much more linear than the M system. This is the case in the sense that for the parvo system, the centre and surround of the receptive fields add linearly, in that when a light covers the whole of the receptive field, the centre and surround effects sum to zero, and there is no firing of the cells (cf. Fig. 2.4). Conversely, the magno system will respond, typically transiently, when a light is switched on to cover the whole of the receptive field.

These five major differences between the magno and parvo neurons in the LGN (and retina) are maintained in the visual cortex because the segregated connectivity of the visual paths is maintained at the cortical level. The magno- and parvo-cellular subdivisions of the LGN are the pre-cortical correlate of the partially segregated processing streams in the visual cortical areas.

The properties of the P stream are overall computationally appropriate for shape perception, in that the responses tend to be linear (which means that the exact difference in brightness can be encoded, and are encoded with high spatial precision, in that the receptive fields are small, and the difference in brightness must be within the visual field); in that colour is encoded, and this can be very helpful in delineating edges in a scene as an aid to shape description; and in that the P system has sustained responses. The centre-surround organization is an efficient way of using the limited bandwidth of the neuronal channel capacity to encode the relative brightness in a scene (which is what is needed for shape perception, in which edges must be detected and steady gradients of illumination across the scene should be discounted). The centre-surround organization is produced by lateral inhibition, and in the spatial frequency domain corresponds to removing low spatial frequencies, that is high-pass filtering. The high-pass filtering is more evident in lateral geniculate neurons than in retinal ganglion cells, in that the lateral geniculate provides an extra stage of lateral inhibition.

Conversely, the properties of the M system are broadly computationally appropriate for motion detection and the detection of any sudden change in illumination, in that they have fast response dynamics, and do not respond well to sustained visual stimulation.

2.3 Striate cortex: Area V1

2.3.1 Classification of V1 neurons

In primates the striate cortex is a 2 mm thick sheet containing some 200 million neurons in an area of a few square centimetres. The term 'striate' refers to the stripe (stria) of Gennari, which is produced in layer 4 by the massive number of fibres received from the lateral geniculate nucleus. The characteristics of the receptive fields of the neurons included in this area (usually called V1) were studied by Hubel and Wiesel (1959), (1962), (1968), (1972), and (1977). They discovered simple cells and complex cells in V1. They discovered that these cells respond optimally to an oriented edge or bar in their receptive field. The discovery was made during testing with circular spots of light generated using slides. These stimuli were relatively ineffective, but Hubel and Wiesel one day observed a large response when the slide was being loaded, and produced a straight edge which passed through the receptive field (Hubel and Wiesel 1959). Their term 'hypercomplex cell' is no longer used, because it is now realized that both simple and complex cells may or may not be endstopped. Endstopped cells respond when an edge or bar terminates within their receptive field, whereas non-endstopped cells respond just as well when the edge or line extends beyond the receptive field.

2.3.1.1 Simple cells

Simple cells have a receptive field that can be mapped just by determining its response, relative to the background firing, to an individual small spot of light located at different positions on the retina[2]. The receptive fields are typically small, usually in the range 0.25–1 degree close to the fovea. Different simple cells are responsive to bars versus edges. The typical orientation tuning is approximately 45 degrees (in that the width of the tuning function of the response versus orientation curve is typically 45 deg at a half-maximal neuronal response). Thus approximately four sets of simple (and complex) cells cover the orientation tuning domain. Hubel and Wiesel (1962) pointed out that such receptive fields can be constructed from the output of a set of conveniently aligned centre/surround LGN neurons. Fig. 2.6 shows as an example of the underlying functional connectivity of LGN cells that could produce the responses of a bar-responsive V1 simple cell.

More recent neurophysiological studies (De Valois and De Valois 1988) revealed a more complicated structure for the receptive fields of simple V1 neurons. De Valois and De Valois (1988) found that the receptive fields of simple cells were not only sensitive to a specific position and orientation of the stimulus, but were also sensitive to the spatial frequency of the visual stimuli. Figure 2.7 shows the typical receptive profiles found in simple cells. The receptive fields have additional lobes of excitation and inhibition, such that the simple neurons are not only sensitive to a specific position and orientation but also to the spatial frequency of the stimuli. In fact, the size of the receptive field is strongly correlated with the spatial frequency (i.e. frequency of extra lobes) at which they preferentially respond. These kinds of profiles can be matched very well with so-called Gabor functions or wavelets. A Gabor function is constructed by multiplying a sinusoidal function by a Gaussian function. A Gabor function is given by an expression of the form

---

[2]This is the reason why Hubel and Wiesel (1959) called these neurons 'simple cells'.

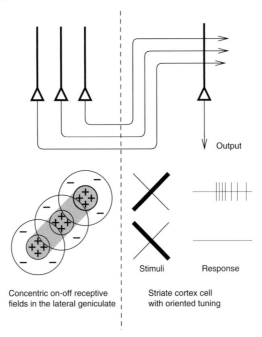

Fig. 2.6 Receptive fields in the lateral geniculate have concentric on-centre off-surround (or vice versa) receptive fields (left). Neurons in the striate cortex, such as the simple cell illustrated on the right, respond to elongated lines or edges at certain orientations. A suggestion that the lateral geniculate neurons might combine their responses in the way shown to produce orientation-selective simple cells was made by Hubel and Wiesel (1962).

$$G(x,y) = \frac{1}{2\pi\sigma\beta}e^{-\pi\left[\frac{(x-x_0)^2}{\sigma^2}+\frac{(y-y_0)^2}{\beta^2}\right]}e^{i[\xi_0 x+\nu_0 y]} \tag{2.1}$$

where (x_0, y_0) is the centre of the receptive field in the spatial domain and (ξ_0, ν_0) is the optimal spatial frequency of the filter in the frequency domain. σ and β are the standard deviations of the elliptical Gaussian along x and y. We present in Chapter 9 the detailed implementation of the neurophysiological constraints (e.g. relationship between size and spatial frequency) that restrict the degree of freedom of Gabor functions for modelling V1 simple cells. In fact, we will use the mathematical framework of Gabor wavelets for the computational models of visual attention that we present in Chapters 9, 10 and 11.

An alternative mathematical formulation, which also produces good fits to the actual receptive fields of simple cells, is a difference of two Gaussian functions (DOG) (Hawken and Parker 1987, Parker 1997). This formulation is described in Chapter 8, and is used in VisNet as described in Chapter 8.

2.3.1.2 Complex cells

Complex cells are similar to simple cells in that they also respond to bars or edges and are orientation tuned, but differ in that they are highly nonlinear, are not sensitive to the exact position of the edge or bar within the receptive field, and are frequently tuned to the direction of motion. Complex neurons do not respond to small spots (and in this sense are non-linear),

(a)

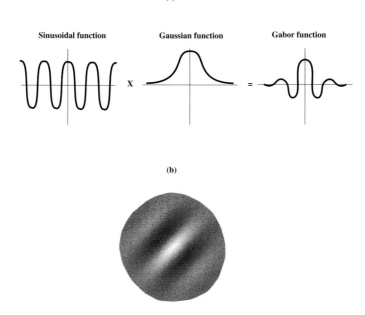

(b)

Fig. 2.7 (a) Gabor receptive field obtained as a multiplicative modulation of a Gaussian function by a sinusoidal function (in one dimension). (b) Example of the typical response of a neuron with a 2-dimensional Gabor receptive field when a small spot stimulus is scanned in the retinal plane. High responses are plotted by high luminance values.

and therefore their receptive fields can not be mapped just by measuring their response to an individual spot of light[3]. The receptive fields of complex cells are larger than the receptive fields of simple cells. They do not show special sensitivity to small variations of the position of the stimuli inside their larger receptive fields. This type of neuron is the most common cell type in area V1. Hubel and Wiesel speculated that complex neurons were constructed by integrating the responses of many simple cells. The underlying assumed wiring for that is shown in Fig. 2.8.

2.3.1.3 Endstopped cells

These cells require an oriented bar or edge to terminate in their receptive field, and can be simple or complex. The complex variety were called hypercomplex by Hubel and Wiesel. These cells could be constructed in the same way as simple or complex cells respectively, but include in addition inhibitory inputs at one end or both ends of the receptive field (Palmer 1999).

2.3.2 Organization of the striate cortex

V1 neurons respond selectively to different spatial features such as position, size and orientation, to different non-spatial features such as colours and motion, and to the eye (left or

---

[3]This is the reason why they are called "complex cells".

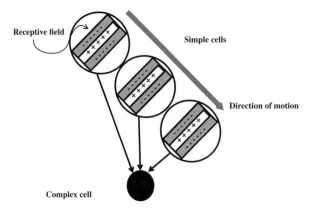

Fig. 2.8 Hypothesized underlying wiring of a complex V1 neuron. Complex neurons receive input synapses from several simple neurons sensitive to the same orientation but with centres shifted to different successive positions. An appropriate time delay from the different simple cell inputs could account for the motion direction sensitivity of many complex cells.

right) where the stimulus is presented. Three main spatial patterns of topological organization of neurons that are tuned to different features are found: retinotopic organization, ocular dominance slabs, and columnar organization with each column being tuned to, for example, a particular orientation, as described next and reviewed by Palmer (1999).

2.3.2.1 The retinotopic map

The primary visual cortex is spatially organized such that retinal topography is approximately preserved. In each hemisphere, the neurons in the striate cortex are organized into a retinotopic map of half of the contralateral visual field. The left (right) visual field is topographically mapped in the right (left) striate cortex. This topographic organization preserves the relative spatial relationship (i.e. adjacent locations in the retina are also adjacent in the striate cortex), but the metric properties are distorted. The main distortion corresponds to the cortical magnification of central retinal areas relative to the periphery (Rolls and Cowey 1970, Cowey and Rolls 1975). Tootell, Silverman, Switkes and De Valois (1982) used autoradiographic techniques to study experimentally the organization of the striate cortex in monkeys. The region that has been activated by visual stimuli can be identified with this technique, which labels areas with high metabolic activity. They showed that topographic relationships were preserved, but the metric properties were distorted. Logarithmically spaced semicircular lines in the stimulated right (left) visual field were projected with equal spacing in the left (right) striate cortex. This corresponds to the fact that the retinal fovea is disproportionately mapped in a large area of the striate cortex, whereas the retinal periphery has a much smaller representation.

2.3.2.2 Ocular dominance

Cells in the striate cortex receive inputs from both eyes in differing proportions: some V1 neurons are more dominated by one that by the other eye. LeVay, Hubel and Wiesel (1975) termed this phenomenon 'ocular dominance'. They discovered that neurons dominated by a particular eye cluster together in 'slabs'. The slabs corresponding to the left and right eye are not regularly organized, but they are also not random. They are interleaved in clear stripes,

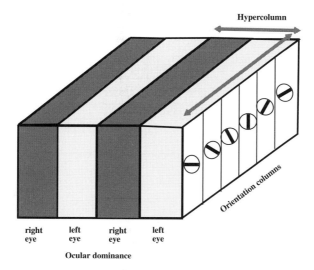

Fig. 2.9 Hypercolumn modular organization of V1. Each hypercolumn is composed of a group of orientation columns. Within a hypercolumn, the neurons are sensitive to the same retinal location. The orientation columns vary systematically along one dimension of the cortical sheet, tangential to the cortical surface. Along the other dimension, also tangential to the cortical surface, different ocular dominance columns are interleaved.

tuned to respond better to one than the other eye, that run tangentially to the cortical surface.

2.3.2.3 Columnar structure

The striate cortex shows at a finer scale a columnar organization. The basic unit is called a hypercolumn. Hypercolumns are small columnar regions of the cortical sheet, around 1 mm by 1 mm on the surface, that extend perpendicularly through all six cortical layers. In one hypercolumn which is concerned with a given part of the visual field and in which the receptive field centres are similar, there is a series of columns (see Fig. 2.9). Each vertical column contain neurons tuned to a particular orientation. Along one dimension of the cortex, tangential to the cortical surface, the orientation tuning of neurons varies systematically, and a set of orientation columns with preferential tuning for one eye forms a hypercolumn. Along the other dimension, also tangential to the cortical surface, different ocular dominance columns are interleaved, forming slabs (see Fig. 2.9). Perpendicular to the orientation dimension, a third dimension may provide a regular progression of different spatial frequency sensitivities, from small to large (De Valois and De Valois 1988).

A computational interpretation of hypercolumns containing computational units sensitive to a common centre position but tuned to different orientations and spatial frequencies has been proposed by Buhmann, Lades and von der Malsburg (1990). They relate the concept of a set of Gabor filters (see Section 2.3.1.1) for one position in the visual field to that of a hypercolumn of cells in V1. They introduced the concept of Gabor jets. A Gabor jet corresponds to a set of Gabor filters sensitive to different orientations and spatial frequencies and with a common centre (see Section 9.3 for details). In other words, a Gabor jet corresponds to a hypercolumn. Some of the underlying processes that lead to the organization of these ocular dominance

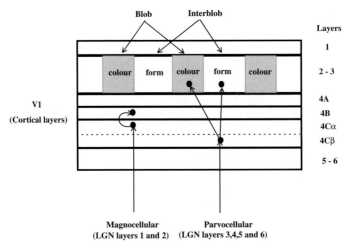

Fig. 2.10 Functional segregation of the primate striate cortex into three visual streams, associated with the processing of stereopsis and movement (layer 4B), colour (blobs in layers 2–3), and form (interblobs in layers 2–3).

and orientation columns are described in Chapter 7, and by Erwin, Obermayer and Schulten (1995), and Erwin and Miller (1998).

2.3.3 Visual streams within the striate cortex

The segregation of the visual pathways from the retina through the LGN into magnocellular and parvocellular systems is further continued in the striate cortex. Neurons in the magnocellular geniculate layers project to layer 4Cα of the striate cortex, which is further synaptically connected with neurons in layer 4B, and from there to visual area V2 and MT (see Figs. 2.10 and 2.1). Neurons in the parvocellular lateral geniculate layers project to layer 4Cβ, and from there neurons project to neurons in layer 2 and 3, and from there neurons project to visual area V2. An extra subdivision of the parvocellular system inside the striate cortex has been shown by staining techniques that stain the mitochondrial enzyme cytochrome oxidase. Cytochrome oxidase is concentrated in regions of high metabolic activity. Application of this staining technique to the striate cortex reveals, mainly in layers 2 and 3, a regular organization of regions which show high and low metabolic activity. Livingstone and Hubel (1988) called these two different 3D stained regions 'blobs' (dark oval regions in sections cut parallel to the surface), and 'interblobs'. Fig. 2.10 shows a diagram of these three separate functional pathways in the striate cortex.

V1 neurons corresponding to these three at least partially segregated systems show different stimulus selectivity associated with different types of perceptual processing.

The magno system within the striate cortex, i.e. neurons in layer 4B, are orientation selective, and also show selectivity for the direction of movement of the corresponding lines and edges in a direction orthogonal to the optimal orientation. (Their responses can only reflect the direction of motion perpendicular to the orientation to which they are tuned because they have small receptive fields. This is an instance of the aperture problem. This problem is solved to obtain the true motion of the edge in area MT, as described in Chapter 3). As usual in the magnocellular system, the neurons are not selective for colour. They seem to be mainly responsible for the encoding of motion, and of stereopsis by computing the retinal disparity between the signals coming from the two eyes (as described further in Section 3.4.2 and Chapters 4 and 5).

Blob V1 neurons of the parvo-related striate system are not orientation–selective, but are either colour or brightness selective. Consequently, they are engaged with the coding of colour and contrast information.

On the other hand, interblob V1 neurons of the parvo striate system are orientation selective, not motion selective, and most of them are not very colour selective. Their receptive field is in general small, so that they are specially suited for the analysis of high spatial resolution. Hence, they are responsible for the coding of form.

In summary, there seem to be partially separated neural pathways within the striate cortex that imply a segregation of the processing channels into three functionally distinct pathways: a stereopsis and motion pathway (magno V1 neurons), a colour pathway (blob V1 neurons), and a form pathway (interblob V1 neurons). Nevertheless, the segregation of pathways is not perfect in the sense that there is cross-talk between them (Van Essen and DeYoe 1995), and indeed in later areas this is computationally important for producing, for example, structure from motion.

2.4 Computational processes that give rise to V1 simple cells

The idea of the nervous system and brain being regulated by an economy principle has been well known since the pioneering work of Zipf (1949) and the ideas of Attneave (1954) about information processing in visual perception. Barlow (1989) related network models of unsupervised learning to the concept of redundancy reduction. The minimum entropy coding method was introduced for the generation of factorial codes (Barlow, Kaushal and Mitchison 1989). Factorial learning or redundancy reduction is defined as an unsupervised extraction of 'features', where each feature corresponds to a set of statistically correlated stimuli in the sensory input space. This type of learning can be formalized in the general frameworks of competitive neural network learning (see Section 7.4), and nonlinear independent component analysis (Deco and Obradovic 1996, Deco and Schürmann 2001).

Linsker (1988) and (1992) has proposed an optimization principle called Infomax, according to which the synaptic weights develop in such a way that the mutual information between the input and output layers of a cortical network is maximized under constrained boundary conditions. Some algorithms were developed for maximizing mutual information by using probabilistic linear neurons (Linsker 1992). In the linear case the Infomax principle is related to principal component analysis. Rubner and Tavan (1989) and Földiák (1989) proved that a network composed of linear neurons with the synaptic modification defined as Hebbian with respect to the connections between the input and output neurons, and anti-Hebbian for the inhibitory lateral synaptic connection between the output units, can be trained to perform principal component analysis (see Section 7.4). The Hebbian and anti-Hebbian form of the learning rules can be derived from first principles using information theoretic concepts (Linsker 1988). Atick and Redlich (1990) demonstrated that statistically salient input features can be optimally extracted from a noisy input by maximizing mutual information. Simultane-

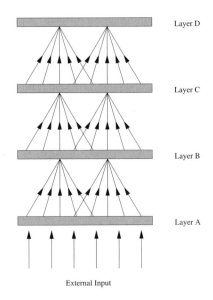

Fig. 2.11 Architecture of Linsker's multilayer Hebbian learning network, showing the receptive fields of a few neurons.

ously, Atick and Redlich (1990) and especially the work of Redlich (1993) concentrated on the original idea of feature extraction by redundancy reduction.

Olshausen and Field (1996) proposed an alternative principle, maximization of sparseness, that accounts for many of the properties of V1 simple cell receptive fields, and is consistent with the idea of efficient coding generation via information maximization and generation of statistically independent or minimal redundancy codes.

We review in the next two subsections, the maximal information transmission approach of Linsker (1988), and the more complete information-theoretic framework of Olshausen and Field (1996).

2.4.1 Linsker's method: Information maximization

One approach to how the processing from the retina to V1 might self-organize in a multilayer neuronal network has been described by Linsker (1986), (1988). In the model architecture he describes, neurons with simple cell-like receptive fields develop. They develop even in the absence of visual input, and require only random firing of the neurons in the system. Visual input would help the network to learn by providing evidence on the statistics of the visual world. The network and its properties are described next, not because this is necessarily a sufficient approach to understanding how the preprocessing stages of the visual system develop (see Rose and Dobson (1985) for alternative models), but because it is very helpful to understand what can develop in even quite a simple multilayer processing system that uses local Hebb-like synaptic modification rules.

The architecture of Linsker's net is shown in Fig. 2.11. It is a multilayer feedforward network. A neuron in one layer receives from a small neighbourhood of the preceding layer, with the density of connections received falling off with a Gaussian profile within that neighbourhood. These limited receptive fields are crucial, for they enable units to respond to spatial correlations in the previous layer. Layer A receives the input, and is analogous to the receptors of the retina, for example cones. The neurons are linear (that is, they have linear activation functions, see Fig. 1.3). This means that the multilayer network could be replaced by a single-layer network, since a set of linear transforms can be replaced by a single linear transform, the product of the set of transforms (see Appendix A). This does not of course imply that any local unsupervised learning rule would be able in a one-layer net to produce the equivalent transform. What is of interest in this network is the type of receptive fields that self-organize with the architecture and a local Hebb-like learning rule, and this is what Linsker (1986), (1988) has investigated. The modified Hebb rule he used (for the jth synaptic weight w_j from an axon with firing rate x_j on a neuron with firing rate y) is

$$\delta w_j = k(y x_j + b x_j + c y + d) \tag{2.2}$$

where the first product inside the brackets is the Hebbian term, and k is the learning rate. The weights in Linsker's network are clipped at maximum and minimum values. The second to fourth terms in the brackets alter how the network behaves within these constraints, with separate constants b, c and d for each of these terms. The rule acts, as does Oja's (see Chapter 7), to maximize the variance of the neuron's outputs given a set of inputs, that is it finds the first principal component of the inputs (Linsker 1988, Hertz, Krogh and Palmer 1991).

Random firing of the input neurons was used to train the network model. (A parallel might be the conditions when the visual system is set up before birth, so that at birth there is some organization of the receptive fields already present.) For a range of parameters, it was found that neurons in layer B simply averaged the input activity over their receptive fields. Because the receptive fields of neighbouring neurons in layer B overlapped considerably and they therefore received correlated inputs, the activity of neighbouring neurons in layer B was highly correlated. Correspondingly, the activity of neurons further apart in layer B was not highly correlated. This led to the emergence of a new type of receptive field in layer C, with centre-surround organization, in which a neuron might respond to inputs in the centre of its receptive field, and be inhibited by inputs towards the edge of its receptive field, responding maximally, for example, to a bright spot with a dark surround (Fig. 2.12a). Other cells in layer C developed off-centre / on-surround receptive fields.

Within layer C, nearby units were positively correlated, while further away there was a ring of negative correlation. (This is again related to the geometry of the connections, with nearby neurons receiving from overlapping and therefore correlated regions of layer B.) The result of the neighbourhood correlations in layer C, which had a spatial profile like that of a Mexican hat (see Fig. 7.13), was that later layers of the simulation (D–F) also had centre-surround receptive fields, which had sharper and sharper Mexican hat correlations.

For layer G the parameters were altered, and it was found that many of the neurons in the model were no longer circularly symmetric, but had elongated receptive fields, which responded best to bars or edges of a particular orientation (see Fig. 2.12b). Thus, symmetry had been broken, and cells with similarities to simple cells in V1 with oriented bar or edge selectivity had developed.

Cells in layer G developed the orientation to which they responded best independently of their neighbours. In the architecture described so far, there were no lateral connections. If lateral connections were introduced (for example, short-range excitatory connections) in layer G, then neighbouring cells developed nearby orientation preferences, and there was a transition of orientation-sensitivity across the cell layer, reminiscent of orientation columns

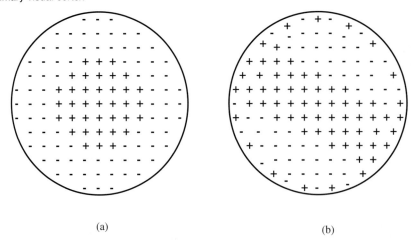

(a)   (b)

Fig. 2.12 Diagram of positive and negative connection strengths within the receptive fields of units in Linsker's network: (a) a centre-surround cell in layer C; (b) an orientation-selective cell in layer G, which would respond well to a light bar.

in V1 (Linsker 1986). This is an example of the formation of topological maps described in Chapter 7.

The exact ways in which different types of receptive fields develop in this architecture depending on the parameters in the learning rule have been analyzed in detail by Kammen and Yuille (1988), Yuille, Kammen and Cohen (1989), and MacKay and Miller (1990). Perhaps the most important point from these results in relation to the network just described is that a multilayer network with local feedforward connectivity, and using a local, modified Hebb, learning rule, can, even without structured inputs, develop quite complicated receptive fields that would be biologically useful as preprocessors, and that have similarities to the receptive fields of neurons in the mammalian visual system.

There are a number of issues that are not addressed by Linsker's network that are of fundamental importance, and that we now go on to address. One is that neighbouring neurons in Linsker's network develop within a layer independently of any information from their neighbours about how they are responding, so that the layer as a whole cannot learn to remove redundancy from the representations it forms, and in this way produce efficient encoding (cf. Barlow (1989)). (The lateral connections added as an option to layer G of Linsker's network assist in the formation of maps, but not directly in redundancy removal.)

A second issue is that Linsker's net was trained on random input firings. What now needs exploration is how networks can learn the statistics of the visual world they are shown, to set up analyzers appropriate for efficiently encoding that world, and removing the redundancies present in the visual scenes we see. If the visual world consisted of random pixels, then this would not be an issue. But because our visual world is far from random, with, for example, many elongated edges (at least over the short range), this implies that it is useful to extract these elements as features.

A third issue is that higher–order representations are not explicitly addressed by the network just described. A higher–order feature might consist of a horizontal line or bar above a vertical line or bar, which if occurring in combination might signify a letter 'T'. It is the co-occurrence, that is the pairwise correlation, of these two features which is significant in detecting the T, and which is a specific higher–order (in this case second–order) combination of the two line elements. Such higher–order spatial representations are crucial for distinguishing real objects in the world. (Form cannot be encoded by neurons sensitive to edges or bars if nothing is specified about the relative spatial positions of the edges or bars.)

A fourth issue is that nothing of what has been considered so far deals with or solves the problem of invariant representation, that is how objects can be identified irrespective of their position, size, etc., on the retina, and even independently of view.

A fifth issue is that Linsker's net is relevant to early visual processing, as far as V1, but does not consider how processing beyond that might occur, including object identification.

A sixth issue is that, for reasons of mathematical tractability, Linsker's network was linear. We know that many mappings cannot be performed in a single-layer network (to which Linsker's is equivalent, in terms of how it maps inputs to outputs), and it is therefore appropriate to consider how non–linearities in the network may help in the solution of some of the issues just raised.

These issues are all addressed in Chapter 8 on visual processing in multilayer networks which model some of the aspects of visual processing in the primate visual system between V1 and the anterior inferior temporal visual cortex (IT).

2.4.2   Olshausen and Field's method: Sparseness maximization

Olshausen and Field (1996) show that a learning algorithm that attempts to find sparse linear codes for natural scenes will develop a complete family of Gabor-like receptive fields similar to those found in the striate cortex. They assume that an image $R(x, y)$ can be decomposed as a linear superposition of basis function $\phi_i(x, y)$:

$$R(x, y) = \sum_i a_i \phi_i(x, y) \tag{2.3}$$

The aim of efficient coding is to find a complete set of basis functions $\phi_i(x, y)$, such that the resulting coefficient values a_i are as statistically independent as possible over a collection of natural scenes. A theory for the generation using learning of statistically independent or non-redundant codes has been developed in the framework of nonlinear independent component analysis (Deco and Obradovic 1996). If the information from the input image is conserved in the output code, statistical dependencies in the output code can be reduced by minimizing each individual entropy $H(a_i)$. In fact, the output redundancy is given by the following mutual information:

$$I(a_1, ..., a_i) = \sum_i H(a_i) - H(a_1, ..., a_i) \tag{2.4}$$

with the joint entropy always less than or equal to the sum of the individual entropies, i.e. $H(a_1, ..., a_i) \leq \sum_i H(a_i)$. Consequently, if the total information $H(a_1, ..., a_i)$ is preserved, the redundancy $I(a_1, ..., a_i)$ can be reduced by minimizing each individual entropy $H(a_i)$. This is the formulation of independent component analysis (ICA). Bell and Sejnowski (1997) have shown that the formation of edge-filter receptive fields can be accounted for by just applying linear ICA.

The theory of Olshausen and Field goes even further. They assumed that natural images have a sparse structure, i.e. any image can be described by a small number of active coefficients

a_i. They demonstrated that biologically realistic Gabor-like receptive fields can be described as conserving the transmitted information and as achieving a particular low-redundancy code in which the probability distribution of each coefficient's activity is unimodal and peaked around zero, i.e. a sparse code. From this, they derive a learning rule that minimizes a cost function that guarantees information maximization and sparseness[4]. This cost function can be written as follows:

$$E = \sum_{x,y} \left[R(x,y) - \sum_i a_i \phi_i(x,y) \right]^2 + \lambda \sum_i S\left(\frac{a_i}{\sigma}\right) \tag{2.5}$$

The first term in the right hand side (r.h.s.) of equation 2.5 accounts for the conservation of transmitted information, whereas the second term assesses the sparseness of the generated output code. The constant λ determines the relative weight between these optimization terms. The choice of the function $S(x) = log(1 + x^2)$, which is a function with high kurtosis, promotes the generation of sparse codes. For each image presentation, E is minimized with respect to the a_i according to the equilibrium solution to the differential equation

$$\frac{\partial a_i}{\partial t} = \sum_{x,y} \phi_i(x,y) R(x,y) - \sum_j a_j \sum_{x,y} \phi_i(x,y) \phi_j(x,y) - \frac{\lambda}{\sigma} S'\left(\frac{a_i}{\sigma}\right) . \tag{2.6}$$

The $\phi_i(x,y)$ then evolve by gradient descent on E averaged over presentations of many images. The learning rule for updating ϕ is then

$$\Delta\phi_i(x_m, y_m) = \eta \langle a_i \left[R(x_m, y_m) - \hat{R}(x_m, y_m) \right] \rangle \tag{2.7}$$

where $\hat{R}(x_m, y_m) = \sum_i a_i \phi_i(x_m, y_m)$, η is the learning rate, and $\langle \rangle$ means the average over many images.

With this learning rule Olshausen and Field (1996) demonstrated that localized, oriented, bandpass receptive fields emerge as the result of generating a non-redundant sparse code that maximizes information transmission. The result indicates that the receptive fields of simple cells are efficient in maximizing information transmission when a sparse code is used, but do not of course give an account for how the receptive fields of simple cells actually develop.

We think it likely that the actual development of simple cells is guided by a few genetic parameters of the connectivity and properties of the neurons (Rolls and Stringer 2000) which specify a Hebbian rule for the forward connectivity to V1 from the LGN, and mutual (lateral) feedback inhibition between neurons implemented by inhibitory neurons, to make the network operate like a competitive network (see Section 7.4) to produce a sparse representation. Such a network, working on the statistical distribution of the inputs actually received from the visual world, and utilizing the non-linear properties of the neuronal activation function, can help to produce quite useful firing rate distributions (Treves, Panzeri, Rolls, Booth and Wakeman 1999). Some of the possible underlying neural circuits involved in producing simple cells are described by Ferster and Miller (2000).

---

[4] An equivalent but purer statistical Bayesian formulation was produced by Lewicki and Sejnowski (2000).

2.5 The computational role of V1 for form processing

In addition to their functions in edge detection described in Section 2.3, V1 neurons also may function more collectively to incorporate contextual information from outside their classical receptive fields. These contextual influences may serve pre-attentive visual segmentation by causing relatively higher neural responses to important or conspicuous image locations, making them more salient for perceptual pop-out. These locations include boundaries between regions, smooth contours, and pop-out targets against backgrounds. The mark of these locations is the breakdown of spatial homogeneity in the input, for instance, at the border between two texture regions of equal mean luminance. This breakdown causes changes in contextual influences, often resulting in higher responses at the border than at surrounding locations. These processes may contribute to texture segmentation, figure-ground segregation, target-distractor asymmetry, and contour enhancement, and have been modelled by Grossberg and Mingolla (1985a), Grossberg and Mingolla (1985b), and Li (2000). These models are examples of the application of dynamical models to intermediate-level visual processing. We illustrate the application of dynamical models to understanding high-level perceptual processing in Chapters 9–11.

The theory developed in Chapter 8 of invariant object representation and recognition predicts and requires that neurons in early cortical processing stages should respond to combinations of features in a particular spatial configuration. Evidence for this is now starting to appear. Some V1 neurons in cats respond to two lines that cross each other, or join in 'L' or 'T' arrangements (Shevelev, Novikova, Lazareva, Tikhomirov and Sharaev 1995). Some of these neurons have responses to single bars, but the orientation tuning shows two peaks that correspond to the orientations of the two lines to which the neuron is most responsive. Others of these neurons respond well to a combination of two lines forming a spatial feature combination, but respond very little to single lines. This emphasizes that non-linearity is a property of at least some V1 cells. These neurons require the two lines to be in the correct spatial arrangement for maximal responses, and have small receptive fields. More neurons were found to be sensitive to combinations of lines that to single oriented lines of bars. Sillito, Grieve, Jones, Cudeiro and Davis (1995) stimulated V1 cells at their optimal orientation and then introduced a surrounding field at an orthogonal orientation and found that this enhanced neuronal responses by both disinhibitory and an active facilitatory mechanism. Further evidence consistent with the hypothesis that responses to single bars or edges is not all that is encoded in V1 is that the majority (80%) of V1 neurons in macaques have responses that are greater to complex stimuli (such as Walsh patterns) than to the best oriented bar or edge (B.J.Richmond, personal communication, 2001.) Other features that are useful descriptors of low-order feature combinations that occur in natural and man-made scenes include 'Y' and arrow features, and it is likely that these, and further related descriptors, will be found to be very effective stimuli for some neurons throughout early and intermediate visual cortical processing areas such as V1, V2, and V4.

2.6 Backprojections from the primary visual cortex, V1, to the lateral geniculate nucleus

One aspect of V1 architecture that has led to much speculation is the enormous number (perhaps 15–30 million) of backprojecting fibres from V1 to the lateral geniculate nucleus.

This is much greater than the number of fibres from the lateral geniculate to V1 (approximately 1 million, figures for Old World monkeys). Many functions have been suggested. One very simple possible explanation for the much larger number of backprojecting axons in this V1-to-geniculate architecture is that deep cortical pyramidal cells have a genetic specification to have backprojecting axons, and because there are many more V1 cells than lateral geniculate cells (15–30 million in layer 6, the origin of the backprojecting axons), there are more backprojecting than forward projecting fibres in this particular part of the system. The suggestion is that this simple genetic specification works well for most cortico-cortical systems, producing approximately equal numbers of forward and backprojecting fibres, because of the approximately similar numbers of cells in connected cortical regions that are adjacent in the hierarchy. Use of this simple rule economizes on special genetic specification for the particular case of backprojections from V1 to the lateral geniculate (cf. Rolls and Stringer (2000)).

3 Extrastriate visual areas

3.1 Introduction

In the previous chapter, we described the early segregation of the visual pathways from the retina to the lateral geniculate nuclei (LGN), and then to the striate cortex as constituting at least three partially segregated visual streams: the magnocellular system; the parvocellular-interblob system; and the parvocellular-blob system. We discuss in this Chapter, the anatomical and functional continuation of these three parallel visual pathways in cortical areas beyond the striate cortex (V1), namely into the extrastriate cortical areas V2, V4 and V5. At the perceptual level, these different parallel pathways mediate the processing of information for, respectively, motion and depth; form; and colour. The unity achieved by the visual system which allows us to experience the world as a whole, is not realized by a pure serial feedforward hierarchical system, but by the interaction between partly separate, parallel, functionally specialized systems. (As a matter of fact, conscious visual perception may be more associated with the operation of the ventral than the dorsal processing stream (Milner and Goodale 1995) (see Section 12.9 and Fig. 1.10), and this may be related to the fact that we can perform long-term planning about objects, a process that when helped by higher–order thoughts used to correct the plans, may be closely related to consciousness (Rolls 1999a). We will see in further Chapters that visual attentional mechanisms allow these processing streams to interact, and that these attentional effects are implemented via top-down feedback parallel interactions.

3.2 Visual pathways in extrastriate cortical areas

Before we analyze the continuation of the three parallel early visual processing streams in extrastriate areas, let us review briefly the architecture and pattern of connections of the different areas in the visual cortex. Figure 3.1 shows a schematic connectivity diagram of the cortical visual areas in the macaque. There are around 30 cortical areas in the primate brain that are known to process visual information (Felleman and Van Essen 1991). As we have already described in Chapter 2, the striate cortex, or area V1, is the first projection region of the lateral geniculate neurons into the cortex. The extrastriate cortical streams V1–V2–V4–TEO–TE and V1–V2–MT(or V5)–MST (shown in Fig. 3.1) are not connected in a purely hierarchical form, but, especially in the dorsal stream, may have connections that jump one stage, and some cross-connections between them (see Chapters 5 and 9 for more details). Even more, in general the connections between cortical areas are reciprocal (i.e. if one area projects forward to a second area, this second area also projects back to the first area, as described in Section 1.11). In the ventral path (V1–V2–V4–TEO–TE), the size of the receptive fields of the corresponding neurons increases from 0.5 deg in V1 to 40 or more deg in the inferior temporal cortex (IT) (architectonic area TE) (see Chapters 5 and 8), and the receptive fields also become larger along the dorsal stream. The larger receptive fields enable some computations to be performed that cannot be performed with small receptive fields,

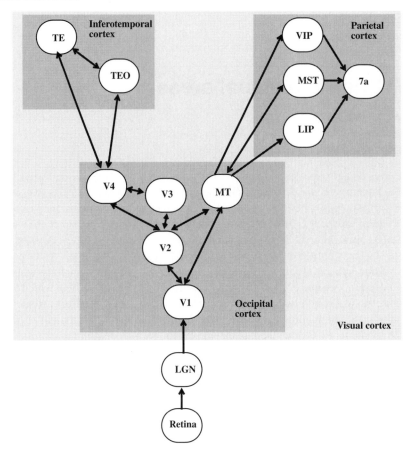

Fig. 3.1 Main visual processing areas in the macaque cortex. The 'what' or ventral pathway computing shape, colour, etc., runs from the occipital lobe down to the inferior temporal lobe (areas V1, V2, V4, and inferior temporal (IT) areas TEO (posterior inferior temporal cortex) and TE (anterior inferior temporal cortex). This pathway is involved in the identification of objects or parts of objects. The second pathway, associated with the extraction of spatial properties like location, motion, etc., is called the 'where' or dorsal pathway and runs from the occipital lobe up to the parietal lobe (areas V1, V2, V3, MT, MST, and further areas in the inferior parietal cortex and the cortex in the posterior part of the superior temporal sulcus).

such as colour constancy in V4, translation (shift) invariance for objects across the retina in the inferior temporal visual cortex, and the computation of global motion when the size of the 'aperture' or receptive field is no longer small, in MT, as described below.

The characteristics of the stimuli to which neurons respond become more complex along the ventral stream. The striate and early extrastriate areas are topographically organized, so that they form (at least) six retinotopic maps, namely in areas V1, V2, V3, V3a, V4 and V5. With respect to Brodmann's nomenclature based on the microscopically defined architecture of the cortex, area V1 is Brodmann's area 17, V2 and V3 are in Brodmann's area 18, and V3a, V4, and V5 are in Brodmann's area 19. V5 is also called MT (middle temporal area), and is located on the posterior bank of the cortex in the posterior (upper) part of the superior temporal sulcus. On the anterior bank of the same sulcus also posteriorly lies area V5a or

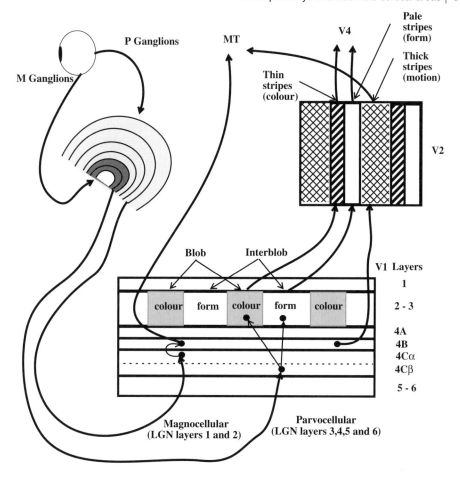

Fig. 3.2 Schematic diagram of the anatomical segregation of the visual pathways in extrastriate areas.

MST (medial superior temporal). These extrastriate areas show different retinotopic precision and their neurons respond to different features.

Visual area V1 projects mainly to area V2. Tootell, Silverman, De Valois and Jacobs (1983) discovered a distinctive pattern in area V2 when stained with cytochrome oxidase. Instead of finding small dark blobs as in area V1, they found in V2 much coarser tangential stripes. They distinguished between three different type of stripes, namely two types of alternating dark thick and dark thin stripes separated by a third type of pale stripes. Livingstone and Hubel (1988) analyzed how the three different segregated systems in the striate cortex project into V2 and higher areas. From tracer injections into the three kinds of stripes in V2, they found that the blobs of layers 2–3 of V1 are reciprocally connected to the thin stripes, the interblobs to the pale stripes, and layer 4B to the thick stripes (see Fig. 3.2). Furthermore, analysing the types of responses of neurons in the three different kinds of stripes and in MT, they identified four different segregated visual pathways in the extrastriate cortex, which are associated with four different perceptual subfunctions. Figure 3.2 shows schematically the underlying connectivity associated with these four segregated visual pathways.

The first segregated visual pathway is mainly responsible for motion processing and its anatomical substrate is the connection from the retinal M cells, to the LGN magno system, to V1 layer 4Cα, to V1 layer 4B, to MT, and thus to MST.

The second segregated visual pathway processes binocular information to extract depth and its anatomical substrate is the connection from the retinal M cells, to the LGN magno system, to V1 layer 4Cα, to V1 layer 4B, to the V2 thick stripes, and thus to MT.

The third segregated visual pathway is engaged with the analysis of form and its anatomical substrate is the connection from the retinal P cells, to the LGN parvo system, to V1 layer 4Cβ, to the V1 interblobs, to the V2-pale stripes, to V4, and thus to IT.

The fourth segregated visual pathway is responsible for the processing of colour and its anatomical substrate is the connection from the retinal P cells, to the LGN parvo system, to V1 layer 4Cβ, to the V1 blobs, to the V2 thin stripes, to V4, and thus to IT.

In fact, Livingstone and Hubel (1988) found that neurons in the thin stripes of visual area **V2** showed no orientation selectivity, and over half were colour coded, just as in the blobs of V1. As in the blobs, these cells were doubly opponent, i.e. with two antagonistic inputs to their centres (e.g. excitatory to red and inhibitory to green), and surround antagonism for both of these centre inputs (e.g. inhibitory to red and excitatory to green). Neurons in the pale stripes are orientation selective but insensitive to motion and colour. Over 50% of the pale stripe neurons are end-stopped, which makes these cells especially suitable for form processing. Neurons in the thick stripes are orientation sensitive and insensitive to colour. In contrast to neurons in the pale stripes, neurons in the thick stripes are seldom end-stopped. The most important feature of these cells is their binocular sensitivity (i.e. most of these cells respond best when both eyes are simultaneously stimulated). Even more, they show great sensitivity to stereoscopic depth and retinal disparity.

Neurons in higher visual areas, are also very specialized. Zeki (1976) (see also Zeki (1993)) discovered that for example area V4 is particularly sensitive to colour and orientation, whereas area MT is sensitive to motion and depth analysis.

Clinical neuropsychological evidence (Farah 1990, Farah 2000, Zihl 2000) confirms the functional segregation of the visual system into anatomically separated subsystems. Sigmund Freud introduced the term 'agnosias' to characterize the inability of certain patients to recognize visual objects due to a localized cortical injury, when low-level visual function as measured for example by visual acuity may be normal. Some patients, for example, cannot process colour information, saying that there is no colour in objects (achromatopsia), because of localized damage to a part of the temporal cortex, which is the region that in humans contains what may be a homologue of visual area V4. Other patients are not able to process motion information (movement agnosia), due to bilateral damage to visual areas MT or MST. There also exist visual agnosias related to shape processing in which patients cannot identify objects, and these agnosias can be related to damage to the ventral visual system, especially equivalents in humans of the inferior temporal visual cortical areas. In one type of object agnosia, **apperceptive agnosia**, subjects have normal acuity, but cannot match or draw objects. This type of agnosia thus reflects damage to intermediate-level shape analysis. In a second type of agnosia, **associative visual agnosia**, patients can match or draw objects (showing that intermediate-level shape analysis can still operate), but cannot identify or name objects (Lissauer 1890, Farah 1990, Farah 2000, Kolb and Whishaw 1996).

We conclude this Section by emphasizing the relevance of reciprocal interactions between the different visual subsystems that provide an essential component of the mechanisms by which the different localized and partly specialized modular subcomponents can be integrated to provide a single and coherent visual percept. Analysis of the dynamical interactions between neural modules, which leads to a quantitative understanding of how visual attention works, is one of the main issues treated in later Chapters of this book, and we believe that computational neuroscience offers an appropriate theoretical framework for analysing the neurodynamical and parallel integration and interaction of the different segregated subfunctions that result in human behaviour.

3.3   Colour processing

3.3.1   Trichromacy theory

It is well known that human subjects with normal colour vision are able to match single colours by a combination of three colours in different parts of the spectrum, e.g. blue, green and red. In 1802, the British scientist Thomas Young proposed a trichromatic theory of colour vision. This theory postulates that colour experiences originate from the simultaneous activation of three types of photoreceptors with different but overlapping spectral responses. In 1964, Edward MacNichol and his collaborators and George Wald and Paul Brown (see Kandel, Schwartz and Jessel (2000)) confirmed the existence of three different types of visual pigments of single cones in the retina. Individual cones contain only one of these three pigments, giving origin to the three different types of cone already mentioned in Section 2.2. One type of cone, the short or blue cone, is mainly sensitive to short wavelengths and therefore related to the perception of the colour blue. The second type of cone, the middle or green cone, is selective to middle wavelengths and primarily associated with the colour green; and the third type of cone, the long or red cone, is sensitive to long wavelengths which make a strong contribution to the perception of the colour red. These three cone types cover the spectral range of wavelengths from 400 to 700 nm, which is the range to which the human eye is sensitive. At any particular wavelength, all three photoreceptors are simultaneously activated to different extents. A perceived colour is determined by the pattern of activation of these three different photoreceptors. Note that each single cone type cannot extract colour information, because each type reacts to a broad range of different wavelengths. Only the simultaneous combination of the activation of the three different cones allows the visual system to extract normal information about colour. In fact, colour blindness is caused by a defect in the cone pigments. Dichromacy results from the absence of one type of cone and monochromacy results from the simultaneous absence of two types of cones. In the case of red or green blindness, it is even known that the defects in the red or green cones can be genetic, with the main colour genes present on the X chromosome leading to sex-linked colour deficits that occur mainly in males (see Kandel, Schwartz and Jessel (2000)).

3.3.2   Colour opponency, and colour contrast: Opponent cells

The existence of three different types of photoreceptors in the retina rationalize several aspect of colour perception, such as the trichromacy property just described. Nevertheless, in general colour processing requires computational neural mechanisms that are much more complex than the single analysis of a pattern of activation of three different photoreceptors. This is shown particularly by analysis of three important properties of colour perception as described next: colour opponency, colour contrast, and colour constancy.

3.3.2.1 Colour opponency

Colour opponency refers to the fact that the pairs red and green, yellow and blue, and white and black antagonize each other in contributing to colour perception when they come from the same region of the retina. This is the reason why the subjective experience of yellow when an object reflects green and red does not appear to be a mixture of green and red, in the way that purples appear to be a combination of red and blue. Ewald Hering (1878 (republished 1964)) formulated the opponent process theory in order to explain this phenomenon. The opponent process theory proposes that each antagonistic colour pair is processed in a separate neural channel in the retina. Each channel responds with excitation or inhibition to one of the colours of the pair (e.g. red) and in an antagonistic way, with inhibition or excitation, to the other opponent colour (e.g. green). When both opponent colours stimulate the corresponding channel in a balanced form, the output activity of the neural channel vanishes. The neurophysiological substrate corresponding to these colour-opponent neural channels has been described in both the retina and in the lateral geniculate nucleus of primates (Kandel, Schwartz and Jessel 2000). Retinal ganglion cells and neurons in the lateral geniculate nucleus of primates can be classified into: broad-band cells, concentric single-opponent cells, and co-extensive single-opponent cells.

Broad-band cells show the typical centre-surround antagonism of retinal ganglion cells (see Section 2.2) but they do not show antagonism between cone mechanisms. Both the centre and the surround receive input from red and green cones, so that they are not colour specific[5]. In fact, broad-band cells do not contribute to the processing of colours and they respond only to brightness. Figure 3.3 shows the underlying receptive field structure of this type of neuron.

Specific colour processing is performed by colour-opponent retinal and lateral geniculate cells. The concentric single-opponent cells present not only the typical centre/surround antagonism but they show also antagonism between red and green cone inputs. As shown in Fig. 3.3, the centres of these neurons receive excitatory or inhibitory input from red or green cones, and the surround receives the opposite antagonistic input from the opposite type of cone. Concentric single-opponent cells do not respond only to chromatic stimuli. They process information about both colour and achromatic brightness contrast.

The third type of cell, coextensive single-opponent cells, transmit information from the blue cones. They have a uniform receptive field without centre/surround antagonism, but with inputs from blue cones that antagonize the combined inputs of red and green cones (see Fig. 3.3).

3.3.2.2 Colour contrast

The outputs of lateral geniculate single-opponent cells are combined in the cortex in the so-called double opponent cells. This type of neuron is concentrated in the blob zones of layers 2 and 3 of the primary visual cortex (V1), which is part of the parvocellular processing system. The outputs of the blob cells are transmitted to the thin stripes of V2 and from there to the colour-specific neurons of V4. The receptive fields of double-opponent cells have a centre/surround antagonistic organization, but each type of cone provides inputs to both parts of the receptive field, eliciting opposite effects in a given part of the receptive field, as shown in Fig. 3.4. Figure 3.4 also shows the possible functional connectivity from LGN cells to

---

[5]The blue cones do not connect with broad-band cells, meaning that short wavelengths are not used for the processing of form.

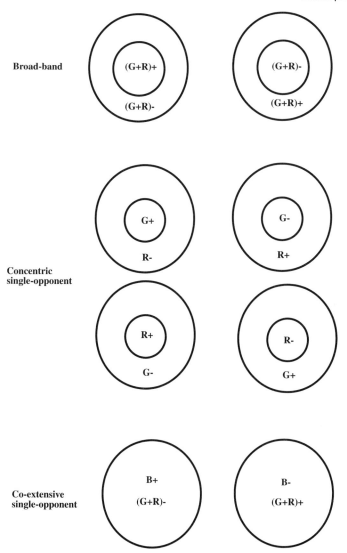

Fig. 3.3 Receptive field structure of broad-band cells, concentric single-opponent cells, and co-extensive single-opponent cells. Symbols: R (red), G (green), B (blue), + (excitatory), and − (inhibitory).

double opponent cells. The example in the Figure corresponds to a double-opponent neuron that receives excitatory input from red cones in the centre and inhibitory in the surround, and inhibitory input from green cones in the centre and excitatory in the surround. These kind of neurons are quite selective for chromatic stimuli. They support the phenomenon of colour opponency because of the antagonistic effect of opposite colours throughout the receptive field, and they provide an explanation of the phenomenon of colour contrast.

The phenomenon of colour contrast is present perceptually when opposite colours come from different positions, for example from different sides of a boundary. For example, a grey object in a red background looks greenish, whereas the same grey object in a green background

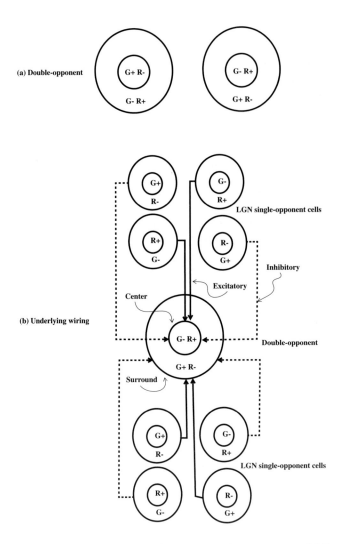

Fig. 3.4 (a) Double-opponent cells responding preferentially to red-green contrast. (b) Hypothetical underlying wiring of a red-green contrast double-opponent cell. On- and off-centre single-opponent cells from the lateral geniculate contribute to both the centre and surround of double-opponent cells in the cortex.

looks reddish. In contrast to colour-opponency in which opposite colour pairs cancel each other if they are in the same spatial location, with colour contrast they seem to facilitate each other when they are in different neighbouring spatial locations. A double-opponent neuron that is stimulated by red and inhibited by green in its centre will respond the same either when a small red spot light excites its centre, or a green light excites its surround. This explains why a green background may cause a central grey object to have a red tinge.

3.3.2.3 Colour constancy

Colour constancy refers to the fact that we perceive the colour of an object as relatively constant independently of the global wavelength composition of the illuminating light, even though this causes dramatic changes in the spectral composition of the reflected light captured by the photoreceptors of the retina. For example, we have the experience that a banana is always yellow whether seen under natural daylight, or at sunset when the illuminating light is biassed towards the longer wavelengths, or in artificial illumination.

Double-opponent cells contribute to the phenomenon of colour constancy because an increase in the long wavelength component of global illumination has little effect on the output of these cells. This is because the increase in wavelength is the same for both the centre and the surround of the cell's receptive fields. Effectively, subtracting the input to the surround from the input to the centre enables double-opponent cells to code for the difference in the spectra of the light in the centre and the surround, and thus to discount the wavelength composition of the illuminating light. This enables the cells to provide evidence about the true colour of each patch in the visual field, and thus to implement colour constancy.

Zeki (1993) has shown that further on in the system, neurons in macaque area V4 respond in a way which reflects colour constancy and not the wavelength at the centre of the receptive field. These neurons contribute to discounting the wavelength composition of the illuminating light, and their role in this is likely to be facilitated by their larger receptive fields than in earlier processing areas. Their larger receptive fields, up to several degrees across, enable them to obtain, in their surround, a reasonable estimate of the illuminant, because in a large region, objects with different colours are more likely to be present than in small areas of the visual field. A possible homologue in humans of this V4 area, or a related region, has been shown with neuroimaging techniques to have a strong response to chromatic stimuli (Zeki, Watson, Lueck, Friston, Kennard and Frackowiak 1991, McKeefry and Zeki 1997, Wandell 2000).

3.4 Motion and depth processing

3.4.1 The motion pathway

In Section 3.2 we have seen that motion processing in the visual cortex is segregated into a pathway that starts at the retina and reaches higher cortical areas in the parietal cortex. The motion pathway (see Figs. 2.1 and 3.2) starts in the M-type ganglion cells of the retina which project through the magnocellular layers of the lateral geniculate nucleus to layer $4C\alpha$ in V1, from there to layer 4B, then to the thick stripes of V2, then both directly and through V3 on to MT (V5), and finally from there to area MST (V5a) in the parietal cortex. The retinal M-type ganglion cells are not motion sensitive, but they respond more rapidly to change in stimulation than P-type cells, they are strongly sensitive to contrast and contrast variation, have large receptive fields, and have low spatial resolution. These make M-type retinal ganglion cells specially suitable for pre-processing for motion. In the primary visual cortex (V1), neurons are explicitly selective to a particular direction of motion. Figure 3.5 shows the response of a motion-sensitive cell in V1. Its directional selectivity can be plotted as a graph in polar coordinates, in which each point's distance from the centre corresponds to the firing rate of the neuron in each motion direction that is measured.

In areas MT and MST, motion processing is further elaborated. Neurons in these areas shown high selectivity to direction of motion and speed. In particular, Movshon and his

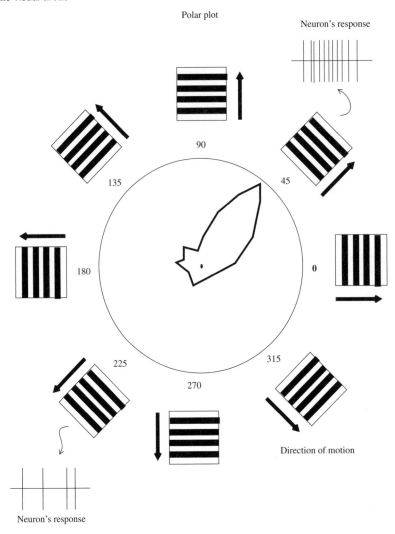

Fig. 3.5 Directional tuning of a V1 direction-selective cell. A grating was moved through the receptive field of this cell in various directions. Directional selectivity can be plotted as a graph in polar coordinates, in which each point's distance from the centre corresponds to the firing rate of the neuron in each motion direction that is measured.

colleagues (Movshon, Adelson, Gizzi and Newsome 1985) found that there are two differ-ent types of motion-sensitive neurons in area MT: component direction-selective neurons, and (global) pattern direction-selective neurons. Component direction-selective neurons are similar to motion-sensitive neurons in V1, in the sense that they respond only to motion per-pendicular to their axis of orientation. Thus, their responses provide information about only one-dimensional local components of motion of a global pattern or object, which of course can be ambiguous with respect to the real state of global motion of the whole pattern. This ambiguity can be understood by considering the so-called aperture problem. Figure 3.6 shows this phenomenon. A large grating moving in three different directions can produce the same

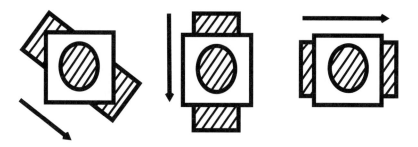

Fig. 3.6 The aperture problem. A large grating moving in three different directions can produce the same physical stimulus when viewed with a small circular aperture, so that the direction of motion within the aperture appears in all three cases to be the same.

physical stimulus when viewed with a small circular aperture, so that the direction of motion within the aperture appears in all three cases to be the same. Movshon and his colleagues, used a similar set up for distinguishing the detection of global motion versus component motion. They used plaid images, produced by overlapping two moving gratings (see Fig. 3.7), which make the aperture problem explicit. The global motion of the plaid pattern is different from the motion of the two component gratings. The responses of V1 neurons, and of component direction-selective neurons in MT, detect only the motion of each grating component and not the global motion of the plaid.

Movshon, Adelson, Gizzi and Newsome (1985) found that a small population of neurons in MT (about 20%), which they called pattern direction-selective neurons, responds to the perceived global motion of the plaid. These neurons integrate information about two- (or three-) dimensional global object motion by combining the output responses of component direction-selective neurons, and of V1 neurons that extract just the motion of the components in different one-dimensional directions.

Further, Newsome, Britten and Movshon (1989) demonstrated experimentally that the responses of MT motion-sensitive neurons correlate with the perceptual judgements about motion performed by a monkey that has to report the direction of motion in a random dot display.

These results are consistent with neuropsychological evidence on human motion perception. Zihl, Von Cramon and Mai (1983) describe a patient with damage in an extrastriate brain region homologous to the MT and MST areas of monkeys who was unable to perceive motion. She reported that the world appeared to her as a series of frozen snapshots. Her primary visual cortex was intact, and this was the reason why she was able to detect simple one-dimensional motion in the plane, but unable to integrate this information to infer global motion.

3.4.2 Depth perception

The convergence of our eyes when we fixate on a point on an object less than 30 metres away produces stimulation for the point fixated that falls on identical regions of each retina. On the other hand, points that are nearer or farther than the fixation point stimulate slightly different portions of the retina of each eye. This phenomenon is known as binocular disparity, and offers a physical cue for stereopsis, i.e. the perception of depth information based on differences in the images in the two eyes.

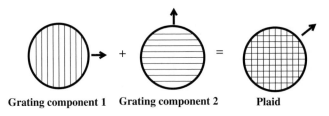

Grating component 1   Grating component 2   Plaid

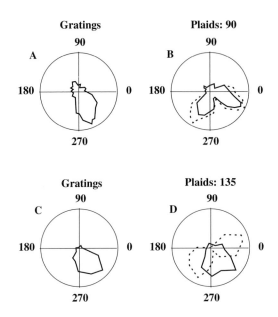

Fig. 3.7 Directional selectivity of component direction-selective neurons and (global) pattern direction-selective neurons in area MT to single versus plaid gratings. The combination of two gratings to produce a single plaid showing global motion is shown in the top part of the diagram. (A) Polar graph of a component direction-selective neuron responding to a single grating moving in different directions. (B) Polar graph of the same component direction-selective neuron responding to a plaid grating (solid line), together with how it should respond if it was responding to each single component (dashed line). (C) and (D) the same as (A) and (B) but for a pattern (or global) direction-selective neuron. (Adapted from Movshon, Adelson, Gizzi, and Newsome, 1985.)

Hubel and Wiesel (1962) showed that the fusion of information coming from the two eyes occurs first in V1. They found this in V1 binocular cells, i.e. cells that responded preferentially when both eyes were stimulated. Nevertheless, stereopsis requires not only binocular neurons but cells that are sensitive to binocular disparity. Barlow, Blakemore and Pettigrew (1967) described such cells in V1. More modern studies of Poggio and Fischer (1977) demonstrated that neurons in V1 are highly sensitive to small or zero disparity, whereas neurons in V2 (in the thick stripes) are highly sensitive to large disparity. In areas V3, MT, and MST, binocular neurons sensitive to disparity were also found.

In V1, MT and MST, depth-tuned neurons can respond to anticorrelated depth images, that is, where white in one eye corresponds to black in the other (Parker, Cumming and Dodd 2000, DeAngelis, Cumming and Newsome 2000), and this may be suitable for eye movement control, but much less suitable for shape and object discrimination and identification. Neurons that respond to 3D cues for shape based on disparity cues have also been found in the inferior temporal cortical visual areas (especially TEa and TEm, see Janssen, Vogels and Orban (1999), Janssen, Vogels and Orban (2000) and Chapter 5). Interestingly, the inferior temporal neurons respond only when the black and white regions in the two eyes are correlated (that is, when a white patch in one eye corresponds to a white patch in the other eye) (P. Janssen, personal communication, June 2001). This corresponds to what we see perceptually and consciously. Moreover, the inferior temporal cortex neurons tend to be tuned to quite small disparities, of for example 0.25 degrees, and this is appropriate for a system that utilizes depth information from stereopsis as a guide to the 3D structure of objects. The coarser tuning of MT and MST neurons may be appropriate for a system that must correct convergence eye movements when disparities become excessive and out of the normal working range.

In summary, the magnocellular pathway (retinal M cells – LGN-magno cell – V1-4Cα – V1-4B – V2-thick stripes – MT) forms a segregated pathway responsible for much of the processing of binocular depth information, and in these areas which eventually reach parietal visual areas, the processing may be used for helping to execute actions in 3D egocentric space. In addition, some binocular depth information reaches the inferior temporal cortex areas, especially TEa and TEm, which receive inputs from the cortex in the intraparietal sulcus, and there it may be useful for structure from depth, that is in helping object recognition by providing some evidence about the 3D structure of objects.

4 The parietal cortex

4.1 Introduction

Humans have the capacity to recognize objects in natural visual scenes with high efficiency despite the complexity of such scenes, which usually contain multiple, partly covered, and often distorted, object features. Visual attention provides a mechanism for the selection of behaviourally relevant information from such complex natural scenes. The limited processing capacity of the visual system does not allow the simultaneous analysis of many different objects[6]. Selective visual attention facilitates therefore the processing of that limited portion of the input associated with relevant information and it suppresses the remaining irrelevant information. If we consider spatial attention, for example, then attention is guided by top-down processing to a certain spatial location in the visual field. Information coming from the attended location will be facilitated whereas information coming from the other unattended locations will be reduced. Thus, in addition to attentional mechanisms, a representation of space is also required for spatial attention. Both components, attention and a spatial representation, constitute some of the basic ingredients of spatial cognition, helping us to recognize objects, and to move and act in a complex and changing world.

In this Chapter, we outline the contribution of the parietal cortex to spatial cognition. In the first Chapters of this book, we have seen that the visual system processes information in distinct brain areas that are hierarchically organized. Within this hierarchical system, there are two parallel streams of areas: the ventral or 'what' stream, and the dorsal or 'where' stream. The posterior region of the parietal cortex contains the higher order cortical areas of the dorsal pathway where extrastriate area MT projects. We discuss in this Chapter the structure and function in spatial attention and spatial representation of these parietal higher order areas of the dorsal visual stream. The structure and role of the higher order areas of the ventral stream progressing to the temporal lobe that are involved in object recognition are described in Chapters 5 and 8.

4.2 Spatial information processing in the parietal lobe

The parietal lobe is divided into superior and inferior lobules. In both humans and monkeys, the anterior region of the parietal cortex, corresponding roughly to the superior lobule and containing areas 5 and 7b, is mainly engaged with the processing of somatosensory information (see Fig. 1.8), whereas the posterior region, corresponding roughly to the inferior lobule, is related primarily with the processing of visuospatial information. In this Chapter, we focus on the posterior parietal cortex. In monkeys, the posterior parietal cortex can be further

---

[6] In fact, in the human visual system, the amount of information transferred from the retina to the brain is estimated to be in the range of 10^8–10^9 bits per second, which is far in excess of what the brain is capable of fully processing and assimilating into conscious experience.

subdivided into functionally segregated regions, including the lateral intraparietal area (LIP), the ventral intraparietal area (VIP), Brodmann area MST, and area 7a (see Figs. 1.9 and 1.10). The homologous regions in humans are still not established.

We consider in the next subsections the role of areas LIP, VIP, MST and 7a in the monkey for the processing of spatial information. Through well-designed experiments with trained monkeys that perform eye saccades to visual targets, it has been possible to differentiate between the responses of single neurons in LIP and VIP during spatial tasks, and how they provide different spatial representations.

4.2.1 Area LIP

Much experimental evidence is starting to provide an understanding of the functional role of LIP neurons in spatial representation and spatial attention (Robinson, Goldberg and Stanton 1978, Bushnell, Goldberg and Robinson 1981, Gnadt and Andersen 1988, Goldberg, Colby and Duhamel 1990, Andersen, Bracewell, Barash, Gnadt and Fogassi 1990b, Duhamel, Colby and Goldberg 1992a, Duhamel, Goldberg, Fitzgibbon, Sirigu and Grafman 1992b, Andersen, Snyder, Bradley and Xing 1997, Andersen, Batista, Snyder, Buneo and Cohen 2000). The responses of LIP neurons implement the transformation of retinotopic spatial coordinates into what is in part a head-based spatial representation. This transformation allows mapping between auditory inputs, which are of course in head-based coordinates, and visual space, which may be useful for example in enabling attention and eye movements to be directed to where a sound is heard (Andersen, Batista, Snyder, Buneo and Cohen 2000). However, the head-based auditory reference frame is converted into a partly eye-centred coordinate frame in LIP, in that the responses to auditory stimuli of some LIP neurons are modulated by eye position (Andersen, Batista, Snyder, Buneo and Cohen 2000). The remapping of visual inputs into at least partially head-based coordinates may also be a factor in the stability of the visual world. In addition, the responses of LIP neurons reflect the allocation of spatial attention to specific task-relevant locations. A brief historical summary of the experiments that lead to these conclusions is provided next.

Robinson, Goldberg and Stanton (1978) showed that LIP neurons have receptive fields at specific retinal locations, in that the neurons respond to the onset of a visual stimulus presented in a region within the neuronal receptive field. Andersen, Bracewell, Barash, Gnadt and Fogassi (1990b) showed that the responses of these neurons are modulated by the position of the eye in the orbit. These neurons effectively respond to a particular combination of visual eccentricity on the retina *and* eye position, and thus reflect where the visual stimulus is located based in head-based coordinates. Although different neurons do not respond to all combinations of retinal position of a stimulus and eye position that together correspond to a particular location in head-centred space, the neurons may nevertheless represent the type of coding that is at an intermediate stage of processing in such a coordinate transform network (Zipser and Andersen 1988). More recently, it has been found that the visual receptive fields of some LIP neurons are also gain-modulated by head position, and together with the eye position 'gain field' effects could provide a representation that is at least partly body-centred (Snyder, Grieve, Brotchie and Andersen 1998, Andersen, Batista, Snyder, Buneo and Cohen 2000). The evidence thus suggests that LIP visual (and auditory) neurons (and reach-related neurons in the nearby parietal reach region) encode information in an eye-based coordinate frame, which effectively comes close to being a head-based or even body-based coordinate frame because of the 'gain field' modulation by eye position and head position (Andersen, Batista,

Snyder, Buneo and Cohen 2000).

In addition, the responses of LIP neurons to a stimulus within their receptive field are modulated by attention. Bushnell, Goldberg and Robinson (1981) observed an enhanced response of LIP neurons when the stimulus within their receptive fields becomes relevant, independently of the particular response the monkey makes to the stimulus. In other words, attentional modulation is observed whether the monkey has to notice the stimulus by performing a hand movement, or an eye saccade, or even by avoiding making a movement to the stimulus.

Another interesting property of LIP neurons was revealed by Gnadt and Andersen (1988), who showed that the typically tonic responses of LIP neurons last even for seconds after the stimulus has disappeared from the corresponding receptive field. This supports the idea that LIP neurons implement a kind of spatial memory trace for the location of absent targets.

Another remarkable property of LIP neurons is their anticipatory response just before an eye saccade is performed toward the position of their receptive field (Goldberg, Colby and Duhamel 1990).

Perhaps the most revealing property of LIP neurons is observed in the context of remapping of visual memory trace activity. As we mentioned above, a remapping of the visual world around us may be necessary for generating a head-centered internal image of the world that is independent of our eye movements. The experiments of Duhamel, Colby and Goldberg (1992a) provide a hint that LIP neurons directly contribute to the update of this stable internal representation. This experiment shows that the memory trace of a previous stimulus event is shifted after an eye movement to match the new eye position. Figure 4.1 illustrates the experimental set up and results of this experiment.

The firing rate activity of a single LIP neuron was recorded under three different conditions. In the first condition, the monkey has to fixate on a particular point (the cross in the top part of the Figure), and a stimulus (the black square in the top part of the Figure) is presented on the receptive field of the neuron. As expected, high activity after stimulus onset was observed. The second condition involves a saccade that brings the receptive field onto a stationary persisting visual stimulus. Initially the receptive field of the neuron is on a blank region of the screen. After the saccade, the receptive field of the neuron has been shifted to a position where a stimulus is present. When the receptive field is aligned with the location of the stimulus after the saccade, the neuron responds strongly. The third condition is similar to the second condition, but now the stimulus presented at the new position disappears before the receptive field of the neuron reaches the final location, so that the stimulus is never physically present in the receptive field. Nevertheless, the neuron responds with high activity after the saccade, as though there was a stimulus in the receptive field.

This result leads to the conclusion that LIP neurons respond to the memory trace of a previous stimulus and that the representation of the memory trace is updated at the time of the saccade. Before the saccade and when the stimulus appears, neurons with the receptive field at the location of the stimulus are highly activated. When the stimulus disappears, a memory trace maintains activation of those neurons. After the saccade, the eye movement information is processed in the parietal cortex and the memorized location is remapped for the new eye position. This produces a spatial head-centred representation of the world that is stable despite eye movements. Sustaining the location of a target object in eye-centered coordinates tells the monkey not just where the stimulus was on the retina before the saccade, but where it would be on the retina after the eyes have moved. Further experimental results (Duhamel, Goldberg, Fitzgibbon, Sirigu and Grafman 1992b) seem to confirm that humans also rely on a similar

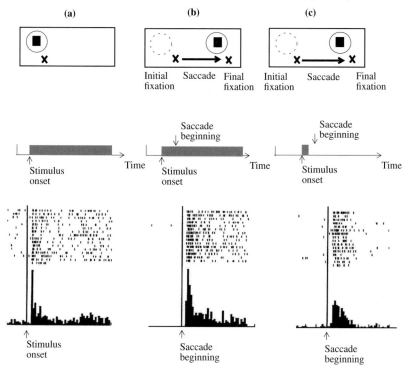

Fig. 4.1 Responses of a LIP neuron that shows remapping of visual memory trace activity. In the top part of the Figure, the **x** is the fixation point, the black square is the presented stimulus, the dashed circle is the location of the receptive field when the monkey was fixating at the left fixation point (before a saccade), and the continuous circle is the location of the receptive field when the monkey was fixating at the right point (after a saccade). At the bottom, the raster display shows the spiking activity of the neuron in 16 successive trials. In these rasters, each dot represents the time at which a spike is elicited. The histogram at the bottom shows the average spiking activity from all trials as a function of time. (a) Stimulus presented in the receptive field of the measured neuron while the monkey is fixating. (b) The monkey performing a saccade that brings the receptive field onto the location of a stationary stimulus. (c) The response following a saccade that brings the receptive field onto a location where a stimulus was previously presented.

kind of remapping of spatial representations.

The head-centred spatial representation is especially useful for guiding eye movements. In fact, LIP neurons participate in the control of saccades by projecting to regions engaged with the generation of eye movements, i.e. to both the superior colliculus and the frontal eye fields. In addition, some LIP neurons show shape selectivity (Sereno and Maunsell 1998), and more generally the shape selectivity of parietal neurons may be related to the utility this has in preparing to make actions to objects (Murata, Gallese, Luppino, Kaseda and Sakata 2000).

4.2.2 Area VIP

VIP neurons also seem to encode spatial information in a head-centred reference frame. Duhamel, Colby and Goldberg (1991) found that VIP neurons are also sensitive to tactile stimulation. A light touch on the face of a monkey elicits a large response in some VIP neurons. Even more, the visual receptive fields of VIP neurons match in location the tactile receptive fields. For example, if a certain VIP neuron responded to visual stimuli in an upper right region with respect to the head of the monkey, then the same neuron would respond to tactile stimulation of the upper right region of the face. Both receptive fields are matched not only with respect to location but also with respect to size.

As in the case of MT and MST neurons, VIP neurons are also selective for motion. They show selectivity to both the direction and speed of motion. Also, moving tactile stimuli are selectively encoded by VIP neurons. In the two modalities, visual and tactile, the direction of motion selectivity is matched.

The head-centred spatial representation is especially useful for guiding head movements. In fact, as expected, anatomical studies reveal that the outputs of VIP neurons project to the region of the prefrontal cortex primarily engaged in the control of head movement.

4.2.3 Area MST

In Chapter 2, we noted that MST neuronal activity is related to global pattern motion processing. One can divide MST into two subareas: MSTd (dorsal) and MSTl (lateral). The dorsal part MSTd is the site of visuo- and oculo-motor integration concerning motion detection, determination of eye position, and the control of slow eye movements. In addition, MSTd is responsible for the processing of complex global motion information like the extraction of rotation and translation, and expansion or contraction. On the other hand, MSTl neurons have smaller receptive fields and are involved in the selection of targets for smooth-pursuit eye movements. By using 'optic flow' information, that is, the motion of surrounding objects on the retina during our own movement, MST neurons can identify self-movement. It has been suggested that area 7a, using data from optic-flow analysis performed by MT and MST, constructs a motion-based representation of extrapersonal space.

4.2.4 Area 7a

Neurons in area 7a are sensitive to both visual input and eye position. Neurons in this area possess large receptive fields that cover both sides of visual space. It has been hypothesized that neurons in area 7a encode positions and motion trajectories of stimuli in egocentric space. In fact, Motter and Mountcastle (1981) observed that a subpopulation of area 7a neurons responded preferentially to the axis and direction of visual motion relative to a fixation point in space. In addition, visual responses in area 7a are modulated by eye position (Andersen, Asanuma, Essick and Siegel 1990a).

Recently, Siegel and Read (1997) discovered a third visuo-spatial signal, the pattern of optic flow, to which neurons in area 7a are sensitive. Selective neuronal responses to simple optic flow patterns including rotatory, radial motion and axis of translation were found in area 7a. Optic flow information could be passed to area 7a via a projection from MST or LIP. Similarly to neurons in MST, neurons in 7a show selective responses to complex motion flow patterns. But contrary to MST neurons, 7a neurons show strong selectivity to the locus of the optic flow. Furthermore, responses to optic flow are modulated by eye position. Convergence of visual and extra-retinal signals occurs onto parietal neurons with optic flow pattern selectivity. These three sensory signals may be used to derive information about the location of objects in the environment relative to the viewer.

4.3   The neuropsychology of the parietal lobe

Patients with lesions in the parietal lobe show a set of symptoms that can be interpreted generally as a failure of spatial cognitive functions. In order to have a better understanding of the functions of the parietal cortex in spatial cognitive functions (e.g. spatial representation and spatial attention) and the parietal cortex in humans, it is essential to consider the neuropsychological syndromes that result from parietal lesions, such as visual neglect, Balint's syndrome, and Gerstmann's syndrome.

4.3.1   Unilateral neglect

One of the most common attentional deficits observed in neurological patients with parietal lesions is unilateral neglect or hemineglect. The main feature that characterizes the neglect syndrome is the systematic failure of patients to notice objects or events in the hemispace opposite to their lesion. For example, patients with right parietal damage will tend to ignore food on the left side of their plate or ignore the presence of friends in their left visual field. A weaker form of this syndrome that typically results from an acute neglect that subsides, is extinction. Extinction refers to the failure to perceive or to respond to a stimulus or event contralateral to the lesion when presented simultaneously with a stimulus ipsilateral to the lesion.

There are several clinical neuropsychological studies suggesting that neglect is primarily due to an impairment in the spatial attentional processing system. Bisiach and Luzzatti (1978) demonstrate that the problem in unilateral neglect is not sensory but is more consistent with an attention-based explanation. They analyzed the performance of unilateral neglect patients in a visual imagery experiment that referred to visual scenes that were well known by the patients. The patients had a perfect memory of the main landmarks characterizing the Piazza del Duomo in Milan. Nevertheless, when they were asking to imagine the Piazza from a vantage point on one side of the Piazza, they systematically failed to report the buildings and landmarks that would have been on the left side of their visual image. On the other hand, when they were then asked to imagine the same Piazza from a vantage point on the opposite side of the Piazza, they reported items that had been previously neglected, and now neglected the side of the Piazza they previously reported (see Fig. 4.2).

Thus, neglect seems to be caused by the incapacity to pay attention to the contralesional side, and this deficit is independent of whether the visual scene to be analyzed comes from imagery or perception. Neglect operates at a high level of spatial attention, and not at a perceptually low level of sensory processing. Therefore, one can conclude that the parietal lobe in humans (especially the right parietal lobe) is mainly involved in high-level spatial attention-related processing.

A further significant indication that neglect is not due to impaired low-level sensory processing is the fact that unilateral neglect patients with parietal lesions sometimes show impaired attention in object-based coordinates. This is consistent with theories of object-based attention (see Chapter 11). Behrman and Tipper (1994) demonstrated this effect by analysing the performance of unilateral neglect patients that are asked to report a target flashed within the confines of the two circular ends of a dumbbell-shaped stimulus (see Fig. 4.3). As expected they fail to notice the contralesional side. But, when the dumbbell is rotated around the centre of the visual field, the targets that appear on the end of the dumbbell that was first on the neglected side are still not noticed, although the end is rotated to the normal ipsilesional visual

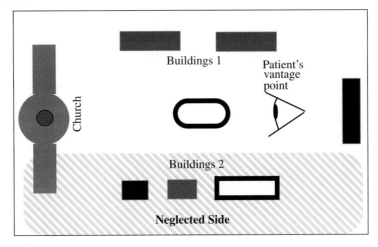

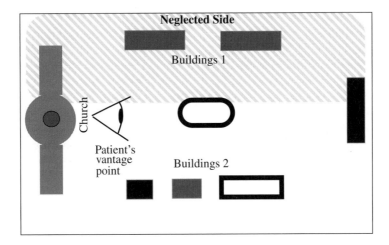

Fig. 4.2 Schematic diagram of the visual imagery experiment of Bisiach and Luzzati (1978). A unilateral neglect patient has to report the different landmarks of the Piazza when they imagine seeing the Piazza from two different opposite vantage points. They neglect always the contralesional side (in this case the left hemifield).

hemifield.

Also, Driver and Halligan (1991) demonstrated that unilateral neglect patients do not perceive parts of an object on the contralateral side to their lesion, even if the entire object is presented within the normal ipsilesional visual hemifield. This kind of phenomenon can only be understood in the framework of impaired object-based attention.

Visual neglect can be diagnosed by neuropsychological tests. The most popular and simple is the so-called line cancellation test. Subjects are given a sheet of paper containing many horizontal lines and they are asked under free-viewing conditions to bisect the lines precisely in the middle by drawing a vertical line. Patients with visual neglect, for example with lesions

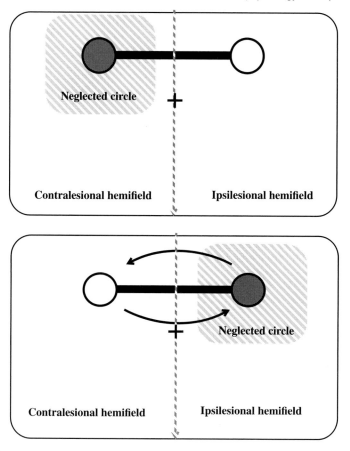

Fig. 4.3 Dumbbell experiment of Behrmann and Tipper (1994) showing object-based neglect.

of the right hemisphere, tend to bisect the lines to the right of their centre. They may also miss lines altogether in the visual field contralateral to the impaired hemisphere. Figure 4.4 shows a typical example of this test. In Chapter 11, we will analyze computationally the underlying attentional mechanisms that cause this syndrome.

4.3.2 Balint's syndrome

In general, unilateral neglect is caused by a unilateral lesion of the posterior parietal cortex. A more severe syndrome due to bilateral lesion of the posterior parietal cortex is Balint's syndrome. Balint (1909) described a patient with a bilateral parietal lesion. This patient was able to see, recognize and name objects, photos and colours normally. Nevertheless, the patient behaved as functionally blind because she was only able to see single fixated visual objects. Balint's patients are incapable of attending simultaneously to multiple objects. In fact, one of the typical symptoms, simultanagnosia, alludes to the inability to perceive more than one object at a time during a single fixation. They can report only one object even when more than one object is located at the fixated position. For example, patients shown a person wearing glasses may report the face of the person or the glasses, but not a person wearing glasses.

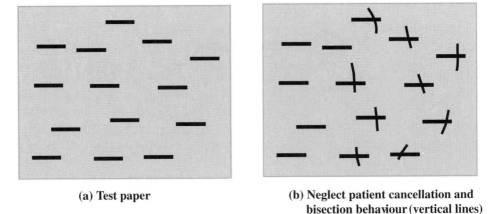

(a) Test paper   (b) Neglect patient cancellation and
bisection behaviour (vertical lines)

Fig. 4.4 Neuropsychological line cancellation and bisection test for diagnosing the visual neglect syndrome. Patients suffering from neglect are given a sheet of paper (a) containing many horizontal lines and are asked under free-viewing conditions to bisect the lines precisely in the middle with a vertical line. (b). They tend to bisect the lines to the right (for a right hemisphere lesion) of their centre, and may completely miss many lines on the left half of space due to neglect for the contralesional space.

Simultanagnosia is usually interpreted as being caused by a deficit in the attentional system which makes the person unable to attend simultaneously to more than one object and even to disengage attention from this object to another (see Kolb and Whishaw (1996) and Farah (1990)).

A second symptom that can also be interpreted as an underlying inability to disengage attention is ocular ataxia. Ocular ataxia refers to the inability to change visual fixation from one object to another. Balint's patients fixate rigidly on one object and are unable to perform a task that requires multiple fixations, which are required by most things we do in daily life. Rehabilitation of this kind of syndrome involves trying to develop conscious strategies, such as closing the eyes, to disengage attention from one fixated object to another. In addition to ocular ataxia, Balint's patients show optic ataxia, which refers to the inability to grasp objects. Another typical symptom observed in Balint's syndrome is spatial disorientation. This causes serious problems for navigation.

Balint's syndrome can be interpreted as a failure of the attentional system that shows specific object-based characteristics. Humphreys and Riddoch (1992) proved that simultanagnosia for example can be associated with object-based attention. They asked patients to report the colours that appeared on a screen of coloured circles (see Fig. 4.5). They defined three different conditions. In the first condition, all circles had the same colour, namely all green or all red. The patients always reported the same correct colour. Probably they were fixating and perceiving only one stimulus, but due to the fact that all stimuli had the same colours, they performed well under this condition. In the second condition, the items presented on the screen had different colours (green or red). The patients did not perform well under this condition, because they probably were fixating randomly on only one item and reported only the corresponding colour, being unable to report the presence of the other colours. The third condition is the essential one, where again items of both colours were presented on the screen, but now items of different colours were linked with a line forming one dumbbell

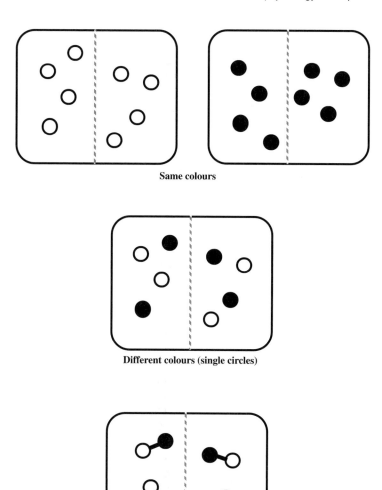

Fig. 4.5 Object-based character of Balint's syndrome demonstrated by the colour reporting experiment of Humphreys and Riddoch (1992). See text for details.

object. In this case, the patients were able to report both colours, because they were fixating and perceiving a single dumbbell object that contained both colours.

4.3.3 Gerstmann's syndrome

Gerstmann's syndrome is associated with a lesion of the left inferior region of the parietal cortex. Patients with Gerstmann's syndrome show four typical symptoms. The first symptom is a left–right ataxia, i.e. an inability to distinguish left from right. The second symptom is a finger agnosia. Patients with Gerstmann's syndrome cannot recognize or name their own finger. This is related to the third symptom, dysgraphia, a writing disability, which occurs in the absence of motor or sensory impairments. Finally, the fourth symptom is dyscalculia, i.e. the inability to perform mathematical calculations. All these symptoms are related to a failure in the spatial representation system, which is necessary for establishing mathematical relationships, and for identifying parts or orientations in a body-centred frame of reference.

All these clinical neuropsychological results provide clues about how spatial functions, and their interaction with object vision, normally occur in the brain. These findings provide evidence that the computational models of brain function developed in later Chapters seek to explain, and evidence on which these models build.

5   Inferior temporal cortical visual areas

5.1   Introduction

There is now good evidence that neural systems in temporal cortical visual areas process information about faces. Because a large number of neurons are devoted to this class of stimuli, these systems have proved amenable to experimental analysis. Face recognition and the identification of face expression are important in primate social behaviour, and analysis of the neural systems involved is important for understanding the effects of damage to these systems in humans. Damage to these or related systems can lead to prosopagnosia, an impairment in recognizing individuals from the sight of their face; or in difficulty in identifying the expression on a face. It turns out that the temporal cortical visual areas also have similar neuronal populations that code for objects, and study of both sets of neurons is helping to unravel the enormous computational problem being solved of invariant visual object recognition. The neurophysiological recordings are made mainly in non-human primates, macaques, firstly because the temporal lobe, in which this processing occurs, is much more developed than in non-primates; and secondly because the findings are relevant to understanding the effects of brain damage in patients.

5.2   Neuronal responses found in different temporal lobe cortex visual areas

While recording in the temporal lobe cortical visual areas of macaques, Charles Gross and colleagues found some neurons that appeared to respond best to complex visual stimuli such as faces (Desimone and Gross 1979, Bruce, Desimone and Gross 1981, Desimone 1991). It was soon found that while some of these neurons could respond to parts of faces, other neurons required several parts of the face to be present in the correct spatial arrangement; and that many of these neurons did not just respond to any face that was shown, but responded differently to different faces (Perrett, Rolls and Caan 1982, Desimone, Albright, Gross and Bruce 1984, Rolls 1984, Gross, Desimone, Albright and Schwartz 1985). By responding differently to different faces, these neurons potentially encode information useful for identifying individual faces. It also appears that there is some specialization of function of different temporal cortical visual areas, as described next.

The visual pathways project from the primary visual cortex to the temporal lobe visual cortical areas by a number of intervening cortical stages (Seltzer and Pandya 1978, Maunsell and Newsome 1987, Baizer, Ungerleider and Desimone 1991). The inferior temporal visual cortex, area TE, is divided on the basis of cytoarchitecture, myeloarchitecture, and afferent input, into areas TEa, TEm, TE3, TE2 and TE1. In addition, there is a set of different areas in the cortex in the superior temporal sulcus (Seltzer and Pandya 1978, Baylis, Rolls and Leonard 1987) (see Fig. 5.1). Of these latter areas, TPO receives inputs from temporal,

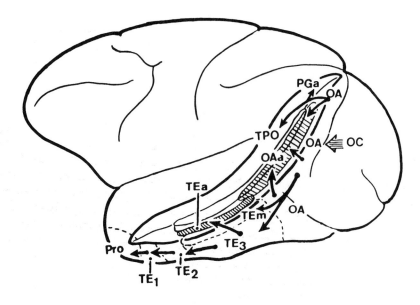

Fig. 5.1 Lateral view of the macaque brain (left hemisphere) showing the different architectonic areas (e.g. TEm, TPO) in and bordering the anterior part of the superior temporal sulcus (STS) of the macaque (see text).

parietal and occipital cortices; PGa and IPa from parietal and temporal cortices; and TS and TAa primarily from auditory areas.

There is considerable specialization of function in these architectonically defined areas (Baylis, Rolls and Leonard 1987). Areas TPO, PGa and IPa are multimodal, with neurons that respond to visual, auditory and/or somatosensory inputs. The more ventral areas in the inferior temporal gyrus (areas TE3, TE2, TE1, TEa and TEm) are primarily unimodal visual areas. Areas in the cortex in the anterior and dorsal part of the superior temporal sulcus (e.g. TPO, IPa and IPg) have neurons specialized for the analysis of moving visual stimuli. Neurons responsive primarily to faces are found more frequently in areas TPO, TEa and TEm, where they comprise approximately 20% of the visual neurons responsive to stationary stimuli, in contrast to the other temporal cortical areas in which they comprise 4–10%. The stimuli that activate other cells in these TE regions include simple visual patterns such as gratings, and combinations of simple stimulus features (Gross, Desimone, Albright and Schwartz 1985, Tanaka, Saito, Fukada and Moriya 1990). Due to the fact that face-selective neurons have a wide distribution, it might be expected that only large lesions, or lesions that interrupt outputs of these visual areas, would produce readily apparent face-processing deficits. Moreover, neurons with responses related to facial expression, movement, and gesture are more likely to be found in the cortex in the superior temporal sulcus, whereas neurons with activity related to facial identity are more likely to be found in the TE areas (see below and Hasselmo, Rolls and Baylis (1989a)). Another specialization is that areas TEa and TEm, which receive inter alia from the cortex in the intraparietal sulcus, have neurons that are tuned to binocular disparity, so that information derived from stereopsis about the 3D structure of objects is represented in the inferior temporal cortical visual areas (Janssen, Vogels and

Orban 1999, Janssen, Vogels and Orban 2000). Interestingly, these neurons respond only when the black and white regions in the two eyes are correlated (that is, when a white patch in one eye corresponds to a white patch in the other eye) (P. Janssen, personal communication, June 2001). This corresponds to what we see perceptually and consciously. In contrast, in V1, MT and MST, depth-tuned neurons can respond to anticorrelated depth images, that is where white in one eye corresponds to black in the other (Parker, Cumming and Dodd 2000, DeAngelis, Cumming and Newsome 2000), and this may be suitable for eye movement control, but much less for shape and object discrimination and identification. It may be expected that other depth cues, such as perspective, surface shading, and occlusion, affect the response properties of some neurons in inferior temporal cortex visual areas. Binocular disparity, and information from these other depth cues, may be used to compute the absolute size of objects, which is represented independently of the distance of the object for a small proportion of inferior temporal cortex neurons (Rolls and Baylis 1986). Knowing the absolute size of an object is useful evidence to include in the identification of an object. Although cues from binocular disparity can thus drive some temporal cortex visual neurons, and there is a small proportion of inferior temporal that responds better to real faces and objects than to 2D representations on a monitor, it is found that the majority of TE neurons respond as well or almost as well to 2D images on a video monitor as to real faces or objects (Perrett, Rolls and Caan 1982). Moreover, the tuning of inferior temporal cortex neurons to images of faces or objects on a video monitor is similar to that to real objects (personal observations of E. T. Rolls).

5.3   The selectivity of one population of neurons for faces

Neurons with responses selective for faces respond 2–20 times more to faces than to a wide range of gratings, simple geometrical stimuli, or complex 3D objects (Rolls 1984, Rolls 1992a, Baylis, Rolls and Leonard 1985, Baylis, Rolls and Leonard 1987). The responses to faces are excitatory with firing rates often reaching 100 spikes/s, are sustained, and have typical latencies of 80–100 ms. The neurons are typically unresponsive to auditory or tactile stimuli and to the sight of arousing or aversive stimuli. These findings indicate that explanations in terms of arousal, emotional or motor reactions, and simple visual feature sensitivity, are insufficient to account for the selective responses to faces and face features observed in this population of neurons (Perrett, Rolls and Caan 1982, Baylis, Rolls and Leonard 1985, Rolls and Baylis 1986). Observations consistent with these findings have been published by Desimone, Albright, Gross and Bruce (1984), who described a similar population of neurons located primarily in the cortex in the superior temporal sulcus that responded to faces but not to simpler stimuli such as edges and bars or to complex non-face stimuli (see also Gross, Desimone, Albright and Schwartz (1985)).

These neurons are specialized to provide information about faces in that they provide much more information (on average, 0.4 bits in a 400 ms epoch) about which (of 20) face stimuli is being seen than about which (of 20) non-face stimuli is being seen (on average 0.07 bits) (Rolls and Tovee 1995b, Rolls, Treves, Tovee and Panzeri 1997d). These information theoretic procedures provide an objective and quantitative way to show what is 'represented' by a particular population of neurons.

5.4 The selectivity of these neurons for individual face features or for combinations of face features

Masking out or presenting parts of the face (e.g. eyes, mouth, or hair) in isolation reveal that different cells respond to different features or subsets of features. For some cells, responses to the normal organization of cut-out or line-drawn facial features are significantly larger than to images in which the same facial features are jumbled (Perrett, Rolls and Caan 1982, Rolls, Tovee, Purcell, Stewart and Azzopardi 1994b). These findings are consistent with the hypotheses developed below that by competitive self-organization some neurons in these regions respond to parts of faces by responding to combinations of simpler visual properties received from earlier stages of visual processing, and that other neurons respond to combinations of parts of faces and thus respond only to whole faces. Moreover, the finding that for some of these latter neurons, the parts must be in the correct spatial configuration shows that the combinations formed can reflect not just the features present, but also their spatial arrangement. This provides a way in which binding can be implemented in neural networks (see further Elliffe, Rolls and Stringer (2001)). Further evidence that neurons in these regions respond to combinations of features in the correct spatial configuration was found by Tanaka and colleagues (see for example Tanaka, Saito, Fukada and Moriya (1990)) using combinations of features that are used by comparable neurons to define objects.

5.5 Distributed encoding of object and face identity

How is information encoded in the inferior temporal visual cortical areas? Can we read the code being used by the cortex? What are the advantages of the encoding scheme used for the neuronal network computations being performed in different areas of the cortex? These are some of the key issues considered in this Section. Because information is exchanged between the computing elements of the cortex (the neurons) by their spiking activity, which is conveyed by their axon to synapses onto other neurons, the appropriate level of analysis is how single neurons, and populations of single neurons, encode information in their firing. More global measures that reflect the averaged activity of large numbers of neurons (for example, PET (positron emission tomography) and fMRI (functional magnetic resonance imaging), EEG (electroencephalographic recording), and ERPs (event-related potentials)) cannot reveal how the information is represented, or how the computation is being performed.

We summarize some of the types of representation that might be found at the neuronal level next (cf. Section 1.6 and Chapter 7). A **local representation** is one in which all the information that a particular stimulus or event occurred is provided by the activity of one of the neurons. This is sometimes called a grandmother cell representation, because in a famous example, a single neuron might be active only if one's grandmother was being seen (see Barlow (1995)). A **fully distributed representation** is one in which all the information that a particular stimulus or event occurred is provided by the activity of the full set of neurons. If the neurons are binary (for example, either active or not), the most distributed encoding is when half the neurons are active for any one stimulus or event. A **sparse distributed representation** is a distributed representation in which a small proportion of the neurons is active at any one time. Equation 5.1 defines a measure of the sparseness, a:

$$a = \frac{(\sum\limits_{s=1}^{S} y_s/S)^2}{(\sum\limits_{s=1}^{S} y_s^2)/S} \qquad (5.1)$$

where y_s is the mean firing rate of the neuron to stimulus s in the set of S stimuli (Rolls and Treves (1998). For a binary representation, a is 0.5 for a fully distributed representation, and $1/S$ if a neuron responds to one of a set of S stimuli. Another measure of sparseness is the kurtosis of the distribution, which is the fourth moment of the distribution. It reflects the length of the tail of the distribution. (An actual distribution of the firing rates of a neuron to a set of 65 stimuli is shown in Fig. 5.3. The sparseness a for this neuron was 0.69, see Rolls, Treves, Tovee and Panzeri (1997d).)

5.5.1 Distributed representations evident in the firing rate distributions

Barlow (1972) proposed a single neuron doctrine for perceptual psychology. He proposed that sensory systems are organized to achieve as complete a representation as possible with the minimum number of active neurons. He suggested that at progressively higher levels of sensory processing, fewer and fewer cells are active, and that each represents a more and more specific happening in the sensory environment. He suggested that 1,000 active neurons (which he called cardinal cells) might represent the whole of a visual scene. An important principle involved in forming such a representation was the reduction of redundancy (see Appendix B for an introduction to information theory and redundancy). The implication of Barlow's (1972) approach was that when an object is being recognized, there are, towards the end of the visual system, a small number of neurons (the cardinal cells) that are so specifically tuned that the activity of these neurons encodes the information that one particular object is being seen. (He thought that an active neuron conveys something of the order of complexity of a word.) The encoding of information in such a system is described as local, in that knowing the activity of just one neuron provides evidence that a particular stimulus (or, more exactly, a given 'trigger feature') is present. Barlow (1972) eschewed 'combinatorial rules of usage of nerve cells', and believed that the subtlety and sensitivity of perception results from the mechanisms determining when a single cell becomes active. In contrast, with distributed or ensemble encoding, the activity of several or many neurons must be known in order to identify which stimulus is present, that is, to read the code. It is the relative firing of the different neurons in the ensemble that provides the information about which object is present.

At the time Barlow (1972) wrote, there was little actual evidence on the activity of neurons in the higher parts of the visual and other sensory systems. There is now considerable evidence, which is now described.

First, it has been shown that the representation of which particular object (face) is present is actually rather distributed. Baylis, Rolls and Leonard (1985) showed this with the responses of temporal cortical neurons that typically responded to several members of a set of five faces, with each neuron having a different profile of responses to each face (see examples in Fig. 5.2). It would be difficult for most of these single cells to tell which of even five faces, let alone which of hundreds of faces, had been seen. (At the same time, the neurons discriminated between the faces reliably, as shown by the values of d', taken, in the case of the neurons, to be the number of standard deviations of the neuronal responses that separated the response to the best face in the set from that to the least effective face in the set. The values of d' were typically in the range 1–3.)

Second, the distributed nature of the representation can be further understood by the finding that the firing rate distribution of single neurons, when a wide range of natural visual stimuli are being viewed, is approximately exponentially distributed, with rather few stimuli producing high firing rates, and increasingly large numbers of stimuli producing lower and lower firing rates (Rolls and Tovee 1995b, Baddeley, Abbott, Booth, Sengpiel, Freeman, Wakeman and Rolls 1997, Treves, Panzeri, Rolls, Booth and Wakeman 1999).

For example, in a recent study, the responses of another set of temporal cortical neurons to 23 faces and 42 non-face natural images were measured, and again a distributed representation was found (Rolls and Tovee 1995b). The tuning was typically graded, with a range of different firing rates to the set of faces, and very little response to the non-face stimuli (see Fig. 5.3). The spontaneous firing rate of the neuron in Fig. 5.3 was 20 spikes/s, and the histogram bars indicate the change of firing rate from the spontaneous value produced by each stimulus. Stimuli that are faces are marked F, or P if they are in profile. B refers to images of scenes that included either a small face within the scene, sometimes as part of an image that included a whole person, or other body parts, such as hands (H) or legs. The non-face stimuli are unlabelled. The neuron responded best to three of the faces (profile views), had some response to some of the other faces, and had little or no response, and sometimes had a small decrease of firing rate below the spontaneous firing rate, to the non-face stimuli. The sparseness value a for this cell across all 68 stimuli was 0.69, and the response sparseness a_r (based on the evoked responses minus the spontaneous firing of the neuron) was 0.19. It was found that the sparseness of the representation of the 68 stimuli by each neuron had an average across all neurons of 0.65 (Rolls and Tovee 1995b). This indicates a rather distributed representation. (If neurons had a continuum of firing rates equally distributed between zero and maximum rate, a would be 0.75, while if the probability of each response decreased linearly, to reach zero at the maximum rate, a would be 0.67). If the spontaneous firing rate was subtracted from the firing rate of the neuron to each stimulus, so that the changes of firing rate, that is the active responses of the neurons, were used in the sparseness calculation, then the 'response sparseness' a_r had a lower value, with a mean of 0.33 for the population of neurons, or 0.60 if calculated over the set of faces rather than over all the face and non-face stimuli. Thus the representation was rather distributed. (It is, of course, important to remember the relative nature of sparseness measures, which (like the information measures to be discussed below) depend strongly on the stimulus set used.) Thus we can reject a cardinal cell representation. As shown below, the readout of information from these cells is actually much better in any case than would be obtained from a local representation, and this makes it unlikely that there is a further population of neurons with very specific tuning that use local encoding.

These data provide a clear answer to whether these neurons are grandmother cells: they are not, in the sense that each neuron has a graded set of responses to the different members of a set of stimuli, with the prototypical distribution similar to that of the neuron illustrated in Fig. 5.3. On the other hand, each neuron does respond very much more to some stimuli than to many others, and in this sense is tuned to some stimuli.

The large set of 68 stimuli used by Rolls and Tovee (1995b) was chosen to produce an approximation to a set of stimuli that might be found to natural stimuli in a natural environment, and thus to provide evidence about the firing rate distribution of neurons to natural stimuli. Another approach to the same fundamental question was taken by Baddeley,

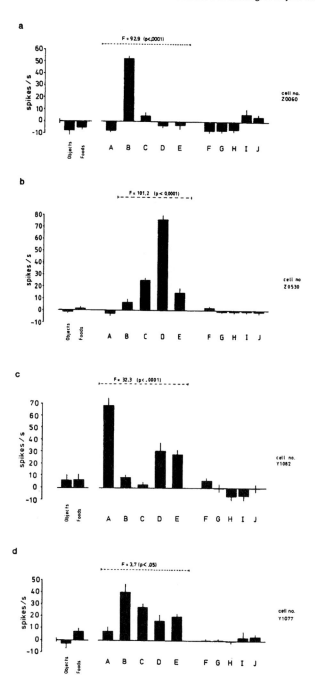

Fig. 5.2 Responses of four different temporal cortex visual neurons to a set of five faces (A–E), and, for comparison, to a wide range of non-face objects and foods. F–J are non-face stimuli. The means and standard errors of the responses computed over 8–10 trials are shown. (From Baylis, Rolls and Leonard, 1985.)

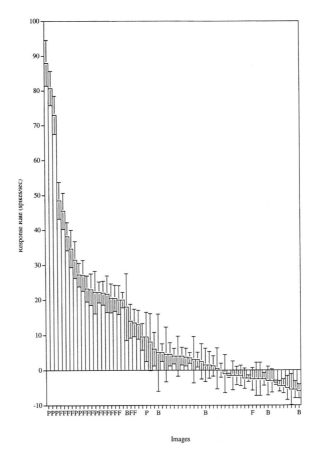

Images

Fig. 5.3 Firing rate distribution of a single neuron in the temporal visual cortex to a set of 23 face (F) and 45 non-face images of natural scenes. The firing rate to each of the 68 stimuli is shown. The neuron does not respond to just one of the 68 stimuli. Instead, it responds to a small proportion of stimuli with high rates, to more stimuli with intermediate rates, and to many stimuli with almost no change of firing. This is typical of the distributed representations found in temporal cortical visual areas. (After Rolls and Tovee, 1995a).

Abbott, Booth, Sengpiel, Freeman, Wakeman, and Rolls (1997) who measured the firing rates over short periods of individual inferior temporal cortex neurons while monkeys watched continuous videos of natural scenes. They found that the firing rates of the neurons were again approximately exponentially distributed (see Fig. 5.4), providing further evidence that this type of representation is characteristic of inferior temporal cortex (and indeed also V1) neurons.

The actual distribution of the firing rates to a wide set of natural stimuli is of interest, because it has a rather stereotypical shape, typically following a graded unimodal distribution with a long tail extending to high rates (see for example Figs. 5.3 and 5.4). The mode of the distribution is close to the spontaneous firing rate, and sometimes it is at zero firing. If the number of spikes recorded in a fixed time window is taken to be constrained by a fixed

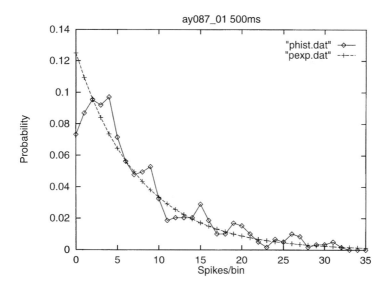

Fig. 5.4 The probability of different firing rates measured in short (e.g. 100 ms or 500 ms) time windows of a temporal cortex neuron calculated over a 5 min period in which the macaque watched a video showing natural scenes, including faces. An exponential fit (+) to the data (diamonds) is shown. (After Baddeley, Abbott, Booth, Sengpiel, Freeman, Wakeman and Rolls, 1997.)

maximum rate, one can try to interpret the distribution observed in terms of optimal information transmission (Shannon 1948), by making the additional assumption that the coding is noiseless. An exponential distribution, which maximizes entropy (and hence information transmission for noiseless codes) is the most efficient in terms of energy consumption if its mean takes an optimal value that is a decreasing function of the relative metabolic cost of emitting a spike (Levy and Baxter 1996). This argument would favour sparser coding schemes the more energy expensive neuronal firing is (relative to rest). Although the tail of actual firing rate distributions is often approximately exponential (see for example Figs. 5.3 and 5.4; Baddeley, Abbott, Booth, Sengpiel, Freeman, Wakeman and Rolls (1997); Rolls, Treves, Tovee and Panzeri (1997d)), the maximum entropy argument cannot apply as such, because noise is present and the noise level varies as a function of the rate, which makes entropy maximization different from information maximization. Moreover, a mode at low but non-zero rate, which is often observed, is inconsistent with the energy efficiency theorem.

A simpler explanation for the characteristic firing rate distribution arises by appreciating that the value of the activation of a neuron across stimuli, reflecting a multitude of contributing factors, will typically have a Gaussian distribution; and by considering a physiological input-output transform (i.e. activation function), and realistic noise levels. In fact, an input-output transform that is supralinear in a range above threshold results from a fundamentally linear transform and fluctuations in the activation, and produces a variance in the output rate, across repeated trials, that increases with the rate itself, consistent with common observations. At the same time, such a supralinear transform tends to convert the Gaussian tail of the activation distribution into an approximately exponential tail, without implying a fully exponential distribution with the mode at zero. Such basic assumptions yield excellent fits with observed distributions (Treves, Panzeri, Rolls, Booth and Wakeman 1999), which often differ from exponential in that there are too few very low rates observed, and too many low rates (Rolls, Treves, Tovee and Panzeri 1997d).

This peak at low but non-zero rates may be related to the low firing rate spontaneous activity that is typical of many cortical neurons. Keeping the neurons close to threshold in this way may maximize the speed with which a network can respond to new inputs (because time is not required to bring the neurons from a strongly hyperpolarized state up to threshold). The advantage of having low spontaneous firing rates may be a further reason why a curve such as an exponential cannot sometimes be exactly fitted to the experimental data.

A conclusion of this analysis was that the firing rate distribution may arise from the threshold non-linearity of neurons combined with short-term variability in the responses of neurons (Treves, Panzeri, Rolls, Booth and Wakeman 1999).

However, given that the firing rate distribution is approximately exponential, some properties of this type of representation are worth elucidation. The sparseness of such an exponential distribution of firing rates is 0.5. This has interesting implications, for to the extent that the firing rates are exponentially distributed, this fixes an important parameter of cortical neuronal encoding to be close to 0.5. Indeed, only one parameter specifies the shape of the exponential distribution, and the fact that the exponential distribution is at least a close approximation to the firing rate distribution of real cortical neurons implies that the sparseness of the cortical representation of stimuli is kept under precise control. The utility of this may be to ensure that any neuron receiving from this representation can perform a dot product operation between its inputs and its synaptic weights that produces similarly distributed outputs; and that the information being represented by a population of cortical neurons is kept high. It is interesting to realize that the representation that is stored in an associative network (see Chapter 7) may be more sparse than the 0.5 value for an exponential firing rate distribution, because the non-linearity of learning introduced by the voltage dependence of the NMDA receptors (see Chapters 1 and 7) effectively means that synaptic modification in, for example, an autoassociative network will occur only for the neurons with relatively high firing rates, i.e. for those that are strongly depolarized.

5.5.2 The representation of information in the responses of single neurons to a set of stimuli

The use of an information theoretic approach (see Appendix B) makes it clear that there is considerable information in this distributed encoding of these temporal visual cortex neurons about which face is being seen. Fig. 5.5 shows typical firing rate changes on different trials to each of several different faces. This makes it clear that from the firing rate on any one trial, information is available about which stimulus was shown. In order to clarify the representation of individual stimuli by individual cells, Fig. 5.6 shows the information $I(s,R)$[7] available in the neuronal response about each of 20 face stimuli calculated for the neuron (am242) whose firing rate response profile to the set of 65 stimuli is shown in Fig. 5.3. Unless otherwise stated, the information measures given are for the information available on a single trial from the firing rate of the neuron in a 500 ms period starting 100 ms after the onset of the stimuli. It is shown in Fig. 5.6 that 2.2, 2.0, and 1.5 bits of information were present about the three face

---

[7] $I(s, R)$ has more recently been called the stimulus-specific surprise, see DeWeese and Meister (1999).

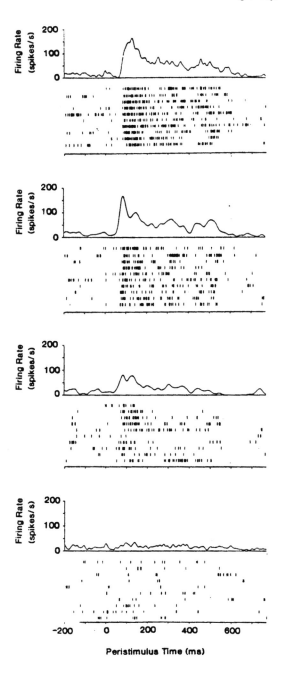

Fig. 5.5 Peristimulus time histograms and rastergrams showing the responses on different trials (originally in random order) of a face-selective neuron to four different faces. (In the rastergrams each vertical line represents one spike from the neuron, and each row is a separate trial.)

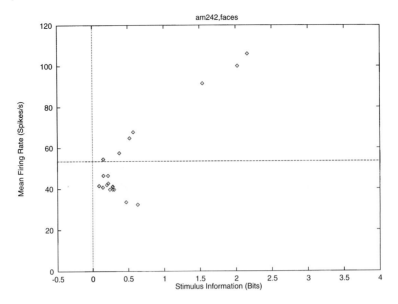

Fig. 5.6 The information I(s,R) available in the response of the same single neuron as in Fig. 5.3 about each of the stimuli in the set of 20 face stimuli (abscissa), with the firing rate of the neuron to the corresponding stimulus plotted as a function of this on the ordinate. (From Rolls, Treves, Tovee and Panzeri, 1997.)

stimuli to which the neuron had the highest firing rate responses. The neuron conveyed some but smaller amounts of information about the remaining face stimuli. The average information I(S,R) about this set (S) of 20 faces for this neuron was 0.55 bits. The average firing rate of this neuron to these 20 face stimuli was 54 spikes/s. It is clear from Fig. 5.6 that little information was available from the responses of the neuron to a particular face stimulus if that response was close to the average response of the neuron across all stimuli. At the same time, it is clear from Fig. 5.6 that information was present depending on how far the firing rate to a particular stimulus was from the average response of the neuron to the stimuli. Of particular interest, it is evident that information is present from the neuronal response about which face was shown if that neuronal response was below the average response, as well as when the response was greater than the average response.

One intuitive way to understand the data shown in Fig. 5.6 is to appreciate that low probability firing rate responses, whether they are greater than or less than the mean response rate, convey much information about which stimulus was seen. This is of course close to the definition of information (see Appendix B). Given that the firing rates of neurons are always positive, and follow an asymmetric distribution about their mean, it is clear that deviations above the mean have a different probability to occur than deviations by the same amount below the mean. One may attempt to capture the relative likelihood of different firing rates above and below the mean by computing a z score obtained by dividing the difference between the mean response to each stimulus and the overall mean response by the standard deviation of the response to that stimulus. The greater the number of standard deviations (i.e. the greater the z score) from the mean response value, the greater the information might be expected to be. We therefore show in Fig. 5.7 the relation between the z score and $I(s, R)$. (The z score

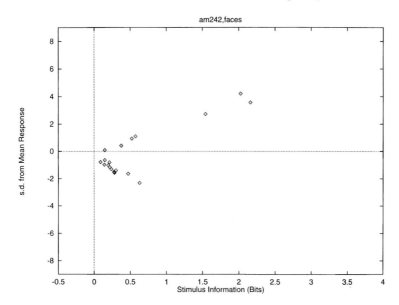

Fig. 5.7 The relation for a single cell between the number of standard deviations the response to a stimulus was from the average response to all stimuli (see text, z score) plotted as a function of $I(s, R)$, the information available about the corresponding stimulus, s. (From Rolls, Treves, Tovee and Panzeri, 1997, Fig. 2c)

was calculated by obtaining the mean and standard deviation of the response of a neuron to a particular stimulus s, and dividing the difference of this response from the mean response to all stimuli by the calculated standard deviation for that stimulus.) This results in a C-shaped curve in Figs. 5.6 and 5.7, with more information being provided by the cell the further its response to a stimulus is in spikes per second or in z scores either above or below the mean response to all stimuli (which was 54 spikes/s). The specific C-shape is discussed further in Section B.3.2.

The information $I(s, R)$ about each stimulus in the set of 65 stimuli is shown in Fig. 5.8 for the same neuron, am242. The 23 face stimuli in the set are indicated by a diamond, and the 42 non-face stimuli by a cross. Using this much larger and more varied stimulus set, which is more representative of stimuli in the real world, a C-shaped function again describes the relation between the information conveyed by the cell about a stimulus and its firing rate to that stimulus. In particular, this neuron reflected information about most, but not all, of the faces in the set, that is those faces that produced a higher firing rate than the overall mean firing rate to all the 65 stimuli, which was 31 spikes/s. In addition, it conveyed information about the majority of the 42 non-face stimuli by responding at a rate below the overall mean response of the neuron to the 65 stimuli. This analysis usefully makes the point that the information available in the neuronal responses about which stimulus was shown is relative to (dependent upon) the nature and range of stimuli in the test set of stimuli.

This evidence makes it clear that a single cortical visual neuron tuned to faces conveys information not just about one face, but about a whole set of faces, with the information conveyed on a single trial related to the difference in the firing rate response to a particular stimulus compared to the average response to all stimuli.

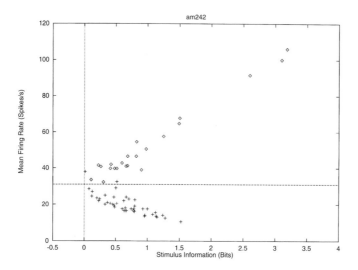

Fig. 5.8 The information $I(s, R)$ available in the response of the same neuron about each of the stimuli in the set of 23 face and 42 non-face stimuli (abscissa), with the firing rate of the neuron to the corresponding stimulus plotted as a function of this on the ordinate. The 23 face stimuli in the set are indicated by a diamond, and the 42 non-face stimuli by a cross. (From Rolls, Treves, Tovee and Panzeri, 1997, Fig. 2d.)

5.5.3 The representation of information in the responses of a population of inferior temporal visual cortex neurons

Complementary evidence comes from applying information theory to analyze how information is represented by a population of these neurons. The underlying idea is indicated in Fig. 5.9, which shows that if we know the average firing rate of each cell in a population to each stimulus, then on any single trial we can guess the stimulus that was present by taking into account the response of all the cells. What we wish to know is how the percentage correct, or better still the information, based on the evidence from any single trial about which stimulus was shown, increases as the number of cells in the population sampled increases. We can expect that the more cells there are in the sample, the more accurate the estimate of the stimulus is likely to be. If the encoding was local, the number of stimuli encoded by a population of neurons would be expected to rise approximately linearly with the number of neurons in the population. In contrast, with distributed encoding, provided that the neuronal responses are sufficiently independent, and are sufficiently reliable (not too noisy), information from the ensemble would be expected to rise linearly with the number of cells in the ensemble, and (as information is a log measure) the number of stimuli encodable by the population of neurons might be expected to rise exponentially as the number of neurons in the sample of the population was increased.

In the first series of experiments we conducted on this issue, we measured the information present in the firing rates of inferior temporal cortex neurons about which of 20 faces had been shown on the video monitor. The neuronal responses were measured in a 500 ms period starting 100 ms after stimulus onset while the macaques performed a visual discrimination task (Rolls, Treves and Tovee 1997b, Abbott, Rolls and Tovee 1996). (The responses of this

How well can one predict which stimulus was
shown on a single trial from the mean responses
of different neurons to each stimulus?

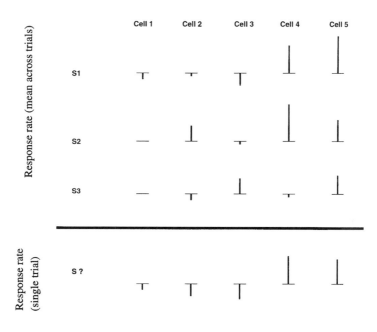

Fig. 5.9 This diagram shows the average response for each of several cells (Cell 1, etc.) to each of several stimuli (S1, etc.). The change of firing rate from the spontaneous rate is indicated by the vertical line above or below the horizontal line, which represents the spontaneous rate. We can imagine guessing or predicting from such a table the stimulus S? that was present on any one trial (see text and Section B.2.3).

population of neurons is similar in visual fixation and visual discrimination tasks, as shown by E. T. Rolls and M. C. A. Booth in unpublished experiments in 1997. What is important for the responses of these neurons is that the monkey is looking alertly at the images on the screen, and this was ensured by the visual fixation task being performed.) To measure the information from the neuronal responses, a decoding procedure was used in which the probability that it was each of the set of stimuli was estimated from the neuronal response on any one trial, given the responses to each stimulus on every other trial. This decoding procedure is one way to reduce the dimensionality of the space, which is reduced by the decoding procedure to the probability table between the stimulus s' guessed from the neuronal response and the real stimulus s that was shown on that trial. The decoding procedure used was either a probability estimation (PE) procedure, or a very neuronally plausible dot product decoding in which the vector of cell firing rates on the single trial is compared using the dot product calculation with the dot products obtained to each of the stimuli on all the other trials. Details of the method are

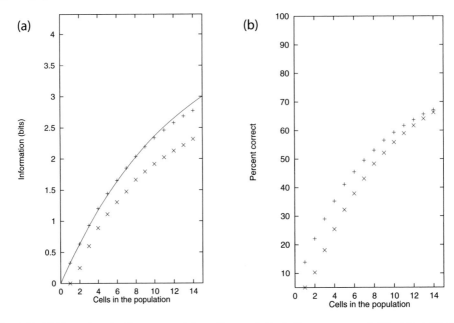

Fig. 5.10 (a) The values for the average information available in the responses of different numbers of these neurons on each trial, about which of a set of 20 face stimuli has been shown. The decoding method was Dot Product (DP, ×) or Probability Estimation (PE, +). The full line indicates the amount of information expected from populations of increasing size, when assuming random correlations within the constraint given by the ceiling (the information in the stimulus set, I = 4.32 bits). (b) The percent correct for the corresponding data to those shown in (a). (After Rolls, Treves and Tovee, 1997)

described by Rolls, Treves and Tovee (1997b) and in Appendix B. The information available about which of 20 equiprobable faces had been shown that was available from the responses of different numbers of these neurons is shown in Fig. 5.10. First, it is clear that some information is available from the responses of just one neuron: on average approximately 0.34 bits. Thus, knowing the activity of just one neuron in the population does provide some evidence about which stimulus was present. This evidence that information is available in the responses of individual neurons in this way, without having to know the state of all the other neurons in the population, indicates that information is made explicit in the firing of individual neurons in a way that will allow neurally plausible decoding, involving computing a sum of input activities each weighted by synaptic strength, to work (see below, Section 5.5.4.2). Second, it is clear (Fig. 5.10) that the information rises approximately linearly, and the number of stimuli encoded thus rises approximately exponentially (see Fig. 5.11), as the number of cells in the sample increases (Rolls, Treves and Tovee 1997b).

It has also been shown that there are neurons in the inferior temporal visual cortex that encode view invariant representations of objects, and for these neurons the same type of representation is found, namely distributed encoding with independent information conveyed by different neurons (Booth and Rolls 1998) (see Section 5.6.4).

This direct neurophysiological evidence thus demonstrates that the encoding is distributed, and the responses are sufficiently independent and reliable that the representational capacity increases exponentially. The consequence of this is that large numbers of stimuli, and fine

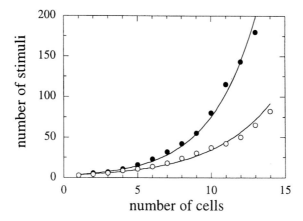

Fig. 5.11 The number of stimuli (in this case from a set of 20 faces) that are encoded in the responses of different numbers of neurons in the temporal lobe visual cortex, based on the results shown in Fig. 5.10. (After Rolls, Treves and Tovee, 1997; Abbott, Rolls and Tovee, 1996.)

discriminations between them, can be represented without having to measure the activity of an enormous number of neurons. Although the information rises approximately linearly with the number of neurons when this number is small, gradually each additional neuron does not contribute as much as the first. In the sample analyzed by Rolls, Treves and Tovee (1997b), the first neuron contributed 0.34 bits, on average, with 3.23 bits available from the 14 neurons analyzed. This reduction is, however, exactly what could be expected to derive from a simple ceiling effect, in which the ceiling is just the information in the stimulus set, or $\log_2(20) = 4.32$ bits, as discussed further in Appendix B (cf. Fig. 5.10). This indicates that, on the one hand, each neuron does not contribute independently to the sum, and there is some overlap or redundancy in what is contributed by each neuron; and that, on the other hand, the degree of redundancy is not a property of the neuronal representation, but just a contingent feature dependent on the particular set of stimuli used in probing that representation. The data available are consistent with the hypothesis, explored by Abbott, Rolls and Tovee (1996) through simulations, that if the ceiling provided by the limited number of stimuli that could be presented were at a much higher level, each neuron would continue to contribute as much as the first few, up to much larger neuronal populations, so that the number of stimuli that could be encoded would still continue to increase exponentially even with larger numbers of neurons (Fig. 5.11).

The redundancy observed could be characterized as flexible, in that it is the task that determines the degree to which large neuronal populations need to be sampled. If the task requires discriminations with very fine resolution in between many different stimuli (i.e. in a high-dimensional space), then the responses of many neurons must be taken into account. If very simple discriminations are required (requiring little information), small subsets of neurons or even single neurons may be sufficient (see Section 5.5.5). The importance of this type of flexible redundancy in the representation is discussed below.

We believe that the same type of ensemble encoding of what stimulus is present (i.e. stimulus identity) is likely to be used in other sensory systems, and have evidence that this is the case for the primate taste and olfactory systems, in particular for the cortical taste and olfactory areas in the orbitofrontal cortex (Rolls, Critchley and Treves 1996a). This type of ensemble encoding is also used in the primate hippocampus, in that the information about which spatial view is being seen rises approximately linearly with the number of hippocampal neurons in the sample (Rolls, Treves, Robertson, Georges-François and Panzeri 1998).

The analyses just described were obtained with neurons that were not simultaneously recorded, but similar results have now been obtained with simultaneously recorded neurons. That is, the information about which stimulus was shown increases approximately linearly with the number of neurons, showing that the neurons convey information that is nearly independent (Panzeri, Schultz, Treves and Rolls 1999a, Rolls, Webb and Booth 2000). (Consistently, Gawne and Richmond (1993) showed that even adjacent pairs of neurons recorded simultaneously from the same electrode carried information that was approximately 80% independent.)

Panzeri, Schultz, Treves and Rolls (1999a) developed a method for measuring the information in the relative time of firing of simultaneously recorded neurons (see Section B.2.4), which might be significant if the neurons became synchronized to some but not other stimuli in a set, as postulated by W. Singer and colleagues (e.g. Engel, Konig, Kreiter, Schillen and Singer (1992)). We found that for the set of inferior temporal cortex neurons currently available, almost all the information was available in the firing rates of the cells, and almost no information was available about which static image was shown in the relative time of firing of different simultaneously recorded neurons (Panzeri, Schultz, Treves and Rolls 1999a, Rolls, Webb and Booth 2000) (see Section B.3.4). Consistently, there were no significant cross-correlations between the spikes of many pairs of these simultaneously recorded inferior temporal cortex neurons. Thus the evidence is that most of the information is available in the firing rates of the neurons and not in synchronization, for representations of faces and objects in the inferior temporal visual cortex (see Section B.3.4) (and is also the case for space in the hippocampus, see Panzeri, Schultz, Treves and Rolls (1999a)).

It is unlikely that there are further processing areas beyond those described where ensemble coding changes into grandmother cell (local) encoding. Anatomically, there does not appear to be a whole further set of visual processing areas present in the brain; and outputs from the temporal lobe visual areas such as those described are taken to limbic and related regions such as the amygdala and orbitofrontal cortex, and via the entorhinal cortex to the hippocampus, where associations between the visual stimuli and other sensory representations are formed (see Chapter 12 and Rolls and Treves (1998), Rolls (1999a)). Indeed, tracing this pathway onwards, we have found a population of neurons with face-selective responses in the amygdala (Leonard, Rolls, Wilson and Baylis 1985, Rolls 1992a, Rolls 2000a) and orbitofrontal cortex (Rolls, Critchley, Browning and Inoue 2002b), and in the majority of these neurons, different responses occur to different faces, with ensemble (not local) coding still being present. The amygdala in turn projects to another structure that may be important in other behavioural responses to faces, the ventral striatum, and comparable neurons have also been found in the ventral striatum (Williams, Rolls, Leonard and Stern 1993).

5.5.4 Advantages for brain processing of the distributed representation of objects and faces

Three key types of evidence that the visual representation provided by neurons in the temporal cortical areas, and the olfactory and taste representations in the orbitofrontal cortex, are distributed, have been provided, and reviewed above. One is that the coding is not sparse

(Baylis, Rolls and Leonard 1985, Rolls and Tovee 1995b). The second is that different neurons have different response profiles to a set of stimuli, and thus have at least partly independent responses (Baylis, Rolls and Leonard 1985, Rolls and Tovee 1995b, Rolls, Treves and Tovee 1997b, Rolls 2000a). The third is that the capacity of the representations rises exponentially with the number of neurons (Rolls, Treves and Tovee 1997b, Abbott, Rolls and Tovee 1996). The advantages of the distributed encoding actually found are as follows. (These advantages do not apply to local, that is grandmother cell, encoding schemes, nor to all distributed encoding schemes (see Sections 1.6 and 5.5.4.2)).

5.5.4.1 Exponentially high coding capacity

This property arises from two factors: first, the encoding by the different neurons is sufficiently close to independent (i.e. factorial); and second the encoding is sufficiently distributed (Rolls, Treves and Tovee 1997b, Abbott, Rolls and Tovee 1996) (see further Rolls and Treves (1998)). Part of the biological significance of the exponential encoding capacity is that a receiving neuron or neurons can obtain information about which one of a very large number of stimuli is present by receiving the activity of relatively small numbers of inputs (in the order of hundreds) from each of the neuronal populations from which it receives. In particular, the characteristics of the actual visual cells described here indicate that the activity of 15 cells could encode 192 face stimuli (at 50% accuracy), 20 neurons could encode 768 stimuli, 25 neurons could encode 3,072 stimuli, 30 neurons could encode 12,288 stimuli, and 35 neurons could encode 49,152 stimuli (Abbott, Rolls and Tovee (1996); the values are for the optimal decoding case). Given that most neurons receive a limited number of synaptic contacts, in the order of several thousand, this type of encoding is ideal. (The capacity of the distributed representations was calculated from ensembles of neurons each already shown to provide information about faces. If inferior temporal cortex neurons were chosen at random, 20 times as many neurons would be needed in the sample if face-selective neurons comprised 5% of the population. This brings the number of inputs to a receiving neuron required from an ensemble of sending inferior temporal cortex neurons up to a reasonable number given the several thousand synapses typically received by each neuron.) This type of encoding (in contrast to local encoding) would enable for example neurons in the amygdala and orbitofrontal cortex to learn associations of visual stimuli with reinforcers such as the taste of food when each neuron received a reasonable number, perhaps in the order of hundreds, of inputs from the visually responsive neurons in the temporal cortical visual areas that specify which visual stimulus or object is being seen (see Rolls (1989b), Rolls (1992a), Rolls (1992c), Rolls and Treves (1998)). This type of representation is also appropriate for interfacing to the hippocampal system, to allow an episodic memory to be formed, that for example a particular visual object was seen in a particular place in the environment (Treves and Rolls 1994, Rolls 1996b, Rolls and Treves 1998).

It is useful to realize that although the sensory representation may have exponential encoding capacity, this does not mean that the associative networks in brain regions such as the amygdala, orbitofrontal cortex, and hippocampus that receive the information can store such large numbers of different patterns. Indeed, there are strict limits on the number of memories that associative networks can store (Rolls and Treves 1990, Treves and Rolls 1991, Rolls and Treves 1998) (see Chapter 7). The particular value of the exponential encoding capacity of sensory representations is that very fine discriminations can be made as there is much information in the representation, and that the representation can be decoded if the activity of even a limited number of neurons in the representation is known.

One of the underlying themes here is the neural representation of objects. How would one know that one had found a neuronal representation of objects in the brain? The criterion suggested (Rolls and Treves 1998) is that when receiving neurons can identify (with neuronally plausible decoding, such as the synaptically weighted sum of inputs described above) the object or stimulus that is present (from a large set of stimuli, thousands or more) from a realistic number of sending neurons, say in the order of 100, then the sending neurons provide a useful representation of the object.

The properties of the representation of faces, of objects (Booth and Rolls 1998), and of olfactory and taste stimuli, have been evident when the readout of the information was by measuring the firing rate of the neurons, typically over a 20, 50 or 500 ms period. Thus, at least where objects are represented in the visual, olfactory, and taste systems (e.g. individual faces, odours, and tastes), information can be read rather efficiently by the receiving neurons without taking into account any aspects of the possible temporal synchronization between neurons (Engel, Konig, Kreiter, Schillen and Singer 1992, Rolls, Treves and Tovee 1997b, Panzeri, Schultz, Treves and Rolls 1999a), or temporal encoding within a spike train (Tovee, Rolls, Treves and Bellis 1993). Thus rate coding carries an enormous amount of information. The challenge for those who believe that synchronization is important in neuronal coding is to quantify how much extra information it may add (see Panzeri, Schultz, Treves and Rolls (1999a)). In addition, correlations between the firing of different neurons to a set of stimuli do not appear to have any major impact in decreasing what is encoded by small ensembles of neurons in the situation in which there are many objects to be coded for, that is, in which the stimulus space is high dimensional (Panzeri, Schultz, Treves and Rolls 1999a).

5.5.4.2 Ease with which the code can be read by receiving neurons

For a code to be plausible, it is a requirement that neurons should be able to read the code. This is why when we have estimated the information from populations of neurons, we have used, in addition to a probability estimating measure (PE, optimal, in the Bayesian sense), also a dot product measure, which is a way of specifying that all that is required of decoding neurons would be the property of adding up postsynaptic potentials produced through each synapse as a result of the activity of each incoming axon (Rolls, Treves and Tovee 1997b, Abbott, Rolls and Tovee 1996) (see Fig. 5.12). More formally, the way in which the activation h of a neuron would be produced is by the following principle:

$$h = \sum_j x_j w_j \tag{5.2}$$

where $\sum_j$ indicates that the sum is over the C input axons (or connections) indexed by j to each neuron, and w_j is the strength of the jth synapse. The firing y of the neuron is a function of the activation

$$y = \mathrm{f}(h) \tag{5.3}$$

This activation function f may be linear, sigmoid, binary threshold, etc. (see Fig. 1.3).

It was found that with such a neuronally plausible decoding algorithm (the Dot Product, DP, algorithm), the same generic results were obtained, with only at most a 40% reduction of information compared to the more efficient (PE) algorithm. This is an indication that the brain could utilize the exponentially increasing capacity for encoding stimuli as the number of neurons in the population increases. For example, by using the representation provided by

Dot Product Multiple Cell Decoding

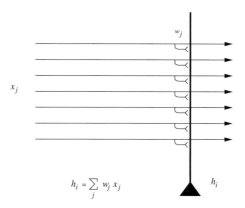

$$h_i = \sum_j w_j \, x_j \qquad h_i$$

Fig. 5.12 Dot product decoding or readout of information from a population of sending neurons x_j specifies that what is required of a decoding neuron i is that it adds up postsynaptic potentials produced through each synapse w_{ij} as a result of the activity of each incoming axon x_j.

the neurons described here as the input to an associative or autoassociative memory, which computes effectively the dot product on each neuron between the input vector and the synaptic weight vector, most of the information available would in fact be extracted (see Rolls and Treves (1990), Rolls and Treves (1998), Treves and Rolls (1991)). (The 40% reduction with dot product as compared to optimal decoding is a minor cost of using this simple type of decoding. The important point is that the number of stimuli still rises exponentially with the number of neurons, so that the number of stimuli that can be represented by a given sample of these neurons is still very large.)

Such a dot product decoding algorithm requires that the representation to be decoded is formed of linearly separable patterns. (Such a decoder might consist of neurons in associative networks in the next brain region such as the orbitofrontal cortex, amygdala, or hippocampus.) Moreover, the representation should be in a form in which approximately equal numbers of neurons are active for each pattern (i.e. a constant sparseness of the representation), and similar patterns should be represented by similar subsets of active neurons. This implies that the representation must not be too compact. Consider, for example, the binary encoding of numbers $(0 = 0000; 1 = 0001; 2 = 0010; 3 = 0011; 4 = 0100; ...; 7 = 0111; 8 = 1000;$ etc.). Now take the encoding of the numbers 7 (0111) and 8 (1000). Although these numbers are very similar, the nodes that represent these numbers are completely different, and the sparseness of the representation fluctuates wildly. So although such binary encoding is optimal in terms of the number of bits used, and has exponential encoding capacity, dot product decoding could not be used to decode it, and the number of neurons active for any stimulus fluctuates wildly. Also, for such binary encoding, a receiving neuron would need to receive from all the neurons in the representation; and generalization and graceful degradation would not occur. This indicates that a neuronal code must not be too compact in order to possess all these properties of generalization, constant sparseness, dot product decoding, and exponential capacity.

The use of this type of neuronal code by the brain is one of the factors that enables single neuron recording to be so useful in understanding brain function, as a correlation can frequently be found between the activity of even a single neuron and a subset of the stimuli being shown, of the motor responses being made, etc.

The fundamental point being made here is that the actual code being used in these brain areas can be read efficiently with the simplest type of decoding that neurons are thought to use, taking a sum of the input firings to the neuron each weighted by the synaptic weight connecting the input to the neuron. It is from the fact that a weighted sum of the firings can be used to read the code about which stimulus was shown that the interesting properties of generalization etc described next arise. (It may be emphasized that not all distributed codes have this property, so that the finding being discussed is important. If for example binary encoding of numbers as in computers were being used (with the bits representing from the lowest 1, 2, 4, 8, 16, 32, 64, etc.), then the code would not provide for generalization, completion, etc.)

5.5.4.3 Generalization, completion, graceful degradation, and higher resistance to noise

Because the decoding of a distributed representation involves assessing the activity of a whole population of neurons, and computing a dot product or correlation between the set (or vector) of inputs and the synaptic weights, a distributed representation provides more resistance to variation in individual components than does a local encoding scheme, and this provides for higher resistance to noise (Panzeri, Biella, Rolls, Skaggs and Treves 1996), and for graceful (in that it is gradual) degradation of performance when synapses or input axons are lost. The dot product decoding coupled with the type of representation actually found also enables the receiving neuron(s) to generalize to similar stimuli (where the similarity is measured by the number of inputs that correspond), and to complete an incomplete pattern when a network with recurrent collateral connections is used (see Section 7.3). Although these are well-known properties of some neural networks (Willshaw, Buneman and Longuet-Higgins 1969, Kohonen 1977, Kohonen 1989, Hopfield 1982, Bishop 1995, Rolls and Treves 1998) (see Chapter 7), the important point being made here is that the encoding being used by the inferior temporal visual cortex (and by the olfactory cortex and hippocampus, see Rolls, Critchley and Treves (1996a), Rolls, Treves, Robertson, Georges-François and Panzeri (1998)) is of a type that allows these properties to arise when simple neuronally plausible dot product decoding is being used by the receiving neurons.

A point being made here is that the sparse distributed representation found in the inferior temporal cortex as an output stage of the visual system has all the properties required for an input to pattern associators in receiving structures such as the orbitofrontal cortex and amygdala, and autoassociators in structures such as the hippocampus, as the sparseness of the representation allows these associative networks to operate with high memory capacity, and the distributed nature of the representation allows for the properties of generalization, completion etc just described. The argument here is completely different to that of Olshausen and Field (1996) who suggested that the code in the primary visual cortex, V1, was sparse because of the sparse feature structure of visual images (see Section 2.4.2).

5.5.4.4 Speed of readout of the information

The information available in a distributed representation can be decoded by an analyzer more quickly than can the information from a local representation, given comparable firing rates. Within a fraction of an interspike interval, with a distributed representation, much information can be extracted (Treves 1993, Rolls, Treves and Tovee 1997b, Treves, Rolls and Simmen 1997, Panzeri, Treves, Schultz and Rolls 1999b). An example of how rapidly

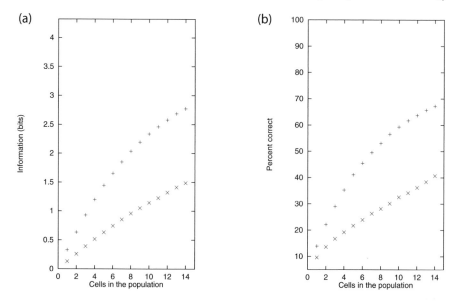

Fig. 5.13 (a) The information available about which of 20 faces had been seen that is available from the responses measured by the firing rates in a time period of 500 ms (+) or a shorter time period of 50 ms (x) of different numbers of temporal cortex cells. (b) The corresponding percentage correct from different numbers of cells. (From Rolls, Treves and Tovee, 1997.)

information can be made available in short time periods from an ensemble of neurons is shown in Fig. 5.13. In effect, spikes from many different neurons can contribute to calculating the angle between a neuronal population and a synaptic weight vector within an interspike interval. In contrast, with local encoding, the speed of information readout depends on the exact model considered, but if the rate of firing needs to be taken into account, this will necessarily take time, because of the time needed for several spikes to accumulate in order to estimate the firing rate.

Rolls and Treves (1998) note that with integrate-and-fire neurons (see their Appendix A5, and Section 7.6 of this book), there will not be a great delay in reading the information, provided that the neurons are kept close to threshold, so that even a small additional input will cause them to fire earlier than otherwise. The fact that many cortical neurons show spontaneous activity may reflect the fact that cortical neurons are indeed kept close to threshold, so that at least some of them can respond quickly when an input is received.

5.5.5 Should one neuron be as discriminative as the whole organism, in object encoding systems?

In the analysis of random dot motion with a given level of correlation among the moving dots, single neurons in area MT in the dorsal visual system of the primate can be approximately as sensitive or discriminative as the psychophysical performance of the whole animal (Zohary, Shadlen and Newsome 1994). The arguments and evidence presented here suggest that this is not the case for the ventral visual system, concerned with object identification. Why should there be this difference?

Rolls and Treves (1998) suggest that the dimensionality of what is being computed may account for the difference. In the case of visual motion (at least in the study referred to), the problem was effectively one-dimensional, in that the direction of motion of the stimulus along a line in 2D space was extracted from the activity of the neurons. In this low-dimensional stimulus space, the neurons may each perform one of a few similar computations on a particular (local) portion of 2D space, with the side effect that, by averaging, one can extract a signal of a global nature. Indeed, the average of the neuronal activity can be used to obtain the average direction of motion, but it is clear that most of the parallel computational power in this system has been devoted to sampling visual space and analyzing local motion, rather than being geared to an effective solution of the particular problem of detecting the global average of random dot motion.

In contrast, in the higher dimensional space of objects, in which there are very many different objects to represent as being different from each other, and in a system that is not concerned with location in visual space but on the contrary tends to be relatively invariant with respect to location, the goal of the representation is to reflect the many aspects of the input information in a way that enables many different objects to be represented, in what is effectively a very high dimensional space. This is achieved by allocating cells, each with an intrinsically limited discriminative power, to sample as thoroughly as possible the many dimensions of the space. Thus the system is geared to use efficiently the parallel computations of all its units precisely for tasks such as that of face discrimination, which was used as an experimental probe. Moreover, object representation must be kept higher dimensional, in that it may have to be decoded by dot product decoders in associative memories, in which the input patterns must be in a space that is as high-dimensional as possible (i.e. the activity on different input axons should not be too highly correlated). In this situation, each neuron should act somewhat independently of its neighbours, so that each provides its own separate contribution that adds together with that of the other neurons (in a linear manner, see above and Figs. 5.10, 5.13 and 5.11) to provide in toto sufficient information to specify which out of perhaps several thousand visual stimuli was seen. The computation involves in this case not an average of neuronal activity (cf. (Robertson, Rolls, Georges-François and Panzeri 1999)), but instead comparing the dot product of the activity of the population of neurons with a previously learned vector, stored in, for example, associative memories as the weight vector on a receiving neuron or neurons.

Zohary, Shadlen and Newsome (1994) put forward another argument which suggested to them that the brain could hardly benefit from taking into account the activity of more than a very limited number of neurons. The argument was based on their measurement of a small (0.12) correlation between the activity of simultaneously recorded neurons in area MT. They suggested that there would because of this be decreasing signal-to-noise ratio advantages as more neurons were included in the population, and that this would limit the number of neurons that it would be useful to decode to approximately 100. However, a measure of correlations in the activity of different neurons depends entirely on the way the space of neuronal activity is sampled, that is on the task chosen to probe the system. Among face cells in the temporal cortex, for example, much higher correlations would be observed when the task is a simple two-way discrimination between a face and a non-face, than when the task involves finer identification of several different faces. (It is also entirely possible that some face cells could be found that perform as well in a given particular face / non-face discrimination as the whole animal.) Moreover, their argument depends on the type of decoding of the activity of the population that is envisaged. It implies that the average of the neuronal activity must be estimated accurately. If a set of neurons uses dot product decoding, and then the activity of the decoding population is scaled or normalized by some negative feedback through inhibitory interneurons, then the effect of such correlated firing in the sending population is reduced, for the decoding effectively measures the relative firing of the different neurons in the population to be decoded. This is equivalent to measuring the angle between the current vector formed by the population of neurons firing, and a previously learned vector, stored in synaptic weights. Thus, with for example this biologically plausible decoding, it is not clear whether the correlation Zohary, Shadlen and Newsome (1994) describe would place a severe limit on the ability of the brain to utilize the information available in a population of neurons.

The main conclusion from this Section is that the information available from a set or ensemble of temporal cortex visual neurons increases approximately linearly as more neurons are added to the sample. This is powerful evidence that distributed encoding is used by the brain; and the code can be read just by knowing the firing rates in a short time of the population of neurons. The fact that the code can be read off from the firing rates, and by a principle as simple and neuron–like as dot product decoding, provides strong support for the general approach taken in this book to brain function. It is possible that more information would be available in the relative time of occurrence of the spikes, either within the spike train of a single neuron, or between the spike trains of different neurons, and it is to this that we now turn.

5.5.6 Temporal encoding in the spike train of a single neuron

It is possible that there is information contained in the relative time of firing of spikes within the spike train of a single neuron. For example, stimulus 1 might regularly elicit a whole burst of action potentials in the period 100–150 ms after the stimulus is shown. Stimulus 2 might regularly elicit a whole burst of action potentials between say 150–200 ms, and 350–400 ms. If one took into account when the action potentials occurred, then one might have more information from the spike train than if one took only the mean number of action potentials over the same 100–500 ms period. This possibility was investigated in a pioneering set of investigations by Optican and Richmond (1987) (see also Richmond and Optican (1987)) for inferior temporal cortex neurons. They used a set of black and white (Walsh) patterns as the stimuli. The way they assessed temporal encoding was by setting up the spikes to each stimulus as a time series with, for example, 64 bins each 6 ms long. With the response to each stimulus on each trial as such a time series, they calculated the first few principal components of the variance of spike trains across trials. The first principal component of the variance was in fact similar to the average response to all stimuli, and the second and higher components had waveforms that reflected more detailed differences in the response to the different stimuli. They found that several principal components were needed to capture most of the variance in their response set. They went on to calculate the information about which stimulus was shown that was available from the first principal component (which was similar to that in the firing rate), and if the second, third, etc. principal components were added. Their results appeared to indicate that considerably more information was available if the second principal component was added (50% more), and more if the third was added (37% more). They interpreted this as showing that temporal encoding in the spike train was used. In subsequent papers, the same authors (Richmond and Optican 1990, Optican, Gawne, Richmond and Joseph 1991, Eskandar, Richmond and Optican 1992) realized the need for a refinement of the analysis procedure.

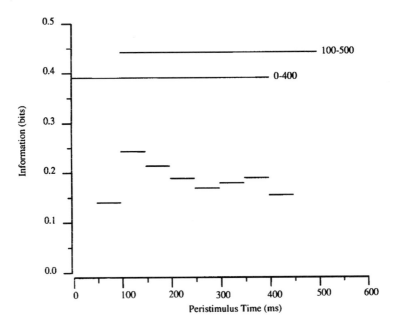

Fig. 5.14 The average information I(S,R) available in short temporal epochs (50 ms as compared to 400 ms) of the spike trains of single neurons about which face had been shown. (From Tovee and Rolls, 1995.)

Tovee, Rolls, Treves and Bellis (1993) independently reinvestigated this issue, using different faces as stimuli while recording from temporal cortical face-selective neurons. They showed that a correction needed to be made in the information analysis routine to correct for the limited number of trials (see Appendix B). When the correction was incorporated, Tovee, Rolls, Treves and Bellis (1993) found that rather little additional information was available if the second (19% more) and third (8% more) principal components were used in addition to the first. Moreover, they showed that if information was available in the second principal component in their data, it did not necessarily reflect interesting evidence about which stimulus had been seen, but frequently reflected latency differences between the different stimuli. (For example, some of their stimuli were presented parafoveally, and the longer onset latency that this produced was often reflected in the second principal component.)

The conclusion at this stage was that little additional evidence about which stimulus had been presented was available if temporal aspects of the spike train were taken into account. To pursue the point further, Rolls and colleagues next reasoned that if information was encoded in the time of arrival of spikes over a 400 ms period, much less information would be available if a short temporal epoch was taken. There would simply not be time within a short temporal epoch for the hypothesized characteristic pattern of firing to each stimulus to be evident. Moreover, it might well matter when the short temporal epoch was taken. Tovee, Rolls, Treves and Bellis (1993) and Tovee and Rolls (1995) therefore analyzed the information available in short temporal epochs (for example 100, 50, and 20 ms) of the spike train. They found that considerable information was present even with short epochs. For example, in periods of 20 ms, 30% of the information present in 400 ms using temporal encoding with the first three principal components was available. Moreover, the exact time when the epoch was taken was not crucial, with the main effect being that rather more information was available if information was measured near the start of the spike train, when the firing rate of the neuron tended to be highest (see Fig. 5.14).

The conclusion was that much information was available when temporal encoding could not be used easily, that is in very short time epochs of 20 or 50 ms. [It is also useful to note from Figs. 5.14 and 5.5 the typical time course of the responses of many temporal cortex visual neurons in the awake behaving primate. Although the firing rate and availability of information is highest in the first 50–100 ms of the neuronal response, the firing is overall well sustained in the 500 ms stimulus presentation period. Cortical neurons in the temporal lobe visual system, in the taste cortex, and in the olfactory cortex, do not in general have rapidly adapting neuronal responses to sensory stimuli. This may be important for associative learning: the outputs of these sensory systems can be maintained for sufficiently long while the stimuli are present for synaptic modification to occur. Although rapid synaptic adaptation within a spike train is seen in some experiments in brain slices (Markram and Tsodyks 1996, Abbott, Rolls and Tovee 1996), it is not a very marked effect in at least some brain systems in vivo, when they operate in normal physiological conditions with normal levels of acetylcholine, etc.[8]]

To pursue this issue even further, Rolls, Tovee, Purcell, Stewart and Azzopardi (1994b) and Rolls and Tovee (1994) limited the period for which visual cortical neurons could respond by using backward masking. In this paradigm, a short (16 ms) presentation of the test stimulus (a face) was followed after a delay of 0, 20, 40, 60, etc. ms by a masking stimulus (which was a high contrast set of letters) (see Fig. 5.15). They showed that the mask did actually interrupt the neuronal response, and that at the shortest interval between the stimulus and the mask (a delay of 0 ms, or a 'Stimulus Onset Asynchrony' of 20 ms), the neurons in the temporal cortical areas fired for approximately 30 ms (see Fig. 5.16). Under these conditions, the subjects could identify which of five faces had been shown much better than chance. Thus when the possibility of temporal encoding within a spike train was minimized by allowing temporal cortex visual neurons to fire for only approximately 30 ms, visual identification was still possible (though perception was certainly not perfect).

Thus most of the evidence indicates that temporal encoding within the spike train of a single neuron does not add much to the information that is present in the firing rate of the neuron. Most of the information is available from the firing rate, even when short temporal epochs are taken (Rolls, Tovee and Panzeri 1999b). Thus a neuron in the next cortical area would obtain considerable information within 20–50 ms of measuring the firing rate of a single neuron. Moreover, if it took a short sample of the firing rate of many neurons in the preceding area, then very much information is made available, as shown in the preceding Section.

---

[8] A situation in which synaptic adaptation might be very interesting is if the time-course of adaptation was different for different inputs to a neuron. If for example the synaptic inputs from forward afferents to cortical neurons showed some partial adaptation over the first 30–50 ms of strong firing, then this might not only account for the small decline in firing seen soon after the onset of the response in Fig. 5.5, but might also allow the recurrent collaterals and backprojections, if their synapses showed less adaptation, to become relatively more important in the neuronal responses soon after the initial feed-forward mediated firing of a neuron. This would enable context-dependent effects to be given more influence at later times of the neuronal response.

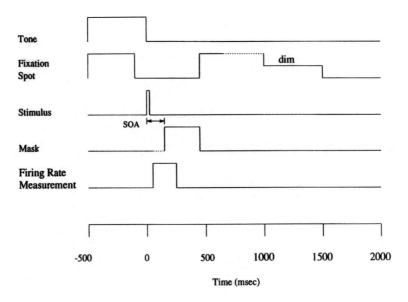

Fig. 5.15 Backward masking paradigm. The visual stimulus appeared at time 0 for 16 ms. The time between the start of the visual stimulus and the masking image is the Stimulus Onset Asynchrony (SOA). A visual fixation task was being performed to ensure correct fixation of the stimulus. In the fixation task, the fixation spot appeared in the middle of the screen at time −500 ms, was switched off 100 ms before the test stimulus was shown, and was switched on again at the end of the mask stimulus. Then when the fixation spot dimmed after a random time, fruit juice could be obtained by licking. No eye movements could be performed after the onset of the fixation spot. (From Rolls and Tovee, 1994.)

5.5.7   Temporal synchronization of the responses of different cortical neurons

Von der Malsburg (1973) (see also Malsburg (1990) and Malsburg and Schneider (1986)) has proposed that information might be encoded by the relative time of firing of different neurons. In particular, he suggested this to help with the feature binding problem, as discussed in Sections 8.5 and 8.4.5. Temporal synchronization of one subset of neurons, and separate temporal synchronization of a different subset, might allow the activity of different subsets of neurons to be kept apart. The decoding process might be performed by neurons with a strong sensitivity to temporal co-occurrence of inputs (which is a property of neurons). Singer, Engel, Konig and colleagues (Engel, Konig and Singer 1991, Engel, Konig, Kreiter, Schillen and Singer 1992, Gray, Konig, Engel and Singer 1989, Gray, Engel and Singer 1992, Singer 1999, Singer 2000) have obtained some evidence that when features must be bound, synchronization of neuronal populations can occur. At first the evidence suggested that oscillations might occur in the visual system during object recognition. However, these oscillations were most evident in the primary visual cortex of the anaesthetized cat, and were best evoked by moving visual stimuli. Indeed, the frequency of the oscillations was a function of the velocity of the moving stimulus. In the auditory cortex, the action potentials of neurons can become synchronized (deCharms and Merzenich 1996). Tovee and Rolls (1992) found no evidence, however, for oscillations in neurons or groups of neurons in the temporal cortical visual areas of awake

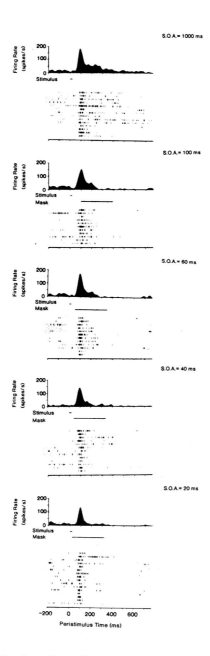

Fig. 5.16 Firing of a temporal cortex cell to a 20 ms presentation of a face stimulus when the face was followed with different Stimulus Onset Asynchronies (SOAs) by a masking visual stimulus. At an SOA of 20 ms, when the mask immediately followed the face, the neuron fired for only approximately 30 ms, yet identification of the face at this SOA by human observers was possible, as shown in Fig. 5.25 (Rolls and Tovee, 1994; Rolls, Tovee, Purcell et al., 1994).

behaving macaques performing a visual fixation task while being shown sets of stationary face and non-face stimuli. Thus at least for form perception of static stimuli in primates, oscillations may not be necessary. Oscillations are likely to occur in any circuit with positive feedback. One possibility is that in evolution the feedback systems for maintaining cortical stability have improved, and that in line with this, oscillations are less evident in the awake behaving monkey than in the anaesthetized cat. Oscillations may also be less likely to occur in systems processing static stimuli, such as what an object is that is presented for at least 0.5 s, than in systems concerned with processing dynamically changing stimuli, such as the motion of visual stimuli.

Engel, Konig and Singer (1991), Engel, Konig, Kreiter, Schillen and Singer (1992) and Singer (1999) do, however, present impressive evidence that under some circumstances, in some animals, synchronization between different subsets of neurons may occur. It remains an open issue whether this is an important part of the code in some parts of the visual system (see further Ferster and Spruston (1995); Singer (1995)). It may perhaps be used under conditions where ambiguous stimuli such as a Necker cube or with figure-ground ambiguity are presented, and there is plenty of time (perhaps hundreds of ms) to code, and decode, the different possibilities by different synchronization of different subsets of neurons. The evidence described above and in Chapter 8 suggests though that normally each cortical area can operate fast, in 20–50 ms, and that much of the information is available from the firing rates of ensembles of neurons, without taking into account the relative time of firing. The challenge for those who believe that synchronization is important in neuronal coding is to quantify how much extra information it may add.

Panzeri, Schultz, Treves and Rolls (1999a), have developed information theoretic methods to quantify the absolute amounts of information available in the firing rates and in the relative times of firing of simultaneously recorded neurons (see Section B.2.4). Using those methods, we found that for the set of inferior temporal cortex neurons currently available, almost all the information was available in the firing rates of the cells, and almost no information was available about which static image was shown in the relative time of firing of different simultaneously recorded neurons (Panzeri, Schultz, Treves and Rolls 1999a, Rolls, Webb and Booth 2000). Consistently, there were no significant cross-correlations between the spikes of these simultaneously recorded inferior temporal cortex neurons. In addition, the methods showed that correlations between the firing of different neurons to a set of stimuli do not appear to have any major impact in decreasing what is encoded by small ensembles of neurons in the situation in which there are many objects to be coded for, that is in which the stimulus space is high dimensional (Panzeri, Schultz, Treves and Rolls 1999a) (see Section B.3.4). Thus the evidence is that most of the information is available in the firing rates of the neurons and not in synchronization, for representations of faces and objects in the inferior temporal visual cortex (and that this is also the case for space in the hippocampus, see Panzeri, Schultz, Treves and Rolls (1999a)).

The neuronal synchronization scheme does has some disadvantages and incongruities. One disadvantage of the scheme is that it may not be sufficiently fast in operation (both to become set up, and to allow appropriate decoding) for the whole process to be sufficiently complete in 20–30 ms of cortical processing time per cortical area that the neurophysiological results above show is sufficient for object recognition. Another argument that the scheme may not at least be used in many parts of the visual system is that the information about which object is present can be read off with an enormous capacity from the firing rates of a small number (for example 10–50) of neurons, as shown above. (The information capacity calculations of neuronal populations performed by Rolls, Treves and Tovee (1997b) and Abbott, Rolls and Tovee (1996) used only measurements of firing rates.) Third, the scheme is computationally very powerful. In a two-layer network, von der Malsburg has shown that it could provide the necessary feature linking to perform object recognition with relatively few neurons, because they can be reused again and again, linked differently for different objects (Malsburg 1999). In contrast, the primate uses a considerable part of its brain, perhaps 50% in monkeys, for visual processing, with therefore what must be in the order of 10^9 neurons and 10^{13} synapses involved, so that the solution adopted by the real visual system may be one which relies on many neurons with simpler processing than arbitrary syntax implemented by synchronous firing of separate assemblies suggests. Instead, many neurons are used at each stage of, for example, the visual system, and the reason suggested for this is that syntax or binding is represented by low-order combinations of what is represented in previous layers of the cortical architecture (see Section 8.4.5). The low-order combinations (i.e. combinations of a small number of inputs) are suggested to occur because otherwise there would be a massive combinatorial explosion of the number of neurons required. This is why it is suggested there are so many neurons devoted to vision, which would not be required if a more powerful system utilizing temporal binding were actually implemented (Singer 1995). It will be fascinating to see how research on these different approaches to processing in the primate visual system develops. For the development of both approaches, the use of well-defined neuronal network models will be very important.

Abeles and colleagues (1990), (1993) (see also Abeles (1991)) have some evidence that occasionally (perhaps every few seconds or trials) groups of neurons emit action potentials with a precise (in the order of 1 ms) timing relationship to each other. They have argued that such synchrony could reflect a binding mechanism for different cell assemblies. They have termed these events Synfire chains. However, such events are relatively rare, and whether they convey significant information remains to be shown.

5.5.8 Conclusions on cortical encoding

The exponential rise in the number of stimuli that can be decoded when the firing rates of different numbers of neurons are analyzed shows that the encoding of information using firing rates is a very powerful coding scheme used by the cerebral cortex. Quantitatively, it is likely to be far more important than temporal encoding, in terms of the number of stimuli that can be encoded. Moreover, the information available from an ensemble of cortical neurons when only the firing rates are read, that is with no temporal encoding within or between neurons, is made available very rapidly (see Fig. 5.14). Further, the neuronal responses in most brain areas of behaving monkeys show sustained firing rate differences to different stimuli (see for example Fig. 5.5), so that it may not usually be necessary to invoke temporal encoding for the information about the stimulus to be present. Temporal encoding may, however, be used as well as rate coding, for example to help solve the binding problem (although there are alternatives to solve the binding problem, as shown in Chapter 8).

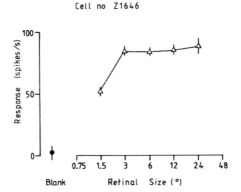

Cell no Z1646

Fig. 5.17 Typical response of an inferior temporal cortex face-selective neuron to faces of different sizes. The size subtended at the retina in degrees is shown. (From Rolls and Baylis, 1986.)

5.6   Invariance in the neuronal representation of stimuli

One of the major problems that must be solved by a visual system is the building of a representation of visual information that allows recognition to occur relatively independently of size, contrast, spatial frequency, position on the retina, angle of view, etc. This is required so that if the receiving regions such as the amygdala, orbitofrontal cortex and hippocampus learn about one view, position, or size of the object, the animal generalizes correctly to other positions, views and sizes of the object.

5.6.1   Size and spatial frequency invariance

The majority of face-selective neurons in the inferior temporal cortex have responses that are relatively invariant with respect to the size of the stimulus (Rolls and Baylis 1986). An example of the responses of an inferior temporal cortex face-selective neurons to faces of different sizes is shown in Fig. 5.17. The median size change tolerated with a response of greater than half the maximal response was 12 times. Also, the neurons typically responded to a face when the information in it had been reduced from 3D to a 2D representation in grey on a monitor, with a response that was on average 0.5 of that to a real face. Another transform over which recognition is relatively invariant is spatial frequency. For example, a face can be identified when it is blurred (when it contains only low spatial frequencies), and when it is high-pass spatial frequency filtered (when it looks like a line drawing). If the face images to which these neurons respond are low-pass filtered in the spatial frequency domain (so that they are blurred), then many of the neurons still respond when the images contain frequencies only up to 8 cycles per face. Similarly, the neurons still respond to high-pass filtered images (with only high spatial frequency edge information) when frequencies down to only 8 cycles per face are included (Rolls, Baylis and Leonard 1985). Face recognition shows similar invariance with respect to spatial frequency (see Rolls, Baylis and Leonard (1985)). Further analysis of these neurons with narrow (octave) bandpass spatial frequency filtered face stimuli shows that the responses of these neurons to an unfiltered face can not be predicted from a linear combination of their responses to the narrow band stimuli (Rolls, Baylis and Hasselmo 1987). This lack of linearity of these neurons, and their responsiveness to a wide range of spatial frequencies, indicate that in at least this part of the primate visual system recognition does not occur using Fourier analysis of the spatial frequency components of images.

5.6.2 Translation (shift) invariance

Inferior temporal visual cortex neurons also often show considerable translation (shift) invariance, not only under anesthesia (see Gross, Desimone, Albright and Schwartz (1985)), but also in the awake behaving primate (Tovee, Rolls and Azzopardi 1994). In most cases the responses of the neurons were little affected by which part of the face was fixated, and the neurons responded (with a greater than half-maximal response) even when the monkey fixated 2–5 degrees beyond the edge of a face that subtended 8–17 degrees at the retina. Moreover, the stimulus selectivity between faces was maintained this far eccentrically within the receptive field.

5.6.3 Reduced translation invariance in natural scenes

Until recently, research on translation invariance considered the case in which there is only one object in the visual field. What happens in a cluttered, natural, environment? Do all objects that can activate an inferior temporal neuron do so whenever they are anywhere within the large receptive fields of inferior temporal neurons (cf. Sato (1989))? If so, the output of the visual system might be confusing for structures that receive inputs from the temporal cortical visual areas. In an investigation of this, it was found that the mean firing rate across all cells to a fixated effective face with a non-effective face in the parafovea (centred 8.5 degrees from the fovea) was 34 spikes/s. On the other hand, the average response to a fixated non-effective face with an effective face in the periphery was 22 spikes/s (Rolls and Tovee 1995a). Thus these cells gave a reliable output about which stimulus is actually present at the fovea, in that their response was larger to a fixated effective face than to a fixated non-effective face, even when there were other parafoveal stimuli effective for the neuron. Thus the neurons provide information biassed towards what is present at the fovea, and not equally about what is present anywhere in the visual field. This makes the interface to action simpler, in that what is at the fovea can be interpreted (e.g. by an associative memory) partly independently of the surroundings, and choices and actions can be directed if appropriate to what is at the fovea (cf. Ballard (1993)). These findings are a step towards understanding how the visual system functions in a normal environment (see also Gallant, Connor and Van-Essen (1998), Stringer and Rolls (2000)).

To investigate further how information is passed from the inferior temporal cortex (IT) to other brain regions to enable stimuli to be selected from natural scenes for action, Rolls, Webb and Booth (2000) analyzed the responses of single and simultaneously recorded IT neurons to stimuli presented in complex natural backgrounds. In one situation, a visual fixation task was performed in which the monkey fixated at different distances from the effective stimulus. In another situation the monkey had to search for two objects on a screen, and a touch of one object was rewarded with juice, and of another object was punished with saline (see Fig. 5.18). In both situations neuronal responses to the effective stimuli for the neurons were compared when the objects were presented in the natural scene or on a plain background. It was found that the overall response of the neuron to objects was sometimes somewhat reduced when they were presented in natural scenes, though the selectivity of the neurons remained. However,

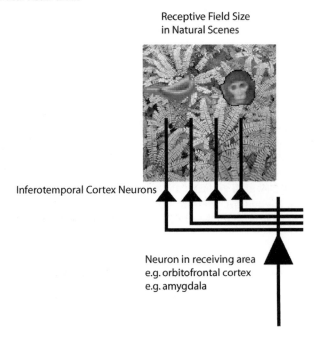

Receptive Field Size
in Natural Scenes

Inferotemporal Cortex Neurons

Neuron in receiving area
e.g. orbitofrontal cortex
e.g. amygdala

Fig. 5.18 Objects shown in a natural scene, in which the task was to search for and touch one of the stimuli. The objects in the task as run were smaller. The diagram shows that if the receptive fields of inferior temporal cortex neurons are large in natural scenes with multiple objects, then any receiving neuron in structures such as the orbitofrontal and amygdala would receive information from many stimuli in the field of view, and would not be able to provide evidence about each of the stimuli separately.

the main finding was that the magnitudes of the responses of the neurons typically became much less in the real scene the further the monkey fixated in the scene away from the object (see Fig. 5.19). It is proposed that this reduced translation invariance in natural scenes helps an unambiguous representation of an object which may be the target for action to be passed to the brain regions which receive from the primate inferior temporal visual cortex. It helps with the binding problem, by reducing in natural scenes the effective receptive field of at least some inferior temporal cortex neurons to approximately the size of an object in the scene. The computational utility and basis for this is considered in Chapters 8 (e.g. Section 8.4.9), 9, 12 and 13.

It is also found that in natural scenes, the effect of object-based attention on the response properties of inferior temporal cortex neurons is relatively small, as illustrated in Fig. 5.20 (Rolls, Zheng and Aggelopoulos 2001b). The results summarized in Fig. 5.20 show that the receptive fields were large (65 deg) with a single stimulus in a blank background (top left), and were greatly reduced in size (to 36.6 deg) when presented in a complex natural scene (top right). The results also show that there was little difference in receptive field size or firing rate in the complex background when the effective stimulus was selected for action (bottom right), and when it was not (middle right). The computational basis for these relatively minor effects of object-based attention when objects are viewed in natural scenes is also considered in Section 8.4.9 and Chapter 9.

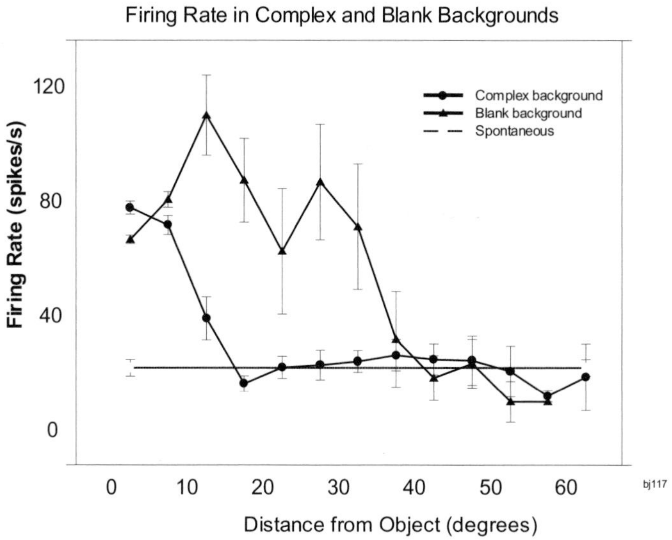

Firing Rate in Complex and Blank Backgrounds

Fig. 5.19 Firing of a temporal cortex cell to an effective stimulus presented either in a blank background or in a natural scene, as a function of the angle in degrees at which the monkey was fixating away from the effective stimulus. The task was to search for and touch the stimulus. (After Rolls, Webb and Booth, 2000.)

5.6.4   A view–independent representation of objects and faces

Some temporal cortical neurons reliably responded differently to the faces of two different individuals independently of viewing angle, although in most cases (16/18 neurons) the response was not perfectly view–independent (Hasselmo, Rolls, Baylis and Nalwa 1989b). Mixed together in the same cortical regions there are neurons with view–dependent responses. Such neurons might respond for example to a view of a profile of a monkey but not to a full–face view of the same monkey (Perrett, Smith, Mistlin, Chitty, Head, Potter, Broennimann, Milner and Jeeves 1985a). These findings, of view–dependent, partially view–independent, and view–independent representations in the same cortical regions are consistent with the hypothesis discussed below that view–independent representations are being built in these regions by associating together neurons that respond to different views of the same individual.

Further evidence that some neurons in the temporal cortical visual areas have object-based rather than view-based responses comes from a study of a population of neurons that responds to moving faces (Hasselmo, Rolls, Baylis and Nalwa 1989b). For example, four neurons responded vigorously to a head undergoing ventral flexion, irrespective of whether the view of the head was full face, of either profile, or even of the back of the head. These different views could only be specified as equivalent in object-based coordinates. Further, the movement specificity was maintained across inversion, with neurons responding for example to ventral flexion of the head irrespective of whether the head was upright or inverted. In this procedure, retinally encoded or viewer-centered movement vectors are reversed, but the object-based description remains the same.

Also consistent with object-based encoding is the finding of a small number of neurons that respond to images of faces of a given absolute size, irrespective of the retinal image size or distance (Rolls and Baylis 1986).

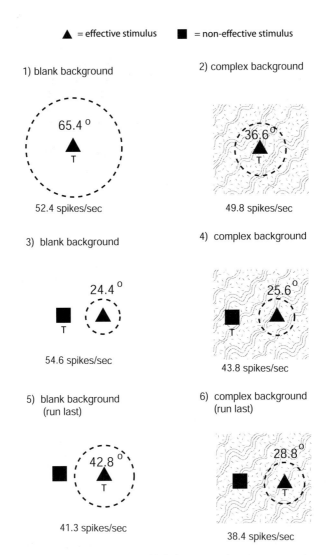

Fig. 5.20 Summary of the receptive field sizes of inferior temporal cortex neurons to an effective stimulus presented with and without a non-effective stimulus for the cell in either a blank background (blank screen) or in a natural scene (complex background). The stimulus that was a target for action in the different experimental conditions is marked by T. When the target stimulus was touched, a reward was obtained. The mean receptive field size of the population of neurons analyzed, and the mean firing rate in spikes/s, is shown. The stimuli subtended 9 deg × 7 deg at the retina, and occurred on each trial in a random position in the 70 deg × 55 deg screen. (After Rolls, Zheng and Aggelopoulos, 2001.)

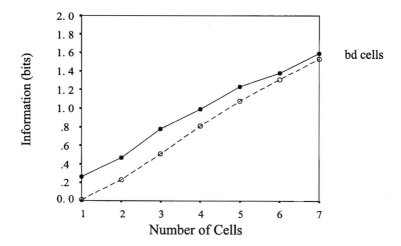

Fig. 5.21 View-independent object encoding: information in a population of different numbers of inferior temporal cortex cells available from a single trial in which one view was shown about which of 10 objects each with 4 different views in the stimulus set had been seen. The solid symbols show the information decoded with a Bayesian probability estimator algorithm, and the open symbols with dot product decoding. (From Booth and Rolls, 1998.)

Neurons with view invariant responses of objects seen naturally by macaques have also been described recently (Booth and Rolls 1998). The stimuli were presented for 0.5 s on a colour video monitor while the monkey performed a visual fixation task. The stimuli were images of 10 real plastic objects that had been in the monkey's cage for several weeks, to enable him to build view invariant representations of the objects. Control stimuli were views of objects that had never been seen as real objects. The neurons analyzed were in the TE cortex in and close to the ventral lip of the anterior part of the superior temporal sulcus. Many neurons were found that responded to some views of some objects. However, for a smaller number of neurons, the responses occurred only to a subset of the objects (using ensemble encoding), irrespective of the viewing angle. Moreover, the firing of a neuron on any one trial, taken at random and irrespective of the particular view of any one object, provided information about which object had been seen, and this information increased approximately linearly with the number of neurons in the sample (see Fig. 5.21). This is strong quantitative evidence that some neurons in the inferior temporal cortex provide an invariant representation of objects. Moreover, the results of Booth and Rolls (1998) show that the information is available in the firing rates, and has all the desirable properties of distributed representations described above, including exponentially high coding capacity, and rapid speed of readout of the information.

Further evidence consistent with these findings is that some studies have shown that the responses of some visual neurons in the inferior temporal cortex do not depend on the presence or absence of critical features for maximal activation (Perrett, Rolls and Caan 1982, Tanaka 1993, Tanaka 1996). For example, Mikami, Nakamura and Kubota (1994) have shown that some TE cells respond to partial views of the same laboratory instrument(s), even when these partial views contain different features. In a different approach, Logothetis, Pauls, Bulthoff and Poggio (1994) have reported that in monkeys extensively trained (over thousands of trials) to treat different views of computer generated wire-frame 'objects' as the same, a small population of neurons in the inferior temporal cortex did respond to different views of the same wire-frame object (see also Logothetis and Sheinberg (1996)). However, extensive training is not necessary for invariant representations to be formed, and indeed no explicit training in invariant object recognition was given in the experiment by Booth and Rolls (1998), as Rolls' hypothesis (1992a) is that view invariant representations can be learned by associating together the different views of objects as they are moved and inspected naturally in a period that may be in the order of a few seconds.

5.7 Different neural systems are specialized for face identification and for face expression decoding

Some neurons respond to face identity, and others to face expression (Hasselmo, Rolls and Baylis 1989a). The neurons responsive to expression were found primarily in the cortex in the superior temporal sulcus, while the neurons responsive to identity were found in the inferior temporal gyrus. Information about facial expression is of potential use in social interactions. Indeed, damage to this population may contribute to the deficits in emotional and social behavior such as tameness and social withdrawal that are part of the Kluver-Bucy syndrome produced by temporal lobe damage in monkeys (see Rolls (1984), Rolls (1999a), Leonard, Rolls, Wilson and Baylis (1985)).

A further way in which some of these neurons in the cortex in the superior temporal sulcus may be involved in social interactions is that some of them respond to gestures, for example to a face undergoing ventral flexion (Perrett, Smith, Potter, Mistlin, Head, Milner and Jeeves 1985b, Hasselmo, Rolls, Baylis and Nalwa 1989b). The interpretation of these neurons as being useful for social interactions is that in some cases these neurons respond not only to ventral head flexion, but also to the eyes lowering and the eyelids closing (Hasselmo, Rolls, Baylis and Nalwa 1989b). These movements (turning the head away, breaking eye contact, and eyelid lowering) often occur together when a monkey is breaking social contact with another, and neurons that respond to these components could be built by associative synaptic modification. It is also important when decoding facial expression to retain some information about the direction of the head relative to the observer, for this is very important in determining whether a threat is being made in your direction. The presence of view–dependent, head and body gesture (Hasselmo, Rolls, Baylis and Nalwa 1989b), and eye gaze (Perrett, Smith, Potter, Mistlin, Head, Milner and Jeeves 1985b), representations in some of these cortical regions where face expression is represented is consistent with this requirement. In contrast, the TE areas (more ventral, mainly in the macaque inferior temporal gyrus), in which neurons tuned to face identity (Hasselmo, Rolls and Baylis 1989a) and with view–independent responses (Hasselmo, Rolls, Baylis and Nalwa 1989b) are more likely to be found, may be more related to an object–based representation of identity. Of course, for appropriate social and emotional responses, both types of subsystem would be important, for it is necessary to know both the direction of a social gesture, and the identity of the individual, in order to make the correct social or emotional response.

As we have seen, outputs from the temporal cortical visual areas reach the amygdala and the orbitofrontal cortex, and evidence is accumulating that these brain areas are involved in social and emotional responses to faces (Rolls 1989b, Rolls 1992b, Rolls 1992a, Rolls 1999a,

Rolls 2000d). For example, lesions of the amygdala in monkeys disrupt social and emotional responses to faces, and we have identified a population of neurons with face-selective responses in the primate amygdala (Leonard, Rolls, Wilson and Baylis 1985), some of which respond to facial and body gesture (Brothers, Ring and Kling 1990). Rolls, Critchley, Browning and Inoue (2002b) have found a number of face-responsive neurons in the orbitofrontal cortex, and such neurons are also present in adjacent prefrontal cortical areas (Wilson, O'Sclaidhe and Goldman-Rakic 1993, O'Sclaidhe, Wilson and Goldman-Rakic 1999).

We have applied this research to the study of humans with frontal lobe damage, to try to develop a better understanding of the social and emotional changes that may occur in these patients. Impairments in the identification of facial and vocal emotional expression were demonstrated in a group of patients with ventral frontal lobe damage who had behavioural problems such as disinhibited or socially inappropriate behaviour (Hornak, Rolls and Wade 1996). A group of patients with lesions outside this brain region, without these behavioural problems, was unimpaired on the expression identification tests. These findings suggest that some of the social and emotional problems associated with ventral frontal lobe or amygdala damage may be related to a difficulty in identifying correctly facial (and vocal) expression, and in learning associations involving such visual stimuli (Hornak, Rolls and Wade 1996, Rolls 1999a, Rolls 1999b, Rolls, Hornak, Wade and McGrath 1994a).

Neuroimaging data, while not being able to address the details of what is encoded in a brain area or of how it is encoded, does provide evidence consistent with the neurophysiology that there are different face processing systems in the human brain. For example, Kanwisher, McDermott and Chun (1997) and Ishai, Ungerleider, Martin, Schouten and Haxby (1999) have shown activation by faces of an area in the fusiform gyrus; Hoffman and Haxby (2000) have shown that distinct areas are activated by eye gaze and face identity; Dolan, Fink, Rolls, Booth, Holmes, Frackowiak and Friston (1997) have shown that a fusiform gyrus area becomes activated after humans learn to identify faces in complex scenes; and the amygdala (Morris, Fritch, Perrett, Rowland, Young, Calder and Dolan 1996) and orbitofrontal cortex (Blair, Morris, Frith, Perrett and Dolan 1999) may become activated particularly by certain face expressions.

Different inferior temporal cortex neurons in macaques not only provide different types of information about different aspects of faces, as just described, but also respond differently to faces and objects, with some neurons conveying information primarily about faces, and others about objects (Rolls and Tovee 1995b, Rolls, Treves, Tovee and Panzeri 1997d, Booth and Rolls 1998). In fact, when recording in the inferior temporal cortex, one finds small clustered groups of neurons, with neurons within a cluster responding to somewhat similar attributes of stimuli, and different clusters responding to different types of visual stimuli. For example, within the set of neurons responding to moving faces, there are some neuronal clusters that respond to for example ventral and dorsal flexion of the head on the body; others that respond to axial rotation of the head; others that respond to the sight of mouth movements; and others that respond to the gaze direction in a face being viewed (Hasselmo, Rolls, Baylis and Nalwa 1989b, Perrett, Smith, Potter, Mistlin, Head, Milner and Jeeves 1985b). Within the domain of objects, some neuronal clusters respond based on the shapes of the object (e.g. responding to elongated and not round objects), others respond based on the texture of the object being viewed, and others are sensitive to what the object is doing (Hasselmo, Rolls, Baylis and Nalwa 1989b, Perrett, Smith, Mistlin, Chitty, Head, Potter, Broennimann, Milner and Jeeves 1985a). Within a cluster, each neuron is tuned differently, and may combine at least partly with information predominant in other clusters, so that at least some of the neurons in a cluster can become quite selective between different objects, with the preference for individual objects showing the usual exponential firing rate distribution (see Figs. 5.3 and 5.4). Using optical imaging, Tanaka and colleagues (Tanaka 1996) have seen comparable evidence for different localized clusters of neuronal activation produced by different types of stimuli.

The principle that Rolls proposes underlies this local clustering is the representation of the high dimensional space of objects and their features into a two-dimensional cortical sheet using the local self-organizing mapping principles described in Section 7.4.6. In this situation, these principles produce maps that have many fractures, but nevertheless include local clusters of similar neurons. These local clusters minimize the wiring length between neurons, if the computations involve exchange of information within neurons of a similar general type, as will usually be the case. It is consistent with this general conceptual background that Krieman, Koch and Fried (2000) have described some neurons in the human high-order visual cortical areas that seem to respond to categories of object. This is consistent with the principles just described, although in humans backprojections from language or other cognitive areas concerned for example with tool use might also influence the categories represented in high order cortical areas, as described in Sections 7.4.5 and 1.11, and by Farah, Meyer and McMullen (1996) and Farah (2000).

5.8   Learning of new representations in the temporal cortical visual areas

To investigate the idea that visual experience might guide the formation of the responsiveness of neurons so that they provide an economical and ensemble-encoded representation of items actually present in the environment, the responses of inferior temporal cortex face-selective neurons have been analyzed while a set of new faces were shown. Some of the neurons studied in this way altered the relative degree to which they responded to the different members of the set of novel faces over the first few (1–2) presentations of the set (Rolls, Baylis, Hasselmo and Nalwa 1989a) (see examples in Fig. 5.22). If in a different experiment a single novel face was introduced when the responses of a neuron to a set of familiar faces were being recorded, the responses to the set of familiar faces were not disrupted, while the responses to the novel face became stable within a few presentations. Alteration of the tuning of individual neurons in this way may result in a good discrimination over the population as a whole of the faces known to the monkey. This evidence is consistent with the categorization being performed by self-organizing competitive neuronal networks, as described in Section 7.4 and elsewhere (Rolls and Treves 1998).

Further evidence that these neurons can learn new representations very rapidly comes from an experiment in which binarized black and white (two-tone) images of faces that blended with the background were used (see Fig. 5.23, left). These did not activate face-selective neurons. Full grey-scale images of the same photographs were then shown for ten 0.5s presentations (Fig. 5.23, middle). In a number of cases, if the neuron happened to be responsive to that face, when the binarized version of the same face was shown next (Fig. 5.23, right), the neurons responded to it (Tovee, Rolls and Ramachandran 1996). This is a direct parallel to the same phenomenon that is observed psychophysically, and provides dramatic evidence that these neurons are influenced by only a very few seconds (in this case 5 s) of experience with a visual stimulus. We have shown a neural correlate of this effect using similar stimuli and a

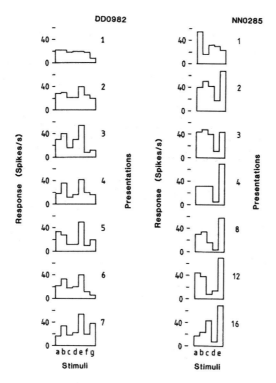

Fig. 5.22 Learning in the responses of inferior temporal cortex neurons. Results for two face-selective neurons are shown. On presentation 1 a set of 7 completely novel faces was shown when recording from neuron DD0982. The firing rates elicited on 1-sec presentations of each stimulus (a-g) are shown. After the first presentation, the set of stimuli was repeated in random sequence for a total of 7 iterations. After the first two presentations, the relative responses of the neuron to the different stimuli settled down to a fairly reliable response profile to the set of stimuli (as shown by statistical analysis). NN0285 – data from a similar experiment with another neuron, and another completely new set of face stimuli. (From Rolls, Baylis, Hasselmo and Nalwa, 1989.)

similar paradigm in a PET (positron emission tomography) neuroimaging study in humans, with a region showing an effect of the learning found for faces in the right temporal lobe, and for objects in the left temporal lobe (Dolan, Fink, Rolls, Booth, Holmes, Frackowiak and Friston 1997).

Such rapid learning of representations of new objects appears to be a major type of learning in which the temporal cortical areas are involved. Ways in which this learning could occur are considered in Chapter 8. In addition, some of these neurons may be involved in a short term memory for whether a particular familiar visual stimulus (such as a face) has been seen recently. The evidence for this is that some of these neurons respond differently to recently seen stimuli in short term visual memory tasks (Baylis and Rolls 1987, Miller and Desimone 1994, Xiang and Brown 1998). In the inferior temporal visual cortex proper, neurons respond more to novel than to familiar stimuli, but treat the stimuli as novel if more than one other stimuli intervene between the first (novel) and second (familiar) presentations of a particular stimulus (Baylis and Rolls 1987). More ventrally, in what is in or close to

Objects

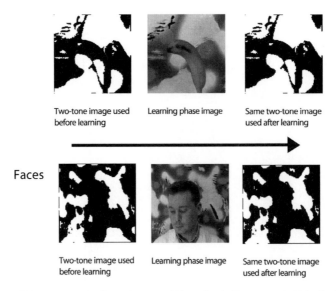

Faces

Fig. 5.23 Images of the type used to investigate rapid learning in the neurophysiological experiments of Tovee, Rolls and Ramachandran (1996) and the PET imaging study of Dolan, Fink, Rolls et al. (1997). When the black and white (two-tone) images at the left are shown, the objects or faces are not generally recognized. After the full grey scale images (middle) have been shown for a few seconds, humans and inferior temporal cortex neurons respond to the faces or objects in the black and white images (right).

the perirhinal cortex, these memory spans may hold for several intervening stimuli in the same task (Xiang and Brown 1998). Some neurons in these areas respond more when a sample stimulus reappears in a delayed match to sample task with intervening stimuli (Miller and Desimone 1994), and the basis for this using a short term memory implemented in the prefrontal cortex is described in Section 12.1. Neurons in the more ventral (perirhinal) cortical area respond during the delay in a match to sample task with a delay between the sample stimulus and the to-be-matched stimulus (Miyashita 1993, Renart, Parga and Rolls 2000) (see Section 12.1).

5.9 The speed of processing in the temporal cortical visual areas

Given that there is a whole sequence of visual cortical processing stages including V1, V2, V4, and the posterior inferior temporal cortex to reach the anterior temporal cortical areas, and that the response latencies of neurons in V1 are approximately 40–50 ms, and in the anterior inferior temporal cortical areas approximately 80–100 ms, each stage may need to perform processing for only 15–30 ms before it has performed sufficient processing to start influencing the next stage. Consistent with this, response latencies between V1 and the inferior temporal cortex increase from stage to stage (Thorpe and Imbert 1989). In a first approach to the very rapid processing apparently being performed by each cortical stage, we

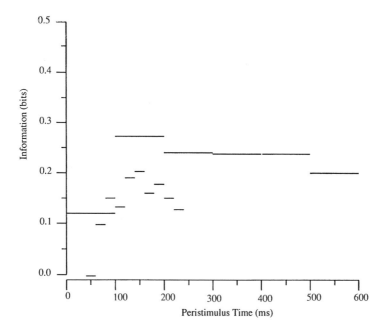

Fig. 5.24 The average information I(S,R) available in 20 ms temporal epochs of the spike trains of single neurons about which of 20 faces had been shown. (From Tovee and Rolls, 1995.)

measured the information available in short temporal epochs of the responses of temporal cortical face-selective neurons about which face had been seen. If a period of the firing rate of 50 ms was taken, then this contained 84.4% of the information available in a much longer period of 400 ms about which of four faces had been seen. If the epoch was as little as 20 ms, then there was as much as 65% if the information available from the firing rate in the 400 ms period (Tovee, Rolls, Treves and Bellis 1993). These high information yields were obtained with the short epochs taken near the start of the neuronal response, for example in the post-stimulus period 100–120 ms. Moreover, the firing rate in short periods taken near the start of the neuronal response was highly correlated with the firing rate taken over the whole response period, so that the information available was stable over the whole response period of the neurons (Tovee, Rolls, Treves and Bellis 1993). A comparable result is also found with a much larger set of stimuli: with 20 faces the information available in short (e.g. 50 ms) epochs was a considerable proportion (e.g. 65%) of that available in a 400 ms long firing rate analysis period (Tovee and Rolls 1995) (see e.g. Fig. 5.14), and considerable amounts were available in epochs as short as 20 ms (see e.g. Fig. 5.24). This analysis shows that there is considerable information about which stimulus has been seen in short time epochs of the responses of temporal cortex visual neurons.

The next type of experiment showed that very short periods of firing are sufficient for a cortical stage to perform its computation. This experiment was the backward masking experiment described in Section 5.5.6, in which there is a brief presentation of a test stimulus that is rapidly followed (within 1–100 ms) by the presentation of a second stimulus (the mask). The mask stimulus impairs or masks the perception of the test stimulus. When there

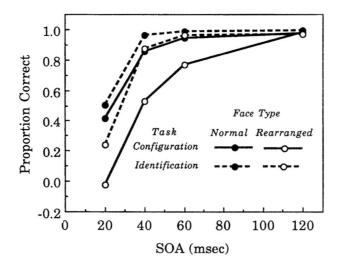

Fig. 5.25 The percentage correct in discriminating which of 6 faces had been shown with different Stimulus Onset Asynchronies (SOA) by human observers in the backward masking paradigm used in neurophysiological experiments.(Data are shown separately for identifying which stimulus was shown and for judging whether the stimulus as configured had jumbled features; and for normal faces, and for faces with jumbled (rearranged) features.) (From Rolls, Tovee, Purcell, Stewart, and Azzopardi, 1994. The data are corrected for guessing, that is chance performance is at 0.)

is no mask, inferior temporal cortex neurons respond to a 16 ms presentation of the test stimulus for 200–300 ms, far longer than the presentation time (Rolls and Tovee 1994). This reflects the operation of a short term memory system implemented in the cortical circuitry (see Fig. 5.16). If the pattern mask followed the onset of the test face stimulus by 20 ms (a Stimulus Onset Asynchrony of 20 ms), face-selective neurons in the inferior temporal cortex of macaques responded for a period of 20–30 ms before their firing was interrupted by the mask (Rolls and Tovee 1994, Rolls, Tovee and Panzeri 1999b) (see Fig. 5.16). Under these conditions (a test-to-mask Stimulus Onset Asynchrony of 20 ms), human observers looking at the same displays could just identify which of 6 faces was shown (Rolls, Tovee, Purcell, Stewart and Azzopardi 1994b) (see Fig. 5.25). Interestingly, the subjects did not have full conscious perception of the faces, and felt, as in blindsight (Weiskrantz 1998), that they were guessing, even though their performance was well above chance. It may be that a rather strong signal from a sensory system is needed before it is sufficient to overcome a threshold for access to the conscious processing system. This might have adaptive value, by preventing the conscious processing system being interrupted by what could be just noise in a sensory system (see Chapter 9 of Rolls (1999a)).

These results provide evidence that a cortical area can perform the computation it contributes to the identification of a visual stimulus in 20–30 ms, and emphasizes just how rapidly cortical circuitry can operate. Although this speed of operation does seem fast for a network with recurrent connections (mediated for example by recurrent collateral connections between pyramidal cells or by inhibitory interneurons), recent analyses of integrate-and-fire networks with biophysically modelled neurons that integrate their inputs and have spontaneous activity to keep the neurons close to the firing threshold, show that such networks can settle very rapidly (Treves 1993, Treves, Rolls and Simmen 1997, Rolls and Treves 1998) (see Section 7.6). This approach has been extended to multilayer networks such as those found in the visual system, and again very rapid propagation (in 50–60 ms) of information through such a 4-layer network, with recurrent collaterals operating at each stage, has been found (Panzeri, Rolls, Battaglia and Lavis 2001) (see Section 7.6.5).

5.10 Conclusions

Neurophysiological investigations of the inferior temporal cortex are revealing at least part of the way in which neuronal firing encodes information about faces and objects, and are showing that the representation implements several types of invariance. The representation found has clear utility for the receiving networks. The invariant representations are ideal as inputs to associative networks used for recognition, stimulus-reward association, episodic memory including object-place memory, etc. (see Chapter 12), and are also ideal because they do not themselves reflect the reward association of objects (Rolls, Judge and Sanghera 1977) (see Section 12.3). The fact that inferior temporal cortex neurons are neutral with respect to reward, motivational state such as whether one is hungry, etc, is appropriate for a representation that must be used for many different purposes. For example, it may be highly adaptive to learn about the location of foods even when one is not hungry, and this would not be possible if reward value were represented in the output of the visual system, for then neurons might not be responding to the sight of food. Instead, the use of the visual information for purposes such as memory, emotion, and motivation is left to the areas that receive from the inferior temporal visual cortex, which are typically multimodal areas that use the multimodal including visual inputs for the memory, emotional, and motivational functions that they perform, as described in Chapter 12.

These neurophysiological findings have stimulated the development of and provide constraints for the computational neuronal network models described in Chapter 8, which suggest that part of the cellular processing involves the operation of a modified associative learning rule with a short term memory trace to help the system learn invariances from the statistical properties of the inputs it receives. It is an interesting issue about what cellular or network processes might implement this short term memory trace, but the continuing firing of inferior temporal cortex neurons for 100–300 ms after a 20 ms presentation of a test stimulus (see Fig. 5.16) does suggest that there is a network process implemented, probably by the recurrent collateral cortical connections, that contributes to providing a short term memory trace.

6 Visual attentional mechanisms

6.1 Introduction

Up to a few years ago, the investigation of attentional mechanisms was of interest mainly to psychophysicists and cognitive psychologists trying to understand the operating principles and computational constraints of visual perception. Especially over the last twenty years, visual search tasks have been widely used as a tool to explore the functional role and operating principles of attention in visual perception. The classical view of attention that has emerged from this research refers to two functionally distinct stages of visual processing. One stage, termed the pre-attentive stage, implies an unlimited-capacity system capable of processing the information contained in the entire visual field in parallel. The other stage is termed the attentive or focal stage, and is characterized by the serial processing of visual information corresponding to local spatial regions.

Recently, the merging of neurobiology and psychology in the study of visual perception is defining the cognitive neuroscience of attention. Cognitive neuroscience has started to explore more directly the neural mechanisms underlying visual attention in humans and primates. New observations from a number of cognitive neuroscience experiments led to an alternative account of attention termed the "**biased competition hypothesis**", which aims to explain the computational algorithms governing visual attention and their implementation in the brain's neural circuits and neural systems. According to this hypothesis, attentional selection operates in parallel by biasing an underlying competitive interaction between multiple stimuli in the visual field. Moreover, this hypothesis implies a theory of visual perception that is very distinct in nature from the classical purely feedforward analyses such as Marr's analysis of vision (Marr 1982), in that it is based in an essential way on a dynamical relaxation between feedforward and feedback processes. The massively feedforward and feedback connections that exist in the anatomy and physiology of the cortex (described in Section 1.11) implement the attentional mutual biasing interactions that result in a neurodynamical account of visual perception consistent with the seminal constructivist theories of Helmholtz (1867) and Gregory (1970).

In this Chapter, we review and discuss the classical psychophysical view of selective visual attention, and the physiological findings that suggest an alternative neurodynamical view based on the biased competition hypothesis.

6.2 The classical view

6.2.1 The spotlight metaphor and feature integration theory

Lower animals possess only limited sensory capabilities and a restricted behaviour repertoire. On the other hand, highly developed animals, like primates, have a much more efficient sensory system and, consequently, greater possibilities in their behaviour. Expansion of the sensory and motor capabilities increases the problem of stimulus selection, i.e. the question as to which information is relevant to react to. Paradoxically, in spite of the massively parallel character of computations performed by the brain, it seems that biological systems employ a selection processing strategy for managing the enormous amount of information resulting from their interaction with the environment. This selection of relevant information is referred to as visual attention. The concept of attention implies that we can concentrate on certain portions of the sensory input, motor programs, memory contents, or internal representations to be processed preferentially, shifting the processing focus from one location to another or from one object to another in a serial fashion. This mechanism is commonly known by psychologists as selective or focal attention (Broadbent 1958, Kahneman 1973, Neisser 1967).

In the case of the visual system, only a small fraction of the information received reaches a level of processing that can be voluntarily reported or directly used to influence behaviour. The psychophysical work of Helmholtz (1867) provided the origin of a commonly employed metaphor for focal attention in terms of a spotlight (Treisman 1982, Crick 1984). This metaphor postulates the existence of a spotlight of attention that 'illuminates' a portion of the field of view where stimuli are processed in higher detail and brings them to a higher level of processing (Eriksen and Hoffmann 1973). In other words, information outside the spotlight is filtered out. Experimental maps of the attentional spotlight have been introduced by Sagi and Julesz (1986) by asking observers to detect the presence or absence of a peripheral probe dot while carrying out a concurrent letter discrimination task. Sperling and Weichselgartner (1995) demonstrated experimentally that the spotlight does not sweep continuously across the visual field, but rather fades in one place while increasing in strength in another. Pylyshyn and Storm (1988) have also shown that the focal selection performed by the spotlight does not necessarily involve contiguous parts of the visual field. Since Helmholtz it has been clear that visual selective attention can diverge from the direction of gaze and that attention can be voluntarily focused on a peripheral part of the visual field, i.e. the spotlight can be moved about the scene with or without eye movements. This entails the distinction of shifts of attention into overt and covert shifts of attention (Helmholtz 1867, Pashler 1996). In the case of **overt attention**, eye movements (saccades) occur, whereas **covert attention** changes the focus selected without any movement of the eyes.

The Helmholtzian spotlight paradigm alludes to a serial processing mode responsible for the complete scanning of the visual display. A complementary aspect of visual attention originates from James (1890), who generalized Helmholtz's attentional theory by introducing the concept of dispersed attention. Dispersed attention is a parallel process that operates across the entire visual field. James (1890) proposed that focused and dispersed attention are extremes in the spectrum of attentional states. Based on these notions, Neisser (1967) formulated a view in which visual search is done in two stages (Shaw and Shaw 1977, Shaw 1978). The first preattentive part comprises processes that are fast, parallel and involuntary. The second attentive part comprises processes that are slower, serial and voluntary.

An influential refinement of this latter approach is the Feature Integration Theory of visual selective attention (Treisman and Gelade 1980). This concept explains numerous psychophysical experiments on visual search and offers an interpretation of the binding problem. The binding problem concerns the question of which mechanisms are involved in the fusion of features, such as colour, form and motion, that compose an object. The revised version of feature integration theory (Treisman 1988) is in many aspects different from the original theory by Treisman and Gelade (1980). In the feature integration theory, the first preattentive process runs in parallel across the complete visual field extracting single primitive features without

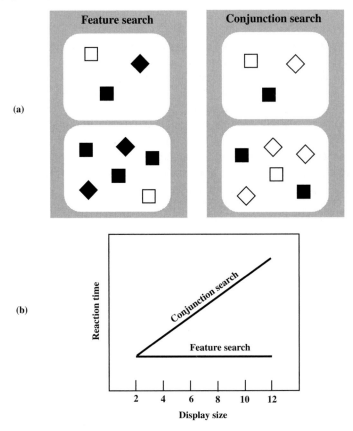

Fig. 6.1 Visual search experiments: feature and conjunction search. (a) Typical sample displays in visual feature and conjunction search tasks. The target (white square) 'pops out' in the case of feature search despite the variation in the number of distractors, whereas in the case of conjunction search the target is harder to find, especially when there are many distractors. (b) Feature search compared with conjunction search times. In feature searches, the subjects' reaction times do not increase as a function of the display size as they do in conjunction searches.

integrating them. The second attentive stage corresponds to the serial specialized integration of information from a limited part of the field at any one time.

Evidence for these two stages of attentional visual processing comes from psychophysical experiments using visual search tasks where subjects examine a display containing randomly positioned items in order to detect an a-priori defined target (see Fig. 6.1). All other items in the display which are different from the target serve the role of distractors. The number of items in the display is called the display size. The relevant variable typically measured is search time as a function of the display size.

Much work has been based on these two kinds of search paradigms: feature and conjunction search (Quinlan and Humphreys 1987, Wolfe, Cave and Franzel 1989, Treisman and Sato 1990). In a feature search task the target differs from the distractors in a single feature, e.g. only in its colour. In a conjunction search task the target is defined by a conjunction of features and each distractor shares at least one of those features with the target. One can distinguish between standard and triple conjunction search tasks. In a standard conjunction search, there are two distractor groups, each sharing one feature with the target, and sharing one feature with the other distractor group. In a triple conjunction search, the groups share one or two features with the target and two or one feature with the other distractor group. So far, models based on the above two-stage processes have been able to distinguish feature from conjunction search on the basis of the observed slopes of the function relating search time to display size. The conjunction search experiments show that search time increases linearly with the display size, implying a serial process. On the other hand, search times in feature search can be independent of the display size. The result is consistent with the activation of only parallel processes, which generate 'pop out' of the target. Figure 6.1 gives a schematic account of such results. Feature integration theory achieves dynamic binding by selectively gating each of the separate feature maps so that only those features lying within the attentional spotlight are passed to higher level recognition systems.

The feature integration theory has not been developed within a computational framework, but it involves clearly articulated processing stages: parallel coding of independent visual features, followed by serial coding of conjunctions (see Fig. 6.2). Feature integration theory essentially assumes the involvement of attention for visual conjunction search, based on an explicit serial scanning mechanism.

More recently, Nakayama and Silverman (1986) and Treisman (1988) have shown that there are some conjunction search tasks that can be accomplished in parallel. For example, Nakayama and Silverman (1986) showed that targets defined by colour and motion can be searched in parallel, and McLeod, Driver and Crisp (1988) showed that visual search for a conjunction of movement and form is parallel. On the other hand, targets defined by colour and orientation, or shape and orientation, (Posner and Dehaene 1994) yielded large slopes in the reaction time versus display size graphs. Furthermore, conjunction search can yield reaction times that range continuously from roughly 0 msec per item, i.e. parallel, to 30–50 msec per item, contradicting the simple distinction between parallel and serial search. A modified theory (Duncan and Humphreys 1989) proposes that focal attention does not necessarily need to be constrained to spatial dimensions, but can also be involved in feature dimensions like colour, motion, etc. as well. Thus, the objects possessing common features with the target may be prioritized for selection. Let us briefly review in Section 6.2.2 the existing computational models of visual selective attention based on this classical view.

6.2.2 Computational models of visual attention

In order to have a more concrete specification of the mechanism underlying visual attention and the binding of features, a number of computational models have been proposed aiming to explain psychophysical findings. In general, all models postulate the existence of a **saliency or priority map** for registering the potentially interesting areas of the retinal input, and a gating mechanism for reducing the amount of incoming visual information, so that the limited computational resources of the brain can handle it.

The aim of a priority map is to represent topographically the relevance of the different parts of the visual field, in order to have an *a posteriori* mechanism for guiding the attentional focus onto salient regions of the retinal input. The focused region will be gated, so that only the information within it will be passed further to the higher level and will be processed. There are several implementations of priority maps. Koch and Ullman (1985) proposed to build such a map by bottom-up information (e.g. how different is a particular stimulus relative

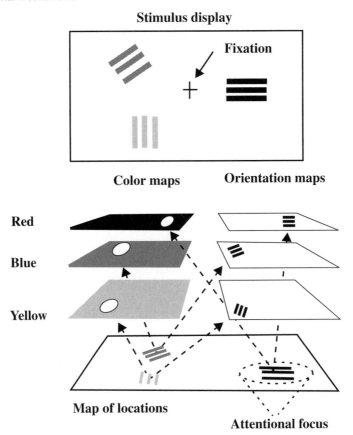

Fig. 6.2 Psychological model of feature integration for visual search experiments. The dimensions of colour and orientation are portrayed by a series of feature maps tuned to specific values. Stimuli activate neurons adjusted to their features on these maps. Attention is focused on particular locations and integrates the features at that location. (Adapted from Treisman, 1988.)

to its neighbourhood), which in combination with a winner-takes-all mechanism selects the currently most salient feature in the map and directs attention to its location via a gating strategy. Van de Laar, Heskes and Gielen (1997) generated a priority map in a task-dependent way in order to incorporate top-down information. They use this approach for handling visual search. The position on the priority map associated with the target of the search is maximized, i.e. the position is selected where all input features are equal or similar to the ones defining the target. Due to the fact that the target can change in each task, this approach defines a modifiable task-dependent priority map.

From a neurophysiological point of view it is still unclear whether such a functional saliency map is most likely to be implemented in a highly distributed manner across different cortical and subcortical areas, or whether saliency is implemented directly into the individual cortical feature maps. Gottlieb, Kusunoki and Goldberg (1998) have published electrophysiological evidence of neurons in area LIP (lateral intraparietal, a subdivision of the posterior parietal cortex PP) that directly encode stimulus salience. They observed that these neurons show a very strong response to the sharp onset of a stimulus, but do not respond when a steady stimulus is brought inside their receptive field. In both cases, the stimulus in the receptive field is the same, but is only salient ('interesting') in the first case. However, Desimone and Duncan (1995) claim that saliency-based attention can be realized without a priority map.

A more refined account of visual search is the 'Guided Search Model' of Wolfe, Cave and Franzel (1989) (see also Wolfe (1994)). It is formulated in a computational framework and is used to explain psychophysical data. The basic idea is that a serial visual attention stage can be guided by a parallel feature-computation stage. The latter generates a priority map that is used for guiding the localized control of the focal attention mechanism in an explicit serial stage of processing. When saliency values along each dimension are summed to calculate the order of serial inspection, the target should be the first candidate. However, because of noise in the parallel processing stage, the target will not always be given the highest activation values. Consequently, some distractors will be inspected before conjunction targets, producing effects of display size and visual search. It is interesting to remark that in the 'Guided Search Model' the priority map is generated after the conjunction of the features, so that competition mechanisms are only involved at this high level.

Recently, Olshausen, Anderson and Van Essen (1993) presented a model of how visual attention might solve the object recognition problem of position and scale invariance. The model relies on a set of control neurons to dynamically modify (in what is effectively a multiplication operation) the synaptic strengths of intracortical connections so that information from a windowed region of primary visual cortex (V1) is selectively routed to higher cortical areas. This corresponds to a dynamical routing of the information flow from the sensory input to the higher levels by a gating mechanism. They also hypothesize that the pulvinar nucleus of the thalamus may provide the control signals for routing information through the cortex and that the control neurons modify intracortical connection strengths via multiplicative dendrite interactions (e.g., via the NMDA receptor channel). This is consistent with work of Zihl and von Cramon (1979) who on the basis of observations in brain-injured patients proposed that the pulvinar nucleus of the thalamus is involved in the control of attention. Other neural network based models of attention have also utilized the concept of control neurons for directing information flow (Niebur and Koch 1994, Ahmad 1992, Tsotsos 1991). On the other hand, the explicit neural implementation of the Olshausen gating mechanism is in fact implausible, given the need to multiply and thus map large parts of the retinal input with shift- and scale-modifying connections to a particular set of outputs (see further Section 8.6).

It is important to remark that models utilizing the concept of priority maps and dynamical routing of information flow are compatible with Treisman's feature integration theory. In fact, the routing of information lying within the attentional window during each attentional step offers a mechanism for the binding of the involved features. This avoids the combinatorial explosion of computational resources that might be required if the brain utilized a hard-wired neural representation for each of the possible feature conjunctions that could occur in the visual world (though see Chapter 8 for a hypothesis about how this problem can be made tractable by representing only low-order combinations of features). The SERR (for SEarch via Recursive Rejection) model of Humphreys and Müller (1993) is based on the recursive rejection of areas of the visual field where stable and unambiguous grouping has been achieved. Spatially parallel grouping plays an important role in visual search and can generate both flat and linear reaction time functions. Search is easy if identical distractors group separately from targets. On the other hand, search is hard if distractors tend not to group between themselves while also grouping to some degree with the target, i.e. when the distractors are heterogeneous and share features with the target. The SERR model is especially suitable for describing such grouping effects in visual search. Furthermore, as suggested by Humphreys and Müller (1993), grouping can operate in different ways at different levels of image processing, in the sense that there may be inhibitory or facilitatory interaction at the single feature level or at the level after conjoining the features. Other authors postulate as a binding and attentional mechanism the synchronous firing of neurons in order to change connection strengths (Crick 1984, Malsburg and Bienenstock 1986, Crick and Koch 1990, Gray and Singer 1987, Gray and Singer 1989). The synchronization specifies just a possible physical implementation of binding and is not directly associated with the computational description of attention.

6.3 The biased competition hypothesis in single-cell studies

Recently, the dichotomy between parallel and serial operations in visual search has been challenged by other alternative psychological models suggesting that all kinds of search tasks can be solved by a parallel competitive mechanism. Duncan (1980) and Duncan and Humphreys (1989) have proposed a scheme that integrates both attentional modes (i.e. parallel and serial) as an instantiation of a common principle. They explain both single feature and conjunction searches on the basis of the same operations involving, on the one hand, grouping between items in the field and, on the other hand, matching of those items or groups to a memory template of the target. This matching process supports items with features consistent with the template and inhibits those with different features. This would be the same for all the features comprising the stimuli: colour, shape, location, etc. This process of feature selection suggests that subjects utilize top-down information (from the template) independent of localization of the stimuli in space. The attentional theory of Duncan and Humphreys (1989) proposed that there is both parallel activation of a target template (from multiple items in the field) and competition between items (and between the template and non-matching items) so that, ultimately, only one object is selected. In particular, there is evidence in the literature suggesting that parallel competitive processes in the brain are responsible for human performance in visual selective attention tasks (Duncan, Humphreys and Ward 1997, Mozer and Sitton 1998, Phaf, Van der Heijden and Hudson 1990, Tsotsos 1990).

Duncan and Humphreys (1989) used the biased competition mechanism to explain serial visual search as a top-down template influencing the competitions between neurons coding for different object attributes in the early visual areas, leading to the selection of one object in the visual field. Subsequently, this biased competition mechanism has also been used in several models for explaining attentional effects in neural responses observed in the inferotemporal cortex (Usher and Niebur 1996), and in V2 and V4 (Reynolds, Chelazzi and Desimone 1999). Deco and Zihl (2001b) extended Usher and Niebur's model to simulate some psychophysical results of visual search experiments in humans. The hypothesis for this mechanism can be traced back to the 'adaptive resonance' model (Grossberg 1987) and the 'interactive activation' model (McClelland and Rumelhart 1981) in the neural network and connectionist literature.

A challenging question is therefore: Is the linearly increasing search time observed in some visual search tasks necessarily due to a serial mechanism? Or, could it also be explained by the dynamical time-consuming latency of a parallel process? In other words, are priority maps and spotlight mechanisms required? A second interesting question is to determine the characteristics of top-down aided competition in feature space. Clarification is needed of whether this competition is achieved independently in each feature dimension or whether it occurs after binding the feature dimensions of each item. In order to answer these fundamental questions about the nature of attentional operations in visual perception, the underlying microscopic and intermediate scale mechanisms have to be assessed using cognitive neuroscience techniques, including especially recordings of the activity of single neurons from the brain of behaving macaques, as well as functional brain imaging. In this section we review the electrophysiological facts that suggested the formulation of the biased competition hypothesis. The evidence on the biased competition hypothesis provided by functional brain imaging experiments will be described in Section 6.4. Chapter 9 describes computational simulations of these single-cell recording experiments based on our model of the biased competition hypothesis.

6.3.1 Neurophysiology of attention

Recently, several neurophysiological experiments (Moran and Desimone 1985, Spitzer, Desimone and Moran 1988, Sato 1989, Motter 1993, Miller, Gochin and Gross 1993a, Chelazzi, Miller, Duncan and Desimone 1993, Motter 1994, Reynolds and Desimone 1999, Chelazzi 1998) have been performed suggesting biased competition neural mechanisms that are consistent with the theory of Duncan and Humphreys (1989) (i.e., with the role for a top-down memory target template in visual search). The biased competition hypothesis proposes that multiple stimuli in the visual field activate populations of neurons that engage in competitive interactions. Attending to a stimulus at a particular location or with a particular feature biases this competition in favour of neurons that respond to the feature or location of the attended stimulus. This attentional effect is produced by generating signals within areas outside the visual cortex that are then fed back to extrastriate areas, where they bias the competition such that when multiple stimuli appear in the visual field, the cells representing the attended stimulus 'win', thereby suppressing cells representing distracting stimuli (Duncan and Humphreys 1989, Desimone and Duncan 1995, Duncan 1996).

Single-cell recording studies in monkeys from extrastriate areas seem to support the biased competition theory. Moran and Desimone (1985) showed that the firing activity of visually tuned neurons in the cortex was modulated if monkeys were instructed to attend to the location of the target stimulus. In these studies, Moran and Desimone first identified the V4 neuron's classic receptive field and then determined which stimulus was effective for exciting a response from the neuron (e.g. a vertical white bar) and which stimulus was ineffective (e.g. a horizontal black bar). In other words, the effective stimulus made the neuron fire whereas the ineffective stimulus did not. They presented these two stimuli simultaneously within the neuron's receptive field, which for V4 neurons may extend over several degrees of visual angle. The task required attending covertly to a cued spatial location. When spatial attention was directed to the effective stimulus, the pair elicited a strong response. On the other hand, when spatial attention was directed to the ineffective stimulus, the identical pair elicited a weak response, even though the effective stimulus was still in its original location (see Fig. 6.3).

Based on results of this type, the spatial attentional modulation could be described as a shrinkage of the classical receptive field around the attended location. Similar studies of Luck, Chelazzi, Hillyard and Desimone (1997) and Reynolds, Chelazzi and Desimone (1999) replicated this result in area V4, and showed similar attentional modulation effects in area V2

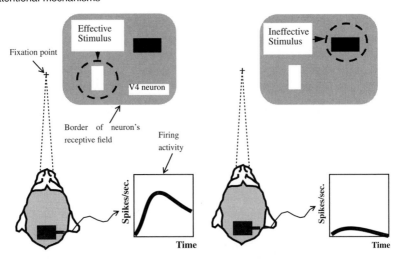

Fig. 6.3 Spatial attention and shrinking receptive fields in single-cell recordings of V4 neurons from the brain of a behaving monkey. The areas that are circled indicate the attended locations. When the monkey attended to effective sensory stimuli, the V4 neuron produced a good response, whereas a poor response was observed when the monkey attended to the ineffective sensory stimulus. The point fixated was the same in both conditions. Adapted from Moran and Desimone (1985).

as well. Even in area V1 a weak attentional modulation has been described (McAdams and Maunsell 1999).

Maunsell (1995) and many others (Duhamel, Colby and Goldberg 1992a, Colby, Duhamel and Goldberg 1993, Gnadt and Andersen 1988) have shown that the modification of neural activity by attentional and behavioral states of the animal is not only true for the extrastriate areas in the ventral stream, but is also true for the extrastriate areas of the dorsal stream as well. In particular, Maunsell (1995) demonstrated that the biased competitive interaction due to spatial and object attention exists not only for objects within the same receptive field, but also for objects in spatially distant receptive fields. This suggests that mechanisms exist to provide biased competition of a more global nature. Furthermore, Connor, Gallant and Van Essen (1993) and Connor, Gallant, Preddie and Van Essen (1996) showed that the locus of spatial attention can modulate the structure of the receptive fields of V4 neurons (see Chapter 9). By asking the monkeys to discriminate subtle changes in features in particular points in visual space, they managed to shift the hot spot of the receptive field of nearby V4 neurons towards the spatial focus of attention. One question then is whether there is a general mechanism, for example projections from the pulvinar as Olshausen, Anderson and Van Essen (1993) suggest, that modulates spatial attention and attentional gating in both the dorsal stream and the ventral stream. Alternatively, spatial attentional gating in the ventral stream could be modulated by the parietal cortex. In fMRI (functional magnetic resonance imaging) experiments that involve object discrimination, Corbetta and Shulman (1998) suggested that the dorsal frontal-parietal network is active concurrently with the ventral occipital-temporal regions. However, the time resolution of fMRI is insufficient to establish this, and studies with single and multiple single neuron recording studies are the only way to address this, in that they operate at the speed of neural computation (which their firing implements), and in that they also provide evidence on what the computational elements, the neurons, are actually encoding in different areas. We note for example that the speed of the interactions between the dorsal and ventral visual streams is likely to take less than 100 ms, given the speed with which networks of real neurons in the brain interact, as described in Section 7.6. Studies of this type, and experimental anatomical and disconnection studies, are needed to address the issue of the speed of interaction between the dorsal stream and the ventral stream, and whether this takes place through direct connections, or through their interaction with the early visual cortex, in order to implement these attentional processes. We will investigate this hypothesis thoroughly in Chapter 9.

6.3.2 The role of competition

In order to test at the neuronal level the biased competition hypothesis more directly, Reynolds, Chelazzi and Desimone (1999) performed single-cell recordings of V2 and V4 neurons in a behavioural paradigm that explicitly separated sensory processing mechanisms from attentional effects. They first examined the presence of competitive interactions in the absence of attentional effects by having the monkey attend to a location far outside the receptive field of the neuron that they were recording. They compared the firing activity response of the neuron when a single reference stimulus was within the receptive field, with the response when a probe stimulus was added to the field. When the probe was added to the field, the activity of the neuron was shifted toward the activity level that would have been evoked if the probe had appeared alone. This effect is shown in Fig. 6.4. When the reference was an effective stimulus (high response) and the probe was an ineffective stimulus (low response), the firing activity was suppressed after adding the probe (Fig. 6.4a). On the other hand, the response of the cell increased when an effective probe stimulus was added to an ineffective reference stimulus (Fig. 6.4b).

These results are explained in Chapter 9 by assuming that V2 and V1 neurons coding the different stimuli engage in competitive interactions, mediated for example through intermediary inhibitory neurons (Fig. 6.5).

The neurons in the extrastriate area receiving input from the competing neurons respond according to their input activity after the competition; that is, the response of a V4 neuron to two stimuli in its field is not the sum of its responses to both, but rather is a weighted average of the responses to each stimulus alone.

Attentional modulatory effects have been independently tested by repeating the same experiment, but now having the monkey attend to the reference stimulus within the receptive field of the recorded neuron. The effect of attention on the response of the V2 or V4 neuron was to almost compensate for the suppressive or excitatory effect of the probe. That is, if the probe caused a suppression of the neuronal response to the reference when attention was outside the receptive field, then attending to the reference restored the neuron's activity to the level corresponding to the response of the neuron to the reference stimulus alone (Fig. 6.4a). Symmetrically, if the probe stimulus had increased the neuron's level of activity, attending to the reference stimulus compensated the response by shifting the activity to the level that had been recorded when the reference was presented alone (Fig. 6.4b). As is shown in Fig. 6.5, this attentional modulation can be understood by assuming that attention biases the competition between V1 neurons in favour of a specific location or feature.

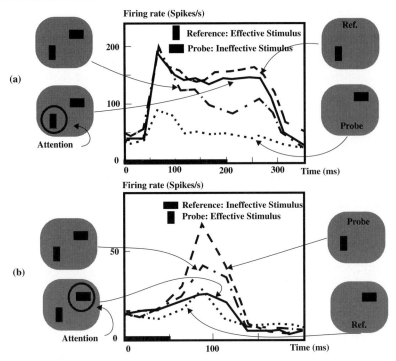

Fig. 6.4 Competitive interactions and attentional modulation of responses of single neurons in area V2. (a) Inhibitory suppression by the probe and attentional compensation. (b) Excitatory reinforcement by the probe and attentional compensation. The black horizontal bar at the bottom indicates stimulus duration. (Adapted from Reynolds, Chelazzi and Desimone, 1999.)

6.3.3 Evidence of attentional bias

Direct physiological evidence for the existence of an attentional bias was provided by Luck, Chelazzi, Hillyard and Desimone (1997) and by Spitzer, Desimone and Moran (1988). Luck, Chelazzi, Hillyard and Desimone (1997) discovered that attending to a location within the receptive field of a V2 or V4 neuron increased its spontaneous firing activity. They reported that in the absence of stimuli, the spontaneous firing activity of V2 and V4 neurons increased by 30–40% when the monkey attended to a location within the receptive field of the recorded neuron, compared with when attention was directed outside the field.

When a single stimulus is presented in the receptive field of a V4 neuron, Spitzer, Desimone and Moran (1988) observed an increase of the neuronal response to that stimulus when the monkey directed attention inside the receptive field, compared to when attention was directed outside the field.

6.3.4 Non-spatial attention

Additional evidence for the biased competition hypothesis as a mechanism for the selection of non-spatial attributes has been put forward by Chelazzi, Miller, Duncan and Desimone (1993). Chelazzi, Miller, Duncan and Desimone (1993) measured the responses of inferior temporal cortex (IT) neurons in monkeys while the animals were looking at a display containing a

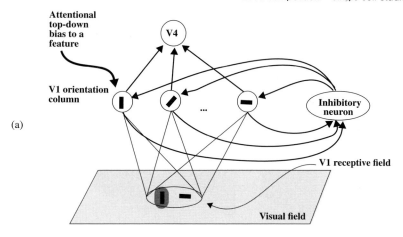

(a)

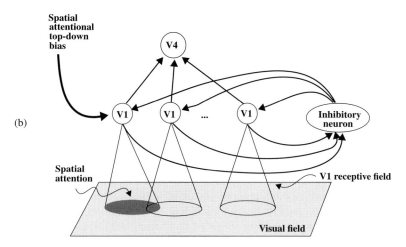

(b)

Fig. 6.5 Biased competition hypothesis at the neuronal level. (a) Attention to a particular top-down pre-specified feature. Intermediate neurons provide an inhibitory signal to V1 neurons in an orientation column (with receptive fields at the same spatial location) that are sensitive to an orientation different from the to-be-attended orientation. (b) Attention to a particular top-down prespecified spatial location. Intermediate neurons provide an inhibitory signal to V1 neurons with receptive fields corresponding to regions that are outside the to-be-attended location. In this schematic representation, we assume for simplicity direct connections from V1 to V4 neurons without intermediate V2 neurons.

target and a distractor. Figure 6.6 illustrates schematically the task design and results in the experiment of Chelazzi, Miller, Duncan and Desimone (1993).

At the start of a trial, the macaque was cued with what was to be the target stimulus. After a delay period of 1.5 s, during which the display remained blank, the target and the distractor were simultaneously presented. The monkey had to initiate an eye movement towards the location of the target. Chelazzi, Miller, Duncan and Desimone (1993) recorded activity in IT neurons using two kinds of stimuli: a highly effective stimulus for activating the corresponding IT neuron, and an ineffective stimulus, whose presence suppressed that neuron's response. In some trials, the effective stimulus was the target and the distractor was the ineffective

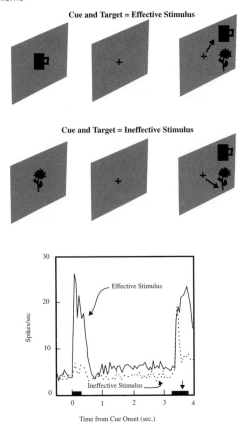

Fig. 6.6 Schematic illustration of the task design and results in the experiment of Chelazzi et al. (1993). Top: Each trial starts with presentation of a cue stimulus at the fixation point, which the monkey has to retain in short-term memory. After a blank delay interval, an array appeared in random peripheral locations of the visual field. The animal was trained to make a saccadic eye movement to the stimulus array that matched the initial cue. Arrays of 1, 2, 3, and 5 stimuli were used. (The Figure shows an array of 2 stimuli). Bottom: peristimulus time histograms of the firing rates of 20 recorded IT cells. On some trials, the cue and the target corresponded to the effective stimulus (e.g. the cup in the Figure) of the cell, and in other trials to the ineffective stimulus (e.g. the flower in the Figure). The black horizontal bar at the bottom left and right of the histograms indicates the cue and the array durations, respectively. The arrow indicates the average saccade latency. In both conditions the receptive field of each of the neurons included both items (target and distractor). See text for explanation. (Modified from Chelazzi et al., 1993).

stimulus, while in other trials the opposite arrangement held. In both conditions, the receptive field of each of the recorded neurons included both items (target and distractor). During the delay period, the neuron sensitive to the target showed a higher firing rate, presumably reflecting a top-down projection from a working memory system. After the delay period, the recordings of the average firing rates showed that the neuron sensitive to the target and the neuron sensitive to the distractor became active, but only the activity of the neuron sensitive to the target remained high (or even increased), while the activity of the neuron sensitive to the distractor decreased. These results clearly support the idea of two phases in the response of IT neurons. During the first phase, the activity of the neuron is enhanced due to the presence of the effective stimulus in the display, independently of the attentional state of the monkey, i.e. independently of whether the stimulus is the target. During the second phase, the response is maintained or increased only if the effective stimulus is the target, and the firing rate decreases if the stimulus is not a target. Thus, the activity of IT neurons depends on the internal attentional state of the monkey. Usher and Niebur (1996) formulated a detailed neural model that explains these data assuming parallel dynamic processing driven by a competition mechanism.

6.3.5   High-resolution buffer hypothesis

Attentional effects have also been observed in the early visual areas V1 and V2 under particular experimental conditions in which competing objects are present simultaneously in the visual field (Motter 1993, Roelfsema, Lamme and Spekreijse 1998, Ito and Gilbert 1999). The magnitude of such attention-related enhancement effects was about 15–25% of the response magnitude. However, Lee (1996) (see also Lee, Mumford, Romero and Lamme (1998)) have argued that the state-dependent modulation of the V1 responses is not simply an attentional gain-control mechanism, but reflects a more general role of V1 in visual processing. In particular, they proposed that V1 is a unique high resolution buffer available to the cortex for calculations, that can be used by any computation, high or low level, that requires high resolution image details and spatial precision. While the early response of the V1 neurons could be considered as these neurons operating as filters on the visual input (see Chapter 2), the late responses of V1 neurons might reflect the consequence of an elaborate interactive and concurrent computation involving the whole visual hierarchy via the feedfoward/feedback loops between the different cortical areas. This conjecture was triggered in part by the 'figure-highlighting' effect that Lamme (1995) observed in V1, i.e. that V1 neurons' activity was higher when their receptive fields were located inside the figure than when their receptive fields were in the background. In a series of experiments, Lee and his colleagues (Lee, Mumford, Romero and Lamme 1998, Lee, Romero and Mumford 2000) showed that the figure-highlighting effect could be observed not only in texture figures, but also in luminance figures, in equi-luminance colour figures, and in figures defined by shape from shading (Lee, Romero and Mumford 2000). In particular, the recent research of Lee's group furnished two further important pieces of evidence in support of the high resolution buffer hypothesis, and more generally, that perceptual computation is an emergent phenomenon that arises from the interactive and distributed computations between many cortical areas. First, they (Lee, Romero and Mumford 2000) showed that the figure-highlighting effect in V1 due to shape from shading, a higher order visual construct, can be observed only after the stimulus has been made relevant to the monkeys' behaviour; and that the figure-highlighting effect is a top-down object-specific effect that depends on what the monkeys are looking for behaviourally. Second, they (Lee and Nguyen 2001) demonstrated that the illusory contours of Kanizsa squares could be observed in V1, but occurred 35 ms after V2 activation. This evidence, together with the curve-tracing experiment in V1 of Roelfsema, Lamme and Spekreijse (1998), support the notion that higher order modulation effects in V1 are very spatially- and object-specific.

  The basic premise of the high-resolution buffer hypothesis is that different cortical areas may interact through V1 in order to fulfil the specific computations required by the task. In Chapter 9, we will see that when the animal is searching for a specific object (visual search), or recognizing objects at a specific spatial location (object recognition), the early visual cortex, including V1, is actively involved in these processes, and serves as one of the sites of interaction between the dorsal and the ventral streams (Deco and Lee 2001).

6.4 The biased competition hypothesis in functional brain imaging

In recent functional magnetic resonance imaging (fMRI) studies, additional evidence for similar mechanisms in human extrastriate cortex has been obtained (Kastner, De Weerd, Desimone and Ungerleider 1998, Kastner, Pinsk, De Weerd, Desimone and Ungerleider 1999). In line with the biased competition hypothesis, these studies have shown that multiple stimuli in the visual field interact in a mutually suppressive way when presented simultaneously, but not when presented sequentially, and that spatially directed attention to one stimulus location reduces the mutually suppressive effect. These studies also revealed increased activity in extrastriate cortex in the absence of visual stimulation, when subjects covertly directed attention to a peripheral location at which they expected the onset of visual stimuli. This increased activity in visual cortex was related to a top-down bias of neural signals, deriving from frontal and parietal areas, to favour the attended location. We will analyze these results in more detail in the next two subsections. Chapter 9 presents simulations of these fMRI experiments based on our particular computational theory and model of biased competition.

6.4.1 Neuroimaging of attention

Using fMRI in humans, Kastner, De Weerd, Desimone and Ungerleider (1998) in a first experiment tested for the presence of suppressive interactions between visual stimuli presented simultaneously within the visual field in the absence of directed attention. In a second experiment they considered the role of attentional modulation on these suppressive interactions. Figure 6.7 shows the design of the experiment. Complex images were presented in randomized order in four nearby locations within the right upper quadrant of the visual field under two different conditions: sequential (SEQ) and simultaneous (SIM). In the sequential condition, each stimulus appeared successively alone in one of the four locations, whereas in the simultaneous condition all four stimuli appeared at the same time in all four locations. Under the two conditions, the physical stimulation parameters integrated over time were identical. Suppressive interactions between stimuli can be demonstrated if the fMRI activity during the simultaneous condition is smaller than during the sequential presentations. In this case, the mutual suppression can be explained by means of competitive interactions between the simultaneously presented stimuli. In fact, they observed this effect in several visual areas. Figure 6.7 shows these suppressive effect in area V1, V2, V4, and TEO by the reduction of the fMRI signal in the unattended simultaneous condition (SIM blocks without shading) compared to the unattended sequential condition (SEQ blocks without shading). The difference in activation between sequential and simultaneous presentations increased from V1 to V4 and TEO (Fig. 6.7).

In a second experiment, Kastner, De Weerd, Desimone and Ungerleider (1998) studied the influence of attention on the suppressive interactions between simultaneously presented stimuli. In this case, during each scan, four blocks of visual stimulation (SEQ-SIM-SIM-SEQ) were tested in an unattended and an attended condition. During the attended condition the subjects were instructed to covertly attend specifically to one of the four locations where the stimuli could appear. Consistent with the single-cell experiments reported in the previous

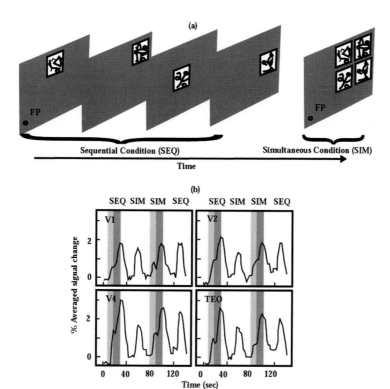

Fig. 6.7 fMRI signals in visual cortex averaged over all subjects. (a) Task. Subjects fixated a spot while stimuli appeared either asynchronously (left images, SEQ) or simultaneously (right images, SIM). The total amount of time each stimulus appeared was constant. Subjects either performed an attentionally demanding task at the fixation point or attended to one of the four stimuli. (b) fMRI signals. The light shading indicates the expectation period, the dark shading the attended presentations, and blocks without shading correspond to the unattended condition. (Adapted from Kastner et al., 1999.)

section, Fig. 6.7 shows an attentional modulatory compensation of the otherwise suppressive interactions between stimuli. Spatially directed attention reduces these interactions by partially cancelling out their suppressive effects, as is shown in Fig. 6.7 (dark shading). A significantly greater compensatory influence of attention on the fMRI signal elicited by simultaneously presented stimuli, compared with to that elicited by sequentially presented stimuli, can also be seen in Fig. 6.7.

6.4.2 Attentional effects in the absence of visual stimulation

In order to study in more detail the nature of the top-down signal that modulates the response to an attended versus an unattended stimulus in human visual cortical areas, Kastner, Pinsk, De Weerd, Desimone and Ungerleider (1999) extended their previous experiment and considered the influence and origin of top-down attentional bias in the absence of visual stimulation. This experiment relates to the increase in spontaneous firing activity demonstrated in monkey extrastriate cortex during attention when no stimulus is present, as described in Section 6.3.3.

They modified the previous experiment by including a new block in which the subject had to attend to a specific location in the absence of any stimuli. They implemented this block by marking the attended condition 10 s before the onset of the visual presentation (SIM or SEQ). They found increased activity related to attention in the absence of visual stimulation in extrastriate cortex when subjects covertly directed attention to a location where the onset of a visual stimulus was expected (light shading in Fig. 6.7).

Even more, during this expectation period they observed a strong signal increase in frontal and parietal areas, suggesting that the biasing attentional top-down signal to the extrastriate visual cortex could be derived from the fronto-parietal network.

In conclusion, these human neuroimaging experiments extend the validity of the biased competition hypothesis from the neuronal level to the intermediate level of cortical areas, and suggest candidate areas from where the biasing signal may originate.

6.5   The computational role of top-down feedback connections

The biased competition hypothesis suggests that a massively parallel neurodynamical mechanism underlies phenomena of visual attention and perception. On the other hand, the anatomy and physiology of the cortex reveal a complex architecture that is fundamentally based on separated modules interacting with each other via feedforward and feedback connections. This design is not only compatible with, but is ideal for, the implementation of biased competition-based neurodynamics. Competitive mechanisms act within each module, as the result of intra-modular lateral inhibitory connections implemented through inhibitory interneurons, while the inter-modular feedforward and feedback connections between different modules implement the attention-related mutual interactions that bias the competition within each module.

These ideas are consistent with the constructivist theories of visual perception initially set forth by the pioneering work of Hermann von Helmholtz (1867). In contrast to the convictions of Gibson (1950) who assumed that optical retinal information is the foundation of vision, constructivism postulates that vision goes beyond the optical information by constructing a model of what environmental situation might have produced the observed pattern of sensory stimulation. In fact, visual illusions (non-veridical perceptions) show that our models are sometimes inaccurate, and ambiguous figures (simple images that can give rise to two or more distinct perceptions) show that they are sometimes not unique. The essential contribution of Helmholtz to the understanding of visual perception is his proposal that perception depends on a process of *unconscious inference*. Helmholtz bridged the gap between optical retinal information and perceptual knowledge by postulating hidden *assumptions* in conjunction with retinal images to derive perceptual *conclusions* about the environment. He called this process *inference*. The process of perceptual inference is unconscious in that, unlike the normal process of cognitive inference, humans have no awareness that they are making inferences at the purely perceptual level.

A modernized version of Helmholtz's ideas was elaborated by Gregory (1970) in the framework of statistical theory. This proposal is associated with the maximum likelihood principle, and hypothesis testing, in probability theory. The visual system performs inference by hypothesizing the interpretation with the highest probability, given the retinal stimulation. In the Bayesian view, the visual system infers the most likely interpretation by calculating the conditional probability of many different 3D scenes (I_i) given the particular retinal evidence (R) with the following expression:

$$P(I_i|R) = \frac{P(R|I_i)P(I_i)}{P(R)}, \tag{6.1}$$

which corresponds to the well-known Bayes' theorem. In equation 6.1, $P(I_i|R)$ is the probability of a particular interpretation of a scene among N possible interpretations (i.e. $i = 1,...,N$), $P(R|I_i)$ is the probability of the retinal optical evidence given a concrete interpretation of the scene, $P(I_i)$ is the prior probability of the different scene interpretations, and $P(R)$ is the prior probability of the evidence. The quantities on the right-hand side of the equation can be computed from prior experience in viewing the world, so that the probability of each particular interpretation given the optical evidence on the retina $P(I_i|R)$ can be calculated. Alternatively, without using Bayes theorem, the maximum likelihood principle is used to select the interpretation with maximal probability given the retinal evidence (i.e. $\arg\max_{I_i} P(R|I_i)$) (Rieke, Warland, de Ruyter van Steveninck and Bialek 1996).

In particular, a concrete realization of this philosophy was carried out by the computational theory of vision formulated by David Mumford (1991, 1992) and usually known as 'Pattern Theory' (a term introduced in the pioneering work of Grenander (1976)). Pattern Theory is based on the interactions between top-down and bottom-up processes. This theory presupposes that sensory signals are coded versions of what is really going on in the world, and that the goal of sensory information processing is to synthesize, i.e. to reconstruct, with minimum error, the state of the world. In other words, in order to successfully reconstruct the world variables, one must synthesize the coded observed signals, so that hypothetical reconstructions of the world variables can be compared with the actual observed signal. Therefore, Mumford postulates that the underlying computational architecture is not purely feedforward, bottom-up, but fundamentally recursive, combining feedforward bottom-up actions with feedback top-down processing. He assumes that iterative cortical feedback loops accomplish the postulated analysis-synthesis loop required for sensory information processing. For example, in the case of visual perception, in a feedforward step a recognizer draws on its database of prototypes to synthesize a standard instantiation of the hypothetical object being seen. In subsequent iterations, the hypothesis will be refined, adding details of size, orientation, missing parts, etc., so that the synthesized image agrees with the observed visual information, up to a minimum error. This theory is consistent with the layered structure of the cortex. In particular, there are two kinds of anatomical pathways that connect two different cortical areas, as described in Section 1.11. The forward pathway has connections that terminate in layer 4, the standard *input* layer for bottom-up cortical processing, and the route from raw sensory information to higher association areas. Mumford associated these pathways with pattern analysis. The top-down pathways terminate usually in layer 1 (and 6) (see section 1.11), and were related by Mumford to top-down hypothesis testing and pattern synthesis.

In Chapter 9, we propose that the ventral and the dorsal streams in the visual system interact to mediate spatial and object attention, and that the early visual areas such as V1 and V2 provide a possible locus for mediating this interaction through the inter-cortical feedforward/feedback connections with these areas. This proposal is inspired in part by the experiments presented in this Chapter that provided evidence for the biased competition hypothesis, by some of the recent experiments that demonstrated top-down effects in the early visual cortex (Haenny and Schiller 1988, Haenny, Maunsell and Schiller 1988, Motter 1993, Lamme 1995, Ito and Gilbert 1999, Roelfsema, Lamme and Spekreijse 1998, Hupe, James, Payne, Lomber, Girard and Bullier 1998, Lee and Nguyen 2001), and by the high-resolution buffer hypothesis of Lee, Mumford, Romero and Lamme (1998) for the primary visual cortex (V1). Using a physiologically constrained model, we demonstrate in Chapter 9 the computational viability of this proposal: the early visual areas can provide an appropriate topologically organized representation to allow the dorsal stream and the ventral stream to interact in the organization of attention.

7 Neural network models

7.1 Introduction

Formal models of neural networks are needed in order to provide a basis for understanding the processing performed by real neuronal networks in the brain. The formal models included in this Chapter all describe fundamental types of network found in different brain regions and the computations they perform. Each of the types of network described can be thought of as providing one of the fundamental building blocks that the brain uses. Often these building blocks are combined within a brain area to perform a particular computation.

The aim of this Chapter is to describe a set of fundamental networks used by the brain, including the parts of the brain involved in vision. As each type of network is introduced, we will point briefly to parts of the visual system in which each network is found. Understanding these models provides a basis for understanding the theories of how different types of visual information processing are performed that are presented in later Chapters. The descriptions of these networks are kept relatively concise in this Chapter. More detailed descriptions starting at a simpler level of many of the networks described in this Chapter are provided in Rolls and Treves (1998) *Neural Networks and Brain Function*, and readers who wish to start at a simpler level, or to find a fuller description, or who are interested in how these networks contribute to processing in many different brain systems, are encouraged to consult that book. Another book that provides a clear and quantitative introduction to some of these networks is Hertz, Krogh and Palmer (1991) *Introduction to the Theory of Neural Computation*.

The network models on which we focus in this Chapter utilize a local learning rule, that is a rule for synaptic modification in which the signals needed to alter the synaptic strength are present in the pre- and post-synaptic neurons. We focus on these networks because use of a local learning rule is biologically plausible. We discuss the issue of biological plausibility of the networks described, and show how they differ from less biologically plausible networks such as multilayer backpropagation of error networks, in Section 7.12.

7.2 Pattern association memory

A fundamental operation of most nervous systems is to learn to associate a first stimulus with a second that occurs at about the same time, and to retrieve the second stimulus when the first is presented. The first stimulus might be the sight of food, and the second stimulus the taste of food. After the association has been learned, the sight of food would enable its taste to be retrieved. In classical conditioning, the taste of food might elicit an unconditioned response of salivation, and if the sight of the food is paired with its taste, then the sight of that food would by learning come to produce salivation. Pattern associators are thus used where the outputs of the visual system interface to learning systems in the orbitofrontal cortex and amygdala that learn associations between the sight of objects and their taste or touch in what is called stimulus-reinforcement association learning (see Section 12.3). Pattern association is also

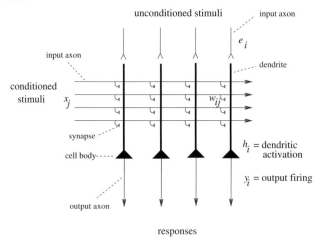

Fig. 7.1 A pattern association memory. An unconditioned stimulus has activity or firing rate e_i for the ith neuron, and produces firing y_i of the ith neuron. An unconditioned stimulus may be treated as a vector, across the set of neurons indexed by i, of activity $\mathbf{e}$. The firing rate response can also be thought of as a vector of firing $\mathbf{y}$. The conditioned stimuli have activity or firing rate x_j for the jth axon, which can also be treated as a vector $\mathbf{x}$.

used throughout the visual processing cortical areas, as it is the architecture that describes the backprojection connections from one cortical area to the preceding cortical area (see Section 1.11). Pattern association thus contributes to implementing top-down influences in vision, including the effects of attention from higher to lower cortical areas, and thus between the object and spatial processing streams (see Chapters 9–11); the effects of mood on memory and visual information processing (see Section 12.3.6); the recall of visual memories (see Section 12.2); and the operation of visual short term memory (see Section 12.1).

7.2.1 Architecture and operation

The essential elements necessary for pattern association, forming what could be called a prototypical pattern associator network, are shown in Fig. 7.1. What we have called the second or unconditioned stimulus pattern is applied through unmodifiable synapses generating an input to each neuron, which, being external with respect to the synaptic matrix we focus on, we can call the external input e_i for the ith neuron. (We can also treat this as a vector, $\mathbf{e}$, as indicated in the legend to Fig. 7.1. Vectors and simple operations performed with them are summarized in Appendix A). This unconditioned stimulus is dominant in producing or forcing the firing of the output neurons (y_i for the ith neuron, or the vector $\mathbf{y}$). At the same time, the first or conditioned stimulus pattern consisting of the set of firings on the horizontally running input axons in Fig. 7.1 (x_j for the jth axon) (or equivalently the vector $\mathbf{x}$) is applied through modifiable synapses w_{ij} to the dendrites of the output neurons. The synapses are modifiable in such a way that if there is presynaptic firing on an input axon x_j paired during learning with postsynaptic activity on neuron i, then the strength or weight w_{ij} between that axon and the dendrite increases. This simple learning rule is often called the Hebb rule, after Donald Hebb who in 1949 formulated the hypothesis that if the firing of one neuron was regularly associated with another, then the strength of the synapse or synapses between the neurons should increase[9]. After learning, presenting the pattern **x** on the input axons will activate the dendrite through the strengthened synapses. If the cue or conditioned stimulus pattern is the same as that learned, the postsynaptic neurons will be activated, even in the absence of the external or unconditioned input, as each of the firing axons produces through a strengthened synapse some activation of the postsynaptic element, the dendrite. The total activation h_i of each postsynaptic neuron i is then the sum of such individual activations. In this way, the 'correct' output neurons, that is those activated during learning, can end up being the ones most strongly activated, and the second or unconditioned stimulus can be effectively recalled. The recall is best when only strong activation of the postsynaptic neuron produces firing, that is if there is a threshold for firing, just like real neurons. The reasons for this arise when many associations are stored in the memory, as will soon be shown.

Next we introduce a more precise description of the above by writing down explicit mathematical rules for the operation of the simple network model of Fig. 7.1, which will help us to understand how pattern association memories in general operate. (In this description we introduce simple vector operations, and, for those who are not familiar with these, have provided in Appendix A a concise summary for later reference.) We have denoted above a conditioned stimulus input pattern as **x**. Each of the axons has a firing rate, and if we count or index through the axons using the subscript j, the firing rate of the first axon is x_1, of the second x_2, of the jth x_j, etc. The whole set of axons forms a vector, which is just an ordered (1, 2, 3, etc.) set of elements. The firing rate of each axon x_j is one element of the firing rate vector **x**. Similarly, using i as the index, we can denote the firing rate of any output neuron as y_i, and the firing rate output vector as **y**. With this terminology, we can then identify any synapse onto neuron i from neuron j as w_{ij} (see Fig. 7.1). In this book, the first index, i, always refers to the receiving neuron (and thus signifies a dendrite), while the second index, j, refers to the sending neuron (and thus signifies a conditioned stimulus input axon in Fig. 7.1). We can now specify the learning and retrieval operations as follows:

7.2.1.1 Learning

The firing rate of every output neuron is forced to a value determined by the unconditioned (or external or forcing stimulus) input. In our simple model this means that for any one neuron i,

$$y_i = f(e_i) \tag{7.1}$$

which indicates that the firing rate is a function of the dendritic activation, taken in this case to reduce essentially to that resulting from the external forcing input (see Fig. 7.1). The function f is called the activation function (see Fig. 1.3), and its precise form is irrelevant, at least during this learning phase. For example, the function at its simplest could be taken to be linear, so that the firing rate would be just proportional to the activation.

The Hebb rule can then be written as follows:

$$\delta w_{ij} = \alpha y_i x_j \tag{7.2}$$

[9]In fact, the terms in which Hebb put the hypothesis were a little different from an association memory, in that he stated that if one neuron regularly comes to elicit firing in another, then the strength of the synapses should increase. He had in mind the building of what he called cell assemblies. In a pattern associator, the conditioned stimulus need not produce before learning any significant activation of the output neurons. The connections must simply increase if there is associated pre- and post-synaptic firing when, in pattern association, most of the postsynaptic firing is being produced by a different input.

where δw_{ij} is the change of the synaptic weight w_{ij} that results from the simultaneous (or conjunctive) presence of presynaptic firing x_j and postsynaptic firing or activation y_i, and α is a learning rate constant that specifies how much the synapses alter on any one pairing.

The Hebb rule is expressed in this multiplicative form to reflect the idea that both presynaptic and postsynaptic activity must be present for the synapses to increase in strength. The multiplicative form also reflects the idea that strong pre- and post-synaptic firing will produce a larger change of synaptic weight than smaller firing rates. It is also assumed for now that before any learning takes place, the synaptic strengths are small in relation to the changes that can be produced during Hebbian learning. We will see that this assumption can be relaxed later when a modified Hebb rule is introduced that can lead to a reduction in synaptic strength under some conditions.

7.2.1.2 Recall

When the conditioned stimulus is present on the input axons, the total activation h_i of a neuron i is the sum of all the activations produced through each strengthened synapse w_{ij} by each active neuron x_j. We can express this as

$$h_i = \sum_{j=1}^{C} x_j w_{ij} \tag{7.3}$$

where $\sum_{j=1}^{C}$ indicates that the sum is over the C input axons (or connections) indexed by j to each neuron.

The multiplicative form here indicates that activation should be produced by an axon only if it is firing, and only if it is connected to the dendrite by a strengthened synapse. It also indicates that the strength of the activation reflects how fast the axon x_j is firing, and how strong the synapse w_{ij} is. The sum of all such activations expresses the idea that summation (of synaptic currents in real neurons) occurs along the length of the dendrite, to produce activation at the cell body, where the activation h_i is converted into firing y_i. This conversion can be expressed as

$$y_i = f(h_i) \tag{7.4}$$

where the function f is again the activation function. The form of the function now becomes more important. Real neurons have thresholds, with firing occurring only if the activation is above the threshold. A threshold linear activation function is shown in Fig. 1.3b. This has been useful in formal analysis of the properties of neural networks. Neurons also have firing rates that become saturated at a maximum rate, and we could express this as the sigmoid activation function shown in Fig. 1.3c. Yet another simple activation function, used in some models of neural networks, is the binary threshold function (Fig. 1.3d), which indicates that if the activation is below threshold, there is no firing, and that if the activation is above threshold, the neuron fires maximally. Whatever the exact shape of the activation function, some non-linearity is an advantage, for it enables small activations produced by interfering memories to be minimized, and it can enable neurons to perform logical operations, such as to fire or respond only if two or more sets of inputs are present simultaneously.

Examples of these learning and recall operations are provided in a very simple form by Rolls and Treves (1998). They illustrate how a number of different associations can be stored in such a pattern associator, and retrieved correctly. The illustration also shows the value of some threshold non-linearity in the activation function of the neurons, which can remove cross-talk or interference between the different pattern associations stored in the network.

7.2.2   The vector interpretation

The way in which recall is produced, equation 7.3, consists for each output neuron i of multiplying each input firing rate x_j by the corresponding synaptic weight w_{ij} and summing the products to obtain the activation h_i. Now we can consider the firing rates x_j where j varies from 1 to N', the number of axons, to be a vector. (A vector is simply an ordered set of numbers — see Appendix A.) Let us call this vector $\mathbf{x}$. Similarly, on a neuron i, the synaptic weights can be treated as a vector, $\mathbf{w}_i$. (The subscript i here indicates that this is the weight vector on the ith neuron.) The operation we have just described to obtain the activation of an output neuron can now be seen to be a simple multiplication operation of two vectors to produce a single output value (called a scalar output). This is the inner product or dot product of two vectors, and can be written

$$h_i = \mathbf{x} \cdot \mathbf{w}_i. \tag{7.5}$$

The inner product of two vectors indicates how similar they are. If two vectors have corresponding elements the same, then the dot product will be maximal. If the two vectors are similar but not identical, then the dot product will be high. If the two vectors are completely different, the dot product will be 0, and the vectors are described as orthogonal. (The term orthogonal means at right angles, and arises from the geometric interpretation of vectors, which is summarized in Appendix A.) Thus the dot product provides a direct measure of how similar two vectors are.

It can now be seen that a fundamental operation many neurons perform is effectively to compute how similar an input pattern vector $\mathbf{x}$ is to their stored weight vector $\mathbf{w}_i$. The similarity measure they compute, the dot product, is a very good measure of similarity, and indeed, the standard (Pearson product-moment) correlation coefficient used in statistics is the same as a normalized dot product with the mean subtracted from each vector, as shown in Appendix A. (The normalization used in the correlation coefficient results in the coefficient varying always between $+1$ and -1, whereas the actual scalar value of a dot product clearly depends on the length of the vectors from which it is calculated.)

With these concepts, we can now see that during learning, a pattern associator adds to its weight vector a vector $\delta\mathbf{w}_i$ that has the same pattern as the input pattern $\mathbf{x}$, if the postsynaptic neuron i is strongly activated. Indeed, we can express equation 7.2 in vector form as

$$\delta\mathbf{w}_i = \alpha y_i \mathbf{x}. \tag{7.6}$$

We can now see that what is recalled by the neuron depends on the similarity of the recall cue vector $\mathbf{x}_r$ to the originally learned vector $\mathbf{x}$. The fact that during recall the output of each neuron reflects the similarity (as measured by the dot product) of the input pattern $\mathbf{x}_r$ to each of the patterns used originally as $\mathbf{x}$ inputs (conditioned stimuli in Fig. 7.1) provides a simple way to appreciate many of the interesting and biologically useful properties of pattern associators, as described next.

7.2.3 Properties

7.2.3.1 Generalization

During recall, pattern associators generalize, and produce appropriate outputs if a recall cue vector $\mathbf{x}_r$ is similar to a vector that has been learned already. This occurs because the recall operation involves computing the dot (inner) product of the input pattern vector $\mathbf{x}_r$ with the synaptic weight vector $\mathbf{w}_i$, so that the firing produced, y_i, reflects the similarity of the current input to the previously learned input pattern $\mathbf{x}$. (Generalization will occur to input cue or conditioned stimulus patterns $\mathbf{x}_r$ that are incomplete versions of an original conditioned stimulus $\mathbf{x}$, although the term completion is usually applied to the autoassociation networks described in Section 7.3.)

This is an extremely important property of pattern associators, for input stimuli during recall will rarely be absolutely identical to what has been learned previously, and automatic generalization to similar stimuli is extremely useful, and has great adaptive value in biological systems. This property is illustrated in the simple model described in Section 2.2 of Rolls and Treves (1998).

7.2.3.2 Graceful degradation or fault tolerance

If the synaptic weight vector $\mathbf{w}_i$ (or the weight matrix, which we can call $\mathbf{W}$) has synapses missing (e.g. during development), or loses synapses, then the activation h_i or $\mathbf{h}$ is still reasonable, because h_i is the dot product (correlation) of $\mathbf{x}$ with $\mathbf{w}_i$. The result, especially after passing through the activation function, can frequently be perfect recall. The same property arises if for example one or some of the conditioned stimulus (CS) input axons are lost or damaged. This is a very important property of associative memories, and is not a property of conventional computer memories, which produce incorrect data if even only 1 storage location (for 1 bit or binary digit of data) of their memory is damaged or cannot be accessed. This property of graceful degradation is of great adaptive value for biological systems.

7.2.3.3 The importance of distributed representations for pattern associators

A distributed representation is one in which the firing or activity of all the elements in the vector is used to encode a particular stimulus. For example, in a conditioned stimulus vector CS1 that has the value 101010, we need to know the state of all the elements to know which stimulus is being represented. Another stimulus, CS2, is represented by the vector 110001. We can represent many different events or stimuli with such overlapping sets of elements, and because in general any one element cannot be used to identify the stimulus, but instead the information about which stimulus is present is distributed over the population of elements or neurons, this is called a distributed representation (see Section 1.6). If, for binary neurons, half the neurons are in one state (e.g. 0), and the other half are in the other state (e.g. 1), then the representation is described as fully distributed. The CS representations above are thus fully distributed. If only a smaller proportion of the neurons is active to represent a stimulus, as in the vector 100001, then this is a sparse representation. For binary representations, we can quantify the sparseness by the proportion of neurons in the active (1) state.

In contrast, a local representation is one in which all the information that a particular stimulus or event has occurred is provided by the activity of one of the neurons, or elements in the vector. One stimulus might be represented by the vector 100000, another stimulus by the vector 010000, and a third stimulus by the vector 001000. The activity of neuron or element

1 would indicate that stimulus 1 was present, and of neuron 2, that stimulus 2 was present. The representation is local in that if a particular neuron is active, we know that the stimulus represented by that neuron is present. In neurophysiology, if such cells were present, they might be called 'grandmother cells' (cf. Barlow (1972), (1995)), in that one neuron might represent a stimulus in the environment as complex and specific as one's grandmother. Where the activity of a number of cells must be taken into account in order to represent a stimulus (such as an individual taste), then the representation is sometimes described as using ensemble encoding.

The properties just described for associative memories, generalization, and graceful degradation, are only implemented if the representation of the CS or x vector is distributed. This occurs because the recall operation involves computing the dot (inner) product of the input pattern vector $\mathbf{x}_r$ with the synaptic weight vector $\mathbf{w}_i$. This allows the activation h_i to reflect the similarity of the current input pattern to a previously learned input pattern $\mathbf{x}$ only if several or many elements of the $\mathbf{x}$ and $\mathbf{x}_r$ vectors are in the active state to represent a pattern. If local encoding were used, e.g. 100000, then if the first element of the vector (which might be the firing of axon 1, i.e. x_1, or the strength of synapse $i1$, w_{i1}) is lost, the resulting vector is not similar to any other CS vector, and the activation is 0. In the case of local encoding, the important properties of associative memories, generalization and graceful degradation, do not thus emerge. Graceful degradation and generalization are dependent on distributed representations, for then the dot product can reflect similarity even when some elements of the vectors involved are altered. If we think of the correlation between Y and X in a graph, then this correlation is affected only a little if a few X, Y pairs of data are lost (see Appendix A).

7.2.4 Prototype extraction, extraction of central tendency, and noise reduction

If a set of similar conditioned stimulus vectors $\mathbf{x}$ are paired with the same unconditioned stimulus e_i, the weight vector $\mathbf{w}_i$ becomes (or points towards) the sum (or with scaling, the average) of the set of similar vectors $\mathbf{x}$. This follows from the operation of the Hebb rule in equation 7.2. When tested at recall, the output of the memory is then best to the average input pattern vector denoted $< \mathbf{x} >$. If the average is thought of as a prototype, then even though the prototype vector $< \mathbf{x} >$ itself may never have been seen, the best output of the neuron or network is to the prototype. This produces 'extraction of the prototype' or 'central tendency'. The same phenomenon is a feature of human memory performance (see McClelland and Rumelhart (1986) Chapter 17, and Section 7.3.3.4), and this simple process with distributed representations in a neural network accounts for the phenomenon.

If the different exemplars of the vector $\mathbf{x}$ are thought of as noisy versions of the true input pattern vector $< \mathbf{x} >$ (with incorrect values for some of the elements), then the pattern associator has performed 'noise reduction', in that the output produced by any one of these vectors will represent the true, noiseless, average vector $< \mathbf{x} >$.

7.2.5 Speed

Recall is very fast in a real neuronal network, because the conditioned stimulus input firings x_j ($j = 1, C$ axons) can be applied simultaneously to the synapses w_{ij}, and the activation h_i can be accumulated in one or two time constants of the dendrite (e.g. 10–20 ms). Whenever the threshold of the cell is exceeded, it fires. Thus, in effectively one step, which takes the brain no more than 10–20 ms, all the output neurons of the pattern associator can be firing with rates that reflect the input firing of every axon. This is very different from a conventional digital computer, in which computing h_i in equation 7.3 would involve C multiplication and addition operations occurring one after another, or $2C$ time steps.

The brain performs parallel computation in at least two senses in even a pattern associator. One is that for a single neuron, the separate contributions of the firing rate x_j of each axon j multiplied by the synaptic weight w_{ij} are computed in parallel and added in the same time step. The second is that this can be performed in parallel for all neurons $i = 1, N$ in the network, where there are N output neurons in the network. It is these types of parallel processing that enable these classes of neuronal network in the brain to operate so fast, in effectively so few steps.

Learning is also fast ('one-shot') in pattern associators, in that a single pairing of the conditioned stimulus **x** and the unconditioned stimulus (UCS) **e** which produces the unconditioned output firing **y** enables the association to be learned. There is no need to repeat the pairing in order to discover over many trials the appropriate mapping. This is extremely important for biological systems, in which a single co-occurrence of two events may lead to learning that could have life-saving consequences. (For example, the pairing of a visual stimulus with a potentially life-threatening aversive event may enable that event to be avoided in future.) Although repeated pairing with small variations of the vectors is used to obtain the useful properties of prototype extraction, extraction of central tendency, and noise reduction, the essential properties of generalization and graceful degradation are obtained with just one pairing. The actual time scales of the learning in the brain are indicated by studies of associative synaptic modification using long-term potentiation paradigms (LTP, see Section 1.5). Co-occurrence or near simultaneity of the CS and UCS is required for periods of as little as 100 ms, with expression of the synaptic modification being present within typically a few seconds.

7.2.6 Local learning rule

The simplest learning rule used in pattern association neural networks, a version of the Hebb rule, is, as shown in equation 7.2 above,

$$\delta w_{ij} = \alpha y_i x_j.$$

This is a local learning rule in that the information required to specify the change in synaptic weight is available locally at the synapse, as it is dependent only on the presynaptic firing rate x_j available at the synaptic terminal, and the postsynaptic activation or firing y_i available on the dendrite of the neuron receiving the synapse (see Fig. 7.2b). This makes the learning rule biologically plausible, in that the information about how to change the synaptic weight does not have to be carried from a distant source, where it is computed, to every synapse. Such a non-local learning rule would not be biologically plausible, in that there are no appropriate connections known in most parts of the brain to bring in the synaptic training or teacher signal to every synapse.

Evidence that a learning rule with the general form of equation 7.2 is implemented in at least some parts of the brain comes from studies of long-term potentiation, described in Chapter 1. Long-term potentiation (LTP) has the synaptic specificity defined by equation 7.2, in that only synapses from active afferents, not those from inactive afferents, become

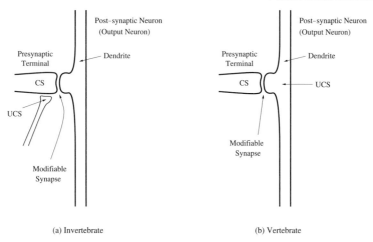

Fig. 7.2 (b) In vertebrate pattern association learning, the unconditioned stimulus (UCS) may be made available at all the conditioned stimulus (CS) terminals onto the output neuron because the dendrite of the postsynaptic neuron is electrically short, so that the effect of the UCS spreads for long distances along the dendrite. (a) In contrast, in at least some invertebrate association learning systems, the unconditioned stimulus or teaching input makes a synapse onto the presynaptic terminal carrying the conditioned stimulus.

strengthened. Synaptic specificity is important for a pattern associator, and most other types of neuronal network, to operate correctly. The number of independently modifiable synapses on each neuron is a primary factor in determining how many different memory patterns can be stored in associative memories (see Sections 7.2.6.1 and 7.3.3.7).

Another useful property of real neurons in relation to equation 7.2 is that the postsynaptic term, y_i, is available on much of the dendrite of a cell, because the electrotonic length of the dendrite is short. Thus if a neuron is strongly activated with a high value for y_i, then any active synapse onto the cell will be capable of being modified. This enables the cell to learn an association between the pattern of activity on all its axons and its postsynaptic activation, which is stored as an addition to its weight vector $\mathbf{w}_i$. Then later on, at recall, the output can be produced as a vector dot product operation between the input pattern vector $\mathbf{x}$ and the weight vector $\mathbf{w}_i$, so that the output of the cell can reflect the correlation between the current input vector and what has previously been learned by the cell.

It is interesting that at least many invertebrate neuronal systems may operate very differently from those described here, as described by Rolls and Treves (1998) (see Fig. 7.2a).

7.2.6.1 Capacity

The question of the storage capacity of a pattern associator is considered in detail in Appendix A3 of Rolls and Treves (1998). It is pointed out there that, for this type of associative network, the number of memories that it can hold simultaneously in storage has to be analyzed together with the retrieval quality of each output representation, and then only for a given quality of the representation provided in the input. This is in contrast to autoassociative nets (Section 7.3), in which a critical number of stored memories exists (as a function of various parameters of the network), beyond which attempting to store additional memories results in it becoming impossible to retrieve essentially anything. With a pattern associator, instead, one will always retrieve something, but this something will be very small (in information or correlation terms) if too many associations are simultaneously in storage and/or if too little is provided as input.

The conjoint quality-capacity input analysis can be carried out, for any specific instance of a pattern associator, by using formal mathematical models and established analytical procedures (see e.g. Treves (1995)). This, however, has to be done case by case. It is anyway useful to develop some intuition for how a pattern associator operates, by considering what its capacity would be in certain well-defined simplified cases.

Linear associative neuronal networks These networks are made up of units with a linear activation function, which appears to make them unsuitable to represent real neurons with their positive-only firing rates. However, even purely linear units have been considered as provisionally relevant models of real neurons, by assuming that the latter operate sometimes in the linear regime of their transfer function. (This implies a high level of spontaneous activity, and may be closer to conditions observed early on in sensory systems rather than in areas more specifically involved in memory.) As usual, the connections are trained by a Hebb (or similar) associative learning rule. The capacity of these networks can be defined as the total number of associations that can be learned independently of each other, given that the linear nature of these systems prevents anything more than a linear transform of the inputs. This implies that if input pattern C can be written as the weighted sum of input patterns A and B, the output to C will be just the same weighted sum of the outputs to A and B. If there are N' input axons, then there can be only at most N' mutually independent input patterns (i.e. none able to be written as a weighted sum of the others), and therefore the capacity of linear networks, defined above, is just N', or equal to the number of inputs to each neuron. In general, a random set of less than N' vectors (the CS input pattern vectors) will tend to be mutually independent but not mutually orthogonal (at 90 deg to each other) (see Appendix A). If they are not orthogonal (the normal situation), then the dot product of them is not 0, and the output pattern activated by one of the input vectors will be partially activated by other input pattern vectors, in accordance with how similar they are (see equations 7.5 and 7.6). This amounts to interference, which is therefore the more serious the less orthogonal, on the whole, is the set of input vectors.

Since input patterns are made of elements with positive values, if a simple Hebbian learning rule like the one of equation 7.2 is used (in which the input pattern enters directly with no subtraction term), the output resulting from the application of a stored input vector will be the sum of contributions from all other input vectors that have a non-zero dot product with it (see Appendix A), and interference will be disastrous. The only situation in which this would not occur is when different input patterns activate completely different input lines, but this is clearly an uninteresting circumstance for networks operating with distributed representations. A solution to this issue is to use a modified learning rule of the following form:

$$\delta w_{ij} = \alpha y_i (x_j - x) \tag{7.7}$$

where x is a constant, approximately equal to the average value of x_j. This learning rule includes (in proportion to y_i) increasing the synaptic weight if $(x_j - x) > 0$ (long-term potentiation), and decreasing the synaptic weight if $(x_j - x) < 0$ (heterosynaptic long-term depression). It is useful for x to be roughly the average activity of an input axon x_j across patterns, because then the dot product between the various patterns stored on the weights and the input vector will tend to cancel out with the subtractive term, except for the pattern equal to (or correlated with) the input vector itself. Then up to N' input vectors can still be learned by the network, with only minor interference (provided of course that they are mutually independent, as they will in general tend to be). This modified learning rule can

Table 7.1 Effects of pre- and post-synaptic activity on synaptic modification

		Post-synaptic activation	
		0	high
	0	No change	Heterosynaptic LTD
Presynaptic firing			
	high	Homosynaptic LTD	LTP

also be described in terms of a contingency table (Table 7.1) showing the synaptic strength modifications produced by different types of learning rule, where LTP indicates an increase in synaptic strength (called Long-Term Potentiation in neurophysiology), and LTD indicates a decrease in synaptic strength (called Long-Term Depression in neurophysiology). Heterosynaptic long-term depression is so-called because it is the decrease in synaptic strength that occurs to a synapse that is other than that through which the postsynaptic cell is being activated. This heterosynaptic long-term depression is the type of change of synaptic strength that is required (in addition to LTP) for effective subtraction of the average presynaptic firing rate, in order, as it were, to make the CS vectors appear more orthogonal to the pattern associator. The rule is sometimes called the Singer-Stent rule, after work by Singer (1987) and Stent (1973), and was discovered in the brain by Levy (Levy (1985); Levy and Desmond (1985); see Brown, Kairiss and Keenan (1990b)). Homosynaptic long-term depression is so-called because it is the decrease in synaptic strength that occurs to a synapse which is (the same as that which is) active. For it to occur, the postsynaptic neuron must simultaneously be inactive, or have only low activity. (This rule is sometimes called the BCM rule after the paper of Bienenstock, Cooper and Munro (1982); see Section 7.4 on competitive networks).

Associative neuronal networks with non-linear neurons With non-linear neurons, that is with at least a threshold in the activation function so that the output firing y_i is 0 when the activation h_i is below the threshold, the capacity can be measured in terms of the number of different clusters of output pattern vectors that the network produces. This is because the non-linearities now present (one per output neuron) result in some clustering of all possible (conditioned stimulus) input patterns **x**. Input patterns that are similar to a stored input vector can result due to the non-linearities in output patterns even closer to the stored output; and vice versa sufficiently dissimilar inputs can be assigned to different output clusters thereby increasing their mutual dissimilarity. As with the linear counterpart, in order to remove the correlation that would otherwise occur between the patterns because the elements can take only positive values, it is useful to use a modified Hebb rule of the form shown in equation 7.7.

   With fully distributed output patterns, the number p of associations that leads to different clusters is of order C, the number of input lines (axons) per output unit (that is, of order

N' for a fully connected network), as shown in Appendix A3 of Rolls and Treves (1998). If sparse patterns are used in the output, or alternatively if the learning rule includes a non-linear postsynaptic factor that is effectively equivalent to using sparse output patterns, the coefficient of proportionality between p and C can be much higher than one, that is, many more patterns can be stored than inputs per unit (see Appendix A3 of Rolls and Treves (1998)). Indeed, the number of different patterns or prototypes p that can be stored can be derived for example in the case of binary units (Gardner 1988) to be

$$p \approx C/[a_o log(1/a_o)] \tag{7.8}$$

where a_o is the sparseness of the output firing pattern **y** produced by the unconditioned stimulus. p can in this situation be much larger than C (see Rolls and Treves (1990), and Appendix A3 of Rolls and Treves (1998)). This is an important result for encoding in pattern associators, for it means that provided that the activation functions are non-linear (which is the case with real neurons), there is a very great advantage to using sparse encoding, for then many more than C pattern associations can be stored. Sparse representations may well be present in brain regions involved in associative memory (see Chapter 12) for this reason.

The non-linearity inherent in the NMDA receptor-based Hebbian plasticity present in the brain may help to make the stored patterns more sparse than the input patterns, and this may be especially beneficial in increasing the storage capacity of associative networks in the brain by allowing participation in the storage of especially those relatively few neurons with high firing rates in the exponential firing rate distributions typical of neurons in sensory systems (see Sections 7.4.8.3 and 5.5.1).

7.2.6.2 Interference

Interference occurs in linear pattern associators if two vectors are not orthogonal, and is simply dependent on the angle between the originally learned vector and the recall cue or CS vector (see Appendix A), for the activation of the output neuron depends simply on the dot product of the recall vector and the synaptic weight vector (equation 7.5). Also in non-linear pattern associators (the interesting case for all practical purposes), interference may occur if two CS patterns are not orthogonal, though the effect can be controlled with sparse encoding of the UCS patterns, effectively by setting high thresholds for the firing of output units. In other words, the CS vectors need not be strictly orthogonal, but if they are too similar, some interference will still be likely to occur.

The fact that interference is a property of neural network pattern associator memories is of interest, for interference is a major property of human memory. Indeed, the fact that interference is a property of human memory and of network association memories is entirely consistent with the hypothesis that human memory is stored in associative memories of the type described here, or at least that network associative memories of the type described represent a useful exemplar of the class of parallel distributed storage network used in human memory. It may also be suggested that one reason that interference is tolerated in biological memory is that it is associated with the ability to generalize between stimuli, which is an invaluable feature of biological network associative memories, in that it allows the memory to cope with stimuli that will almost never be identical on different occasions, and in that it allows useful analogies that have survival value to be made.

Input A	1	0	1
Input B	0	1	1

Required Output	1	1	0

Fig. 7.3 A non-linearly separable mapping.

7.2.6.3 Expansion recoding

If patterns are too similar to be stored in associative memories, then one solution that the brain seems to use repeatedly is to expand the encoding to a form in which the different stimuli are less correlated, that is, more orthogonal, before they are presented as CS stimuli to a pattern associator. The problem can be highlighted by a non-linearly separable mapping (which captures part of the eXclusive OR (XOR) problem), in which the mapping that is desired is as follows. The neuron has two inputs, A and B (see Fig. 7.3).

This is a mapping of patterns that is impossible for a one-layer network, because the patterns are not linearly separable[10]. A solution is to remap the two input lines A and B to three input lines 1–3, that is to use expansion recoding, as shown in Fig. 7.4. This can be performed by a competitive network (see Section 7.4). The synaptic weights on the dendrite

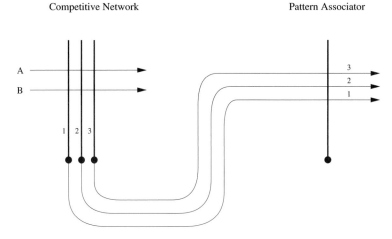

Fig. 7.4 Expansion recoding. A competitive network followed by a pattern associator that can enable patterns that are not linearly separable to be learned correctly.

of the output neuron could then learn the following values using a simple Hebb rule, equation 7.2, and the problem could be solved as in Fig. 7.5. The whole network would look like that shown in Fig. 7.4.

Expansion encoding which maps vectors with N' inputs to a set of neurons that is larger

[10]See Appendix A. There is no set of synaptic weights in a one-layer net that could solve the problem shown in Fig. 7.3. Two classes of patterns are not linearly separable if no hyperplane can be positioned in their N-dimensional space so as to separate them (see Appendix A). The XOR problem has the additional constraint that $A = 0, B = 0$ must be mapped to Output $= 0$.

	Synaptic weight
Input 1 (A=1, B=0)	1
Input 2 (A=0, B=1)	1
Input 3 (A=1, B=1)	0

Fig. 7.5 Synaptic weights on the dendrite of the output neuron in Fig. 7.4.

than N' appears to be present in several parts of the brain, and to precede networks that perform pattern association. Marr (1969) suggested that one such expansion recoding was performed in the cerebellum by the mapping from the mossy fibres to the granule cells, which provide the associatively modifiable inputs to the Purkinje cells in the cerebellum (see Chapter 9 of Rolls and Treves (1998)). The expansion in this case is in the order of 1000 times. Marr (1969) suggested that the expansion recoding was performed by each cerebellar granule cell responding to a low-order combination of mossy fibre activity, implemented by each granule cell receiving 5–7 mossy fibre inputs. Another example is in the hippocampus, in which there is an expansion from the perforant path fibres originating in the entorhinal cortex to the dentate granule cells, which are thought to decorrelate the patterns that are stored by the CA3 cells, which form an autoassociative network (see Chapter 6 of Rolls and Treves (1998)). The suggestion is that this expansion, performed by the dentate granule cells, helps to separate patterns so that overlapping patterns in the entorhinal cortex are made separate in CA3, to allow separate episodic memories with overlapping information to be stored and recalled separately. A similar principle was probably being used as a preprocessor in Rosenblatt's original perceptron (see Chapter 5 of Rolls and Treves (1998)).

7.2.7 Implications of different types of coding for storage in pattern associators

Throughout this Section, we have made statements about how the properties of pattern associators — such as the number of patterns that can be stored, and whether generalization and graceful degradation occur — depend on the type of encoding of the patterns to be associated. (The types of encoding considered, local, sparse distributed, and fully distributed, are described in Chapter 1.) We draw together these points in Table 7.2. The amount of information

Table 7.2 Coding in associative memories*

	Local	Sparse distributed	Fully distributed
Generalization, Completion, Graceful degradation	No	Yes	Yes
Number of patterns that can be stored	N (large)	of order $C/[a_o \log(1/a_o)]$ (can be larger)	of order C (usually smaller than N)
Amount of information in each pattern (values if binary)	Minimal ($\log(N)$ bits)	Intermediate ($Na_o \log(1/a_o)$ bits)	Large (N bits)

* N refers here to the number of output units, and C to the average number of inputs to each output unit. a_o is the sparseness of output patterns, or roughly the proportion of output units activated by a UCS pattern. Note: logs are to the base 2.

that can be stored in each pattern in a pattern associator is considered in Appendix A3 of Rolls and Treves (1998).

7.3 Autoassociation memory

Autoassociative memories, or attractor neural networks, store memories, each one of which is represented by a pattern of neural activity. The memories are stored in the recurrent synaptic connections between the neurons of the network, for example in the recurrent collateral connections between cortical pyramidal cells. Autoassociative networks can then recall the appropriate memory from the network when provided with a fragment of one of the memories. This is called completion. Many different memories can be stored in the network and retrieved correctly. A feature of this type of memory is that it is content addressable; that is, the information in the memory can be accessed if just the contents of the memory (or a part of the contents of the memory) are used. This is in contrast to a conventional computer, in which the address of what is to be accessed must be supplied, and used to access the contents of the memory. Content addressability is an important simplifying feature of this type of memory, which makes it suitable for use in biological systems. The issue of content addressability will be amplified below.

An autoassociation memory can be used as a short term memory, in which iterative processing round the recurrent collateral connection loop keeps a representation active by continuing neuronal firing. The short term memory reflected in continuing neuronal firing for several hundred ms after a visual stimulus is removed which is present in visual cortical areas such as the inferior temporal visual cortex (see Chapter 5) is probably implemented in this way. This short term memory is one possible mechanism that contributes to the implementation of the trace memory learning rule which can help to implement invariant object recognition as described in Chapter 8. Autoassociation memories also appear to be used in a short term memory role in the prefrontal cortex. In particular, the temporal visual cortical areas have connections to the ventrolateral prefrontal cortex which help to implement the short term memory for visual stimuli (in for example delayed match to sample tasks, and visual search tasks, as described in Section 12.1, and in Chapters 6 and 8 of Rolls and Treves (1998)). In an analogous way the parietal cortex has connections to the dorsolateral prefrontal cortex for the short term memory of spatial responses (see Section 12.1, and Chapter 10 of Rolls and Treves (1998)). These short term memories provide a mechanism that enables attention to be maintained through backprojections from prefrontal cortex areas to the temporal and parietal areas that send connections to the prefrontal cortex, as described in Chapters 6, 9 and 10 and Section 12.1. Autoassociation networks implemented by the recurrent collateral synapses between cortical pyramidal cells also provide a mechanism for constraint satisfaction and also noise reduction whereby the firing of neighbouring neurons can be taken into account in enabling the network to settle into a state that reflects all the details of the inputs activating the population of connected neurons, as well as the effects of what has been set up during developmental plasticity as well as later experience. Attractor networks are also effectively implemented by virtue of the forward and backward connections between cortical areas (see Sections 1.11 and 12.1). An autoassociation network with rapid synaptic plasticity can learn each memory in one trial. Because of its 'one-shot' rapid learning, and ability to complete, this type of network is well suited for episodic memory storage, in which each past episode must

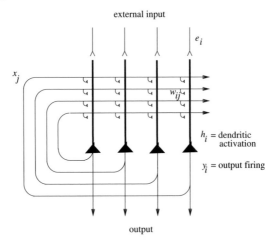

external input

e_i

x_j

w_{ij}

h_i = dendritic activation

y_i = output firing

output

Fig. 7.6 The architecture of an autoassociative neural network.

be stored and recalled later from a fragment, and kept separate from other episodic memories (see Section 12.2, and Chapter 6 of Rolls and Treves (1998)).

7.3.1   Architecture and operation

The prototypical architecture of an autoassociation memory is shown in Fig. 7.6. The external input e_i is applied to each neuron i by unmodifiable synapses. This produces firing y_i of each neuron, or a vector of firing on the output neurons **y**. Each output neuron i is connected by a recurrent collateral connection to the other neurons in the network, via modifiable connection weights w_{ij}. This architecture effectively enables the output firing vector **y** to be associated during learning with itself. Later on, during recall, presentation of part of the external input will force some of the output neurons to fire, but through the recurrent collateral axons and the modified synapses, other neurons in **y** can be brought into activity. This process can be repeated a number of times, and recall of a complete pattern may be perfect. Effectively, a pattern can be recalled or recognized because of associations formed between its parts. This of course requires distributed representations.

Next we introduce a more precise and detailed description of the above, and describe the properties of these networks. Ways to analyze formally the operation of these networks are introduced in Appendix A4 of Rolls and Treves (1998) and by Amit (1989).

7.3.1.1   Learning

The firing of every output neuron i is forced to a value y_i determined by the external input e_i. Then a Hebb-like associative local learning rule is applied to the recurrent synapses in the network:

$$\delta w_{ij} = \alpha y_i y_j. \tag{7.9}$$

It is notable that in a fully connected network, this will result in a symmetric matrix of synaptic weights, that is the strength of the connection from neuron 1 to neuron 2 will be the same as the strength of the connection from neuron 2 to neuron 1 (both implemented via recurrent collateral synapses).

It is a factor that is sometimes overlooked that there must be a mechanism for ensuring that during learning y_i does approximate e_i, and must not be influenced much by activity in the recurrent collateral connections, otherwise the new external pattern e will not be stored in the network, but instead something will be stored that is influenced by the previously stored memories. It is thought that in some parts of the brain, such as the hippocampus, there are processes that help the external connections to dominate the firing during learning (see Chapter 6 and (Rolls 1989c, Rolls 1989b, Rolls 1989e, Rolls 1989a, Treves and Rolls 1992)).

7.3.1.2 Recall

During recall, the external input e_i is applied, and produces output firing, operating through the non-linear activation function described below. The firing is fed back by the recurrent collateral axons shown in Fig. 7.6 to produce activation of each output neuron through the modified synapses on each output neuron. The activation h_i produced by the recurrent collateral effect on the ith neuron is, in the standard way, the sum of the activations produced in proportion to the firing rate of each axon y_j operating through each modified synapse w_{ij}, that is,

$$h_i = \sum_j y_j w_{ij} \tag{7.10}$$

where $\sum_j$ indicates that the sum is over the C input axons to each neuron, indexed by j.

The output firing y_i is a function of the activation produced by the recurrent collateral effect (internal recall) and by the external input (e_i):

$$y_i = \mathrm{f}(h_i + e_i) \tag{7.11}$$

The activation function should be nonlinear, and may be for example binary threshold, linear threshold, sigmoid, etc. (see Fig. 1.3). The threshold at which the activation function operates is set in part by the effect of the inhibitory neurons in the network (not shown in Fig. 7.6). The connectivity is that the pyramidal cells have collateral axons that excite the inhibitory interneurons, which in turn connect back to the population of pyramidal cells to inhibit them by a mixture of shunting (divisive) and subtractive inhibition using GABA (gamma-amino-butyric acid) terminals, as described in Section 7.6. There are many fewer inhibitory neurons than excitatory neurons (in the order of 5–10%, see Table 1.1) and of connections to and from inhibitory neurons (see Table 1.1), and partly for this reason the inhibitory neurons are considered to perform generic functions such as threshold setting, rather than to store patterns by modifying their synapses. Similar inhibitory processes are assumed for the other networks described in this Chapter. The non-linear activation function can minimize interference between the pattern being recalled and other patterns stored in the network, and can also be used to ensure that what is a positive feedback system remains stable. The network can be allowed to repeat this recurrent collateral loop a number of times. Each time the loop operates, the output firing becomes more like the originally stored pattern, and this progressive recall is usually complete within 5–15 iterations.

7.3.2 Introduction to the analysis of the operation of autoassociation networks

With complete connectivity in the synaptic matrix, and the use of a Hebb rule, the matrix of synaptic weights formed during learning is symmetric. The learning algorithm is fast, 'one-shot', in that a single presentation of an input pattern is all that is needed to store that pattern.

During recall, a part of one of the originally learned stimuli can be presented as an external input. The resulting firing is allowed to iterate repeatedly round the recurrent collateral system, gradually on each iteration recalling more and more of the originally learned pattern. Completion thus occurs. If a pattern is presented during recall that is similar but not identical to any of the previously learned patterns, then the network settles into a stable recall state in which the firing corresponds to that of the previously learned pattern. The network can thus generalize in its recall to the most similar previously learned pattern. The activation function of the neurons should be non-linear, since a purely linear system would not produce any categorization of the input patterns it receives, and therefore would not be able to effect anything more than a trivial (i.e. linear) form of completion and generalization.

Recall can be thought of in the following way, relating it to what occurs in pattern associators. The external input $\mathbf{e}$ is applied, produces firing $\mathbf{y}$, which is applied as a recall cue on the recurrent collaterals as $\mathbf{y}^{\mathrm{T}}$. (The notation $\mathbf{y}^{\mathrm{T}}$ signifies the transpose of $\mathbf{y}$, which is implemented by the application of the firing of the neurons $\mathbf{y}$ back via the recurrent collateral axons as the next set of inputs to the neurons.) The activity on the recurrent collaterals is then multiplied with the synaptic weight vector stored during learning on each neuron to produce the new activation h_i which reflects the similarity between $\mathbf{y}^{\mathrm{T}}$ and one of the stored patterns. Partial recall has thus occurred as a result of the recurrent collateral effect. The activations h_i after thresholding (which helps to remove interference from other memories stored in the network, or noise in the recall cue) result in firing y_i, or a vector of all neurons $\mathbf{y}$, which is already more like one of the stored patterns than, at the first iteration, the firing resulting from the recall cue alone, $\mathbf{y} = \mathrm{f}(\mathbf{e})$. This process is repeated a number of times to produce progressive recall of one of the stored patterns.

Autoassociation networks operate by effectively storing associations between the elements of a pattern. Each element of the pattern vector to be stored is simply the firing of a neuron. What is stored in an autoassociation memory is a set of pattern vectors. The network operates to recall one of the patterns from a fragment of it. Thus, although this network implements recall or recognition of a pattern, it does so by an association learning mechanism, in which associations between the different parts of each pattern are learned. These memories have sometimes been called autocorrelation memories (Kohonen 1977), because they learn correlations between the activity of neurons in the network, in the sense that each pattern learned is defined by a set of simultaneously active neurons. Effectively each pattern is associated by learning with itself. This learning is implemented by an associative (Hebb-like) learning rule.

The system formally resembles spin glass systems of magnets analyzed quantitatively in statistical mechanics. This has led to the analysis of (recurrent) autoassociative networks as dynamical systems made up of many interacting elements, in which the interactions are such as to produce a large variety of basins of attraction of the dynamics. Each basin of attraction corresponds to one of the originally learned patterns, and once the network is within a basin it keeps iterating until a recall state is reached that is the learned pattern itself or a pattern closely similar to it. (Interference effects may prevent an exact identity between the recall state and a learned pattern). This type of system is contrasted with other, simpler, systems of magnets (e.g. ferromagnets), in which the interactions are such as to produce only a limited number of related basins, since the magnets tend to be, for example, all aligned with each other. The states reached within each basin of attraction are called attractor states, and the analogy between autoassociator neural networks and physical systems with multiple attractors was drawn by Hopfield (1982) in a very influential paper. He was able to show that the recall state can be thought of as the local minimum in an energy landscape, where the energy would be defined as

$$E = -\frac{1}{2}\sum_{i,j} w_{ij}(y_i - <y>)(y_j - <y>).$$

(7.12)

This equation can be understood in the following way. If two neurons are both firing above their mean rate (denoted by $<y>$), and are connected by a weight with a positive value, then the firing of these two neurons is consistent with each other, and they mutually support each other, so that they contribute to the system's tendency to remain stable. If across the whole network such mutual support is generally provided, then no further change will take place, and the system will indeed remain stable. If, on the other hand, either of our pair of neurons was not firing, or if the connecting weight had a negative value, the neurons would not support each other, and indeed the tendency would be for the neurons to try to alter ('flip' in the case of binary units) the state of the other. This would be repeated across the whole network until a situation in which most mutual support, and least 'frustration', was reached. What makes it possible to define an energy function and for these points to hold is that the matrix is symmetric (see Hopfield (1982), Hertz, Krogh and Palmer (1991), Amit (1989)).

Physicists have generally analyzed a system in which the input pattern is presented and then immediately removed, so that the network then 'falls' without further assistance (in what is referred to as the unclamped condition) towards the minimum of its basin of attraction. A more biologically realistic system is one in which the external input is left on contributing to the recall during the fall into the recall state. In this clamped condition, recall is usually faster, and more reliable, so that more memories may be usefully recalled from the network. The approach using methods developed in theoretical physics has led to rapid advances in the understanding of autoassociative networks, and its basic elements are described in Appendix A4 of Rolls and Treves (1998) and by Hertz, Krogh and Palmer (1991) and Amit (1989).

7.3.3 Properties

The internal recall in autoassociation networks involves multiplication of the firing vector of neuronal activity by the vector of synaptic weights on each neuron. This inner product vector multiplication allows the similarity of the firing vector to previously stored firing vectors to be provided by the output (as effectively a correlation), if the patterns learned are distributed. As a result of this type of 'correlation computation' performed if the patterns are distributed, many important properties of these networks arise, including pattern completion (because part of a pattern is correlated with the whole pattern), and graceful degradation (because a damaged synaptic weight vector is still correlated with the original synaptic weight vector). Some of these properties are described next.

7.3.3.1 Completion

Perhaps the most important and useful property of these memories is that they complete an incomplete input vector, allowing recall of a whole memory from a small fraction of it. The memory recalled in response to a fragment is that stored in the memory that is closest in pattern similarity (as measured by the dot product, or correlation). Because the recall is iterative and progressive, the recall can be perfect.

This property and the associative property of pattern associator neural networks are very similar to the properties of human memory. This property may be used when we recall a part of a recent memory of a past episode from a part of that episode. The way in which this could be implemented in the hippocampus is summarized in Section 12.2 and described in Chapter 6 of Rolls and Treves (1998).

7.3.3.2 Generalization

The network generalizes in that an input vector similar to one of the stored vectors will lead to recall of the originally stored vector, provided that distributed encoding is used. The principle by which this occurs is similar to that described for a pattern associator.

7.3.3.3 Graceful degradation or fault tolerance

If the synaptic weight vector $\mathbf{w}_i$ on each neuron (or the weight matrix) has synapses missing (e.g. during development), or loses synapses (e.g. with brain damage or aging), then the activation h_i (or vector of activations $\mathbf{h}$) is still reasonable, because h_i is the dot product (correlation) of $\mathbf{y}^T$ with $\mathbf{w}_i$. The same argument applies if whole input axons are lost. If an output neuron is lost, then the network cannot itself compensate for this, but the next network in the brain is likely to be able to generalize or complete if its input vector has some elements missing, as would be the case if some output neurons of the autoassociation network were damaged.

7.3.3.4 Prototype extraction, extraction of central tendency, and noise reduction

These arise when a set of similar input pattern vectors $\{\mathbf{e}\}$ (which induce firing of the output neurons $\{\mathbf{y}\}$) are learned by the network. The weight vectors $\mathbf{w}_i$ (or strictly $\mathbf{w}_i^T$) become (or point towards) the average $\{< \mathbf{y} >\}$ of that set of similar vectors. This produces 'extraction of the prototype' or 'extraction of the central tendency', and 'noise reduction'. This process can result in better recognition or recall of the prototype than of any of the exemplars, even though the prototype may never itself have been presented. The general principle by which the effect occurs is similar to that by which it occurs in pattern associators. It of course only occurs if each pattern uses a distributed representation.

Related to outputs of the visual system to long term memory systems (see Section 12.2), there has been intense debate about whether when human memories are stored, a prototype of what is to be remembered is stored, or whether all the instances or the exemplars are each stored separately so that they can be individually recalled (McClelland and Rumelhart (1986), Chapter 17, p. 172). Evidence favouring the prototype view is that if a number of different examples of an object are shown, then humans may report more confidently that they have seen the prototype before than any of the different exemplars, even though the prototype has never been shown (Posner and Keele 1968, Rosch 1975). Evidence favouring the view that exemplars are stored is that in categorization and perceptual identification tasks the responses made are often sensitive to the congruity between particular training stimuli and particular test stimuli (Brooks 1978, Medin and Schaffer 1978, Jacoby 1983a, Jacoby 1983b, Whittlesea 1983). It is of great interest that both types of phenomena can arise naturally out of distributed information storage in a neuronal network such as an autoassociator. This can be illustrated by the storage in an autoassociation memory of sets of stimuli that are all somewhat different examples of the same pattern. These can be generated, for example, by randomly altering each of the input vectors from the input stimulus. After many such randomly altered exemplars have been learned by the network, recall can be tested, and it is found that the network responds best to the original input vector, with which it has never been presented. The reason for this is that the autocorrelation components that build up in the synaptic matrix with repeated presentations of the exemplars represent the average correlation between the different elements of the vector, and this is highest for the prototype. This effect also gives the storage some noise immunity, in that variations in the input that are random noise average out, while the signal that is constant builds up with repeated learning.

7.3.3.5 Speed

The recall operation is fast on each neuron on a single iteration, because the pattern $\mathbf{y}^T$ on the axons can be applied simultaneously to the synapses $\mathbf{w}_i$, and the activation h_i can be accumulated in one or two time constants of the dendrite (e.g. 10–20 ms). If a simple implementation of an autoassociation net such as that described by Hopfield (1982) is simulated on a computer, then 5–15 iterations are typically necessary for completion of an incomplete input cue $\mathbf{e}$. This might be taken to correspond to 50–200 ms in the brain, rather too slow for any one local network in the brain to function. However, recent work (see Section 7.6 and Treves (1993), Battaglia and Treves (1998a) and Appendix A5 of Rolls and Treves (1998)) has shown that if the neurons are treated not as McCulloch-Pitts neurons which are simply 'updated' at each iteration, or cycle of time steps (and assume the active state if the threshold is exceeded), but instead are analyzed and modelled as 'integrate-and-fire' neurons in real continuous time, then the network can effectively 'relax' into its recall state very rapidly, in one or two time constants of the synapses. This corresponds to perhaps 20 ms in the brain. One factor in this rapid dynamics of autoassociative networks with brain-like 'integrate-and-fire' membrane and synaptic properties is that with some spontaneous activity, some of the neurons in the network are close to threshold already before the recall cue is applied, and hence some of the neurons are very quickly pushed by the recall cue into firing, so that information starts to be exchanged very rapidly (within 1–2 ms of brain time) through the modified synapses by the neurons in the network. The progressive exchange of information starting early on within what would otherwise be thought of as an iteration period (of perhaps 20 ms, corresponding to a neuronal firing rate of 50 spikes/s), is the mechanism accounting for rapid recall in an autoassociative neuronal network made biologically realistic in this way. Further analysis of the fast dynamics of these networks if they are implemented in a biologically plausible way with 'integrate-and-fire' neurons, is provided in Section 7.6, in Appendix A5 of Rolls and Treves (1998), and by Treves (1993). *The general approach applies to other networks with recurrent connections, not just autoassociators, and the fact that such networks can operate much faster than it would seem from simple models that follow discrete time dynamics, is probably a major factor in enabling these networks to provide some of the building blocks of brain function.*

Learning is fast, 'one-shot', in that a single presentation of an input pattern $\mathbf{e}$ (producing $\mathbf{y}$) enables the association between the activation of the dendrites (the post-synaptic term h_i) and the firing of the recurrent collateral axons $\mathbf{y}^T$, to be learned. Repeated presentation with small variations of a pattern vector is used to obtain the properties of prototype extraction, extraction of central tendency, and noise reduction, because these arise from the averaging process produced by storing very similar patterns in the network.

7.3.3.6 Local learning rule

The simplest learning used in autoassociation neural networks, a version of the Hebb rule, is (as in equation 7.9)

$$\delta w_{ij} = \alpha y_i y_j.$$

The rule is a local learning rule in that the information required to specify the change in synaptic weight is available locally at the synapse, as it is dependent only on the presynaptic firing rate y_j available at the synaptic terminal, and the postsynaptic activation or firing y_i available on the dendrite of the neuron receiving the synapse. This makes the learning rule biologically plausible, in that the information about how to change the synaptic weight does not have to be carried to every synapse from a distant source where it is computed. As with pattern associators, since firing rates are positive quantities, a potentially interfering correlation is induced between different pattern vectors. This can be removed by subtracting the mean of the presynaptic activity from each presynaptic term, using a type of long-term depression. This can be specified as

$$\delta w_{ij} = \alpha y_i (y_j - z) \tag{7.13}$$

where α is a learning rate constant. This learning rule includes (in proportion to y_i) increasing the synaptic weight if $(y_j - z) > 0$ (long-term potentiation), and decreasing the synaptic weight if $(y_j - z) < 0$ (heterosynaptic long-term depression). This procedure works optimally if z is the average activity $< y_j >$ of an axon across patterns.

Evidence that a learning rule with the general form of equation 7.9 is implemented in at least some parts of the brain comes from studies of long-term potentiation, described in Section 1.5. One of the important potential functions of heterosynaptic long-term depression is its ability to allow in effect the average of the presynaptic activity to be subtracted from the presynaptic firing rate (see Appendix A3 of Rolls and Treves (1998) and Rolls and Treves (1990)).

Autoassociation networks can be trained with the error-correction or delta learning rule described in Chapter 5. Although a delta rule is less biologically plausible than a Hebb-like rule, a delta rule can help to store separately patterns that are very similar (see McClelland and Rumelhart (1988), Hertz, Krogh and Palmer (1991)).

7.3.3.7 Capacity

One measure of storage capacity is to consider how many orthogonal patterns could be stored, as with pattern associators. If the patterns are orthogonal, there will be no interference between them, and the maximum number p of patterns that can be stored will be the same as the number N of output neurons in a fully connected network. Although in practice the patterns that have to be stored will hardly be orthogonal, this is not a purely academic speculation, since it was shown how one can construct a synaptic matrix that effectively orthogonalizes any set of (linearly independent) patterns (Kohonen 1977, Kohonen 1989, Personnaz, Guyon and Dreyfus 1985, Kanter and Sompolinsky 1987). However, this matrix cannot be learned with a local, one-shot learning rule, and therefore its interest for autoassociators in the brain is limited. The more general case of random non-orthogonal patterns, and of Hebbian learning rules, is considered next.

With non-linear neurons used in the network, the capacity can be measured in terms of the number of input patterns $\mathbf{y}$ (produced by the external input $\mathbf{e}$, see Fig. 7.6) that can be stored in the network and recalled later whenever the network settles within each stored pattern's basin of attraction. The first quantitative analysis of storage capacity (Amit, Gutfreund and Sompolinsky 1987) considered a fully connected Hopfield (1982) autoassociator model, in which units are binary elements with an equal probability of being 'on' or 'off' in each pattern, and the number C of inputs per unit is the same as the number N of output units. (Actually it is equal to $N - 1$, since a unit is taken not to connect to itself.) Learning is taken to occur by clamping the desired patterns on the network and using a modified Hebb rule, in which the mean of the presynaptic and postsynaptic firings is subtracted from the firing on any one learning trial (this amounts to a covariance learning rule, and is described more fully in Appendix A4 of Rolls and Treves (1998)). With such fully distributed random patterns, the number of patterns that can be learned is (for C large) $p \approx 0.14C = 0.14N$, hence well below what could be achieved with orthogonal patterns or with an 'orthogonalizing' synaptic matrix. Many variations of this 'standard' autoassociator model have been analyzed subsequently.

Treves and Rolls (1991) have extended this analysis to autoassociation networks that are much more biologically relevant in the following ways. First, some or many connections between the recurrent collaterals and the dendrites are missing (this is referred to as diluted connectivity, and results in a non-symmetric synaptic connection matrix in which w_{ij} does not equal w_{ji}, one of the original assumptions made in order to introduce the energy formalism in the Hopfield model). Second, the neurons need not be restricted to binary threshold neurons, but can have a threshold linear activation function (see Fig. 1.3). This enables the neurons to assume real continuously variable firing rates, which are what is found in the brain (Rolls and Tovee 1995b, Treves, Panzeri, Rolls, Booth and Wakeman 1999). Third, the representation need not be fully distributed (with half the neurons 'on', and half 'off'), but instead can have a small proportion of the neurons firing above the spontaneous rate, which is what is found in parts of the brain such as the hippocampus that are involved in memory (see Treves and Rolls (1994), and Chapter 6 of Rolls and Treves (1998)). Such a representation is defined as being sparse, and the sparseness a of the representation can be measured, by extending the binary notion of the proportion of neurons that are firing, as

$$a = \frac{(\sum\limits_{i=1}^{N} y_i/N)^2}{\sum\limits_{i=1}^{N} y_i^2/N} \tag{7.14}$$

where y_i is the firing rate of the ith neuron in the set of N neurons. Treves and Rolls (1991) have shown that such a network does operate efficiently as an autoassociative network, and can store (and recall correctly) a number of different patterns p as follows

$$p \approx \frac{C^{\mathrm{RC}}}{a \ln(\frac{1}{a})} k \tag{7.15}$$

where C^{RC} is the number of synapses on the dendrites of each neuron devoted to the recurrent collaterals from other neurons in the network, and k is a factor that depends weakly on the detailed structure of the rate distribution, on the connectivity pattern, etc., but is roughly in the order of 0.2–0.3.

The main factors that determine the maximum number of memories that can be stored in an autoassociative network are thus the number of connections on each neuron devoted to the recurrent collaterals, and the sparseness of the representation. For example, for $C^{RC} = 12,000$ and $a = 0.02$, p is calculated to be approximately $36,000$. This storage capacity can be realized, with little interference between patterns, if the learning rule includes some form of heterosynaptic Long-Term Depression that counterbalances the effects of associative Long-Term Potentiation (Treves and Rolls (1991); see Appendix A4 of Rolls and Treves (1998)). It should be noted that the number of neurons N (which is greater than C^{RC}, the number of recurrent collateral inputs received by any neuron in the network from the other neurons in the network) is not a parameter that influences the number of different memories that can be stored in the network. The implication of this is that increasing the number of neurons (without increasing the number of connections per neuron) does not increase the number of different patterns that can be stored (see Rolls and Treves (1998) Appendix A4), although it may enable simpler encoding of the firing patterns, for example more orthogonal encoding, to be used. This latter point may account in part for why there are generally in the brain more neurons in a recurrent network than there are connections per neuron (see e.g. Section 12.2).

The non-linearity inherent in the NMDA receptor-based Hebbian plasticity present in the brain may help to make the stored patterns more sparse than the input patterns, and this may be especially beneficial in increasing the storage capacity of associative networks in the brain by allowing participation in the storage of especially those relatively few neurons with high firing rates in the exponential firing rate distributions typical of neurons in sensory systems (see Sections 7.4.8.3 and 5.5.1).

7.3.3.8 Context

The environmental context in which learning occurs can be a very important factor that affects retrieval in humans and other animals. Placing the subject back into the same context in which the original learning occurred can greatly facilitate retrieval.

Context effects arise naturally in association networks if some of the activity in the network reflects the context in which the learning occurs. Retrieval is then better when that context is present, for the activity contributed by the context becomes part of the retrieval cue for the memory, increasing the correlation of the current state with what was stored. (A strategy for retrieval arises simply from this property. The strategy is to keep trying to recall as many fragments of the original memory situation, including the context, as possible, as this will provide a better cue for complete retrieval of the memory than just a single fragment.)

The effects that mood has on memory including visual memory retrieval may be accounted for by backprojections from brain regions such as the amygdala in which the current mood, providing a context, is represented, to brain regions involved in memory such as the perirhinal cortex, and in visual representations such as the inferior temporal visual cortex (see Rolls and Stringer (2001b) and Section 12.3.6). The very well-known effects of context in the human memory literature could arise in the simple way just described. An implication of the explanation is that context effects will be especially important at late stages of memory or information processing systems in the brain, for there information from a wide range of modalities will be mixed, and some of that information could reflect the context in which the learning takes place. One part of the brain where such effects may be strong is the hippocampus, which is implicated in the memory of recent episodes, and which receives inputs derived from most of the cortical information processing streams, including those involved in space (see Section 12.2 and Chapter 6 of Rolls and Treves (1998)).

7.3.3.9   Mixture states

If an autoassociation memory is trained on pattern vectors $\mathbf{A}$, $\mathbf{B}$, and $\mathbf{A} + \mathbf{B}$ (i.e. $\mathbf{A}$ and $\mathbf{B}$ are both included in the joint vector $\mathbf{A} + \mathbf{B}$; that is if the vectors are not linearly independent), then the autoassociation memory will have difficulty in learning and recalling these three memories as separate, because completion from either $\mathbf{A}$ or $\mathbf{B}$ to $\mathbf{A} + \mathbf{B}$ tends to occur during recall. (This is referred to as configurational learning in the animal learning literature, see e.g. Sutherland and Rudy (1991).) This problem can be minimized by re-representing $\mathbf{A}$, $\mathbf{B}$, and $\mathbf{A} + \mathbf{B}$ in such a way that they are different vectors before they are presented to the autoassociation memory. This can be performed by recoding the input vectors to minimize overlap using, for example, a competitive network, and possibly involving expansion recoding, as described for pattern associators (see Section 7.2, Fig. 7.4). It is suggested that this is a function of the dentate granule cells in the hippocampus, which precede the CA3 recurrent collateral network (Treves and Rolls 1992, Treves and Rolls 1994).

7.3.3.10   Memory for sequences

One of the first extensions of the standard autoassociator paradigm that has been explored in the literature is the capability to store and retrieve not just individual patterns, but whole sequences of patterns. Hopfield, in the same 1982 paper, suggested that this could be achieved by adding to the standard connection weights, which associate a pattern with itself, a new, asymmetric component, that associates a pattern with the next one in the sequence. In practice this scheme does not work very well, unless the new component is made to operate on a slower time scale that the purely autoassociative component (Kleinfeld 1986, Sompolinsky and Kanter 1986). With two different time scales, the autoassociative component can stabilize a pattern for a while, before the heteroassociative component moves the network, as it were, into the next pattern. The heteroassociative retrieval cue for the next pattern in the sequence is just the previous pattern in the sequence. A particular type of 'slower' operation occurs if the asymmetric component acts after a delay τ. In this case, the network sweeps through the sequence, staying for a time of order τ in each pattern.

One can see how the necessary ingredient for the storage of sequences is only a minor departure from purely Hebbian learning: in fact, the (symmetric) autoassociative component of the weights can be taken to reflect the Hebbian learning of strictly simultaneous conjunctions of pre- and post-synaptic activity, whereas the (asymmetric) heteroassociative component can be implemented by Hebbian learning of each conjunction of postsynaptic activity with presynaptic activity shifted a time τ in the past. Both components can then be seen as resulting from a generalized Hebbian rule, which increases the weight whenever postsynaptic activity is paired with presynaptic activity occurring within a given time range, which may extend from a few hundred milliseconds in the past up to include strictly simultaneous activity. This is similar to a trace rule (see Chapter 8), which itself matches very well the observed conditions for induction of Long-Term Potentiation, and appears entirely plausible. The learning rule necessary for learning sequences, though, is more complex than a simple trace rule in that the time-shifted conjunctions of activity that are encoded in the weights must in retrieval produce activations that are time-shifted as well (otherwise one falls back into the Hopfield (1982) proposal, which does not quite work). The synaptic weights should therefore keep separate 'traces' of what was simultaneous and what was time-shifted during the original experience, and this is not very plausible. Levy and colleagues (Levy, Wu and Baxter 1995, Wu, Baxter and Levy 1996) have investigated these issues further, and the temporal asymmetry that may be present in LTP (see Section 1.5) has been suggested as a mechanism that might provide some of the temporal properties that are necessary for the brain to store and recall sequences (Minai and Levy 1993, Abbott and Blum 1996, Markram, Pikus, Gupta and Tsodyks 1998, Abbott and Nelson 2000). A problem with this suggestion is that, given that the temporal dynamics of attractor networks are inherently very fast when the networks have continuous dynamics (see Section 7.6), and that the temporal asymmetry in LTP may be in the order of only milliseconds to a few tens of milliseconds (see Section 1.5), the recall of the sequences would be very fast, perhaps 10–20 ms per step of the sequence, with every step of a 10–step sequence effectively retrieved and gone in a quick-fire session of 100–200 ms.

Another way in which a delay could be inserted in a recurrent collateral path in the brain is by inserting another cortical area in the recurrent path. This could fit in with the cortico-cortical backprojection connections described in Section 1.11, which would introduce some conduction delay (see Panzeri, Rolls, Battaglia and Lavis (2001)).

7.3.4   Use of autoassociation networks in the brain

Because of its 'one-shot' rapid learning, and ability to complete, this type of network is well suited for episodic memory storage, in which each episode must be stored and recalled later from a fragment, and kept separate from other episodic memories. It does not take a long time (the 'many epochs' of backpropagation networks) to train this network, because it does not have to 'discover the structure' of a problem. Instead, it stores information in the form in which it is presented to the memory, without altering the representation. An autoassociation network may be used for this function in the CA3 region of the hippocampus (see Section 12.2 and Chapter 6 of Rolls and Treves (1998)).

An autoassociation memory can also be used as a short term memory, in which iterative processing round the recurrent collateral loop keeps a representation active until another input cue is received. This may be used to implement many types of short term memory in the brain (see Section 12.1). For example, it may be used in the perirhinal cortex and adjacent temporal lobe cortex to implement short term visual object memory (Miyashita and Chang 1988, Amit 1995); in the dorsolateral prefrontal cortex to implement a short term memory for spatial responses (Goldman-Rakic 1996); and in the prefrontal cortex to implement a short term memory for where eye movements should be made in space. Such an autoassociation memory in the temporal lobe visual cortical areas may be used to implement the firing that continues for often 300 ms after a very brief (16 ms) presentation of a visual stimulus (Rolls and Tovee 1994), and may be one way in which a short memory trace is implemented to facilitate invariant learning about visual stimuli (see Chapter 8). In all these cases, the short term memory may be implemented by the recurrent excitatory collaterals that connect nearby pyramidal cells in the cerebral cortex. The connectivity in this system, that is the probability that a neuron synapses on a nearby neuron, may be in the region of 10% (Braitenberg and Schuz 1991, Abeles 1991).

The recurrent connections between nearby neocortical pyramidal cells may also be important in defining the response properties of cortical cells, which may be triggered by external inputs (from for example the thalamus or a preceding cortical area), but may be considerably dependent on the synaptic connections received from nearby cortical pyramidal cells.

The cortico-cortical backprojection connectivity described in Chapters 1 and 12 can be interpreted as a system that allows the forward-projecting neurons in one cortical area to be linked autoassociatively with the backprojecting neurons in the next cortical area (see Section

1.11 and Chapters 9, 10 and 12). This particular architecture may be especially important in constraint satisfaction (as well as recall), that is it may allow the networks in the two cortical areas to settle into a mutually consistent state. This would effectively enable information in higher cortical areas, which would include information from more divergent sources, to influence the response properties of neurons in earlier processing stages.

7.4 Competitive networks, including self-organizing maps

7.4.1 Function

Competitive neural networks learn to categorize input pattern vectors. Each category of inputs activates a different output neuron (or set of output neurons — see below). The categories formed are based on similarities between the input vectors. Similar, that is correlated, input vectors activate the same output neuron. In that the learning is based on similarities in the input space, and there is no external teacher that forces classification, this is an unsupervised network. The term categorization is used to refer to the process of placing vectors into categories based on their similarity. The term classification is used to refer to the process of placing outputs in particular classes as instructed or taught by a teacher. Examples of classifiers are pattern associators, one-layer delta-rule perceptrons, and multilayer perceptrons taught by error backpropagation (see Sections 7.2, 7.3, 7.10 and 7.11). In supervised networks there is usually a teacher for each output neuron.

The categorization produced by competitive nets is of great potential importance in perceptual systems including the whole of the visual cortical processing hierarchies, as described in Chapters 2–5 and modelled in Chapter 8. Each category formed reflects a set or cluster of active inputs x_j which occur together. This cluster of coactive inputs can be thought of as a feature, and the competitive network can be described as building feature analyzers, where a feature can now be defined as a correlated set of inputs. During learning, a competitive network gradually discovers these features in the input space, and the process of finding these features without a teacher is referred to as self-organization. Another important use of competitive networks is to remove redundancy from the input space, by allocating output neurons to reflect a set of inputs that co-occur. Another important aspect of competitive networks is that they separate patterns that are somewhat correlated in the input space, to produce outputs for the different patterns that are less correlated with each other, and may indeed easily be made orthogonal to each other. This has been referred to as orthogonalization. Another important function of competitive networks is that partly by removing redundancy from the input information space, they can produce sparse output vectors, without losing information. We may refer to this as sparsification.

7.4.2 Architecture and algorithm

7.4.2.1 Architecture

The basic architecture of a competitive network is shown in Fig. 7.7. It is a one-layer network with a set of inputs that make modifiable excitatory synapses w_{ij} with the output neurons. The output cells compete with each other (for example by mutual inhibition) in such a way that the most strongly activated neuron or neurons win the competition, and are left firing strongly. The synaptic weights, w_{ij}, are initialized to random values before learning starts. If

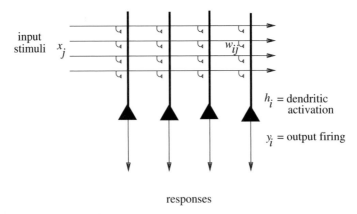

input
stimuli x_j

w_{ij}

h_i = dendritic
activation

y_i = output firing

responses

Fig. 7.7 The architecture of a competitive network.

some of the synapses are missing, that is if there is randomly diluted connectivity, that is not a problem for such networks, and can even help them (see below).

In the brain, the inputs arrive through axons, which make synapses with the dendrites of the output or principal cells of the network. The principal cells are typically pyramidal cells in the cerebral cortex. In the brain, the principal cells are typically excitatory, and mutual inhibition between them is implemented by inhibitory interneurons, which receive excitatory inputs from the principal cells. The inhibitory interneurons then send their axons to make synapses with the pyramidal cells, typically using GABA (gamma-aminobutyric acid) as the inhibitory transmitter.

7.4.2.2 Algorithm

1. Apply an input vector **x** and calculate the activation h_i of each neuron

$$h_i = \sum_j x_j w_{ij} \tag{7.16}$$

where the sum is over the C input axons, indexed by j. (It is useful to normalize the length of each input vector **x**. In the brain, a scaling effect is likely to be achieved both by feedforward inhibition, and by feedback inhibition among the set of input cells (in a preceding network) that give rise to the axons conveying **x**.)

The output firing y_i^1 is a function of the activation of the neuron

$$y_i^1 = f(h_i). \tag{7.17}$$

The function f can be linear, sigmoid, monotonically increasing, etc. (see Fig. 1.3).

2. Allow competitive interaction between the output neurons by a mechanism such as lateral or mutual inhibition (possibly with self-excitation), to produce a contrast-enhanced version of the firing rate vector

$$y_i = g(y_i^1). \tag{7.18}$$

Function g is typically a non-linear operation, and in its most extreme form may be a winner-take-all function, in which after the competition one neuron may be 'on', and the others 'off'. Algorithms that produce softer competition without a single winner are described in Section

7.4.8.4 below.

3. Apply an associative Hebb-like learning rule

$$\delta w_{ij} = \alpha y_i x_j. \tag{7.19}$$

4. Normalize the length of the synaptic weight vector on each dendrite to prevent the same few neurons always winning the competition:

$$\sum_j (w_{ij})^2 = 1. \tag{7.20}$$

(A less efficient alternative is to scale the sum of the weights to a constant, e.g. 1.0)

5. Repeat steps 1–4 for each different input stimulus $\mathbf{x}$, in random sequence, a number of times.

7.4.3 Properties

7.4.3.1 Feature discovery by self-organization

Each neuron in a competitive network becomes activated by a set of consistently coactive, that is correlated, input axons, and gradually learns to respond to that cluster of coactive inputs. We can think of competitive networks as discovering features in the input space, where features can now be defined by a set of consistently coactive inputs. Competitive networks thus show how feature analyzers can be built, with no external teacher. The feature analyzers respond to correlations in the input space, and the learning occurs by self-organization in the competitive network. Competitive networks are therefore well suited to the analysis of sensory inputs. Ways in which they may form fundamental building blocks of sensory systems are described in Chapter 8.

The operation of competitive networks can be visualized with the help of Fig. 7.8. The input patterns are represented as dots on the surface of a sphere. (The patterns are on the surface of a sphere because they are normalized to the same length.) The directions of the weight vectors of the three neurons are represented by '×'s. The effect of learning is to move the weight vector of each of the neurons to point towards the centre of one of the clusters of inputs. If the neurons are winner-take-all, the result of the learning is that although there are correlations between the input stimuli, the outputs of the three neurons are orthogonal. In this sense, orthogonalization is performed. At the same time, given that each of the patterns within a cluster produces the same output, the correlations between the patterns within a cluster become higher. In a winner-take-all network, the within-pattern correlation becomes 1, and the patterns within a cluster have been placed within the same category.

7.4.3.2 Removal of redundancy

In that competitive networks recode sets of correlated inputs to one or a few output neurons, then redundancy in the input representation is removed. Identifying and removing redundancy in sensory inputs is an important part of processing in sensory systems (cf. Barlow (1989)), in part because a compressed representation is more manageable as an output of sensory systems. The reason for this is that neurons in the receiving systems, for example pattern associators in the orbitofrontal cortex or autoassociation networks in the hippocampus, can

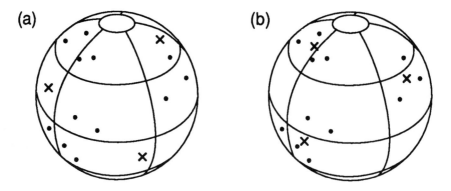

Fig. 7.8 Competitive learning. The dots represent the directions of the input vectors, and the '×'s the weights for each of three neurons. (a) Before learning. (b) After learning. (After Rumelhart and Zipser, 1986.)

then operate with the limited numbers of inputs that each neuron can receive. For example, although the information that a particular face is being viewed is present in the 10^6 fibres in the optic nerve, the information is unusable by associative networks in this form, and is compressed through the visual system until the information about which of many hundreds of faces is present can be represented by less than 100 neurons in the temporal cortical visual areas (Rolls, Treves and Tovee 1997b, Abbott, Rolls and Tovee 1996). (Redundancy can be defined as the difference between the maximum information content of the input data stream (or channel capacity) and its actual content; see Appendix B.)

The recoding of input pattern vectors into a more compressed representation that can be conveyed by a much reduced number of output neurons of a competitive network is referred to in engineering as vector quantization. With a winner-take-all competitive network, each output neuron points to or stands for one of or a cluster of the input vectors, and it is more efficient to transmit the states of the few output neurons than the states of all the input elements. (It is more efficient in the sense that the information transmission rate required, that is the capacity of the channel, can be much smaller.) Vector quantization is of course possible when the input representation contains redundancy.

7.4.3.3 Orthogonalization and categorization

Fig. 7.8 shows visually how competitive networks reduce the correlation between different clusters of patterns, by allocating them to different output neurons. This is described as orthogonalization. It is a process that is very usefully applied to signals before they are used as inputs to associative networks (pattern associators and autoassociators) trained with Hebbian rules (see Sections 7.2 and 7.3), because it reduces the interference between patterns stored in these memories. The opposite effect in competitive networks, of bringing closer together very similar input patterns, is referred to as categorization.

These two processes are also illustrated in Fig. 7.9, which shows that in a competitive network, very similar input patterns (with correlations higher in this case than approximately 0.8) produce more similar outputs (close to 1.0), whereas the correlations between pairs of

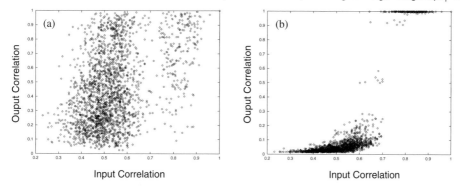

Fig. 7.9 Orthogonalization and categorization in a competitive network: (a) before learning; (b) after learning. The correlations between pairs of output vectors (abscissa) are plotted against the correlations of the corresponding pairs of input vectors that generated the output pair, for all possible pairs in the input set. The competitive net learned for 16 cycles. One cycle consisted of presenting the complete input set of stimuli in a renewing random sequence. The correlation measure shown is the cosine of the angle between two vectors (i.e. the normalized dot product). The network used had 64 input axons to each of 8 output neurons. The net was trained with 64 stimuli, made from 8 initial random binary vectors with each bit having a probability of 0.5 of being 1, from each of which 8 noisy exemplars were made by randomly altering 10% of the 64 elements. Soft competition was used between the output neurons. (A normalized exponential activation function described in Section 7.4.8.4 was used to implement the soft competition.) The sparseness a of the input patterns thus averaged 0.5; and the sparseness a of the output firing vector after learning was close to 0.17 (i.e. after learning, primarily one neuron was active for each input pattern. Before learning, the average sparseness of the output patterns produced by each of the inputs was 0.39).

input patterns that are smaller than approximately 0.7 become much smaller in the output representation. (This simulation used soft competition between neurons with graded firing rates.)

7.4.3.4 Sparsification

Competitive networks can produce more sparse representations than those that they receive, depending on the degree of competition. With the greatest competition, winner-take-all, only one output neuron remains active, and the representation is at its most sparse. This effect can be understood further using Figs. 7.8 and 7.9. This sparsification is useful to apply to representations before input patterns are applied to associative networks, because sparse representations allow many different pattern associations or memories to be stored in these networks (see Sections 7.2 and 7.3).

7.4.3.5 Capacity

In a competitive net with N output neurons and a simple winner-take-all rule for the competition, it is possible to learn up to N output categories, in that each output neuron may be allocated a category. When the competition acts in a less rudimentary way, the number of categories that can be learned becomes a complex function of various factors, including the number of modifiable connections per cell and the degree of dilution, or incompleteness, of the connections. Such a function has not yet been described analytically in general, but an upper bound on it can be deduced for the particular case in which the learning is fast, and can be achieved effectively in one shot, or one presentation of each pattern. In that case, the number of categories that can be learned (by the self-organizing process) will at most be equal to the number of associations that can be formed by the corresponding pattern associators, a process that occurs with the additional help of the driving inputs, which effectively determine the categorization in the pattern associator.

Separate constraints on the capacity result if the output vectors are required to be strictly orthogonal. Then, if the output firing rates can assume only positive values, the maximum number p of categories arises, obviously, in the case when only one output neuron is firing for any stimulus, so that up to N categories are formed. If ensemble encoding of output neurons is used (soft competition), again under the orthogonality requirement, then the number of output categories that can be learned will be reduced according to the degree of ensemble encoding. The p categories in the ensemble-encoded case reflect the fact that the between-cluster correlations in the output space are lower than those in the input space. The advantages of ensemble encoding are that dendrites are more evenly allocated to patterns (see Section 7.4.8.5), and that correlations between different input stimuli can be reflected in correlations between the corresponding output vectors, so that later networks in the system can generalize usefully. This latter property is of crucial importance, and is utilized for example when an input pattern is presented that has not been learned by the network. The relative similarity of the input pattern to previously learned patterns is indicated by the relative activation of the members of an ensemble of output neurons. This makes the number of different representations that can be reflected in the output of competitive networks with ensemble encoding much higher than with winner-take-all representations, even though with soft competition all these representations cannot strictly be learned.

7.4.3.6 Separation of non-linearly separable patterns

A competitive network can not only separate (e.g. by activating different output neurons) pattern vectors that overlap in almost all elements, but can also help with the separation of vectors that are not linearly separable. An example is that three patterns $\mathbf{A}$, $\mathbf{B}$, and $\mathbf{A} + \mathbf{B}$ will lead to three different output neurons being activated (see Fig. 7.10). For this to occur, the length of the synaptic weight vectors must be normalized (to for example unit length), so that they lie on the surface of a sphere or hypersphere (see Fig. 7.8). (If the weight vectors of each neuron are scaled to the same sum, then the weight vectors do not lie on the surface of a hypersphere, and the ability of the network to separate patterns is reduced.)

The property of pattern separation makes a competitive network placed before an autoassociator network very valuable, for it enables the autoassociator to store the three patterns separately, and to recall $\mathbf{A} + \mathbf{B}$ separately from $\mathbf{A}$ and $\mathbf{B}$. This is referred to as the configuration learning problem in animal learning theory (Sutherland and Rudy 1991). Placing a competitive network before a pattern associator will enable a linearly inseparable problem to be solved. For example, three different output neurons of a two-input competitive network could respond to the patterns 01, 10, and 11, and a pattern associator can learn different outputs for neurons 1–3, which are orthogonal to each other (see Fig. 7.4). This is an example of expansion recoding (cf. Marr (1969), who used a different algorithm to obtain the expansion). The sparsification that can be produced by the competitive network can also be advantageous in preparing patterns for presentation to a pattern associator or autoassociator, because the sparsification can increase the number of memories that can be associated or stored.

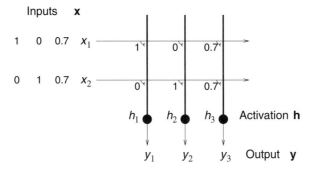

Fig. 7.10 Separation of linearly dependent patterns by a competitive network. The network was trained on patterns 10, 01 and 11, applied on the inputs x_1 and x_2. After learning, the network allocated output neuron 1 to pattern 10, neuron 2 to pattern 01, and neuron 3 to pattern 11. The weights in the network produced during the learning are shown. Each input pattern was normalized to unit length, and thus for pattern 11, x_1=0.7 and x_2=0.7, as shown. Because the weight vectors were also normalized to unit length, $w_{31} = 0.7$ and w_{32}=0.7.

7.4.3.7 Stability

These networks are generally stable if the input statistics are stable. If the input statistics keep varying, then the competitive network will keep following the input statistics. If this is a problem, then a critical period in which the input statistics are learned, followed by stabilization, may be useful. This appears to be a solution used in developing sensory systems, which have critical periods beyond which further changes become more difficult. An alternative approach taken by Carpenter and Grossberg in their 'Adaptive Resonance Theory' is to allow the network to learn only if it does not already have categorizers for a pattern (see Hertz, Krogh and Palmer (1991), p. 228).

Diluted connectivity can help stability, by making neurons tend to find inputs to categorize in only certain parts of the input space, and then making it difficult for the neuron to wander randomly throughout the space later.

7.4.3.8 Frequency of presentation

If some stimuli are presented more frequently than others, then there will be a tendency for the weight vectors to move more rapidly towards frequently presented stimuli, and more neurons may become allocated to the frequently presented stimuli. If winner-take-all competition is used, the result is that the neurons will tend to become allocated during the learning process to the more frequently presented patterns. If soft competition is used, the tendency of neurons to move from patterns that are infrequently or never presented can be reduced by making the competition fairly strong, so that only a few neurons show any learning when each pattern is presented. Provided that the competition is moderately strong (see Section 7.4.8.4), the result is that more neurons are allocated to frequently presented patterns, but one or some neurons are allocated to infrequently presented patterns. These points can all be easily demonstrated in simulations.

7.4.3.9 Comparison to principal component analysis (PCA) and cluster analysis

Although competitive networks find clusters of features in the input space, they do not perform hierarchical cluster analysis as typically performed in statistics. In hierarchical cluster analysis, input vectors are joined starting with the most correlated pair, and the level of the joining of vectors is indicated. Competitive nets produce different outputs (i.e. activate different output neurons) for each cluster of vectors (i.e. perform vector quantization), but do not compute the level in the hierarchy, unless the network is redesigned (see Hertz, Krogh and Palmer (1991)).

The feature discovery can also be compared to principal component analysis (PCA). (In PCA, the first principal component of a multidimensional space points in the direction of the vector that accounts for most of the variance, and subsequent principal components account for successively less of the variance, and are mutually orthogonal.) In competitive learning with a winner-take-all algorithm, the outputs are mutually orthogonal, but are not in an ordered series according to the amount of variance accounted for, unless the training algorithm is modified. The modification amounts to allowing each of the neurons in a winner-take-all network to learn one at a time, in sequence. The first neuron learns the first principal component. (Neurons trained with a modified Hebb rule learn to maximize the variance of their outputs — see Hertz, Krogh and Palmer (1991).) The second neuron is then allowed to learn, and because its output is orthogonal to the first, it learns the second principal component. This process is repeated. Details are given by Hertz, Krogh and Palmer (1991), but as this is not a biologically plausible process, it is not considered in detail here. We note that simple competitive learning is very helpful biologically, because it can separate patterns, but that a full ordered set of principal components as computed by PCA would probably not be very useful in biologically plausible networks. Our point here is that biological neuronal networks may operate well if the variance in the input representation is distributed across many input neurons, whereas principal component analysis would tend to result in most of the variance being allocated to a few neurons, and the variance being unevenly distributed across the neurons.

7.4.4 Utility of competitive networks in information processing by the brain

7.4.4.1 Feature analysis and preprocessing

Neurons that respond to correlated combinations of their inputs can be described as feature analyzers. Neurons that act as feature analyzers perform useful preprocessing in many sensory systems (see e.g. Chapter 8 of Rolls and Treves (1998)). The power of competitive networks in multistage hierarchical processing to build combinations of what is found at earlier stages, and thus effectively to build higher-order representations, is also described in Chapter 8 of this book.

7.4.4.2 Removal of redundancy

The removal of redundancy by competition is thought to be a key aspect of how sensory systems, including the ventral cortical visual system, operate. Competitive networks can also be thought of as performing dimension reduction, in that a set of correlated inputs may be responded to as one category or dimension by a competitive network. The concept of redundancy removal can be linked to the point that individual neurons trained with a modified Hebb rule point their weight vector in the direction of the vector that accounts for most of the variance in the input, that is (acting individually) they find the first principal component of the input space (see Section 7.4.3.9 and Hertz, Krogh and Palmer (1991)). Although networks with anti-Hebbian synapses between the principal cells (in which the anti-Hebbian learning forces neurons with initially correlated activity to effectively inhibit each other) (Földiák 1991), and networks that perform Independent Component Analysis (Bell and Sejnowski 1995), could in principle remove redundancy more effectively, it is not clear that they are implemented biologically. In contrast, competitive networks are more biologically plausible, and illustrate redundancy reduction. The more general use of an unsupervised competitive preprocessor is discussed below (see Fig. 7.16).

7.4.4.3 Orthogonalization

The orthogonalization performed by competitive networks is very useful for preparing signals for presentation to pattern associators and autoassociators, for this re-representation decreases interference between the patterns stored in such networks. Indeed, this can be essential if patterns are overlapping and not linearly independent, e.g. 01, 10, and 11. If three such binary patterns were presented to an autoassociative network, it would not form separate representations of them, because either of the patterns 01 or 10 would result by completion in recall of the 11 pattern. A competitive network allows a separate neuron to be allocated to each of the three patterns, and this set of orthogonal representations can be learned by associative networks (see Fig. 7.10).

7.4.4.4 Sparsification

The sparsification performed by competitive networks is very useful for preparing signals for presentation to pattern associators and autoassociators, for this re-representation increases the number of patterns that can be associated or stored in such networks (see Sections 7.2 and 7.3).

7.4.4.5 Brain systems in which competitive networks may be used for orthogonalization and sparsification

One system is the hippocampus, in which the dentate granule cells are believed to operate as a competitive network in order to prepare signals for presentation to the CA3 autoassociative network (see Chapter 6 of Rolls and Treves (1998)). In this case, the operation is enhanced by expansion recoding, in that (in the rat) there are approximately three times as many dentate granule cells as there are cells in the preceding stage, the entorhinal cortex. This expansion recoding will itself tend to reduce correlations between patterns (cf. Marr (1970), Marr (1969)).

Also in the hippocampus, the CA1 neurons are thought to act as a competitive network that recodes the separate representations of each of the parts of an episode that must be separately represented in CA3, into a form more suitable for the recall using pattern association performed by the backprojections from the hippocampus to the cerebral cortex (see Section 12.2 and Chapter 6 of Rolls and Treves (1998)).

The granule cells of the cerebellum may perform a similar function, but in this case the principle may be that each of the very large number of granule cells receives a very small random subset of inputs, so that the outputs of the granule cells are decorrelated with respect to the inputs (Marr (1969); see Chapter 9 of Rolls and Treves (1998)).

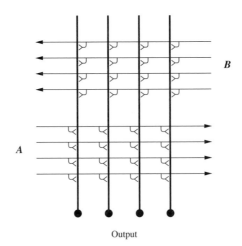

Output

Fig. 7.11 Competitive net receiving a normal set of inputs A, but also another set of inputs B that can be used to influence the categories formed in response to A inputs.

7.4.5 Guidance of competitive learning

Although competitive networks are primarily unsupervised networks, it is possible to influence the categories found by supplying a second input, as follows (Rolls 1989a). Consider a competitive network as shown in Fig. 7.11 with the normal set of inputs A to be categorized, and with an additional set of inputs B from a different source. Both sets of inputs work in the normal way for a competitive network, with random initial weights, competition between the output neurons, and a Hebb-like synaptic modification rule that normalizes the lengths of the synaptic weight vectors onto each neuron. The idea then is to use the B inputs to influence the categories formed by the A input vectors. The influence of the B vectors works best if they are orthogonal to each other. Consider any two A vectors. If they occur together with the same B vector, then the categories produced by the A vectors will be more similar than they would be without the influence of the B vectors. The categories will be pulled closer together if soft competition is used, or will be more likely to activate the same neuron if winner-take-all competition is used. Conversely, if any two A vectors are paired with two different, preferably orthogonal, B vectors, then the categories formed by the A vectors will be drawn further apart than they would be without the B vectors. The differences in categorization remain present after the learning when just the A inputs are used.

This guiding function of one of the inputs is one way in which the consequences of sensory stimuli could be fed back to a sensory system to influence the categories formed when the A inputs are presented. This could be one function of backprojections in the cerebral cortex (Rolls 1989c, Rolls 1989a). In this case, the A inputs of Fig. 7.11 would be the forward inputs from a preceding cortical area, and the B inputs backprojecting axons from the next cortical area, or from a structure such as the amygdala or hippocampus. If two A vectors were both associated with positive reinforcement that was fed back as the same B vector from another part of the brain, then the two A vectors would be brought closer together in the representational space provided by the output of the neurons. If one of the A vectors was associated with positive reinforcement, and the other with negative reinforcement, then the output representations of the two A vectors would be further apart. This is one way in which

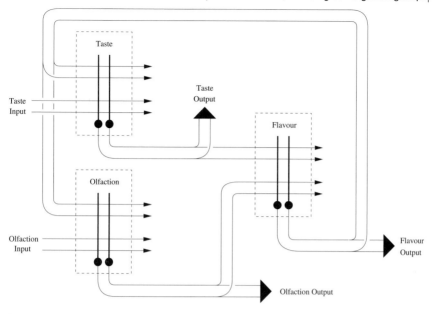

Fig. 7.12 A two-layer set of competitive nets in which feedback from layer 2 can influence the categories formed in layer 1. Layer 2 could be a higher cortical visual area with convergence from earlier cortical visual areas (see Chapter 8). In the example, taste and olfactory inputs are received by separate competitive nets in layer 1, and converge into a single competitive net in layer 2. The categories formed in layer 2 (which may be described as representing 'flavour') may be dominated by the relatively orthogonal set of a few tastes that are received by the net. When these layer 2 categories are fed back to layer 1, they may produce in layer 1 categories in, for example, the olfactory network that reflect to some extent the flavour categories of layer 2, and are different from the categories that would otherwise be formed to a large set of rather correlated olfactory inputs. A similar principle may operate in any multilayer hierarchical cortical processing system, such as the ventral visual system, in that the categories that can be formed only at later stages of processing may help earlier stages to form categories relevant to what can be identified at later stages.

external signals could influence in a mild way the categories formed in sensory systems. Another is that if any B vector only occurred for important sensory A inputs (as shown by the immediate consequences of receiving those sensory inputs), then the A inputs would simply be more likely to have any representation formed than otherwise, due to strong activation of neurons only when combined A and B inputs are present.

A similar architecture could be used to provide mild guidance for one sensory system (e.g. olfaction) by another (e.g. taste), as shown in Fig. 7.12. (Another example of where this architecture could be used is convergence in the visual system at the next cortical stage of processing, with guiding feedback to influence the categories formed in the different regions of the preceding cortical area, as illustrated in Section 1.11.) The idea is that the taste inputs would be more orthogonal to each other than the olfactory inputs, and that the taste inputs would influence the categories formed in the olfactory input categorizer in layer 1, by feedback from a convergent net in layer 2. The difference from the previous architecture is that we now have a two-layer net, with unimodal or separate networks in layer 1, each feeding forward to a single competitive network in layer 2. The categories formed in layer 2 reflect the co-occurrence of a particular taste with particular odours (which together form flavour in layer 2). Layer 2 then provides feedback connections to both the networks in layer 1. It can be shown in such a network that the categories formed in, for example, the olfactory net in layer 1 are influenced by the tastes with which the odours are paired. The feedback signal is built only in layer 2, after there has been convergence between the different modalities. This architecture captures some of the properties of sensory systems, in which there are unimodal processing cortical areas followed by multimodal cortical areas. The multimodal cortical areas can build representations that represent the unimodal inputs that tend to co-occur, and the higher level representations may in turn, by the highly developed cortico-cortical backprojections, be able to influence sensory categorization in earlier cortical processing areas (Rolls 1989a). Another such example might be the effect by which the phonemes heard are influenced by the visual inputs produced by seeing mouth movements (cf. McGurk and MacDonald (1976)). This could be implemented by auditory inputs coming together in the cortex in the superior temporal sulcus onto neurons activated by the sight of the lips moving (recorded during experiments of Baylis, Rolls and Leonard (1987), and Hasselmo, Rolls, Baylis and Nalwa (1989b)), using Hebbian learning with co-active inputs. Backprojections from such multimodal areas to the early auditory cortical areas could then influence the responses of auditory cortex neurons to auditory inputs (see Section 7.9 and cf. Calvert, Bullmore, Brammer, Campbell, Williams, McGuire, Woodruff, Iversen and David (1997)).

A similar principle may operate in any multilayer hierarchical cortical processing system, such as the ventral visual system, in that the categories that can be formed only at later stages of processing may help earlier stages to form categories relevant to what can be identified at the later stages as a result of the operation of backprojections (Rolls 1989a). The idea that the statistical correlation between the inputs received by neighbouring processing streams can be used to guide unsupervised learning within each stream has also been developed by Becker and Hinton (1992) and others (see Phillips, Kay and Smyth (1995)). The networks considered by these authors self-organize under the influence of collateral connections, such as may be implemented by cortico-cortical connections between parallel processing systems in the brain. They use learning rules that, although somewhat complex, are still local in nature, and tend to optimize specific objective functions. The locality of the learning rule, and the simulations performed so far, raise some hope that, once the operation of these types of networks is better understood, they might achieve similar computational capabilities to backpropagation networks (see Section 7.11) while retaining biological plausibility.

7.4.6 Topographic map formation

A simple modification to the competitive networks described so far enables them to develop topological maps. In such maps, the closeness in the map reflects the similarity (correlation) between the features in the inputs. The modification that allows such maps to self-organize is to add short-range excitation and long-range inhibition between the neurons. The function to be implemented has a spatial profile that is described as having a Mexican hat shape (see Fig. 7.13). The effect of this connectivity between neurons, which need not be modifiable, is to encourage neurons that are close together to respond to similar features in the input space, and to encourage neurons that are far apart to respond to different features in the input space. When these response tendencies are present during learning, the feature analyzers that are built by modifying the synapses from the input onto the activated neurons tend to be similar if they are close together, and different if far apart. This is illustrated in Figs. 7.14

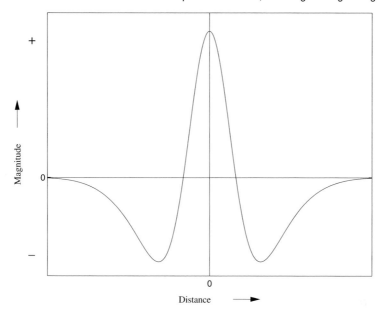

Fig. 7.13 Mexican hat lateral spatial interaction profile.

and 7.15. Feature maps built in this way were described by von der Malsburg (1973) and Willshaw and von der Malsburg (1976). It should be noted that the learning rule needed is simply the modified Hebb rule described above for competitive networks, and is thus local and biologically plausible. (For computational convenience, the algorithm that Kohonen (Kohonen 1982, Kohonen 1989, Kohonen 1995) has mainly used does not use Mexican hat connectivity between the neurons, but instead arranges that when the weights to a winning neuron are updated, so to a smaller extent are those of its neighbours – see further Hertz, Krogh and Palmer (1991)).

A very common characteristic of connectivity in the brain, found for example throughout the neocortex, consists of short-range excitatory connections between neurons, with inhibition mediated via inhibitory interneurons. The density of the excitatory connectivity even falls gradually as a function of distance from a neuron, extending typically a distance in the order of 1 mm from the neuron (Braitenberg and Schuz 1991), contributing to a spatial function quite like that of a Mexican hat. (Longer-range inhibitory influences would form the negative part of the spatial response profile.) This supports the idea that topological maps, though in some cases probably seeded by chemoaffinity, could develop in the brain with the assistance of the processes just described. It is noted that some cortico-cortical connections even within an area may be longer, skipping past some intermediate neurons, and then making connections after some distance with a group of neurons. Such longer-range connections are found for example between different columns with similar orientation selectivity in the primary visual cortex. The longer range connections may play a part in stabilizing maps, and again in the exchange of information between neurons performing related computations, in this case about features with the same orientations.

If a low-dimensional space, for example the orientation sensitivity of cortical neurons in the primary visual cortex (which is essentially one-dimensional, the dimension being angle),

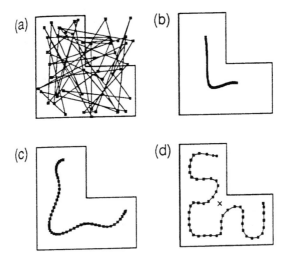

Fig. 7.14 Kohonen feature mapping from a two-dimensional L-shaped region to a linear array of 50 units. Each unit has 2 inputs. The input patterns are the X,Y coordinates of points within the L-shape shown. In the diagrams, each point shows the position of a weight vector. Lines connect adjacent units in the 1-D (linear) array of 50 neurons. The weights were initialized to random values within the unit square (a). During feature mapping training, the weights evolved through stages (b) and (c) to (d). By stage (d) the weights have formed so that the positions in the original input space are mapped to a 1-D vector in which adjacent points in the input space activate neighbouring units in the linear array of output units. (Reproduced with permission from Hertz, Krogh and Palmer, 1991, Fig. 9.13.)

is mapped to a two-dimensional space such as the surface of the cortex, then the resulting map can have long spatial runs where the value along the dimension (in this case orientation tuning) alters gradually, and continuously. Such self-organization can account for many aspects of the mapping of orientation tuning, and of ocular dominance columns, in V1 (Miller 1994, Harris, Ermentrout and Small 1997). If a high-dimensional information space is mapped to the two-dimensional cortex, then there will be only short runs of groups of neurons with similar feature responsiveness, and then the map must fracture, with a different type of feature mapped for a short distance after the discontinuity. This is exactly what Rolls suggests is the type of topology found in the inferior temporal visual cortex, with the individual groupings representing what can be self-organized by competitive networks combined with a trace rule as described in Section 8.4. Here, visual stimuli are not represented with reference to their position on the retina, because here the neurons are relatively translation invariant. Instead, when recording here, small clumps of neurons with similar responses may be encountered close together, and then one moves into a group of neurons with quite different feature selectivity. This topology will arise naturally, given the anatomical connectivity of the cortex with its short-range excitatory connections, because there are very many different objects in the world and different types of features that describe objects, with no special continuity between the different combinations of features possible. Rolls' hypothesis contrasts with the view of Tanaka (1996), who has claimed that the inferior temporal cortex provides an alphabet of visual features arranged in discrete modules. The type of mapping found in higher cortical

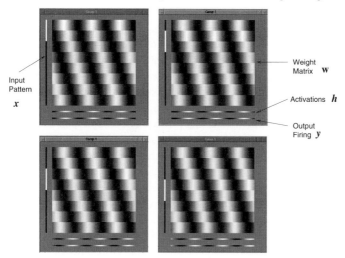

Fig. 7.15 Example of a one-dimensional topological map that self-organized from inputs in a low-dimensional space. The network has 64 neurons (vertical elements in the diagram) and 64 inputs per neuron (horizontal elements in the diagram). The four different diagrams represent the net tested with different input patterns. The input patterns x are displayed at the left of each diagram, with white representing firing and black not firing for each of the 64 inputs. The central square of each diagram represents the synaptic weights of the neurons, with white representing a strong weight. The row vector below each weight matrix represents the activations of the 64 output neurons, and the bottom row vector the output firing y. The network was trained with a set of 8 binary input patterns, each of which overlapped in 8 of its 16 'on' elements with the next pattern. The diagram shows that as one moves through correlations in the input space (top left to top right to bottom left to bottom right), so the output neurons activated move steadily across the output array of neurons. Closely correlated inputs are represented close together in the output array of neurons. The way in which this occurs can be seen by inspection of the weight matrix. The network architecture was the same as for a competitive net, except that the activations were converted linearly into output firings, and then each neuron excited its neighbours and inhibited neurons further away. This lateral inhibition was implemented for the simulation by a spatial filter operating on the output firings with the following filter weights (cf. Fig. 7.13): 5, 5, 5, 5, 5, 5, 5, 5, 5, 10, 10, 10, 10, 10, 10, 10, 10, 10, 5, 5, 5, 5, 5, 5, 5, 5 which operated on the 64-element firing rate vector.

visual areas as proposed by Rolls, implies that topological self-organization is an important way in which maps in the brain are formed, for it seems most unlikely that the locations in the map of different types of object seen in an environment could be specified genetically (Rolls and Stringer 2000).

The biological utility of developing such topology-preserving feature maps may be that if the computation requires neurons with similar types of response to exchange information more than neurons involved in different computations (which is more than reasonable), then the total length of the connections between the neurons is minimized if the neurons that need to exchange information are close together (cf. Cowey (1979), Durbin and Mitchison (1990)). Examples of this include the separation of colour constancy processing in V4 from global motion processing in MT as follows (see also Chapter 3). In V4, to compute colour constancy, an estimate of the illuminating wavelengths can be obtained by summing the outputs of the pyramidal cells in the inhibitory interneurons over several degrees of visual space, and subtracting this from the excitatory central ON colour-tuned region of the receptive field by (subtractive) feedback inhibition. This enables the cells to discount the illuminating wavelength, and thus compute colour constancy. For this computation, no inputs from motion-selective cells (which in the dorsal stream are colour insensitive) are needed. In MT, to compute global motion (e.g. the motion produced by the average flow of local motion elements, exemplified for example by falling snow), the computation can be performed by averaging in the larger (several degrees) receptive fields of MT the local motion inputs received by neurons in earlier cortical areas (V1 and V2) with small receptive fields (see Chapter 3). For this computation, no input from colour cells is useful. Having separate areas (V4 and MT) because these different computations minimizes the wiring lengths, for having intermingled colour and motion cells in a single cortical area would increase the average connection length between the neurons that need to be connected for the computations being performed. Minimizing the total connection length between neurons in the brain is very important in order to keep the size of the brain relatively small.

Placing close to each other neurons that need to exchange information, or that need to receive information from the same source, or which need to project towards the same destination, may also help to minimize the complexity of the rules required to specify cortical (and indeed brain) connectivity (Rolls and Stringer 2000). For example, in the case of V4 and MT, the connectivity rules can be simpler (e.g. connect to neurons in the vicinity, rather than look for colour-, or motion-marked cells, and connect only to the cells with the correct genetically specified label specifying that the cell is either part of motion or of colour processing). Further, the V4 and MT example shows that how the neurons are connected can be specified quite simply, but of course it needs to be specified to be different for different computations. Specifying a general rule for the classes of neurons in a given area also provides a useful simplification to the genetic rules needed to specify the functional architecture of a given cortical area (Rolls and Stringer 2000). In our V4 and MT example, the genetic rules would need to specify the rules separately for different populations of inhibitory interneurons if the computations performed by V4 and MT were performed with intermixed neurons in a single brain area. Together, these two principles, of minimization of wiring length, and allowing simple genetic specification of wiring rules, may underlie the separation of cortical visual information processing into different (e.g. ventral and dorsal) processing streams. The same two principles operating within each brain processing stream may underlie (taken together with the need for hierarchical processing to enable the computations to be biologically plausible in terms of the number of connections per neuron, and the need for local learning rules, see Section 7.12) much of the overall architecture of visual cortical processing, and of information processing and its modular architecture throughout the cortex more generally.

The rules of information exchange just described could also tend to produce more gross topography in cortical regions. For example, neurons that respond to animate objects may have certain visual feature requirements in common, and may need to exchange information about these features. Other neurons that respond to inanimate objects might have somewhat different visual feature requirements for their inputs, and might need to exchange information strongly. (For example, selection of whether an object is a chisel or a screwdriver may require competition by mutual (lateral) inhibition to produce the contrast enhancement necessary to result in unambiguous neuronal responses.) The rules just described would account for neurons with responsiveness to inanimate and animate objects tending to be grouped in separate parts of a cortical map or representation, and thus separately susceptible to brain damage (see e.g. Farah (1990), Farah (2000)).

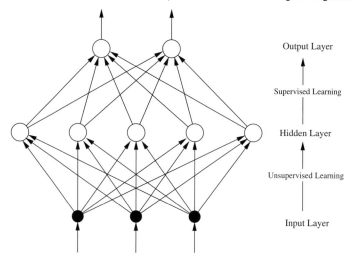

Output Layer

Supervised Learning

Hidden Layer

Unsupervised Learning

Input Layer

Fig. 7.16 A hybrid network, in which for example unsupervised learning rapidly builds relatively orthogonal representations based on input differences, and this is followed by a one-layer supervised network (taught for example by the delta rule) that learns to classify the inputs based on the categorizations formed in the hidden/intermediate layer.

7.4.7 Radial Basis Function networks

As noted above, a competitive network can act as a useful preprocessor for other networks. In the neural examples above, competitive networks were useful preprocessors for associative networks. Competitive networks are also used as preprocessors in artificial neural networks, for example in hybrid two-layer networks such as that illustrated in Fig. 7.16. The competitive network is advantageous in this hybrid scheme, because as an unsupervised network, it can relatively quickly (with a few presentations of each stimulus) discover the main features in the input space, and code for them. This leaves the second layer of the network to act as a supervised network (taught for example by the delta rule, see Section 7.10), which learns to map the features found by the first layer into the output required. This learning scheme is very much faster than that of a (two-layer) backpropagation network, which learns very slowly because it takes it a long time to perform the credit assignment to build useful feature analyzers in layer one (the hidden layer) (see Section 7.11).

The general scheme shown in Fig. 7.16 is used in radial basis function (RBF) neural networks. The main difference from what has been described is that in an RBF network, the hidden neurons do not use a winner-take-all function (as in some competitive networks), but instead use a normalized activation function in which the measure of distance from a weight vector of the neural input is (instead of the dot product $\mathbf{x} \cdot \mathbf{w}_i$ used for most of the networks described in this book), a Gaussian measure of distance:

$$y_i = \frac{\exp[-(\mathbf{x} - \mathbf{w}_i)^2/2\sigma_i^2]}{\sum_k \exp[-(\mathbf{x} - \mathbf{w}_k)^2/2\sigma_k^2]}. \tag{7.21}$$

The effect is that the response y_i of neuron i is a maximum if the input stimulus vector $\mathbf{x}$ is centred at $\mathbf{w}_i$, the weight vector of neuron i (this is the upper term in equation 7.21). The magnitude is normalized by dividing by the sum of the activations of all the k neurons in the network. If the input vector $\mathbf{x}$ is not at the centre of the receptive field of the neuron, then the response is decreased according to how far the input vector is from the weight vector $\mathbf{w}_i$ of the neuron, with the weighting decreasing as a Gaussian function with a standard deviation of σ. The idea is like that implemented with soft competition, in that the relative response of different neurons provides an indication of where the input pattern is in relation to the weight vectors of the different neurons. The rapidity with which the response falls off in a Gaussian radial basis function neuron is set by σ_i, which is adjusted so that for any given input pattern vector, a number of RBF neurons are activated. The positions in which the RBF neurons are located (i.e. the directions of their weight vectors, $\mathbf{w}$) are determined usually by unsupervised learning, e.g. the vector quantization that is produced by the normal competitive learning algorithm. The first layer of a RBF network is not different in principle from a network with soft competition, and it is not clear how biologically a Gaussian activation function would be implemented, so the treatment is not developed further here (see Hertz, Krogh and Palmer (1991), Poggio and Girosi (1990a), and Poggio and Girosi (1990b) for further details).

7.4.8 Further details of the algorithms used in competitive networks

7.4.8.1 Normalization of the inputs

Normalization is useful because in step 1 of the training algorithm described in Section 7.4.2.2, the neuronal activations, formed by the inner product of the pattern and the normalized weight vector on each neuron, are scaled in such a way that they have a maximum value of 1.0. This helps different input patterns to be equally effective in the learning process. A way in which this normalization could be achieved by a layer of input neurons is given by Grossberg (1976a). In the brain, a number of factors may contribute to normalization of the inputs. One factor is that a set of input axons to a neuron will come from another network in which the firing is controlled by inhibitory feedback, and if the numbers of axons involved is large (hundreds or thousands), then the inputs will be in a reasonable range. Second, there is increasing evidence that the different classes of input to a neuron may activate different types of inhibitory interneuron (e.g. Buhl, Halasy and Somogyi (1994)), which terminate on separate parts of the dendrite, usually close to the site of termination of the corresponding excitatory afferents. This may allow separate feedforward inhibition for the different classes of input. In addition, the feedback inhibitory interneurons also have characteristic termination sites, often on or close to the cell body, where they may be particularly effective in controlling firing of the neuron by shunting (divisive) inhibition, rather than by scaling a class of input (see Section 7.6).

7.4.8.2 Normalization of the length of the synaptic weight vector on each dendrite

This is necessary to ensure that one or a few neurons do not always win the competition. (If the weights on one neuron were increased by simple Hebbian learning, and there was no normalization of the weights on the neuron, then it would tend to respond strongly in the future to patterns with some overlap with patterns to which that neuron has previously learned, and gradually that neuron would capture a large number of patterns.) A biologically plausible way to achieve this weight adjustment is to use a modified Hebb rule:

$$\delta w_{ij} = \alpha y_i (x_j - w_{ij}) \tag{7.22}$$

where α is a constant, and x_j and w_{ij} are in appropriate units. In vector notation,

$$\delta\mathbf{w}_i = \alpha y_i(\mathbf{x} - \mathbf{w}_i) \tag{7.23}$$

where $\mathbf{w}_i$ is the synaptic weight vector on neuron i. This implements a Hebb rule that increases synaptic strength according to conjunctive pre- and post-synaptic activity, and also allows the strength of each synapse to decrease in proportion to the firing rate of the postsynaptic neuron (as well as in proportion to the existing synaptic strength). This results in a decrease in synaptic strength for synapses from weakly active presynaptic neurons onto strongly active postsynaptic neurons. Such a modification in synaptic strength is termed heterosynaptic long-term depression in the neurophysiological literature, referring to the fact that the synapses that weaken are other than those that activate the neuron. This is an important computational use of heterosynaptic long-term depression (LTD). In that the amount of decrease of the synaptic strength depends on how strong the synapse is already, the rule is compatible with what is frequently reported in studies of LTD (see Section 1.5). This rule can maintain the sums of the synaptic weights on each dendrite to be very similar without any need for explicit normalization of the synaptic strengths, and is useful in competitive nets. This rule was used by Willshaw and von der Malsburg (1976). As is made clear with the vector notation above, the modified Hebb rule moves the direction of the weight vector $\mathbf{w}_i$ towards the current input pattern vector $\mathbf{x}$ in proportion to the difference between these two vectors and the firing rate y_i of neuron i.

If explicit weight (vector length) normalization is needed, the appropriate form of the modified Hebb rule is:

$$\delta w_{ij} = \alpha y_i(x_j - y_i w_{ij}). \tag{7.24}$$

This rule, formulated by Oja (1982), makes weight decay proportional to y_i^2, normalizes the synaptic weight vector (see Hertz, Krogh and Palmer (1991)), is still a local learning rule, and is known as the Oja rule.

7.4.8.3 Non-linearity in the learning rule

Non-linearity in the learning rule can assist competition (Rolls 1989b, Rolls 1996b). For example, in the brain, long-term potentiation typically occurs only when strong activation of a neuron has produced sufficient depolarization for the voltage-dependent NMDA receptors to become unblocked, allowing Ca^{2+} to enter the cell (see Section 1.5). This means that synaptic modification occurs only on neurons that are strongly activated, effectively assisting competition to select few winners. The learning rule can be written:

$$\delta w_{ij} = \alpha m_i x_j \tag{7.25}$$

where m_i is a (e.g. threshold) non-linear function of the post-synaptic firing y_i which mimics the operation of the NMDA receptors in learning. (It is noted that in associative networks the same process may result in the stored pattern being more sparse than the input pattern, and that this may be beneficial, especially given the exponential firing rate distribution of neurons, in helping to maximize the number of patterns stored in associative networks (see Sections 7.2, 7.3, and 5.5.1).

7.4.8.4 Competition

In a simulation of a competitive network, a single winner can be selected by searching for the neuron with the maximum activation. If graded competition is required, this can be achieved by an activation function that increases greater than linearly. In some of the networks we have simulated (Rolls 1989b, Rolls 1989a, Wallis and Rolls 1997), raising the activation to a fixed power, typically in the range 2–5, and then rescaling the outputs to a fixed maximum (e.g. 1) is simple to implement. In a real neuronal network, winner-take-all competition can be implemented using mutual (lateral) inhibition between the neurons with non-linear activation functions, and self-excitation of each neuron (see e.g. Grossberg (1976a), Grossberg (1988), Hertz, Krogh and Palmer (1991)).

Another method to implement soft competition in simulations is to use the normalized exponential or 'softmax' activation function for the neurons (Bridle (1990); see Bishop (1995)):

$$y = \exp(h)/\sum_i \exp(h_i) . \tag{7.26}$$

This function specifies that the firing rate of each neuron is an exponential function of the activation, scaled by the whole vector of activations h_i, $i = 1, N$. The exponential function (in increasing supralinearly) implements soft competition, in that after the competition the faster firing neurons are firing relatively much faster than the slower firing neurons. In fact, the strength of the competition can be adjusted by using a 'temperature' T greater than 0 as follows:

$$y = \exp(h/T)/\sum_i \exp(h_i/T). \tag{7.27}$$

Very low temperatures increase the competition, until with $T \to 0$, the competition becomes 'winner-take-all'. At high temperatures, the competition becomes very soft. (When using the function in simulations, it may be advisable to prescale the firing rates to for example the range 0–1, both to prevent machine overflow, and to set the temperature to operate on a constant range of firing rates, as increasing the range of the inputs has an effect similar to decreasing T.)

The softmax function has the property that activations in the range $-\infty$ to $+\infty$ are mapped into the range 0 to 1.0, and the sum of the firing rates is 1.0. This facilitates interpretation of the firing rates under certain conditions as probabilities, for example that the competitive network firing rate of each neuron reflects the probability that the input vector is within the category or cluster signified by that output neuron (see Bishop (1995)).

7.4.8.5 Soft competition

The use of graded (continuous valued) output neurons in a competitive network, and soft competition rather than winner-take-all competition, has the value that the competitive net generalizes more continuously to an input vector that lies between input vectors that it has learned. Also, with soft competition, neurons with only a small amount of activation by any of the patterns being used will nevertheless learn a little, and move gradually towards the patterns that are being presented. The result is that with soft competition, the output neurons all tend to become allocated to one of the input patterns or one of the clusters of input patterns.

7.4.8.6 Untrained neurons

In competitive networks, especially with winner-take-all or finely tuned neurons, it is possible that some neurons remain unallocated to patterns. This may be useful, in case patterns in the unused part of the space occur in future. Alternatively, unallocated neurons can be made to move towards the parts of the space where patterns are occurring by allowing such losers in the competition to learn a little. Another mechanism is to subtract a bias term μ_i from y_i, and to use a 'conscience' mechanism that raises μ_i if a neuron wins frequently, and lowers μ_i if it wins infrequently (Grossberg 1976b, Bienenstock and Munro 1982, De Sieno 1988).

7.4.8.7 Large competitive nets: further aspects

If a large neuronal network is considered, with the number of synapses on each neuron in the region of 10,000, as occurs on large pyramidal cells in some parts of the brain, then there is a potential disadvantage in using neurons with synaptic weights that can take on only positive values. This difficulty arises in the following way. Consider a set of positive normalized input firing rates and synaptic weight vectors (in which each element of the vector can take on any value between 0.0 and 1.0). Such vectors of random values will on average be more highly aligned with the direction of the central vector (1, 1, 1, ..., 1) than with any other vector. An example can be given for the particular case of vectors evenly distributed on the positive 'quadrant' of a high-dimensional hypersphere: the average overlap (i.e. normalized dot product) between two binary random vectors with half the elements on and thus a sparseness of 0.5 (e.g. a random pattern vector and a random dendritic weight vector) will be approximately 0.5, while the average overlap between a random vector and the central vector will be approximately 0.707. A consequence of this will be that if a neuron begins to learn towards several input pattern vectors it will get drawn towards the average of these input patterns which will be closer to the 1,1,1,...,1 direction than to any one of the patterns. As a dendritic weight vector moves towards the central vector, it will become more closely aligned with more and more input patterns so that it is more rapidly drawn towards the central vector. The end result is that in large nets of this type, many of the dendritic weight vectors will point towards the central vector. This effect is not seen so much in small systems, since the fluctuations in the magnitude of the overlaps are sufficiently large that in most cases a dendritic weight vector will have an input pattern very close to it and thus will not learn towards the centre. In large systems, the fluctuations in the overlaps between random vectors become smaller by a factor of $\frac{1}{\sqrt{N}}$ so that the dendrites will not be particularly close to any of the input patterns.

One solution to this problem is to allow the elements of the synaptic weight vectors to take negative as well as positive values. This could be implemented in the brain by feedforward inhibition. A set of vectors taken with random values will then have a reduced mean correlation between any pair, and the competitive net will be able to categorize them effectively. A system with synaptic weights that can be negative as well as positive is not physiologically plausible, but we can instead imagine a system with weights lying on a hypersphere in the positive quadrant of space but with additional inhibition that results in the cumulative effects of some input lines being effectively negative. This can be achieved in a network by using positive input vectors, positive synaptic weight vectors, and thresholding the output neurons at their mean activation. A large competitive network of this general nature does categorize well, and has been described more fully elsewhere (Bennett 1990). In a large network with inhibitory feedback neurons, and principal cells with thresholds, the network could achieve at least in part an approximation to this type of thresholding useful in large competitive networks.

A second way in which nets with positive-only values of the elements could operate is by making the input vectors sparse and initializing the weight vectors to be sparse, or to have a reduced contact probability. (A measure a of sparseness is defined (as before) in equation 7.28:

$$a = \frac{(\sum\limits_{i=1}^{N} y_i/N)^2}{\sum\limits_{i=1}^{N} y_i^2/N} \tag{7.28}$$

where y_i is the firing rate of the ith neuron in the set of N neurons.) For relatively small net sizes simulated ($N = 100$) with patterns with a sparseness a of, for example, 0.1 or 0.2, learning onto the average vector can be avoided. However, as the net size increases, the sparseness required does become very low. In large nets, a greatly reduced contact probability between neurons (many synapses kept identically zero) would prevent learning of the average vector, thus allowing categorization to occur. Reduced contact probability will, however, prevent complete alignment of synapses with patterns, so that the performance of the network will be affected.

7.5 Continuous attractor networks

7.5.1 Introduction

Single-cell recording studies have shown that some neurons represent the current position along a continuous physical dimension or space even when no inputs are available, for example in darkness (see Section 12.2). Examples include neurons that represent the positions of the eyes (i.e. eye direction with respect to the head), the place where the animal is looking in space, head direction, and the place where the animal is located. In particular, examples of such classes of cells include head direction cells in rats (Ranck 1985, Taube, Muller and Ranck 1990, Taube, Goodridge, Golob, Dudchenko and Stackman 1996, Muller, Ranck and Taube 1996) and primates (Robertson, Rolls, Georges-François and Panzeri 1999), which respond maximally when the animal's head is facing in a particular preferred direction; place cells in rats (O'Keefe and Dostrovsky 1971, McNaughton, Barnes and O'Keefe 1983, O'Keefe 1984, Muller, Kubie, Bostock, Taube and Quirk 1991, Markus, Qin, Leonard, Skaggs, McNaughton and Barnes 1995) that fire maximally when the animal is in a particular location; and spatial view cells in primates that respond when the monkey is looking towards a particular location in space (Rolls, Robertson and Georges-François 1997a, Georges-Francois, Rolls and Robertson 1999, Robertson, Rolls and Georges-François 1998). In the parietal cortex there are many spatial representations, in several different coordinate frames (see Chapter 4 and Andersen, Batista, Snyder, Buneo and Cohen (2000)), and they have some capability to remain active during memory periods when the stimulus is no longer present. Even more than this, the dorsolateral prefrontal cortex networks to which the parietal networks project have the capability to maintain spatial representations active for many seconds or minutes during short term memory tasks, when the stimulus is no longer present (see Section 12.1). In this section, we describe how such networks representing continuous physical space could operate. The locations of such spatial networks in the brain are the parietal areas (see Chapters 6 and 9–11), the prefrontal areas that implement short term spatial memory and receive from the parietal cortex (see Section 12.1), and the hippocampal system which combines information about objects from the inferior temporal visual cortex with spatial information (see Section 12.2).

A class of network that can maintain the firing of its neurons to represent any location along a continuous physical dimension such as spatial position, head direction, etc is a 'Continuous Attractor' neural network (CANN). It uses excitatory recurrent collateral connections between

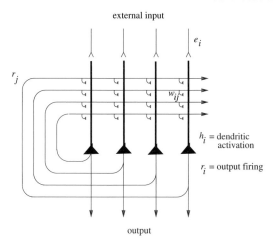

external input

e_i

r_j

w_{ij}

h_i = dendritic activation

r_i = output firing

output

Fig. 7.17 The architecture of a continuous attractor neural network (CANN).

the neurons to reflect the distance between the neurons in the state space of the animal (e.g. head direction space). These networks can maintain the bubble of neural activity constant for long periods wherever it is started to represent the current state (head direction, position, etc) of the animal, and are likely to be involved in many aspects of spatial processing and memory, including spatial vision. Global inhibition is used to keep the number of neurons in a bubble or packet of actively firing neurons relatively constant, and to help to ensure that there is only one activity packet. Continuous attractor networks can be thought of as very similar to autoassociation or discrete attractor networks (described in Section 7.3), and have the same architecture, as illustrated in Fig. 7.17. The main difference is that the patterns stored in a CANN are continuous patterns, with each neuron having broadly tuned firing which decreases with for example a Gaussian function as the distance from the optimal firing location of the cell is varied, and with different neurons having tuning that overlaps throughout the space. Such tuning is illustrated in Fig. 7.18. For comparison, the autoassociation networks described in Section 7.3 have discrete (separate) patterns (each pattern implemented by the firing of a particular subset of the neurons), with no continuous distribution of the patterns throughout the space (see Fig. 7.18). A consequent difference is that the CANN can maintain its firing at any location in the trained continuous space, whereas a discrete attractor or autoassociation network moves its population of active neurons towards one of the previously learned attractor states, and thus implements the recall of a particular previously learned pattern from an incomplete or noisy (distorted) version of one of the previously learned patterns. The energy landscape of a discrete attractor network (see equation 7.12) has separate energy minima, each one of which corresponds to a learned pattern, whereas the energy landscape of a continuous attractor network is flat, so that the activity packet remains stable with continuous firing wherever it is started in the state space. (The state space refers to set of possible spatial states of the animal in its environment, e.g. the set of possible head directions.)

In Section 7.5.2, we first describe the operation and properties of continuous attractor networks, which have been studied by for example Amari (1977), Zhang (1996), and Taylor (1999), and then, following Stringer, Trappenberg, Rolls and de Araujo (2002c), address four key issues about the biological application of continuous attractor network models.

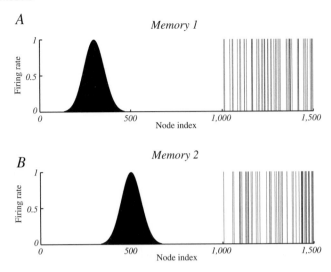

Fig. 7.18 The types of firing patterns stored in continuous attractor networks are illustrated for the patterns present on neurons 1–1000 for Memory 1 (when the firing is that produced when the spatial state represented is that for location 300), and for Memory 2 (when the firing is that produced when the spatial state represented is that for location 500). The continuous nature of the spatial representation results from the fact that each neuron has a Gaussian firing rate that peaks at its optimal location. This particular mixed network also contains discrete representations that consist of discrete subsets of active binary firing rate neurons in the range 1001–1500. The firing of these latter neurons can be thought of as representing the discrete events that occur at the location. Continuous attractor networks by definition contain only continuous representations, but this particular network can store mixed continuous and discrete representations, and is illustrated to show the difference of the firing patterns normally stored in separate continuous attractor and discrete attractor networks. For this particular mixed network, during learning, Memory 1 is stored in the synaptic weights, then Memory 2, etc, and each memory contains part that is continuously distributed to represent physical space, and part that represents a discrete event or object.

One key issue in such Continuous Attractor neural networks is how the synaptic strengths between the neurons in the continuous attractor network could be learned in biological systems (Section 7.5.3).

A second key issue in such Continuous Attractor neural networks is how the bubble of neuronal firing representing one location in the continuous state space should be updated based on non-visual cues to represent a new location in state space (Section 7.5.5). This is essentially the problem of path integration: how a system that represents a memory of where the agent is in physical space could be updated based on idiothetic (self-motion) cues such as vestibular cues (which might represent a head velocity signal), or proprioceptive cues (which might update a representation of place based on movements being made in the space, during for example walking in the dark).

A third key issue is how stability in the bubble of activity representing the current location can be maintained without much drift in darkness, when it is operating as a memory system (Section 7.5.6).

A fourth key issue is considered in Section 7.5.8 in which we describe networks that store both continuous patterns and discrete patterns (see Fig. 7.18), which can be used to store for example the location in (continuous, physical) space where an object (a discrete item) is present.

7.5.2 The generic model of a continuous attractor network

The generic model of a continuous attractor is as follows. (The model is described in the context of head direction cells, which represent the head direction of rats (Taube, Goodridge, Golob, Dudchenko and Stackman 1996, Muller, Ranck and Taube 1996) and macaques (Robertson, Rolls, Georges-François and Panzeri 1999), and can be reset by visual inputs after gradual drift in darkness.) The model is a recurrent attractor network with global inhibition. It is different from a Hopfield attractor network primarily in that there are no discrete attractors formed by associative learning of discrete patterns. Instead there is a set of neurons that are connected to each other by synaptic weights w_{ij} that are a simple function, for example Gaussian, of the distance between the states of the agent in the physical world (e.g. head directions) represented by the neurons. Neurons that represent similar states (locations in the state space) of the agent in the physical world have strong synaptic connections, which can be set up by an associative learning rule, as described in Section 7.5.3. The network updates its firing rates by the following 'leaky-integrator' dynamical equations. The continuously changing activation h_i^{HD} of each head direction cell i is governed by the Equation

$$\frac{dh_i^{\mathrm{HD}}(t)}{dt} = -h_i^{\mathrm{HD}}(t) + \frac{\phi_0}{C^{\mathrm{HD}}} \sum_j (w_{ij} - w^{\mathrm{inh}}) r_j^{\mathrm{HD}}(t) + I_i^V, \tag{7.29}$$

where r_j^{HD} is the firing rate of head direction cell j, w_{ij} is the excitatory (positive) synaptic weight from head direction cell j to cell i, w^{inh} is a global constant describing the effect of inhibitory interneurons, and τ is the time constant of the system[11]. The term $-h_i^{\mathrm{HD}}(t)$ indicates the amount by which the activation decays (in the leaky integrator neuron) at time t. (The network is updated in a typical simulation at much smaller timesteps than the time constant of the system, τ.) The next term in equation 7.29 is the input from other neurons in the network r_j^{HD} weighted by the recurrent collateral synaptic connections w_{ij} (scaled by a constant ϕ_0 and C^{HD} which is the number of synaptic connections received by each head direction cell from other head direction cells in the continuous attractor). The term I_i^V represents a visual input to head direction cell i. Each term I_i^V is set to have a Gaussian response profile in most continuous attractor networks, and this sets the firing of the cells in the continuous attractor to have Gaussian response profiles as a function of where the agent is located in the state space (see e.g. Fig. 7.18), but the Gaussian assumption is not crucial. (It is known that the firing rates of head direction cells in both rats (Taube, Goodridge, Golob, Dudchenko and Stackman 1996, Muller, Ranck and Taube 1996) and macaques (Robertson, Rolls, Georges-François and Panzeri 1999) is approximately Gaussian.) When the agent is operating without visual input, in memory mode, then the term I_i^V is set to zero. The firing rate r_i^{HD} of cell i is determined from the activation h_i^{HD} and the sigmoid function

$$r_i^{\mathrm{HD}}(t) = \frac{1}{1 + e^{-2\beta(h_i^{\mathrm{HD}}(t) - \alpha)}}, \tag{7.30}$$

where α and β are the sigmoid threshold and slope, respectively.

---

[11] Note that for this Section, we use r rather than y to refer to the firing rates of the neurons in the network, remembering that, because this is a recurrently connected network (see Fig. 7.6), the output from a neuron y_i might be the input x_j to another neuron.

7.5.3 Learning the synaptic strengths between the neurons that implement a continuous attractor network

So far we have said that the neurons in the continuous attractor network are connected to each other by synaptic weights w_{ij} that are a simple function, for example Gaussian, of the distance between the states of the agent in the physical world (e.g. head directions, spatial views etc) represented by the neurons. In many simulations, the weights are set by formula to have weights with these appropriate Gaussian values. However, Stringer, Trappenberg, Rolls and de Araujo (2002c) showed how the appropriate weights could be set up by learning. They started with the fact that since the neurons have broad tuning that may be Gaussian in shape, nearby neurons in the state space will have overlapping spatial fields, and will thus be co-active to a degree that depends on the distance between them. They postulated that therefore the synaptic weights could be set up by associative learning based on the co-activity of the neurons produced by external stimuli as the animal moved in the state space. For example, head direction cells are forced to fire during learning by visual cues in the environment that produce Gaussian firing as a function of head direction from an optimal head direction for each cell. The learning rule is simply that the weights w_{ij} from head direction cell j with firing rate r_j^{HD} to head direction cell i with firing rate r_i^{HD} are updated according to an associative (Hebb) rule

$$\delta w_{ij} = k r_i^{\mathrm{HD}} r_j^{\mathrm{HD}} \tag{7.31}$$

where δw_{ij} is the change of synaptic weight and k is the learning rate constant. During the learning phase, the firing rate r_i^{HD} of each head direction cell i might be the following Gaussian function of the displacement of the head from the optimal firing direction of the cell

$$r_i^{\mathrm{HD}} = e^{-s_{\mathrm{HD}}^2 / 2\sigma_{\mathrm{HD}}^2}, \tag{7.32}$$

where s_{HD} is the difference between the actual head direction x (in degrees) of the agent and the optimal head direction x_i for head direction cell i, and σ_{HD} is the standard deviation.

Stringer, Trappenberg, Rolls and de Araujo (2002c) showed that after training at all head directions, the synaptic connections develop strengths that are an almost Gaussian function of the distance between the cells in head direction space, as shown in Fig. 7.19 (left). Interestingly if a non-linearity is introduced into the learning rule that mimics the properties of NMDA receptors by allowing the synapses to modify only after strong postsynaptic firing is present, then the synaptic strengths are still close to a Gaussian function of the distance between the connected cells in head direction space (see Fig. 7.19, left). They showed that after training, the continuous attractor network can support stable activity packets in the absence of visual inputs (see Fig. 7.19, right) provided that global inhibition is used to prevent all the neurons becoming activated. (The exact stability conditions for such networks have been analyzed by Amari (1977)). Thus Stringer, Trappenberg, Rolls and de Araujo (2002c) demonstrated biologically plausible mechanisms for training the synaptic weights in a continuous attractor using a biologically plausible local learning rule.

Stringer, Trappenberg, Rolls and de Araujo (2002c) went on to show that if there was a short term memory trace built into the operation of the learning rule, then this could help to produce smooth weights in the continuous attractor if only incomplete training was available, that is if the weights were trained at only a few locations. The same rule can take advantage in training the synaptic weights of the temporal probability distributions of firing when they happen to reflect spatial proximity. For example, for head direction cells the agent will necessarily

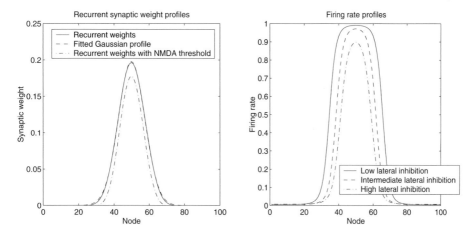

Fig. 7.19 Training the weights in a continuous attractor network with an associative rule (equation 7.31). Left: The trained recurrent synaptic weights from head direction cell 50 to the other head direction cells in the network arranged in head direction space (solid curve). The dashed line shows a Gaussian curve fitted to the weights shown in the solid curve. The dash-dot curve shows the recurrent synaptic weights trained with rule equation (7.31), but with a non-linearity introduced that mimics the properties of NMDA receptors by allowing the synapses to modify only after strong postsynaptic firing is present. Right: The stable firing rate profiles forming an activity packet in the continuous attractor network during the testing phase when the training (visual) inputs are no longer present. The firing rates are shown after the network has been initially stimulated by visual input to initialize an activity packet, and then allowed to settle to a stable activity profile without visual input. The three graphs show the firing rates for low, intermediate and high values of the lateral inhibition parameter w^{inh}. For both left and right plots, the 100 head direction cells are arranged according to where they fire maximally in the head direction space of the agent when visual cues are available. After Stringer, Trappenberg, Rolls and de Araujo (2002).

move through similar head directions before reaching quite different head directions, and so the temporal proximity with which the cells fire can be used to set up the appropriate synaptic weights. This new proposal for training continuous attractor networks can also help to produce broadly tuned spatial cells even if the driving (e.g. visual) input (I_i^V in equation 7.29) during training produces rather narrowly tuned neuronal responses. The learning rule with such temporal properties is a memory trace learning rule that strengthens synaptic connections between neurons, based on the temporal probability distribution of the firing. There are many versions of such rules (Rolls and Milward 2000, Rolls and Stringer 2001a), which are described more fully in Chapter 8, but a simple one that works adequately is

$$\delta w_{ij} = k\overline{r}_i^{\mathrm{HD}}\overline{r}_j^{\mathrm{HD}} \tag{7.33}$$

where δw_{ij} is the change of synaptic weight, and $\overline{r}^{\mathrm{HD}}$ is a local temporal average or trace value of the firing rate of a head direction cell given by

$$\overline{r}^{\mathrm{HD}}(t+\delta t) = (1-\eta)r^{\mathrm{HD}}(t+\delta t) + \eta\overline{r}^{\mathrm{HD}}(t) \tag{7.34}$$

where η is a parameter set in the interval [0,1] which determines the contribution of the current firing and the previous trace. For $\eta = 0$ the trace rule (7.33) becomes the standard Hebb rule (7.31), while for $\eta > 0$ learning rule (7.33) operates to associate together patterns of activity that occur close together in time. The rule might allow temporal associations to influence the synaptic weights that are learned over times in the order of 1 s. The memory trace required for operation of this rule might be no more complicated than the continuing firing that is an inherent property of attractor networks, but it could also be implemented by a number of biophysical mechanisms, discussed in Chapter 8. Finally, we note that some long term depression (LTD) in the learning rule could help to maintain the weights of different neurons equally potent (see Section 7.4.8.2 and equation 7.22), and could compensate for irregularity during training in which the agent might be trained much more in some than other locations in the space (see Stringer, Trappenberg, Rolls and de Araujo (2002c)).

7.5.4   The capacity of a continuous attractor network

The capacity of a continuous attractor network can be approached on the following bases. First, as there are no discrete attractor states, but instead a continuous physical space is being represented, some concept of spatial resolution must be brought to bear, that is the number of different positions in the space that can be represented. Second, the number of connections per neuron in the continuous attractor will directly influence the number of different spatial positions (locations in the state space) that can be represented. Third, the sparseness of the representation can be thought of as influencing the number of different spatial locations (in the continuous state space) that can be represented, in a way analogous to that described for discrete attractor networks in equation 7.14 (Battaglia and Treves 1998b). That is, if the tuning of the neurons is very broad, then fewer locations in the state space may be represented. Fourth, and very interestingly, if representations of different continuous state spaces, for example maps or charts of different environments, are stored in the same network, there may be little cost of adding extra maps or charts. The reason for this is that the large part of the interference between the different memories stored in such a network arises from the correlations between the different positions in any one map, which are typically relatively high because quite broad tuning of individual cells is common. In contrast, there are in general low correlations between the representations of places in different maps or charts, and so many different maps can be simultaneously stored in a continuous attractor network (Battaglia and Treves 1998b).

7.5.5   Continuous attractor models: moving the activity packet of neuronal activity

So far, we have considered how spatial representations could be stored in continuous attractor networks, and how the activity can be maintained at any location in the state space in a form of short term memory when the external (e.g. visual) input is removed. However, many networks with spatial representations in the brain can be updated by internal, self-motion (i.e. idiothetic), cues even when there is no external (e.g. visual) input. Examples are head direction cells in the presubiculum of rats and macaques, place cells in the rat hippocampus, and spatial view cells in the primate hippocampus (see Section 12.2). The major question arises about how such idiothetic inputs could drive the activity packet in a continuous attractor network, and in particular, how such a system could be set up biologically by self-organizing learning.

   One approach to simulating the movement of an activity packet produced by idiothetic cues (which is a form of path integration whereby the current location is calculated from recent movements) is to employ a look-up table that stores (taking head direction cells as an example), for every possible head direction and head rotational velocity input generated by the vestibular

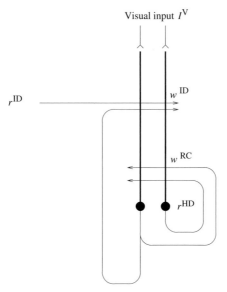

Visual input I^V

r^{ID}

w^{ID}

w^{RC}

r^{HD}

Fig. 7.20 General network architecture for a one-dimensional continuous attractor model of head direction cells which can be updated by idiothetic inputs produced by head rotation cell firing r^{ID}. The head direction cell firing is r^{HD}, the continuous attractor synaptic weights are w^{RC}, the idiothetic synaptic weights are w^{ID}, and the external visual input is I^V.

system, the corresponding new head direction (Samsonovich and McNaughton 1997). Another approach involves modulating the strengths of the recurrent synaptic weights in the continuous attractor on one but not the other side of a currently represented position, so that the stable position of the packet of activity, which requires symmetric connections in different directions from each node, is lost, and the packet moves in the direction of the temporarily increased weights, although no possible biological implementation was proposed of how the appropriate dynamic synaptic weight changes might be achieved (Zhang 1996). Another mechanism (for head direction cells) (Skaggs, Knierim, Kudrimoti and McNaughton 1995) relies on a set of cells, termed (head) rotation cells, which are co-activated by head direction cells and vestibular cells and drive the activity of the attractor network by anatomically distinct connections for clockwise and counter-clockwise rotation cells, in what is effectively a look-up table. However, no proposal was made about how this could be achieved by a biologically plausible learning process, and this has been the case until recently for most approaches to path integration in continuous attractor networks, which rely heavily on rather artificial pre-set synaptic connectivities.

Stringer, Trappenberg, Rolls and de Araujo (2002c) introduced a proposal with more biological plausibility about how the synaptic connections from idiothetic inputs to a continuous attractor network can be learned by a self-organizing learning process. The essence of the hypothesis is described with Fig. 7.20. The continuous attractor synaptic weights w^{RC} are set up under the influence of the external visual inputs I^V as described in Section 7.5.3. At the same time, the idiothetic synaptic weights w^{ID} (in which the ID refers to the fact that they are in this case produced by idiothetic inputs, produced by cells that fire to represent the velocity of clockwise and anticlockwise head rotation), are set up by associating the change of head direction cell firing that has just occurred (detected by a trace memory mechanism described below) with the current firing of the head rotation cells r^{ID}. For example, when the trace memory mechanism incorporated into the idiothetic synapses w^{ID} detects that the head direction cell firing is at a given location (indicated by the firing r^{HD}) and is moving clockwise (produced by the altering visual inputs I^{V}), and there is simultaneous clockwise head rotation cell firing, the synapses w^{ID} learn the association, so that when that rotation cell firing occurs later without visual input, it takes the current head direction firing in the continuous attractor into account, and moves the location of the head direction attractor in the appropriate direction.

For the learning to operate, the idiothetic synapses onto head direction cell i with firing r_i^{HD} need two inputs: the memory traced term from other head direction cells $\bar{r}_j^{\mathrm{HD}}$ (given by equation 7.34), and the head rotation cell input with firing r_k^{ID}; and the learning rule can be written

$$\delta w_{ijk}^{\mathrm{ID}} = \tilde{k}\, r_i^{\mathrm{HD}}\, \bar{r}_j^{\mathrm{HD}}\, r_k^{\mathrm{ID}}, \tag{7.35}$$

where $\tilde{k}$ is the learning rate associated with this type of synaptic connection. The head rotation cell firing (r_k^{ID}) could be as simple as one set of cells that fire for clockwise head rotation (for which k might be 1), and a second set of cells that fire for anticlockwise head rotation (for which k might be 2).

After learning, the firing of the head direction cells would be updated in the dark (when $I_i^V = 0$) by idiothetic head rotation cell firing r_k^{ID} as follows

$$\tau \frac{dh_i^{\mathrm{HD}}(t)}{dt} = -h_i^{\mathrm{HD}}(t) + \frac{\phi_0}{C^{\mathrm{HD}}} \sum_j (w_{ij} - w^{inh}) r_j^{\mathrm{HD}}(t) + I_i^V$$

$$+ \phi_1 \left(\frac{1}{C^{\mathrm{HD} \times \mathrm{ID}}} \sum_{j,k} w_{ijk}^{\mathrm{ID}} r_j^{\mathrm{HD}} r_k^{\mathrm{ID}} \right). \tag{7.36}$$

Equation 7.36 is similar to equation 7.29, except for the last term, which introduces the effects of the idiothetic synaptic weights w_{ijk}^{ID}, which effectively specify that the current firing of head direction cell i, r_i^{HD}, must be updated by the previously learned combination of the particular head rotation now occurring indicated by r_k^{ID}, and the current head direction indicated by the firings of the other head direction cells r_j^{HD} indexed through j[12]. This makes it clear that the idiothetic synapses operate using combinations of inputs, in this case of two inputs. Neurons that sum the effects of such local products are termed Sigma-Pi neurons. Although such synapses are more complicated than the two-term synapses used throughout the rest of this book, such three-term synapses appear to be useful to solve the computational problem of updating representations based on idiothetic inputs in the way described. Synapses that operate according to Sigma-Pi rules might be implemented in the brain by a number of mechanisms described by Koch (1999) (Section 21.1.1), Jonas and Kaczmarek (1999), and Stringer, Trappenberg, Rolls and de Araujo (2002c), including having two inputs close together on a thin dendrite, so that local synaptic interactions would be emphasized.

Simulations demonstrating the operation of this self-organizing learning to produce movement of the location being represented in a continuous attractor network were described by Stringer, Trappenberg, Rolls and de Araujo (2002c), and one example of the operation is

---

[12]The term $\phi_1/C^{\mathrm{HD} \times \mathrm{ID}}$ is a scaling factor that reflects the number $C^{\mathrm{HD} \times \mathrm{ID}}$ of inputs to these synapses, and enables the overall magnitude of the idiothetic input to each head direction cell to remain approximately the same as the number of idiothetic connections received by each head direction cell is varied.

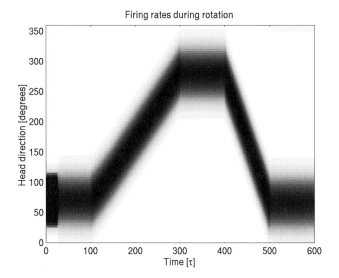

Fig. 7.21 Idiothetic update of the location represented in a continuous attractor network. The firing rate of the cells with optima at different head directions (organized according to head direction on the ordinate) is shown by the blackness of the plot, as a function of time. The activity packet was initialized to a head direction of 75 degrees, and the packet was allowed to settle without visual input. For $t = 0$ to $t = 100$ there was no rotation cell input, and the activity packet in the continuous attractor remained stable at 75 degrees. For $t = 100$ to $t = 300$ the clockwise rotation cells were active with a firing rate of 0.15 to represent a moderate angular velocity, and the activity packet moved clockwise. For $t = 300$ to $t = 400$ there was no rotation cell firing, and the activity packet immediately stopped, and remained still. For $t = 400$ to $t = 500$ the anti-clockwise rotation cells had a high firing rate of 0.3 to represent a high velocity, and the activity packet moved anti-clockwise with a greater velocity. For $t = 500$ to $t = 600$ there was no rotation cell firing, and the activity packet immediately stopped.

shown in Fig. 7.21. They also showed that, after training with just one value of the head rotation cell firing, the network showed the desirable property of moving the head direction being represented in the continuous attractor by an amount that was proportional to the value of the head rotation cell firing. Stringer, Trappenberg, Rolls and de Araujo (2002c) also describe a related model of the idiothetic cell update of the location represented in a continuous attractor, in which the rotation cell firing directly modulates in a multiplicative way the strength of the recurrent connections in the continuous attractor in such a way that clockwise rotation cells modulate the strength of the synaptic connections in the clockwise direction in the continuous attractor, and vice versa.

It should be emphasized that although the cells are organized in Fig. 7.21 according to the spatial position being represented, there is no need for cells in continuous attractors that represent nearby locations in the state space to be close together, as the distance in the state space between any two neurons is represented by the strength of the connection between them, not by where the neurons are physically located. This enables continuous attractor networks to represent spaces with arbitrary topologies, as the topology is represented in the connection strengths (Stringer, Trappenberg, Rolls and de Araujo 2002c, Stringer, Rolls, Trappenberg and de Araujo 2002b, Stringer, Rolls and Trappenberg 2002a, Stringer and Rolls 2002b). Indeed, it is this that enables many different charts each with its own topology to be represented in a single continuous attractor network (Battaglia and Treves 1998b).

7.5.6 Stabilization of the activity packet within the continuous attractor network when the agent is stationary

With irregular learning conditions (in which identical training with high precision of every node cannot be guaranteed), the recurrent synaptic weights between nodes in the continuous attractor will not be of the perfectly regular and symmetric form normally required in a continuous attractor neural network. This can lead to drift of the activity packet within the continuous attractor network of e.g. head direction cells when no visual cues are present, even when the agent is not moving. This drift is a common property of the short term memories of spatial position implemented in the brain, which emphasizes the computational problems which can arise in continuous attractor networks if the weights between nodes are not balanced in different directions in the space being represented. We now describe an approach to stabilizing the activity packet when it should not be drifting in real nervous systems, which does help to minimize the drift that can occur.

The activity packet may be stabilized within a continuous attractor network, when the agent is stationary and the network is operating in memory mode without an external stabilizing input, by enhancing the firing of those cells that are already firing. In biological systems this might be achieved through mechanisms for short term synaptic enhancement (Koch 1999). Another way is to take advantage of the non-linearity of the activation function of neurons with NMDA receptors, which only contribute to neuronal firing once the neuron is sufficiently depolarized (Wang 1999). The effect is to enhance the firing of neurons that are already reasonably well activated. The effect has been utilized in a model of a network with recurrent excitatory synapses that can maintain active an arbitrary set of neurons that are initially sufficiently strongly activated by an external stimulus (see Lisman, Fellous and Wang (1998), and, for a discussion on whether these networks could be used to implement short term memories, see Kesner and Rolls (2001). We have incorporated this non-linearity into a model of a head direction continuous attractor network by adjusting the sigmoid threshold α_i (see equation 7.30) for each head direction cell i as follows (Stringer, Trappenberg, Rolls and de Araujo 2002c). If the head direction cell firing rate r_i^{HD} is lower than a threshold value, γ, then the sigmoid threshold α_i is set to a relatively high value α^{high}. Otherwise, if the head direction cell firing rate r_i^{HD} is greater than or equal to the threshold value, γ, then the sigmoid threshold α_i is set to a relatively low value α^{low}. It was shown that this procedure has the effect of enhancing the current position of the activity packet within the continuous attractor network, and so prevents the activity packet drifting erratically due to the noise in the recurrent synaptic weights produced for example by irregular learning. An advantage of using the nonlinearity in the activation function of a neuron (produced for example by the operation of NMDA receptors) is that this tends to enable packets of activity to be kept active without drift even when the packet is not in one of the energy minima that can result from irregular learning (or from diluted connectivity in the continuous attractor as described below). Thus use of this non-linearity increases the number of locations in the continuous physical state space at which a stable activity packet can be maintained (Stringer, Trappenberg, Rolls and de Araujo 2002c).

7.5.7 Continuous attractor networks in two or more dimensions

Some types of spatial representation used by the brain are of spaces that exist in two or more dimensions. Examples are the two- (or three-) dimensional space representing where one is looking at in a spatial scene. Another is the two- (or three-) dimensional space representing where one is located. It is possible to extend continuous attractor networks to operate in higher dimensional spaces than the one-dimensional spaces considered so far (Taylor 1999, Stringer, Rolls, Trappenberg and de Araujo 2002b). Indeed, it is also possible to extend the analyses of how idiothetic inputs could be used to update two-dimensional state spaces, such as the locations represented by place cells in rats (Stringer, Rolls, Trappenberg and de Araujo 2002b) and the location at which one is looking represented by primate spatial view cells (Stringer, Rolls and Trappenberg 2002a, Stringer and Rolls 2002b). Interestingly, the number of terms in the synapses implementing idiothetic update do not need to increase beyond three (as in Sigma-Pi synapses) even when higher dimensional state spaces are being considered (Stringer, Rolls, Trappenberg and de Araujo 2002b). Also interestingly, a continuous attractor network can in fact represent the properties of very high dimensional spaces, because the properties of the spaces are captured by the connections between the neurons of the continuous attractor, and these connections are of course, as in the world of discrete attractor networks, capable of representing high dimensional spaces (Stringer, Rolls, Trappenberg and de Araujo 2002b). With these approaches, continuous attractor networks have been developed of the two-dimensional representation of rat hippocampal place cells with idiothetic update by movements in the environment (Stringer, Rolls, Trappenberg and de Araujo 2002b), and of primate hippocampal spatial view cells with idiothetic update by eye and head movements (Stringer, Rolls and Trappenberg 2002a, Stringer and Rolls 2002b).

7.5.8 Mixed continuous and discrete attractor networks

It has now been shown that attractor networks can store both continuous patterns and discrete patterns, and can thus be used to store for example the location in (continuous, physical) space where an object (a discrete item) is present (see Fig. 7.18 and Rolls, Stringer and Trappenberg (2002c)). In this network, when events are stored that have both discrete (object) and continuous (spatial) aspects, then the whole place can be retrieved later by the object, and the object can be retrieved by using the place as a retrieval cue. Such networks are likely to be present in parts of the brain that receive and combine inputs both from systems that contain representations of continuous (physical) space, and from brain systems that contain representations of discrete objects, such as the inferior temporal visual cortex. One such brain system is the hippocampus, which appears to combine and store such representations in a mixed attractor network in the CA3 region, which thus is able to implement episodic memories which typically have a spatial component, for example where an item such as a key is located (see Section 12.2). This network thus shows that in brain regions where the spatial and object processing streams are brought together, then a single network can represent and learn associations between both types of input. Indeed, in brain regions such as the hippocampal system, it is essential that the spatial and object processing streams are brought together in a single network, for it is only when both types of information are in the same network that spatial information can be retrieved from object information, and vice versa, which is a fundamental property of episodic memory (see Section 12.2). It may also be the case that in the prefrontal cortex, attractor networks can store both spatial and discrete (e.g. object-based) types of information in short term memory (see Section 12.1).

7.6 Network dynamics: the integrate-and-fire approach

The concept that attractor (autoassociation) networks can operate very rapidly if implemented with neurons that operate dynamically in continuous time was introduced in Section 7.3.3.5. The result described was that the principal factor affecting the speed of retrieval is the time constant of the synapses between the neurons that form the attractor ((Treves 1993, Rolls and Treves 1998, Battaglia and Treves 1998a, Panzeri, Rolls, Battaglia and Lavis 2001). This was shown analytically by Treves (1993), and described by Rolls and Treves (1998) Appendix 5. We now describe in more detail the approaches that produce these results, and the actual results found on the speed of processing.

The networks described so far in this Chapter, and analyzed in Appendices 3 and 4 of Rolls and Treves (1998), were described in terms of the steady-state activation of networks of neuron-like units. Those may be referred to as 'static' properties, in the sense that they do not involve the time dimension. In order to address 'dynamical' questions, the time dimension has to be reintroduced into the formal models used, and the adequacy of the models themselves has to be reconsidered in view of the specific properties to be discussed.

Consider for example a real network whose operation has been described by an autoassociative formal model that acquires, with learning, a given attractor structure. How does the state of the network approach, in real time during a retrieval operation, one of those attractors? How long does it take? How does the amount of information that can be read off the network's activity evolve with time? Also, which of the potential steady states is indeed a stable state that can be reached asymptotically by the net? How is the stability of different states modulated by external agents? These are examples of dynamical properties, which to be studied require the use of models endowed with some dynamics.

7.6.1 From discrete to continuous time

Already at the level of simple models in which each unit is described by an input-output relation, one may introduce equally simple 'dynamical' rules, in order both to fully specify the model, and to simulate it on computers. These rules are generally formulated in terms of 'updatings': time is considered to be discrete, a succession of time steps, and at each time step the output of one or more of the units is set, or updated, to the value corresponding to its input variable. The input variable may reflect the outputs of other units in the net as updated at the previous time step or, if delays are considered, the outputs as they were at a prescribed number of time steps in the past. If all units in the net are updated together, the dynamics is referred to as parallel; if instead only one unit is updated at each time step, the dynamics is sequential. (One main difference between the Hopfield (1982) model of an autoassociator and a similar model considered earlier by Little (1974) is that the latter was based on parallel rather than sequential dynamics.) Many intermediate possibilities obviously exist, involving the updating of groups of units at a time. The order in which sequential updatings are performed may for instance be chosen at random at the beginning and then left the same in successive cycles across all units in the net; or it may be chosen anew at each cycle; yet a third alternative is to select at each time step a unit, at random, with the possibility that a particular unit may be selected several times before some of the other ones are ever updated. The updating may also be made probabilistic, with the output being set to its new value only with a certain probability, and otherwise remaining at the current value.

Variants of these dynamical rules have been used for decades in the analysis and computer simulation of physical systems in statistical mechanics (and field theory). They can reproduce in simple but effective ways the stochastic nature of transitions among discrete quantum states, and they have been subsequently considered appropriate also in the simulation of neural network models in which units have outputs that take discrete values, implying that a change from one value to another can only occur in a sudden jump. To some extent, different rules are equivalent, in that they lead, in the evolution of the activity of the net along successive steps and cycles, to the same set of possible steady states. For example, it is easy to realize that when no delays are introduced, states that are stable under parallel updating are also stable under sequential updating. The reverse is not necessarily true, but on the other hand states that are stable when updating one unit at a time are stable irrespective of the updating order. Therefore, static properties, which can be deduced from an analysis of stable states, are to some extent robust against differences in the details of the dynamics assigned to the model. (This is a reason for using these dynamical rules in the study of the thermodynamics of physical systems.) Such rules, however, bear no relation to the actual dynamical processes by which the activity of real neurons evolves in time, and are therefore inadequate for the discussion of dynamical issues in neural networks.

A first step towards realism in the dynamics is the substitution of discrete time with continuous time. This somewhat parallels the substitution of the discrete output variables of the most rudimentary models with continuous variables representing firing rates. Although continuous output variables may evolve also in discrete time, and as far as static properties are concerned differences are minimal, with the move from discrete to continuous outputs the main raison d'etre for a dynamics in terms of sudden updatings ceases to exist, since continuous variables can change continuously in continuous time. A paradox arises immediately, however, if a continuous time dynamics is assigned to firing rates. The paradox is that firing rates, although in principle continuous if computed with a generic time-kernel, tend to vary in jumps as new spikes — essentially discrete events — come to be included in the kernel. To avoid this paradox, a continuous time dynamics can be assigned, instead, to instantaneous continuous variables like membrane potentials. Hopfield (1984), among others, has introduced a model of an autoassociator in which the output variables represent membrane potentials and evolve continuously in time, and has suggested that under certain conditions the stable states attainable by such a network are essentially the same as for a network of binary units evolving in discrete time. If neurons in the central nervous system communicated with each other via the transmission of graded membrane potentials, as they do in some peripheral and invertebrate neural systems, this model could be an excellent starting point. The fact that, centrally, transmission is primarily via the emission of discrete spikes makes a model based on membrane potentials as output variables inadequate to correctly represent spiking dynamics.

7.6.2 Continuous dynamics with discontinuities

In principle, a solution would be to keep the membrane potential as the basic dynamical variable, evolving in continuous time, and to use as the output variable the spike emission times, as determined by the rapid variation in membrane potential corresponding with each spike. A point-neuron-like processing unit in which the membrane potential V is capable of undergoing spikes is the one described by equations of the Hodgkin-Huxley type:

$$C\frac{dV}{dt} = g_0(V_{\text{rest}} - V) + g_{Na}mh^3(V_{Na} - V) + g_K n^4(V_K - V) + I \tag{7.37}$$

$$\tau_m \frac{dm}{dt} = m_\infty(V) - m \tag{7.38}$$

$$\tau_h \frac{dh}{dt} = h_\infty(V) - h \tag{7.39}$$

$$\tau_n \frac{dn}{dt} = n_\infty(V) - n \tag{7.40}$$

in which changes in the membrane potential, driven by the input current I, interact with the opening and closing of intrinsic conductances (here a sodium conductance, whose channels are gated by the 'particles' m and h and a potassium conductance, whose channels are gated by n; Hodgkin and Huxley (1952)). These equations provide an effective description, phenomenological but broadly based on physical principles, of the conductance changes underlying action potentials, and they are treated in any standard neurobiology text. From the point of view of formal models of neural networks, this level of description is too complicated to be the basis for an analytical understanding of the operation of networks, and it must be simplified. The most widely used simplification is the so-called integrate-and-fire model (see for example MacGregor (1987)), which is legitimized by the observation that (sodium) action potentials are typically brief and self-similar events. If, in particular, the only relevant variable associated with the spike is its time of emission (at the soma, or axon hillock), which essentially coincides with the time the potential V reaches a certain threshold level V_{thr}, then the conductance changes underlying the rest of the spike can be omitted from the description, and substituted with the ad hoc prescription that (i) a spike is emitted, with its effect on receiving units and on the unit itself, and (ii) after a brief time corresponding to the duration of the spike plus a refractory period, the membrane potential is reset and resumes its integration of the input current I. After a spike the membrane potential is taken to be reset to a value V_{ahp} (for after-hyperpolarization). This type of simplified dynamics of the membrane potentials is thus in continuous time with added discontinuities: continuous in between spikes, with discontinuities occurring at different times for each unit in a population, every time a unit emits a spike.

Although the essence of the simplification is in omitting the description of the evolution in time of intrinsic conductances, in order to model the phenomenon of adaptation in the firing rate, prominent especially with pyramidal cells, it is necessary to include at least an intrinsic (potassium-like) conductance (Brown, Gähwiler, Griffith and Halliwell 1990a). This can be done in a rudimentary way with minimal complication, by specifying that this conductance, which if open tends to shunt the membrane and thus to prevent firing, opens by a fixed amount with the potential excursion associated with each spike, and then relaxes exponentially to its closed state. In this manner sustained firing driven by a constant input current occurs at lower rates after the first few spikes, in a way similar, if the relevant parameters are set appropriately, to the behaviour observed in vitro of many pyramidal cells (for example, Lanthorn, Storn and Andersen (1984), Mason and Larkman (1990)).

The equations for the dynamics of each unit, which replace the input-output transduction function, are then

$$C\frac{dV(t)}{dt} = g_0(V_{\text{rest}} - V(t)) + g_K(t)(V_K - V(t)) + I(t) \tag{7.41}$$

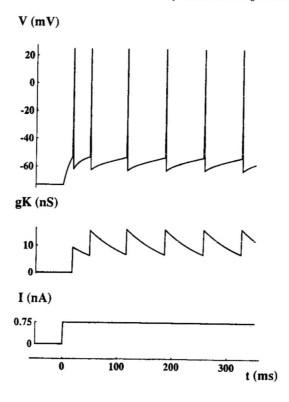

Fig. 7.22 Model behaviour of an integrate-and-fire unit: the membrane potential and adaptation-producing potassium conductance in response to a step of injected current. The spikes were added to the graph by hand, as they do not emerge from the simplified voltage equation. (From Treves, 1993.)

$$\frac{\mathrm{d}g_K(t)}{\mathrm{d}t} = -\frac{g_K(t)}{\tau_K} + \sum_k \Delta g_K \delta(t - t_k) \tag{7.42}$$

supplemented by the prescription that when at time $t = t_{k+1}$ the potential reaches the level V_{thr}, a spike is emitted, and hence included also in the sum of the second equation, and the potential resumes its evolution according to the first equation from the reset level V_{ahp}. The resulting behaviour is exemplified in Fig. 7.22, while Fig. 7.23b shows the input-output transform (current to frequency transduction) operated by an integrate-and-fire unit of this type. One should compare this with the transduction operated by real cells, as exemplified for example in Fig. 1.4.

7.6.3 Conductance dynamics for the input current

Besides the input–output transduction, the neuron–like units of the static formulations are characterized by the weighted summation of inputs, which models the quasi-linear (Rall and Segev 1987) summation of currents of synaptic origin flowing into the soma. This description can be made dynamical by writing each current, at the synapse, as the product of a conductance, which varies in time, with the difference between the membrane potential and the equilibrium potential characteristic of each synapse

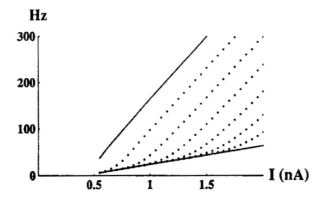

Fig. 7.23 Current-to-frequency transduction in a pyramidal cell modelled as an integrate-and-fire unit. The top solid curve is the firing frequency in the absence of adaptation, $\Delta g_K = 0$. The dotted curves are the instantaneous frequencies computed as the inverse of the ith interspike interval (top to bottom, $i = 1, ..., 6$). The bottom solid curve is the adapted firing curve ($i \rightarrow \infty$). With or without adaptation, the input–output transform is close to threshold–linear. (From Treves, 1993.)

$$I(t) = \sum_{\alpha} I_{\alpha}(t) = \sum_{\alpha} g_{\alpha}(t)(V_{\alpha} - V(t)) . \tag{7.43}$$

The opening of each synaptic conductance is driven by the arrival of spikes at the presynaptic terminal, and its closing can often be described as a simple exponential process. A simplified equation for the dynamics of $g_{\alpha}(t)$ is then

$$\frac{dg_{\alpha}(t)}{dt} = -\frac{g_{\alpha}(t)}{\tau_{\alpha}} + \Delta g_{\alpha} \sum_{l} \delta(t - \Delta t - t_l) \tag{7.44}$$

According to the above equation, each conductance opens instantaneously by a fixed amount Δg a time Δt after the emission of the presynaptic spike at t_l. Δt summarizes delays (axonal, synaptic, and dendritic), and each opening superimposes linearly, without saturating, on previous openings. An alternative equation that incorporates also a finite synaptic opening time and that is fairly accurate at least for certain classes of synapses, is the one that leads to the so-called α-function for the fraction of open channels:

$$\tau_{\alpha} \frac{d^2 g_{\alpha}(t)}{dt^2} + 2 \frac{dg_{\alpha}(t)}{dt} = -\frac{g_{\alpha}(t)}{\tau_{\alpha}} + \Delta g_{\alpha} \sum_{l} \delta(t - \Delta t - t_l) . \tag{7.45}$$

The main feature of this second equation is that it yields a finite opening time without introducing a second time parameter alongside τ_{α}. More complex descriptions of synaptic dynamics may be more appropriate especially, for example, when the interaction between transmitter receptor and channel is mediated by second messengers, but they introduce the complication of additional parameters, and often additional dynamical variables as well (such as calcium concentration).

It should be noted that conductance dynamics is not always included in integrate-and-fire models: sometimes it is substituted with current dynamics, which essentially amounts to neglecting non-linearities due to the appearance of the membrane potential in the driving force for synaptic action (see for example, Amit and Tsodyks (1991), Gerstner (1995)); and sometimes it is simplified altogether by assuming that the membrane potential undergoes small sudden jumps when it receives instantaneous pulses of synaptic current (see the review in Gerstner (1995)). The latter simplification is quite drastic and changes the character of the dynamics markedly; whereas the former can be a reasonable simplification in some circumstances, but it produces serious distortions in the description of inhibitory $GABA_A$ currents, which, having an equilibrium (Cl^-) synaptic potential close to the operating range of the membrane potential, are quite sensitive to the instantaneous value of the membrane potential itself.

7.6.4 The speed of processing of one-layer attractor networks with integrate-and-fire neurons

Given that the analytic approach to the rapidity of the dynamics of attractor networks with integrate-and-fire dynamics (Treves 1993, Rolls and Treves 1998) applies mainly when the state is close to the attractor basin, it is of interest to check the performance of such networks by simulation when the completion of partial patterns that may be towards the edge of the attractor basin can be tested. Simmen, Rolls and Treves (1996) and Treves, Rolls and Simmen (1997) made a start with this, and showed that retrieval could indeed be fast, within 1–2 time constants of the synapses. However, they found that they could not load the systems they simulated with many patterns, and the firing rates during the retrieval process tended to be unstable. The cause of this turned out to be that the inhibition they used to maintain the activity level during retrieval was subtractive, and it turns out that divisive (shunting) inhibition is much more effective in such networks, as described by Rolls and Treves (1998) in Appendix 5. Divisive inhibition is likely to be organized by inhibitory inputs that synapse close to the cell body (where the reversal potential is close to that of the channels opened by GABA receptors), in contrast to synapses on dendrites, where the different potentials result in opening of the same channels producing hyperpolarization (that is an effectively subtractive influence with respect to the depolarizing currents induced by excitatory (glutamate-releasing) terminals). Battaglia and Treves (1998a) therefore went on to study networks with neurons where the inhibitory neurons could be made to be divisive by having them synapse close to the cell body in neurons modelled with multiple (ten) dendritic compartments. The excitatory inputs terminated on the compartments more distant from the cell body in the model. They found that with this divisive inhibition, the neuronal firing during retrieval was kept under much better control, and the number of patterns that could be successfully stored and retrieved was much higher. Some details of their simulation follow.

Battaglia and Treves (1998a) simulated a network of 800 excitatory and 200 inhibitory cells in its retrieval of one of a number of memory patterns stored in the synaptic weights representing the excitatory-to-excitatory recurrent connections. The memory patterns were assigned at random, drawing the value of each unit in each of the patterns from a binary distribution with sparseness 0.1, that is a probability of 1 in 10 for the unit to be active in the pattern. No baseline excitatory weight was included, but the modifiable weights were instead constrained to remain positive (by clipping at zero synaptic modifications that would make a synaptic weight negative), and a simple exponential decay of the weight with successive modifications was also applied, to prevent runaway synaptic modifications within a rudimentary model of forgetting. Both excitation on inhibitory units and inhibition were mediated by non-modifiable uniform synaptic weights, with values chosen so as to satisfy stability conditions of the type shown in equations A5.13 of Rolls and Treves (1998). Both inhibitory and excitatory neurons were of the general integrate-and-fire type, but excitatory units had in addition an extended dendritic cable, and they received excitatory inputs only at the more distal end of the cable, and inhibitory inputs spread along the cable. In this way, inhibitory inputs reached the soma of excitatory cells with variable delays, and in any case earlier than synchronous excitatory inputs, and at the same time they could shunt the excitatory inputs, resulting in a largely multiplicative form of inhibition (Abbott 1991). The uniform connectivity was not complete, but rather each type of unit could contact units of the other type with a probability of 0.5, and the same was true for inhibitory-to-inhibitory connections.

After 100 (simulated) ms of activity evoked by external inputs uncorrelated with any of the stored patterns, a cue was provided that consisted of the external input becoming correlated with one of the patterns, at various levels of correlation, for 300 ms. After that, external inputs were removed, but when retrieval operated successfully the activity of the units remained strongly correlated with the memory pattern, or even reached a higher level of correlation if a rather corrupted cue had been used, so that, if during the 300 ms the network had stabilized into a state rather distant from the memory pattern, it got much closer to it once the cue was removed. All correlations were quantified using information measures (see Appendix B), both in terms of mutual information between the firing rate pattern across units and the particular memory pattern being retrieved, or in terms of mutual information between the firing rate of one unit and the set of patterns, or, finally, in terms of mutual information between the decoded firing rates of a subpopulation of 10 excitatory cells and the set of memory patterns. The same algorithms were used to extract information measures as were used for example by Rolls, Treves, Tovee and Panzeri (1997d) with real neuronal data from inferior temporal cortex neurons. The firing rates were measured over sliding windows of 30 ms, after checking that shorter windows produced noisier measures. The effect of using a relatively long window, 30 ms, for measuring rates is an apparent linear early rise in information values with time. Nevertheless, in the real system the activity of these cells is 'read' by other cells receiving inputs from them, and that in turn have their own membrane capacitance-determined characteristic time for integrating input activity, a time broadly in the order of 30 ms. Using such a time window for integrating firing rates did not therefore artificially slow down the read-out process.

It was shown that the time course of different information measures did not depend significantly on the firing rates prevailing during the retrieval state, nor on the Resistance-Capacitance-determined membrane time constants of the units. Figure 7.24 shows that the rise in information after providing the cue at time = 100 ms followed a roughly exponential approach to its steady-state value, which continued until the steady state switched to a new value when the retrieval cue was removed at time = 400 ms. The time constant of the approach to the first steady state was a linear function, as shown in Fig. 7.24, of the time constant for excitatory conductances, as predicted by the analysis. (The proportionality factor in the Figure is 2.5, or a collective time constant 2.5 times longer than the synaptic time constant.) The approach to the second steady-state value was more rapid, and the early apparent linear rise prevented the detection of a consistent exponential mode. Therefore, it appears that the cue leads to the basin of attraction of the correct retrieval state by activating transient modes, whose time constant is set by that of excitatory conductances; once the network is in the correct basin, its subsequent reaching the 'very bottom' of the basin after the removal of the

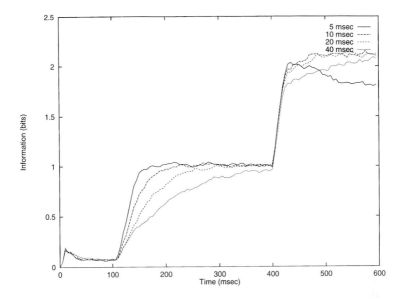

Fig. 7.24 Time course of the transinformation about which memory pattern had been selected, as decoded from the firing rates of 10 randomly selected excitatory units. Excitatory conductances closed exponentially with time constants of 5, 10, 20 and 40 ms (curves from top to bottom). A cue of correlation 0.2 with the memory pattern was presented from 100 to 400 ms, uncorrelated external inputs with the same mean strength and sparseness as the cue were applied at earlier times, and no external inputs were applied at later times. (After Battaglia and Treves, 1998b.)

cue is not accompanied by any prominent transient mode (see further Battaglia and Treves (1998a)).

Overall, these simulations confirm that recurrent networks, in which excitation is mediated mainly by fast (AMPA, i.e. kainate/quisqualate, see Section 1.5) channels, can reach asynchronous steady firing states very rapidly, over a few tens of milliseconds, and the rapid approach to steady state is reflected in the relatively rapid rise of information quantities that measure the speed of the operation in functional terms.

An analysis based on integrate-and-fire model units thus indicates that recurrent dynamics can be so fast as to be practically indistinguishable from purely feedforward dynamics, in contradiction with what simple intuitive arguments would suggest. This makes it hazardous to draw conclusions on the underlying circuitry on the basis of the experimentally observed speed with which selective neuronal responses arise, as attempted by Thorpe and Imbert (1989). The results also show that networks that implement feedback processing can settle into a global retrieval state very rapidly, and that rapid processing is not just a feature of feedforward networks.

We return to the intuitive understanding of this rapid processing. The way in which networks with continuous dynamics (such as networks made of real neurons in the brain, and networks modelled with integrate-and-fire neurons) can be conceptualized as settling so fast into their attractor states is that spontaneous activity in the network ensures that some neurons are close to their firing threshold when the retrieval cue is presented, so that the firing of these neurons is influenced within 1–2 ms by the retrieval cue. These neurons then influence other neurons within milliseconds (given the point that some other neurons will be close to threshold) through the modified recurrent collateral synapses that store the information. In this way, the neurons in networks with continuous dynamics can influence each other within a fraction of the synaptic time constant, and retrieval can be very rapid.

7.6.5 The speed of processing of a four-layer hierarchical network with integrate-and-fire attractor dynamics in each layer

Given that the visual system has a whole series of cortical areas organized predominantly hierarchically (e.g. V1 to V2 to V4 to inferior temporal cortex), the issue arises of whether the rapid information processing that can be performed for object recognition is predominantly feedforward, or whether there is sufficient time for feedback processing within each cortical area implemented by the local recurrent collaterals to contribute to the visual information processing being performed. Some of the constraints are as follows.

An analysis of response latencies indicates that there is sufficient time for only 10–20 ms per processing stage in the visual system. In the primate cortical ventral visual system the response latency difference between neurons in layer $4C\beta$ of V1 and Inferior Temporal cortical cells is approximately 60 ms (Bullier and Nowak 1995, Nowak and Bullier 1997, Schmolesky, Wang, Hanes, Thompson, Leutgeb, Schall and Leventhal 1998). For example, the latency of the responses of neurons in V1 is approximately 30–40 ms (Celebrini, Thorpe, Trotter and Imbert 1993), and in the temporal cortex visual areas approximately 80–110 ms (Baylis, Rolls and Leonard 1987, Sugase, Yamane, Ueno and Kawano 1999). Given that there are 4–6 stages of processing in the ventral visual system from V1 to the anterior inferior temporal cortex, the difference in latencies between each ventral cortical stage is on this basis approximately 10 ms (Rolls 1992a, Oram and Perrett 1994). Information theoretic analyses of the responses of single visual cortical cells in primates reveal that much of the information that can be extracted from neuronal spike trains is often found to be present in periods as short as 20–30 ms (Tovee, Rolls, Treves and Bellis 1993, Tovee and Rolls 1995, Heller, Hertz, Kjaer and Richmond 1995, Rolls, Tovee and Panzeri 1999b). Backward masking experiments indicate that each cortical area needs to fire for only 20–30 ms to pass information to the next stage (Rolls and Tovee 1994, Rolls, Tovee, Purcell, Stewart and Azzopardi 1994b, Kovacs, Vogels and Orban 1995, Rolls, Tovee and Panzeri 1999b) (see Sections 5.5.6 and Appendix B.3.2). Rapid serial visual presentation of image sequences shows that cells in the temporal visual cortex are still face selective when faces are presented at the rate of 14 ms/image (Keysers, Xiao, Foldiak and Perrett 2001). Finally, event-related potential studies in humans provide strong evidence that the visual system is able to complete some analyses of complex scenes in less than 150 ms (Thorpe, Fize and Marlot 1996).

To investigate whether feedback processing within each layer could contribute to information processing in such a multilayer system in times as short as 10–20 ms per layer, Panzeri, Rolls, Battaglia and Lavis (2001) simulated a four-layer network with attractor networks in each layer. The network architecture is shown schematically in Fig. 7.25. All the neurons realized integrate-and-fire dynamics, and indeed the individual layers and neurons were implemented very similarly to the implementation used by Battaglia and Treves (1998a). In particular, the current flowing from each compartment of the multicompartment neurons to the external medium was expressed as:

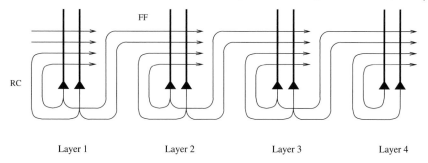

FF

RC

Layer 1   Layer 2   Layer 3   Layer 4

Fig. 7.25 The structure of the excitatory connections in the network. There are feedforward (FF) connections between each layer and the next, and excitatory recurrent collaterals (RC) in each layer. Inhibitory connections are also present within each layer, but they are not shown in this Figure.

$$I(t) = g_{\text{leak}}(V(t) - V^0) + \sum_j g_j(t)(V(t) - V_j), \qquad (7.46)$$

where g_{leak} is a constant passive leakage conductance, V^0 the membrane resting potential, $g_j(t)$ the value of the jth synapse conductance at time t, and V_j the reversal potential of the jth synapse. $V(t)$ is the potential in the compartment at time t. The most important parameter in the simulation, the AMPA inactivation time constant, was set to 10 ms. The recurrent collateral (RC) integration time constant of the membrane of excitatory cells was 20 ms long for the simulations presented. The synaptic conductances decayed exponentially in time, obeying the equation

$$\frac{dg_j}{dt} = -\frac{g_j}{\tau_j} + \Delta g_j \sum_k \delta(t - \Delta t - t_k^j), \qquad (7.47)$$

where τ_j is the synaptic decay time constant, Δt is a delay term summarizing axonal and synaptic delays, and Δg_j is the amount that the conductance is increased when the presynaptic unit fires a spike. Δg_j thus represents the (unidirectional) coupling strength between the pre-synaptic and the post-synaptic cell. t_k^j is the time at which the pre-synaptic unit fires its kth spike.

An example of the rapid information processing of the system is shown in Fig. 7.26, obtained under conditions in which the local recurrent collaterals can contribute to correct performance because the feedforward (FF) inputs from the previous stage are noisy. (The noise implemented in these simulations was some imperfection in the FF signals produced by some alterations to the FF synaptic weights.) Fig. 7.26 shows that, when the FF carry an incomplete signal, some information is still transmitted successfully in the 'No RC' condition (in which the recurrent collateral connections in each layer are switched off), and with a relatively short latency. However, the noise term in the FF synaptic strengths makes the retrieval fail more and more layer by layer. When in contrast the recurrent collaterals (RC) are present and operating after Hebbian training, the amount of information retrieved is now much higher, because the RC are able to correct a good part of the erroneous information injected into the neurons by the noisy FF synapses. In Layer 4, 66 ms after cue injection in Layer 1, the information in the Hebbian RC case is 0.2 bits higher than that provided by the FF connections in the 'No RC' condition. This shows that the RC are able to retrieve information in Layer 4 that is not available by any other purely FF mechanism after only roughly 50–55 ms from the time when Layer 1 responds. (This corresponds to 17–18 ms per layer.)

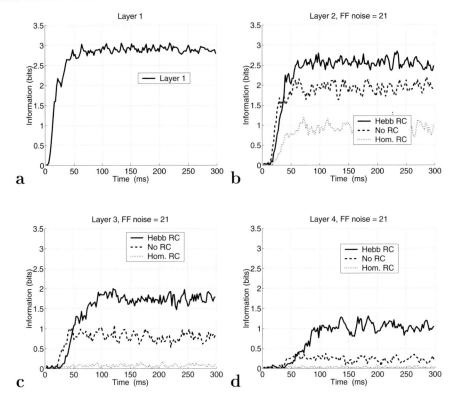

Fig. 7.26 The speed of information processing in a 4-layer network with integrate-and-fire neurons. The information time course of the average information carried by the responses of a population of 30 excitatory neurons in each layer. In the simulations considered here, there is noise in the feedforward (FF) synapses. Layer 1 was tested in just one condition. Layers 2–4 are tested in three different conditions: No RC (in which the recurrent collateral synaptic effects do not operate), Hebbian RC (in which the recurrent collaterals have been trained by as associative rule and can help pattern retrieval in each layer), and a control condition named Homogeneous RC (in which the recurrent collaterals could inject current into the neurons, but no useful information was provided by them because they were all set to the same strength). (After Panzeri, Rolls, Battaglia and Lavis, 2001.)

A direct comparison of the latency differences in layers 1–4 of the integrate-and-fire network simulated by Panzeri, Rolls, Battaglia and Lavis (2001) is shown in Fig. 7.27. The results are shown for the Hebbian condition illustrated in Fig. 7.26, and separate curves are shown for each of the layers 1–4. The Figure shows that, with the time constant of the synapses set to 10 ms, the network can operate with full utilization of and benefit from recurrent processing within each layer in a time which enables the signal to propagate through the 4–layer system with a time course of approximately 17 ms per layer.

The overall results of Panzeri, Rolls, Battaglia and Lavis (2001) were as follows. Through the implementation of continuous dynamics, latency differences were found in information retrieval of only 5 ms per layer when local excitation was absent and processing was purely feedforward. However, information latency differences increased significantly when non-associative local excitation (simulating spontaneous firing or unrelated inputs present in the brain) was included. It was also found that local recurrent excitation through associatively

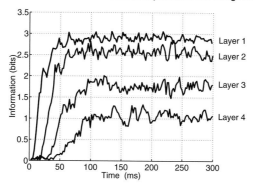

Fig. 7.27 The speed of information processing in a 4-layer network with integrate-and-fire neurons. The information time course of the average information carried by the responses of a population of 30 excitatory neurons in each layer. The results are shown for the Hebbian condition illustrated in Fig. 7.26, and separate curves are shown for each of the layers 1–4. The Figure shows that, with the time constant of the synapses set to 10 ms, the network can operate with full utilization of and benefit from recurrent processing within each layer in a time in the order of 17 ms per layer. (After Panzeri, Rolls, Battaglia and Lavis, 2001.)

modified synapses can contribute significantly to processing in as little as 15 ms per layer, including the feedforward and local feedback processing. Moreover, and in contrast to purely feed-forward processing, the contribution of local recurrent feedback was useful and approximately this rapid even when retrieval was made difficult by noise. These findings provide evidence that cortical information processing can be very fast even when local recurrent circuits are critically involved. The time cost of this recurrent processing is minimal when compared with a feedforward system with spontaneous firing or unrelated inputs already present, and the performance is better than that of a purely feedforward system when noise is present.

It is concluded that local feedback loops within each cortical area can contribute to fast visual processing and cognition.

7.6.6 Spike response model

In this section, we describe another mathematical model that models the activity of single spiking neurons. This model captures the principal effects of real neurons in a realistic way and is simple enough to permit analytical calculations (Gerstner, Ritz and Van Hemmen 1993). In contrast to some integrate-and-fire models (Tuckwell 1988), which are essentially given by differential equations, the spike-response model is based on response kernels that describe the integrated effect of spike reception or emission on the membrane potential. In this model, spikes are generated by a threshold process (i.e. the firing time t' is given by the condition that the membrane potential reaches the firing threshold θ; that is, $h(t') = \theta$). Figure 7.28 (bottom) shows schematically the spike-generating mechanism.

The membrane potential is given by the integration of the input signal weighted by a kernel defined by the equations

$$h(t') = h^{\text{refr}}(t') + h^{\text{syn}}(t') \tag{7.48}$$

$$h^{\text{refr}}(t') = \int_0^\infty \eta^{\text{refr}}(z)\delta(t' - z - t'_{\text{last}})dz \tag{7.49}$$

$$h^{\text{syn}}(t') = \sum_j J_j \int_0^\infty \Lambda(z', t' - t'_{\text{last}}) s(t' - z') dz' . \tag{7.50}$$

The kernel $\eta^{\text{refr}}(z)$ is the refractory function. If we consider only absolute refractoriness, $\eta^{\text{refr}}(z)$ is given by:

$$\eta^{\text{refr}}(z) = \begin{cases} -\infty & \text{for } 0 < z \leq \tau^{\text{refr}} \\ 0 & \text{for } z \geq \tau^{\text{refr}} \end{cases} \tag{7.51}$$

where τ^{refr} is the absolute refractory time. The time t'_{last} corresponds to the last postsynaptic spike (i.e. the most recent firing of the particular neuron). The second response function is the synaptic kernel $\Lambda(z', t' - t'_{\text{last}})$. It describes the effect of an incoming spike on the membrane potential at the soma of the postsynaptic neuron, and it eventually includes also the dependence on the state of the receiving neuron through the difference $t' - t'_{\text{last}}$ (i.e. through the time that has passed since the last postsynaptic spike). The input spike train yields $s(t' - z') = \sum_i \delta(t' - z' - t_{ij})$, t_{ij} being the ith spike at presynaptic input j. In order to simplify the discussion and without losing generality, let us consider only a single synaptic input, and therefore we can remove the subindex j. In addition, we assume that the synaptic strength J is positive (i.e. excitatory). Integrating equations 7.49 and 7.50, we obtain

$$h(t') = \eta^{\text{refr}}(t' - t'_{\text{last}}) + J \sum_i \Lambda(t' - t_i, t' - t'_{\text{last}}) \tag{7.52}$$

Synaptic kernels are of the form

$$\Lambda(t' - t_i, t' - t'_{\text{last}}) = \text{H}(t' - t_i)\text{H}(t_i - t'_{\text{last}})\Psi(t' - t_i) \tag{7.53}$$

where $\text{H}(s)$ is the step (or Heaviside) function which vanishes for $s \leq 0$ and takes a value of 1 for $s > 0$. After firing, the membrane potential is reset according to the renewal hypothesis.

This spike-response model is not used in the models described in this book, but is presented to show alternative approaches to modelling the dynamics of network activity.

7.7 Network dynamics: introduction to the mean field approach

Units whose potential and conductances follow the equations above can be assembled together in a model network of any composition and architecture. It is convenient to imagine that units are grouped into classes, such that the parameters quantifying the electrophysiological properties of the units are uniform, or nearly uniform, within each class, while the parameters assigned to synaptic connections are uniform or nearly uniform for all connections from a given presynaptic class to another given postsynaptic class. The parameters that have to be set in a model at this level of description are quite numerous, as listed in Tables 7.3 and 7.4.

In the limit in which the parameters are constant within each class or pair of classes, a mean-field treatment can be applied to analyze a model network, by summing equations that describe the dynamics of individual units to obtain a more limited number of equations that describe the dynamical behaviour of groups of units (Frolov and Medvedev 1986). The treatment is exact in the further limit in which very many units belong to each class, and is

Table 7.3 Cellular parameters (chosen according to the class of each unit)

V_{rest}	Resting potential
V_{thr}	Threshold potential
V_{ahp}	Reset potential
V_K	Potassium conductance equilibrium potential
C	Membrane capacitance
τ_K	Potassium conductance time constant
g_0	Leak conductance
Δg_K	Extra potassium conductance following a spike
Δt	Overall transmission delay

Table 7.4 Synaptic parameters (chosen according to the classes of presynaptic and postsynaptic units)

V_α	Synapse equilibrium potential
τ_α	Synaptic conductance time constant
Δg_α	Conductance opened by one presynaptic spike
Δt_α	Delay of the connection

an approximation if each class includes just a few units. Suppose that N_C is the number of classes defined. Summing equations 7.41 and 7.42 across units of the same class results in N_C functional equations describing the evolution in time of the fraction of cells of a particular class that at a given instant have a given membrane potential. In other words, from a treatment in which the evolution of the variables associated with each unit is followed separately, one moves to a treatment based on density functions, in which the common behaviour of units of the same class is followed together, keeping track solely of the portion of units at any given value of the membrane potential. Summing equation 7.44 or 7.45 across connections with the same class of origin and destination results in $N_C \times N_C$ equations describing the dynamics of the overall summed conductance opened on the membrane of a cell of a particular class by all the cells of another given class. A more explicit derivation of mean-field equations is given by Treves (1993) and in Section 7.8.

The system of mean-field equations can have many types of asymptotic solutions for long times, including chaotic, periodic, and stationary ones. The stationary solutions are stationary in the sense of the mean fields, but in fact correspond to the units of each class firing tonically at a certain rate. They are of particular interest as the dynamical equivalent of the steady states analyzed by using non-dynamical model networks. In fact, since the neuronal current-to-frequency transfer function resulting from the dynamical equations is rather similar to a threshold linear function (see Fig. 7.23), and since each synaptic conductance is constant in time, the stationary solutions are essentially the same as the states described using model networks made up of threshold linear, non-dynamical units. Thus the dynamical formulation reduces to the simpler formulation in terms of steady-state rates when applied to asymptotic stationary solutions; but, among simple rate models, it is equivalent only to those that allow description of the continuous nature of neuronal output, and not to those, for example based on binary units, that do not reproduce this fundamental aspect. The advantages of the dynamical formulation are that (i) it enables one to describe the character and prevalence of other types of asymptotic solutions, and (ii) it enables one to understand how the network reaches, in time, the asymptotic behaviour.

The development of this mean-field approach, and the foundations for its application to models of cortical visual processing and attention, are described in Section 7.8.

7.8 Mean-field based neurodynamics

A model of brain functions requires the choice of an appropriate theoretical framework, which permits the investigation and simulation of large-scale biologically realistic neural networks. Starting from the mathematical models of biologically realistic single neurons (i.e. spiking neurons), one can derive models that describe the joint activity of pools of equivalent neurons. This kind of neurodynamical model at the neuronal assembly level is motivated by the experimental observation that cortical neurons of the same type that are near to each other tend to receive similar inputs. As described in the previous Section, it is convenient in this simplified approach to neural dynamics to consider all neurons of the same type in a small cortical volume as a computational unit of a neural network. This computational unit is called a neuronal pool or assembly. The mathematical description of the dynamical evolution of neuronal pool activity in multimodular networks, associated with different cortical areas, establishes the roots of the dynamical approach that we will use in Chapters 9, 10 and 11. In this Section, we introduce the mathematical fundamentals utilized for a neurodynamical description of pool activity. Beginning at the microscopic level and using single spiking neurons to form the pools of a network, we derive the mathematical formulation of the neurodynamics of cell assemblies. Further, we introduce the basic architecture of neuronal pool networks that fulfil the basic mechanisms consistent with the biased competition hypothesis. Each of these networks corresponds to cortical areas that also communicate with each other. We describe therefore the dynamical interaction between different modules or networks, which will be the basis for the implementation of attentional top-down bias.

7.8.1 Population activity

We now introduce thoroughly the concept of a neuronal pool and the differential equations representing the neurodynamics of pool activity.

Starting from individual spiking neurons one can derive a differential equation that describes the dynamical evolution of the averaged activity of a pool of extensively many equivalent neurons. Several areas of the brain contain groups of neurons that are organized in populations of units with (somewhat) similar properties (though in practice the neurons convey independent information, as described in Chapter 5). These groups for mean-field modelling purposes are usually called pools of neurons and are constituted by a large and similar population of identical spiking neurons that receive similar external inputs and are mutually coupled by synapses of similar strength. Assemblies of motor neurons (Kandel, Schwartz and Jessel 2000) and the columnar organization in the visual and somatosensory cortex (Hubel and Wiesel 1962, Mountcastle 1957) are examples of these pools. Each single cell in a pool can be described by a spiking model, e.g. the spike response model presented in Section 7.6.6. Due to the fact that for large-scale cortical modelling, neuronal pools form a relevant computational unit, we adopt a population code. We take the activity level of each pool of neurons as the relevant dependent variable rather than the spiking activity of individual neurons. We therefore derive a dynamical model for the mean activity of a neural population.

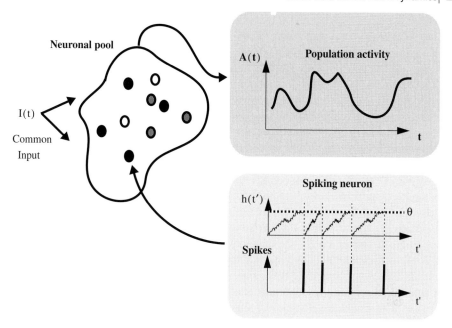

Fig. 7.28 Population averaged rate of a neuronal pool of spiking neurons (top) and the action potential generating mechanism of single neurons (bottom). In a neuronal pool, the mean activity $A(t)$ is determined by the proportion of active neurons by counting the number of spikes in a small time interval Δt and dividing by the number of neurons in the pool and by Δt. Spikes are generated by a threshold process. The firing time t' is given by the condition that the membrane potential $h(t')$ reaches the firing threshold θ. The membrane potential $h(t')$ is given by the integration of the input signal weighted by a given kernel (see text for details).

In a population of M neurons, the mean activity $A(t)$ is determined by the proportion of active neurons by counting the number of spikes $n_{\text{spikes}}(t, t + \Delta t)$ in a small time interval Δt and dividing by M and by Δt (Gerstner 2000), i.e. formally

$$A(t) = \lim_{\Delta t \to 0} \frac{n_{\text{spikes}}(t, t + \Delta t)}{M \Delta t} . \tag{7.54}$$

As indicated by Gerstner (2000), and as depicted in Fig. 7.28, the concept of pool activity is quite different from the definition of the average firing rate of a single neuron. Contrary to the concept of temporal averaging over many spikes of a single cell, which requires that the input is slowly varying compared with the size of the temporal averaging window, a coding scheme based on pool activity allows rapid adaptation to real-world situations with quickly changing inputs. It is possible to derive dynamical equations for pool activity levels by utilizing the mean-field approximation (Wilson and Cowan 1972, Abbott 1991, Amit and Tsodyks 1991). The mean-field approximation consists of replacing the temporally averaged discharge rate of a cell with an equivalent momentary activity of a neural population (ensemble average) that corresponds to the assumption of ergodicity. According to this approximation, we categorize each cell assembly by means of its activity $A(t)$. A pool of excitatory neurons without external input can be described by the dynamics of the pool activity given by

$$\tau \frac{\partial A(t)}{\partial t} = -A(t) + qF(A(t)) \tag{7.55}$$

where the first term on the right hand side is a decay term and the second term takes into account the excitatory stimulation between the neurons in the pool. In the previous equation, the non-linearity

$$F(x) = \frac{1}{T_r - \tau \log(1 - \frac{1}{\tau x})} \tag{7.56}$$

is the response function (transforming current into discharge rate) for a spiking neuron with deterministic input, membrane time constant τ, and absolute refractory time T_r. Equation 7.55 was derived by Gerstner (2000) assuming adiabatic conditions. Gerstner (2000) has shown that the population activity in a homogeneous population of neurons can be described by an integral equation. A systematic reduction of the integral equation to a single differential equation of the form 7.55 always supposes that the activity changes only slowly compared with the typical interval length. In other words, the mean-field approach described in the above equations and utilized in Chapters 9–11 generates a dynamics that neglects fast, transient behaviour. This means that we are assuming that rapid oscillations (and synchronization) do not play a computational role at least for the brain functions that we will consider. Rapid oscillations of neural activity could have a relevant functional role, namely of dynamical cooperation between pools in the same or different brain areas. It is well known in the theory of dynamical systems that the synchronization of oscillators is a cooperative phenomenon. Cooperative mechanisms might complement the competitive mechanisms on which our computational cortical model is based (see Chapter 13).

7.8.2 A basic computational module based on biased competition

We are interested in the neurodynamics of modules composed of several pools that implement a competitive mechanism[13]. This can be achieved by connecting the pools of a given module with a common inhibitory pool, as is schematically shown in Fig. 7.29.

In this way, the more pools of the module that are active, the more active the common inhibitory pool will be, and consequently, the more feedback inhibition will affect the pools in the module, such that only the most excited group of pools will survive the competition. On the other hand, external top-down bias could shift the competition in favour of a specific group of pools. This basic computational module implements therefore the biased competition hypothesis described in Chapter 6. Let us assume that there are m pools in a given module. The system of differential equations describing the dynamics of such a module is given by two differential equations, both of the type of equation 7.55. The first differential equation describes the dynamics of the activity level of the excitatory pools (pyramidal neurons) and is mathematically expressed by

$$\tau \frac{\partial A_i(t)}{\partial t} = -A_i(t) + aF(A_i(t)) - bF(A^I(t)) +$$
$$I_0 + I_i^E(t) + I_i^A(t) + \nu \quad ; \qquad \text{for } i = 1, ..., m \tag{7.57}$$

and the second one describes the dynamics of the activity level of the common inhibitory pool for each feature dimension (stellate neurons)

---

[13] These neurodynamics are used in Chapters 9–11.

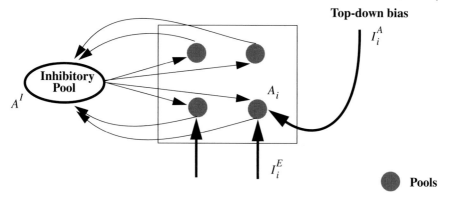

Fig. 7.29 Basic computational module for biassed competition: a competitive network with external top-down bias. Excitatory pools with activity A_i for the ith pool are connected with a common inhibitory pool with activity A^I in order to implement a competition mechanism. I_i^E is the external sensory input to the cells in pool i, and I_i^A attentional top-down bias, an external input coming from higher modules. The external top-down bias can shift the competition in favour of a specific pool or group of pools. This architecture is similar to that shown in Fig. 7.11, but with competition between pools of similar neurons.

$$\tau_p \frac{\partial A^I(t)}{\partial t} = -A^I(t) + c \sum_{i=1}^{m} \mathrm{F}(A_i(t)) - d\mathrm{F}(A^I(t)) \tag{7.58}$$

where $A_i(t)$ is the activity for pool i, $A^I(t)$ is the activity in the inhibitory pool, I_0 is a diffuse spontaneous background input, $I^E(t)$ is the external sensory input to the cells in pool i, and ν is additive Gaussian noise. The attentional top-down bias $I_i^A(t)$ is defined as an external input coming from higher modules that is not explicitly modelled.

A qualitative description of the main fixed point attractors of the system of differential equations 7.57 and 7.58 was provided by Usher and Niebur (1996). Basically, we will be interested in the fixed points corresponding to zero activity and larger activation. The parameters will therefore be fixed such that the dynamics evolves to these attractors.

7.8.3 Multimodular neurodynamical architectures

In order to model complex psychophysically and neuropsychologically relevant brain functions like visual search or object recognition, we must take into account the computational role of individual brain areas and their mutual interaction. The macroscopic phenomenological behaviour will therefore be the result of the mutual interaction of several computational modules[14].

The dynamical coupling of different basic modules in a multimodular architecture can be described, in our neurodynamical framework, by allowing mutual interaction between pools belonging to different modules. Figure 7.30 shows schematically this idea. The system of differential equations describing the global dynamics of such a multimodular system is given by a set of equations of the type of equation 7.55.

---

[14]In this sense, our approach is a functional approach that integrates elements of Gall's localizationist view and Flourens' idea of aggregate fields (Kandel, Schwartz and Jessel 2000). Gall's localizationist view is consistent with our identification of specific local brain areas with individual computational modules, whereas Flourens' aggregate

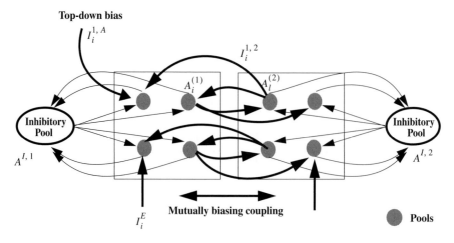

Fig. 7.30 Two competitive networks mutually biased through intermodular connections. The activity $A_i^{(1)}$ of the ith excitatory pool in module 1 (on the left) and of the lth excitatory pool in module 2 (on the right) are connected by the mutually biasing coupling $I_i^{1,2}$. The architecture could implement top-down feedback originating from the interaction between brain areas that are explicitly modelled in the system. (Module 2 might be the higher module.) The external top-down bias $I_i^{1,A}$ corresponds to the coupling to pool i of module 1 from brain area A that is not explicitly modelled in the system.

The excitatory pools belonging to a module obey the following equations:

$$\tau \frac{\partial A_i^{(j)}(t)}{\partial t} = -A_i^{(j)}(t) + aF(A_i^{(j)}(t)) - bF(A^{I,j}(t)) +$$
$$I_i^{j,k} + I_0 + I_i^E(t) + I_i^A(t) + \nu \qquad \text{for } i = 1, ..., m \qquad (7.59)$$

and the corresponding inhibitory pools evolve according to

$$\tau_p \frac{\partial A^{I,j}(t)}{\partial t} = -A^{I,j}(t) + c \sum_{i=1}^{m} F(A_i^{(j)}(t)) - dF(A^{I,j}(t)). \qquad (7.60)$$

The mutual coupling $I_i^{j,k}$ between module (j) and (k) is given by

$$I_i^{j,k} = \sum_l W_{il} F(I_l^{(k)}(t)) \qquad (7.61)$$

where W_{il} is the synaptic coupling strength between the pool i of module (j) and pool l of module (k). This mutual coupling term can be interpreted as a top-down bias originating from the interaction between brain areas that is explicitly modelled in our system. On the other hand, the external top-down bias I_i^A corresponds to the coupling with brain areas A that are not explicitly modelled in our system.

Additionally, it is interesting to note that the top-down bias in this kind of architecture modulates the response of the pool activity in a multiplicative manner. Responses of neurons

fields approach is analogous in our approach to the global dynamical interactions between different modules, which explain the observed macroscopic behaviour.

(a)

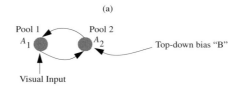

(b)

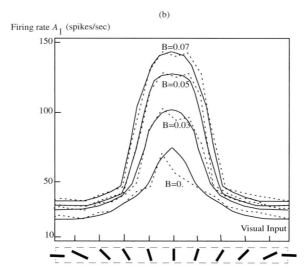

Fig. 7.31 (a) The basic building block of the system utilizes non-specific competition between pools within the same module and specific mutual facilitation between pools in different modules. Excitatory neuronal pools within the same module compete with each other through one or more inhibitory neuronal pool(s) I_I with activity A^{I_I}. Excitatory pool 1 (with activity A_1) receives a bottom-up (visual) input I_1^V, and excitatory pool 2 receives a top-down ('attentional') bias input I_2^A. Excitatory neuronal pools in the two different modules can excite each other via mutually biased coupling. (b) The effect of altering the bias input B to pool 2 on the responses or activity of pool 1 to its orientation-tuned visual input (see text).

in parietal area 7a are modulated by combined eye and head movement, exhibiting a multiplicative gain modulation that modifies the amplitude of the neural responses to retinal input but does not change the preferred retinal location of a cell, nor in general the width of the receptive field (Brotchie, Andersen, Snyder and Goodman 1995). It has also been suggested that multiplicative gain modulation might play a role in translation invariant object representation (Salinas and Abbott 1997). We will use this multiplicative effect for formulating an architecture for attentional gain modulation, which can contribute to correct translation invariant object recognition in ways to be described, in Chapters 9–11.

We show now that multiplicative-like responses can arise from the top-down biased mutual interaction between pools. Another alternative architecture that can also perform product operations on additive synaptic inputs was proposed by Salinas and Abbott (1996). Our basic architecture for showing this multiplicative effect is presented schematically in Fig. 7.31a.

Two pools are mutually connected via fixed weights. The first pool or unit receives a bottom-up visual input I_1^V, modelled by the response of a vertically-oriented complex cell. The second pool receives a top-down attentional bias $I_2^A = B$. The two pools are mutually coupled with unity weight. The equations describing the activities of the two pools are given by:

$$\tau \frac{\partial A_1(t)}{\partial t} = -A_1(t) + \alpha F(A_1(t)) + I_o + I_1^V + \nu \tag{7.62}$$

$$\tau \frac{\partial A_2(t)}{\partial t} = -A_2(t) + \alpha F(A_2(t)) + I_o + I_2^A + \nu \tag{7.63}$$

where A_1 and A_2 are the activities of pool 1 and pool 2 respectively, $\alpha = 0.95$ is the coefficient of recurrent self-excitation of the pool, ν is the noise input to the pool drawn from a normal distribution $N(\mu = 0, \sigma = 0.02)$, and $I_o = 0.025$ is a direct current biasing input to the pool.

Simulation of the above dynamical equations produces the results shown in Fig. 7.31b. The orientation tuning curve of unit (or pool) 1 was modulated by a top-down bias B introduced to unit (pool) 2. The gain modulation was transmitted through the coupling from pool 2 to pool 1 after a few steps of evolution of the dynamical equations. Without the feedback from unit 2, unit 1 exhibits the orientation tuning curve shown as $(B = 0)$. As B increased, the increase in pool 1's response to the vertical bar was significantly greater than the increase in its response to the horizontal bar. Therefore, the attentional gain modulation produced in pool 1 through the mutual coupling was not a simple additive effect, but had a strong multiplicative component. The net effect was that the width of the orientation tuning curve of the cell was roughly preserved under attentional modulation. This was due to the non-linearity in the activation function.

This finding is basically consistent with McAdams and Maunsell's observation (1999) on the effect of attention on the orientation tuning curves of neurons in V4.

Summarizing, the neurodynamics of the competitive mechanisms between neuronal pools, and their mutual gain modulation, are the two main ingredients that we will use in Chapters 9, 10 and 11 for proposing a cortical architecture that models attention and different kinds of object search in a visual scene.

7.9 Interacting attractor networks

It is prototypical of the cerebral neocortical areas that there are recurrent collateral connections between the neurons within an area or module, and forward connections to the next cortical area in the hierarchy, which in turn sends backprojections (see Section 1.11). This architecture, made explicit in Fig. 12.2, immediately suggests, given that the recurrent connections within a module, and the forward and backward connections, are likely to be associatively modifiable, that the operation incorporates at least to some extent, interactions between coupled attractor (autoassociation) networks. For these reasons, it is important to analyze the rules that govern the interactions between coupled attractor networks. This has been done using the formal type of model described in Section 12.1.2.

One boundary condition is when the coupling between the networks is so weak that there is effectively no interaction. This holds when the coupling parameter g between the networks is less than approximately 0.002, where the coupling parameter indicates the relative strength of the intermodular to the intramodular connections, and measures effectively the relative strengths of the currents injected into the neurons by the inter-modular relative to the intramodular (recurrent collateral) connections (Renart, Parga and Rolls 1999b). At the other extreme, if the coupling parameter is strong, all the networks will operate as a single

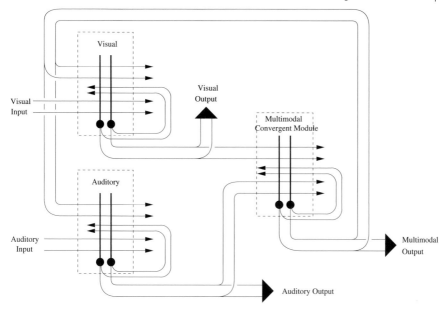

Fig. 7.32 A two-layer set of attractor nets in which feedback from layer 2 can influence the states reached in layer 1. Layer 2 could be a higher cortical visual area with convergence from earlier cortical visual areas (see Chapter 8). Layer 2 could also be a multimodal area receiving inputs from unimodal visual and auditory cortical areas, as labelled. Each of the 3 modules has recurrent collateral synapses that are trained by an associative synaptic learning rule, and also inter-modular synaptic connections in the forward and backward direction that are also associatively trained. Attractors are formed within modules, the different modules interact, and attractors are also formed by the forward and backward intermodular connections. The higher area may affect not only the states reached during attractor settling in the input layers, but may also, as a result of this, influence the representations that are learned in earlier cortical areas. A similar principle may operate in any multilayer hierarchical cortical processing system, such as the ventral visual system, in that the categories that can be formed only at later stages of processing may help earlier stages to form categories relevant to what can be diagnosed at later stages.

attractor network, together able to represent only one state (Renart, Parga and Rolls 1999b). This critical value of the coupling parameter (at least for reciprocally connected networks with symmetric synaptic strengths) is relatively low, in the region of 0.024 (Renart, Parga and Rolls 1999b). This is one reason why cortico-cortical backprojections are predicted to be quantitatively relatively weak, and for this reason it is suggested end on the apical parts of the dendrites of cortical pyramidal cells (see Section 1.11). In the strongly coupled regime when the system of networks operates as a single attractor, the total storage capacity (the number of patterns that can be stored and correctly retrieved) of all the networks will be set just by the number of synaptic connections received from other neurons in the network, a number in the order of a few thousand. This is one reason why connected cortical networks are thought not to act in the strongly coupled regime, because the total number of memories that could be represented in the whole of the cerebral cortex would be so small, in the order of a few thousand, depending on the sparseness of the patterns (see equation 7.15) (O'Kane and Treves 1992).

Between these boundary conditions, that is in the region where the inter-modular coupling parameter g is in the range 0.002–0.024, it has been shown that interesting interactions can occur (Renart, Parga and Rolls 1999b, Renart, Parga and Rolls 1999a). In a bimodular architecture, with forward and backward connections between the modules, the capacity of one module can be increased, and an attractor is more likely to be found under noisy conditions, if there is a consistent pattern in the coupled attractor. By consistent we mean a pattern that during training was linked associatively by the forward and backward connections, with the pattern being retrieved in the first module. This provides a quantitative model for understanding some of the effects that backprojections can produce by supporting particular states in earlier cortical areas (Renart, Parga and Rolls 1999b). The total storage capacity of the two networks is however, in line with O'Kane and Treves (1992), not a great deal greater than the storage capacity of one of the modules alone. Thus the help provided by the attractors in falling into a mutually compatible global retrieval state (in e.g. the scenario of a hierarchical system) is where the utility of such coupled attractor networks must lie. Another interesting application of such weakly coupled attractor networks is in coupled perceptual and short term memory systems in the brain, described in Section 12.1.

In a trimodular attractor architecture shown in Fig. 7.32 (which is similar to the architecture of the multilayer competitive net illustrated in Fig. 7.12 but has recurrent collateral connections within each module), further interesting interactions occur that account for effects such as the McGurk effect, in which what is seen affects what is heard (Renart, Parga and Rolls 1999a). The effect was originally demonstrated with the perception of auditory syllables, which were influenced by what is seen (McGurk and MacDonald 1976). The trimodular architecture (studied using similar methods to those used by Renart, Parga and Rolls (1999b) and frequently utilizing scenarios in which first a stimulus was presented to a module, then removed during a memory delay period in which stimuli were applied to other modules) showed a phase with $g < 0.005$ in which the modules operated in an isolated way. With g in the range 0.005–0.012, an 'independent' regime existed in which each module could be in a separate state to the others, but in which interactions between the modules occurred, which could assist or hinder retrieval in a module depending on whether the states in the other modules were consistent or inconsistent. It is in this 'independent' regime that a module can be in a continuing attractor that can provide other modules with a persistent external modulatory input that is helpful for tasks such as making comparisons between stimuli processed sequentially (as in delayed match-to-sample tasks and visual search tasks) (see Section 12.1). In this regime, if the modules are initially quiescent, then application of a stimulus to one input module propagates to the central module, and from it to the non-stimulated input module as well (see Fig. 7.32). When g grows beyond 0.012, the picture changes and the independence between the modules is lost. The delay activity states found in this region (of the phase space) *always* involve the three modules in attractors correlated with consistent features associated in the synaptic connections. Also, since g is now larger, changes in the properties of the external stimuli have more impact on the delay activity states. The general trend seen in this phase under the change of stimulus after a previous consistent attractor has been reached is that, first, if the second stimulus is not effective enough (it is weak or brief), it is unable to move any of the modules from their current delay activity states. If the stimulus is made more effective, then as soon as it is able to change the state of the stimulated input module, the internal and non-stimulated input modules follow, and the whole network moves into the new consistent attractor selected by the second stimulus. In this case, the interaction between the modules is so large that it does not allow contradictory local delay activity states to coexist, and the network is described as being in a 'locked' state.

The conclusion is that the most interesting scenario for coupled attractor networks is when they are weakly coupled (in the trimodular architecture $0.005 < g < 0.012$), for then interactions occur whereby how well one module responds to its own inputs can be influenced by the states of the other modules, but it can retain partly independent representations. This emphasizes the importance of weak interactions between coupled modules in the brain (Renart, Parga and Rolls 1999b, Renart, Parga and Rolls 1999a, Renart, Parga and Rolls 2000).

These generally useful interactions between coupled attractor networks can be useful in implementing top-down constraint satisfaction (see Section 1.11) and short term memory (see Section 12.1). One type of constraint satisfaction in which they are also probably important is cross-modal constraint satisfaction, which occurs for example when the sight of the lips moving assists the hearing of syllables. If the experimenter mismatches the visual and auditory inputs, then auditory misperception can occur, as in the McGurk effect. In such experiments (McGurk and MacDonald 1976) the subject receives one stimulus through the auditory pathway (e.g. the syllables *ga-ga*) and a *different* stimulus through the visual pathway (e.g. the lips of a person performing the movements corresponding to the syllables *ba-ba* on a video monitor). These stimuli are such that their acoustic waveforms as well as the lip motions needed to pronounce them are rather different. One can then assume that although they share the same vowel 'a', the internal representation of the syllables is dominated by the consonant, so that the representations of the syllables *ga-ga* and *ba-ba* are not correlated either in the primary visual cortical areas or in the primary auditory ones. At the end of the experiment, the subject is asked to repeat what he heard. When this procedure is repeated with many subjects, it is found that roughly 50% of them claim to have heard either the auditory stimulus (*ga-ga*), or the visual one (*ba-ba*). The rest of the subjects report to have heard neither the auditory nor the visual stimuli, but actually a combination of the two (e.g. *gabga*) or even something else including phonemes not presented auditorally or visually (e.g. *gagla*).

Renart, Parga and Rolls (1999a) were able to show that the McGurk effect can be accounted for by the operation of coupled attractor networks of the form shown in Fig. 7.32. One input module is for the auditory input, the second is for the visual input, and both converge into a higher area which represents the syllable formed on the evidence of combination of the two inputs. There are backprojections from the convergent module back to the input modules. Persistent (continuing) inputs were applied to both the inputs, and during associative training of all the weights the visual and auditory inputs corresponded to the same syllable. When tested with inconsistent visual and auditory inputs, it was found for g between ~ 0.10 and ~ 0.11, the convergent module can either remain in a symmetric state in which it represents a mixture of the two inputs, or choose between one of the inputs, with either situation being stable. For lower g the convergent module always settles into a state corresponding to the input in one of the input modules. It is the random fluctuations produced during the convergence to the attractor that determine the pattern selected by the convergent module. When the convergent module becomes correlated with *one* of its stored patterns, the signal back-projected to the input module stimulated with the feature associated to that pattern becomes stronger and the overlap in this module is increased. Thus, with low values of the inter-module coupling parameter g, situations are found in which sometimes the input to one module dominates, and sometimes the input to the other module dominates what is represented in the convergent module, and sometimes mixture states are stable in the convergent module. This model can thus account for the influences that visual inputs can have on what is heard, in for example the McGurk effect.

The interactions between coupled attractor networks can lead to the following effects. Facilitation can occur in a module if its external input is matched by an input from another module, whereas suppression in a module of its response to an external input can occur if the two inputs mismatch. This type of interaction can be used in imaging studies to identify brain regions where different signals interact with each other. One example is to locate brain regions where multimodal inputs converge. If the inputs in two sensory modalities are consistent based on previous experience, then facilitation will occur, whereas if they are inconsistent, suppression of the activity in a module can occur. This is one of the effects described in the bimodular and trimodular architectures investigated by Renart, Parga and Rolls (1999b), Renart, Parga and Rolls (1999a) and Rolls and Stringer (2001b), and found in architectures such as that illustrated in Fig. 7.32.

If a multimodular architecture is trained with each of many patterns (which might be visual stimuli) in one module associated with one of a few patterns (which might be mood states) in a connected module, then interesting effects due to this asymmetry are found, as described in Section 12.3.6 and by Rolls and Stringer (2001b).

An interesting issue that arises is how rapidly a system of interacting attractor networks such as that illustrated in Fig. 7.32 settles into a stable state. Is it sufficiently rapid for the interacting attractor effects described to contribute to cortical information processing? It is likely that the settling of the whole system is quite rapid, if it is implemented (as it is in the brain) with synapses and neurons that operate with continuous dynamics, where the time constant of the synapses dominates the retrieval speed, and is in the order of 15 ms for each module, as described in Section 7.6 and by Panzeri, Rolls, Battaglia and Lavis (2001). In that Section, it is shown that a multimodular attractor network architecture can process information in approximately 15 ms per module (assuming an inactivation time constant for the synapses of 10 ms), and similarly fast settling may be expected of a system of the type shown in Fig. 7.32.

7.10 Perceptrons and one-layer error correction networks

The networks described in the next two Sections are capable of mapping a set of inputs to a set of required outputs using correction when errors are made. Although some of the networks are very powerful in the types of mapping they can perform, the power is obtained at the cost of learning algorithms that do not use local learning rules. A local learning rule specifies that synaptic strengths should be altered on the basis of information available locally at the synapse, for example the activity of the presynaptic and the post-synaptic neurons. Because the networks described here do not use local learning rules, their biological plausibility remains at present uncertain. One of the aims of future research must be to determine whether comparably difficult problems to those solved by the networks described in Sections 7.10 and 7.11 can be solved by biologically plausible neuronal networks.

We now describe one-layer networks taught by an error-correction algorithm. The term *perceptron* refers strictly to networks with binary threshold activation functions. The outputs might take the values only 1 or 0 for example. The term perceptron arose from networks designed originally to solve perceptual problems (Rosenblatt 1961, Minsky and Papert 1969), and these networks are referred to briefly below. If the output neurons have continuous-valued firing rates, then a more general error-correcting rule called the delta rule is used, and is

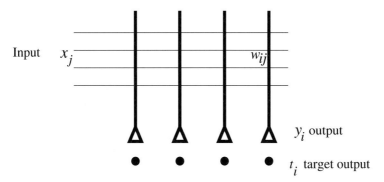

Fig. 7.33 One-layer perceptron.

introduced in this Chapter. For such networks, the activation function may be linear, or it may be non-linear but monotonically increasing, without a sharp threshold, as in the sigmoid activation function (see Fig. 1.3).

7.10.1 Architecture and general description

The one-layer error-correcting network has a set of inputs that it is desired to map or classify into a set of outputs (see Fig. 7.33). During learning, an input pattern is selected, and produces output firing by activating the output neurons through modifiable synapses, which then fire as a function of their typically non-linear activation function. The output of each neuron is then compared with a target output for that neuron given that input pattern, an error between the actual output and the desired output is determined, and the synaptic weights on that neuron are then adjusted to minimize the error. This process is then repeated for all patterns until the average error across patterns has reached a minimum. A one-layer error-correcting network can thus produce output firing for each pattern in a way that has similarities to a pattern associator. It can perform more powerful mappings than a pattern associator, but requires an error to be computed for each neuron, and for that error to affect the synaptic strength in a way that is not altogether local. A more detailed description follows.

These one-layer networks have a target for each output neuron (for each input pattern). They are thus an example of a supervised network. With the one-layer networks taught with the delta rule or perceptron learning rule described next, there is a separate teacher for each output neuron.

7.10.2 Generic algorithm (for a one-layer network taught by error correction)

For each input pattern and desired target output:

1. Apply an input pattern to produce input firing $\mathbf{x}$, and obtain the activation of each neuron in the standard way by computing the dot product of the input pattern and the synaptic weight vector. The synaptic weight vector can be initially zero, or have random values.

$$h_i = \sum_j x_j w_{ij} \qquad (7.64)$$

where $\sum\limits_{j}$ indicates that the sum is over the C input axons (or connections) indexed by j to each neuron.

2. Apply an activation function to produce the output firing y_i:

$$y_i = f(h_i) . \tag{7.65}$$

This activation function f may be sigmoid, linear, binary threshold, linear threshold, etc. If the activation function is non-linear, this helps to classify the inputs into distinct output patterns, but a linear activation function may be used if an optimal linear mapping is desired (see Adaline and Madaline, below).

3. Calculate the difference for each cell i between the target output t_i and the actual output y_i produced by the input, which is the error Δ_i

$$\Delta_i = t_i - y_i. \tag{7.66}$$

4. Apply the following learning rule, which corrects the (continuously variable) weights according to the error and the input firing x_j

$$\delta w_{ij} = k(t_i - y_i)x_j \tag{7.67}$$

where k is a constant that determines the learning rate. This is often called the delta rule, the Widrow-Hoff rule, or the LMS (Least Mean Squares) rule (see below).

5. Repeat steps 1–4 for all input pattern—output target pairs until the root mean square error becomes zero or reaches a minimum.

In general, networks taught by the delta rule may have linear, binary threshold, or non-linear but monotonically increasing (e.g. sigmoid) activation functions, and may be taught with binary or continuous input patterns (see Rolls and Treves (1998), Chapter 5). The properties of these variations are made clear next.

7.10.3 Capability and limitations of single-layer error-correcting networks

Perceptrons perform pattern classification. That is, each neuron classifies the input patterns it receives into classes determined by the teacher. This is thus an example of a supervised network, with a separate teacher for each output neuron. The classification is most clearly understood if the output neurons are binary, or are strongly non-linear, but the network will still try to obtain an optimal mapping with linear or near-linear output neurons.

When each neuron operates as a binary classifier, we can consider how many input patterns p can be classified by each neuron, and the classes of pattern that can be correctly classified. The result is that the maximum number of patterns that can be correctly classified by a neuron with C inputs is

$$p_{\max} = 2C \tag{7.68}$$

when the inputs have random continuous-valued inputs, but the patterns must be linearly separable (see Section A.2.1, and Hertz, Krogh and Palmer (1991)). For a one-layer network, no set of weights can be found that will perform the XOR (exclusive OR), or any other non-linearly separable function (see Appendix A).

Although the inability of one-layer networks with binary neurons to solve non-linearly separable problems is a limitation, it is not in practice a major limitation on the processing that can be performed in a neural network for a number of reasons. First, if the inputs can take continuous values, then if the patterns are drawn from a random distribution, the one-layer network can map up to $2C$ of them. Second, as described for pattern associators, the perceptron could be preceded by an expansion recoding network such as a competitive network with more output than input neurons. This effectively provides a two-layer network for solving the problem, and multilayer networks are in general capable of solving arbitrary mapping problems. Ways in which such multilayer networks might be trained are discussed later in this Chapter.

We now return to the issue of the capacity of one-layer perceptrons, that is, how many patterns p can be correctly mapped to correct binary outputs if the input patterns are linearly separable.

7.10.3.1 Output neurons with continuous values, and random patterns

Before treating this case, we note that if the inputs are orthogonal, then just as in the pattern associator, C patterns can be correctly classified, where there are C inputs, x_j, $(j = 1, C)$, per neuron. The argument is the same as for a pattern associator.

We consider next the capacity of a one-layer error-correcting network that learns patterns drawn from a random distribution. For neurons with continuous output values, whether the activation function is linear or not, the capacity (for fully distributed inputs) is set by the criterion that the set of input patterns must be linearly independent (see Hertz, Krogh and Palmer (1991)). (Three patterns are linearly independent if any one cannot be formed by addition (with scaling allowed) of the other two patterns — see Appendix A.) Given that there can be a maximum of C linearly independent patterns in a C-dimensional space (see Appendix A), the capacity of the perceptron with such patterns is C patterns. If we choose p random patterns with continuous values, then they will be linearly independent for $p \leq C$ (except for cases with very low probability when the randomly chosen values may not produce linearly independent patterns). (With random continuous values for the input patterns, it is very unlikely that the addition of any two, with scaling allowed, will produce a third pattern in the set.) Thus with continuous valued input patterns,

$$p_{\max} = C. \qquad (7.69)$$

If the inputs are not linearly independent, networks trained with the delta rule produce a Least Mean Squares (LMS) error (optimal) solution (see below).

7.10.3.2 Output neurons with binary threshold activation functions

Let us consider here strictly defined perceptrons, that is, networks with (binary) threshold output neurons, and taught by the perceptron learning procedure.

Capacity with fully distributed output patterns

The condition here for correct classification is that described in Appendix A for the AND and XOR functions, that the patterns must be linearly separable. If we consider random continuous-valued inputs, then the capacity is

$$p_{\max} = 2C \qquad (7.70)$$

(see Cover (1965), Hertz, Krogh and Palmer (1991); this capacity is the case with C large, and the number of output neurons small). The interesting point to note here is that, even with fully distributed inputs, a perceptron is capable of learning more (fully distributed) patterns than there are inputs per neuron. This formula is in general valid for large C, but happens to hold also for the AND function illustrated in Appendix A.2.1.

Sparse encoding of the patterns
    If the output patterns **y** are sparse (but still distributed), then just as with the pattern associator, it is possible to map many more than C patterns to correct outputs. Indeed, the number of different patterns or prototypes p that can be stored is

$$p \approx C/a \qquad (7.71)$$

where a is the sparseness of the target pattern **t**. p can in this situation be much larger than C (cf. Rolls and Treves (1990), and Rolls and Treves (1998) Appendix A3).

Perceptron convergence theorem
    It can be proved that such networks will learn the desired mapping in a finite number of steps (Block 1962, Minsky and Papert 1969, Hertz, Krogh and Palmer 1991). (This of course depends on there being such a mapping, the condition for this being that the input patterns are linearly separable.) This is important, for it shows that single-layer networks can be proved to be capable of solving certain classes of problem.
    As a matter of history, Minsky and Papert (1969) went on to emphasize the point that no one-layer network can correctly classify non-linearly separable patterns. Although it was clear that multilayer networks can solve such mapping problems, Minsky and Papert were pessimistic that an algorithm for training such a multilayer network would be found. Their emphasis that neural networks might not be able to solve general problems in computation, such as computing the XOR, which is a non-linearly separable mapping, resulted in a decline in research activity in neural networks. In retrospect, this was unfortunate, for humans are rather poor at solving parity problems such as the XOR (Thorpe, O'Regan and Pouget 1989), yet can perform many other useful neural network operations very fast. Algorithms for training multilayer perceptrons were gradually discovered by a number of different investigators, and became widely known after the publication of the algorithm described by Rumelhart, Hinton and Williams (1986b), and Rumelhart, Hinton and Williams (1986a). Even before this, interest in neural network pattern associators, autoassociators and competitive networks was developing (see Hinton and Anderson (1981), Kohonen (1977), Kohonen (1988)), but the acceptance of the algorithm for training multilayer perceptrons led to a great rise in interest in neural networks, partly for use in connectionist models of cognitive function (McClelland and Rumelhart 1986, McLeod, Plunkett and Rolls 1998), and partly for use in applications (see Bishop (1995)).
    In that perceptrons can correctly classify patterns provided only that they are linearly separable, but pattern associators are more restricted, perceptrons are more powerful learning devices than Hebbian pattern associators.

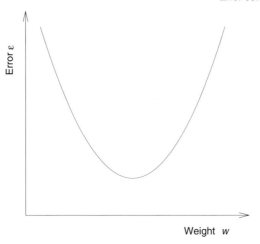

Fig. 7.34 The error function ϵ for a neuron in the direction of a particular weight w.

7.10.3.3 Gradient descent for neurons with continuous-valued outputs

We now consider networks trained by the delta (error correction) rule 7.67, and having continuous-valued outputs. The activation function may be linear or non-linear, but provided that it is differentiable (in practice, does not include a sharp threshold), the network can be thought of as gradually decreasing the error on every learning trial, that is as performing some type of gradient descent down a continuous error function. The concept of gradient descent arises from defining an error ϵ for a neuron as

$$\epsilon = \sum_{\mu}(t^{\mu} - y^{\mu})^2 \tag{7.72}$$

where μ indexes the patterns learned by the neuron. The error function for a neuron in the direction of a particular weight would have the form shown in Fig. 7.34. The delta rule can be conceptualized as performing gradient descent of this error function, in that for the jth synaptic weight on the neuron

$$\delta w_j = -k\partial\epsilon/\partial w_j \tag{7.73}$$

where $\partial\epsilon/\partial w_j$ is just the slope of the error curve in the direction of w_j in Fig. 7.34. This will decrease the weight if the slope is positive and increase the weight if the slope is negative. Given equation 7.66, and recalling that $h = \sum_j x_j w_j$, equation 7.34 becomes

$$\delta w_j = -k\partial/\partial w_j \sum_{\mu}[(t^{\mu} - \mathrm{f}(h^{\mu}))^2] \tag{7.74}$$

$$= 2k\sum_{\mu}[(t^{\mu} - y^{\mu})]\mathrm{f}'(h)x_j$$

where $\mathrm{f}'(h)$ is the derivative of the activation function. Provided that the activation function is monotonically increasing, its derivative will be positive, and the sign of the weight change will only depend on the mean sign of the error. Equation 7.75 thus shows one way in which, from a gradient descent conceptualization, equation 7.67 can be derived.

With linear output neurons, this gradient descent is proved to reach the correct mapping (see Hertz, Krogh and Palmer (1991)). (As with all single-layer networks with continuous-valued output neurons, a perfect solution is only found if the input patterns are linearly independent. If they are not, an optimal mapping is achieved, in which the sum of the squares of the errors is a minimum.) With non-linear output neurons (for example with a sigmoid activation function), the error surface may have local minima, and is not guaranteed to reach the optimal solution, although typically a near-optimal solution is achieved. Part of the power of this gradient descent conceptualization is that it can be applied to multilayer networks with neurons with non-linear but differentiable activation functions, for example with sigmoid activation functions (see Hertz, Krogh and Palmer (1991)).

7.10.4 Properties

The properties of single-layer networks trained with a delta rule (and of perceptrons) are similar to those of pattern associators trained with a Hebbian rule in many respects (see Section 7.2). In particular, the properties of generalization and graceful degradation are similar, provided that (for both types of network) distributed representations are used. The main differences are in the types of pattern that can be separated correctly, the learning speed (in that delta-rule networks can take advantage of many training trials to learn to separate patterns that could not be learned by Hebbian pattern associators), and in that the delta-rule network needs an error term to be supplied for each neuron, whereas an error term does not have to be supplied for a pattern associator, just an unconditioned or forcing stimulus. Given these overall similarities and differences, the properties of one-layer delta-rule networks are considered here briefly.

7.10.4.1 Generalization

During recall, delta-rule one-layer networks with non-linear output neurons produce appropriate outputs if a recall cue vector x_r is similar to a vector that has been learned already. This occurs because the recall operation involves computing the dot (inner) product of the input pattern vector x_r with the synaptic weight vector w_i, so that the firing produced, y_i, reflects the similarity of the current input to the previously learned input pattern x. Distributed representations are needed for this property. If two patterns that a delta-rule network has learned to separate are very similar, then the weights of the network will have been adjusted to force the different outputs to occur correctly. At the same time, this will mean that the way in which the network generalizes in the space between these two vectors will be very sharply defined. (Small changes in the input vector will force it to be classified one way or the other.)

7.10.4.2 Graceful degradation or fault tolerance

One-layer delta-rule networks show graceful degradation provided that the input patterns x are distributed.

7.10.4.3 Prototype extraction, extraction of central tendency, and noise reduction

These occur as for pattern autoassociators.

7.10.4.4 Speed

Recall is very fast in a one-layer pattern associator or perceptron, because it is a feedforward network (with no recurrent or lateral connections). Recall is also fast if the neuron has cell-like properties, because the stimulus input firings x_j ($j = 1, C$ axons) can be applied simultaneously to the synapses w_{ij}, and the activation h_i can be accumulated in one or two time constants of the synapses and dendrite (e.g. 10–20 ms) (see Section 7.6.5). Whenever the threshold of the cell is exceeded, it fires. Thus, in effectively one time step, which takes the brain no more than 10–20 ms, all the output neurons of the delta-rule network can be firing with rates that reflect the input firing of every axon.

Learning is as fast ('one-shot') in perceptrons as in pattern associators if the input patterns are orthogonal. If the patterns are not orthogonal, so that the error-correction rule has to work in order to separate patterns, then the network may take many trials to achieve the best solution (which will be perfect under the conditions described above).

7.10.4.5 Non-local learning rule

The learning rule is not truly local, as it is in pattern associators, autoassociators, and competitive networks, in that with one-layer delta-rule networks, the information required to change each synaptic weight is not available in the presynaptic terminal (reflecting the presynaptic rate) and the postsynaptic activation. Instead, an error for the neuron must be computed, possibly by another neuron, and then this error must be conveyed back to the postsynaptic neuron to provide the postsynaptic error term, which together with the presynaptic rate determines how much the synapse should change, as in equation 7.67,

$$\delta w_{ij} = k(t_i - y_i)x_j \tag{7.75}$$

where $(t_i - y_i)$ is the error.

A rather special architecture would be required if the brain were to utilize delta-rule error-correcting learning. One such architecture might require each output neuron to be supplied with its own error signal by another neuron. The possibility (Albus 1971) that this is implemented in one part of the brain, the cerebellum, is introduced in Rolls and Treves (1998) Chapter 9. Another functional architecture would require each neuron to compute its own error by subtracting its current activation by its **x** inputs from another set of afferents providing the target activation for that neuron. A neurophysiological architecture and mechanism for this is not currently known.

7.10.4.6 Interference

Interference is less of a property of single-layer delta rule networks than of pattern autoassociators and autoassociators, in that delta rule networks can learn to separate patterns even when they are highly correlated. However, if patterns are not linearly independent, then the delta rule will learn a least mean squares solution, and interference can be said to occur.

7.10.4.7 Expansion recoding

As with pattern associators and autoassociators, expansion recoding can separate input patterns into a form that makes them learnable, or that makes learning more rapid with only a few trials needed, by delta rule networks. It has been suggested that this is the role of the granule cells in the cerebellum, which provide for expansion recoding by 1,000:1 of the mossy fibre inputs before they are presented by the parallel fibres to the cerebellar Purkinje cells (Marr 1969, Albus 1971, Rolls and Treves 1998).

7.10.4.8 Utility of single-layer error-correcting networks in information processing by the brain

In the cerebellum, each output cell, a Purkinje cell, has its own climbing fibre, that distributes from its inferior olive cell its terminals throughout the dendritic tree of the Purkinje cell. It is this climbing fibre that controls whether learning of the **x** inputs supplied by the parallel fibres onto the Purkinje cell occurs, and it has been suggested that the function of this architecture is for the climbing fibre to bring the error term to every part of the postsynaptic neuron (see Rolls and Treves (1998) Chapter 9). This rather special arrangement with each output cell apparently having its own teacher is probably unique in the brain, and shows the lengths to which the brain might need to go to implement a teacher for each output neuron. The requirement for error-correction learning is to have the neuron forced during a learning phase into a state that reflects its error while presynaptic afferents are still active, and rather special arrangements are needed for this.

7.11 Multilayer perceptrons: backpropagation of error networks

7.11.1 Introduction

So far, we have considered how error can be used to train a one-layer network using a delta rule. Minsky and Papert (1969) emphasized the fact that one-layer networks cannot solve certain classes of input-output mapping problems (as described above). It was clear then that these restrictions would not apply to the problems that can be solved by feedforward multilayer networks, if they could be trained. A multilayer feedforward network has two or more connected layers, in which connections allow activity to be projected forward from one layer to the next, and in which there are no lateral connections within a layer. Such a multilayer network has an output layer (which can be trained with a standard delta rule using an error provided for each output neuron), and one or more hidden layers, in which the neurons do not receive separate error signals from an external teacher. (Because they do not provide the outputs of the network directly, and do not directly receive their own teaching error signal, these layers are described as hidden.) To solve an arbitrary mapping problem (in which the inputs are not linearly separable), a multilayer network could have a set of hidden neurons that would remap the inputs in such a way that the output layer can be provided with a linearly separable problem to solve using training of its weights with the delta rule. The problem was: how could the synaptic weights into the hidden neurons be trained in such a way that they would provide an appropriate representation? Minsky and Papert (1969) were pessimistic that such a solution would be found, and partly because of this, interest in computations in neural networks declined for many years. Although some work in neural networks continued in the following years (e.g. (Marr 1969, Marr 1970, Marr 1971, Willshaw and Longuet-Higgins 1969, Willshaw 1981, Malsburg 1973, Grossberg 1976a, Grossberg 1976b, Arbib 1964, Amari 1982, Amari, Yoshida and Kanatani 1977)), widespread interest in neural networks was revived by the type of approach to associative memory and its relation to human memory taken by the work described in the volume edited by Hinton and Anderson (1981), and by Kohonen (Kohonen 1977, Kohonen 1989). Soon after this, a solution to training a multilayer perceptron using backpropagation of error became widely known (Rumelhart, Hinton and Williams 1986b, Rumelhart, Hinton and Williams 1986a)

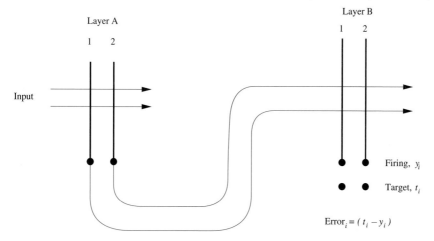

Fig. 7.35 A two-layer perceptron. Inputs are applied to layer A through modifiable synapses. The outputs from layer A are applied through modifiable synapses to layer B. Layer B can be trained using a delta rule to produce firing y_i which will approach the target t_i. It is more difficult to modify the weights in layer A, because appropriate error signals must be backpropagated from layer B.

(although earlier solutions had been found), and very great interest in neural networks and also in neural network approaches to cognitive processing (connectionism) developed (Rumelhart and McClelland 1986, McClelland and Rumelhart 1986, McLeod, Plunkett and Rolls 1998).

7.11.2 Architecture and algorithm

An introduction to the way in which a multilayer network can be trained by backpropagation of error is described next. Then we consider whether such a training algorithm is biologically plausible. A more formal account of the training algorithm for multilayer perceptrons (sometimes abbreviated MLP) is given by Rumelhart, Hinton and Williams (1986b), Rumelhart, Hinton and Williams (1986a) and Hertz, Krogh and Palmer (1991).

Consider the two-layer network shown in Fig. 7.35. Inputs to the hidden neurons in layer A feed forward activity to the output neurons in layer B. The neurons in the network have a sigmoid activation function. One reason for such an activation function is that it is non-linear, and non-linearity is needed to enable multilayer networks to solve difficult (non-linearly separable) problems. (If the neurons were linear, the multilayer network would be equivalent to a one-layer network, which cannot solve such problems.) Neurons B1 and B2 of the output layer, B, are each trained using a delta rule and an error computed for each output neuron from the target output for that neuron when a given input pattern is being applied to the network. Consider now the error that needs to be used to train neuron A1 by a delta rule. This error clearly influences the error of neuron B1 in a way that depends on the magnitude of the synaptic weight from neuron A1 to B1; and on the error of neuron B2 in a way that depends on the magnitude of the synaptic weight from neuron A1 to B2. In other words, the error for neuron A1 depends on:

the weight from A1 to B1 (w_{11}) · error of neuron B1
+ the weight from A1 to B2 (w_{21}) · error of neuron B2.

In this way, the error calculation can be propagated backwards through the network to any neuron in any hidden layer, so that each neuron in the hidden layer can be trained, once its error is computed, by a delta rule (which uses the computed error for the neuron and the presynaptic firing at the synapse to correct each synaptic weight). For this to work, the way in which each neuron is activated and sends a signal forward must be continuous (not binary), so that the extent to which there is an error in, for example, neuron B1 can be related back in a graded way to provide a continuously variable correction signal to previous stages. This is one of the requirements for enabling the network to descend a continuous error surface. The activation function must be non-linear (e.g. sigmoid) for the network to learn more than could be learned by a single-layer network. (Remember that a multilayer linear network can always be made equivalent to a single-layer linear network, and that there are some problems that cannot be solved by single-layer networks.) For the way in which the error of each output neuron should be taken into account to be specified in the error-correction rule, the position at which the output neuron is operating on its activation function must also be taken into account. For this, the slope of the activation function is needed, and because the slope is needed, the activation function must be differentiable. Although we indicated use of a sigmoid activation function, other activation functions that are non-linear and monotonically increasing (and differentiable) can be used. (For further details, see Rumelhart, Hinton and Williams (1986b), Rumelhart, Hinton and Williams (1986a) and Hertz, Krogh and Palmer (1991)).

7.11.3 Properties of multilayer networks trained by error backpropagation

7.11.3.1 Arbitrary mappings

Arbitrary mappings of non-linearly separable patterns can be achieved. For example, such networks can solve the XOR problem, and parity problems in general of which XOR is a special case. (The parity problem is to determine whether the sum of the (binary) bits in a vector is odd or even.) Multilayer feedforward backpropagation of error networks are not guaranteed to converge to the best solution, and may become stuck in local minima in the error surface. However, they generally perform very well.

7.11.3.2 Fast operation

The network operates as a feedforward network, without any recurrent or feedback processing. Thus (once it has learned) the network operates very fast, with a time proportional to the number of layers.

7.11.3.3 Learning speed

The learning speed can be very slow, taking many thousands of trials. The network learns to gradually approximate the correct input-output mapping required, but the learning is slow because of the credit assignment problem for neurons in the hidden layers. The credit assignment problem refers to the issue of how much to correct the weights of each neuron in the hidden layer. As the example above shows, the error for a hidden neuron could influence the errors of many neurons in the output layers, and the error of each output neuron reflects the error from many hidden neurons. It is thus difficult to assign credit (or blame) on any single trial to any particular hidden neuron, so an error must be estimated, and the net run until the weights of the crucial hidden neurons have become altered sufficiently to allow good performance of the network. Another factor that can slow learning is that if a neuron operates close to a horizontal part of its activation function, then the output of the neuron will depend rather little on its activation, and correspondingly the error computed to backpropagate will depend rather little on the activation of that neuron, so learning will be slow.

More general approaches to this issue suggest that the number of training trials for such a network will (with a suitable training set) be of the same order of magnitude as the number of synapses in the network (see Cortes, Jaeckel, Solla, Vapnik and Denker (1996)).

7.11.3.4   Number of hidden neurons and generalization

Backpropagation networks are generally intended to discover regular mappings between the input and output, that is mappings in which generalization will occur usefully. If there were one hidden neuron for every combination of inputs that had to be mapped to an output, then this would constitute a look-up table, and no generalization between similar inputs (or inputs not yet received) would occur. The best way to ensure that a backpropagation network learns the structure of the problem space is to set the number of neurons in the hidden layers close to the minimum that will allow the mapping to be implemented. This forces the network not to operate as a look-up table. A problem is that there is no general rule about how many hidden neurons are appropriate, given that this depends on the types of mappings required. In practice, these networks are sometimes trained with different numbers of hidden neurons, until the minimum number required to perform the required mapping has been approximated.

7.12   Biologically plausible networks

Given that the error for a hidden neuron in an error backpropagation network is calculated by propagating backwards information based on the errors of all the output neurons to which a hidden neuron is connected, and all the relevant synaptic weights, and the activations of the output neurons to define the part of the activation function on which they are operating, it is implausible to suppose that the correct information to provide the appropriate error for each hidden neuron is propagated backwards between real neurons. A hidden neuron would have to 'know', or receive information about, the errors of all the neurons to which it is connected, and its synaptic weights to them, and their current activations. If there were more than one hidden layer, this would be even more difficult.

To expand on the difficulties: first, there would have to be a mechanism in the brain for providing an appropriate error signal to each output neuron in the network. With the possible exception of the cerebellum, an architecture where a separate error signal could be provided for each output neurons is difficult to identify in the brain. Second, any retrograde passing of messages across multiple-layer forward-transmitting pathways in the brain that could be used for backpropagation seems highly implausible, not only because of the difficulty of getting the correct signal to be backpropagated, but also because retrograde signals passed by axonal transport in a multilayer net would take days to arrive, long after the end of any feedback given in the environment indicating a particular error. Third, as noted in Section 1.11, the backprojection pathways that are present in the cortex seem suited to perform recall, and this would make it difficult for them also to have the correct strength to carry the correct error signal.

A problem with the backpropagation of error approach in a biological context is thus that in order to achieve their competence, backpropagation networks use what is almost certainly a learning rule that is much more powerful than those that could be implemented biologically, and achieves its excellent performance by performing the mapping though a minimal number of hidden neurons. In contrast, real neuronal networks in the brain probably use much less powerful learning rules, in which errors are not propagated backwards, and at the same time have very large numbers of hidden neurons, without the bottleneck that helps to provide backpropagation networks with their good performance. A consequence of these differences between backpropagation and biologically plausible networks may be that the way in which biological networks solve difficult problems may be rather different from the way in which backpropagation networks find mappings. Thus the solutions found by connectionist systems may not always be excellent guides to how biologically plausible networks may perform on similar problems. Part of the challenge for future work is to discover how more biologically plausible networks than backpropagation networks can solve comparably hard problems, and then to examine the properties of these networks, as a perhaps more accurate guide to brain computation.

As stated above, it is a major challenge for brain research to discover whether there are algorithms that will solve comparably difficult problems to backpropagation, but with a local learning rule. Such algorithms may be expected to require many more hidden neurons than backpropagation networks, in that the brain does not appear to use information bottlenecks to help it solve difficult problems. The issue here is that much of the power of backpropagation algorithms arises because there is a minimal number of hidden neurons to perform the required mapping using a final one-layer delta-rule network. Useful generalization arises in such networks because with a minimal number of hidden neurons, the net sets the representation they provide to enable appropriate generalization. The danger with more hidden neurons is that the network becomes a look-up table, with one hidden neuron for every required output, and generalization when the inputs vary becomes poor. The challenge is to find a more biologically plausible type of network that operates with large numbers of neurons, and yet that still provides useful generalization.

7.13 Reinforcement learning

One issue with perceptrons and multilayer perceptrons that makes them generally biologically implausible for many brain regions is that a separate error signal must be supplied for each output neuron. When operating in an environment, usually a simple binary or scalar signal representing success or failure is received. Partly for this reason, there has been some interest in networks that can be taught with such a single reinforcement signal. In this Section, we introduce one such approach to such networks. We note that such networks are classified as supervised networks in which there is a single teacher, and that these networks attempt to perform an optimal mapping between an input vector and an output neuron or set of neurons. They thus solve the same class of problems as single and (if multilayer) multilayer perceptrons. They should be distinguished from pattern association networks in the brain, which might learn associations between previously neutral stimuli and primary reinforcers such as taste (signals that might be interpreted appropriately by a subsequent part of the brain), but do not attempt to produce arbitrary mappings between an input and an output, using a single reinforcement signal. A class of problems to which such networks might be applied are motor control problems. It was to such a problem that Barto and Sutton (Barto 1985, Sutton and Barto 1981) applied their reinforcement learning algorithm. The algorithm can in principle be applied to multilayer networks, and the learning is relatively slow. The algorithm is summarized by Rolls and Treves (1998) and Hertz, Krogh and Palmer (1991), and more recent developments in reinforcement learning are described by Sutton and Barto (1998).

7.14 Contrastive Hebbian learning: the Boltzmann machine

In a move towards a learning rule that is more local than in backpropagation networks, yet that can solve similar mapping problems in a multilayer architecture, we describe briefly contrastive Hebbian learning. The multilayer architecture has forward connections through the network to the output layer, and a set of matching backprojections from the output layer through each of the hidden layers to the input layer. The forward connection strength between any pair of neurons has the same value as the backward connection strength between the same two neurons, resulting in a symmetric set of forward and backward connection strengths. An input pattern is applied to the multilayer network, and an output is computed using normal feedforward activation processing with neurons with a sigmoid (non-linear and monotonically increasing) activation function. The output firing then via the backprojections is used to create firing of the input neurons. This process is repeated until the firing rates settle down, in an iterative way (which is similar to the settling of the autoassociative nets described in Section 7.3). After settling, the correlations between any two neurons are remembered, for this type of unclamped operation, in which the output neurons fire at the rates that the process just described produces. The correlations reflect the normal presynaptic and postsynaptic terms used in the Hebb rule, e.g. $(x_j y_i)^{\mathrm{uc}}$, where uc refers to the unclamped condition, and as usual x_j is the firing rate of the input neuron, and y_i is the activity of the receiving neuron. The output neurons are then clamped to their target values, and the iterative process just described is repeated, to produce for every pair of synapses in the network $(x_j y_i)^{\mathrm{c}}$, where the c refers now to the clamped condition. An error-correction term for each synapse is then computed from the difference between the remembered correlation of the unclamped and the clamped conditions, to produce a synaptic weight correction term as follows:

$$\delta w_{ij} = k[(x_j y_i)^{\mathrm{c}} - (x_j y_i)^{\mathrm{uc}}], \tag{7.76}$$

where k is a learning rate constant. This process is then repeated for each input pattern to output pattern to be learned. The whole process is then repeated many times with all patterns until the output neurons fire similarly in the clamped and unclamped conditions, that is until the errors have become small. Further details are provided by Hinton and Sejnowski (1986). The version described above is the mean field (or deterministic) Boltzmann machine (Peterson and Anderson 1987, Hinton 1989). More traditionally, a Boltzmann machine updates one randomly chosen neuron at a time, and each neuron fires with a probability that depends on its activation (Ackley, Hinton and Sejnowski 1985, Hinton and Sejnowski 1986). The latter version makes fewer theoretical assumptions, while the former may operate an order of magnitude faster (Hertz, Krogh and Palmer 1991).

In terms of biological plausibility, it certainly is the case that there are backprojections between adjacent cortical areas (see Chapter 1). Indeed, there are as many backprojections between adjacent cortical areas as there are forward projections. The backward projections seem to be more diffuse than the forward projections, in that they connect to a wider region of the preceding cortical area than the region that sends the forward projections. If the backward and the forward synapses in such an architecture were Hebb-modifiable, then there is a possibility that the backward connections would be symmetric with the forward connections. Indeed, such a connection scheme would be useful to implement top-down recall, as summarized in Section 12.2 and described by Rolls and Treves (1998) in their Chapter 6. What seems less biologically plausible is that after an unclamped phase of operation, the correlations between all pairs of neurons would be remembered, there would then be a clamped phase of operation with each output neuron clamped to the required rate for that particular input pattern, and then the synapses would be corrected by an error-correction rule that would require a comparison of the correlations between the neuronal firing of every pair of neurons in the unclamped and clamped conditions.

Although this algorithm has the disadvantages that it is not very biologically plausible, and does not operate as well as standard backpropagation, it has been made use of by O'Reilly and Munakata (2000) in approaches to connectionist modelling in cognitive neuroscience.

8 Models of invariant object recognition

8.1 Introduction

One of the major problems that is solved by the visual system in the cerebral cortex is the building of a representation of visual information which allows recognition to occur relatively independently of size, contrast, spatial frequency, position on the retina, angle of view, etc. It is important to realize that this type of generalization is not a simple property of one-layer neural networks. Although neural networks do generalize well, the type of generalization they show naturally is to vectors which have a high dot product or correlation with what they have already learned. To make this clear, Fig. 8.1 is a reminder that the activation h_i of each neuron is computed as

$$h_i = \sum_j x_j w_{ij} \tag{8.1}$$

where the sum is over the C input axons, indexed by j. Now consider translation (or shift) of the input pattern vector by one position. The dot product will now drop to a low level, and the neuron will not respond, even though it is the same pattern, just shifted by one location. This makes the point that special processes are needed to compute invariant representations. Network approaches to such invariant pattern recognition are described in this Chapter. Once an invariant representation has been computed by a sensory system, it is in a form which is suitable for presentation to a pattern association or autoassociation neural network (see Chapter 7).

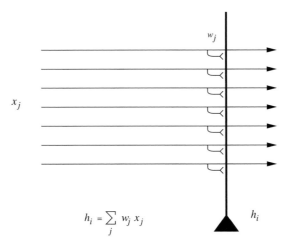

Fig. 8.1 A neuron which computes a dot product of the input pattern with its synaptic weight vector generalizes well to other patterns based on their similarity measured in terms of dot product or correlation, but shows no translation (or size, etc.) invariance.

Before considering networks that can perform invariant pattern recognition, we note that these build on the preprocessing performed in V1, the striate visual cortex of primates, which is described in Chapter 2, with an outline of the overall architecture of the visual pathways shown in Fig. 2.1. The model to be described in Section 8.4 also builds on what is known about visual processing in extrastriate visual areas, described in Chapter 3, and offers solutions to many of the issues raised at the end of Section 2.4.1. Before considering VisNet in Section 8.4 and other models, we next consider the different general approaches that have been taken to invariant object recognition in the brain.

8.2 Approaches to invariant object recognition

A number of different computational approaches that have been taken both in artificial vision systems and as suggestions for how the brain performs invariant object recognition are described in this Section. This places in context the approach that appears to be taken in the brain and that forms the basis for VisNet.

8.2.1 Feature spaces

One very simple possibility for performing object classification is based on feature spaces, which amount to lists of (the extent to which) different features are present in a particular object. The features might consist of textures, colours, areas, ratios of length to width, etc. The spatial arrangement of the features is not taken into account. If n different properties are used to characterize an object, each viewed object is represented by a set of n real numbers. It then becomes possible to represent an object by a point R^n in an n-dimensional space (where R is the resolution of the real numbers used). Such schemes have been investigated (Tou and Gonzalez 1974, Gibson 1950, Gibson 1979, Bolles and Cain 1982, Mundy and Zisserman 1992, Selfridge 1959, Mel 1997), but, because the relative positions of the different parts are not implemented in the object recognition scheme, are not sensitive to spatial jumbling of the features. For example, if the features consisted of nose, mouth, and eyes, such a system would respond to faces with jumbled arrangements of the eyes, nose and mouth, which does not match human vision, nor the responses of macaque inferior temporal cortex neurons, which are sensitive to the spatial arrangement of the features in a face (Rolls, Tovee, Purcell, Stewart and Azzopardi 1994b). Similarly, such an object recognition system might not distinguish a normal car from a car with the back wheels removed and placed on the roof. Such systems do not therefore perform shape recognition (where shape implies something about the spatial arrangement of features within an object, see further Ullman (1996)), and something more is needed, and is implemented in the primate visual system. However, I note that the features that are present in objects, e.g. a furry texture, are useful to incorporate in object recognition systems, and the brain may well use, and the model VisNet in principle can use, evidence from which features are present in an object as part of the evidence for identification of a particular object. I note that the features might consist also of for example the pattern of movement that is characteristic of a particular object (such as a buzzing fly), and use this as part of the input to final object identification.

The capacity to use shape in invariant object recognition is fundamental to primate vision, but may not be used or fully implemented on the visual systems of some other animals with less developed visual systems. For example, pigeons may correctly identify pictures containing

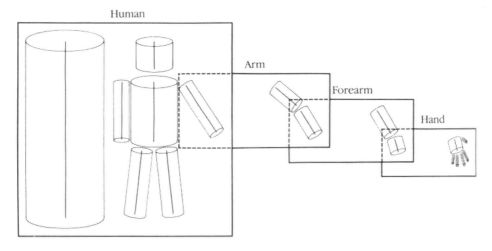

Fig. 8.2 A 3D structural description of an object based on generalized cone parts. Each box corresponds to a 3D model, with its model axis on the left side of the box and the arrangement of its component axes on the right. In addition, some component axes have 3D models associated with them, as indicated by the way the boxes overlap. (After Marr and Nishihara, 1978.)

people, a particular person, trees, pigeons etc, but may fail to distinguish a figure from a scrambled version of a figure (Herrnstein 1984, Cerella 1986). Thus their object recognition may be based more on a collection of parts than on a direct comparison of complete figures in which the relative positions of the parts are important. Even if the details of the conclusions reached from this research are revised (Wasserman, Kirkpatrick-Steger and Biederman 1998), it nevertheless does appear that at least some birds may use computationally simpler methods than those needed for invariant shape recognition. For example, it may be that when some birds are trained to discriminate between images in a large set of pictures, they tend to rely on some chance detail of each picture (such as a spot appearing by mistake on the picture), rather than on recognition of the shapes of the object in the picture (Watanabe, Lea and Dittrich 1993).

8.2.2 Structural descriptions and syntactic pattern recognition

A second approach to object recognition is to decompose the object or image into parts, and to then produce a structural description of the relations between the parts. The underlying assumption is that it is easier to capture object invariances at a level where parts have been identified. This is the type of scheme for which Marr and Nishihara (1978) and Marr (1982) opted. The particular scheme (Binford 1981) they adopted consists of generalized cones, series of which can be linked together to form structural descriptions of some, especially animate, stimuli (see Fig. 8.2). Such schemes assume that there is a 3D internal model (structural description) of each object. Perception of the object consists of parsing or segmenting the scene into objects, and then into parts, then producing a structural description of the object, and then testing whether this structural description matches that of any known object stored in the system. Other examples of structural description schemes include those of Winston (1975), Sutherland (1968), and Milner (1974). The relations in the structural description may need to be quite complicated, for example 'connected together', 'inside of', 'larger than' etc. Perhaps the most developed model of this type is the recognition by components (RBC) model of Biederman (1987), implemented in a computational model by Hummel and Biederman (1992). His small set (less than 50) of primitive parts named 'geons' include simple 3D shapes such as boxes, cylinders and wedges. Objects are described by a syntactically linked list of the relations between each of the geons of which they are composed. Describing a table in this way (as a flat top supported by three or four legs) seems quite economical. Other schemes use 2D surface patches as their primitives (Dane and Bajcsy 1982, Brady, Ponce, Yuille and Asada 1985, Faugeras and Hebert 1986, Faugeras 1993). When 3D objects are being recognized, the implication is that the structural description is a 3D description. This is in contrast to feature hierarchical systems, in which recognition of a 3D object from any view might be accomplished by storing a set of associated 2D views (see below, Section 8.2.5).

There are a number of difficulties with schemes based on structural descriptions, some general, and some with particular reference to the potential difficulty of their implementation in the brain. First, it is not always easy to decompose the object into its separate parts, which must be performed before the structural description can be produced. For example, it may be difficult to produce a structural description of a cat curled up asleep from separately identifiable parts. Identification of each of the parts is also frequently very difficult when 3D objects are seen from different viewing angles, as key parts may be invisible or highly distorted. This is particularly likely to be difficult in 3D shape perception. It appears that being committed to producing a correct description of the parts before other processes can operate is making too strong a commitment early on in the recognition process.

A second difficulty is that many objects or animals that can be correctly recognized have rather similar structural descriptions. For example, the structural description of many four-legged animals is rather similar. Rather more than a structural description seems necessary to identify many objects and animals.

A third difficulty, which applies especially to biological systems, is the difficulty of implementing the syntax needed to hold the structural description as a 3D model of the object, of producing a syntactic structural description on the fly (in real time, and with potentially great flexibility of the possible arrangement of the parts), and of matching the syntactic description of the object in the image to all the stored representations in order to find a match. An example of a structural description for a limb might be body > thigh > shin > foot > toes. In this description > means 'is linked to', and this link must be between the correct pair of descriptors. If we had just a set of parts, without the syntactic or relational linking, then there would be no way of knowing whether the toes are attached to the foot or to the body. In fact, worse than this, there would be no evidence about what was related to what, just a set of parts. Such syntactical relations are difficult to implement in neuronal networks, because if the representations of all the features or parts just mentioned were active simultaneously, how would the spatial relations between the features also be encoded? (How would it be apparent just from the firing of neurons that the toes were linked to the rest of foot but not to the body?) It would be extremely difficult to implement this 'on the fly' syntactic binding in a biologically plausible network (though cf. Hummel and Biederman (1992)), and the only suggested mechanism for flexible syntactic binding, temporal synchronization of the firing of different neurons, is not well supported as a quantitatively important mechanisms for information encoding in the ventral visual system, and would have major difficulties in implementing correct. relational, syntactic binding (see Sections 5.5.7 and 13.7).

A fourth difficulty of the structural description approach is that segmentation into objects must occur effectively before object recognition, so that the linked structural description list can be of one object. Given the difficulty of segmenting objects in typical natural cluttered scenes (Ullman 1996), and the compounding problem of overlap of parts of objects by other objects, segmentation as a first necessary stage of object recognition adds another major difficulty for structural description approaches.

A fifth difficulty is that metric information, such as the relative size of the parts that are linked syntactically, needs to be specified in the structural description (Stankiewicz and Hummel 1994), which complicates the parts that have to be syntactically linked.

It is because of these difficulties that even in artificial vision systems implemented on computers, where almost unlimited syntactic binding can easily be implemented, the structural description approach to object recognition has not yet succeeded in producing a scheme which actually works in more than an environment in which the types of objects are limited, and the world is far from the natural world, consisting for example of 2D scenes (Mundy and Zisserman 1992).

Although object recognition in the brain is unlikely to be based on the structural description approach, for the reasons given above, and the fact that the evidence described in Chapter 5 supports a feature hierarchy rather than the structural description implementation in the brain, it is certainly the case that humans can provide verbal, syntactic, descriptions of objects in terms of the relations of their parts, and that this is often a useful type of description. Humans may therefore it is suggested supplement a feature hierarchical object recognition system built into their ventral visual system with the additional ability to use the type of syntax that is necessary for language to provide another level of description of objects. This is of course useful in, for example, engineering applications.

8.2.3 Template matching and the alignment approach

Another approach is template matching, comparing the image on the retina with a stored image or picture of an object. This is conceptually simple, but there are in practice major problems. One major problem is how to align the image on the retina with the stored images, so that all possible images on the retina can be compared with the stored template or templates of each object.

The basic idea of the alignment approach (Ullman 1996) is to compensate for the transformations separating the viewed object and the corresponding stored model, and then compare them. For example, the image and the stored model may be similar, except for a difference in size. Scaling one of them will remove this discrepancy and improve the match between them. For a 2D world, the possible transforms are translation (shift), scaling, and rotation. Given for example an input letter of the alphabet to recognize, the system might, after segmentation (itself a very difficult process if performed independently of (prior to) object recognition), compensate for translation by computing the centre of mass of the object, and shifting the character to a 'canonical location'. Scale might be compensated for by calculating the convex hull (the smallest envelope surrounding the object), and then scaling the image. Of course how the shift and scaling would be accomplished is itself a difficult point — easy to perform on a computer using matrix multiplication as in simple computer graphics, but not the sort of computation that could be performed easily or accurately by any biologically plausible network. Compensating for rotation is even more difficult (Ullman 1996). All this has to happen before the segmented canonical representation of the object is compared to the stored object templates with the same canonical representation. The system of course becomes vastly more complicated when the recognition must be performed of 3D objects seen in a 3D world, for now the particular view of an object after segmentation must be placed into a canonical form, regardless of which view, or how much of any view, may be seen in a natural scene with occluding contours. However, this process is helped, at least in computers that can perform high precision matrix multiplication, by the fact that (for many continuous transforms such as 3D rotation, translation, and scaling) all the possible views of an object transforming in 3D space can be expressed as the linear combination of other views of the same object (see Chapter 5 of Ullman (1996); Koenderink and van Doorn (1991); and Koenderink (1990)).

This alignment approach is the main theme of the book by Ullman (1996), and there are a number of computer implementations (Lowe 1985, Grimson 1990, Huttenlocher and Ullman 1990, Shashua 1995). However, as noted above, it seems unlikely that the brain is able to perform the high precision calculations needed to perform the transforms required to align any view of a 3D object with some canonical template representation. For this reason, and because the approach also relies on segmentation of the object in the scene before the template alignment algorithms can start, and because key features may need to be correctly identified to be used in the alignment (Edelman 1999), this approach is not considered further here.

We may note here in passing that some animals with a less computationally-developed visual system appear to attempt to solve the alignment problem by actively moving their heads or eyes to see what template fits, rather than starting with an image on the eye and attempting to transform it into canonical coordinates. This 'active vision' approach used for example by some invertebrates has been described by Land (1999) and Land and Collett (1997).

8.2.4 Invertible networks that can reconstruct their inputs

Hinton, Dayan, Frey and Neal (1995) and Hinton and Ghahramani (1997) have argued that cortical computation is invertible, so that for example the forward transform of visual information from V1 to higher areas loses no information, and there can be a backward transform from the higher areas to V1. A comparison of the reconstructed representation in V1 with the actual image from the world might in principle be used to correct all the synaptic weights between the two (in both the forward and the reverse directions), in such a way that there are no errors in the transform. This suggested reconstruction scheme would seem to involve non-local synaptic weight correction (though see Section 7.14 for a suggested, although still biologically implausible, neural implementation), or other biologically implausible operations. The scheme also does not seem to provide an account for why or how the responses of inferior temporal cortex neurons become the way they are (providing information about which object is seen relatively independently of position on the retina, size, or view). The whole forward transform performed in the brain seems to lose information about the size, position and view of the object, as it is evidence about which object is present invariant of its size, view, etc. that is useful to the stages of processing about objects that follow (as described in Chapter 12). (It could be that it is just that object information is made explicit in the firing of inferior temporal cortex single neurons and ensembles of neurons (see Chapter 5), and that position, size, and view information is all somehow represented in a very distributed way across the whole population of inferior temporal cortex neurons that is difficult to measure, but this seems to be conjecture rather than what is indicated by the data.) Because of these difficulties, and because the backprojections are needed for processes such as recall (see Section 1.11),

Top layer

Intermediate
layers

Input (V1)

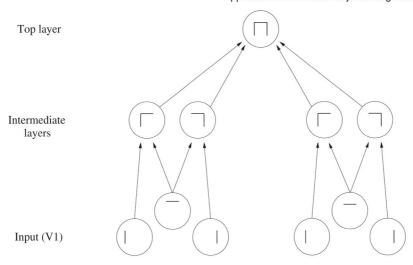

Fig. 8.3 The feature hierarchy approach to object recognition. The inputs may be neurons tuned to oriented straight line segments. In early intermediate layers neurons respond to a combination of these inputs in the correct spatial position with respect to each other. In further intermediate layers, of which there may be several, neurons respond with some invariance to the feature combinations represented early, and form higher order feature combinations. Finally, in the top layer, neurons respond to combinations of what is represented in the preceding intermediate layer, and thus provide evidence about objects in a position (and scale and even view) invariant way. Convergence through the network is designed to provide top layer neurons with information from across the entire input retina, as part of the solution to translation invariance, and other types of invariance are treated similarly.

this approach is not considered further in this book.

In the context of recall, if the visual system were to perform a reconstruction in V1 of a visual scene from what is represented in the inferior temporal visual cortex, then it might be supposed that remembered visual scenes might be as information-rich (and subjectively as full of rich detail) as seeing the real thing. This is not the case for most humans, and indeed this point suggests that at least what reaches consciousness from the inferior temporal visual cortex (which is activated during the recall of visual memories) is the identity of the object (as made explicit in the firing rate of the neurons), and not the low-level details of the exact place, size, and view of the object in the recalled scene, even though, according to the reconstruction argument, that information should be present in the inferior temporal visual cortex.

8.2.5 Feature hierarchies and 2D view-based object recognition

Another approach, and one that is much closer to what appears to be present in the primate ventral visual system as described in Chapter 5 and for most of this Chapter, is a feature hierarchy system (see Fig. 8.3).

In this approach, the system starts with some low-level description of the visual scene, in terms for example of oriented straight line segments of the type that are represented in the responses of primary visual cortex (V1) neurons, and then builds in repeated hierarchical layers features based on what is represented in previous layers. A feature may thus be defined as a combination of what is represented in the previous layer. For example, after V1, features might consist of combinations of straight lines, which might represent longer curved lines (Zucker, Dobbins and Iverson 1989), or terminated lines (in fact represented in V1 as end-stopped cells), corners, 'T' junctions which are characteristic of obscuring edges, and (at least in humans) the arrow and 'Y' vertices which are characteristic properties of man-made environments. As one ascends the hierarchy, neurons might respond to more complex trigger features (such as two parts of a complex figure in the correct spatial arrangement with respect to each other, as shown by Tanaka (1996) for V4 and posterior inferior temporal cortex neurons). Further on, neurons might respond to combinations of several such intermediate-level feature combination neurons, and thus come to respond systematically differently to different objects, and thus to convey information about which object is present. This approach received neurophysiological support early on from the results of Hubel and Wiesel (1962) and Hubel and Wiesel (1968) in the cat and monkey, and much of the data described in Chapter 5 is consistent with this scheme.

A number of problems need to be solved for such feature hierarchy visual systems to provide a useful model of object recognition in the primate visual system.

First, some way needs to be found to keep the number of feature combination neurons realistic at each stage, without undergoing a combinatorial explosion. If a separate feature combination neuron was needed to code for every possible combination of n types of feature each with a resolution of 2 levels (binary encoding) in the preceding stage, then 2^n neurons would be needed. The suggestion that is made in Section 8.3 is that by forming neurons that respond to low-order combinations of features (neurons that respond to just say 2–4 features from the preceding stage), the number of actual feature analysing neurons can be kept within reasonable numbers. By reasonable we mean the number of neurons actually found at any one stage of the visual system, which, for V4 might be in the order of 60×10^6 neurons (assuming a volume for macaque V4 of approximately 2,000 mm^3, and a cell density of 20,000–40,000 neurons per mm^3, see Table 1.1). This is certainly a large number; but the fact that a large number of neurons is present at each stage of the primate visual system is in fact consistent with the hypothesis that feature combination neurons are part of the way in which the brain solves object recognition. A factor which also helps to keep the number of neurons under control is the statistics of the visual world, which contain great redundancies. The world is not random, and indeed the statistics of natural images are such that many regularities are present (Field 1994), and not every possible combination of pixels on the retina needs to be separately encoded. A third factor which helps to keep the number of connections required onto each neuron under control is that in a multilayer hierarchy each neuron can be set up to receive connections from only a small region of the preceding layer. Thus an individual neuron does not need to have connections from all the neurons in the preceding layer. Over multiple layers, the required convergence can be produced so that the same neurons in the top layer can be activated by an image of an effective object anywhere on the retina (see Fig. 8.4).

A second problem of feature hierarchy approaches is how to map all the different possible images of an individual object through to the same set of neurons in the top layer by modifying the synaptic connections (see Fig. 8.4). The solution discussed in Sections 8.3, 8.4.1.1 and 8.4.4 is the use of a synaptic modification rule with a short term memory trace of the previous activity of the neuron, to enable it to learn to respond to the now transformed version of what was seen very recently, which, given the statistics of looking at the visual world, will probably be an input from the same object.

A third problem of feature hierarchy approaches is how they can learn in just a few seconds of inspection of an object to recognize it in different transforms, for example in different positions on the retina in which it may never have been presented during training. A solution to this problem is provided in Section 8.4.5, in which it is shown that this can be a natural property of feature hierarchy object recognition systems, if they are trained first for all locations on the intermediate level feature combinations of which new objects will simply be a new combination, and therefore requiring learning only in the upper layers of the hierarchy.

A fourth potential problem of feature hierarchy systems is that when solving translation invariance they need to respond to the same local spatial arrangement of features (which are needed to specify the object), but to ignore the global position of the whole object. It is shown in Section 8.4.5 that feature hierarchy systems can solve this problem by forming feature combination neurons at an early stage of processing (e.g. V1 or V2 in the brain) that respond with high spatial precision to the local arrangement of features. Such neurons would respond differently for example to L, +, and T if they receive inputs from two line-responding neurons. It is shown in Section 8.4.5 that at later layers of the hierarchy, where some of the intermediate level feature combination neurons are starting to show translation invariance, then correct object recognition may still occur because only one object contains just those sets of intermediate level neurons in which the spatial representation of the features is inherent in the encoding.

The type of representation developed in a hierarchical object recognition system, in the brain, and by VisNet as described in the rest of this Chapter would be suitable for recognition of an object, and for linking associative memories to objects, but would be less good for making actions in 3D space to particular parts of, or inside, objects, as the 3D coordinates of each part of the object would not be explicitly available. It is therefore proposed that visual fixation is used to locate in foveal vision part of an object to which movements must be made, and that local disparity and other measurements of depth (made explicit in the dorsal visual system) then provide sufficient information for the motor system to make actions relative to the small part of space in which a local, view-dependent, representation of depth would be provided (cf. Ballard (1990)).

One advantage of feature hierarchy systems is that they can operate fast (see Sections 7.4 and 7.6.5).

A second advantage is that the feature analyzers can be built out of the rather simple competitive networks described in Section 7.4 which use a local learning rule, and have no external teacher, so that they are rather biologically plausible. Another advantage is that, once trained on subset features common to most objects, the system can then learn new objects quickly.

A related third advantage is that, if implemented with competitive nets as in the case of VisNet (see Section 8.4), then neurons are allocated by self-organization to represent just the features present in the natural statistics of real images (cf. Field (1994)), and not every possible feature that could be constructed by random combinations of pixels on the retina.

A related fourth advantage of feature hierarchy networks is that because they can utilize competitive networks, they can still produce the best guess at what is in the image under non-ideal conditions, when only parts of objects are visible because for example of occlusion by other objects, etc. The reasons for this are that competitive networks assess the evidence for the presence of certain 'features' to which they are tuned using a dot product operation on their inputs, so that they are inherently tolerant of missing input evidence; and reach a state that reflects the best hypothesis or hypotheses (with soft competition) given the whole set of inputs, because there are competitive interactions between the different neurons (see Section 7.4).

A fifth advantage of a feature hierarchy system is that, as shown in Section 8.4.6, the system does not need to perform segmentation into objects as part of pre-processing, nor does it need to be able to identify parts of an object, and can also operate in cluttered scenes in which the object may be partially obscured. The reason for this is that once trained on objects, the system then operates somewhat like an associative memory, mapping the image properties forward onto whatever it has learned about before, and then by competition selecting just the most likely output to be activated. Indeed, the feature hierarchy approach provides a mechanism by which processing at the object recognition level could feed back using backprojections to early cortical areas to provide top-down guidance to assist segmentation. Although backprojections are not built into VisNet, they are in principle part of the architecture that could easily be added, are present in the brain, and are incorporated into some of the models described in Chapters 9–11 and in Section 13.27.

A sixth advantage of feature hierarchy systems is that they can naturally utilize features in the images of objects which are not strictly part of a shape description scheme, such as the fact that different objects have different textures, colours etc. Feature hierarchy systems, because they utilize whatever is represented at earlier stages in forming feature combination neurons at the next stage, naturally incorporate such 'feature list' evidence into their analysis, and have the advantages of that approach (see Section 8.2.1 and also Mel (1997)). Indeed, the feature space approach can utilize a hybrid representation, some of whose dimensions may be discrete and defined in structural terms, while other dimensions may be continuous and defined in terms of metric details, and others may be concerned with non-shape properties such as texture and colour (cf. Edelman (1999)).

A seventh advantage of feature hierarchy systems is that they do not need to utilize 'on-the-fly' or run-time arbitrary binding of features. Instead, the spatial syntax is effectively hardwired into the system when it is trained, in that the feature combination neurons have learned to respond to their set of features when they are in a given spatial arrangement on the retina.

An eighth advantage of feature hierarchy systems is that they can self-organize (given the right functional architecture, trace synaptic learning rule, and the temporal statistics of the normal visual input from the world), with no need for an external teacher to specify that the neurons must learn to respond to objects. The correct, object, representation self-organizes itself given rather economically specified genetic rules for building the network (cf. Rolls and Stringer (2000)).

Ninth, it is also noted that hierarchical visual systems may recognize 3D objects based on a limited set of 2D views of objects, and that the same architectural rules just stated and implemented in VisNet will correctly associate together the different views of an object. It is part of the concept (see below), and consistent with neurophysiological data (Tanaka 1996), that the neurons in the upper layers will generalize correctly within a view (see Section 8.4.7).

After the immediately following description of early models of a feature hierarchy approach implemented in the Cognitron and Neocognitron, we turn for the remainder of this Chapter to analyses of how a feature hierarchy approach to invariant visual object recognition might be implemented in the brain, and how key computational issues could be solved by such a system. The analyses are developed with and tested with a model, VisNet, which will shortly be described. Much of the data we have on the operation of the high order visual cortical areas suggests that they implement a feature hierarchy approach to visual object recognition, as made evident in the remainder of this Chapter and in Chapter 5.

8.2.5.1 The Cognitron and Neocognitron

An early computational model of a hierarchical feature-based approach to object recognition, joining other early discussions of this approach (Selfridge 1959, Sutherland 1968, Barlow 1972, Milner 1974), was proposed by Fukushima (1975), (1980), (1989), and (1991). His model used two types of cell within each layer to approach the problem of invariant representations. In each layer, a set of 'simple cells', with defined position, orientation, etc. sensitivity for the stimuli to which they responded, was followed by a set of 'complex cells', which generalized a little over position, orientation, etc. This simple cell—complex cell pairing within each layer provided some invariance. When a neuron in the network using competitive learning with its stimulus set, which was typically letters on a 16 x 16 pixel array, learned that a particular feature combination had occurred, that type of feature analyzer was replicated in a non-local manner throughout the layer, to provide further translation invariance. Invariant representations were thus learned in a different way from VisNet. Up to eight layers were used. The network could learn to differentiate letters, even with some translation, scaling, or distortion. Although internally it is organized and learns very differently to VisNet, it is an independent example of the fact that useful invariant pattern recognition can be performed by multilayer hierarchical networks. A major biological implausibility of the system is that once one neuron within a layer learned, other similar neurons were set up throughout the layer by a non-local process. A second biological limitation was that no learning rule or self-organizing process was specified as to how the complex cells can provide translation invariant representations of simple cell responses — this was simply handwired in by lateral replication. Solutions to both these issues are provided by VisNet.

8.3 Hypotheses about the computational mechanisms in the visual cortex for object recognition

The neurophysiological findings described in Chapter 5, and wider considerations on the possible computational properties of the cerebral cortex (Rolls 1989b, Rolls 1989f, Rolls 1992a, Rolls 1994, Rolls 1995b, Rolls 1997b, Rolls 2000a, Rolls and Treves 1998), lead to the following outline working hypotheses on object recognition by visual cortical mechanisms (see Rolls (1992a)). The principles underlying the processing of faces and other objects may be similar, but more neurons may become allocated to represent different aspects of faces because of the need to recognize the faces of many different individuals, that is to identify many individuals within the category faces.

Cortical visual processing for object recognition is considered to be organized as a set of hierarchically connected cortical regions consisting at least of V1, V2, V4, posterior inferior temporal cortex (TEO), inferior temporal cortex (e.g. TE3, TEa and TEm), and anterior temporal cortical areas (e.g. TE2 and TE1). (This stream of processing has many connections with a set of cortical areas in the anterior part of the superior temporal sulcus, including area TPO.) There is convergence from each small part of a region to the succeeding region (or layer in the hierarchy) in such a way that the receptive field sizes of neurons (e.g. 1 degree near the fovea in V1) become larger by a factor of approximately 2.5 with each succeeding stage (and the typical parafoveal receptive field sizes found would not be inconsistent with

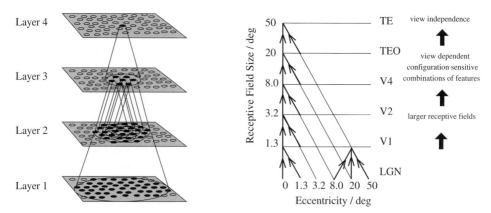

Fig. 8.4 Convergence in the visual system. Right – as it occurs in the brain. V1: visual cortex area V1; TEO: posterior inferior temporal cortex; TE: inferior temporal cortex (IT). Left – as implemented in VisNet. Convergence through the network is designed to provide fourth layer neurons with information from across the entire input retina.

the calculated approximations of e.g. 8 degrees in V4, 20 degrees in TEO, and 50 degrees in the inferior temporal cortex (Boussaoud and Ungerleider 1991)) (see Fig. 8.4). Such zones of convergence would overlap continuously with each other (see Fig. 8.4). This connectivity would be part of the architecture by which translation invariant representations are computed.

Each layer is considered to act partly as a set of local self-organizing competitive neuronal networks with overlapping inputs. (The region within which competition would be implemented would depend on the spatial properties of inhibitory interneurons, and might operate over distances of 1–2 mm in the cortex.) These competitive nets operate by a single set of forward inputs leading to (typically non-linear, e.g. sigmoid) activation of output neurons; of competition between the output neurons mediated by a set of feedback inhibitory interneurons which receive from many of the principal (in the cortex, pyramidal) cells in the net and project back (via inhibitory interneurons) to many of the principal cells and serve to decrease the firing rates of the less active neurons relative to the rates of the more active neurons; and then of synaptic modification by a modified Hebb rule, such that synapses to strongly activated output neurons from active input axons strengthen, and from inactive input axons weaken (see Section 7.4). A biologically plausible form of this learning rule that operates well in such networks is

$$\delta w_{ij} = \alpha y_i (x_j - w_{ij}) \tag{8.2}$$

where α is a learning rate constant, and x_j and w_{ij} are in appropriate units (see Section 7.4). Such competitive networks operate to detect correlations between the activity of the input neurons, and to allocate output neurons to respond to each cluster of such correlated inputs. These networks thus act as categorizers. In relation to visual information processing, they would remove redundancy from the input representation, and would develop low entropy representations of the information (cf. Barlow (1985), Barlow, Kaushal and Mitchison (1989)). Such competitive nets are biologically plausible, in that they utilize Hebb-modifiable forward excitatory connections, with competitive inhibition mediated by cortical inhibitory neurons. The competitive scheme I suggest would not result in the formation of 'winner-take-all' or 'grand-mother' cells, but would instead result in a small ensemble of active neurons representing each input (Rolls 1989b, Rolls 1989f, Rolls and Treves 1998). The scheme has the advantages that the output neurons learn better to distribute themselves between the input patterns (cf. Bennett (1990)), and that the sparse representations formed have utility in maximizing the number of memories that can be stored when, towards the end of the visual system, the visual representation of objects is interfaced to associative memory (Rolls 1989b, Rolls 1989f, Rolls and Treves 1998)[15].

Translation invariance would be computed in such a system by utilizing competitive learning to detect regularities in inputs when real objects are translated in the physical world. The hypothesis is that because objects have continuous properties in space and time in the world, an object at one place on the retina might activate feature analyzers at the next stage of cortical processing, and when the object was translated to a nearby position, because this would occur in a short period (e.g. 0.5 s), the membrane of the postsynaptic neuron would still be in its 'Hebb-modifiable' state (caused for example by calcium entry as a result of the voltage dependent activation of NMDA receptors), and the presynaptic afferents activated with the object in its new position would thus become strengthened on the still-activated postsynaptic neuron. It is suggested that the short temporal window (e.g. 0.5s) of Hebb-modifiability helps neurons to learn the statistics of objects moving in the physical world, and at the same time to form different representations of different feature combinations or objects, as these are physically discontinuous and present less regular correlations to the visual system. Földiák (1991) has proposed computing an average activation of the postsynaptic neuron to assist with the same problem. One idea here is that the temporal properties of the biologically implemented learning mechanism are such that it is well suited to detecting the relevant continuities in the world of real objects. Another suggestion is that a memory trace for what has been seen in the last 300 ms appears to be implemented by a mechanism as simple as continued firing of inferior temporal neurons after the stimulus has disappeared, as was found in the masking experiments described in Sections 5.5.6 and 5.9 (see also Rolls and Tovee (1994), Rolls, Tovee, Purcell, Stewart and Azzopardi (1994b)). I also suggest that other invariances, for example size, spatial frequency, and rotation invariance, could be learned by a comparable process. (Early processing in V1 which enables different neurons to represent inputs at different spatial scales would allow combinations of the outputs of such neurons to be formed at later stages. Scale invariance would then result from detecting at a later stage which neurons are almost conjunctively active as the size of an object alters.) It is suggested that this process takes place at each stage of the multiple-layer cortical processing hierarchy, so that invariances are learned first over small regions of space, and then over successively

---

[15] In that each neuron has graded responses centred about an optimal input, the proposal has some of the advantages with respect to hypersurface reconstruction described by Poggio and Girosi (1990a). However, the system I propose learns differently, in that instead of using perhaps non biologically-plausible algorithms to optimally locate the centres of the receptive fields of the neurons, the neurons use graded competition to spread themselves throughout the input space, depending on the statistics of the inputs received, and perhaps with some guidance from backprojections (see below). In addition, the competitive nets I propose use as a distance function the dot product between the input vector to a neuron and its synaptic weight vector, whereas radial basis function networks use a Gaussian measure of distance (see Section 7.4.7). Both systems benefit from the finite width of the response region of each neuron which tapers from a maximum, and is important for enabling the system to generalize smoothly from the examples with which it has learned (cf. Poggio and Girosi (1990b), Poggio and Girosi (1990a)), to help the system to respond for example with the correct invariances as described below.

larger regions. This limits the size of the connection space within which correlations must be sought.

Increasing complexity of representations could also be built in such a multiple layer hierarchy by similar mechanisms. At each stage or layer the self-organizing competitive nets would result in combinations of inputs becoming the effective stimuli for neurons. In order to avoid the combinatorial explosion, it is proposed, following Feldman (1985), that low-order combinations of inputs would be what is learned by each neuron. (Each input would not be represented by activity in a single input axon, but instead by activity in a set of active input axons.) Evidence consistent with this suggestion that neurons are responding to combinations of a few variables represented at the preceding stage of cortical processing is that some neurons in V1 respond to combinations of bars or edges (see Section 2.5); V2 and V4 respond to end-stopped lines, to tongues flanked by inhibitory subregions, or to combinations of colours (see Chapter 3); in posterior inferior temporal cortex to stimuli which may require two or more simple features to be present (Tanaka, Saito, Fukada and Moriya 1990); and in the temporal cortical face processing areas to images that require the presence of several features in a face (such as eyes, hair, and mouth) in order to respond (Perrett, Rolls and Caan 1982, Yamane, Kaji and Kawano 1988) (see Chapter 5). (Precursor cells to face-responsive neurons might, it is suggested, respond to combinations of the outputs of the neurons in V1 that are activated by faces, and might be found in areas such as V4.) It is an important part of this suggestion that some local spatial information would be inherent in the features which were being combined. For example, cells might not respond to the combination of an edge and a small circle unless they were in the correct spatial relation to each other. (This is in fact consistent with the data of Tanaka, Saito, Fukada and Moriya (1990), and with our data on face neurons, in that some faces neurons require the face features to be in the correct spatial configuration, and not jumbled, Rolls, Tovee, Purcell, Stewart and Azzopardi (1994b).) The local spatial information in the features being combined would ensure that the representation at the next level would contain some information about the (local) arrangement of features. Further low-order combinations of such neurons at the next stage would include sufficient local spatial information so that an arbitrary spatial arrangement of the same features would not activate the same neuron, and this is the proposed, and limited, solution which this mechanism would provide for the feature binding problem (Elliffe, Rolls and Stringer 2001) (cf. Malsburg (1990)). By this stage of processing a view-dependent representation of objects suitable for view-dependent processes such as behavioural responses to face expression and gesture would be available.

It is suggested that view-independent representations could be formed by the same type of computation, operating to combine a limited set of views of objects. The plausibility of providing view-independent recognition of objects by combining a set of different views of objects has been proposed by a number of investigators (Koenderink and Van Doorn 1979, Poggio and Edelman 1990, Logothetis, Pauls, Bulthoff and Poggio 1994, Ullman 1996). Consistent with the suggestion that the view-independent representations are formed by combining view-dependent representations in the primate visual system, is the fact that in the temporal cortical areas, neurons with view-independent representations of faces are present in the same cortical areas as neurons with view-dependent representations (from which the view-independent neurons could receive inputs) (Hasselmo, Rolls, Baylis and Nalwa 1989b, Perrett, Smith, Potter, Mistlin, Head, Milner and Jeeves 1985b, Booth and Rolls 1998). This solution to 'object-based' representations is very different from that traditionally proposed for artificial vision systems, in which the coordinates in 3D space of objects are stored in a database, and general-purpose algorithms operate on these to perform transforms such as translation, rotation, and scale change in 3D space (e.g. Marr (1982)). In the present, much more limited but more biologically plausible scheme, the representation would be suitable for recognition of an object, and for linking associative memories to objects, but would be less good for making actions in 3D space to particular parts of, or inside, objects, as the 3D coordinates of each part of the object would not be explicitly available. It is therefore proposed that visual fixation is used to locate in foveal vision part of an object to which movements must be made, and that local disparity and other measurements of depth then provide sufficient information for the motor system to make actions relative to the small part of space in which a local, view-dependent, representation of depth would be provided (cf. Ballard (1990)).

The computational processes proposed above operate by an unsupervised learning mechanism, which utilizes statistical regularities in the physical environment to enable representations to be built. In some cases it may be advantageous to utilize some form of mild teaching input to the visual system, to enable it to learn for example that rather similar visual inputs have very different consequences in the world, so that different representations of them should be built. In other cases, it might be helpful to bring representations together, if they have identical consequences, in order to use storage capacity efficiently. It is proposed elsewhere (Rolls 1989b, Rolls 1989f, Rolls and Treves 1998) (see Section 1.11 that the backprojections from each adjacent cortical region in the hierarchy (and from the amygdala and hippocampus to higher regions of the visual system) play such a role by providing guidance to the competitive networks suggested above to be important in each cortical area. This guidance, and also the capability for recall, are it is suggested implemented by Hebb-modifiable connections from the backprojecting neurons to the principal (pyramidal) neurons of the competitive networks in the preceding stages (Rolls 1989b, Rolls 1989f, Rolls and Treves 1998) (see Section 7.4).

The computational processes outlined above use sparse distributed coding with relatively finely tuned neurons with a graded response region centred about an optimal response achieved when the input stimulus matches the synaptic weight vector on a neuron. The distributed nature of the coding but with fine tuning would help to limit the combinatorial explosion, to keep the number of neurons within the biological range. The graded response region would be crucial in enabling the system to generalize correctly to solve for example the invariances. However, such a system would need many neurons, each with considerable learning capacity, to solve visual perception in this way. This is fully consistent with the large number of neurons in the visual system, and with the large number of, probably modifiable, synapses on each neuron (e.g. 5000). Further, the fact that many neurons are tuned in different ways to faces is consistent with the fact that in such a computational system, many neurons would need to be sensitive (in different ways) to faces, in order to allow recognition of many individual faces when all share a number of common properties.

8.4 The feature hierarchy approach to invariant object recognition: computational issues

The feature hierarchy approach to invariant object recognition was introduced in Section 8.3, and advantages and disadvantages of it were discussed. Hypotheses about how object recognition could be implemented in the brain which are consistent with much of the neurophysiology discussed in Chapter 5 and earlier Chapters were set out in Section 8.3. These hypotheses effectively incorporate a feature hierarchy system while encompassing much of the neurophysiological evidence. In this Section (8.4), we consider the computational issues that arise in such feature hierarchy systems, and in the brain systems that implement visual object recognition. The issues are considered with the help of a particular model, VisNet, which requires precise specification of the hypotheses, and at the same time enables them to be explored and tested numerically and quantitatively. However, we emphasize that the issues to be covered in Section 8.4 are key and major computational issues for architectures of this feature hierarchical type, and are very relevant to understanding how invariant object recognition is implemented in the brain.

VisNet is a model of invariant object recognition based on Rolls' (1992a) hypotheses. It is a computer simulation that allows hypotheses to be tested and developed about how multilayer hierarchical networks of the type believed to be implemented in the visual cortical pathways operate. The architecture captures a number of aspects of the architecture of the visual cortical pathways, and is described next. The model of course, as with all models, requires precise specification of what is to be implemented, and at the same time involves specified simplifications of the real architecture, as investigations of the fundamental aspects of the information processing being performed are more tractable in a simplified and at the same time quantitatively specified model. First the architecture of the model is described, and this is followed by descriptions of key issues in such multilayer feature hierarchical models, such as the issue of feature binding, the optimal form of training rule for the whole system to self-organize, the operation of the network in natural environments and when objects are partly occluded, how outputs about individual objects can be read out from the network, and the capacity of the system.

8.4.1   The architecture of VisNet

Fundamental elements of Rolls' (1992a) theory for how cortical networks might implement invariant object recognition are described in Section 8.3. They provide the basis for the design of VisNet, and can be summarized as:

- A series of competitive networks, organized in hierarchical layers, exhibiting mutual inhibition over a short range within each layer. These networks allow combinations of features or inputs occurring in a given spatial arrangement to be learned by neurons, ensuring that higher order spatial properties of the input stimuli are represented in the network.
- A convergent series of connections from a localized population of cells in preceding layers to each cell of the following layer, thus allowing the receptive field size of cells to increase through the visual processing areas or layers.
- A modified Hebb-like learning rule incorporating a temporal trace of each cell's previous activity, which, it is suggested, will enable the neurons to learn transform invariances.

The first two elements of Rolls' theory are used to constrain the general architecture of a network model, VisNet, of the processes just described that is intended to learn invariant representations of objects. The simulation results described in this Chapter using VisNet show that invariant representations can be learned by the architecture. It is moreover shown that successful learning depends crucially on the use of the modified Hebb rule. The general architecture simulated in VisNet, and the way in which it allows natural images to be used as stimuli, has been chosen to enable some comparisons of neuronal responses in the network and in the brain to similar stimuli to be made.

8.4.1.1 The trace rule

The learning rule implemented in the VisNet simulations utilizes the spatio-temporal constraints placed upon the behaviour of 'real-world' objects to learn about natural object transformations. By presenting consistent sequences of transforming objects the cells in the network can learn to respond to the same object through all of its naturally transformed states, as described by Földiák (1991) and Rolls (1992a). The learning rule incorporates a decaying trace of previous cell activity and is henceforth referred to simply as the 'trace' learning rule. The learning paradigm we describe here is intended in principle to enable learning of any of the transforms tolerated by inferior temporal cortex neurons (Rolls 1992a, Rolls 1994, Rolls 1995b, Rolls 1997b, Rolls 2000a).

To clarify the reasoning behind this point, consider the situation in which a single neuron is strongly activated by a stimulus forming part of a real world object. The trace of this neuron's activation will then gradually decay over a time period in the order of 0.5s. If, during this limited time window, the net is presented with a transformed version of the original stimulus then not only will the initially active afferent synapses modify onto the neuron, but so also will the synapses activated by the transformed version of this stimulus. In this way the cell will learn to respond to either appearance of the original stimulus. Making such associations works in practice because it is very likely that within short time periods different aspects of the same object will be being inspected. The cell will not, however, tend to make spurious links across stimuli that are part of different objects because of the unlikelihood in the real world of one object consistently following another.

Various biological bases for this temporal trace have been advanced[16] :

- The persistent firing of neurons for as long as 100–400ms observed after presentations of stimuli for 16 ms (Rolls and Tovee 1994) could provide a time window within which to associate subsequent images. Maintained activity may potentially be implemented by recurrent connections between cortical areas (O'Reilly and Johnson 1994, Rolls 1994, Rolls 1995b)[17].

- The binding period of glutamate in the NMDA channels, which may last for 100 or more ms, may implement a trace rule by producing a narrow time window over which the *average* activity at each presynaptic site affects learning (Rolls 1992a, Rhodes 1992, Földiák 1992).

- Chemicals such as nitric oxide may be released during high neural activity and gradually decay in concentration over a short time window during which learning could be enhanced (Földiák 1992, Montague, Gally and Edelman 1991).

The trace update rule used in the baseline simulations of VisNet (Wallis and Rolls 1997) is equivalent to both Földiák's used in the context of translation invariance and to the earlier

---

[16]The precise mechanisms involved may alter the precise form of the trace rule which should be used. Földiák (1992) describes an alternative trace rule which models individual NMDA channels. Equally, a trace implemented by extended cell firing should be reflected in representing the trace as an external firing rate, rather than an internal signal.

[17]The prolonged firing of inferior temporal cortex neurons during memory delay periods of several seconds, and associative links reported to develop between stimuli presented several seconds apart (Miyashita 1988) are on too long a time scale to be immediately relevant to the present theory. In fact, associations between visual events occurring several seconds apart would, under *normal* environmental conditions, be detrimental to the operation of a network of the type described here, because they would probably arise from different objects. In contrast, the system described benefits from associations between visual events which occur close in time (typically within 1 s), as they are likely to be from the same object.

rule of Sutton and Barto (1981) explored in the context of modelling the temporal properties of classical conditioning, and can be summarized as follows:

$$\delta w_j = \alpha \bar{y}^\tau x_j \qquad (8.3)$$

where

$$\bar{y}^\tau = (1 - \eta)y^\tau + \eta \bar{y}^{\tau-1} \qquad (8.4)$$

and

x_j:	j^{th} input to the neuron.	y:	Output from the neuron.
$\bar{y}^\tau$:	Trace value of the output of the neuron at time step τ.	α:	Learning rate. Annealed between unity and zero.
w_j:	Synaptic weight between j^{th} input and the neuron.	η:	Trace value. The optimal value varies with presentation sequence length.

To bound the growth of each neuron's synaptic weight vector, $\mathbf{w}_i$ for the ith neuron, its length is explicitly normalized (a method similarly employed by von der Malsburg (1973) which is commonly used in competitive networks, see Section 7.4). An alternative, more biologically relevant implementation, using a local weight bounding operation which utilizes a form of heterosynaptic long-term depression (see Section 1.5), has in part been explored using a version of the Oja (1982) rule (see Wallis and Rolls (1997)).

8.4.1.2 The network implemented in VisNet

The network itself is designed as a series of hierarchical, convergent, competitive networks, in accordance with the hypothesis advanced above. The actual network consists of a series of four layers, constructed such that the convergence of information from the most disparate parts of the network's input layer can potentially influence firing in a single neuron in the final layer — see Fig. 8.4. This corresponds to the scheme described by many researchers (Van Essen, Anderson and Felleman 1992, Rolls 1992a, for example) as present in the primate visual system — see Fig. 8.4. The forward connections to a cell in one layer are derived from a topologically related and confined region of the preceding layer. The choice of whether a connection between neurons in adjacent layers exists or not, is based upon a Gaussian distribution of connection probabilities which roll off radially from the focal point of connections for each neuron. (A minor extra constraint precludes the repeated connection of any pair of cells.) In particular, the forward connections to a cell in one layer come from a small region of the preceding layer defined by the radius in Table 8.1 which will contain approximately 67% of the connections from the preceding layer. Figure 8.4 shows the general

Table 8.1 VisNet Dimensions

	Dimensions	# Connections	Radius
Layer 4	32x32	100	12
Layer 3	32x32	100	9
Layer 2	32x32	100	6
Layer 1	32x32	272	6
Retina	128x128x32	-	-

convergent network architecture used. Localization and limitation of connectivity in the network is intended to mimic cortical connectivity, partially because of the clear retention of retinal topology through regions of visual cortex. This architecture also encourages the gradual combination of features from layer to layer which has relevance to the binding problem, as described in Section 8.4.5[18].

8.4.1.3 Competition and lateral inhibition

In order to act as a competitive network some form of mutual inhibition is required within each layer, which should help to ensure that all stimuli presented are evenly represented by the neurons in each layer. This is implemented in VisNet by a form of lateral inhibition. The idea behind the lateral inhibition, apart from this being a property of cortical architecture in the brain, was to prevent too many neurons that received inputs from a similar part of the preceding layer responding to the same activity patterns. The purpose of the lateral inhibition was to ensure that different receiving neurons coded for different inputs. This is important in reducing redundancy (Rolls and Treves 1998). The lateral inhibition is conceived as operating within a radius that was similar to that of the region within which a neuron received converging inputs from the preceding layer (because activity in one zone of topologically organized processing within a layer should not inhibit processing in another zone in the same layer, concerned perhaps with another part of the image)[19].

The lateral inhibition and contrast enhancement just described is actually implemented in VisNet2 (Rolls and Milward 2000) in two stages, to produce filtering of the type illustrated in Fig. 8.5. This lateral inhibition is implemented by convolving the activation of the neurons in a layer with a spatial filter, I, where δ controls the contrast and σ controls the width, and a and b index the distance away from the centre of the filter

$$
I_{a,b} = \begin{cases} -\delta e^{-\frac{a^2+b^2}{\sigma^2}} & \text{if } a \neq 0 \text{ or } b \neq 0, \\ 1 - \sum_{a\neq 0, b\neq 0} I_{a,b} & \text{if } a = 0 \text{ and } b = 0. \end{cases} \tag{8.5}
$$

[18]Modelling topological constraints in connectivity leads to an issue concerning neurons at the edges of the network layers. In principle these neurons may either receive no input from beyond the edge of the preceding layer, or have their connections repeatedly sample neurons at the edge of the previous layer. In practice either solution is liable to introduce artificial weighting on the few active inputs at the edge and hence cause the edge to have unwanted influence over the development of the network as a whole. In the real brain such edge-effects would be naturally smoothed by the transition of the locus of cellular input from the fovea to the lower acuity periphery of the visual field. However, it poses a problem here because we are in effect only simulating the small high acuity foveal portion of the visual field in our simulations. As an alternative to the former solutions Wallis and Rolls (1997) elected to form the connections into a toroid, such that connections wrap back onto the network from opposite sides. This wrapping happens at all four layers of the network, and in the way an image on the 'retina' is mapped to the input filters. This solution has the advantage of making all of the boundaries effectively invisible to the network. Further, this procedure does not itself introduce problems into evaluation of the network for the problems set, as many of the critical comparisons in VisNet involve comparisons between a network with the same architecture trained with the trace rule, or with the Hebb rule, or not trained at all. In practice, it is shown below that only the network trained with the trace rule solves the problem of forming invariant representations.

[19]Although the extent of the lateral inhibition actually investigated by Wallis and Rolls (1997) in VisNet operated over adjacent pixels, the lateral inhibition introduced by Rolls and Milward (2000) in what they named VisNet2 and which has been used in subsequent simulations operates over a larger region, set within a layer to approximately half of the radius of convergence from the preceding layer. Indeed, Rolls and Milward (2000) showed in a problem in which invariant representations over 49 locations were being used with a 17 face test set, that the best performance was with intermediate range lateral inhibition, using the parameters for σ shown in Table 8.3. These values of σ set the lateral inhibition radius within a layer to be approximately half that of the spread of the excitatory connections from the preceding layer.

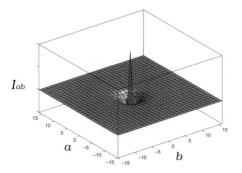

Fig. 8.5 Contrast enhancing filter, which has the effect of local lateral inhibition. The parameters δ and σ are variables used to modify the amount and extent of inhibition respectively.

Table 8.2 Sigmoid parameters for the runs with 25 locations by Rolls and Milward 2000

Layer	1	2	3	4
Percentile	99.2	98	88	91
Slope β	190	40	75	26

This is a filter which leaves the average activity unchanged.

The second stage involves contrast enhancement. In VisNet (Wallis and Rolls 1997), this was implemented by raising the neuronal activations to a fixed power and normalizing the resulting firing within a layer to have an average firing rate equal to 1.0. In VisNet2 (Rolls and Milward 2000) and in subsequent simulations a more biologically plausible form of the activation function, a sigmoid, was used:

$$y = f^{\text{sigmoid}}(r) = \frac{1}{1 + e^{-2\beta(r-\alpha)}} \qquad (8.6)$$

where r is the activation (or firing rate) of the neuron after the lateral inhibition, y is the firing rate after the contrast enhancement produced by the activation function, and β is the slope or gain and α is the threshold or bias of the activation function. The sigmoid bounds the firing rate between 0 and 1 so global normalization is not required. The slope and threshold are held constant within each layer. The slope is constant throughout training, whereas the threshold is used to control the sparseness of firing rates within each layer. The sparseness of the firing within a layer is defined (Rolls and Treves 1998) as:

$$a = \frac{\left(\sum_i y_i/n\right)^2}{\sum_i y_i^2/n} \qquad (8.7)$$

where n is the number of neurons in the layer. To set the sparseness to a given value, e.g. 5%, the threshold is set to the value of the 95th percentile point of the activations within the layer. (Unless otherwise stated here, the neurons used the sigmoid activation function as just described.)

In most simulations with VisNet2 and later, the sigmoid activation function was used with parameters (selected after a number of optimization runs) as shown in Table 8.2.

Table 8.3 Lateral inhibition parameters for the 25-location runs

Layer	1	2	3	4
Radius, σ	1.38	2.7	4.0	6.0
Contrast, δ	1.5	1.5	1.6	1.4

In addition, the lateral inhibition parameters normally used in VisNet2 simulations are as shown in Table 8.3[20].

8.4.1.4 The input to VisNet

VisNet is provided with a set of input filters which can be applied to an image to produce inputs to the network which correspond to those provided by simple cells in visual cortical area 1 (V1). The purpose of this is to enable within VisNet the more complicated response properties of cells between V1 and the inferior temporal cortex (IT) to be investigated, using as inputs natural stimuli such as those that could be applied to the retina of the real visual system. This is to facilitate comparisons between the activity of neurons in VisNet and those in the real visual system, to the same stimuli. In VisNet no attempt is made to train the response properties of simple cells, but instead we start with a defined series of filters to perform fixed feature extraction to a level equivalent to that of simple cells in V1, as have other researchers in the field (Hummel and Biederman 1992, Buhmann, Lange, von der Malsburg, Vorbrüggen and Würtz 1991, Fukushima 1980), because we wish to simulate the more complicated response properties of cells between V1 and the inferior temporal cortex (IT). The elongated orientation-tuned input filters used accord with the general tuning profiles of simple cells in V1 (Hawken and Parker 1987) and are computed by weighting the difference of two Gaussians by a third orthogonal Gaussian as described by Wallis, Rolls and Foldiak (1993) and Wallis and Rolls (1997). Each individual filter is tuned to spatial frequency (0.0625 to 0.5 cycles pixels^{-1} over four octaves); orientation (0° to 135° in steps of 45°); and sign (± 1). Of the 272 layer 1 connections, the number to each group is as shown in Table 8.4. In the current

Table 8.4 VisNet layer 1 connectivity. The frequency is in cycles per pixel

Frequency	0.5	0.25	0.125	0.0625
# Connections	201	50	13	8

model only even symmetric — 'bar detecting' — filter shapes are used, which take the form of a Gaussian shape along the axis of orientation tuning for the filter, and a difference of Gaussians along the perpendicular axis.

This filter is referred to as an oriented difference of Gaussians, or DOG filter[21]. It was chosen for VisNet in preference to the often used Gabor filter on the grounds of its better fit to available neurophysiological data including its zero D.C. response (Hawken and Parker 1987, Wallis, Rolls and Foldiak 1993). Any zero D.C. filter can of course produce a negative as well as positive output, which would mean that this simulation of a simple cell would permit negative as well as positive firing. In contrast to some other models the response of

---

[20] Where a power activation function was used in the simulations of Wallis and Rolls (1997), the power for layer 1 was 6, and for the other layers was 2.

[21] Professor R. Watt, of Stirling University, is thanked for assistance with the implementation of this filter scheme.

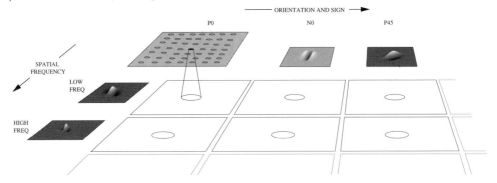

Fig. 8.6 The filter sampling paradigm. Here each square represents the retinal image presented to the network after being filtered by a Difference Of Gaussian filter of the appropriate orientation sign and frequency. The circles represent the consistent retinotopic coordinates used to provide input to a layer one cell. The filters double in spatial frequency towards the reader. Left to right the orientation tuning increases from 0° in steps of 45°, with segregated pairs of positive (P) and negative (N) filter responses.

each filter is zero thresholded and the negative results used to form a separate anti-phase input to the network. The filter outputs are also normalized across scales to compensate for the low frequency bias in the images of natural objects.

Cells of layer one receive a topologically consistent, localized, random selection of the filter responses in the input layer, under the constraint that each cell samples every filter spatial frequency and receives a constant number of inputs. Figure 8.6 shows pictorially the general filter sampling paradigm, and Fig. 8.7 the typical connectivity to a layer 1 cell from the filters of the input layer. The blank squares indicate that no connection exists between the layer one cell chosen and the filters of that particular orientation, sign, and spatial frequency.

8.4.1.5 Measures for network performance

A neuron can be said to have learnt an invariant representation if it discriminates one set of stimuli from another set, across all transformations. For example, a neuron's response is translation invariant if its response to one set of stimuli irrespective of presentation is consistently higher than for all other stimuli irrespective of presentation location. Note that we state 'set of stimuli' since neurons in the inferior temporal cortex are not generally selective for a single stimulus but rather a sub-population of stimuli (see Chapter 5) (Baylis, Rolls and Leonard 1985, Abbott, Rolls and Tovee 1996, Rolls, Treves and Tovee 1997b, Rolls and Treves 1998). The measure of network performance used in VisNet, the 'Fisher metric' (referred to in some Figure Labels as the Discrimination Factor), reflects how well a neuron discriminates between stimuli, compared to how well it discriminates between different locations (or more generally the images used rather than the objects, each of which is represented by a set of images, over which invariant stimulus or object representations must be learned). The Fisher measure is very similar to taking the ratio of the two F values in a two-way ANOVA, where one factor is the stimulus shown, and the other factor is the position in which a stimulus is shown. The measure takes a value greater than 1.0 if a neuron has more different responses to the stimuli than to the locations. That is, values greater than 1 indicate invariant representations when this measure is used in the following Figures. Further details of how the measure is calculated are given by Wallis and Rolls (1997).

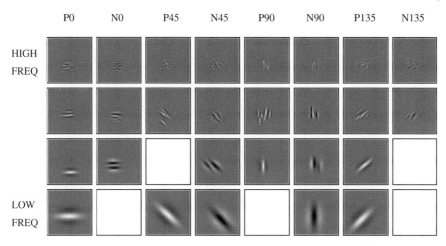

Fig. 8.7 Typical connectivity before training between a single cell in the first layer of the network and the input layer, represented by plotting the receptive fields of every input layer cell connected to the particular layer one cell. Separate input layer cells have activity that represents a positive (P) or negative (N) output from the bank of filters which have different orientations in degrees (the columns) and different spatial frequencies (the rows). Here the overall receptive field of the layer one cell is centred just below the centre-point of the retina. The connection scheme allows for relatively fewer connections to lower frequency cells than to high frequency cells in order to cover a similar region of the input at each frequency. A blank square indicates that there is no connection to the layer one neuron from an input neuron with that particular filter type.

Measures of network performance based on information theory and similar to those used in the analysis of the firing of real neurons in the brain (see Appendix B) were introduced by Rolls and Milward (2000) for VisNet2, and are used in later papers. A single cell information measure was introduced which is the maximum amount of information the cell has about any one stimulus / object independently of which transform (e.g. position on the retina) is shown. Because the competitive algorithm used in VisNet tends to produce local representations (in which single cells become tuned to one stimulus or object), this information measure can approach $\log_2 N_S$ bits, where N_S is the number of different stimuli. Indeed, it is an advantage of this measure that it has a defined maximal value, which enables how well the network is performing to be quantified. Rolls and Milward (2000) showed that the Fisher and single cell information measures were highly correlated, and given the advantage just noted of the information measure, it was adopted in Rolls and Milward (2000) and subsequent papers. Rolls and Milward (2000) also introduced a multiple cell information measure, which has the advantage that it provides a measure of whether all stimuli are encoded by different neurons in the network. Again, a high value of this measure indicates good performance.

For completeness, we provide further specification of the two information theoretic measures, which are described in detail by Rolls and Milward (2000) (see Appendix B for introduction of the concepts). The measures assess the extent to which either a single cell, or a population of cells, responds to the same stimulus invariantly with respect to its location, yet responds differently to different stimuli. The measures effectively show what one learns about which stimulus was presented from a single presentation of the stimulus at any randomly chosen location. Results for top (4th) layer cells are shown. High information measures thus show that cells fire similarly to the different transforms of a given stimulus (object), and differently to the other stimuli. The single cell stimulus-specific information, $I(s, R)$, is the amount of information the set of responses, R, has about a specific stimulus, s (see Rolls, Treves, Tovee and Panzeri (1997d) and Rolls and Milward (2000)). $I(s, R)$ is given by

$$I(s, R) = \sum_{r \in R} P(r|s) \log_2 \frac{P(r|s)}{P(r)} \tag{8.8}$$

where r is an individual response from the set of responses R of the neuron. For each cell the performance measure used was the maximum amount of information a cell conveyed about any one stimulus. This (rather than the mutual information, $I(S, R)$ where S is the whole set of stimuli s), is appropriate for a competitive network in which the cells tend to become tuned to one stimulus[22].

If all the output cells of VisNet learned to respond to the same stimulus, then the information about the set of stimuli S would be very poor, and would not reach its maximal value of $\log_2$ of the number of stimuli (in bits). The second measure that is used here is the information provided by a set of cells about the stimulus set, using the procedures described by Rolls, Treves and Tovee (1997b) and Rolls and Milward (2000). The multiple cell information is the mutual information between the whole set of stimuli S and of responses R calculated using a decoding procedure in which the stimulus s' that gave rise to the particular firing rate response vector on each trial is estimated. (The decoding step is needed because the high dimensionality of the response space would lead to an inaccurate estimate of the information if the responses were used directly, as described by Rolls, Treves and Tovee (1997b) and Rolls and Treves (1998)). A probability table is then constructed of the real stimuli s and the decoded stimuli s'. From this probability table, the mutual information between the set of actual stimuli S and the decoded estimates S' is calculated as

$$I(S, S') = \sum_{s, s'} P(s, s') \log_2 \frac{P(s, s')}{P(s)P(s')} \tag{8.9}$$

This was calculated for the subset of cells which had as single cells the most information about which stimulus was shown. In particular, in Rolls and Milward (2000) and subsequent papers, the multiple cell information was calculated from the first five cells for each stimulus that had maximal single cell information about that stimulus, that is from a population of 35 cells if there were seven stimuli (each of which might have been shown in for example 9 or 25 positions on the retina).

8.4.2 Initial experiments with VisNet

Having established a network model Wallis and Rolls (1997) (following a first report by Wallis, Rolls and Foldiak (1993)) described four experiments in which the theory of how invariant representations could be formed was tested using a variety of stimuli undergoing a number of natural transformations. In each case the network produced neurons in the final layer whose responses were largely invariant across a transformation and highly discriminating between stimuli or sets of stimuli.

---

[22] $I(s, R)$ has more recently been called the stimulus-specific surprise, see DeWeese and Meister (1999). Its average across stimuli is the mutual information $I(S, R)$.

Fig. 8.8 The three stimuli used in the first two experiments.

8.4.2.1 'T','L' and '+' as stimuli: Learning translation invariance

One of the classical properties of inferior temporal cortex face cells is their invariant response to face stimuli translated across the visual field (Tovee, Rolls and Azzopardi 1994). In this first experiment, the learning of translation invariant representations by VisNet was investigated.

In order to test the network a set of three stimuli, based upon probable 3D edge cues — consisting of a 'T', 'L' and '+' shape — was constructed[23]. The actual stimuli used are shown in Fig. 8.8. These stimuli were chosen partly because of their significance as form cues, but on a more practical note because they each contain the same fundamental features - namely a horizontal bar conjoined with a vertical bar. In practice this means that the oriented simple cell filters of the input layer can not distinguish these stimuli on the basis of which features are present. As a consequence of this, the representation of the stimuli received by the network is non-orthogonal and hence considerably more difficult to classify than was the case in earlier experiments involving the trace rule described by Földiák (1991). The expectation is that layer one neurons would learn to respond to spatially selective combinations of the basic features thereby helping to distinguish these non-orthogonal stimuli. The trajectory followed by each stimulus consisted of sweeping left to right horizontally across three locations in the top row, and then sweeping back, right to left across the middle row, before returning to the right hand side across the bottom row — tracing out a 'Z' shape path across the retina. Unless stated otherwise this pattern of nine presentation locations was adopted in all image translation experiments described by Wallis and Rolls (1997).

Training was carried out by permutatively presenting all stimuli in each location a total of 800 times. The sequence described above was followed for each stimulus, with the sequence start point and direction of sweep being chosen at random for each of the 800 training trials.

Figures 8.9 and 8.10 show the response after training of a first layer neuron selective for the 'T' stimulus. The weighted sum of all filter inputs reveals the combination of horizontally and vertically tuned filters in identifying the stimulus. In this case many connections to the lower frequency filters have been reduced to zero by the learning process, except at the relevant orientations. This contrasts strongly with the random wiring present before training, as seen previously in Fig. 8.7.

Likewise, Fig. 8.11 depicts two neural responses, but now from the two intermediate layers of the network, taken from the top 30 most highly invariant cells, not merely the top two or three. The gradual increase in the discrimination indicates that the tolerance to shifts of the preferred stimulus gradually builds up through the layers.

---

[23]Chakravarty (1979) describes the application of these shapes as cues for the 3D interpretation of edge junctions, and Tanaka et al (1991) have demonstrated the existence of cells responsive to such stimuli in IT.

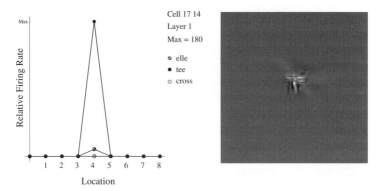

Fig. 8.9 The left graph shows the response of a layer one neuron to the three training stimuli for the nine training locations. Alongside this are the results of summating all the filter inputs to the neuron. The discrimination factor for this cell was 1.04.

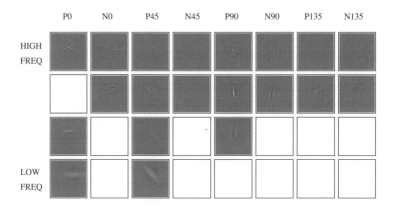

Fig. 8.10 The same cell as in the previous Figure and the same input reconstruction results but separated into four rows of differing spatial frequency, and eight columns representing the four filter tuning orientations in positive and negative complementary pairs.

The results for layer four neurons are illustrated in Fig. 8.12. By this stage translation-invariant, stimulus-identifying, cells have emerged. The response profiles confirm the high level of neural selectivity for a particular stimulus irrespective of location.

Figure 8.13 contrasts the measure of invariance, or discrimination factor, achieved by cells in the four layers, averaged over five separate runs of the network. Translation invariance clearly increases through the layers, with a considerable increase in translation invariance between layers three and four. This sudden increase may well be a result of the geometry of the network, which enables cells in layer 4 to receive inputs from any part of the input layer.

Having established that invariant cells have emerged in the final layer, we now consider the role of the trace rule, by assessing the network tested under two new conditions. Firstly, the performance of the network was measured before learning occurs, that is with its initially random connection weights. Secondly, the network was trained with η in the trace rule set to 0, which causes learning to proceed in a traceless, standard Hebbian, fashion.

Figure 8.14 shows the results under the three training conditions. The results show that

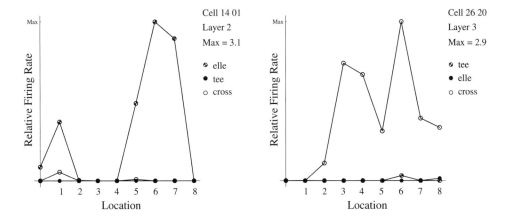

Fig. 8.11 Response profiles for two intermediate layer neurons — discrimination factors 1.34 and 1.64 — in the L, T and + experiment.

the trace rule is the decisive factor in establishing the invariant responses in the layer four neurons. It is interesting to note that the Hebbian learning results are actually *worse* than those achieved by chance in the untrained net. In general, with Hebbian learning, the most highly discriminating cells barely rate higher than 1. This value of discrimination corresponds to the case in which a cell responds to only one stimulus and in only one location. The poor performance with the Hebb rule comes as a direct consequence of the presentation paradigm being employed. If we consider an image as representing a vector in multidimensional space, a particular image in the top left hand corner of the input retina will tend to look more like any other image in that same location than the same image presented elsewhere. A simple competitive network using just Hebbian learning will thus tend to categorize images by *where* they are rather than what they are — the exact opposite of what the net was intended to learn. This comparison thus indicates that a small memory trace acting in the standard Hebbian learning paradigm can radically alter the normal vector averaging, image classification, performed by a Hebbian based competitive network.

One question which emerges about the representation in the final layer of the network relates to how evenly the network divides up its resources to represent the learnt stimuli. It is conceivable that one stimulus stands out among the set of stimuli by containing very distinctive features which would make it easier to categorize. This may produce an unrepresentative number of neurons with high discrimination factors which are in fact all responding to the same stimulus. It is important that at least some cells code for (or provide information about) each of the stimuli. As a simple check on this, the preferred stimulus of each cell was found and the associated measure of discrimination added to a total for each stimulus. This measure in practice never varied by more than a factor of 1.3 : 1 for all stimuli. The multiple cell information measure used in some later Figures addresses the same issue, with similar results.

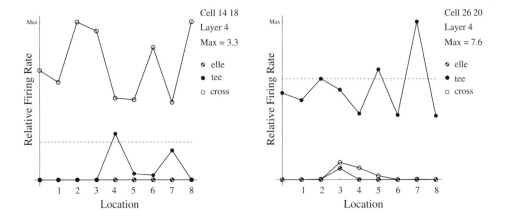

Fig. 8.12 Response profiles for two fourth layer neurons - discrimination factors 4.07 and 3.62 - in the L, T and + experiment.

8.4.2.2 'T','L' and '+' as stimuli: Optimal network parameters

The second series of investigations described by Wallis and Rolls (1997) using the 'T','L' and '+' stimuli, centred upon finding optimal parameters for elements of the network, such as the optimal trace time constant η, which controls the relative effect of previous activities on current learning as described above. The network performance was gauged using a single 800 epoch training run of the network with the median discrimination factor (with the upper and lower quartile values) for the top sixteen cells of the fourth layer being displayed at each parameter value.

   Figure 8.15 displays the effect of varying the value of η for the nine standard presentation locations. The optimal value of η might conceivably change with the alteration of the number of training locations, and indeed one might predict that it would be smaller if the number of presentation locations was reduced. To confirm this, network performance was also measured for presentation sweeps over only five locations. Figure 8.16 shows the results of this experiment, which confirm the expected shift in the general profile of the curve towards shorter time constant values. Of course, the optimal value of η derived is in effect a compromise between optimal values for the three layers in which the trace operates. Since neurons in each layer have different effective receptive field sizes, one would expect each layer's neurons to be exposed to different portions of the full sweep of a particular stimulus. This would in turn suggest that the optimal value of η will grow through the layers.

8.4.2.3 Faces as stimuli: Translation invariance

The aim of the next set of experiments described by Wallis and Rolls (1997) was to start to address the issues of how the network operates when invariant representations must be learned for a larger number of stimuli, and whether the network can learn when much more complicated, real biological stimuli (faces) are used. The set of face images used appears in Fig. 8.17. In practice, to equalize luminance the dc component of the images was removed. In

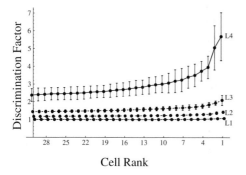

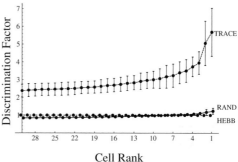

Fig. 8.13 Variation in neural discrimination factors as a measure of performance for the top 30 most highly discriminating cells through the four layers of the network, averaged over five runs of the network in the L, T and + experiment.

Fig. 8.14 Variation in neural discrimination factors as a measure of performance for the top 30 most highly discriminating cells in the fourth layer for the three training regimes, averaged over five runs of the network.

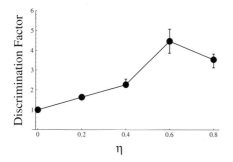

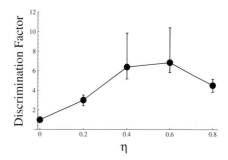

Fig. 8.15 Variation in network performance as a function of the trace rule parameter η in neurons of layers two to four — over nine locations in the L, T and + experiment.

Fig. 8.16 Variation in network performance as a function of the trace rule parameter η in neurons of layers two to four — over five presentation locations in the L, T and + experiment.

addition, so as to minimize the effect of cast shadows, an oval Hamming window was applied to the face image which also served to remove any hard edges of the image relative to the plain background upon which they were set.

The results of training in the translation invariance paradigm with 7 faces each in 9 locations are shown in Figs. 8.18, 8.19 and 8.20. The network produces neurons with high discrimination factors, and this only occurs if it is trained with the trace rule. Some layer 4 neurons showed a somewhat distributed representation, as illustrated in the examples of layer 4 neurons shown in Fig. 8.18.

In order to check that there was an invariant representation in layer 4 of VisNet that could

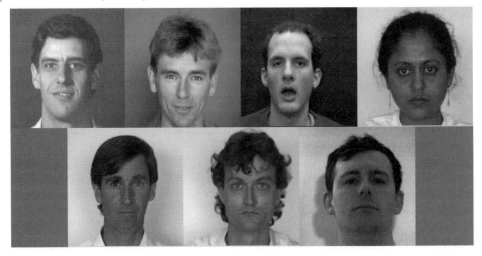

Fig. 8.17 Seven faces used as stimuli in the face translation experiment.

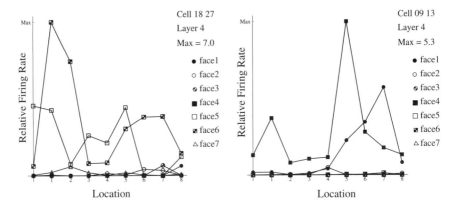

Fig. 8.18 Response profiles for two neurons in the fourth layer — discrimination factors 2.64 and 2.10. The net was trained on 7 faces each in 9 locations.

be read by a receiving population of neurons, a fifth layer was added to the net which fully sampled the fourth layer cells. This layer was in turn trained in a supervised manner using gradient descent or with a Hebbian associative learning rule. Wallis and Rolls (1997) showed that the object classification performed by the layer 5 network was better if the network had been trained with the trace rule than when it was untrained or was trained with a Hebb rule.

8.4.2.4 Faces as stimuli: View Invariance

Now that the network has been shown to be able to operate usefully with a more difficult translation invariance problem, we now address the question of whether the network can solve other types of transform invariance, as we had intended. The next experiment addresses this question, by training the network on the problem of 3D stimulus rotation, which produces non-isomorphic transforms, to determine whether the network can build a view-invariant categorization of the stimuli. The trace rule learning paradigm should, in conjunction with

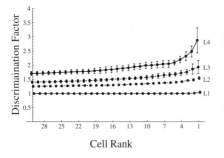

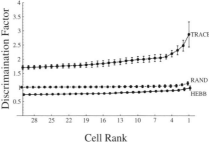

Fig. 8.19 Variation in network performance for the top 30 most highly discriminating cells through the four layers of the network, averaged over five runs of the network. The net was trained on 7 faces each in 9 locations.

Fig. 8.20 Variation in network performance for the top 30 most highly discriminating cells in the fourth layer for the three training regimes, averaged over five runs of the network. The net was trained on 7 faces each in 9 locations.

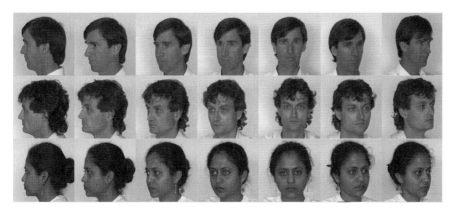

Fig. 8.21 Three faces in seven different views used as stimuli in an experiment by Wallis and Rolls (1997).

the architecture we describe here, prove capable of learning any of the transforms tolerated by IT neurons, so long as each stimulus is presented in short sequences during which the transformation occurs and can be learned. This experiment continues with the use of faces but now presents them centrally in the retina in a sequence of different views of a face. The images used are shown in Fig. 8.21. The faces were again smoothed at the edges to erase the harsh image boundaries, and the dc term was removed. During the 800 epochs of learning, each stimulus was chosen at random, and a sequence of preset views of it was shown, rotating the face either to the left or right.

Although the actual number of images being presented is smaller, some 21 views in all, there is good reason to think that this problem may be harder to solve than the previous translation experiments. This is simply due to the fact that all 21 views exactly overlap with one another. The net was indeed able to solve the invariance problem, with examples of invariant layer four neuron response profiles appearing in Fig. 8.22.

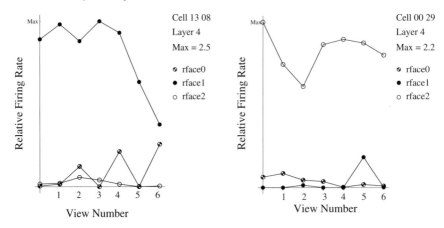

Fig. 8.22 Response profiles for cells in the last two layers of the network - discrimination factors 11.12 and 12.40 - in the experiment with seven different views of each of three faces.

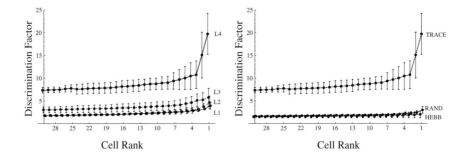

Fig. 8.23 Variation in network performance for the top 30 most highly discriminating cells through the four layers of the network, averaged over five runs of the network in the experiment with seven different views of each of three faces.

Fig. 8.24 Variation in network performance for the top 30 most highly discriminating cells in the fourth layer for the three training regimes, averaged over five runs of the network in the experiment with seven different views of each of three faces.

Figure 8.23 confirms the improvement in invariant stimulus representation found through the layers, and that layer four provides a considerable improvement in performance over the previous layers. Figure 8.24 shows the Hebb trained and untrained nets performing equally poorly, whilst the trace trained net shows good invariance across the entire 30 cells selected.

8.4.3 The optimal parameters for the temporal trace used in the learning rule

The trace used in VisNet enables successive features that, based on the natural statistics of the visual input, are likely to be from the same object or feature complex to be associated together. For good performance, the temporal trace needs to be sufficiently long that it covers the period in which features seen by a particular neuron in the hierarchy are likely to come from the same object. On the other hand, the trace should not be so long that it produces associations between features that are parts of different objects, seen when for example the eyes move to another object. One possibility is to reset the trace during saccades between different objects. If explicit trace resetting is not implemented, then the trace should, to optimize the compromise implied by the above, lead to strong associations between temporally close stimuli, and increasingly weaker associations between temporally more distant stimuli. In fact, the trace implemented in VisNet has an exponential decay, and it has been shown that this form is optimal in the situation where the exact duration over which the same object is being viewed varies, and where the natural statistics of the visual input happen also to show a decreasing probability that the same object is being viewed as the time period in question increases (Wallis and Baddeley 1997). Moreover, as is made evident in Figs. 8.16 and 8.15, performance can be enhanced if the duration of the trace does at the same time approximately match the period over which the input stimuli are likely to come from the same object or feature complex. Nevertheless, good performance can be obtained under conditions under which the trace rule allows associations to be formed only between successive items in the visual stream (Rolls and Milward 2000, Rolls and Stringer 2001a).

It is also the case that the optimal value of η in the trace rule is likely to be different for different layers of VisNet, and for cortical processing in the 'what' visual stream. For early layers of the system, small movements of the eyes might lead to different feature combinations providing the input to cells (which at early stages have small receptive fields), and a short duration of the trace would be optimal. However, these small eye movements might be around the same object, and later layers of the architecture would benefit from being able to associate together their inputs over longer times, in order to learn about the larger scale properties that characterize individual objects, including for example different views of objects observed as an object turns or is turned. Thus the suggestion is made that the temporal trace could be effectively longer at later stages (e.g. inferior temporal visual cortex) compared to early stages (e.g. V2 and V4) of processing in the visual system. In addition, as will be shown in Section 8.4.5, it is important to form feature combinations with high spatial precision before invariance learning supported by a temporal place starts, in order that the feature combinations and not the individual features have invariant representations. This leads to the suggestion that the trace rule should either not operate, or be short, at early stages of cortical visual processing such as V1. This is reflected in the operation of VisNet2, which does not use a temporal trace in layer 1.

8.4.4 Different forms of the trace learning rule, and their relation to error correction and temporal difference learning

The original trace learning rule used in the simulations of Wallis and Rolls (1997) took the form

$$\delta w_j = \alpha \overline{y}^\tau x_j^\tau \tag{8.10}$$

where the trace $\overline{y}^\tau$ is updated according to

$$\overline{y}^\tau = (1 - \eta)y^\tau + \eta \overline{y}^{\tau-1}. \tag{8.11}$$

The parameter $\eta \in [0, 1]$ controls the relative contributions to the trace $\overline{y}^\tau$ from the instantaneous firing rate y^τ and the trace at the previous time step $\overline{y}^{\tau-1}$, where for $\eta = 0$ we have $\overline{y}^\tau = y^\tau$ and equation (8.10) becomes the standard Hebb rule

$$\delta w_j = \alpha y^\tau x_j^\tau. \tag{8.12}$$

In the start of a series of investigations of different forms of the trace learning rule, Rolls and Milward (2000) demonstrated that VisNet's performance could be greatly enhanced with a modified Hebbian learning rule that incorporated a trace of activity from the preceding time steps, with no contribution from the activity being produced by the stimulus at the current time step. This rule took the form

$$\delta w_j = \alpha \overline{y}^{\tau-1} x_j^\tau. \tag{8.13}$$

The trace shown in equation 8.13 is in the postsynaptic term, and similar effects were found if the trace was in the presynaptic term, or in both the pre- and the postsynaptic term. The crucial difference from the earlier rule (see equation 8.10) was that the trace should be calculated up to only the preceding timestep, with no contribution to the trace from the firing on the current trial to the current stimulus. How might this be understood?

One way to understand this is to note that the trace rule is trying to set up the synaptic weight on trial τ based on whether the neuron, based on its previous history, is responding to that stimulus (in other positions). Use of the trace rule at $\tau - 1$ does this, that is it takes into account the firing of the neuron on previous trials, with no contribution from the firing being produced by the stimulus on the current trial. On the other hand, use of the trace at time τ in the update takes into account the current firing of the neuron to the stimulus in that particular position, which is not a good estimate of whether that neuron should be allocated to invariantly represent that stimulus. Effectively, using the trace at time τ introduces a Hebbian element into the update, which tends to build position encoded analyzers, rather than stimulus encoded analyzers. (The argument has been phrased for a system learning translation invariance, but applies to the learning of all types of invariance.) A particular advantage of using the trace at $\tau - 1$ is that the trace will then on different occasions (due to the randomness in the location sequences used) reflect previous histories with different sets of positions, enabling the learning of the neuron to be based on evidence from the stimulus present in many different positions. Using a term from the current firing in the trace (i.e. the trace calculated at time τ) results in this desirable effect always having an undesirable element from the current firing of the neuron to the stimulus in its current position.

8.4.4.1 The Modified Hebbian trace rule and its relation to error correction

Rule (8.13) corrects the weights using a postsynaptic trace obtained from the previous firing (produced by other transforms of the same stimulus), with no contribution to the trace from the current postsynaptic firing (produced by the current transform of the stimulus). Indeed, insofar as the current firing y^τ is not the same as $\overline{y}^{\tau-1}$, this difference can be thought of as an error. This leads to a conceptualization of using the difference between the current firing and the preceding trace as an error correction term, as noted in the context of modelling the temporal properties of classical conditioning by Sutton and Barto (1981), and developed next in the context of invariant learning (see Rolls and Stringer (2001a)).

First, we re-express rule (8.13) in an alternative form as follows. Suppose we are at timestep τ and have just calculated a neuronal firing rate y^τ and the corresponding trace $\overline{y}^\tau$ from the trace update equation (8.11). If we assume $\eta \in (0, 1)$, then rearranging equation (8.11) gives

$$\overline{y}^{\tau-1} = \frac{1}{\eta}(\overline{y}^\tau - (1-\eta)y^\tau), \tag{8.14}$$

and substituting equation (8.14) into equation (8.13) gives

$$\delta w_j = \alpha \frac{1}{\eta}(\overline{y}^\tau - (1-\eta)y^\tau)x_j^\tau$$
$$= \alpha \frac{1-\eta}{\eta}(\frac{1}{1-\eta}\overline{y}^\tau - y^\tau)x_j^\tau$$
$$= \hat{\alpha}(\hat{\beta}\overline{y}^\tau - y^\tau)x_j^\tau \tag{8.15}$$

where $\hat{\alpha} = \alpha\frac{1-\eta}{\eta}$ and $\hat{\beta} = \frac{1}{1-\eta}$. The modified Hebbian trace learning rule (8.13) is thus equivalent to equation (8.15) which is in the general form of an error correction rule (Hertz, Krogh and Palmer 1991). That is, rule (8.15) involves the subtraction of the current firing rate y^τ from a target value, in this case $\hat{\beta}\overline{y}^\tau$.

Although above we have referred to rule (8.13) as a modified Hebbian rule, we note that it is only associative in the sense of associating *previous* cell firing with the current cell inputs. In the next Section we continue to explore the error correction paradigm, examining five alternative examples of this sort of learning rule.

8.4.4.2 Five forms of error correction learning rule

Error correction learning rules are derived from gradient descent minimization (Hertz, Krogh and Palmer 1991), and continually compare the current neuronal output to a target value t and adjust the synaptic weights according to the following equation at a particular timestep τ

$$\delta w_j = \alpha(t - y^\tau)x_j^\tau. \tag{8.16}$$

In this usual form of gradient descent by error correction, the target t is fixed. However, in keeping with our aim of encouraging neurons to respond similarly to images that occur close together in time it seems reasonable to set the target at a particular timestep, t^τ, to be some function of cell activity occurring close in time, because encouraging neurons to respond to temporal classes will tend to make them respond to the different variants of a given stimulus (Földiák 1991, Rolls 1992a, Wallis and Rolls 1997). For this reason, Rolls and Stringer (2001a) explored a range of error correction rules where the targets t^τ are based on the trace of neuronal activity calculated according to equation (8.11). We note that although the target is not a fixed value as in standard error correction learning, nevertheless the new learning rules perform gradient descent on each timestep, as elaborated discussed below. Although the target may be varying early on in learning, as learning proceeds the target is expected to become more and more constant, as neurons settle to respond invariantly to particular stimuli. The first set of five error correction rules we discuss are as follows.

$$\delta w_j = \alpha(\beta\overline{y}^{\tau-1} - y^\tau)x_j^\tau, \tag{8.17}$$

$$\delta w_j = \alpha(\beta y^{\tau-1} - y^\tau)x_j^\tau, \tag{8.18}$$

$$\delta w_j = \alpha(\beta\overline{y}^\tau - y^\tau)x_j^\tau, \tag{8.19}$$

$$\delta w_j = \alpha(\beta\overline{y}^{\tau+1} - y^\tau)x_j^\tau, \tag{8.20}$$

$$\delta w_j = \alpha(\beta y^{\tau+1} - y^\tau)x_j^\tau, \tag{8.21}$$

where updates (8.17), (8.18) and (8.19) are performed at time step τ, and updates (8.20) and (8.21) are performed at time step $\tau + 1$. (The reason for adopting this convention is that

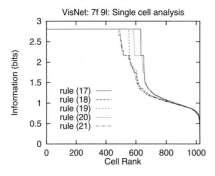
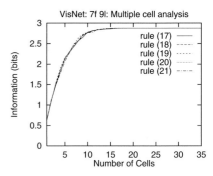

Fig. 8.25 Numerical results with the five error correction rules (8.17), (8.18), (8.19), (8.20), (8.21), (with positive clipping of synaptic weights) trained on 7 faces in 9 locations. On the left are single cell information measures, and on the right are multiple cell information measures.

the basic form of the error correction rule (8.16) is kept, with the five different rules simply replacing the term t.) It may be readily seen that equations (8.18) and (8.21) are special cases of equations (8.17) and (8.20) respectively, with $\eta = 0$.

These rules are all similar except for their targets t^τ, which are all functions of a temporally nearby value of cell activity. In particular, rule (8.19) is directly related to rule (8.15), but is more general in that the parameter $\hat{\beta} = \frac{1}{1-\eta}$ is replaced by an unconstrained parameter β. In addition, we also note that rule (8.17) is closely related to a rule developed in Peng, Sha, Gan and Wei (1998) for view invariance learning. The above five error correction rules are biologically plausible in that the targets t^τ are all local cell variables (Rolls and Treves 1998). In particular, rule (8.19) uses the trace $\overline{y}^\tau$ from the current time level τ, and rules (8.18) and (8.21) do not need exponential trace values $\overline{y}$, instead relying only on the instantaneous firing rates at the current and immediately preceding timesteps. However, all five error correction rules involve decrementing of synaptic weights according to an error which is calculated by subtracting the current activity from a target.

Numerical results with the error correction rules trained on 7 faces in 9 locations are presented in Fig. 8.25. For all the results shown the synaptic weights were clipped to be positive during the simulation, because it is important to test that decrementing synaptic weights purely within the positive interval $w \in [0, \infty)$ will provide significantly enhanced performance. That is, it is important to show that error correction rules do not necessarily require possibly biologically implausible modifiable negative weights. For each of the rules (8.17), (8.18), (8.19), (8.20), (8.21), the parameter β has been individually optimized to the following respective values: 4.9, 2.2, 2.2, 3.8, 2.2. On the left and right are results with the single and multiple cell information measures, respectively. Comparing Fig. 8.25 with Fig. 8.26 shows that all five error correction rules offer considerably improved performance over both the standard trace rule (8.10) and rule (8.13). From the left hand side of Fig. 8.25 it can be seen that rule (8.17) performs best, and this is probably due to two reasons. Firstly, rule (8.17) incorporates an exponential trace $\overline{y}^{\tau-1}$ in its target t^τ, and we would expect this to help neurons to learn more quickly to respond invariantly to a class of inputs that occur close together in time. Hence, setting $\eta = 0$ as in rule (8.18) results in reduced performance. Secondly, unlike rules (8.19) and (8.20), rule (8.17) does not contain any component of y^τ in its target. If we examine rules (8.19), (8.20), we see that their respective targets $\beta\overline{y}^\tau$, $\beta\overline{y}^{\tau+1}$ contain significant components of y^τ.

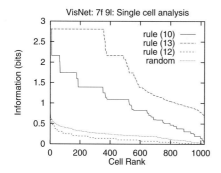

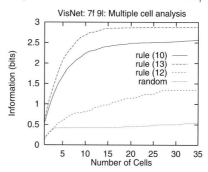

Fig. 8.26 Numerical results with the standard trace rule (8.10), learning rule (8.13), the Hebb rule (8.12), and random weights, trained on 7 faces in 9 locations: single cell information measure (left), multiple cell information measure (right).

8.4.4.3 Relationship to temporal difference learning

Rolls and Stringer (2001a) not only considered the relationship of rule (8.13) to error correction, but also considered how the error correction rules shown in equations (8.17), (8.18), (8.19), (8.20) and (8.21) are related to temporal difference learning (Sutton 1988, Sutton and Barto 1998). Sutton (1988) described temporal difference methods in the context of prediction learning. These methods are a class of incremental learning techniques that can learn to predict final outcomes through comparison of successive predictions from the preceding time steps. This is in contrast to traditional supervised learning, which involves the comparison of predictions only with the final outcome. Consider a series of multi-step prediction problems in which for each problem there is a sequence of observation vectors, $\mathbf{x}^1$, $\mathbf{x}^2$, ..., $\mathbf{x}^m$, at successive time steps, followed by a final scalar outcome z. For each sequence of observations temporal difference methods form a sequence of predictions y^1, y^2, ..., y^m, each of which is a prediction of z. These predictions are based on the observation vectors $\mathbf{x}^\tau$ and a vector of modifiable weights $\mathbf{w}$; i.e. the prediction at time step τ is given by $y^\tau(\mathbf{x}^\tau, \mathbf{w})$, and for a linear dependency the prediction is given by $y^\tau = \mathbf{w}^T\mathbf{x}^\tau$. (Note here that $\mathbf{w}^T$ is the transpose of the weight vector $\mathbf{w}$.) The problem of prediction is to calculate the weight vector $\mathbf{w}$ such that the predictions y^τ are good estimates of the outcome z.

The supervised learning approach to the prediction problem is to form pairs of observation vectors $\mathbf{x}^\tau$ and outcome z for all time steps, and compute an update to the weights according to the gradient descent equation

$$\delta\mathbf{w} = \alpha(z - y^\tau)\nabla_\mathbf{w} y^\tau \tag{8.22}$$

where α is a learning rate parameter and $\nabla_\mathbf{w}$ indicates the gradient with respect to the weight vector $\mathbf{w}$. However, this learning procedure requires all calculation to be done at the end of the sequence, once z is known. To remedy this, it is possible to replace method (8.22) with a temporal difference algorithm that is mathematically equivalent but allows the computational workload to be spread out over the entire sequence of observations. Temporal difference methods are a particular approach to updating the weights based on the values of successive predictions, y^τ, $y^{\tau+1}$. Sutton (1988) showed that the following temporal difference algorithm is equivalent to method (8.22)

$$\delta \mathbf{w} = \alpha(y^{\tau+1} - y^\tau) \sum_{k=1}^{\tau} \nabla_{\mathbf{w}} y^k, \qquad (8.23)$$

where $y^{m+1} \equiv z$. However, unlike method (8.22) this can be computed incrementally at each successive time step since each update depends only on $y^{\tau+1}$, y^τ and the sum of $\nabla_{\mathbf{w}} y^k$ over previous time steps k. The next step taken in Sutton (1988) is to generalize equation (8.23) to the following final form of temporal difference algorithm, known as 'TD(λ)'

$$\delta \mathbf{w} = \alpha(y^{\tau+1} - y^\tau) \sum_{k=1}^{\tau} \lambda^{\tau-k} \nabla_{\mathbf{w}} y^k \qquad (8.24)$$

where $\lambda \in [0, 1]$ is an adjustable parameter that controls the weighting on the vectors $\nabla_{\mathbf{w}} y^k$. Equation (8.24) represents a much broader class of learning rules than the more usual gradient descent based rule (8.23), which is in fact the special case TD(1).

A further special case of equation (8.24) is for $\lambda = 0$, i.e. TD(0), as follows

$$\delta \mathbf{w} = \alpha(y^{\tau+1} - y^\tau) \nabla_{\mathbf{w}} y^\tau. \qquad (8.25)$$

But for problems where y^τ is a linear function of $\mathbf{x}^\tau$ and $\mathbf{w}$, we have $\nabla_{\mathbf{w}} y^\tau = \mathbf{x}^\tau$, and so equation (8.25) becomes

$$\delta \mathbf{w} = \alpha(y^{\tau+1} - y^\tau) \mathbf{x}^\tau. \qquad (8.26)$$

If we assume the prediction process is being performed by a neuron with a vector of inputs $\mathbf{x}^\tau$, synaptic weight vector $\mathbf{w}$, and output $y^\tau = \mathbf{w}^T \mathbf{x}^\tau$, then we see that the TD(0) algorithm (8.26) is identical to the error correction rule (8.21) with $\beta = 1$. In understanding this comparison with temporal difference learning, it may be useful to note that the firing at the end of a sequence of the transformed exemplars of a stimulus is effectively the temporal difference target z. This establishes a link to temporal difference learning. Further, we note that from learning epoch to learning epoch, the target z for a given neuron will gradually settle down to be more and more fixed as learning proceeds.

We now explore in more detail the relation between the error correction rules described above and temporal difference learning. For each sequence of observations with a single outcome the temporal difference method (8.26), when viewed as an error correction rule, is attempting to adapt the weights such that $y^{\tau+1} = y^\tau$ for all successive pairs of time steps — the same general idea underlying the error correction rules (8.17), (8.18), (8.19), (8.20), (8.21). Furthermore, in Sutton and Barto (1998), where temporal difference methods are applied to reinforcement learning, the TD(λ) approach is again further generalized by replacing the target $y^{\tau+1}$ by any weighted average of predictions y from arbitrary future time steps, e.g $t^\tau = \frac{1}{2} y^{\tau+3} + \frac{1}{2} y^{\tau+7}$, including an exponentially weighted average extending forward in time. So a more general form of the temporal difference algorithm has the form

$$\delta \mathbf{w} = \alpha(t^\tau - y^\tau) \mathbf{x}^\tau, \qquad (8.27)$$

where here the target t^τ is an arbitrary weighted average of the predictions y over future time steps. Of course, with standard temporal difference methods the target t^τ is always an average over *future* time steps $k = \tau + 1, \tau + 2$, etc. But in the five error correction rules this is only true for the last exemplar (8.21). This is because with the problem of prediction, for example, the ultimate target of the predictions $y^1,...,y^m$ is a final outcome $y^{m+1} \equiv z$.

However, this restriction does not apply to our particular application of neurons trained to respond to temporal classes of inputs within VisNet. Here we only wish to set the firing rates $y^1,...,y^m$ to the same value, not some final given value z. However, the more general error correction rules clearly have a close relationship to standard temporal difference algorithms. For example, it can be seen that equation (8.18) with $\beta = 1$ is in some sense a temporal mirror image of equation (8.26), particularly if the updates δw_j are added to the weights w_j only at the end of a sequence. That is, rule (8.18) will attempt to set $y^1,...,y^m$ to an *initial* value $y^0 \equiv z$. This relationship to temporal difference algorithms allows us to begin to exploit established temporal difference analyses to investigate the convergence properties of the error correction methods (Rolls and Stringer 2001a).

Although the main aim of Rolls and Stringer (2001a) in relating error correction rules to temporal difference learning was to begin to exploit established temporal difference analyses, they observed that the most general form of temporal difference learning, TD(λ), in fact suggests an interesting generalization to the existing error correction learning rules for which we currently have $\lambda = 0$. Assuming $y^\tau = \mathbf{w}^T \mathbf{x}^\tau$ and $\nabla_{\mathbf{w}} y^\tau = \mathbf{x}^\tau$, the general equation (8.24) for TD(λ) becomes

$$\delta \mathbf{w} = \alpha(y^{\tau+1} - y^\tau) \sum_{k=1}^{\tau} \lambda^{\tau-k} \mathbf{x}^k \tag{8.28}$$

where the term $\sum_{k=1}^{\tau} \lambda^{\tau-k} \mathbf{x}^k$ is a weighted sum of the vectors $\mathbf{x}^k$. This suggests generalizing the original five error correction rules (8.17), (8.18), (8.19), (8.20), (8.21) by replacing the term x_j^τ by a weighted sum $\hat{x}_j^\tau = \sum_{k=1}^{\tau} \lambda^{\tau-k} x_j^k$ with $\lambda \in [0, 1]$. In Sutton (1988) $\hat{x}_j^\tau$ is calculated according to

$$\hat{x}_j^\tau = x_j^\tau + \lambda \hat{x}_j^{\tau-1} \tag{8.29}$$

with $\hat{x}_j^0 \equiv 0$. This gives the following five temporal difference inspired error correction rules

$$\delta w_j = \alpha(\beta \overline{y}^{\tau-1} - y^\tau)\hat{x}_j^\tau, \tag{8.30}$$

$$\delta w_j = \alpha(\beta y^{\tau-1} - y^\tau)\hat{x}_j^\tau, \tag{8.31}$$

$$\delta w_j = \alpha(\beta \overline{y}^\tau - y^\tau)\hat{x}_j^\tau, \tag{8.32}$$

$$\delta w_j = \alpha(\beta \overline{y}^{\tau+1} - y^\tau)\hat{x}_j^\tau, \tag{8.33}$$

$$\delta w_j = \alpha(\beta y^{\tau+1} - y^\tau)\hat{x}_j^\tau, \tag{8.34}$$

where it may be readily seen that equations (8.31) and (8.34) are special cases of equations (8.30) and (8.33) respectively, with $\eta = 0$. As with the trace $\overline{y}^\tau$, the term $\hat{x}_j^\tau$ is reset to zero when a new stimulus is presented. These five rules can be related to the more general TD(λ) algorithm, but continue to be biologically plausible using only local cell variables. Setting $\lambda = 0$ in rules (8.30), (8.31), (8.32), (8.33), (8.34), gives us back the original error correction rules (8.17), (8.18), (8.19), (8.20), (8.21), which may now be related to TD(0).

Numerical results with error correction rules (8.30), (8.31), (8.32), (8.33), (8.34), and $\hat{x}_j^\tau$ calculated according to equation (8.29) with $\lambda = 1$, with positive clipping of weights, trained on 7 faces in 9 locations are presented in Fig. 8.27. For each of the rules (8.30), (8.31), (8.32), (8.33), (8.34), the parameter β has been individually optimized to the following respective values: 1.7, 1.8, 1.5, 1.6, 1.8. On the left and right are results with the single and multiple cell information measures, respectively. Comparing these five temporal difference inspired rules

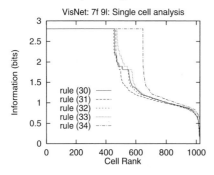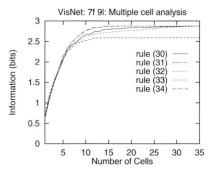

Fig. 8.27 Numerical results with the five temporal difference inspired error correction rules (8.30), (8.31), (8.32), (8.33), (8.34), and $\hat{x}_j^\tau$ calculated according to equation (8.29) (with positive clipping of synaptic weights) trained on 7 faces in 9 locations. On the left are single cell information measures, and on the right are multiple cell information measures.

it can be seen that the best performance is obtained with rule (8.34) where many more cells reach the maximum level of performance possible with respect to the single cell information measure. In fact, this rule offered the best such results. This may well be due to the fact that this rule may be directly compared to the standard TD(1) learning rule, which itself may be related to classical supervised learning for which there are well known optimality results, as discussed further by Rolls and Stringer (2001a).

From the simulations described by Rolls and Stringer (2001a) it appears that the form of optimization described above associated with TD(1) rather than TD(0) leads to better performance within VisNet. Comparing Figs. 8.25 and 8.27 shows that the TD(1)–like rule (8.34) with $\lambda = 1.0$ and $\beta = 1.8$ gives considerably superior results to the TD(0)–like rule (8.21) with $\beta = 2.2$. In fact, the former of these two rules provided the best single cell information results in these studies. We hypothesize that these results are related to the fact that only a finite set of image sequences is presented to VisNet, and so the type of optimization performed by TD(1) for repeated presentations of a finite data set is more appropriate for this problem than the form of optimization performed by TD(0).

8.4.4.4 Discussion of the different training rules

In terms of biological plausibility, we note the following. First, all the learning rules investigated by Rolls and Stringer (2001a) are local learning rules, and in this sense are biologically plausible (see Rolls and Treves (1998)). (The rules are local in that the terms used to modify the synaptic weights are potentially available in the pre- and post-synaptic elements.)

Second we note that all the rules do require some evidence of the activity on one or more previous stimulus presentations to be available when the synaptic weights are updated. Some of the rules, e.g. learning rule (8.19), use the trace $\overline{y}^\tau$ from the current time level, while rules (8.18) and (8.21) do not need to use an exponential trace of the neuronal firing rate, but only the instantaneous firing rates y at two successive time steps. It is known that synaptic plasticity does involve a combination of separate processes each with potentially differing time courses (Koch 1999), and these different processes could contribute to trace rule learning. Another mechanism suggested for implementing a trace of previous neuronal activity is the continuing firing for often 300 ms produced by a short (16 ms) presentation of a visual stimulus (Rolls and Tovee 1994) which is suggested to be implemented by local cortical recurrent attractor networks (Rolls and Treves 1998).

Third, we note that in utilizing the trace in the targets t^τ, the error correction (or temporal difference inspired) rules perform a comparison of the instantaneous firing y^τ with a temporally nearby value of the activity, and this comparison involves a subtraction. The subtraction provides an error, which is then used to increase or decrease the synaptic weights. This is a somewhat different operation from long term depression (LTD) as well as long term potentiation (LTP), which are *associative* changes which depend on the pre- and post-synaptic activity. However, it is interesting to note that an error correction rule which appears to involve a subtraction of current firing from a target might be implemented by a combination of an associative process operating with the trace, and an anti-Hebbian process operating to remove the effects of the current firing. For example, the synaptic updates $\delta w_j = \alpha(t^\tau - y^\tau)x_j^\tau$ can be decomposed into two separate associative processes $\alpha t^\tau x_j^\tau$ and $-\alpha y^\tau x_j^\tau$, that may occur independently. (The target, t^τ, could in this case be just the trace of previous neural activity from the preceding trials, excluding any contribution from the current firing.) Another way to implement an error correction rule using associative synaptic modification would be to force the post-synaptic neuron to respond to the error term. Although this has been postulated to be an effect which could be implemented by the climbing fibre system in the cerebellum (Ito 1989, Ito 1984, Rolls and Treves 1998), there is no similar system known for the neocortex, and it is not clear how this particular implementation of error correction might operate in the neocortex.

In Section 8.4.4.2 we describe five learning rules as error correction rules. We now discuss an interesting difference of these error correction rules from error correction rules as conventionally applied. It is usual to derive the general form of error correction learning rule from gradient descent minimization in the following way (Hertz, Krogh and Palmer 1991). Consider the idealized situation of a single neuron with a number of inputs x_j and output $y = \sum_j w_j x_j$, where w_j are the synaptic weights. We assume that there are a number of input patterns and that for the kth input pattern, $\mathbf{x}^k = [x_1^k, x_2^k, ...]^T$, the output y^k has a target value t^k. Hence an error measure or cost function can be defined as

$$e(\mathbf{w}) = \frac{1}{2}\sum_k (t^k - y^k)^2 = \frac{1}{2}\sum_k (t^k - \sum_j w_j x_j^k)^2. \tag{8.35}$$

This cost function is a function of the input patterns $\mathbf{x}^k$ and the synaptic weight vector $\mathbf{w} = [w_1, w_2, ...]^T$. With a fixed set of input patterns, we can reduce the error measure by employing a gradient descent algorithm to calculate an improved set of synaptic weights. Gradient descent achieves this by moving downhill on the error surface defined in $\mathbf{w}$ space using the update

$$\delta w_j = -\alpha \frac{\partial e}{\partial w_j} = \alpha \sum_k (t^k - y^k)x_j^k. \tag{8.36}$$

If we update the weights after each pattern k, then the update takes the form of an error correction rule

$$\delta w_j = \alpha(t^k - y^k)x_j^k, \tag{8.37}$$

which is also commonly referred to as the delta rule or Widrow-Hoff rule (see Widrow and Hoff (1960) and Widrow and Stearns (1985)). Error correction rules continually compare the neuronal output with its pre-specified target value and adjust the synaptic weights accordingly. In contrast, the way Rolls and Stringer (2001a) introduced of utilizing error correction is to specify the target as the activity trace based on the firing rate at nearby time-steps. Now the actual firing at those nearby time steps is not a pre-determined fixed target, but instead depends on how the network has actually evolved. This effectively means the cost function $e(\mathbf{w})$ that is being minimized changes from timestep to timestep. Nevertheless, the concept of calculating an error, and using the magnitude and direction of the error to update the synaptic weights, is the similarity we wish to draw out to gradient descent learning.

To conclude this discussion, we see the error correction and temporal difference rules explored by Rolls and Stringer (2001a) as providing interesting approaches to help understand invariant pattern recognition learning. Although we do not know whether the full power of these rules is expressed in the brain, we have provided suggestions about how they might be implemented. At the same time, we note that the original trace rule used by Földiák (1991), Rolls (1992a), and Wallis and Rolls (1997) is a simple associative rule, is therefore biologically very plausible, and while not as powerful as many of the other rules introduced here, can nevertheless solve the same class of problem. We would also like to emphasize that although we have demonstrated how a number of new error correction and temporal difference rules might play a role in the context of view invariant object recognition, they may also operate elsewhere where it is important for neurons to learn to respond similarly to temporal classes of inputs that tend to occur close together in time.

8.4.5 The issue of feature binding, and a solution

In this Section we investigate two key issues that arise in hierarchical layered network architectures, such as VisNet, other examples of which have been described and analyzed by Fukushima (1980), Ackley, Hinton and Sejnowski (1985), and Rosenblatt (1961). One issue is whether the network can discriminate between stimuli that are composed of the same basic alphabet of features. The second issue is whether such network architectures can find solutions to the spatial binding problem. These issues are described in the next two paragraphs and by Elliffe, Rolls and Stringer (2001).

The first issue investigated is whether a hierarchical layered network architecture of the type exemplified by VisNet can discriminate stimuli that are composed of a limited set of features and where the different stimuli include cases where the feature sets are subsets and supersets of those in the other stimuli. In many investigations with VisNet, complex stimuli (such as faces) were used where each stimulus might contain unique features not present in the other stimuli. To address this Elliffe, Rolls and Stringer (2001) used stimuli that are composed from a set of four features which are designed so that each feature is spatially separate from the other features, and no unique combination of firing caused for example by overlap of horizontal and vertical filter outputs in the input representation distinguishes any one stimulus from the others. The results described in Section 8.4.5.1 show that VisNet can indeed learn correct invariant representations of stimuli which do consist of feature sets where individual features do not overlap spatially with each other and where the stimuli can be composed of sets of features which are supersets or subsets of those in other stimuli. Fukushima and Miyake (1982) did not address this crucial issue where different stimuli might be composed of subsets or supersets of the same set of features, although they did show that stimuli with partly overlapping features could be discriminated by the Neocognitron.

In Section 8.4.5.2 we address the spatial binding problem in architectures such as VisNet. This computational problem that needs to be addressed in hierarchical networks such as the primate visual system and VisNet is how representations of features can be (e.g. translation) invariant, yet can specify stimuli or objects in which the features must be specified in the correct spatial arrangement. This is the feature binding problem, discussed for example by Malsburg (1990), and arising in the context of hierarchical layered systems (Ackley, Hinton and Sejnowski 1985, Fukushima 1980, Rosenblatt 1961). The issue is whether or not features are bound into the correct combinations, or if alternative combinations of known features would elicit the same responses. Von der Malsburg suggested that one potential solution is the addition of a temporal dimension to the neuronal response, so that features that should be bound together would be linked by temporal binding. There has been considerable neurophysiological investigation of this possibility (Singer, Gray, Engel, Konig, Artola and Brocher 1990, Abeles 1991, Hummel and Biederman 1992, Singer and Gray 1995). We note that a problem with this approach is that temporal binding might enable features 1, 2 and 3, which might define one stimulus to be bound together and kept separate from for example another stimulus consisting of features 2, 3 and 4, but would require a further temporal binding (leading in the end potentially to a combinatorial explosion) to indicate the relative spatial positions of the 1, 2 and 3 in the 123 stimulus, so that it can be discriminated from e.g. 312. Another approach to a binding mechanism is to group spatial features based on local mechanisms that might operate for closely adjacent synapses on a dendrite (Finkel and Edelman 1987, Mel, Ruderman and Archie 1998). A problem for such architectures is how to force one particular neuron to respond to the same feature combination invariantly with respect to all the ways in which that feature combination might occur in a scene.

The approach to the spatial binding problem that is proposed for VisNet is that individual neurons at an early stage of processing are set up (by learning) to respond to low order combinations of input features occurring in a given relative spatial arrangement and position on the retina (Rolls 1992a, Rolls 1994, Rolls 1995b, Wallis and Rolls 1997, Rolls and Treves 1998) (cf. Feldman (1985)). (By low order combinations if input features we mean combinations of a few input features. By forming neurons that respond to combinations of a few features in the correct spatial arrangement the advantages of the scheme for syntactic binding are obtained, yet without the combinatorial explosion that would result if the feature combination neurons responded to combinations of many input features so producing potentially very specifically tuned neurons which very rarely responded.) Then invariant representations are developed in the next layer from these feature combination neurons which already contain evidence on the local spatial arrangement of features. Finally, in later layers, only one stimulus would be specified by the particular set of low order feature combination neurons present, even though each feature combination neuron would itself be somewhat invariant. The overall design of the scheme is shown in Fig. 8.3. Evidence many neurons in V1 respond to combinations of spatial features with the correct spatial configuration is now starting to appear (see Section 2.5).

8.4.5.1 Discrimination between stimuli with super- and sub-set feature combinations

Previous investigations with VisNet (Wallis and Rolls 1997) have involved groups of stimuli that might be identified by some unique feature common to all transformations of a particular stimulus. This might allow VisNet to solve the problem of transform invariance by simply learning to respond to a unique feature present in each stimulus. For example, even in the case where VisNet was trained on invariant discrimination of T, L and +, the representation of the T stimulus at the spatial filter level inputs to VisNet might contain unique patterns of filter outputs where the horizontal and vertical parts of the T join. The unique filter outputs

Feature-Combination Stimulus Set

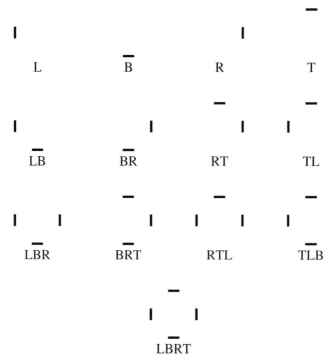

Fig. 8.28 Merged feature objects. All members of the full object set are shown, using a dotted line to represent the central 32x32 square on which the individual features are positioned, with the features themselves shown as dark line segments. Nomenclature is by acronym of the features present, as above.

thus formed might distinguish the T from for example the L.

Elliffe, Rolls and Stringer (2001) tested whether VisNet is able to form transform invariant cells with stimuli that are specially composed from a common alphabet of features, with no stimulus containing any firing in the spatial filter inputs to VisNet not present in at least one of the other stimuli. The limited alphabet enables the set of stimuli to consist of feature sets which are subsets or supersets of those in the other stimuli.

For these experiments the common pool of stimulus features chosen was a set of two horizontal and two vertical 8×1 bars, each aligned with the sides of a 32×32 square. The stimuli can be constructed by arbitrary combination of these base level features. We note that effectively the stimulus set consists of four features, a top bar (T), a bottom bar (B), a left bar (L), and a right bar (R). Figure 8.28 shows the complete set used, containing every possible image feature combination. (Note that the two double-feature combinations where the features are parallel to each other are not included, in the interests of retaining symmetry and equal inter-object overlap within each feature-combination level.) Subsequent discussion will group these objects by the number of features each contain: single-; double-; triple-; and quadruple-feature objects correspond to the respective rows of Fig. 8.28. Stimuli are referred to by the list of features they contain; e.g. 'LBR' contains the left, bottom and right features, while 'TL' contains top and left only. Further details of how the stimuli were prepared is provided by Elliffe, Rolls and Stringer (2001).

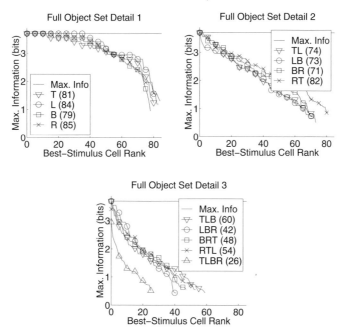

Fig. 8.29 Performance of VisNet 2 on the full set of stimuli shown in Fig. 8.28. Separate graphs showing the information available about the stimulus for cells tuned to respond best to each of the stimuli are shown. The number of cells responding best to each of the stimuli is indicated in parentheses. The information values are shown for the different cells ranked according to how much information about that stimulus they encode. Separate graphs are shown for cells tuned to stimuli consisting of single features, pairs of features, and triples of features as well as the quadruple feature stimulus TLBR.

To train the network a stimulus was presented in a randomized sequence of nine locations in a square grid across the 128×128 input retina. The central location of the square grid was in the centre of the 'retina', and the eight other locations were offset 8 pixels horizontally and/or vertically from this. Two different learning rules were used, 'Hebbian' (8.12), and 'trace' (8.13), and also an untrained condition with random weights. As in earlier work (Wallis and Rolls 1997, Rolls and Milward 2000) only the trace rule led to any cells with invariant responses, and the results shown here are for networks trained with the trace rule.

The results with VisNet trained on the set of stimuli shown in Fig. 8.28 with the trace rule are as follows. Firstly, it was found that single neurons in the top layer learned to differentiate between the stimuli in that the responses of individual neurons were maximal for one of the stimuli and had no response to any of the other stimuli invariantly with respect to location. Secondly, to assess how well every stimulus was encoded for in this way, Fig. 8.29 shows the information available about each of the stimuli consisting of feature singles, feature pairs, feature triples, and the quadruple feature stimulus 'TLBR'. The single cell information available from the 26–85 cells with best tuning to each of the stimuli is shown. The cells in general conveyed translation invariant information about the stimulus to which they responded, with indeed cells which perfectly discriminated one of the stimuli from all others over every testing position (for all stimuli except 'RTL' and 'TLBR').

The results presented show clearly that the VisNet paradigm can accommodate networks that can perform invariant discrimination of objects which have a subset-superset relationship. The result has important consequences for feature binding and for discriminating stimuli for other stimuli which may be supersets of the first stimulus. For example, a VisNet cell which responds invariantly to feature combination TL can genuinely signal the presence of exactly that combination, and will not necessarily be activated by T alone, or by TLB. The basis for this separation by competitive networks of stimuli which are subsets and supersets of each other is described in Chapter 7, Section 7.4, and by Rolls and Treves (1998, Section 4.3.6).

8.4.5.2 Feature binding in a hierarchical network with invariant representations of local feature combinations

In this Section we consider the ability of output layer neurons to learn new stimuli if the lower layers are trained solely through exposure to simpler feature combinations from which the new stimuli are composed. A key question we address is how invariant representations of low order feature combinations in the early layers of the visual system are able to uniquely specify the correct spatial arrangement of features in the overall stimulus and contribute to preventing false recognition errors in the output layer.

The problem, and its proposed solution, can be treated as follows. Consider an object 1234 made from the features 1, 2, 3 and 4. The invariant low order feature combinations might represent 12, 23, and 34. Then if neurons at the next layer respond to combinations of the activity of these neurons, the only neurons in the next layer that would respond would be those tuned to 1234, not to for example 3412, which is distinguished from 1234 by the input of a pair neuron responding to 41 rather than to 23. The argument (Rolls 1992a) is that low-order spatial feature combination neurons in the early stage contain sufficient spatial information so that a particular combination of those low-order feature combination neurons specifies a unique object, even if the relative positions of the low-order feature combination neurons are not known, because they are somewhat invariant.

The architecture of VisNet is intended to solve this problem partly by allowing high spatial precision combinations of input features to be formed in layer 1. The actual input features in VisNet are, as described above, the output of oriented spatial-frequency tuned filters, and the combinations of these formed in layer 1 might thus be thought of in a simple way as for example a T or an L or for that matter a Y. Then in layer 2, application of the trace rule might enable neurons to respond to a T with limited spatial invariance (limited to the size of the region of layer 1 from which layer 2 cells receive their input). Then an 'object' such as H might be formed at a higher layer because of a conjunction of two Ts in the same small region.

To show that VisNet can actually solve this problem, Elliffe, Rolls and Stringer (2001) performed the experiments described next. They trained the first two layers of VisNet with feature pair combinations, forming representations of feature pairs with some translation invariance in layer 2. Then they used feature triples as input stimuli, allowed no more learning in layers 1 and 2, and then investigated whether layers 3 and 4 could be trained to produce invariant representations of the triples where the triples could only be distinguished if the local spatial arrangement of the features within the triple had effectively to be encoded in order to distinguish the different triples. For this experiment, they needed stimuli that could be specified in terms of a set of different features (they chose vertical (1), diagonal (2), and horizontal (3) bars) each capable of being shown at a set of different relative spatial positions (designated A, B and C), as shown in Fig. 8.30. The stimuli are thus defined in terms of what features are present and their precise spatial arrangement with respect to each other. The length of the horizontal and vertical feature bars shown in Fig. 8.30 is 8 pixels. To train the

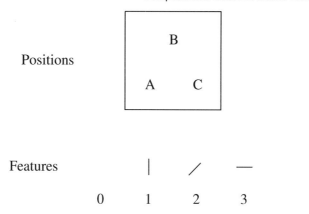

Fig. 8.30 Feature combinations for experiments of Section 8.4.5.2: there are 3 features denoted by 1, 2 and 3 (including a blank space 0) that can be placed in any of 3 positions A, B, and C. Individual stimuli are denoted by three consecutive numbers which refer to the individual features present in positions A, B and C respectively. In the experiments in Section 8.4.5.2, layers 1 and 2 were trained on stimuli consisting of pairs of the features, and layers 3 and 4 were trained on stimuli consisting of triples. Then the network was tested to show whether layer 4 neurons would distinguish between triples, even though the first two layers had only been trained on pairs. In addition, the network was tested to show whether individual cells in layer 4 could distinguish between triples even in locations where the triples were not presented during training.

network a stimulus (that is a pair or triple feature combination) is presented in a randomized sequence of nine locations in a square grid across the 128×128 input retina. The central location of the square grid is in the centre of the 'retina', and the eight other locations are offset 8 pixels horizontally and/or vertically from this. We refer to the two and three feature stimuli as 'pairs' and 'triples', respectively. Individual stimuli are denoted by three numbers which refer to the individual features present in positions A, B and C, respectively. For example, a stimulus with positions A and C containing a vertical and diagonal bar, respectively, would be referred to as stimulus 102, where the 0 denotes no feature present in position B. In total there are 18 pairs (120, 130, 210, 230, 310, 320, 012, 013, 021, 023, 031, 032, 102, 103, 201, 203, 301, 302) and 6 triples (123, 132, 213, 231, 312, 321). This nomenclature not only defines which features are present within objects, but also the spatial relationships of their component features. Then the computational problem can be illustrated by considering the triple 123. If invariant representations are formed of single features, then there would be no way that neurons higher in the hierarchy could distinguish the object 123 from 213 or any other arrangement of the three features. An approach to this problem (see e.g. Rolls (1992a)) is to form early on in the processing neurons that respond to overlapping combinations of features in the correct spatial arrangement, and then to develop invariant representations in the next layer from these neurons which already contain evidence on the local spatial arrangement of features. An example might be that with the object 123, the invariant feature pairs would represent 120, 023, and 103. Then if neurons at the next layer correspond to combinations of these neurons, the only next layer neurons that would respond would be those tuned to 123, not to for example 213. The argument is that the low-order spatial feature combination neurons in the early stage contain sufficient spatial information so that a particular combination of those low-order feature combination neurons specifies a unique object, even if the relative

Table 8.5 The different training regimes used in VisNet experiments 1–4 of Section 8.4.5.2. In the no training condition the synaptic weights were left in their initial untrained random values.

	Layers 1,2	Layers 3,4
Experiment 1	trained on pairs	trained on triples
Experiment 2	no training	no training
Experiment 3	no training	trained on triples
Experiment 4	trained on triples	trained on triples

positions of the low-order feature combination neurons are not known because these neurons are somewhat translation invariant (cf. also Fukushima (1988)).

The stimuli used in the experiments of Elliffe, Rolls and Stringer (2001) were constructed from pre-processed component features as discussed in Section 8.4.5.1. That is, base stimuli containing a single feature were constructed and filtered, and then the pairs and triples were constructed by merging these pre-processed single feature images. In the first experiment layers 1 and 2 of VisNet were trained with the 18 feature pairs, each stimulus being presented in sequences of 9 locations across the input. This led to the formation of neurons that responded to the feature pairs with some translation invariance in layer 2. Then they trained layers 3 and 4 on the 6 feature triples in the same 9 locations, while allowing no more learning in layers 1 and 2, and examined whether the output layer of VisNet had developed transform invariant neurons to the 6 triples. The idea was to test whether layers 3 and 4 could be trained to produce invariant representations of the triples where the triples could only be distinguished if the local spatial arrangement of the features within the triple had effectively to be encoded in order to distinguish the different triples. The results from this experiment were compared and contrasted with results from three other experiments which involved different training regimes for layers 1,2 and layers 3,4. All four experiments are summarized in Table 8.5. Experiment 2 involved no training in layers 1,2 and 3,4, with the synaptic weights left unchanged from their initial random values. These results are included as a baseline performance with which to compare results from the other experiments 1,3 and 4. The model parameters used in these experiments were as described by Rolls and Milward (2000) and Rolls and Stringer (2001a).

In Fig. 8.31 we present numerical results for the four experiments listed in Table 8.5. On the left are the single cell information measures for all top (4th) layer neurons ranked in order of their invariance to the triples, while on the right are multiple cell information measures. To help to interpret these results we can compute the maximum single cell information measure according to

$$\text{Maximum single cell information} = \log_2(\text{Number of triples}), \qquad (8.38)$$

where the number of triples is 6. This gives a maximum single cell information measure of 2.6 bits for these test cases. First, comparing the results for experiment 1 with the baseline performance of experiment 2 (no training) demonstrates that even with the first two layers trained to form invariant responses to the pairs, and then only layers 3 and 4 trained on feature triples, layer 4 is indeed capable of developing translation invariant neurons that can discriminate effectively between the 6 different feature triples. Indeed, from the single cell information measures it can be seen that a number of cells have reached the maximum level of performance in experiment 1. In addition, the multiple cell information analysis presented in Fig. 8.31 shows that all the stimuli could be discriminated from each other by the firing of

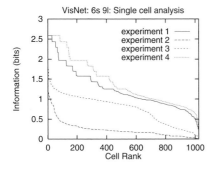
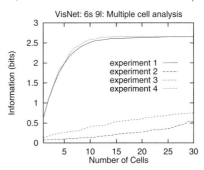

Fig. 8.31 Numerical results for experiments 1–4 as described in Table 8.5, with the trace learning rule (8.13). On the left are single cell information measures, and on the right are multiple cell information measures.

a number of cells. Analysis of the response profiles of individual cells showed that a fourth layer cell could respond to one of the triple feature stimuli and have no response to any other of the triple feature stimuli invariantly with respect to location.

A comparison of the results from experiment 1 with those from experiment 3 (see Table 8.5 and Fig. 8.31) reveals that training the first two layers to develop neurons that respond invariantly to the pairs (performed in experiment 1) actually leads to improved invariance of 4th layer neurons to the triples, as compared with when the first two layers are left untrained (experiment 3).

Two conclusions follow from these results. First, a hierarchical network which seeks to produce invariant representations in the way used by VisNet can solve the feature binding problem. In particular, when feature pairs in layer 2 with some translation invariance are used as the input to later layers, these later layers can nevertheless build invariant representations of objects where all the individual features in the stimulus must occur in the correct spatial position relative to each other. This is possible because the feature combination neurons formed in the first layer (which could be trained just with a Hebb rule) do respond to combinations of input features in the correct spatial configuration, partly because of the limited size of their receptive fields. The second conclusion is that even though early layers can in this case only respond to small feature subsets, these provide, with no further training of layers 1 and 2, an adequate basis for learning to discriminate in layers 3 and 4 stimuli consisting of combinations of larger numbers of features. Indeed, comparing results from experiment 1 with experiment 4 (in which all layers were trained on triples, see Table 8.5) demonstrates that training the lower layer neurons to develop invariant responses to the pairs offers almost as good performance as training all layers on the triples (see Fig. 8.31).

8.4.5.3 Stimulus generalization to new locations

Another important aspect of the architecture of VisNet is that it need not be trained with every stimulus in every possible location. Indeed, part of the hypothesis (Rolls 1992a) is that training early layers (e.g. 1–3) with a wide range of visual stimuli will set up feature analyzers in these early layers which are appropriate later on with no further training of early layers for new objects. For example, presentation of a new object might result in large numbers of low order feature combination neurons in early layers of VisNet being active, but the particular set of feature combination neurons active would be different for the new object. The later layers

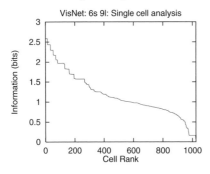
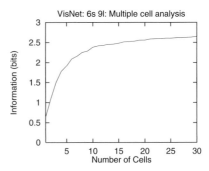

Fig. 8.32 Generalization to new locations: numerical results for a repeat of experiment 1 of Section 8.4.5.2 with the triples presented at only 7 of the original 9 locations during training, and with the trace learning rule (8.13). On the left are single cell information measures, and on the right are multiple cell information measures.

of the network (in VisNet, layer 4) would then learn this new set of active layer 3 neurons as encoding the new object. However, if the new object was then shown in a new location, the same set of layer 3 neurons would be active because they respond with spatial invariance to feature combinations, and given that the layer 3 to 4 connections had already been set up by the new object, the correct layer 4 neurons would be activated by the new object in its new untrained location, and without any further training.

To test this hypothesis, Elliffe, Rolls and Stringer (2001) repeated the general procedure of experiment 1 of Section 8.4.5.2, training layers 1 and 2 with feature pairs, but then instead trained layers 3 and 4 on the triples in only 7 of the original 9 locations. The crucial test was to determine whether VisNet could form top layer neurons that responded invariantly to the 6 triples when presented over all nine locations, not just the seven locations at which the triples had been presented during training. The results are presented in Fig. 8.32, with single cell information measures on the left and multiple cell information measures on the right. VisNet is still able to develop some fourth layer neurons with perfect invariance, that is which have invariant responses over all nine location, as shown by the single cell information analysis. The response profiles of individual fourth layer cells showed that they can continue to discriminate between the triples even in the two locations where the triples were not presented during training. In addition, the multiple cell analysis shown in Fig. 8.32 shows that a small population of cells was able to discriminate between all of the stimuli irrespective of location, even though for two of the test locations the triples had not been trained at those particular locations during the training of layers 3 and 4.

8.4.5.4 Discussion of feature binding in hierarchical layered networks

Elliffe, Rolls and Stringer (2001) thus first showed (see Section 8.4.5.1) that hierarchical feature detecting neural networks can learn to respond differently to stimuli which consist of unique combinations of non-unique input features, and that this extends to stimuli that are direct subsets or supersets of the features present in other stimuli.

Second, Elliffe, Rolls and Stringer (2001) investigated (see Section 8.4.5.2) the hypothesis that hierarchical layered networks can produce identification of unique stimuli even when the feature combination neurons used to define the stimuli are themselves partly translation invariant. The stimulus identification should work correctly because feature combination neurons in which the spatial features are bound together with high spatial precision are formed in the first layer. Then at later layers when neurons with some translation invariance are formed, the neurons nevertheless contain information about the relative spatial position of the original features. There is only then one object which will be consistent with the set of active neurons at earlier layers, which though somewhat translation invariant as combination neurons, reflect in the activity of each neuron information about the original spatial position of the features. We note that the trace rule training used in early layers (1 and 2) in Experiments 1 and 4 would set up partly invariant feature combination neurons, and yet the late layers (3 and 4) were able to produce during training neurons in layer 4 that responded to stimuli that consisted of unique spatial arrangements of lower order feature combinations. Moreover, and very interestingly, Elliffe, Rolls and Stringer (2001) were able to demonstrate that VisNet layer 4 neurons would respond correctly to visual stimuli at untrained locations, provided that the feature subsets had been trained in early layers of the network at all locations, and that the whole stimulus had been trained at some locations in the later layers of the network.

The results described by Elliffe, Rolls and Stringer (2001) thus provide one solution to the feature binding problem. The solution which has been shown to work in the model is that in a multilayer competitive network, feature combination neurons which encode the spatial arrangement of the bound features are formed at intermediate layers of the network. Then neurons at later layers of the network which respond to combinations of active intermediate layer neurons do contain sufficient evidence about the local spatial arrangement of the features to identify stimuli because the local spatial arrangement is encoded by the intermediate layer neurons. The information required to solve the visual feature binding problem thus becomes encoded by self-organization into what become hard-wired properties of the network. In this sense, feature binding is not solved at run time by the necessity to instantaneously set up arbitrary syntactic links between sets of co-active neurons. The computational solution proposed to the superset/subset aspect of the binding problem will apply in principle to other multilayer competitive networks, although the issues considered here have not been explicitly addressed in architectures such as the neocognitron (Fukushima and Miyake 1982).

Consistent with these hypotheses about how VisNet operates to achieve, by layer 4, position-invariant responses to stimuli defined by combinations of features in the correct spatial arrangement, investigations of the effective stimuli for neurons in intermediate layers of VisNet showed as follows. In layer 1, cells responded to the presence of individual features, or to low order combinations of features (e.g. a pair of features) in the correct spatial arrangement at a small number of nearby locations. In layers 2 and 3, neurons responded to single features or to higher order combinations of features (e.g. stimuli composed of feature triples) in more locations. These findings provide direct evidence that VisNet does operate as described above to solve the feature binding problem.

The type of solution investigated by Elliffe, Rolls and Stringer (2001) is thus different to the proposal of von der Malsburg (1990) that feature-selective neurons might be linked by temporal binding. There has been considerable neurophysiological investigation of this possibility (Singer, Gray, Engel, Konig, Artola and Brocher 1990, Abeles 1991, Hummel and Biederman 1992, Singer and Gray 1995). We note that a problem with this approach is that temporal binding might enable features 1, 2 and 3 which might define one stimulus to be bound together and kept separate from for example another stimulus consisting of features 2, 3 and 4, but would require a further temporal binding (leading in the end potentially to a combinatorial explosion) to indicate the relative spatial positions of the 1, 2 and 3 in the 123 stimulus, so that it can be discriminated from e.g. 312 (see Section 13.7). Another approach to a binding mechanism is to group spatial features based on local mechanisms that might operate for closely adjacent synapses on a dendrite (Finkel and Edelman 1987, Mel, Ruderman and Archie 1998).

A further issue with hierarchical multilayer architectures such as VisNet is that false binding errors might occur in the following way (Mozer 1991, Mel and Fiser 2000). Consider the output of one layer in such a network in which there is information only about which pairs are present. How then could a neuron in the next layer discriminate between the whole stimulus (such as the triple 123 in the above experiment) and what could be considered a more distributed stimulus or multiple different stimuli composed of the separated subparts of that stimulus (e.g. the pairs 120, 023, 103 occurring in 3 of the 9 training locations in the above experiment)? The problem here is to distinguish a single object from multiple other objects containing the same component combinations (e.g. pairs). We propose that part of the solution to this general problem in real visual systems is implemented through lateral inhibition between neurons in individual layers, and that this mechanism, implemented in VisNet, acts to reduce the possibility of false recognition errors in the following two ways.

First, consider the situation in which neurons in layer N have learned to represent low order feature combinations with location invariance, and where a neuron n in layer $N+1$ has learned to respond to a particular set Ω of these feature combinations. The problem is that neuron n receives the same input from layer N as long as the same set Ω of feature combinations is present, and cannot distinguish between different spatial arrangements of these feature combinations. The question is how can neuron n respond only to a particular favoured spatial arrangement Ψ of the feature combinations contained within the set Ω. We suggest that as the favoured spatial arrangement Ψ is altered by rearranging the spatial relationships of the component feature combinations, the new feature combinations that are formed in new locations will stimulate additional neurons nearby in layer $N+1$, and these will tend to inhibit the firing of neuron n. Thus, lateral inhibition within a layer will have the effect of making neurons more selective, ensuring neuron n responds only to a single spatial arrangement Ψ from the set of feature combinations Ω, and hence reducing the possibility of false recognition.

The second way in which lateral inhibition may help to reduce binding errors is through limiting the sparseness of neuronal firing rates within layers. In our discussion above the spurious stimuli we suggested that might lead to false recognition of triples were obtained from splitting up the component feature combinations (pairs) so that they occurred in separate training locations. However, this would lead to an increase in the number of features present in the complete stimulus; triples contain 3 features while their spurious counterparts would contain 6 features (resulting from 3 separate pairs). For this trivial example, the increase in the number of features is not dramatic, but if we consider, say, stimuli composed of 4 features where the component feature combinations represented by lower layers might be triples, then to form spurious stimuli we need to use 12 features (resulting from 4 triples occurring in separate locations). But if the lower layers also represented all possible pairs then the number of features required in the spurious stimuli would increase further. In fact, as the size of the stimulus increases in terms of the number of features, and as the size of the component feature combinations represented by the lower layers increases, there is a combinatorial explosion in terms of the number of features required as we attempt to construct spurious stimuli to trigger false recognition. And the construction of such spurious stimuli will then be prevented through setting a limit on the sparseness of firing rates within layers, which will in turn set a limit on the number of features that can be represented. Lateral inhibition is likely to contribute in both these ways to the performance of VisNet when the stimuli consist of subsets and supersets of each other, as described in Section 8.4.5.1.

Another way is which the problem of multiple objects is addressed is by limiting the size of the receptive fields of inferior temporal cortex neurons so that neurons in IT respond primarily to the object being fixated. Multiple objects are then "seen" by virtue of being added to a visuo-spatial scratchpad, as addressed in Section 12.8.

A related issue that arises in this class of network is whether forming neurons that respond to feature combinations in the way described here leads to a combinatorial explosion in the number of neurons required. The solution that is proposed to this issue is to form only low-order combinations of features at any one stage of the network (Rolls (1992a); cf. Feldman (1985)). Using low-order combinations limits the number of neurons required, yet enables the type of computation that relies on feature combination neurons that is analyzed here to still be performed. The actual number of neurons required depends also on the redundancies present in the statistics of real-world images. Even given these factors, it is likely that a large number of neurons would be required if the ventral visual system performs the computation of invariant representations in the manner captured by the hypotheses implemented in VisNet. Consistent with this, a considerable part of the non-human primate brain is devoted to visual information processing. The fact that large numbers of neurons and a multi-layer organization are present in the primate ventral visual system is actually thus consistent with the type of model of visual information processing described here.

8.4.6   Operation in a cluttered environment

In this Section we consider how hierarchical layered networks of the type exemplified by VisNet operate in cluttered environments. Although there has been much work involving object recognition in cluttered environments with artificial vision systems, many such systems typically rely on some form of explicit segmentation followed by search and template matching procedure (see Ullman (1996) for a general review). In natural environments, objects may not only appear against cluttered (natural) backgrounds, but also the object may be partially occluded. Biological nervous systems operate in quite a different manner to those artificial vision systems that rely on search and template matching, and the way in which biological systems cope with cluttered environments and partial occlusion is likely to be quite different also.

One of the factors that will influence the performance of the type of architecture considered here, hierarchically organized series of competitive networks, which form one class of approaches to biologically relevant networks for invariant object recognition (Fukushima 1980, Rolls 1992a, Wallis and Rolls 1997, Poggio and Edelman 1990, Rolls and Treves 1998), is how lateral inhibition and competition are managed within a layer. Even if an object is not obscured, the effect of a cluttered background will be to fire additional neurons, which will in turn to some extent compete with and inhibit those neurons that are specifically tuned to respond to the desired object. Moreover, where the clutter is adjacent to part of the object, the feature analysing neurons activated against a blank background might be different from those activated against a cluttered background, if there is no explicit segmentation process. We consider these issues next, following investigations of Stringer and Rolls (2000).

Fig. 8.33 Cluttered backgrounds used in VisNet simulations: backgrounds 1 and 2 are on the left and right respectively.

8.4.6.1 VisNet simulations with stimuli in cluttered backgrounds

In this Section we show that recognition of objects learned previously against a blank background is hardly affected by the presence of a natural cluttered background. We go on to consider what happens when VisNet is set the task of learning new stimuli presented against cluttered backgrounds.

The images used for training and testing VisNet in the simulations described next performed by Stringer and Rolls (2000) were specially constructed. There were 7 face stimuli approximately 64 pixels in height constructed without backgrounds from those shown in Fig. 8.17. In addition there were 3 possible backgrounds: a blank background (greyscale 127, where the range is 0-255), and two cluttered backgrounds as shown in Fig. 8.33 which are 128×128 pixels in size. Each image presented to VisNet's 128×128 input retina was composed of a single face stimulus positioned at one of 9 locations on either a blank or cluttered background. The cluttered background was intended to be like the background against which an object might be viewed in a natural scene. If a background is used in an experiment described here, the same background is always used, and it is always in the same position, with stimuli moved to different positions on it. The 9 stimulus locations are arranged in a square grid across the background, where the grid spacings are 32 pixels horizontally or vertically. Before images are presented to VisNet's input layer they were pre-processed by the standard set of input filters which accord with the general tuning profiles of simple cells in V1 (Hawken and Parker 1987); full details are given in Rolls and Milward (2000). To train the network a sequence of images is presented to VisNet's retina that corresponds to a single stimulus occurring in a randomized sequence of the 9 locations across a background. At each presentation the activation of individual neurons is calculated, then their firing rates are calculated, and then the synaptic weights are updated. After a stimulus has been presented in all the training locations, a new stimulus is chosen at random and the process repeated. The presentation of all the stimuli across all locations constitutes 1 epoch of training. In this manner the network is trained one layer at a time starting with layer 1 and finishing with layer 4. In the investigations described in this subsection, the numbers of training epochs for layers 1–4 were 50, 100, 100 and 75 respectively.

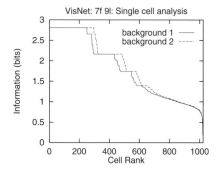
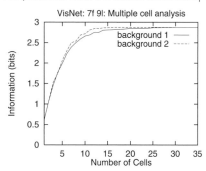

Fig. 8.34 Numerical results for experiment 2, with the 7 faces presented on a blank background during training and a cluttered background during testing. On the left are single cell information measures, and on the right are multiple cell information measures.

In this experiment (see Stringer and Rolls (2000), experiment 2), VisNet was trained with the 7 face stimuli presented on a blank background, but tested with the faces presented on each of the 2 cluttered backgrounds. Figure 8.34 shows results for experiment 2, with single and multiple cell information measures on the left and right respectively. It can be seen that a number of cells have reached the maximum possible single cell information measure of 2.8 bits ($\log_2$ of the number of stimuli) for this test case, and that the multiple cell information measures also reach the 2.8 bits indicating perfect performance. Compared to performance when shown against a blank background, there was very little deterioration in performance when testing with the faces presented on either of the two cluttered backgrounds. This is an interesting result to compare with many artificial vision systems that would need to carry out computationally intensive serial searching and template matching procedures in order to achieve such results. In contrast, the VisNet neural network architecture is able to perform such recognition relatively quickly through a simple feedforward computation. Further results from this experiment are presented in Fig. 8.35 where we show the response profiles of a 4th layer neuron to the 7 faces presented on cluttered background 1 during testing. It can be seen that this neuron achieves excellent invariant responses to the 7 faces even with the faces presented on a cluttered background. The response profiles are independent of location but differentiate between the faces in that the responses are maximal for only one of the faces and minimal for all other faces.

This is an interesting and important result, for it shows that after learning, special mechanisms for segmentation and for attention are not needed in order for neurons already tuned by previous learning to the stimuli to be activated correctly in the output layer. Although the experiments described here tested for position invariance, we predict and would expect that the same results would be demonstrable for size and view invariant representations of objects.

In experiments 3 and 4 of Stringer and Rolls (2000), VisNet was trained with the 7 face stimuli presented on either one of the 2 cluttered backgrounds, but tested with the faces presented on a blank background. Results for this experiment showed poor performance. The results of experiments 3 and 4 suggest that in order for a cell to *learn* invariant responses to different transforms of a stimulus when it is presented during training in a cluttered background, some form of segmentation is required in order to separate the figure (i.e. the stimulus or object) from the background. This segmentation might be performed using evidence in the visual scene about different depths, motions, colours, etc. of the object from its

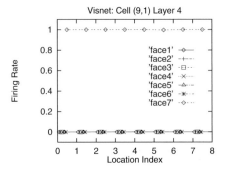

Fig. 8.35 Response profiles of a top layer neuron to the 7 faces from experiment 2 of Stringer and Rolls (2000), with the faces presented against cluttered background 1 during testing.

background. In the visual system, this might mean combining evidence represented in different cortical areas, and might be performed by cross-connections between cortical areas to enable such evidence to help separate the representations of objects from their backgrounds in the form-representing cortical areas. Another mechanism that helps the operation of architectures such as VisNet and the primate visual system to learn about new objects in cluttered scenes is that the receptive fields of inferior temporal cortex neurons become much smaller when objects are seen against natural backgrounds (Sections 5.6.3 and 8.4.9). This will help greatly to learn about new objects that are being fixated, by reducing responsiveness to other features elsewhere in the scene. Another mechanism that might help the learning of new objects in a natural scene is attention. An attentional mechanism might highlight the current stimulus being attended to and suppress the effects of background noise, providing a training representation of the object more like that which would be produced when it is presented against a blank background. The mechanisms that could implement such attentional processes are described in Chapters 9–11. If such an attentional mechanisms do contribute to the development of view invariance, then it follows that cells in the temporal cortex may only develop transform invariant responses to objects to which attention is directed.

Part of the reason for the poor performance in experiments 3 and 4 was probably that the stimuli were always presented against the same fixed background (for technical reasons), and thus the neurons learned about the background rather than the stimuli. Part of the difficulty that hierarchical multilayer competitive networks have with learning in cluttered environments may more generally be that without explicit segmentation of the stimulus from its background, at least some of the features that should be formed to encode the stimuli are not formed properly, because the neurons learn to respond to combinations of inputs which come partly from the stimulus, and partly from the background. To investigate this, Stringer and Rolls (2000) performed experiment 5 in which layers 1–3 were pretrained with stimuli to ensure that good feature combination neurons for stimuli were available, and then allowed learning in only layer 4 when stimuli were presented in the cluttered backgrounds. Layer 4 was then trained in the usual way with the 7 faces presented against a cluttered background. The results for this experiment are shown in Fig. 8.36, with single and multiple cell information measures on the left and right respectively. It was found that prior random exposure to the face stimuli led to much improved performance. Indeed, it can be seen that a number of cells have reached the maximum possible single cell information measure of 2.8 bits for this test case, although the multiple cell information measures do not quite reach the 2.8 bits that would indicate

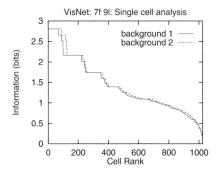

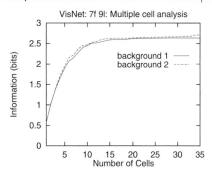

Fig. 8.36 Numerical results for experiment 5 of Stringer and Rolls (2000). In this experiment VisNet is first exposed to a completely random sequence of faces in different positions against a blank background during which layers 1–3 are allowed to learn. This builds general feature detecting neurons in the lower layers that are tuned to the face stimuli, but cannot develop view invariance since there is no temporal structure to the order in which different views of different faces occur. Then layer 4 is trained in the usual way with the 7 faces presented against a cluttered background, where the images are now presented such that different views of the same face occur close together in time. On the left are single cell information measures, and on the right are multiple cell information measures.

perfect performance for the complete face set.

These results demonstrate that the problem of developing position invariant neurons to stimuli occurring against cluttered backgrounds may be ameliorated by the prior existence of stimulus tuned feature detecting neurons in the early layers of the visual system, and that these feature detecting neurons may be set up through previous exposure to the relevant class of objects. When tested in cluttered environments, the background clutter may of course activate some other neurons in the output layer, but at least the neurons that have learned to respond to the trained stimuli are activated. The result of this activity is sufficient for the activity in the output layer to be useful, in the sense that it can be read off correctly by a pattern associator connected to the output layer. Indeed, Stringer and Rolls (2000) tested this by connecting a pattern associator to layer 4 of VisNet. The pattern associator had seven neurons, one for each face, and 1024 inputs, one from each neuron in layer 4 of VisNet. The pattern associator learned when trained with a simple associative Hebb rule (equation 8.12) to activate the correct output neuron whenever one of the faces was shown in any position in the uncluttered environment. This ability was shown to be dependent on invariant neurons for each stimulus in the output layer of VisNet, for the pattern associator could not be taught the task if VisNet had not been previously trained with a trace learning rule to produce invariant representations. Then it was shown that exactly the correct neuron was activated when any of the faces was shown in any position with the cluttered background. This read-off by a pattern associator is exactly what we hypothesize takes place in the brain, in that the inferior temporal visual cortex (where neurons with invariant responses are found) projects to structures such as the orbitofrontal cortex and amygdala, where associations between the invariant visual representations and stimuli such as taste and touch are learned (Rolls and Treves 1998, Rolls 1999a). Thus testing whether the output of an architecture such as VisNet can be used effectively by a pattern associator is a very biologically relevant way to evaluate the performance of this class of architecture.

The results of the experiments just described suggest that in order for a cell to *learn* invariant responses to different transforms of a stimulus when it is presented during training in a cluttered background, some form of segmentation is required in order to separate the figure (i.e. the stimulus or object) from the background. This segmentation might be performed using evidence in the visual scene about different depths, motions, colours, etc. of the object from its background. In the visual system, this might mean combining evidence represented in different cortical areas, and might be performed by cross-connections between cortical areas to enable such evidence to help separate the representations of objects from their backgrounds in the form-representing cortical areas. These processes are also likely to be helped by the reduced receptive field sizes of inferior temporal cortex neurons in natural scenes, to normally represent particularly the object at the fovea, and by attentional mechanisms.

8.4.6.2   VisNet simulations with partially occluded stimuli

In this Section we examine the recognition of partially occluded stimuli. Many artificial vision systems that perform object recognition typically search for specific markers in stimuli, and hence their performance may become fragile if key parts of a stimulus are occluded. However, in contrast we demonstrate that the biologically inspired model discussed here can continue to offer robust performance with this kind of problem, and that the model is able to correctly identify stimuli with considerable flexibility about what part of a stimulus is visible.

In these simulations (Stringer and Rolls 2000), training and testing was performed with a blank background to avoid confounding the two separate problems of occlusion and background clutter. In object recognition tasks, artificial vision systems may typically rely on being able to locate a small number of key markers on a stimulus in order to be able to identify it. This approach can become fragile when a number of these markers become obscured. In contrast, biological vision systems may generalize or complete from a partial input as a result of the use of distributed representations in neural networks, and this could lead to greater robustness in situations of partial occlusion.

In this experiment (6 of Stringer and Rolls, 2000), the network was first trained with the 7 face stimuli without occlusion, but during testing there were two options: either (i) the top halves of all the faces were occluded, or (ii) the bottom halves of all the faces were occluded. Since VisNet was tested with either the top or bottom half of the stimuli no stimulus features were common to the two test options. This ensures that if performance is good with both options, the performance cannot be based on the use of a single feature to identify a stimulus. Results for this experiment are shown in Fig. 8.37, with single and multiple cell information measures on the left and right respectively. When compared with the performance without occlusion (Stringer and Rolls 2000), 8.37 shows that there is only a modest drop in performance in the single cell information measures when the stimuli are partially occluded.

For both options (i) and (ii), even with partially occluded stimuli, a number of cells continue to respond maximally to one preferred stimulus in all locations, while responding minimally to all other stimuli. However, comparing results from options (i) and (ii) shows that the network performance is better when the bottom half of the faces is occluded. This is consistent with psychological results showing that face recognition is performed more easily when the top halves of faces are visible rather than the bottom halves (see Bruce (1988)). The top half of a face will generally contain salient features, e.g. eyes and hair, that are particularly helpful for recognition of the individual, and it is interesting that these simulations appear to further demonstrate this point. Furthermore, the multiple cell information measures confirm that performance is better with the upper half of the face visible (option (ii)) than the lower half (option (i)). When the top halves of the faces are occluded the multiple cell information

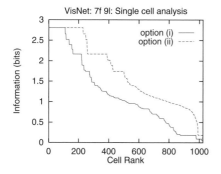

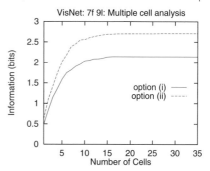

Fig. 8.37 Effects of partial occlusion of a stimulus: numerical results for experiment 6 of Stringer and Rolls, 2000, with the 7 faces presented on a blank background during both training and testing. Training was performed with the whole face. However, during testing there are two options: either (i) the top half of all the faces are occluded, or (ii) the bottom half of all the faces are occluded. On the left are single cell information measures, and on the right are multiple cell information measures.

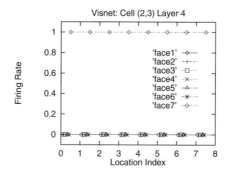

Fig. 8.38 Effects of partial occlusion of a stimulus. Response profiles of a top layer neuron to the 7 faces from experiment 6 of Stringer and Rolls (2000), with the bottom half of all the faces occluded during testing.

measure asymptotes to a sub-optimal value reflecting the difficulty of discriminating between these more difficult images. Further results from experiment 6 are presented in Fig. 8.38 where we show the response profiles of a 4th layer neuron to the 7 faces, with the bottom half of all the faces occluded during testing. It can be seen that this neuron continues to respond invariantly to the 7 faces, responding maximally to one of the faces but minimally for all other faces.

Thus this biologically inspired model offers robust performance with this kind of problem, and the model is able to correctly identify stimuli with considerable flexibility about what part of a stimulus is visible, because it is effectively using distributed representations and associative processing (see Chapter 7).

8.4.7 Learning 3D transforms

In this Section we describe investigations of Stringer and Rolls (2002a) which show that trace learning can in the VisNet architecture solve the problem of in-depth rotation invariant object recognition by developing representations of the transforms which features undergo when they are on the surfaces of 3D objects. Moreover, it is shown that having learned how features on 3D objects transform as the object is rotated in depth, the network can correctly recognize novel 3D variations within a generic view of an object which is composed of previously learned feature combinations.

Rolls' hypothesis of how object recognition could be implemented in the brain postulates that trace rule learning helps invariant representations to form in two ways (Rolls 1992a, Rolls 1994, Rolls 1995b, Rolls 2000a). The first process enables associations to be learned between different generic 3D views of an object where there are different qualitative shape descriptors. One example of this would be the front and back views of an object, which might have very different shape descriptors. Another example is provided by considering how the shape descriptors typical of 3D shapes, such as Y vertices, arrow vertices, cusps, and ellipse shapes, alter when most 3D objects are rotated in 3 dimensions. At some point in the 3D rotation, there is a catastrophic rearrangement of the shape descriptors as a new generic view can be seen (Koenderink 1990). An example of a catastrophic change to a new generic view is when a cup being viewed from slightly below is rotated so that one can see inside the cup from slightly above. The bottom surface disappears, the top surface of the cup changes from a cusp to an ellipse, and the inside of the cup with a whole set of new features comes into view. The second process is that within a generic view, as the object is rotated in depth, there will be no catastrophic changes in the qualitative 3D shape descriptors, but instead the quantitative values of the shape descriptors alter. For example, while the cup is being rotated within a generic view seen from somewhat below, the curvature of the cusp forming the top boundary will alter, but the qualitative shape descriptor will remain a cusp. Trace learning could help with both processes. That is, trace learning could help to associate together qualitatively different sets of shape descriptors that occur close together in time, and describe for example the generically different views of a cup. Trace learning could also help with the second process, and learn to associate together the different quantitative values of shape descriptors that typically occur when objects are rotated within a generic view.

We note that there is evidence that some neurons in the inferior temporal cortex may show the two types of 3D invariance. First, Booth and Rolls (1998) showed that some inferior temporal cortex neurons can respond to different generic views of familiar 3D objects. Second, some neurons do generalize across quantitative changes in the values of 3D shape descriptors while faces (Hasselmo, Rolls, Baylis and Nalwa 1989b) and objects (Tanaka 1996, Logothetis, Pauls and Poggio 1995) are rotated within generic views. Indeed, Logothetis, Pauls and Poggio (1995) showed that a few inferior temporal cortex neurons can generalize to novel (untrained) values of the quantitative shape descriptors typical of within-generic-view object rotation.

In addition to the qualitative shape descriptor changes that occur catastrophically between different generic views of an object, and the quantitative changes of 3D shape descriptors that occur within a generic view, there is a third type of transform that must be learned for correct invariant recognition of 3D objects as they rotate in depth. This third type of transform is that which occurs to the surface features on a 3D object as it transforms in depth. The main aim here is to consider mechanisms that could enable neurons to learn this third type of transform, that is how to generalize correctly over the changes in the surface markings on 3D objects that are typically encountered as 3D objects rotate within a generic view. Examples of the types of perspectival transforms we investigate here are shown in Fig. 8.39. Surface markings on the sphere that consist of combinations of three features in different spatial arrangements undergo characteristic transforms as the sphere is rotated from 0 degrees towards -60 degrees and +60

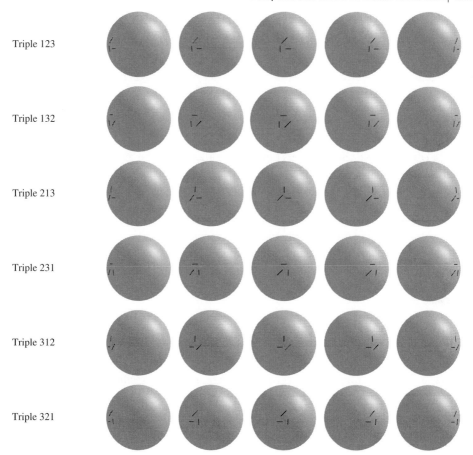

Fig. 8.39 Learning 3D perspectival transforms of features. Representations of the 6 visual stimuli with 3 surface features (triples) presented to VisNet during the simulations described in Section 8.4.7. Each stimulus is a sphere that is uniquely identified by a unique combination of three surface features (a vertical, diagonal and horizontal arc), which occur in 3 relative positions A, B, and C. Each row shows one of the stimuli rotated through the 5 different rotational views in which the stimulus is presented to VisNet. From left to right the rotational views shown are: (i) -60 degrees, (ii) -30 degrees, (iii) 0 degrees (central position), (iv) 30 degrees, and (v) 60 degrees.

degrees. We investigate here whether the class of architecture exemplified by VisNet, and the trace learning rule, can learn about the transforms that surface features of 3D objects typically undergo during 3D rotation in such a way that the network generalizes across the change of the quantitative values of the surface features produced by the rotation, and yet still discriminates between the different objects (in this case spheres). In the cases being considered, each object is identified by surface markings that consist of a different spatial arrangement of the same three features (a horizontal, vertical, and diagonal line, which become arcs on the surface of the object).

We note that it has been suggested that the finding that neurons may offer some degree of 3D rotation invariance after training with a single view (or limited set of views) represents a challenge for existing trace learning models, because these models assume that an initial exposure is required during learning to every transformation of the object to be recognized (Riesenhuber and Poggio 1998). We show here that this is not the case, and that such models can generalize to novel within-generic views of an object provided that the characteristic changes that the features show as objects are rotated have been learned previously for the sets of features when they are present in different objects.

Elliffe, Rolls and Stringer (2001) demonstrated for a 2D system how the existence of translation invariant representations of low order feature combinations in the early layers of the visual system could allow correct stimulus identification in the output layer even when the stimulus was presented in a novel location where the stimulus had not previously occurred during learning. The proposal was that the low-order spatial feature combination neurons in the early stages contain sufficient spatial information so that a particular combination of those low-order feature combination neurons specifies a unique object, even if the relative positions of the low-order feature combination neurons are not known because these neurons are somewhat translation invariant (see Section 8.4.5.2). Stringer and Rolls (2002a) extended this analysis to feature combinations on 3D objects, and indeed in their simulations described in this Section therefore used surface markings for the 3D objects that consisted of triples of features.

The images used for training and testing VisNet were specially constructed for the purpose of demonstrating how the trace learning paradigm might be further developed to give rise to neurons that are able to respond invariantly to novel within-generic view perspectives of an object, obtained by rotations in-depth up to 30 degrees from any perspectives encountered during learning. The stimuli take the form of the surface feature combinations of 3-dimensional rotating spheres, with each image presented to VisNet's retina being a 2-dimensional projection of the surface features of one of the spheres. Each stimulus is uniquely identified by two or three surface features, where the surface features are (1) vertical, (2) diagonal, and (3) horizontal arcs, and where each feature may be centred at three different spatial positions, designated A, B, and C, as shown in Fig. 8.39. The stimuli are thus defined in terms of what features are present and their precise spatial arrangement with respect to each other. We refer to the two and three feature stimuli as 'pairs' and 'triples', respectively. Individual stimuli are denoted by three numbers which refer to the individual features present in positions A, B and C, respectively. For example, a stimulus with positions A and C containing a vertical and diagonal bar, respectively, would be referred to as stimulus 102, where the 0 denotes no feature present in position B. In total there are 18 pairs (120, 130, 210, 230, 310, 320, 012, 013, 021, 023, 031, 032, 102, 103, 201, 203, 301, 302) and 6 triples (123, 132, 213, 231, 312, 321).

To train the network each stimulus was presented to VisNet in a randomized sequence of five orientations with respect to VisNet's input retina, where the different orientations are obtained from successive in-depth rotations of the stimulus through 30 degrees. That is, each stimulus was presented to VisNet's retina from the following rotational views: (i) $-60°$, (ii) $-30°$, (iii) $0°$ (central position with surface features facing directly towards VisNet's retina), (iv) $30°$, (v) $60°$. Fig. 8.39 shows representations of the 6 visual stimuli with 3 surface features (triples) presented to VisNet during the simulations. (For the actual simulations described here, the surface features and their deformations were what VisNet was trained and tested with, and the remaining blank surface of each sphere was set to the same greyscale as the background.) Each row shows one of the stimuli rotated through the 5 different rotational views in which the stimulus is presented to VisNet. At each presentation the activation of individual neurons is calculated, then the neuronal firing rates are calculated, and then the synaptic weights are updated. Each time a stimulus has been presented in all the training orientations, a new stimulus is chosen at random and the process repeated. The presentation of all the stimuli through all 5 orientations constitutes 1 epoch of training. In this manner the network was trained one layer at a time starting with layer 1 and finishing with layer 4. In the investigations described here, the numbers of training epochs for layers 1–4 were 50, 100, 100 and 75, respectively.

In experiment 1, VisNet was trained in two stages. In the first stage, the 18 feature pairs were used as input stimuli, with each stimulus being presented to VisNet's retina in sequences of five orientations as described above. However, during this stage, learning was only allowed to take place in layers 1 and 2. This led to the formation of neurons which responded to the feature pairs with some rotation invariance in layer 2. In the second stage, we used the 6 feature triples as stimuli, with learning only allowed in layers 3 and 4. However, during this second training stage, the triples were only presented to VisNet's input retina in the first 4 orientations (i)–(iv). After the two stages of training were completed, Stringer and Rolls (2002a) examined whether the output layer of VisNet had formed top layer neurons that responded invariantly to the 6 triples when presented in all 5 orientations, not just the 4 in which the triples had been presented during training. To provide baseline results for comparison, the results from experiment 1 were compared with results from experiment 2 which involved no training in layers 1,2 and 3,4, with the synaptic weights left unchanged from their initial random values.

In Fig. 8.40 numerical results are given for the experiments described. On the left are the single cell information measures for all top (4th) layer neurons ranked in order of their invariance to the triples, while on the right are multiple cell information measures. To help to interpret these results we can compute the maximum single cell information measure according to

$$\text{Maximum single cell information } = \log_2(\text{Number of triples}), \qquad (8.39)$$

where the number of triples is 6. This gives a maximum single cell information measure of 2.6 bits for these test cases. The information results from the experiment demonstrate that even with the triples presented to the network in only four of the five orientations during training, layer 4 is indeed capable of developing rotation invariant neurons that can discriminate effectively between the 6 different feature triples in all 5 orientations, that is with correct recognition from all five perspectives. In addition, the multiple cell information for the experiment reaches the maximal level of 2.6 bits, indicating that the network as a whole is capable of perfect discrimination between the 6 triples in any of the 5 orientations. These results may be compared with the very poor baseline performance from the control experiment, where no learning was allowed before testing. Further results from experiment 1 are presented in Fig. 8.41 where we show the response profiles of a top layer neuron to the 6 triples. It can be seen that this neuron has achieved excellent invariant responses to the 6 triples: the response profiles are independent of orientation, but differentiate between triples in that the responses are maximal for triple 132 and minimal for all other triples. In particular, the cell responses are maximal for triple 132 presented in all 5 of the orientations.

Stringer and Rolls (2002a) also performed a control experiment to show that the network really had learned invariant representations specific to the kinds of 3D deformations undergone by the surface features as the objects rotated in-depth. In the control experiment the network

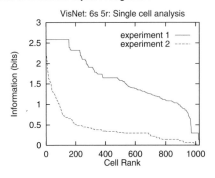
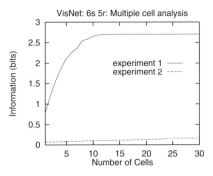

Fig. 8.40 Learning 3D perspectival transforms of features. Numerical results for experiments 1 and 2: On the left are single cell information measures, and on the right are multiple cell information measures.

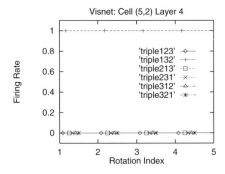

Fig. 8.41 Learning 3D perspectival transforms of features. Numerical results for experiment 1: Response profiles of a top layer neuron to the 6 triples in all 5 orientations.

was trained on 'spheres' with non-deformed surface features; and then as predicted the network failed to operate correctly when it was tested with objects with the features present in the transformed way that they appear on the surface of a real 3D object.

Stringer and Rolls (2002a) were thus able to show how trace learning can form neurons which can respond invariantly to novel rotational within-generic view perspectives of an object, obtained by within-generic view 3D rotations up to 30 degrees from any view encountered during learning. They were able to show in addition that this could occur for a novel view of an object which was not an interpolation from previously shown views. This was possible given that the low order feature combination sets from which an object was composed had been learned about in early layers of VisNet previously. The within-generic view transform invariant object recognition described was achieved through the development of true 3 dimensional representations of objects based on 3-dimensional features and feature combinations, which, unlike 2-dimensional feature combinations, are invariant under moderate in-depth rotations of the object. Thus, in a sense, these rotation invariant representations encode a form of 3-dimensional knowledge with which to interpret the visual input from the real world, that is able provide a basis for robust rotation invariant object recognition with novel perspectives. The particular finding in the work described here was that VisNet can learn how the surface features on 3D objects transform as the object is rotated in depth, and can use knowledge of the characteristics of the transforms to perform 3D object recognition.

The knowledge embodied in the network is knowledge of the 3D properties of objects, and in this sense assists the recognition of 3D objects seen from different views.

The process investigated by Stringer and Rolls (2002a) will only allow invariant object recognition over moderate 3D object rotations, since rotating an object through a large angle may lead to a catastrophic change in the appearance of the object that requires the new qualitative 3D shape descriptors to be associated with those of the former view. In that case, invariant object recognition must rely on the first process referred to at the start of this Section (8.4.7) in order to associate together the different generic views of an object to produce view invariant object identification. For that process, association of a few cardinal or generic views is likely to be sufficient (Koenderink 1990). The process described in this Section of learning how surface features transform is likely to make a major contribution to the within-generic view transform invariance of object identification and recognition.

8.4.8 Capacity of the architecture, and incorporation of a trace rule into a recurrent architecture with object attractors

One issue that has not been considered extensively so far is the capacity of hierarchical feedforward networks of the type exemplified by VisNet that are used for invariant object recognition. One approach to this issue is to note that VisNet operates in the general mode of a competitive network, and that the number of different stimuli that can be categorized by a competitive network is in the order of the number of neurons in the output layer, as described in Section 7.4. Given that the successive layers of the real visual system (V1, V2, V4, posterior inferior temporal cortex, anterior inferior temporal cortex) are of the same order of magnitude[24] VisNet is designed to work with the same number of neurons in each successive layer. The hypothesis is that because of redundancies in the visual world, each layer of the system by its convergence and competitive categorization can capture sufficient of the statistics of the visual input at each stage to enable correct specification of the properties of the world that specify objects. For example, V1 does not compute all possible combinations of a few lateral geniculate inputs, but instead represents linear series of geniculate inputs to form edge-like and bar-like feature analyzers, which are the dominant arrangement of pixels found at the small scale in natural visual scenes. Thus the properties of the visual world at this stage can be captured by a small proportion of the total number of combinations that would be needed if the visual world were random. Similarly, at a later stage of processing, just a subset of all possible combinations of line or edge analyzers would be needed, partly because some combinations are much more frequent in the visual world, and partly because the coding because of convergence means that what is represented is for a larger area of visual space (that is, the receptive fields of the neurons are larger), which also leads to economy and limits what otherwise would be a combinatorial need for feature analyzers at later layers. The hypothesis thus is that the effects of redundancies in the input space of stimuli that result from the statistical properties of natural images (Field 1987), together with the convergent architecture with competitive learning at each stage, produces a system which can perform invariant object recognition for large numbers of objects. Large in this case could be within one or two orders of magnitude of the number of neurons in any one layer of the network (or cortical area in the brain). The extent to which this can be realized can be explored with

---

[24]Of course the details are worth understanding further. V1 is for example somewhat larger than earlier layers, but on the other hand serves the dorsal as well as the ventral stream of visual cortical processing.

simulations of the type implemented in VisNet, in which the network can be trained with natural images which therefore reflect fully the natural statistics of the stimuli presented to the real brain.

We should note that a rich variety of information in perceptual space may be represented by subtle differences in the distributed representation provided by the output of the visual system. At the same time, the actual number of different patterns that may be stored in for example a pattern associator connected to the output of the visual system is limited by the number of input connections per neuron from the output neurons of the visual system (see Section 7.2). One essential function performed by the ventral visual system is to provide an invariant representation which can be read by a pattern associator in such a way that if the pattern associator learns about one view of the object, then the visual system allows generalization to another view of the same object, because the same output neurons are activated by the different view. In the sense that any view can and must activate the same output neurons of the visual system (the input to the associative network), then we can say the invariance is made explicit in the representation. Making some properties of an input representation explicit in an output representation has a major function of enabling associative networks that use visual inputs in for example recognition, episodic memory, emotion and motivation to generalize correctly, that is invariantly with respect to image transforms that are all consistent with the same object in the world (Rolls and Treves 1998).

Another approach to the issue of the capacity of networks that use trace-learning to associate together different instances (e.g. views) of the same object is to reformulate the issue in the context of autoassociation (attractor) networks, where analytic approaches to the storage capacity of the network are well developed (see Section 7.3, Amit (1989), and Rolls and Treves (1998)). This approach to the storage capacity of networks that associate together different instantiations of an object to form invariant representations has been developed by Parga and Rolls (1998) and Elliffe, Rolls, Parga and Renart (2000), and is described next.

In this approach, the storage capacity of a *recurrent* network which performs for example view invariant recognition of objects by associating together different views of the same object which tend to occur close together in time, was studied (Parga and Rolls 1998, Elliffe, Rolls, Parga and Renart 2000). The architecture with which the invariance is computed is a little different to that described earlier. In the model of Rolls ((1992a), (1994), (1995b), Wallis and Rolls (1997), Rolls and Milward (2000)), the postsynaptic memory trace enabled different afferents from the preceding stage to modify onto the same postsynaptic neuron (see Fig. 8.42). In that model there were no recurrent connections between the neurons, although such connections were one way in which it was postulated the memory trace might be implemented, by simply keeping the representation of one view or aspect active until the next view appeared. Then an association would occur between representations that were active close together in time (within e.g. 100–300 ms).

In the model developed by Parga and Rolls (1998) and Elliffe, Rolls, Parga and Renart (2000), there is a set of inputs with fixed synaptic weights to a network. The network itself is a recurrent network, with a trace rule incorporated in the recurrent collaterals (see Fig. 8.43). When different views of the same object are presented close together in time, the recurrent collaterals learn using the trace rule that the different views are of the same object. After learning, presentation of any of the views will cause the network to settle into an attractor which represents all the views of the object, that is which is a view invariant representation

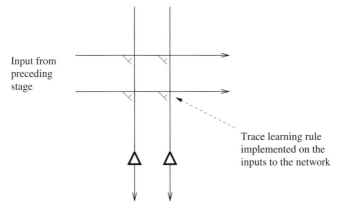

Fig. 8.42 The learning scheme implemented in VisNet. A trace learning rule is implemented in the feedforward inputs to a competitive network.

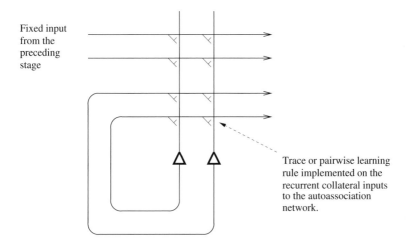

Fig. 8.43 The learning scheme considered by Parga and Rolls (1998) and Elliffe, Rolls, Parga and Renart (2000). There are inputs to the network from the preceding stage via unmodifiable synapses, and a trace or pairwise associative learning rule is implemented in the recurrent collateral synapses of an autoassociative memory to associative together the different exemplars (e.g. views) of the same object.

of an object[25].

We envisage a set of neuronal operations which set up a synaptic weight matrix in the recurrent collaterals by associating together because of their closeness in time the different views of the same object.

In more detail, Parga and Rolls (1998) considered two main approaches. First, one could store in a synaptic weight matrix the s views of an object. This consists in equally associating all the views to each other, including the association of each view with itself. Choosing in

---

[25] In this Section, the different exemplars of an object which need to be associated together are called views, for simplicity, but could at earlier stages of the hierarchy represent for example similar feature combinations (derived from the same object) in different positions in space.

	O_1v_1	O_1v_2	O_1v_3	O_1v_4	O_1v_5	O_2v_1	O_2v_2	O_2v_3	O_2v_4	O_2v_5	$\cdots$
O_1v_1	1	1	1	1	1						
O_1v_2	1	1	1	1	1						
O_1v_3	1	1	1	1	1						
O_1v_4	1	1	1	1	1						
O_1v_5	1	1	1	1	1						
O_2v_1						1	1	1	1	1	
O_2v_2						1	1	1	1	1	
O_2v_3						1	1	1	1	1	
O_2v_4						1	1	1	1	1	
O_2v_5						1	1	1	1	1	

Fig. 8.44 A schematic illustration of the first type of associations contributing to the synaptic matrix considered by Parga and Rolls (1998). Object 1 (O_1) has five views labelled v_1 to v_5, etc. The matrix is formed by associating the pattern presented in the columns with itself, that is with the same pattern presented as rows.

Fig. 8.44 an example such that objects are defined in terms of five different views, this might produce (if each view produced firing of one neuron at a rate of 1) a block of 5×5 pairs of views contributing to the synaptic efficacies each with value 1. Object 2 might produce another block of synapses of value 1 further along the diagonal, and symmetric about it. Each object or memory could then be thought of as a single attractor with a distributed representation involving five elements (each element representing a different view). Then the capacity of the system in terms of the number P_o of objects that can be stored is just the number of separate attractors which can be stored in the network. For random fully distributed patterns this is as shown numerically by Hopfield (1982)

$$P_o = 0.14\,C \tag{8.40}$$

where there are C inputs per neuron (and $N = C$ neurons if the network is fully connected). Now the synaptic matrix envisaged here does not consist of random fully distributed binary elements, but instead we will assume has a sparseness $a = s/N$, where s is the number of views stored for each object, from any of which the whole representation of the object must be recognized. In this case, one can show (Gardner 1988, Tsodyks and Feigel'man 1988, Treves and Rolls 1991) that the number of objects that can be stored and correctly retrieved

$$P_o = \frac{k\,C}{a\,\ln(1/a)} \tag{8.41}$$

where C is the number of synapses on each neuron devoted to the recurrent collaterals from other neurons in the network, and k is a factor that depends weakly on the detailed structure of the rate distribution, on the connectivity pattern, etc., but is approximately in the order of 0.2–0.3. A problem with this proposal is that as the number of views of each object increases to a large number (e.g. > 20), the network will fail to retrieve correctly the internal representation of the object starting from any one view (which is only a fraction $1/s$ of the length of the stored pattern that represents an object).

	O_1v_1	O_1v_2	O_1v_3	O_1v_4	O_1v_5	O_2v_1	O_2v_2	O_2v_3	O_2v_4	O_2v_5	. . .
O_1v_1	1	b	b	b	b						
O_1v_2	b	1	b	b	b						
O_1v_3	b	b	1	b	b						
O_1v_4	b	b	b	1	b						
O_1v_5	b	b	b	b	1						
O_2v_1						1	b	b	b	b	
O_2v_2						b	1	b	b	b	
O_2v_3						b	b	1	b	b	
O_2v_4						b	b	b	1	b	
O_2v_5						b	b	b	b	1	

Fig. 8.45 A schematic illustration of the second and main type of associations contributing to the synaptic matrix considered by Parga and Rolls (1998) and Elliffe, Rolls, Parga and Renart (2000). Object 1 (O_1) has five views labelled v_1 to v_5, etc. The association of any one view with itself has strength 1, and of any one with another view of the same object has strength b.

The second approach, taken by Parga and Rolls (1998) and Elliffe, Rolls, Parga and Renart (2000), is to consider the operation of the network when the associations between pairs of views can be described by a matrix that has the general form shown in Fig. 8.45. Such an association matrix might be produced by different views of an object appearing after a given view with equal probability, and synaptic modification occurring of the view with itself (giving rise to the diagonal term), and of any one view with that which immediately follows it. The same weight matrix might be produced not only by pairwise association of successive views because the association rule allows for associations over the short time scale of e.g. 100–200 ms, but might also be produced if the synaptic trace had an exponentially decaying form over several hundred ms, allowing associations with decaying strength between views separated by one or more intervening views. The existence of a regime, for values of the coupling parameter between pairs of views in a finite interval, such that the presentation of any of the views of one object leads to the same attractor regardless of the particular view chosen as a cue, is one of the issues treated by Parga and Rolls (1998) and Elliffe, Rolls, Parga and Renart (2000). A related problem also dealt with was the capacity of this type of synaptic matrix: how many objects can be stored and retrieved correctly in a view invariant way? Parga and Rolls (1998) and Elliffe, Rolls, Parga and Renart (2000) showed that the number grows linearly with the number of recurrent collateral connections received by each neuron. Some of the groundwork for this approach was laid by the work of Amit and collaborators (Griniasty, Tsodyks and Amit 1993, Amit 1989).

A variant of the second approach is to consider that the remaining entries in the matrix shown in Fig. 8.45 all have a small value. This would be produced by the fact that sometimes a view of one object would be followed by a view of a different object, when for example a large saccade was made, with no explicit resetting of the trace. On average, any one object would follow another rarely, and so the case is considered when all the remaining associations between pairs of views have a low value.

Parga and Rolls (1998) and Elliffe, Rolls, Parga and Renart (2000) were able to show that invariant object recognition is feasible in attractor neural networks in the way described. The system is able to store and retrieve in a view invariant way an extensive number of objects, each defined by a finite set of views. What is implied by extensive is that the number of objects is proportional to the size of the network. The crucial factor that defines this size is the number of connections per neuron. In the case of the fully connected networks considered in this Section, the size is thus proportional to the number of neurons. To be particular, the number of objects that can be stored is $0.081\,N\,/\,5$, when there are five views of each object. The number of objects is $0.073\,N\,/\,11$, when there are eleven views of each object. This is an interesting result in network terms, in that s views each represented by an independent random set of active neurons, can in the network described, be present in the same 'object' attraction basin. It is also an interesting result in neurophysiological terms, in that the number of objects that can be represented in this network scales linearly with the number of recurrent connections per neuron. That is, the number of objects P_o that can be stored is approximately

$$P_o = \frac{k\,C}{s} \tag{8.42}$$

where C is the number of synapses on each neuron devoted to the recurrent collaterals from other neurons in the network, s is the number of views of each object, and k is a factor that is in the region of 0.07–0.09 (Parga and Rolls 1998).

Although the explicit numerical calculation was done for a rather small number of views for each object (up to 11), the basic result, that the network can support this kind of 'object' phase, is expected to hold for any number of views (the only requirement being that it does not increase with the number of neurons). This is of course enough: once an object is defined by a set of views, when the network is presented with a somewhat different stimulus or a noisy version of one of them it will still be in the attraction basin of the object attractor.

Parga and Rolls (1998) thus showed that multiple (e.g. 'view') patterns could be within the basin of attraction of a shared (e.g. 'object') representation, and that the capacity of the system was proportional to the number of synapses per neuron divided by the number of views of each object.

Elliffe, Rolls, Parga and Renart (2000) extended the analysis of Parga and Rolls (1998) by showing that correct retrieval could occur where retrieval 'view' cues were distorted; where there was some association between the views of different objects; and where there was only partial and indeed asymmetric connectivity provided by the associatively modified recurrent collateral connections in the network. The simulations also extended the analysis by showing that the system can work well with sparse patterns, and indeed that the use of sparse patterns increases (as expected) the number of objects that can be stored in the network.

Taken together, the work described by Parga and Rolls (1998) and Elliffe, Rolls, Parga and Renart (2000) introduced the idea that the trace rule used to build invariant representations could be implemented in the recurrent collaterals of a neural network (as well as or as an alternative to its incorporation in the forward connections from one layer to another incorporated in VisNet), and provided a precise analysis of the capacity of the network if it operated in this way. In the brain, it is likely that the recurrent collateral connections between cortical pyramidal cells in visual cortical areas do contribute to building invariant representations, in that if they are associatively modifiable, as seems likely, and because there is continuing firing for typically 100–300 ms after a stimulus has been shown, associations between different exemplars of the same object that occur together close in time would almost necessarily become built into the recurrent synaptic connections between pyramidal cells.

Invariant representation of faces in the context of attractor neural networks has also been discussed by Bartlett and Sejnowski (1997) in terms of a model where different views of faces are presented in a fixed sequence (Griniasty, Tsodyks and Amit 1993). This is not however the general situation; normally any pair of views can be seen consecutively and they will become associated. The model described by Parga and Rolls (1998) treats this more general situation.

We wish to note the different nature of the invariant object recognition problem studied here, and the paired associate learning task studied by Miyashita and Chang (1988), Miyashita (1988), and Sakai and Miyashita (1991). In the invariant object recognition case no particular learning protocol is required to produce an activity of the inferior temporal cortex cells responsible for invariant object recognition that is maintained for 300 ms. The learning can occur rapidly, and the learning occurs between stimuli (e.g. different views) which occur with no intervening delay. In the paired associate task, which had the aim of providing a model of semantic memory, the monkeys must learn to associate together two stimuli that are separated in time (by a number of seconds), and this type of learning can take weeks to train. During the delay period the sustained activity is rather low in the experiments, and thus the representation of the first stimulus that remains is weak, and can only poorly be associated with the second stimulus. However, formally the learning mechanism could be treated in the same way as that used by Parga and Rolls (1998) for invariant object recognition. The experimental difference is just that in the paired associate task used by Miyashita et al, it is the weak memory of the first stimulus that is associated with the second stimulus. In contrast, in the invariance learning, it would be the firing activity being produced by the first stimulus (not the weak memory of the first stimulus), that can be associated together.

The mechanisms described here using an attractor network with a trace associative learning rule would apply most naturally when a small number of representations need to be associated together to represent an object. One example is associating together what is seen when an object is seen from different perspectives. Another example is scale, with respect to which neurons early in the visual system tolerate scale changes of approximately 1.5 octaves, so that the whole scale range could be covered by associating together a limited number of such representations (see Chapter 5 and Fig. 8.4). The mechanism would not be so suitable when a large number of different instances would need to be associated together to form an invariant representation of objects, as might be needed for translation invariance. For the latter, the standard model of VisNet with the associative trace learning rule implemented in the feedforward connections would be more appropriate. However, both types of mechanism, with the trace rule in the feedforward or in recurrent collateral synapses, could contribute (separately or together) to achieve invariant representations. Part of the interest of the attractor approach described in this Section is that it allows analytic investigation.

8.4.9 Vision in natural scenes — effects of background versus attention

The results described in Section 5.6.3 and summarized in Fig. 5.20 show that the receptive fields of inferior temporal cortex neurons were large (65 deg) with a single stimulus in a blank background (top left), and were greatly reduced in size (to 36.6 deg) when presented in a complex natural scene (top right). The results also show that there was little difference in receptive field size or firing rate in the complex background when the effective stimulus was selected for action (bottom right), and when it was not (middle right).

Trappenberg, Rolls and Stringer (2002) have suggested what underlying mechanisms could account for these findings, and simulated a model to test the ideas. The model utilizes an attractor network representing the inferior temporal visual cortex (implemented by the recurrent connections between inferior temporal cortex neurons), and a neural input layer with several retinotopically organized modules representing the visual scene in an earlier visual cortical area such as V4 (see Fig. 8.46). The attractor network aspect of the model produces the property that receptive fields of IT neurons can be large in blank scenes by enabling a weak input in the periphery of the visual field to act as a retrieval cue for the object attractor. On the other hand, when the object is shown in a complex background, the object closest to the fovea tends to act as the retrieval cue for the attractor, because the fovea is given increased weight in activating the IT module because the magnitude of the input activity from objects at the fovea is greatest due to the higher magnification factor of the fovea incorporated into the model. This results in smaller receptive fields of IT neurons in complex scenes, because the object tends to need to be close to the fovea to trigger the attractor into the state representing that object. (In other words, if the object is far from the fovea, then it will not trigger neurons in IT which represent it, because neurons in IT are preferentially being activated by another object at the fovea.) This may be described as an attractor model in which the competition for which attractor state is retrieved is weighted towards objects at the fovea.

Attentional top-down object-based inputs can bias the competition implemented in this attractor model, but have relatively minor effects (in for example increasing receptive field size) when they are applied in a complex natural scene, as then as usual the stronger forward inputs dominate the states reached. In this network, the recurrent collateral connections may be thought of as implementing constraints between the different inputs present, to help arrive at firing in the network which best meets the constraints. In this scenario, the preferential weighting of objects close to the fovea because of the increased magnification factor at the fovea is a useful principle in enabling the system to provide useful output. The attentional object biasing effect is much more marked in a blank scene, or a scene with only two objects present at similar distances from the fovea, which are conditions in which attentional effects have frequently been examined. The results of the investigation (Trappenberg, Rolls and Stringer 2002) thus suggest that attention may be a much more limited phenomenon in complex, natural, scenes than in reduced displays with one or two objects present. The results also suggest that the alternative principle, of providing strong weight to whatever is close to the fovea, is an important principle governing the operation of the inferior temporal visual cortex, and in general of the output of the visual system in natural environments. This principle of operation is very important in interfacing the visual system to action systems, because the effective stimulus in making inferior temporal cortex neurons fire is in natural scenes usually on or close to the fovea. This means that the spatial coordinates of where the object is in the scene do not have to be represented in the inferior temporal visual cortex, nor passed from it to the action selection system, as the latter can assume that the object making IT neurons fire is close to the fovea in natural scenes (see Section 13.14).

There may of course be in addition a mechanism for object selection that takes into account the locus of covert attention when actions are made to locations not being looked at. However, the simulations described in this Section suggest that in any case covert attention is likely to be a much less significant influence on visual processing in natural scenes than in reduced scenes with one or two objects present.

Object bias

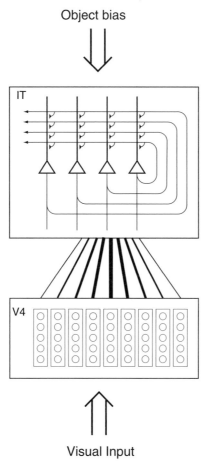

Visual Input

Fig. 8.46 The architecture of the inferior temporal cortex (IT) model of Trappenberg, Rolls and Stringer (2002) operating as an attractor network with inputs from the fovea given preferential weighting by the greater magnification factor of the fovea. The model also has a top-down object-selective bias input. The model was used to analyze how object vision and recognition operate in complex natural scenes.

Given these points, one might question why inferior temporal cortex neurons can have such large receptive fields, which show translation invariance. At least part of the answer to this may be that inferior temporal cortex neurons must have the capability to be large if they are to deal with large objects. A V1 neuron, with its small receptive field, simply could not receive input from all the features necessary to define an object. On the other hand, inferior temporal cortex neurons may be able to adjust their size to approximately the size of objects, using in part the interactive effects described in Chapters 9–11, and need the capability for translation invariance because the actual relative positions of the features of an object could be at different relative positions in the scene. For example, a car can be recognized whichever way it is viewed, so that the parts (such as the bonnet or hood) must be identifiable as parts wherever they happen to be in the image, though of course the parts themselves also have to be in the correct relative positions, as allowed for by the hierarchical feature analysis architecture described in this Chapter.

Some details of the simulations follow. Each independent module within 'V4' in Fig. 8.46 represents a small part of the visual field and receives input from earlier visual areas represented by an input vector for each possible location which is unique for each object. Each module was 6 deg in width, matching the size of the objects presented to the network. For the simulations Trappenberg, Rolls and Stringer (2002) chose binary random input vectors representing objects with $N^{V4}a^{V4}$ components set to ones and the remaining $N^{V4}(1 - a^{V4})$ components set to zeros. N^{V4} is the number of nodes in each module and a^{V4} is the sparseness of the representation which was set to be $a^{V4} = 0.2$ in the simulations.

The structure labelled 'IT' represents areas of visual association cortex such as the inferior temporal visual cortex and cortex in the anterior part of the superior temporal sulcus in which neurons provide distributed representations of faces and objects (Booth and Rolls 1998, Rolls 2000a). Nodes in this structure are governed by leaky integrator dynamics (similar to those used in the mean field approach described in Section 7.8.1) with time constant τ

$$\tau \frac{dh_i^{IT}(t)}{dt} = -h_i^{IT}(t) + \sum_j (w_{ij}^{IT} - c^{IT})y_j^{IT}(t) + \sum_k w_{ik}^{IT-V4}y_k^{V4}(t) + k^{IT\_BIAS}I_i^{OBJ}. \quad (8.43)$$

The firing rate y_i^{IT} of the ith node is determined by a sigmoidal function from the activation h_i^{IT} as follows

$$y_i^{IT}(t) = \frac{1}{1 + \exp[-2\beta(h_i^{IT}(t) - \alpha)]}, \quad (8.44)$$

where the parameters $\beta = 1$ and $\alpha = 1$ represent the gain and the bias, respectively.

The recognition functionality of this structure is modelled as an attractor neural network (ANN) with trained memories indexed by μ representing particular objects. The memories are formed through Hebbian learning on sparse patterns,

$$w_{ij}^{IT} = k^{IT} \sum_\mu (\xi_i^\mu - a^{IT})(\xi_j^\mu - a^{IT}), \quad (8.45)$$

where k^{IT} (set to 1 in the simulations) is a normalization constant that depends on the learning rate, $a^{IT} = 0.2$ is the sparseness of the training pattern in IT, and ξ_i^μ are the components of the pattern used to train the network. The constant c^{IT} in equation 8.43 represents the strength of the activity-dependent global inhibition simulating the effects of inhibitory interneurons. The external 'top-down' input vector I^{OBJ} produces object-selective inputs, which are used as the attentional drive when a visual search task is simulated. The strength of this object bias is modulated by the value of k^{IT_BIAS} in equation 8.43.

The weights w_{ij}^{IT-V4} between the V4 nodes and IT nodes were trained by Hebbian learning of the form

$$w_{ij}^{IT-V4} = k^{IT-V4}(k) \sum_\mu (\xi_i^\mu - a^{V4})(\xi_j^\mu - a^{IT}). \quad (8.46)$$

to produce object representations in IT based on inputs in V4. The normalizing modulation factor $k^{IT-V4}(k)$ allows the gain of inputs to be modulated as a function of their distance from the fovea, and depends on the module k to which the presynaptic node belongs. The model supports translation invariant object recognition of a single object in the visual field if the normalization factor is the same for each module and the model is trained with the objects placed at every possible location in the visual field. The translation invariance of the weight vectors between each 'V4' module and the IT nodes is however explicitly modulated in the model by the module-dependent modulation factor $k^{IT-V4}(k)$ as indicated in Fig. 8.46 by the width of the lines connecting V4 with IT. The strength of the foveal V4 module is strongest, and the strength decreases for modules representing increasing eccentricity. The form of this modulation factor was derived from the parameterization of the cortical magnification factors given by (Dow, Snyder, Vautin and Bauer 1981)[26].

To study the ability of the model to recognize trained objects at various locations relative to the fovea the system was trained on a set of objects. The network was then tested with distorted versions of the objects, and the 'correlation' between the target object and the final state of the attractor network was taken as a measure of the performance. The correlation was estimated from the normalized dot product between the target object vector that was used during training the IT network, and the state of the IT network after a fixed amount of time sufficient for the network to settle into a stable state. The objects were always presented on backgrounds with some noise (introduced by flipping 2% of the bits in the scene which were not the test stimulus) in order to utilize the properties of the attractor network, and because the input to IT will inevitably be noisy under normal conditions of operation.

In the first simulation only one object was present in the visual scene in a plain background at different eccentricities from the fovea. As shown in Fig. 8.47A by the line labelled 'blank background', the receptive fields of the neurons were very large. The value of the object bias $k^{IT-BIAS}$ was set to 0 in these simulations. Good object retrieval (indicated by large correlations) was found even when the object was far from the fovea, indicating large IT receptive fields with a blank background. The reason that any drop is seen in performance as a function of eccentricity is because flipping 2% of the bits outside the object introduces some noise into the recall process. This demonstrates that the attractor dynamics can support translation invariant object recognition even though the translation invariant weight vectors between V4 and IT are explicitly modulated by the modulation factor k^{IT-V4} derived from the cortical magnification factor.

In a second simulation individual objects were placed at all possible locations in a natural and cluttered visual scene. The resulting correlations between the target pattern and the asymptotic IT state are shown in Fig. 8.47A with the line labelled 'natural background'. Many objects in the visual scene are now competing for recognition by the attractor network, and the objects around the foveal position are enhanced through the modulation factor derived from the cortical magnification factor. This results in a much smaller size of the receptive field of IT neurons when measured with objects in natural backgrounds.

In addition to this major effect of the background on the size of the receptive field, which parallels and we suggest may account for the physiological findings outlined above and in Section 5.6.3, there is also a dependence of the size of the receptive fields on the level of object bias provided to the IT network. Examples are shown in Fig. 8.47B where an object bias was used. The object bias biases the IT network towards the expected object with a strength determined by the value of $k^{IT-BIAS}$, and has the effect of increasing the size of the receptive fields in both blank and natural backgrounds (see Fig. 8.47B compared to A). This

---

[26] This parameterization is based on V1 data. However, it was shown that similar forms of the magnification factor hold also in V4 (Gattass, Sousa and Covey 1985). Similar results to the ones presented here can also be achieved with different forms of the modulation factor such as a shifted Gaussian as illustrated by the dashed line in Figure 8.46C.

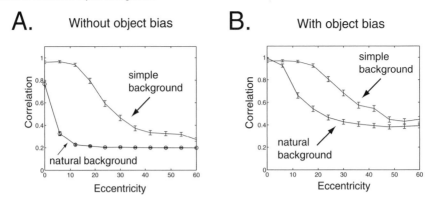

Fig. 8.47 Correlations as measured by the normalized dot product between the object vector used to train IT and the state of the IT network after settling into a stable state with a single object in the visual scene (blank background) or with other trained objects at all possible locations in the visual scene (natural background). There is no object bias included in the results shown in graph A, whereas an object bias is included in the results shown in B with $k^{\mathrm{IT\_BIAS}} = 0.7$ in the experiments with a natural background and $k^{\mathrm{IT\_BIAS}} = 0.1$ in the experiments with a blank background.

models the effect found neurophysiologically (Rolls, Zheng and Aggelopoulos 2001b)[27].

Some of the conclusions are as follows. When single objects are shown in a scene with a blank background, the attractor network helps neurons to respond to an object with large eccentricities of this object relative to the fovea of the agent. When the object is presented in a natural scene, other neurons in the inferior temporal cortex become activated by the other effective stimuli present in the visual field, and these forward inputs decrease the response of the network to the target stimulus by a competitive process. The results found fit well with the neurophysiological data, in that IT operates with almost complete translation invariance when there is only one object in the scene, and reduces the receptive field size of its neurons when the object is presented in a cluttered environment. The model described here provides an explanation of the responses of real IT neurons in natural scenes.

In natural scenes, the model is able to account for the neurophysiological data that the IT neuronal responses are larger when the object is close to the fovea, by virtue of fact that objects close to the fovea are weighted by the cortical magnification factor related modulation $k^{\mathrm{IT}-\mathrm{V4}}$.

The model accounts for the larger receptive field sizes from the fovea of IT neurons in natural backgrounds if the target is the object being selected compared to when it is not selected (Rolls, Zheng and Aggelopoulos 2001b). The model accounts for this by an effect of top-down bias which simply biases the neurons towards particular objects compensating for their decreasing inputs produced by the decreasing magnification factor modulation with increasing distance from the fovea. Such object-based attention signals could originate in the prefrontal cortex and could provide the object bias for the inferotemporal cortex (Renart, Parga and Rolls 2000).

---

[27] $k^{\mathrm{IT\_BIAS}}$ was set to 0.7 in the experiments with a natural background and to 0.1 in blank background, reflecting the fact that more attention may be needed to find objects in natural cluttered environments because of the noise present than in blank backgrounds.

We note that it is possible that a 'spotlight of attention' (Desimone and Duncan 1995) can be moved covertly away from the fovea as described in Chapters 9–10. However, at least during normal visual search tasks in natural scenes, the neurons are sensitive to the object at which the monkey is looking, that is to the object which is on the fovea, as shown by Rolls, Zheng and Aggelopoulos (2001b) and described in Sections 5.6.3 and 13.14.

8.5 Syntactic binding of separate neuronal ensembles by synchronization

The problem of syntactic binding of neuronal representations, in which some features must be bound together to form one object, and other simultaneously active features must be bound together to represent another object, has been addressed by von der Malsburg (see von der Malsburg (1990)). He has proposed that this could be performed by temporal synchronization of those neurons that were temporarily part of one representation in a different time slot from other neurons that were temporarily part of another representation. The idea is attractive in allowing arbitrary relinking of features in different combinations. Singer, Engel, Konig, and colleagues (see Engel, Konig, Kreiter, Schillen and Singer (1992) and Singer (1999), but see Shadlen and Movshon (1999)) have obtained some evidence that when features must be bound, synchronization of neuronal populations can occur. The extent to which it is required for visual object recognition is an open issue, and evidence on the possible role of synchronization in cortical computation is considered further in Sections 5.5.7, by Shadlen and Movshon (1999), and by Riesenhuber and Poggio (1999a).

Synchronization to implement syntactic binding has a number of disadvantages and limitations, as described in Sections 5.5.7 and 13.7. The greatest computational problem is that synchronization does not by itself define the spatial relations between the features being bound, so is not just as a binding mechanism adequate for shape recognition, as described in Section 13.7. In the context of VisNet, and how the real visual system may operate to implement object recognition, the use of synchronization does not appear to match the way in which the visual system is organized. For example, using only a two-layer network, von der Malsburg has shown that synchronization could provide the necessary feature linking to perform object recognition with relatively few neurons, because they can be reused again and again, linked differently for different objects. In contrast, the primate uses a considerable part of its cortex, perhaps 50% in monkeys, for visual processing, with therefore what could be in the order of 6×10^8 neurons and 6×10^{12} synapses involved (estimating from the values given in Table 1.1), so that the solution adopted by the real visual system may be one which relies on many neurons with simpler processing than arbitrary syntax implemented by synchronous firing of separate assemblies suggests. On the other hand, a solution such as that investigated by VisNet which forms low-order combinations of what is represented in previous layers, is very demanding in terms of the number of neurons required, and this matches what is found in the primate visual system. It will be fascinating to see how research on these different approaches to processing in the primate visual system develops. For the development of both approaches, the use of well-defined neuronal network models is proving to be very helpful.

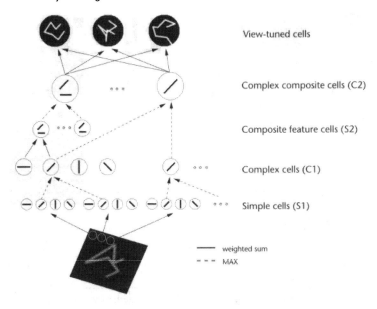

View-tuned cells

Complex composite cells (C2)

Composite feature cells (S2)

Complex cells (C1)

Simple cells (S1)

——— weighted sum

– – – MAX

Fig. 8.48 Sketch of Riesenhuber and Poggio's (1999a,b) model of invariant object recognition. The model includes layers of 'S' cells which perform template matching (solid lines), and 'C' cells (solid lines) which pool information by a non-linear MAX function to achieve invariance (see text). After Riesenhuber and Poggio (1999a,b).

8.6 Further approaches to invariant object recognition

A related approach to invariant object recognition is described by Riesenhuber and Poggio (1999b), and builds on the hypothesis that not just shift invariance (as implemented in the Neocognitron), but also other invariances such as scale, rotation and even view, could be built into a feature hierarchy system, as suggested by Rolls (1992a) (see also Perrett and Oram (1993)). The approach of Riesenhuber and Poggio (1999b) (see also Riesenhuber and Poggio (1999a) and Riesenhuber and Poggio (2000)) is a feature hierarchy approach which uses alternate 'simple cell' and 'complex cell' layers in a way analogous to Fukushima (1980) (see Fig. 8.48). The function of each S cell layer is to build more complicated features from the inputs, and works by template matching. The function of each 'C' cell layer is to provide some translation invariance over the features discovered in the preceding simple cell layer (as in Fukushima (1980)), and operates by performing a MAX function on the inputs. The non-linear MAX function makes a complex cell respond only to whatever is the highest activity input being received, and is part of the process by which invariance is achieved according to this proposal. This C layer process involves 'implicitly scanning over afferents of the same type differing in the parameter of the transformation to which responses should be invariant (for instance, feature size for scale invariance), and then selecting the best-matching afferent' (Riesenhuber and Poggio 1999b). Brain mechanisms by which this computation could be set up are not part of the scheme, and the model is effectively hand-wired, so does not yet provide a biologically plausible model of invariant object recognition. However, the fact that the model sets out to achieve some of the processes specified by Rolls (1992a) (see Section 8.3) and implemented in VisNet (see Section 8.4) does represent useful convergent thinking towards how invariant object recognition might be implemented in the brain.

Further evidence consistent with the approach developed in the investigations of VisNet described in this Chapter comes from psychophysical studies. Wallis and Bulthoff (1999) review psychophysical evidence for learning of view invariant representations by experience, in that the learning can be shown in special circumstances to be affected by the temporal sequence in which different views of objects are seen.

Another approach to the implementation of invariant representations in the brain is the use of neurons with Sigma-Pi synapses. Sigma-Pi synapses, described in Section A.2.3, effectively allow one input to a synapse to be multiplied or gated by a second input to the synapse. The multiplying input might gate the appropriate set of the other inputs to a synapse to produce the shift or scale change required. For example, the x^c input in equation A.14 could be a signal that varies with the shift required to compute translation invariance, effectively mapping the appropriate set of x_j inputs through to the output neurons depending on the shift required (Mel, Ruderman and Archie 1998, Mel and Fiser 2000, Olshausen, Anderson and Van Essen 1993, Olshausen, Anderson and Van Essen 1995). Local operations on a dendrite could be involved in such a process (Mel, Ruderman and Archie 1998). The explicit neural implementation of the gating mechanism seems implausible, given the need to multiply and thus remap large parts of the retinal input depending on shift and scale modifying connections to a particular set of output neurons. Moreover, the explicit control signal to set the multiplication required in V1 has not been identified (see Section 6.2.2). Moreover, if this was the solution used by the brain, the whole problem of shift and scale invariance could in principle be solved in one layer of the system, rather than with the multiple hierarchically organized set of layers actually used in the brain, as shown schematically in Fig. 8.4. The multiple layers actually used in the brain are much more consistent with the type of scheme incorporated in VisNet. Moreover, if a multiplying system of the type hypothesized by Mel, Ruderman and Archie (1998), Olshausen, Anderson and Van Essen (1993) and Olshausen, Anderson and Van Essen (1995) was implemented in a multilayer hierarchy with the shift and scale change emerging gradually, then the multiplying control signal would need to be supplied to every stage of the hierarchy. A further problem with such approaches is how the system is trained in the first place.

8.7 Different processes involved in different types of object identification

To conclude this Chapter, it is proposed that there are (at least) three different types of process that could be involved in object identification. The first is the simple situation where different objects can be distinguished by different non-overlapping sets of features (see Section 8.2.1). An example might be a banana and an orange, where the list of features of the banana might include yellow, elongated, and smooth surface; and of the orange its orange colour, round shape, and dimpled surface. Such objects could be distinguished just on the basis of a list of the properties, which could be processed appropriately by a competitive network, pattern associator, etc. No special mechanism is needed for view invariance, because the list of properties is very similar from most viewing angles. Object recognition of this type may be common in animals, especially those with visual systems less developed than those of primates. However, this approach does not describe the shape and form of objects, and is insufficient to account for primate vision. Nevertheless, the features present in objects are valuable cues to object identity, and are naturally incorporated into the feature hierarchy approach.

A second type of process might involve the ability to generalize across a small range of views of an object, that is within a generic view, where cues of the first type cannot be used to solve the problem. An example might be generalization across a range of views of a cup when looking into the cup, from just above the near lip until the bottom inside of the cup comes into view. Such generalization would work because the neurons are tuned as filters to accept a range of variation of the input within parameters such as relative size and orientation of the components of the features. Generalization of this type would not be expected to work when there is a catastrophic change in the features visible, as for example occurs when the cup is rotated so that one can suddenly no longer see inside it, and the outside bottom of the cup comes into view. This type of process includes the learning of the transforms of the surface markings on 3D objects which occur when the object is rotated, as described in Section 8.4.7.

The third type of process is one that can deal with the sudden catastrophic change in the features visible when an object is rotated to a completely different view, as in the cup example just given (cf. Koenderink (1990)). Another example, quite extreme to illustrate the point, might be when a card with different images on its two sides is rotated so that one face and then the other is in view. This makes the point that this third type of process may involve arbitrary pairwise association learning, to learn which features and views are different aspects of the same object. Another example occurs when only some parts of an object are visible. For example, a red-handled screwdriver may be recognized either from its round red handle, or from its elongated silver coloured blade.

The full view-invariant recognition of objects which occurs even when the objects share the same features, such as colour, texture, etc. is an especially computationally demanding task which the primate visual system is able to perform with its highly developed temporal cortical visual areas. The neurophysiological evidence and the neuronal networks described in this Chapter suggest how the primate visual system may perform this task.

We have seen in this Chapter that the feature hierarchy approach has a number of advantages in performing object recognition over other approaches (see Section 8.2), and that some of the key computational issues that arise in these architectures have solutions (see Sections 8.3 and 8.4). Some further issues that arise with such architectures, such as object selection in cluttered environments, passing the coordinates of the object to the action system from an object recognition system which has little information about where the object is in the scene, and how the outputs are used, are discussed in Chapter 12. The ways in which object representations in feature hierarchy systems operate when attention can be object-based, or spatial, are considered in Chapters 9–11.

9 The cortical neurodynamics of visual attention – a model

9.1 Introduction

Visual attention can function in two distinct modes: spatial focal attention which can be visualized as a spotlight that 'illuminates' a certain location of visual space for focused visual analysis, and object attention which is spatially dispersed and with which a target object can be searched for in parallel over a large area of visual space. Duncan (1980) proposed that both modes of operation are manifestations of a top-down selection process. In spatial attention, the selection is focused in the spatial dimension and spread in the shape feature dimension; while in object attention, the selection is focused in the feature dimension and spread in the spatial dimension. In Chapter 6, we reviewed a number of neurophysiological studies that provide insights into the neural basis of spatial attention and object attention. Experimental observations made with single-cell recording and with functional imaging provide strong evidence for the biased competition hypothesis. This hypothesis is that attention modulates visual processing by enhancing the responses of the neurons representing the features or locations of the attended stimulus and by reducing the suppressive interactions from neurons representing nearby distractors.

In this Chapter, we formulate a neurodynamical theory and model that addresses the issues of how spatial and object attention mechanisms (presumably mediated by the dorsal and the ventral visual stream, respectively) can be integrated, and can function as a unitary system in visual search and visual recognition tasks. The system is built on the biased-competition hypothesis, attentional gain field modulation, and on interacting processes in visual information processing. An important novel idea in this model is that the dorsal stream and the ventral stream interact at multiple points and levels. The locus of intersection is a function of the scale of analysis. For example, detailed spatial and feature integration and attentive analysis would utilize area V1, whereas analysis at a larger scale would involve later stages such as V2 and V4. These early visual computational processing areas interact with the dorsal stream and the ventral stream via the multiple feedforward/feedback inter-cortical loops. These loops, together with the competitive mechanism within each cortical area, enable the system to switch from one mode of operation (e.g. spatial attention) to another (e.g. object attention), depending on the biasing input from regions such as the prefrontal cortex. The model can reproduce the findings of a number of attention-related neurophysiological experiments. It provides a unified conceptual framework to account for several apparently disparate psychological processes such as spatial and object attention, object recognition and localization, and serial and parallel search.

9.2 Physiological constraints

In this Section we make explicit and place in context the three main constraints that are incorporated into the theory and model described in this Chapter, namely: the segregation of the visual system into dorsal and ventral streams, the biased competition hypothesis, and the receptive field structure of neurons in the striate visual cortex. Further evidence on these issues is provided in Chapters 2–6.

9.2.1 The dorsal and ventral paths of the visual cortex

A major portion of the neocortex of primates is dedicated to the processing of visual information. This range of visual cortex has been differentiated into a large number of separate regions, which are organized in a hierarchical fashion (see Figs. 1.10 and 12.1). Around 30 different areas have been identified in the primate brain that are known to process visual information (Felleman and Van Essen 1991, Van Essen 1985, Van Essen, Felleman, DeYoe, Olavarria and Knierim 1990). A first functional distinction between these areas corresponds to the kind of visual processing performed, namely whether low-level or high-level computation is executed (Van Essen 1985). Low-level processing refers to the extraction of edges, textures and colours from sensory inputs. The brain areas involved are typically topographically organized. High-level functions involve the processing of information based on previously stored knowledge, about, for example, objects. The areas associated with such processing are often not topographically organized.

A widely accepted neurophysiological concept describes the visual processing of information as being performed in two main neural streams engaged in the analysis of an object's intrinsic properties, and in spatial analysis (Ungerleider and Mishkin 1982, Maunsell and Newsome 1987). The object pathway computing shape, colour, and other object properties, runs from the occipital lobe down to the inferior temporal lobe (areas V1, V2, V4, and inferotemporal areas TEO and TE). This pathway is commonly called the 'what' – or ventral – path and is involved in the identification of objects or parts of objects. The second pathway, associated with the extraction of spatial properties like location, motion etc. is called the 'where' – or dorsal – pathway, and runs from the occipital lobe up to the parietal lobe (areas V1, V2, V3, middle temporal area MT, medial superior temporal MST, and further stations in the inferior parietal and superior temporal sulcus cortex). This second pathway is sometimes called the 'action' pathway (Goodale and Milner 1992, Milner and Goodale 1995).

Neurons in the ventral processing stream leading to the inferior temporal cortical visual areas are found to process information concerning form, contour and color and are thought to mediate object recognition and awareness (Rolls 2000a, Pasupathy and Connor 1999, Gross 1973, Tovee, Rolls and Azzopardi 1994, Tanaka 1996, Logothetis and Sheinberg 1996, Sheinberg and Logothetis 2001) (see Chapter 5). In fact, neurons in the temporal lobes have large receptive fields that reveal translation invariance and in some cases even view-invariance (Booth and Rolls 1998), but are sensitive to the relative location of features within an object (Perrett, Rolls and Caan 1982, Olson and Gettner 1995). The majority of inferotemporal (IT) cells have distributed representations of shape, objects or faces (Perrett, Rolls and Caan 1982, Desimone, Albright, Gross and Bruce 1984, Tanaka 1996, Rolls 2000a) (see Chapter 5). On the other hand, neurons in posterior parietal areas seem to be concerned with depth, space, motion and coordinate transformations (Bushnell, Goldberg and Robinson 1981, Robinson, Goldberg and Stanton 1978, Robinson, Bowman and Kertzman 1991, Steinmetz, Connor and MacLeod

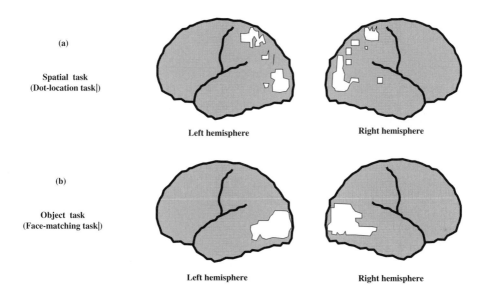

Fig. 9.1 Segregation of the visual system into ventral and dorsal streams as analyzed in the PET study of Haxby et al. (1994). (a) Spatial task: Subjects had to judge which of the two lower figures was a rotated version of the upper figure. (b) Object task: Subjects had to judge which faces were of the same individual. Compared to a control condition, activation in the spatial task was dorsal to that observed in the object task. Adapted from Haxby et al. (1994).

1992, Andersen, Snyder, Bradley and Xing 1997, Colby and Goldberg 1999, Duhamel, Colby and Goldberg 1992a, Galletti, Battaglini and Fattori 1991, Duhamel, Bremmer, BenHamed and Werner 1997) (see Chapter 4). Neurons in the parietal lobes are sensitive to the location of the stimulus with regard to the animal's head (Graziano and Gross 1993). Neurons in the posterior parietal cortex (PP) show an enhanced response to attended targets within their receptive fields, even when no eye movements are made. Bushnell, Goldberg and Robinson (1981), Robinson, Bowman and Kertzman (1991), and Steinmetz, Connor and MacLeod (1992) described a relative enhancement for attended items as opposed to unattended items, which suggests that PP represents the location of potentially attended items. This is in accordance with the studies of Posner, Walker, Friedrich and Rafal (1984) showing that damage to the parietal lobe in humans impairs the ability to move the attentional focus away from the presently attended location to other objects in the visual field.

Haxby, Horwitz, Ungerleider, Maisog, Pietrini and Grady (1994) investigated the what–where dichotomy in normal visual processing by means of positron emission tomography (PET) (Fig. 9.1). Subjects were scanned while performing either an object task, a spatial task, or a control task. The object task consisted of face-matching. Three faces were displayed within three squares arranged in the form of a triangle. Subjects had to judge which of the two lower faces (left or right) was of the same individual in the square at the triangle's apex by pressing one of two keys (left or right). During the spatial task, subjects had to judge which of the two lower figures (left or right) was a rotated version of the upper figure by pressing the corresponding left or right button. A control task presented three empty boxes to subjects who simply responded by alternating between two keys. Activation during the object and spatial task was compared (by subtraction) to activation during the control task. Significant areas of activation during the object task were observed within the inferior and temporal cortex and in the occipital lobe. On the other hand, during the spatial task, significant areas of activation were detected within the parietal and occipital cortex. Figure 9.1 shows these results which illustrate the segregation of the visual system in the dorsal and ventral pathways. The segregation of the 'what' and the 'where' streams in humans has also been demonstrated in functional magnetic resonance imaging (fMRI) studies (Corbetta and Shulman 1998).

We include in our architecture of the visual system this what-where segregation by modelling in different modules the visual areas V1, V4, PP and IT, and by connecting these modules such that the modules V1–V4–IT form a chain of connections corresponding to the ventral stream, and V1–V4–PP form another chain of connected modules associated with the dorsal stream. In the formulation of the model, V4 refers to an area of visual cortex that follows V1 and has connections that lead eventually to both the inferior temporal (IT) module and to the posterior parietal (PP) module. 'V4' in the model could correspond in the brain to regions such as V2, V3 or V4, and this module is also referred to as the 'V2–V4' module.

9.2.2 The biased competition hypothesis

In Sections 6.2 and 6.3 the biased competition hypothesis was introduced. Our model implements this hypothesis at the microscopic level of neuronal pools and at the mesoscopic level of visual areas. In fact, we show in Section 9.3 computational simulations of our architecture, which demonstrate that our model is consistent with experimental observations of biased competition effects, as measured in both single cells and fMRI experiments. At the neuronal pool level, dynamical competition is implemented by using the basic module introduced in Section 7.8.

Intermodular competition and mutual biasing result from the interaction between modules corresponding to different visual areas. In particular, we will show how feature-based attention biases intermodular competition between V4 and IT, whereas spatial attention biases intermodular competition between V1, V4 and PP. There is evidence for different concurrent segregated attentional modulation effects in the two visual 'what–where' streams produced during the performance of object recognition as compared with spatial tasks. (This evidence complements that described in Sections 6.2 and 6.3.) The fMRI study of Corbetta and Shulman (1998) is consistent with the hypothesis that the dorsal fronto-parietal network controls the allocation of spatial attention. More specifically, performance in a spatial attentional task correlates with parietal cortex activation along the post-central and intraparietal sulci, and with frontal cortex activation in the precentral sulcus/gyrus and the posterior tip of the superior frontal sulcus. Further, in tasks which involve object discrimination (Corbetta and Shulman 1998), the dorsal fronto-parietal network is active concurrently with the ventral occipito-temporal regions involved in object analysis. In addition, it is known that objects presented at an attended location are discriminated more accurately, and produce stronger blood flow responses, in ventral regions than targets presented at unattended locations (Vandenberghe, Dupont, Debruyn, Bormans, Michiels, Mortelmans and Orban 1996, Vandenberghe, Duncan, Dupont, Ward, Poline, Bormans, Michiels, Mortelmans and Orban 1997). Further, single cell studies in awake, behaving monkeys have shown that cells in V1, V2, V4 and IT can be modulated by attention (Colby 1991) (see Chapter 6). Finally, the effective receptive field of neurons in V4 can be dynamically modulated by the spatial location of the attentional focus (see Chapter 6 and e.g. Connor, Gallant and Van Essen (1993)).

9.2.3 Neuronal receptive fields

The pioneering work of Hubel and Wiesel (1962) about the organization of the primary visual cortex has given impetus to the theoretical and experimental research of the response properties of the cells in this area.

The theoretical investigations of Daugman (1988) and Marcelja (1980) proposed that simple cells in the primary visual cortex can be modelled by 2D-Gabor functions. There is a trade off between the resolution that can be achieved in the spatial and the spatial frequency domain (Daugman 1997). The 2D-Gabor function achieves the optimal resolution limit in the conjoint spatial and spatial frequency domain. The Gabor receptive fields have five degrees of freedom given essentially by the product of an elliptical Gaussian and a complex plane wave (see Chapter 2 for a further description). The first two degrees of freedom are the 2D-locations of the receptive field's centre; the third is the size of the receptive field; the fourth is the orientation of the boundaries separating excitatory and inhibitory regions; and the fifth is the symmetry. This fifth degree of freedom is given in the standard Gabor transformation by the real and imaginary part, i.e by the phase of the complex function representing it, whereas in a biological context this can be done by combining pairs of neurons with even or odd receptive fields (Daugman 1988). This design is supported by the experimental work of Pollen and Ronner (1981), who found simple cells in quadrature-phase pairs. Figure 9.2 shows examples of 2D-Gabor receptive fields at a specific position, for two different frequencies and eight orientations.

Even more, Daugman (1988) proposed that an ensemble of simple cells is best modelled as a family of 2D-Gabor wavelets sampling the frequency domain in a log-polar manner. Experimental neurophysiological evidence constrains the relation between the free parameters that define a 2D-Gabor receptive field (De Valois and De Valois 1988, Kulikowski and Bishop 1981, Webster and De Valois 1985). There are three constraints fixing the relation between the width, height, orientation, and spatial frequency (Lee 1996). The first constraint posits that the aspect ratio of the elliptical Gaussian envelope is 2:1. The second constraint postulates that the plane wave tends to have its propagating direction along the short axis of the elliptical Gaussian. The third constraint assumes that the half-amplitude bandwidth of the frequency response is about 1 to 1.5 octaves along the optimal orientation. Further, we assume that the mean is zero in order to have an admissible wavelet basis (Lee 1996). The V1 neuronal pools implemented in the model described in this Chapter consist of an ensemble of simple cells whose receptive fields correspond to a 2D-Gabor function sensitive to a particular location, orientation, symmetry, and spatial frequency. We have noted in Chapter 2 that a difference of Gaussian specification of simple cell-like receptive fields (used in the model described in Chapter 8) produces very similar receptive field shapes to those produced by the 2D-Gabor wavelet approach used in the model described in this Chapter.

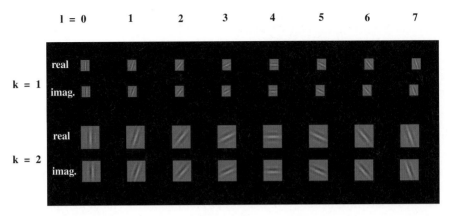

Fig. 9.2 2D-Gabor Wavelet receptive fields for eight different orientations (l) and for two octaves (k). For each octave the real part (top row) and the imaginary (imag.) part (bottom row) are plotted (corresponding to even and odd receptive fields). Simple cells in the primary visual cortex can be modelled by 2D-Gabor functions.

9.3   Neurodynamical architecture of a model of spatial and object based attention

In order to analyze the principles of operation of attentional systems in the brain, we describe in this Section a cortical model of visual attention for object recognition and visual search based on the neurophysiological constraints described in the previous Section. The system is absolutely autonomous and each component of its functional behaviour is explicitly described in a complete mathematical framework.

9.3.1   Overall architecture of the model

The overall systems-level representation of the model is shown in Fig. 9.3. The system is essentially composed of six modules structured such that they resemble the two known main visual paths of the mammalian visual cortex. Information from the retino–geniculo–striate pathway enters the visual cortex through area V1 in the occipital lobe and proceeds into two processing streams. The occipital-temporal stream leads ventrally through V2–V4 and IT (inferotemporal cortex) and the occipito-parietal stream leads dorsally into PP (posterior parietal complex). In this model system, we choose to model with our ventral stream one particular aspect of the inferior temporal cortex function: translation invariant object recognition. (A fuller model of the processing that leads to invariant object representations is provided in Chapter 8). We choose to model with our dorsal stream one particular aspect of parietal cortex function: encoding of visual space in retinotopic coordinates. This is obviously a great simplification of the complex hierarchical architecture and functions of the primate visual system (Felleman and Van Essen 1991) (see Chapters 2–5 and 8), but the model is sufficient to test and demonstrate our fundamental proposals for how the object and spatial streams interact to implement visual attentional processes.

First, we give an overview of the entire architecture. The ventral stream consists of four modules as shown in Fig. 9.3: 1) a V1 module; 2) a V2–V4 module; 3) an IT module;

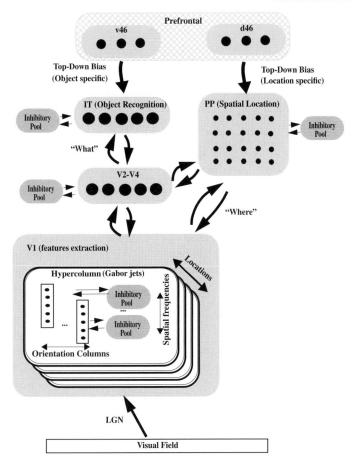

Fig. 9.3 The systems-level architecture of a model of the cortical mechanisms of visual attention. The system is essentially composed of six modules structured such that they resemble the two known main visual pathways of the primate visual cortex. Information from the retino-geniculo-striate pathway enters the visual cortex through area V1 in the occipital lobe and proceeds into two processing streams. The occipital-temporal stream leads ventrally through V2–V4 and IT (inferotemporal), and is mainly concerned with object recognition. The occipito-parietal stream leads dorsally into PP (posterior parietal complex) and is responsible for maintaining a spatial map of an object's location.

4) a module v46 corresponding to the ventral part of area 46 of the prefrontal cortex that maintains a short-term memory of a recognized object or holds a representation of the target object in a visual search task. The V1 module contains $P \times P$ hypercolumns, covering a $N \times N$ pixel scene. Each hypercolumn contains L orientation columns of complex cells with K octave levels corresponding to different spatial frequencies. The complex cells are modelled by the power modulus of Gabor wavelets. There is one inhibitory pool interacting with complex cells of all orientations at each scale. This module sends spatial and feature information up to the dorsal stream and the ventral stream respectively. It also provides a high-resolution representation for the two streams to interact through recurrent feedback. This interaction between the streams may be important for several functions such as binding,

and high-resolution visual analysis. We will first focus our treatment on the possible role for the interaction in mediating spatial and object attention in visual search and object recognition tasks. The V2–V4 module serves primarily to pool and channel the responses of V1 neurons to IT, to achieve a limited degree of translation invariance. It also implements a certain degree of localized competitive interaction between different targets. We use a topologically organized lattice to represent V2 or V4. Each node in this lattice has $L \times K$ cell assemblies as in a hypercolumn in V1. Each cell assembly, however, receives convergent input from the cell assemblies with the same tuning from a $M \times M$ hypercolumn neighbourhood in V1. The feedforward connections from V1 to the V4 module are modelled with convergent Gaussian weight functions, with symmetric recurrent connections. The IT module contains C pools, as the network is trained to search for or recognize C particular objects. Each of these cell assemblies is fully connected to each of the V4 pools. The connection weights from V4 to IT are trained by Hebbian learning rules in a learning phase.

The dorsal stream consists of three modules: 1) the V1 module; 2) a PP module; 3) a module d46 corresponding to the dorsal part of area 46 of prefrontal cortex that maintains a representation of a spatial location in a short term spatial memory. This d46 module not only provides a short term memory function, but can be used to generate an attentional bias for a spatial location. The PP module is responsible for mediating spatial attention modulation and for updating the spatial position of the attended object. A lattice of $N \times N$ nodes represents the topographical organization of module PP. Each node in the lattice corresponds to the spatial position of each pixel in the input image. Each of these assemblies monitors the activities from hypercolumns in V1–V2 via a Gaussian weighting function that connects topologically corresponding locations.

Prefrontal cortex area 46 (modules d46 and v46) is not explicitly simulated in the current version of the model, but is part of the model. We assume that top-down feedback connections from these areas provide the external top-down bias that specifies the processing conditions of earlier modules. Concretely, the feedback connection from area v46 with the IT module specifies the target object in a visual search task; and the feedback connection from area d46 with the PP module generates the bias to a targeted spatial location in a recognition task at a fixed prespecified location.

The system operates in two different modes: the learning mode and the recognition mode. During the learning mode the synaptic connections between V4 and IT are trained by means of Hebbian learning during several presentations of a specific object at changing random positions in the visual field. This is the simple way in which translation invariant representations are produced in IT in this model. During the recognition mode there are two possibilities for running the system, illustrated in Fig. 9.4.

First, in **visual spatial search mode** (Fig. 9.4b), an object can be found in a scene by biasing the system with an external top-down (backprojection) component (from e.g. prefrontal area v46) to the IT module. This drives the competition in IT in favour of the pool associated with the specific object to be searched for. Then, the intermodular backprojection attentional modulation IT–V4–V1 will enhance the activity of the pools in V4 and V1 associated with the component features of the specific object to be searched for. This modulation will add to the visual input being received by V1, resulting in greater local activity where the features in the topologically organized visual input features match the backprojected features being facilitated. Finally, the enhanced firing in a particular part of V1 will lead to increased activity in the forward pathway from V1 to V2–V4 to PP, resulting in increased firing in the PP

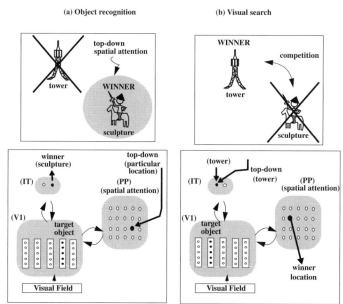

Fig. 9.4 (a) Attentional modulation for finding an object (in visual object identification mode) at a specific spatial location using spatial bias to the PP module from for example a prefrontal module d46. See text for details. (b) Attentional modulation for finding a visual spatial location (in visual spatial search mode) when an object is specified by bias to the IT module from for example a prefrontal module v46.

module in the location that corresponds to where the object being searched for is located. In this way, the architecture automatically finds the location of the object being searched for, and the location found is made explicit by which neurons in the spatially organized PP module are firing.

Second, in **visual object identification mode** (Fig. 9.4a), the PP module receives a top-down (backprojection) input (from e.g. prefrontal area d46) which specifies the location in which to identify an object. The spatially biased PP module then drives by its backprojections the competition in the V2–V4 module in favour of the pool associated with the specified location. This biasing effect in V1 and V2–V4 will bias these modules to have a greater response for the specified location in space. The shape feature representations which happen to be present due to the visual input from the retina at that location in the V1 and V2–V4 modules will therefore be enhanced, and the enhanced firing of these shape feature will by the feedforward pathway V1–V4–IT favour the IT object pool that contains the facilitated features, leading to recognition in IT of the object at the attentional location being specified in the PP module. The operation of these two attentional modes is shown schematically in Fig. 9.4.

9.3.2 Formal description of the model

We now present a formal description of the model (Deco and Lee 2001). We consider a pixelized grey-scale image given by a $N \times N$ matrix $\Gamma_{ij}^{\text{orig}}$. The subindices ij denote the spatial position of the pixel. Each pixel value is given a grey level brightness value coded in a scale between 0 (black) and 255 (white). The first step in the preprocessing consists of removing the DC component of the image (i.e. the mean value of the grey-scale intensity of the pixels). (The equivalent in the brain is the low-pass filtering performed by the retinal ganglion cells and lateral geniculate cells, as described in Chapter 2. The visual representation in the LGN is essentially a contrast invariant pixel representation of the image, i.e. each neuron encodes the relative brightness value at one location in visual space referred to the mean value of the image brightness). We denote this contrast invariant LGN representation by the $N \times N$ matrix Γ_{ij} defined by the equation

$$\Gamma_{ij} = \Gamma_{ij}^{\text{orig}} - \frac{1}{N^2} \sum_{i=1}^{N} \sum_{j=1}^{N} \Gamma_{ij}^{\text{orig}}. \tag{9.1}$$

Feedforward connections to a layer of V1 neurons perform the extraction of simple features like bars at different locations, orientations and sizes. In Section 9.2.3, we argued that realistic receptive fields for V1 neurons that extract these simple features can be represented by 2D-Gabor wavelets. Lee (1996) derived a family of discretized 2D-Gabor wavelets that satisfy the wavelet theory and the neurophysiological constraints for simple cells mentioned in Section 9.2.3. They are given by an expression of the form

$$G_{kpql}(x, y) = a^{-k} \Psi_{\Theta_l}(a^{-k}(x - 2p) - a^{-k}(y - 2q)) \tag{9.2}$$

where

$$\Psi_{\Theta_l} = \Psi(x \cos(l\Theta_0) + y \sin(l\Theta_0), -x \sin(l\Theta_0) + y \cos(l\Theta_0)), \tag{9.3}$$

and the mother wavelet is given by

$$\Psi(x, y) = \frac{1}{\sqrt{2\pi}} e^{-\frac{1}{8}(4x^2 + y^2)} [e^{i\kappa x} - e^{-\frac{\kappa^2}{2}}]. \tag{9.4}$$

In the above equations $\Theta_0 = \pi/L$ denotes the step size of each angular rotation; l the index of rotation corresponding to the preferred orientation $\Theta_l = l\pi/L$; k denotes the octave; and the indices pq the position of the receptive field centre at $c_x = p(N/P)$ and $c_y = q(N/P)$. In this form, the receptive fields at all levels cover the spatial domain in the same way, i.e. by always overlapping the receptive fields in the same fashion. In the model we use $a = 2, b = 1$ and $\kappa = \pi$ corresponding to a spatial frequency bandwidth of one octave.

The neurons in the pools in **V1** have receptive fields performing a Gabor wavelet transform. Let us denote by I_{kpql}^{V1} the sensory input activity to a pool A_{kpql}^{V1} in V1 which is sensitive to a spatial frequency at octave k, to a preferred orientation defined by the rotation index l, and to stimuli at the centre location specified by the indices pq. The sensory input activity to a pool in V1 is therefore defined by the modulus of the convolution between the corresponding receptive fields and the image, i.e.

$$I_{kpql}^{V1} = \sqrt{\|\langle G_{kpql}, \Gamma \rangle\|^2} = \sqrt{\left\| \sum_{i=1}^{N} \sum_{j=1}^{N} G_{kpql}(i, j)\Gamma_{ij} \right\|^2} \tag{9.5}$$

and is normalized to a maximal saturation value of 0.025. Since in our numerical simulations the system needs only to learn a small number of objects (usually 2–4), we temporarily did not include the V2 and V4 module for simplicity in some of the simulations. We did however include an explicit module V2–V4 for the simulations corresponding to measures in brain areas V2 and V4 presented in Section 9.4.1. In the simulations where the V2–V4 module was not present, the V1 and IT cell assemblies were directly connected together, with full connectivity. In fact, the large receptive fields of V2 and V4 can be approximately taken into account by including them in V1 pools with receptive fields corresponding to several octaves of the 2D-Gabor transform wavelets (i.e. not only the typical narrow receptive fields of V1 but also larger receptive fields are included in the modelled V1). The reduced system connects all cell assemblies in V1 with all cell assemblies in IT. In contradistinction to the suggestion of Salinas and Abbott (1997), and to the mechanisms described in Chapter 8 for producing translation invariance, in this particular model translation invariance was implemented by an attentional intermodular biasing interaction between pools in modules V1–V4 and PP. For each V1–V4 neuron, the gain modulation observed decreases as the actual location where attention is focused moves away from the centre of the receptive field in a Gaussian-like way (Connor, Gallant and Van Essen 1993). Consequently, the connections with the pools in the PP module are specified such that the modulation is Gaussian. Let us define in the PP module a pool A_{ij}^{PP} for each location ij in the visual field. The mutual (i.e. forward and back) connections between a pool A_{kpql}^{V1} in V1 (or V4) and a pool A_{ij}^{PP} in PP are therefore defined by

$$
w_{pqij} = A \exp \left\{ -\frac{(i-p)^2 + (j-q)^2}{2S^2} \right\} - B \tag{9.6}
$$

These connections mean that the V1 pool A_{kpql}^{V1} will have maximal amplitude when spatial attention is located at $i = p$ and $j = q$ in the visual field, i.e. when the pool A_{ij}^{PP} in PP is maximally activated and provides an inhibitory contribution $-B$ at the locations not being attended to. In our simulations, we always used $S = 2$, $A = 1.5$ and $B = 0.1$. The V1–PP attentional modulation, in combination with the Hebbian learning that we will define later in this Section, generate translation-invariant recognition pools in the module IT.

Let us now define the neurodynamical equations that regulate the temporal evolution of the whole system. (The dynamics correspond to those described in Section 7.8.1.)

The activity level of the input current in the **V1 module** is given by

$$
\tau \frac{\partial A_{kpql}^{V1}(t)}{\partial t} = -A_{kpql}^{V1} + \alpha F(A_{kpql}^{V1}(t)) - \beta F(A_k^{I,V1}(t)) + I_{kpql}^{V1}(t)
$$
$$
+ I_{pq}^{V1-PP}(t) + I_{kpql}^{V1-IT}(t) + I_0 + \nu \tag{9.7}
$$

where the attentional biasing coupling I_{pq}^{V1-PP} due to the intermodular 'where' connections with the pools in the parietal module PP is given by

$$
I_{pq}^{V1-PP} = \sum_{i,j} W_{pqij} F(A_{ij}^{PP}(t)) \tag{9.8}
$$

and the attentional biasing term I_{kpql}^{V1-IT} due to the intermodular 'what' connections with the pools in the temporal module IT is defined by

$$
I_{kpql}^{V1-IT} = \sum_{c=1}^{C} w_{ckpql} F(A_c^{IT}(t)) \tag{9.9}
$$

w_{ckpql} being the connection strength between the V1 pool A_{kpql}^{V1} and the IT pool A_c^{IT} corresponding to the coding of a specific object category c. We assume that the IT module has C

pools corresponding to different object categories. For each spatial frequency level, a common inhibitory pool (designated with a superscript I) is defined. The current activity of these inhibitory pools obeys the following equations:

$$\tau_P \frac{\partial A_k^{I,V1}(t)}{\partial t} = -A_k^{I,V1}(t) + \gamma \sum_{p,q,l} F(A_{kpql}^{V1}(t)) - \delta F(A_k^{I,V1}(t)) \tag{9.10}$$

Similarly, the current activity of the excitatory pools in the **posterior parietal module PP** is given by

$$\tau \frac{\partial A_{ij}^{PP}(t)}{\partial t} = -A_{ij}^{PP} + \alpha F(A_{ij}^{PP}(t)) - \beta F(A^{I,PP}(t)) + I_{ij}^{PP-V1}(t)$$
$$+ I_{ij}^{PP,A} + I_0 + \nu \tag{9.11}$$

where $I_{ij}^{PP,A}$ denotes an external attentional spatially-specific top-down bias, and the intermodular attentional biasing I_{ij}^{PP-V1} through the connections with the pools in the module V1 is

$$I_{ij}^{PP-V1} = \sum_{k,p,q,l} W_{pqij} F(A_{kpql}^{V1}(t)) \tag{9.12}$$

and the activity current of the common PP inhibitory pool evolves according to

$$\tau_P \frac{\partial A^{I,PP}(t)}{\partial t} = -A^{I,PP}(t) + \gamma \sum_{i,j} F(A_{ij}^{PP}(t)) - \delta F(A^{I,PP}(t)) \ . \tag{9.13}$$

The dynamics of the **inferotemporal module IT** is given by

$$\tau \frac{\partial A_c^{IT}(t)}{\partial t} = -A_c^{IT} + \alpha F(A_c^{IT}(t)) - \beta F(A^{I,IT}(t)) + I_c^{IT-V1}(t)$$
$$+ I_c^{IT,A} + I_0 + \nu \tag{9.14}$$

where $I_c^{IT,A}$ denotes an external attentional object-specific top-down bias, and the intermodular attentional biasing I_c^{IT-V1} between IT and V1 pools is

$$I_c^{IT-V1} = \sum_{k,p,q,l} w_{ckpql} F(A_{kpql}^{V1}(t)) \tag{9.15}$$

and the activity current of the common PP inhibitory pool evolves according to

$$\tau_P \frac{\partial A^{I,IT}(t)}{\partial t} = -A^{I,IT}(t) + \gamma \sum_c F(A_c^{IT}(t)) - \delta F(A^{I,IT}(t)) \ . \tag{9.16}$$

In our simulations, we use $\alpha = 0.95$, $\beta = 0.8$, $\gamma = 1$, $\delta = 0.1$, $I_0 = 0.025$, and the standard deviation of the additive noise ν, $\sigma_\nu = 0.02$. The values of the external bias $I_{ij}^{PP,A}$ and $I_c^{IT,A}$ are equal to 0.07 for the pools that eventually receive an external positive bias and otherwise are equal to zero. The choice of these parameters is uncritical and is based on biological parameters.

During a **learning phase** each object (or object category) is learned. This is done by training the connections between the modules V1 and IT by Hebbian learning. In order to achieve invariant translation the different objects are presented at random locations. The external attentional location-specific bias in PP $I_{ij}^{PP,A}$ is set so that only the pool ij corresponding to the spatial location of the object to be learned receives a positive bias. In this way the spatial attention defines the localization of the object to be learned. The external attentional object-specific bias in IT $I_c^{IT,A}$ is set similarly, so that only the pool c that will identify the object, receives a positive bias. We therefore define the identity of the object in a supervised way. After presentation of a given stimulus, i.e. a specific object at a specific location, and the corresponding external bias, the system evolves until convergence. After convergence, the V1–IT connections w_{ckpql} are trained through the following Hebbian rule

$$w_{ckpql} = w_{ckpql} + \eta F(A_c^{IT}(t))F(A_{kpql}^{V1}(t)) \tag{9.17}$$

(t) being large enough (i.e. after convergence) and η the learning coefficient. This procedure is repeated for all objects at all possible locations until the weights converge.

During the **recognition phase** there are two possibilities, namely to search for a specific object (visual search), or to identify an object at a given spatial location (object recognition), as shown in Fig. 9.4.

In the case of **visual search** the external attentional object-specific bias in IT $I_c^{IT,A}$ is set so that only the pool c corresponding to the category of the object to be searched for receives a positive bias, while the external attentional location-specific bias in PP $I_{ij}^{PP,A}$ is set equal to zero everywhere. The external attentional bias $I_c^{IT,A}$ drives the competition in the IT module so that the pool corresponding to the searched for object wins. The intermodular attentional modulation between IT and V1 biases the competition in V1 so that all features detected from the retinal inputs at different positions that are compatible with the specific object to be searched for will now win. Finally the intermodular attentional bias between V1 and PP drives the competition in V1 and in PP so that only the spatial location in PP and the associated V1 pools compatible with the presented stimulus and with the top-down specific category of the object to be searched for will remain active after convergence. At convergence, the final state in PP shows where the object has been found. In this form, the final activation state is neurodynamically driven by the stimulus, external top-down bias, and intermodular bias, in an entirely parallel way. Attention is not a mechanism involved in the competition but just an emergent effect that supports the dynamical evolution to a state where all constraints are satisfied.

In the case of **object recognition**, the external attentional bias in PP $I_{ij}^{PP,A}$ is set so that only the pool associated with the spatial location where the object to be identified is receives a positive bias, i.e. a spatial region will be 'illuminated'. The other external bias $I_c^{IT,A}$ is zero everywhere. In this case, the dynamics evolves such that in PP only the pool associated with the top-down biased spatial location will win. This fact drives the competition in V1 such that only the pools corresponding to features of the stimulus at that location will win, biasing the dynamics in IT such that only the pool identifying the class of the features at that position will remain active indicating the category of the object at that predefined spatial location.

9.3.3 Performance measures

Deco and Lee (2001) used two measures of neural activities to characterize the temporal evolution of the units in different modules.

The first measure is the population average response of neuronal pools within a certain neighbourhood. This measure has often been used to describe data in neurophysiological experiments (Lee, Mumford, Romero and Lamme 1998).

The second measure is the population *maximum* activity of the neuronal pools associated with the target and with the distractors. To compute the maximum activity, for each point in time, we take the maximum response in any pool within a specified spatial neighbourhood at that time. More precisely, the maximum activity of pools within a spatial neighbourhood associated with the target T is defined by:

$$A_T^{PP} = \max(A_{ij}^{PP}) \text{ for } ij \text{ such that } \text{dist}(ij, xy) < 4$$

$$A_T^{V1} = \max(\sum_{k,l} A_{klpq}^{V1}) \text{ for } pq \text{ such that } \text{dist}(((N/P)p)((N/P)q), xy) < 4$$

$$A_T^{IT} = \max(A_c^{IT}) \text{ for } c \text{ corresponding to the target}$$

where xy denotes the location of the centre of the target, and ij and pq denote the spatial locations of the V1 module pools and the PP module pools in the same spatial coordinates. dist() is the Euclidean distance in the 2D map. The maximum distractor (D) activities are defined by:

$$A_D^{PP} = \max(A_{ij}^{PP}) \text{ for } ij \text{ such that } \text{dist}(ij, xy) \geq 4$$

$$A_D^{V1} = \max(\sum_{m,l} A_{mlpq}^{V1}) \text{ for } pq \text{ such that } \text{dist}(((N/P)p)((N/P)q), xy) \geq 4$$

$$A_D^{IT} = \max(A_c^{IT}) \text{ for } c \text{ corresponding to non-targets}$$

This measure is more sensitive for detecting and isolating the time point at which the neural activity becomes different under different conditions.

9.4 Simulations of basic experimental findings

According to the biased competition hypothesis (see Sections 6.2 and 6.3) multiple stimuli in the visual field activate populations of neurons that engage in competitive interactions. Moreover, attending to a stimulus at a particular location biases this competition in favour of neurons that respond to the location of the attended stimulus. Different experimental results seem to support these ideas. In particular, we described in Section 6.2 the results of Reynolds, Chelazzi and Desimone (1999) from single cell recording in V2 neurons in monkeys, which show inhibitory competitive effects and attentive biasing modulation at the neuronal level. In addition, we reviewed in Section 6.3 the fMRI studies in humans of Kastner, Pinsk, De Weerd, Desimone and Ungerleider (1999), which are consistent with the idea of biased competition at the larger scale (i.e. mesoscopic) level in which the activation of more gross regions of the brain is measured.

In this Section, we show that simulations based on the model introduced in Section 9.3, explain the results observed in both experiments. Consequently, the dynamical behaviour of the multimodular architecture described is Section 9.3 provides an account of attentional processes in vision that is consistent with the biased competition hypothesis at the (microscopic) level of neuronal firing and at the level of the activation of different large regions of cortex.

9.4.1 Simulations of single-cell experiments

In this Section, we describe the simulation results with the model described in Section 9.3 that correspond to the experiments by Reynolds, Chelazzi and Desimone (1999) on single cell recordings from V2 neurons in monkeys. We describe the dynamical behaviour of the model of the cortical architecture of visual attention by solving numerically the system of coupled differential equations in a computer simulation. For this experiment we did use a module of V2 neurons. The input system processed an input image of 66×66 pixels ($N = 66$). The V1 hypercolumns covered the entire image uniformly. They were distributed in 33×33 locations ($P = 33$) and each hypercolumn was sensitive to two spatial frequencies and to eight different orientations (i.e., $K = 2$ and $L = 8$). The V2 module had 2×8 pools receiving convergent input from the pools of the same tuning from a 10×10 (i.e., $M = 10$) hypercolumn in the neighbourhood in V1. The feedforward connections from V1 to V2 are modelled with convergent Gaussian weight functions, with symmetric recurrent connections. We analyzed the firing activity of a single pool in the V2 module which was highly sensitive to a vertical bar presented in its receptive field (the effective stimulus) and poorly sensitive to a 75 degree oriented bar presented in its receptive field (ineffective stimulus). The size of the bars was 2×4 pixels. Following the experimental setup reviewed in Section 6.2.2, we plot in Figs. 9.5a and 9.5b the development of the firing activity of a V2 pool under four different conditions: 1) with a single reference stimulus within the receptive field; 2) with a single probe stimulus within the receptive field; 3) with a reference and a probe stimulus within the receptive field and without attention; and 4) with a reference and a probe stimulus within the receptive field and with attention directed to the spatial location of the reference stimulus. In our simulation, spatial attention was directed to the location of the reference stimulus by setting in module PP the top-down attentional bias coming from prefrontal area d46 equal to 0.05 if i and j corresponded to the location of the reference stimulus and zero elsewhere. In the unattended condition the external top-down bias was zero everywhere. The computational simulations of Fig. 9.5a and 9.5b should be compared with the experimental results shown in Fig. 6.4a and 6.4b, respectively. The same qualitative behaviour is observed in the model and the real neuronal data in all experimental conditions. The competitive interactions in the absence of attention are due to the intramodular competitive dynamics at the level of V1 (i.e., the suppressive and excitatory effect of the probe shown in Fig. 9.5a and 9.5b, respectively). The modulatory biasing corrections in the attended conditions are caused by the intermodular interactions between the V1 and PP pools, and between the PP pools and prefrontal top-down modulation.

We are also able to account for the findings of the experiments by Connor, Gallant and Van Essen (1993) and Connor, Gallant, Preddie and Van Essen (1996), which showed that the locus of spatial attention can modulate the structure of the receptive fields of V4 neurons. By asking the monkeys to discriminate subtle changes in features in a particular point in visual space, they managed to shift the hot spot of the receptive field of V4 neurons in the neighborhood of the location being fixated towards the spatial focus of attention. For these simulations, we used two cell pools in the V4 module. One pool received input from the vertically oriented cells in the V1 module. The other pool received input from the horizontally

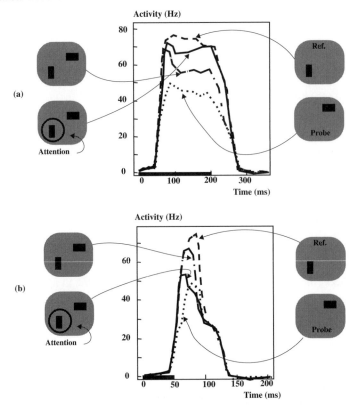

Fig. 9.5 Simulation of the experiment of Reynolds et al. (1999). (a) The stimulus was presented for 200 msec. When an optimal reference stimulus was presented alone, the cell's response (dashed line) was much stronger than its response (dotted line) when a suboptimal probe stimulus was presented alone in the receptive field. Simultaneous presentation of both the reference and the probe stimuli produced an intermediate response (dashed-dotted line), indicating that the probe was producing competitive suppression of the response to the reference stimulus. However, when spatial attention was directed toward to the reference stimulus, the suppression due to the probe was largely eliminated: the neuronal response returned to the level when the reference was presented alone (continuous line). (b) The same manipulation as in (a) except that the stimulus was presented for only 50 ms. In this simulation, we used $I_o = 0$, and slower dynamics, $\tau = 15$ ms.

oriented cells in the V1 module. In our simulation, spatial attention was again specified by introducing a bias to the appropriate cell pool in the PP module. Figure 9.6 shows that the central 'hot spot' peak of the receptive field of V4 neurons was shifted spatially as different PP module pools received the top-down bias.

In conclusion, the dynamical evolution of the firing activity at the neuronal level of our model of the cortical architecture of visual attention is consistent with the single cell experiments of Reynolds, Chelazzi and Desimone (1999) and Connor, Gallant, Preddie and Van Essen (1996). This indicates that the particular computational model we propose, of how biased competition and interaction between the 'what' and 'where' visual processing streams can be used to understand visual attention, is able to account for data at the neuronal level. This represents an advance beyond the biased competition hypothesis, in that it shows how

Attentional Shifting of Receptive Field

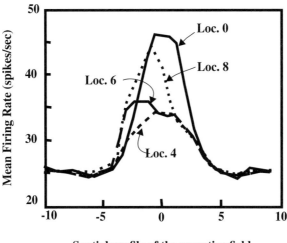

Spatial profile of the receptive field.

Fig. 9.6 Simulation of the experiment by Connor et al. (1993) showing that spatial attention can move the hot spot of the receptive field of a V4 neuron towards the location which is the focus of attention (if the receptive field is close to the focus of attention). The connection between the early V1 module and V4 was modelled with spatially local Gaussian weights centred at location 10, with standard deviation equal to 2.5. The V4 neuron had a more local spatial support than the ventral IT module's neuron in the other simulations in this series. Spatial attention allocated in the dorsal PP module pools corresponding to retinotopic locations 6 and 8 effectively shifted the peak of the receptive field of the V4 neuron to the left. Shift to the right can be produced by biasing locations 12 and 16 (not shown). Location 4 was far away from the centre of the receptive field. Allocating spatial attention to location 4 made the hot spot of the receptive field snap back to its centre position, but the responses of the V4 cell were markedly attenuated.

object and spatial attention can be produced by dynamic interactions between the 'what' and 'where' streams, and in that as a computational model which has been simulated, the details of the model have been made fully explicit and have been defined quantitatively.

9.4.2 Simulations of fMRI experiments

The dynamic evolution of activity at the cortical level, as shown for example by behaviour of fMRI signals in experiments with humans, can be simulated in the framework of our model by integrating the pool of neuronal activity in a given area over space and time (Corchs and Deco 2001). The integration over space yields an average activity of the brain area considered at a given time. (The spatial resolution in most fMRI experiments is worse than 3 mm in any one dimension.) The integration over time is performed in order to simulate the relatively coarse temporal resolution of MRI experiments, which is in the order of a few seconds. In this Section, we simulate fMRI signals from V4 under the experimental conditions defined by Kastner, Pinsk, De Weerd, Desimone and Ungerleider (1999) (see Section 6.3). We use the same parameters as in the last Section with the only difference that the V1 hypercolumns include now three levels of spatial resolution (i.e., $K = 3$). The integrated activity over space at a given time t is given by

$$A^{V4}(t) = \sum_{k,p,q,l} A^{V4}_{kpql}(t) \quad . \tag{9.18}$$

The temporal evolution is plotted at a coarse timescale of 2 seconds that emulates the temporal resolution of fMRI experiments. The integration over time in a time window of 2000 ms is given by

$$A^{V4}(T) = \sum_{t=T-2000}^{T} A^{V4}(t) \tag{9.19}$$

since the differential equations are solved iteratively at the ms scale. Following the experimental paradigm described in Section 6.3, we also use four complex images similar to the ones that Kastner, Pinsk, De Weerd, Desimone and Ungerleider (1999) used in their experiments. They were presented as input images in four nearby locations in the upper right quadrant. The stimuli were shown in the conditions described in Sections 6.3.1 and 6.3.2. The images were shown in randomized order under two presentation conditions: sequential (SEQ) and simultaneous (SIM). In the SEQ condition, each stimulus was shown alone in one of the four locations for 250 ms. In the SIM condition, the stimuli appeared together in all four locations for the first 250 ms, followed by a blank period of 750 ms. This stimulation period of 1 s was repeated for each condition in blocks of 10 sec. interleaved with blank periods of 20 s (BLK). Two attentional conditions were simulated: 1) an unattended condition, during which no external top-down bias from prefrontal areas was present (i.e., $I^{\text{PP},\text{A}}_{ij}$ is zero everywhere); 2) an attended condition which started 10 s before the onset of visual presentation (the expectation period EXP) and continued during the subsequent 10 s block. The attended condition was implemented by setting $I^{\text{PP},\text{A}}_{ij}$ equal to 0.07 for the locations associated with the lower left stimulus and zero elsewhere. Figure 9.7 shows the results of our computational simulations for a sequential stimulation block: BLK–EXP–SEQ(attended)–BLK–SIM–BLK–EXP–SIM(attended)–SEQ–BLK. These results should be contrasted with the experimental fMRI signal evolution shown in Fig. 6.6 for area V4.

As found in the experiments of Kastner, Pinsk, De Weerd, Desimone and Ungerleider (1999), the simulations also showed that the fMRI signals were smaller during the SIM than during the SEQ presentations in the unattended conditions, because of the mutual suppression induced by competitively interacting stimuli. On the other hand, the average fMRI signals with attention increased more strongly for simultaneously presented stimuli than the corresponding ones for sequentially presented stimuli. Thus, the suppressive interactions were partially cancelled out by attention. Finally, during the expectation period activation increased in the absence of visual presentations, and further increased after the onset of visual stimuli.

In summary, these results demonstrate that our cortical attentional architecture shows and explains the typical dynamical competition and attention modulating effects found in attention experiments, even at the level of gross brain area activation as measured with fMRI.

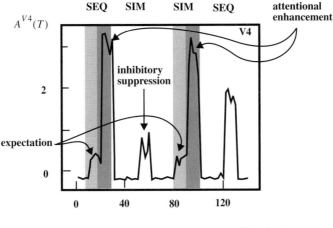

Fig. 9.7 Computer simulations of fMRI signals in module V4 based on the experiments of Kastner et al. (1999). Light shades indicate the expectation period, dark shades the attended presentations, and blocks without shading correspond to the unattended condition. See text and Section 6.3 for details.

9.5 The role of attention in object recognition and visual search

We concentrate now on the macroscopic[28] level of psychophysics. We study the interplay between the microscopic neuronal dynamics and the macroscopic functional behaviour in the particular context of object recognition and visual search. These two different functions of visual perception can be explained in a unifying fashion by our neurodynamical model of the visual cortical system (as described in Section 9.3).

A phenomenological description of these two perceptual functions is schematically presented in Fig. 9.8 in the context of a natural scene.

In the case of object recognition, a particular location in the natural scene is *a priori* specified with the aim of identification of the object which lies at that position. Therefore, object recognition asks for 'what' is at a predefined particular spatial location. In the naive framework of the spotlight metaphor, one can describe the role of attention in object recognition by imagining that the prespecification of the particular spatial location is realized by fixing an attentional window or spotlight at that position. The features inside the fixed attentional window should now be bounded and recognized. On the other hand, in visual search, a given target object (composed of a set of shape features) is *a priori* specified with the goal of finding out whether the target object is present in the scene, and if so, at which location. Consequently, visual search asks for 'where' a predefined set of shape features is located. A naive description of visual search in the framework of the spotlight paradigm considers that during visual search, attentional mechanisms shift a window through the entire scene in

---

[28]Physicists distinguish levels of analysis, and we use microscopic to refer to the neuronal level, mesoscopic to refer to an intermediate level of activation of brain areas as shown for example by fMRI, and macroscopic to refer to the behaviour of the whole organism as measured for example in psychophysical experiments.

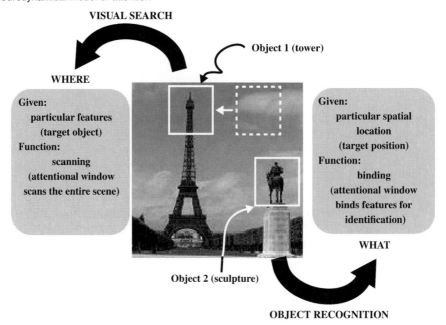

Fig. 9.8 The role of attentional mechanisms in object recognition and visual search in a natural scene. In the case of object recognition, a particular location in the natural scene is *a priori* specified with the aim of identification of the object which lies at that position. On the other hand, in visual search, a given target object (i.e. shape features) is *a priori* specified with the goal of finding out whether the target object is present in the scene, and if so, at which location.

order to serially search for the target object at different positions. The neurodynamical system described in Section 9.3 was tested in these two modes of operation: **object recognition** in an attended spatial location (spatial attention); and **visual search** of a target object (object attention). The input image was the scene shown in Fig. 9.8. The objects to be identified and located were the sculpture, and the top of the tower as indicated in the image. In the following two sub-sections, we will describe the dynamics of object recognition and visual search under these two modes of attention.

The thrust of our computational neuroscience approach to attention, and the particular model we describe in Section 9.3, is that it shows how spatial and object attention mechanisms can be integrated and function as a unitary system in visual search and visual recognition tasks. The dynamical intra- and inter-modular interactions in our cortical model implement attentional top-down feedback mechanisms that embody a physiologically plausible system of active vision unifying in this way different perceptual functions. In our system, attention is a dynamical emergent property, rather than a separate mechanism operating independently of other perceptual and cognitive processes. The dynamical mechanisms that define our system work across the visual field in parallel, but due to the different latencies of settling due to the intra- and intermodular dynamics, actually exhibit temporal properties like those normally described as 'serial' or 'parallel' when used to describe visual search (see Chapter 10).

The model thus accounts for a number of the temporal properties of serial search tasks found psychophysically without either an explicit serial scanning process with an attentional spotlight, or a saliency map. The following two subsections 9.5.1 and 9.5.2 show how these properties arise. In doing this, the model offers new insight into how attention may actually work, and goes beyond the biased competition hypothesis not only by showing how the 'what' and 'where' visual processing streams interact, and how the system operates quantitatively, but also by providing a direct approach to the temporal properties of the system, leading to a new conceptualization of what was described previously as serial and parallel search.

9.5.1 Dynamics of spatial attention and object recognition

In the object recognition task, the system functioned in a spatial attention mode as shown in Fig. 9.4a (see Deco and Lee (2001)). Spatial attention was initiated by introducing a bias to a cell pool coding for a particular location. The input system processed a pixelized 66×66 image ($N = 66$). The V1 hypercolumns covered the entire image uniformly. They are distributed in 33×33 locations ($P = 33$) and each hypercolumn was sensitive to three spatial frequencies and to eight different orientations (i.e., $K = 3$ and $L = 8$). Consequently, the V1 module had 26,136 pools and three inhibitory pools. The IT module utilized had two pools and one common inhibitory pool. Finally, the PP module contained 4,356 pools corresponding to each possible spatial location, i.e. to each of the 66×66 pixels, and a common inhibitory pool. Two objects (also shown in Fig. 9.8) were isolated in order to define two categories to be associated with two different pools in the IT module. During the learning phase (see Section 9.3), these two objects were presented randomly and at random positions in order to learn translation invariant responses. The system required 1,000,000 different presentations for training the IT pools. We used $\tau = \tau_P = 10$ ms and $T_r = 1$ ms.

When the image (Fig. 9.8) was presented, the spatial bias interacted with the visual image inputs provided by the V1 module. If the hypercolumns in the V1 module 'designated' for spatial attention contained sufficiently strong neural activities, the activities in these hypercolumns would interact synergistically with the biased PP cell pool, so that over time that pool would dominate over all the other pools in the PP module, and the activities in the hypercolumns in the V1 module would be enhanced by the top-down bias from the PP module pool. This enhancement of neural activity highlighted the information in the attended location, effectively gating information in that area of the V1 module to the IT module for recognition (and in this way performing a type of shift invariance by using an attentional spatial modulation of early visual cortical processing). When the highlighted image patch contained one of the trained object classes, the activity of the IT cell pools started to polarize, resulting in only one cell pool surviving the competition. The winner indicated the object class being recognized, identifying 'what was where', or 'what from where'.

The actual simulation results just outlined are shown in Fig. 9.9, which shows the neuronal activities in the three modules during object recognition in the spatial attention mode[29]. The 'maximum' population activity of the three modules at the sculpture location was compared against the activity in all other locations when the spatial attention was allocated to the sculpture location by the PP spatial input bias. Spatial attention allocated in the dorsal stream PP module ultimately led to the dominance of the sculpture neurons' activity in the ventral stream IT module. The bifurcation of the maximum activities in different pools showed

---

[29] In the presentation of this and all subsequent simulations, we introduced a delay of 40 msec in the response of the early module neurons relative to stimulus onset. This is because physiologically there is typically a 40 msec delay between the presentation of the stimulus and the onset of V1 neurons' responses (Lee, Mumford, Romero and Lamme 1998) (see Section 7.6.5).

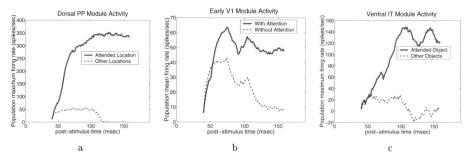

Fig. 9.9 Neuronal activity in the three modules of the trimodular architecture during object recognition in the spatial attention mode. (a) The maximum population activity of the top-down biased neuronal pool in the dorsal PP module was compared against the maximum activity of the other pools corresponding to the other locations in the visual scene. The bias introduced to a particular pool led to a rapid increase in the response of that pool relative to any other pools in the dorsal PP module. (b) The maximum population activity of pools in the highlighted area in the early V1 module was compared against the maximum activity of pools covering all the other locations in the scene. (c) The response of the sculpture neuronal pool (which was where spatial attention was directed) was compared against the maximum activity of all other object pools in the ventral IT module.

the propagation of the spatial attentional effect across the different modules. The effect of attention, as indicated by the polarization of the responses, started at the dorsal stream PP module, then propagated to the early stage V1 module, and then to the ventral stream IT module. The relative time of onset of the attentional effect was not distinguishable between the V1 module and the IT module because the attention-related computation in the V1 module and the IT module was concurrent and interactive. The polarization and stabilization of neuronal pool activity in the IT module corresponded to recognition of the object in the attended location. By moving the attentional spotlight to different spatial locations, the system can gate information at different spatial locations to the IT module realizing a form of translation-invariant object recognition.

Figure 9.10 illustrates the effect of spatial attention on the responses of cell pools in the three modules. The population mean spontaneous firing rate of pools at the sculpture location was plotted under two conditions: when spatial attention was introduced to the PP spatial pool corresponding to the sculpture location, and when no spatial attention was introduced. The population firing rate showed an elevation of the responses in the latter part of the responses of V1 module pools at the attended location. Interestingly, this relatively small elevation in the V1 module was sufficient to bias the IT module cell pools to produce a large polarization in their responses.

It is important to emphasize that no explicit spotlight of attention is forced onto every module in the system in our model. Instead, given just a bias to one of the modules (in this case the spatial PP module), parallel and global dynamical evolution of the entire system converges to a state where a spotlight of attention in PP appears, and is explicitly used (even while it is emerging) to modulate the information processing channel for object recognition. Effectively spatial effects influence the function of the non-spatially organized IT ventral stream module through modulatory effects on information processing implemented through the topologically organized V1 module. An interesting aspect of our theory is that the process of object recognition can be influenced by spatial attention arising from a global dynamical continuing

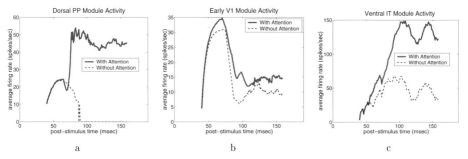

Fig. 9.10 Effect of spatial attention on neural activities in the three modules when performing object recognition. A top-down spatial attentional bias was introduced to the dorsal PP module pool that encodes the location of the sculpture in the scene (curves marked 'With Attention'), or there was no attentional input (curves marked 'Without Attention'). The responses of neuronal pools in the corresponding locations in the dorsal stream PP module and the V1 module, and the response of the neuronal pool encoding the sculpture in the ventral stream IT module were compared with and without spatial attention. (a) The evolution of the neuronal pool response at the dorsal stream PP module with and without the top-down spatial attentional bias. When the top-down bias was absent, the competition in the dorsal stream PP module was entirely determined by the bottom-up signals from V1. In this case, the sculpture shape did not provide the strongest input and was rapidly suppressed by the competitive interactions from the neuronal activities in the other locations in the scene. When there was a top-down spatial bias to the PP neuronal pool, the neuronal pool's activity rose rapidly and was maintained at a sustained level. (b) The increase in the activity of the neuronal pool in the dorsal stream PP module enhanced the activity at the corresponding retinotopic location in the early V1 module, particularly in the latter phase of the response, producing a highlighting effect. (c) The response of the ventral stream IT module neuron pool coding for the sculpture was substantially stronger when the spatial attention bias to the PP module was allocated to the location of the sculpture than when there was no spatial attention. The spatial highlighting gated the sculpture image to be analyzed and recognized by the ventral stream IT module neurons. Without spatial attention, the sculpture did not have bottom-up saliency to enable the domination of the sculpture neuron in the ventral stream IT module.

interaction between all the processing modules involved. Of course each local brain area has a defined functional role, but interesting global attention-related behaviour emerges from the constant cross-talk between the different modules in the ventral and dorsal visual paths, implemented partly through the early processing areas such as V1, using backprojections as well as forward connections.

9.5.2 Dynamics of object attention and visual search

In the visual search task, i.e. when the system was looking for a particular object in a visual scene, the system functioned in an **object attention mode** as shown in Fig. 9.4b. Object attention was created by introducing a top-down bias to a particular cell pool in the IT module corresponding to a particular object class (as will be described in more detail below with Fig. 9.11). This ventral IT module pool backprojected the expected shape activity patterns over all spatial positions in the early V1 visual module through the top-down feedback connections. When the image (Fig. 9.8) containing the target object was presented, the hypercolumns in the early V1 visual module whose activities were closest to the top-down 'template' became more excited because of the interactive activation or resonance between the forward visual inputs and the backprojected activity from the IT module. Over time, these V1 hypercolumns

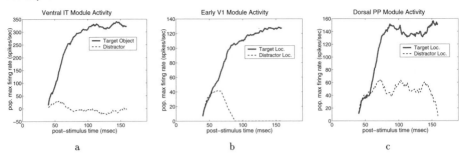

Fig. 9.11 Neuronal activities in the three modules during visual search in object attention mode. The maximum population activity of the neuronal pools corresponding to the identity or location of the sculpture in the scene in the three modules was compared against the maximum activity in pools coding any other locations or objects when sculpture was the target object of attention. The effect of attention, i.e. differentiation of neuronal response between the target and distractor conditions, was observed to start at the ventral stream IT module, and then to propagate to the early V1 module and the dorsal stream PP module. (a) The neuronal activity of the top-down biased 'sculpture' neuronal pool in the ventral-stream module was compared against the maximum activity of all other object pools. The increase in the response of the sculpture pool relative to the response in all other pools was observed to rise rapidly as a result of the top-down object attentional bias applied to the IT module. (b) The maximum population activity at the sculpture location in the early V1 visual module was compared against the maximum activity of pools at all other locations in the scene. (c) The maximum activity of the pools coding in the dorsal stream PP module for the sculpture location was compared against the maximum activity of the pools encoding the other locations.

with neuronal activities best matching the encoded object dominated over all the other hypercolumns, resulting in a spatially localized response peak in the early V1 visual module. Meanwhile, the dorsal PP module was not idle but actively participated by having all its pools engaged in the competitive process to narrow down the location of the target. The simultaneous competition in the spatial domain and in the object domain in the two extrastriate modules as mediated by their reciprocal connections with the early V1 module finally resulted in a localized peak of activation in the spatially mapped dorsal stream PP module, with a corresponding peak of activity for the object mapped in the ventral stream IT module, and corresponding activity in the early V1 module. This corresponds to finding the object's location in the image in a visual search task, or linking 'where is what', or computing 'where' from 'what'.

The actual simulation results just outlined are shown in Fig. 9.11, which compares the responses of the pools in the three modules corresponding to the location or identity of the attended object (the sculpture) to the responses at all other locations or identities. The evolution of the population maximum activity shows the polarization of responses that started in the ventral stream IT module, and then propagated to the other two modules.

Figure 9.12 shows the temporal evolution of the summed pool activity in the space maps of the V1 visual module and the PP module when the system was searching for the sculpture object in the scene in this object attention mode. It shows the gradual emergence of a response peak at the location corresponding to the location of the search object (i.e. the sculpture) in the visual scene as time evolved.

The effect of object attention on the responses of early V1 module neurons is shown in Fig. 9.13. It compares the responses of V1 module neurons when there was a bias imposed

(a) DM: 2 msec · (b) DM: 35 msec · (c) DM: 120 msec

(d) EM: 2 msec · (e) EM: 35 msec · (f) EM: 120 msec

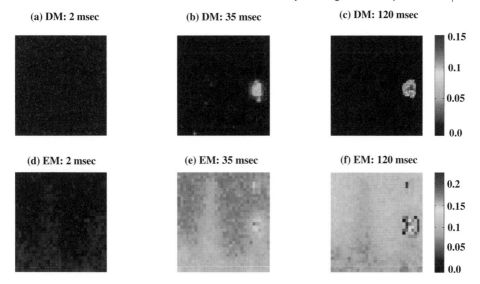

Fig. 9.12 Visual search in object attention mode. Top row: (a–c) show the maximum activity pools in the spatial-position topological map in the Dorsal stream PP Module (DM) at time 2 msec (a), 35 msec (b), and 120 msec (c) after stimulus onset. Bottom row: (d–f) show the maximum neural activity of each hypercolumn in the Early V1 visual Module (EM) at time 2 msec (d), 35 msec (e), 120 msec (f) after stimulus onset. Activities in both the early V1 module and the dorsal stream PP module exhibited a gradual contraction of responses to a localized area in the topological map due to lateral inhibition, corresponding to localizing the searched-for object. This Figure appears in the colour plate Section.

on the IT module sculpture neuron to when there was not. The effect of object attention in this case emerged around 90 msec after the onset of the stimulus. Note that the onset of a significant spatial attention effect was also at around the same time (see Fig. 9.9b).

It should be emphasized that because the V1 visual module interacted with both the PP and the IT modules simultaneously, the attentional effect observed in the later response of the neurons was not purely spatial or featural, but involved both components simultaneously. In the spatial attention mode, the PP module's bias initially highlighted the V1 module's response at a particular location, then the IT module got drawn into the process, and competition in the IT module, combined with the on-going interaction via the reciprocal connections between the PP module and the IT module, produced the enhancement effect in the V1 module.

A similar situation also appeared in the object attention mode. An object attention effect of this magnitude has been observed in V4, but not in V1. We reduced the magnitude of the coupling between the early V1 module and the ventral IT module, and found that the system continues to perform well in the visual search task when the coupling strength is reduced from 1 to 0.4. With this coupling strength, the effect of object attention in V1 is more modest, and yet it can still produce a bias that leads to the emergence of a peak response at the dorsal stream PP module's spatial location map.

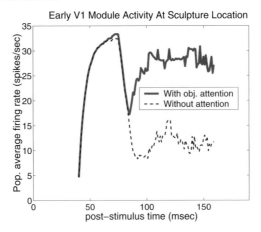

Fig. 9.13 Population average firing rate of neuronal pools in the early V1 module at the sculpture location when the sculpture was the target of object attention and when there was no object attention. Significant and sustained enhancement was observed in the later part of the V1 neuronal pool's response.

9.6 Evaluation of the model

In this Chapter, we demonstrate that spatial and object attention can be mediated by a single unitary system. We show that the early visual areas can provide a high-resolution representation for the dorsal and the ventral streams to interact to integrate abstract spatial and form information. We demonstrate that the system's attentional mechanism is an integral part that enables it to recognize and localize objects in a cluttered scene. In the following subsections, we will highlight and discuss the functional and physiological implications of this model and its simulation results in the context of existing psychological, neurophysiological and computational modelling literature.

9.6.1 Spatial attention and object attention

The concept of biased competition has been used to account for object attention (Usher and Niebur 1996) and spatial attention in the ventral stream (Reynolds, Chelazzi and Desimone 1999). The model described here advances these ideas by bringing in the dorsal stream and the early visual areas to coordinate the organization of attention in a unified system[30]. The simple architecture described in Section 9.3 (see Deco and Lee (2001)) allows spatial attention (Helmholtz 1867) and object attention (James 1890) to be accounted for in a symmetric fashion. The two modes of attention emerge depending simply on whether a top-down bias is introduced to either the dorsal stream PP module or the ventral stream IT module. In this framework, attention is produced by a simple top-down bias communicated from the short term memory systems of the brain which hold the target object or location in memory (e.g. in the prefrontal cortex) to the dorsal stream or the ventral stream. Moreover, this conceptualization offers a way of understanding the 'executive control' that is ascribed to the prefrontal cortex. It appears based on phenomenology to implement 'executive control', but at least a major part of this function we suggest can be understood as providing the short term memory bias to

---

[30] An alternative model is presented by Hamker (1999) and by Tagametz and Horwitz (1998).

posterior (parietal and temporal) perceptual systems to enable them to implement attentional effects as described. Of course, without a short term memory system in the prefrontal cortex to hold the target online in memory while the perceptual systems are processing sensory input (see Section 12.1), the whole organism would appear to an observer to be without 'executive control', and indeed to be displaying a 'dysexecutive syndrome' (Shallice and Burgess 1996).

Spatial attentional effects have been observed in V2, V4 and IT in numerous neurophysiological studies (Moran and Desimone 1985, Motter 1993, Chelazzi, Miller, Duncan and Desimone 1993, Connor, Gallant, Preddie and Van Essen 1996, Luck, Chelazzi, Hillyard and Desimone 1997, Reynolds, Chelazzi and Desimone 1999), but the source of these spatial attentional effects however has not been clear. We demonstrated that a top-down bias introduced to cell pools in the dorsal stream can produce a 'spatial attentional effect' in the early visual module and thus in ventral-stream (IT) modules.

In our simulation of the complete system described in this Chapter, we found that a top-down bias introduced either to the dorsal stream PP module or to the ventral stream IT module can elevate the neural response in the early V1 visual area in the corresponding location, producing a highlighting effect. This elevation in activity in the early V1 module, though relatively small, was sufficient to enable the image patch under the spatial attentional spotlight to be gated to the ventral stream, leading to a large polarization in responses in the relevant object class neurons in the ventral stream IT areas.

Elevated responses of a similar small magnitude in macaque V1 have been observed by Motter (1993) and Ito and Gilbert (1999) while the monkeys were engaging in spatial attention-demanding tasks with stimuli with multiple competitive elements. In the simulation, the elevation was found to become significant at about 90 msec after stimulus onset. This time course is thus very similar to that of the figure-highlighting effect induced texture contrast as observed by Lamme (1995) and Lee, Mumford, Romero and Lamme (1998). This suggests that the figure-highlighting effects observed in these studies might be an effect of attentional modulation driven by bottom-up saliency. We can therefore make the prediction that when the monkeys were asked to do a highly attentionally demanding task to discriminate stimuli at the fovea, the so-called figure-highlighting effect will be dramatically attenuated for neurons covering the peripheral visual field.

Furthermore, provided all three modules are engaged in interaction in either the spatial or object attentional modes, the attention-induced response elevation in the early V1 visual module is a result of interaction with both the ventral stream and the dorsal stream from the perspective of this framework, and thus cannot be considered a purely spatial or purely object attentional effect.

The second simulation described in this Chapter demonstrated that the same architectural model can produce the shifting of the receptive field's hot spot observed in V4 due to spatial attentional modulation (Connor, Gallant and Van Essen 1993, Connor, Gallant, Preddie and Van Essen 1996). The simulation result suggests that the shifting of the hot spot of the receptive field in V4 might be mediated by the highlighting effect that the dorsal stream produces in V1. Interestingly, when the locus of spatial attention was sufficiently far away from the receptive field, the cell's response was markedly attenuated and its hot spot snapped back to its original centre. This is consistent with Maunsell's (1995) observation. Although the shifting of the hot spot can potentially be implemented by known direct connections from parietal areas to V4 (Baizer, Ungerleider and Desimone 1991), our result suggests that the early visual areas also provide an effective mechanism. (Of course, effects similar to those we describe in this

Chapter could be produced by areas other than V1 that connect the dorsal and ventral visual processing streams, provided that the connecting area is topologically organized and has reciprocal connections with both streams.) From this perspective, the shifting of a receptive field produced by spatial attention observed by Connor, Gallant and Van Essen (1993) in V4 is then closely related to the so-called re-mapping phenomenon observed by Duhamel, Colby and Goldberg (1992a) in LIP in the dorsal stream, in that all were modulated by the same spatial attentional mechanism. From this perspective, we would predict presaccadic activities of a smaller magnitude should also be observed in the early visual areas.

The third simulation demonstrated that the spatial attentional effect and competitive interaction effect observed in Moran and Desimone's (1985) and Reynolds et al.'s (1999) experiments can also be accounted for by our model (Fig. 9.5). Reynolds, Chelazzi and Desimone (1999) had simulated a model based on biased competition to explain their data. However, their model required a dynamic modification of synaptic efficacy during attention. Even though dynamic synaptic change is not completely implausible, we think that synaptic plasticity might be more relevant for learning than for moment-to-moment attentional allocation. Attentional mediation in our model is purely mediated by activity alone and hence is simpler and more parsimonious. From the perspective of this model, given that the elevation of the latter part of the responses of V1 neurons is due to interaction with both the ventral and the dorsal streams, one predicts that deactivation of either the dorsal areas such as V3a, MT, LIP, or the ventral areas such as V4, should attenuate the figure-highlighting effect (Lamme 1995) at the level of V1.

9.6.2 Translation-invariant object recognition

By simulating the system in spatial attention mode, we demonstrated that spatial attentional gating is one possible way in which translation-invariant object recognition can be facilitated. In this model, the memory of each object is encoded in the feedforward/feedback loops between the early V1 module and the ventral IT module. Spatial attention from the dorsal stream effectively achieves gating in the ventral stream by simply highlighting the activities at a particular location in the V1 stage.

Although this demonstration is of interest, this property is not the most important aspect of the model, and indeed a more biologically plausible way of achieving translation invariance using feedforward processing is described in Chapter 8. A feedforward system is more plausible in that with the masking experiments described in Sections 5.5.6 and B.3.2, it is found that inferior temporal cortex neurons can still provide object-selective information even when each cortical area can fire for only 30 ms. This time is too short to enable activity once it has reached the inferior temporal cortex to be backprojected to V1, and forward projected again with dynamic settling. One property that may help the real visual cortical system with the feedforward aspect of its processing, especially when operating in cluttered natural scenes, may be the weighting given to whatever is at the fovea, given the larger magnification factor at the fovea (see Section 8.4.6). Of course in natural cluttered scenes when there is sufficient time for attentional processes to operate, good performance may be contributed both by the weighting given to whatever is at the fovea due to its greater magnification (which describes overt attention, where eye position is used to define what object in a scene to process), and to the types of covert attentional process described in this Chapter, where attention can be paid to an object that is not at the fovea. Another reason why the processes described in Chapter 8 are more appropriate for forming invariant representations in general is that they provide an account of view-invariant object recognition, which a spatial attention-based approach to translation invariant object recognition cannot. Further, a spatial attention-based approach to invariant object recognition cannot without special added mechanisms account for the binding of features in which the relative spatial position of the features must be encoded, as discussed in Section 13.7.

The object recognition aspect of our model can be traced back to the Adaptive Resonance Theory of Grossberg (1987) and the interactive activation model of McClelland and Rumelhart (1981). The basic idea is that the recurrent interaction or resonance between the object memory representation and the early representation serves to clarify both representations (see Section 1.11 and Rolls (1989b)). It is also related to the proposal that the feedback connection from the higher area to the lower area carries a reconstructed or synthesized image (template) of what is expected in the earlier area (Rao and Ballard 1999). In this context, because the feedback due to object attention is feature-specific, it is in some sense feeding back a virtual matching template. While the idea of 'resonance' is not new in the neural modelling community, our use of spatial attentional bias from the dorsal stream to facilitate translation-invariant recognition through interaction with an early visual area is novel.

The object recognition capability of the system described in this Chapter may be contrasted with other neural models of object recognition such as the shifter circuit of Olshausen, Anderson and Van Essen (1993), the predictive coding circuit of Rao and Ballard (1999), and the gain field circuit of Salinas and Abbott (1997). All of these models are limited to the restricted domain of template matching. Olshausen et al.'s (1993) shifter-circuit model was primarily a feedforward model and required dynamic modification of synaptic efficacy by a complicated routing circuit that it was presumed could be controlled by the pulvinar. Salinas and Abbott (1997) also proposed a feedforward model and used an enormous number of gain field neurons for each orientation selective cell in V4 in order to achieve translation-invariant object recognition. The system described in this Chapter shows that both of these schemes might be unnecessarily complicated. The spirit of the model described in this Chapter is closest to the interactive feedback models of Rao and Ballard (1999), Grossberg (1987), and McClelland and Rumelhart (1981). The distinction between the model described in this Chapter and the other models is our use of the dorsal stream and the early visual cortex to coordinate with the translation-invariant object recognition process. Some of the differences between the way in which translation invariance is achieved in the model described in this Chapter and the main approach to invariant object recognition described in this book, in Chapter 8, are noted in the second paragraph of this subsection (9.6.2) and in Chapter 13.

9.6.3 Contributions and limitations

The visual system is composed of numerous areas connected in extensive networks (Felleman and Van Essen 1991). The functions of all of these areas have not been clearly elucidated. Therefore, any systems-level model requires simplification and abstraction. Our objective with the architecture described in this Chapter is to investigate possible computational mechanisms with which the two visual processing streams could work together to identify and localize objects in cluttered scenes. In the ventral stream, we have not included many of the intermediate areas required to encode successive abstract representations and to extract invariances necessary for general object recognition as we have done in Chapter 8. In the dorsal stream, we have not included the various coordinate transforms necessary for spatial understanding and planning of action. The system described thus is a simplified prototype, and is aimed to provide an understanding of visual attention, rather than solving the whole problem of vision.

The fundamental contribution of the work with this model is to show that several apparently different dual processes, such as serial versus parallel search, spatial versus object attention, and spatial localization of a target object can be accounted for by a unitary system as proposed in this book. This model is novel in that it suggests that the reciprocal interaction of the early visual areas with the dorsal visual stream and the ventral visual stream can provide an effective means for mediating spatial and object attention, and for organizing visual search, and helping with translation-invariant object recognition. The model includes also a model of the biased competition hypothesis (Duncan and Humphreys 1989), and effectively models many of the functions proposed for the high-resolution buffer hypothesis (Lee, Mumford, Romero and Lamme 1998). Further, in the context of visual search, it shows that Treisman's feature integration can be implemented as an emergent phenomenon arising from the interaction between early visual cortical areas and the various extrastriate areas in the ventral and dorsal visual streams.

Conceptually, this model is much simpler and more parsimonious than other models such as Olshausen et al.'s shifter circuit, and Salinas and Abbott's gain field model, in that it does not require the large number of routing control connections of the former, or the enormous number of gain field neurons of the latter. Instead it simply uses a biased competition mechanism. Earlier models using biased competition mechanisms (Usher and Niebur 1996, Reynolds, Chelazzi and Desimone 1999) were very small scale models using typically 2 or 3 units. Our simulation results show that the system performs stably and well in the object recognition task and the visual search task despite the interaction of the massive number of units in the model. Furthermore, our system can process cluttered realistic images. It has a bottom-up data-driven component (Itti and Koch 2001), as well as a top-down biasing components. The system works by performing constraint satisfaction on the top-down intercortical attentional forces, the intracortical competitive forces, and the bottom-up data-driven input from the retina, to arrive at the desired solution.

The model is based on evidence from functional, neurophysiological, and psychological findings. It is able to successfully reproduce the findings of neurophysiological experiments on spatial attention, and the findings of psychological studies on serial and parallel search (see Chapter 10). It provides insight into some controversial neurophysiological observations in V1 (Lamme 1995, Lee, Mumford, Romero and Lamme 1998, Rossi, Desimone and Ungerleider 2001). It illuminates the potential functional roles in object recognition and visual search of the neural mechanisms described in neurophysiological experiments. Its ability to perform object recognition and visual search in cluttered scenes, and its ability to explain and predict a range of biological, psychological, and neuropsychological observations (as described further in Chapters 10 and 11) provide an indication of the validity of this model in describing the organization of attentive visual processing in the mammalian visual system.

10 Visual search: Attentional neurodynamics at work

10.1 Introduction

Evidence for different temporal behaviours of attention in visual processing comes from psychophysical experiments using visual search tasks where subjects examine a display containing randomly positioned items in order to detect an *a priori* defined target. All other items in the display which are different from the target serve the role of distractors. The relevant variable typically measured is search time as a function of the number of items in the display. Much work has been based on two kinds of covert search paradigm: feature and conjunction search[31]. In a feature search task the target differs from the distractors in one single feature, e.g. only in colour. In a conjunction search task the target is defined by a conjunction of features and each distractor shares at least one of those features with the target. Conjunction search experiments show that search time increases linearly with the number of items in the display, implying a serial process. On the other hand, search times in feature search can be independent of the number of items in the display.

In this Chapter, we show that the attentional architecture described in Chapter 9 performs covert search across the visual field in parallel but, due to the different latencies of its dynamics, can show the two experimentally observed modes of visual attention, namely: serial focal attention, and the parallel spread of attention over space.

Psychophysical investigations also provide evidence for the so-called 'resolution hypothesis', according to which attention can actually enhance spatial resolution, so that one can resolve finer spatial details at the attended location. When a spatial cue directs covert attention to a target location, the observer's performance improves for stimuli designed to measure spatial resolution. Similarly, in visual search tasks, cueing the target location improves performance (reaction time and accuracy) in cases where the target is located at positions with large eccentricity. The resolution hypothesis is usually associated with the so-called 'global precedence effect', which consists of a speed advantage (i.e. a reaction time measure) of global over local processing of hierarchical stimuli (Navon 1977). A hierarchical stimulus is an object that possesses both a global and a local structure (e.g., global letter shapes constructed with local letter elements, as illustrated in Fig. 10.3). This effect comes from the relationship between spatial frequency and selective attention to local and global structures based on psychophysical experiments, in which observers have to respond to the global level of hierarchical stimuli, followed by a detection response to sinusoidal gratings that differ in spatial frequency. In these experiments, high-frequency gratings were better detected when observers had just responded to a local feature of the hierarchical stimulus, whereas low-frequency gratings were better detected when observers had just responded to a global feature

---

[31] Covert attention refers to attention directed at a location that is away from the position being fixated. Overt attention refers to moving the eyes to fixate the location to which attention is being paid.

of the stimulus. In fact, the global precedence effect is lost after a selective removal of low spatial frequencies from the hierarchical stimuli. Similarly, in a neuropsychological context, it has been shown that the global precedence effect can be absent in patients after brain injury.

In this Chapter, we concentrate on the interplay between microscopic neuronal dynamics and systems-level functional behaviour in the context of visual search. We demonstrate that it is possible to build a neural system for visual search, which works across the visual field in parallel, but, due to the different latencies of its dynamics, can show the two experimentally observed modes of visual attention, namely: serial focal attention and the parallel spread of attention over space. Neither explicit serial focal search nor saliency maps need to be assumed. In our system, attention is a dynamical emergent property, rather than a separate mechanism operating independently of other perceptual and cognitive processes. Further, the visual system always works in parallel, but the different latencies associated with different spatial resolutions allow the emergence of an early attentional focus based on coarser features which gradually increases its resolution by processing finer details only at the spatial region to which coarse resolution has been directed early on in processing. The spatial form of the focus that emerges is attached to the coarse structure of an object. When finer details start to modulate the initially activated pools, a focus more attached to the object emerges. Additions to the neurodynamical cortical model we describe in Chapter 9 enable it to explain the underlying massively parallel mechanisms which not only generate the emergence of a kind of attentional 'spotlight', but also its object-based character, and the associated spatially localized enhancement of resolution suggested by the resolution hypothesis.

Finally, we extend our computational multi-area model in order to understand the neurodynamics underlying much more complex cases of visual search tasks that require the binding of different feature dimensions. In fact, we consider different kinds of conjunction search that not only involve the shape of the target and distractors, but also other feature dimensions such as colour, texture, movement, etc. Previous models were unable to explain the variation of the observed slopes of the linear relation between the reaction time and the number of items on the screen for different kinds of conjunction search experiments. In our neurodynamical computational model we postulate independent competition mechanisms along each feature dimension in order to explain the experimental data. This implies the necessity for the independent character of visual search in separated but not integrated feature dimensions. The binding of these feature dimensions is achieved by the attentional dynamical interaction between the posterior parietal spatial module and the separate primary visual modules corresponding to each different independent feature dimension. We show that the posterior parietal module binds the different feature dimensions by representing the spatial information integrated by attention.

10.2   The psychophysics of simple visual search of shapes

In Section 6.1, we reviewed the classical theories of human vision that infer two functionally distinct stages of visual processing. The pre-attentive stage implies an unlimited-capacity system capable of processing the information contained in the entire visual field in parallel. The attentive stage is characterized by the serial processing of visual information corresponding to local spatial regions. This second stage of processing is typically associated with a limited-capacity system which allocates its resources to a single particular location in visual space. In the specific case of visual search, it is usually assumed that these two processes operate sequentially (Posner and Snyder 1975, Treisman and Gelade 1980, Treisman 1988). The second attentive stage has been related to an attentional spotlight. Computational models formulated in the framework of feature integration theory (see Section 6.1) require the existence of a saliency or priority map for registering the potentially interesting areas of the retinal input, and a gating mechanism for reducing the amount of incoming visual information, so that limited computational resources in the system are not overloaded. The priority map serves to represent topographically the relevance of different parts of the visual field, in order to have a mechanism for guiding the attentional focus to salient regions of the retinal input. The focused area will be gated, such that only the information within it will be passed on to higher levels, which are concerned with object recognition and visually guided action.

Evidence for these two stages of attentional visual processing comes from psychophysical experiments using visual search tasks where subjects examine a display containing randomly positioned items in order to detect an a-priori defined target. All other items in the display which are different from the target serve the role of distractors. The number of items in the display is called the display or set size. The relevant variable typically measured is search time as a function of the display size. A challenging question is therefore: is the linearly increasing search time which is observed in some visual search tests necessarily due to a serial mechanism? Or can it be explained by the dynamical time-consuming latency of a parallel process? In other words, are priority maps and spotlight mechanisms required?

Deco and Lee (2001) demonstrate that the system introduced in Chapter 9 for visual search works across the visual field in parallel, but due to its intrinsic dynamics can show the two experimentally observed modes of visual attention, namely: serial focal attention, and the parallel spread of attention over space. The model demonstrates that neither explicit serial focal search nor saliency maps need to be assumed. In the present model the focus of attention is not provided to the system but only emerges after convergence of the dynamical behaviour of the neural networks.

In order to demonstrate these ideas, we simulate the search of a letter located between other distractor letters. We use exactly the same architecture and the same parameters as in Chapter 9. Let us define the task of searching for the target letter 'E'. Figure 10.1 illustrates the basic observations concerning parallel and serial search.

The stimulus in Fig. 10.1a contains shapes E and X. Because the elementary features in E and X are distinct, i.e. their component lines have different orientations, E pops out from X, and its location can be rapidly localized independently of the number of distracting X shapes in the image. On the other hand, the stimulus in Fig. 10.1b contains E and F. Since both letters are composed of vertical and horizontal lines, there is no difference in elementary features to produce a preattentive pop-out, so they can be distinguished from each other only after their features are glued together by attention. It has been thought that because 'attention' is serial, the time required to localize the target in such an image increases linearly with the number of distractors in the image. The serial movement of the attentional spot light has been thought to be governed by a *saliency map* or *priority map* for registering the potentially interesting areas in the retinal input and directing a *gating* mechanism for selecting information for further processing. Does the linear increase in search time observed in visual search tests necessarily imply a serial search process, a saliency map, or a gating mechanism? Could both the serial and the parallel search phenomena be explained by a single parallel neurodynamical process, without the extra serial control mechanism?

To investigate this issue, we presented the stimuli shown in Fig. 10.1a and Fig. 10.1b

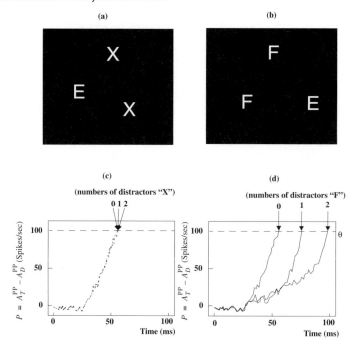

Fig. 10.1 (a) Parallel search example: an image that contains a target E in a field of X distractors. Since the elementary features in E and X are distinct, i.e. their component lines have different orientations, E pops out from X, and its location can be rapidly localized independently of the number of distracting X shapes in the image. This is called parallel search. (b) Serial search example: an image that contains a target E in a field of F distractors. Since both letters are composed of vertical and horizontal lines, there is no difference in the elementary features to produce a preattentive pop-out, so the E and F can be distinguished from each other only after their features are bound or glued by attention. The time required to locate the target in such an image increases linearly with the number of distractors in the image. This is called serial search and has been thought to involve the scanning of the scene with a covert attentional spotlight. (c–d) Simulation result of the network described in Chapter 9 performing visual search on images (a) and (b) respectively. The difference (polarization) between the maximum activity in the neuronal pool corresponding to the target locations and the maximum activity of all other neuronal pools in the dorsal PP module is plotted as a function of time. (c) shows that the difference signal rose to a threshold, corresponding to localization of the object, at about the same time independently of the numbers of distractor items. (d) shows that the time for the polarization signal to rise to a threshold increased linearly with the number of distracting items, with an additional 25 msec required per item.

to the network described in Chapter 9, which had been trained to recognize X, E, and F in a translation invariant manner. The system received a top-down bias for the 'E' pool in the ventral IT module, and then was presented with stimuli containing E in a variable number of X shapes, or E in a variable number of F shapes. Let us use polarization, the difference between the maximal activity of the pools indicating the E location and that indicating the F location, i.e. $P = A_T^{PP} - A_D^{PP}$, as a measure to determine whether detection and localization of the target had been achieved or not. We found that for the E in X case, the time required for the polarization to reach a certain threshold in the dorsal PP module was almost identical whether the number of X shapes was equal to 0, 1 or 2, as shown in Fig. 10.1c. On the other hand, when E and F were presented, the time required for polarization to reach threshold increased

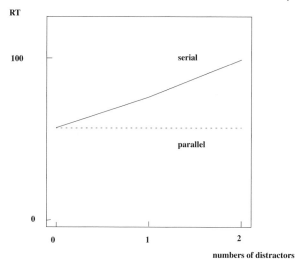

Fig. 10.2 Simulated reaction times for different types and numbers of distractors from the results shown in Fig. 10.1c and 10.1d for 'parallel' and 'serial' search examples.

linearly with the number of distracting items. Although the system was running with the same parallel dynamics, it took an additional 25 msec for each additional distractor added to the stimulus as shown in Fig. 10.1d.

Figure 10.2 shows explicitly the simulated reaction times as a function of the number of distractors for both types of search. The system works across the visual field in parallel, but, due to the different dynamic latencies, resembles the two apparent different modes of visual attention, namely: serial focal search, and parallel search. In the case of serial search, the latency of the dynamics is longer indicating 'apparently' a serial component in the search, although the underlying mechanisms work in parallel. The typical linear increase in the search time with the display size is clearly obtained as the result of a slower convergence (latency) of the dynamics. In this case, the strong competition present in V1 and propagated to PP delays the convergence of the dynamics. The strong competition in the feature extraction module V1 is finally resolved by the feedback received from PP. In other words, stimulus similarity in the feature space is decided by competition mechanisms at the intramodular level of V1 and at the intermodular level of V1–PP.

In this model (described in Chapter 9), the mechanism for object attention is also the mechanism for visual search. A bias to a particular object pool in the ventral IT module starts the system in its search for activity patterns in the early V1 module corresponding to the target object. The feedback connection effectively generates a synthesized image and matches it with activity patterns all over V1. The part of V1 with activity patterns that matches well the feedback signals becomes enhanced, and provides stronger positive input to the corresponding pool in the dorsal PP module. As the biased competition progresses, the neural activities in both the early V1 module and the dorsal PP module will contract to a localized region in their respective spatial maps. In this model, object identification and localization are dual and symmetrical processes being carried out by all the modules concurrently and interactively.

The simulation result shows that the system performs successfully in visual search tasks. It also reveals the possibility that parallel search and serial search might not be two very different independent stages or processes, as was previously thought (Treisman and Gelade 1980). Traditionally, visual processing has been divided into two stages: a preattentive stage in which information in the whole scene is processed in parallel, and an attentive stage in which features in the attentional spotlight are glued together in a process of feature integration for further processing (Posner and Snyder 1975, Treisman and Gelade 1980, Treisman and Sato 1990). In contrast, in the computational model we describe, the two stages of processing involve the same mechanism, and feature integration is accomplished dynamically by the interaction between the ventral IT module and the early V1 module. Feature integration is an emergent phenomenon due to interactive activation among the cortical areas, rather than a separate stage of visual processing, or involving a separate visual area. The linear increase in time observed in 'serial search' in the E–F case reflects simply the fact that when features are similar between objects, the similarity in the stimuli's representations in V1 will require more time for the competition to sort out the target against the distractors. This sorting out at the level of V1 requires a constant interaction with both the ventral stream IT and the dorsal stream PP pathways in the emergent process of the integration and combination of different features to form a shape.

10.3   Visual search of hierarchical patterns

Psychophysical evidence strongly suggests that selective attention can enhance spatial resolution in the input region corresponding to the focus of attention. In this Section, we extend the computational model described in Chapter 9 in order to explain the findings on spatial resolution from psychophysical experiments. We propose that V1 neurons have different time courses of their responses depending on the spatial frequency. To be concrete, we postulate that V1 neurons sensitive to low spatial frequencies have faster dynamics (with e.g. shorter latencies) than V1 neurons sensitive to high spatial frequencies. In this sense, a scene is first predominantly analyzed at a coarse resolution level, and the dynamics then enhances the spatial resolution at the location of an object until the object is identified. We propose and simulate new psychophysical experiments where the effect of the attentional enhancement of spatial resolution can be demonstrated, and predict different reaction times in visual search experiments where the target and distractors are defined with different degrees of spatial resolution.

10.3.1   The spatial resolution hypothesis

The visual system simultaneously analyzes scenes at different spatial scales (Ginsburg 1986, Marr 1982, Wilson 1978). Some channels prefer low spatial frequencies or wide bars, whereas other channels prefer high spatial frequencies or narrow bars (Watson and Robson 1981). Since Navon's (1977) seminal work, a number of psychophysical studies have been devoted to understanding the role of low frequency channels in the processing of global image structure (Lovegrove, Lehmkuhle, Baro and Garzia 1991, Shulman and Wilson 1987, Shulman, Sullivan, Gish and Sakoda 1986, Wong and Weisstein 1998). Navon (1977) developed a task that permits easy manipulation of the global and local features of an image by using hierarchical patterns. A hierarchical pattern refers to a large, global pattern composed of small, local patterns, a classic example of which is a large letter constructed of local letter elements (Kinchla 1974, Navon 1977, Shulman and Wilson 1987, Badcock, Whitworth, Badcock and

Hierarchical Patterns

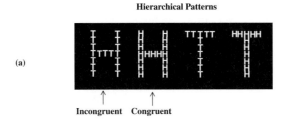

(a)

↑   ↑
Incongruent Congruent

(b)

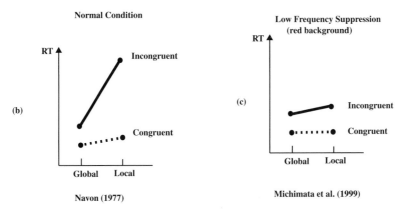

Fig. 10.3 (a) Hierarchical patterns composed of letters. (b) Qualitative results of Navon's (1977) reaction time (RT) experiments showing the global precedence effects. (c) Cancellation of the global precedence effects after suppression of the low-spatial-frequency channel by manipulation of the background color (after Michimata et al. 1999).

Lovegrove 1990). Figure 10.3a shows typical examples of hierarchical patterns formed with letters. The global and local stimulus letters may be either congruent (same letters) or incongruent (different letters). Subjects are required to name either the global letter or the local letter in the display as quickly as possible by pressing a labelled response button. Experimental results demonstrate that the processing of the global level is faster than processing of the local level (i.e. a global reaction time advantage), and that the global level interferes with local processing (i.e., asymmetric global interference). These global precedence effects are qualitatively shown in Fig. 10.3b.

Recently, several studies that manipulate hierarchical patterns have shown that manipulation of low spatial frequencies influences the global task selectively, thus supporting the notion that the global aspect of a stimulus is processed by low spatial frequency mechanisms (Michimata, Okubo and Mugishima 1999, Lovegrove, Lehmkuhle, Baro and Garzia 1991, Badcock, Whitworth, Badcock and Lovegrove 1990, Shulman and Wilson 1987, Shulman, Sullivan, Gish and Sakoda 1986). Shulman and Wilson (1987) studied the relationship between spatial frequency and selective attention to local and global structures by an experiment in which observers had to respond to the global level of hierarchical stimuli, followed by a detection response to sinusoidal gratings that differed in spatial frequency. They reported that high-frequency gratings were better detected when observers had just responded to a local feature of the hierarchical stimulus, whereas low-frequency gratings were better detected when observers had just responded to a global feature of the stimulus. Badcock, Whitworth, Badcock and Lovegrove (1990) also confirmed the role of low spatial frequency information in the processing of global aspects of hierarchical patterns. These authors showed that the global precedence effect was lost after selective removal of low spatial frequencies from the hierarchical stimuli. Similarly, experiments by Rafal and Robertson (1997) have shown that the global precedence effect can be absent in patients with brain injury. Briefly, they argued that mechanisms associated with the left hemisphere, specifically in the region of the temporoparietal junction (TPJ), are biased (in normal subjects) to process local information, whereas mechanisms associated with the right hemisphere are biased to process global information. Consequently, damage of the right hemisphere could cause the cancellation of the global advantage.

An elegant demonstration of the relationship between global processing and low-spatial-frequency mechanisms is offered by the experiments of Michimata, Okubo and Mugishima (1999). They performed an experiment which was designed on the basis of findings about the function and characteristics of the magnocellular pathway in the visual system. The lateral geniculate body in primates and humans has a six-layer structure, which can be divided into two major portions. As described in Chapter 2, the four parvocellular layers contain cells with relatively small receptive fields and have sustained response characteristics. The other two, magnocellular, layers, contain cells with large receptive fields and have transient response characteristics. Both sets of layers provide input to the primary visual cortex (V1) where the segregation is preserved. The magnocellular pathway contains cells of type IV (Livingstone and Hubel 1984) which have receptive fields with a tonic red inhibitory surround mechanism (i.e. imposing diffuse red light causes a tonic suppression of these cells' activity). Breitmeyer and Breier (1994) have provided experimental evidence that the magnocellular pathway tends to act as the low spatial frequency channel of the visual system. Michimata, Okubo and Mugishima (1999) have repeated the reaction-time experiment of Navon (1977) but manipulated the background colour of the stimulus. The control condition with a green background essentially reproduced Navon's results. On the other hand, a red background, which blocked the processing of the low-spatial-frequency magnocellular channel, produced cancellation of the global precedence effects (see Fig. 10.3c).

In the context of attention, the influence of attentional mechanisms on the processing of low and high spatial frequencies is hypothesized in the so-called resolution hypothesis. The resolution hypothesis predicts that attention can actually enhance spatial resolution, so that one can resolve finer details at the attended location (Yeshurun and Carrasco 1998). In the naive framework of the spotlight paradigm, the standard interpretation of selective attention as a spotlight which gates a local region of the visual field to a higher level of processing, the spatial resolution effect can be understood by assuming that the spotlight also gradually enhances the spatial resolution within the 'illuminated' local region for further processing.

Several papers recently provided experimental evidence for the resolution hypothesis by showing that attention can affect performance by enhancing the signal. Yeshurun and Carrasco (1999) demonstrated that when a spatial cue directs covert attention to a future target location, the observer's performance is improved for stimuli designed to measure spatial resolution. Consistently, Yeshurun and Carrasco (1998) also found that for a texture segregation task, attention can improve performance at peripheral locations where spatial resolution was too low, but can impair performance at central locations where spatial resolution was too high for the particular task. Similarly, in visual search tasks, cueing the target location improved performance (reaction time and accuracy) in cases where the target was located at positions with a large eccentricity[32] (Carrasco and Yeshurun 1998). These findings imply that spatial resolution is indeed enhanced at the attended location. Hence, the visual inspection of a scene is initially roughly analyzed at a coarse spatial resolution, and the dynamics of the system then gradually increases the spatial resolution at the location where attention is deployed.

Further, at the level of the responses of single neurons, a dynamic control of visual spatial resolution has been shown (Wörgöter, Suder, Zhao, Kerscher, Eysel and Funke 1998). They found that the shape of the receptive fields of V1 neurons of cats undergo significant modifications which correlate with the behavioural state of the animal. The observed dynamic changes in the sizes of the cortical receptive fields could be a reflection of an attentional process that modulates spatial resolution within the primary visual cortex.

10.3.2 Neurodynamics of the resolution hypothesis

The experimental evidence reviewed in the previous Section suggests that attentional mechanisms actively enhance spatial resolution at the attended location. In Chapter 9, we demonstrated that top-down attentional mechanisms can be understood as the parallel mutual modulation between neuronal pools resulting from intra- and inter-modular dynamical interactions. Hence, a plausible neuronal realization of the resolution hypothesis involves a modification of the underlying dynamics in such a way that the neuronal pools associated with the processing of different spatial resolutions have different response latencies. We implement the resolution hypothesis and the global precedence effect in our theoretical model by assuming that the neuronal pools in module V1 which correspond to different octaves (i.e. different spatial resolutions) have different speeds of temporal evolution. This can be achieved by defining different membrane time constants τ_k for different octaves k. The new update equations for neuronal pools in V1 are similar to equation (9.7) but now with explicitly different τ_k, i.e.:

$$\tau_k \frac{\partial A_{kpql}^{V1}(t)}{\partial t} = -A_{kpql}^{V1} + \alpha F(A_{kpql}^{V1}(t)) - \beta F(A_k^{I,V1}(t)) + I_{kpql}^{V1}(t)$$
$$+ I_{pq}^{V1-PP}(t) + I_{kpql}^{V1-IT}(t) + I_0 + \nu . \tag{10.1}$$

We postulate that the neuronal pools corresponding to low spatial frequency (in the notation used, to higher octaves) evolve faster than the neuronal pools corresponding to high spatial frequency. Consequently, the membrane time constants corresponding to low spatial frequencies are smaller than the ones corresponding to the high spatial frequencies, i.e.

$$\tau_k < \tau_{k'} \text{ for } k > k' . \tag{10.2}$$

In other words, we assume that the processing of coarse spatial information is faster than the processing of finer spatial details. Figure 10.4 shows schematically the implementation of the resolution hypothesis in the cortical architecture introduced in Chapter 9.

The dynamical evolution of the whole visual system is similar to that described before, but with the important difference that now the top-down attentional intra- and inter-modular

---

[32] The lower spatial resolution of the visual system in the periphery of the visual field usually causes the observer's general performance to become slower and less accurate as target eccentricity increases (Carrasco and Frieder 1997).

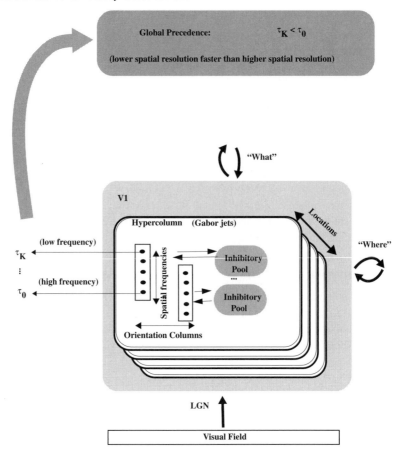

Fig. 10.4 Implementation of the resolution hypothesis in the cortical model for visual attention described in Chapter 9. The neuronal pools in module V1 which correspond to different octaves (i.e. different spatial resolutions) are assumed to also have different speeds of temporal evolution. This can be achieved by defining different membrane time constants τ_k for different octaves k. See text for details.

dynamical modulation especially influences at the beginning the large spatial features in the image because the lower spatial frequencies update much faster. After a while, the slower high spatial frequencies also reach a level of activity capable of influencing the whole dynamical system, which means that finer features in the image are now influential in the visual processing. Due to the dynamical competition existing from the very beginning, the spatial regions already primed by coarser spatial resolution analysis are preferentially analyzed with finer resolution.

The lower spatial resolution features in the image guide the attentional modulation to concentrate on certain spatial regions for a further, finer, analysis later. A naive phenomenological description of the evolution of our system refers to an attentional enhancement of spatial resolution at the attended location. However, in the understanding we provide, the visual system always works in parallel, but the different latencies associated with different spatial resolutions allow the emergence of an early attentional 'spotlight' based on coarser features in the image. This region then has its resolution increased, and finer details are processed mainly at the spatial region to which attention was drawn early on in processing by the low spatial frequencies. Of course, the spatial form of the 'spotlight' that emerges in this way is initially attached to the coarse structure of an object, so that the underlying mechanisms are consistent with the ideas of object-based attention (Behrmann, Zemel and Mozer 1998, Driver and Baylis 1989, Duncan 1984, Kramer and Jacobson 1991, Kramer and Watson 1995, Lavie and Driver 1996, Prinzmetal 1981, Vecera and Farah 1994). When finer details start to modulate the initially activated pools, a 'spotlight' more attached to the object emerges.

In conclusion, our neurodynamical cortical model explains the underlying massively parallel mechanisms that not only generate the emergence of a kind of attentional 'spotlight', but also its object-based character, and the associated spatially localized enhancement of resolution suggested by the resolution hypothesis.

10.3.3   Visual search in the framework of the resolution hypothesis

In order to consider the active role of attentional mechanisms in the enhancement of spatial resolution as we propose and as implemented in our model, we review in this Section a new psychophysical experiment proposed by Deco and Zihl (2001a) where the time course of attentional resolution enhancement can be manipulated. In this form, simulations of the model (Heinke, Deco, Humphreys and Zihl 2001a) allow the generation of specific experimentally testable predictions.

10.3.3.1   A new experimental paradigm

In visual search experiments, the number and the characteristics of the distractors can be manipulated in order to test different dynamical behaviours of attention. Hence, the attentional dynamics that underlies the resolution hypothesis can be demonstrated by combining the processing of multiple spatial resolutions with visual search. To be concrete, we propose a visual search task in the context of the global-local paradigm, i.e. by using hierarchical patterns as target and distractors. As hierarchical patterns we use global letter shapes constructed of local letter elements. Let us denote as 'Xy' an item corresponding to a global letter 'X' and composed by local letters 'y', X and y being arbitrary letters. The aim of our psychophysical paradigm is to find out whether attentional enhancement of spatial resolution has an effect on reaction time during a search experiment which would be predicted by the hypothesis of different update velocities for different spatial frequencies.

We define two different types of visual search experiment. The first experiment (VS1) defines the distractors and the target such that at the global (coarse resolution) level the target can pop-out from the distractors. In this case we say that the display shows global disparity. Figure 10.5a shows a typical screen for this kind of visual search experiment containing 3 items. In this case the target is defined as the pattern Hh, and the distractors are patterns Fh. The global similarity of the patterns (i.e., global H or global F) can be determined reliably at a coarse level of feature extraction using only low frequencies. If the local structure is to be detected in a reliable form as well, then activation of finer levels is required. We measure search time as a function of the number of items (display size). In this case, if the precedence processing of low frequencies holds, we expect a kind of parallel search because the global structure uniquely defines the target and can pop-out from the distractors. The neuronal pools in module V1 extracting the global coarse features cause an early full polarization of spatial

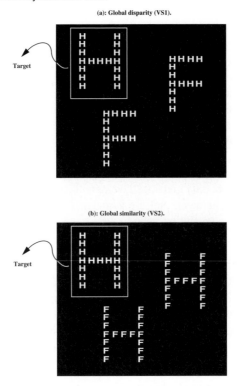

Fig. 10.5 Examples of the visual search displays considered in the context of the global–local paradigm.

attention in PP at the position where the target is since the only item showing the global characteristics of the target is the target itself. The processing of high spatial frequencies (finer details) will only play a role at the spatial position of the target localized early on in processing, and then helps object recognition. On the other hand, if the early processing of low spatial frequencies does not solve the task, then the fine resolution details of the distractors and the target should increase attentional competition from early on in processing, leading to longer overall processing time, and accounting for what otherwise would be interpreted as serial search.

The second experiment (VS2) defines the distractors and the target such that their global characteristics are indistinguishable, and they only differ at the local level. We say in this case that the display shows global similarity. Figure 10.5b shows a typical display for this case. The target is defined as the pattern Hh and the distractors are patterns Hf. To find the target one has to process high spatial frequency information throughout the entire screen because, in this case, the coarse feature information extracted by the low frequencies does not polarize attention, and yields equal levels of activity in PP to each item location (all global forms are H). The competition between the distractors and the target is only solved when the high spatial frequency processing pools are activated. The degree of competition at the high spatial frequency level increases with the number of distractors, and therefore what appears to be a serial kind of search is predicted, even when the fast preferential processing of low spatial frequencies is present.

10.3.3.2 Computational results and predictions

In this Section, we describe simulation results obtained for the visual search experiments on hierarchical patterns introduced in the preceding Section. The input display is given by a pixelized 66×66 image ($N = 66$). The V1 hypercolumns cover the entire image uniformly. They are distributed in 33×33 locations ($P = 33$) and each hypercolumn is sensitive to three spatial frequencies and to eight different orientations (i.e., $K = 3$ and $L = 2$). Consequently, the V1 module has 6,534 pools and three inhibitory pools. The IT module utilized has four pools for recognition of the patterns Hh, Hf, Ff and Fh, and one common inhibitory pool. Finally, the PP module contains 4,356 pools corresponding to each possible spatial location, i.e. to each of the 66×66 pixels, and a common inhibitory pool. Each of the four hierarchical objects (shown in Fig. 10.5) is isolated in order to define four categories to be associated with four different pools in the IT module. During the learning phase, these four objects are presented randomly and at random positions in order to achieve translation invariant responses. The system required 3,000,000 different presentations for training the IT pools. During the search mode, two types of implementations were tested: 1) without the resolution hypothesis, and 2) with the resolution hypothesis. In the case without the resolution hypothesis, the membrane time constants were chosen to be $\tau_k = 30$ ms for all spatial frequencies k. For the implementation with the resolution hypothesis, the membrane time constants for the V1 pools corresponding to the two highest spatial frequencies were $\tau_0 = \tau_1 = 30$ ms, and for the lowest spatial frequency $\tau_2 = 20$ ms. Hence, in this last case, the lowest spatial frequencies produce a temporal processing advantage.

Figure 10.6 shows the dynamical evolution of the polarization P of the pools in the PP module, defined as the difference of the previously defined maximal activity of the target and distractors, i.e. as $P = A_T^{PP} - A_D^{PP}$ (see Section 10.1). The reaction time was defined by determining the point at which the level of polarization of the pools in the PP module P reaches a certain threshold $\theta = 0.05$, meaning that the search target is localized. Figures 10.6a and 10.6b show the dynamical evolution of P for experiments VS1 (global disparity) and VS2 (global similarity), and for different numbers of distractors. Figure 10.6c plots the simulated reaction time as a function of the number of items in the display. When the resolution hypothesis is implemented, VS1 shows a 'parallel' search instead of the 'serial' search observed when the different spatial frequency (octave) latencies in V1 are equalized. On the other hand, in the case of VS2 a 'serial' search is always observed. Only in the case of global disparity does the faster processing of the global structure causes a significant facilitation of the search task, due to an early polarization of the competition towards the location of the target. The implementation of the resolution hypothesis can be described phenomenologically as a kind of 'coarse image' search, in the sense that first the coarse global structure is searched in a scene, and the finer details of the globally preselected spatial regions are analyzed afterwards.

A search of an image based on coarse feature, low spatial frequency, information is only efficient when the global structure of the target is very different from the global appearance of the distractors in the scene. If the scene contains objects which have the same global aspect as a rapid, coarse search, then the low spatial frequency search does not produce any advantage. The simulations explain the neurodynamics underlying this kind of 'impressionist' (low spatial frequency-based) visual processing, which account for the mechanisms through which resolution hypothesis-based processing contributes to visual search.

Summarizing, our cortical model and neurodynamical implementation of the resolution

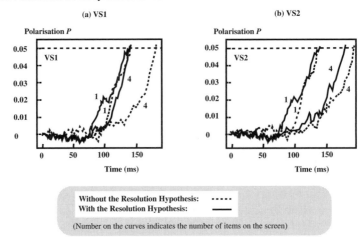

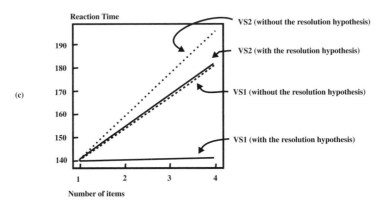

Fig. 10.6 Neurodynamics of visual search of hierarchical patterns. (a) Dynamical evolution of pool activity in the PP module during global disparity visual search VS1; (b) Dynamical evolution of pool activity in the PP module during global similarity visual search VS2; (c) Simulated reaction times for different types and numbers of distractors.

hypothesis predict that visual search of hierarchical patterns takes advantage of the global precedence effect only when the distractors show global differences.

This specific prediction of parallel search for the global differences case, and serial search for the global similarity case, with hierarchical patterns, has now been tested experimentally in human subjects. The preliminary experimental results seem to confirm these predictions (Heinke, Deco, Humphreys and Zihl 2001a). Visual search of hierarchical patterns was tested on 9 normal subjects. Two response buttons of the mouse were interfaced with the computer, and reaction times, contingent on button presses, were recorded. The 17 inch monitor was viewed from a distance of 50 cm. The stimuli consisted of global geometric figures (diamonds or squares) constructed of small letters ('F' or 'E'). Figure 10.7a shows the four kinds of

(a) Hierarchical stimuli

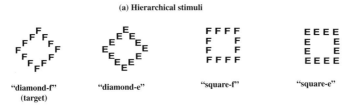

"diamond-f"   "diamond-e"   "square-f"   "square-e"
(target)

(b) Global disparity (VS1)   **(c) Global similarity (VS2)**

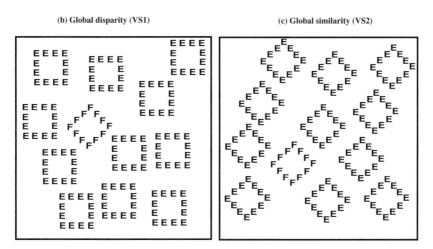

Fig. 10.7 (a) The hierarchical stimuli used as target and distractors for the experimental test in humans of the spatial resolution hypothesis in the framework of a visual search task that involves the global-local paradigm; (b) Typical display for the condition of global disparity (VS1), where the target is defined as a 'diamond-f' and the distractors as 'square-e'; (c) Typical display for the condition of global similarity (VS1), where the target is defined as a 'diamond-f' and the distractors as 'diamond-e'.

stimulus items used as target and distractors for the visual search task. The side length of the global structure consisted of 4 letters. In angles the items covered an area of 1.7 degrees. The letters were 0.28 by 0.28 degrees and the distance between the letters was 0.17 degrees. The stimuli are defined as follows: 1) 'diamond-f': a global diamond constructed of small 'F' letters; 2) 'diamond-e': a global diamond constructed of small 'E' letters; 3) 'square-f': a global square constructed of small 'F' letters; and 4) 'square-e': a global square constructed of small 'E' letters.

Display set sizes of 1, 5, 12, 18 and 25 items occurred equally often on all trials. The items were presented at randomized locations inside a circle within a square of 30 cm. In all search conditions, subjects searched displays in which the target could be present or absent. The target was always the same and was defined *a priori* as a global diamond constructed of small 'F' letters ('diamond-f'). Each display remained until the computer picked up a key press. Timing stopped when the key press was detected. The subjects were instructed to respond as fast and accurately as possible if the target was present (left mouse key) or absent (right mouse key). Each experiment was run as a within-subjects design, with each

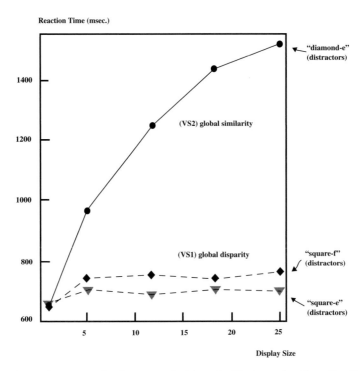

Fig. 10.8 Experimental mean reaction times over all trials and subjects as a function of the display size for a visual search task involving hierarchical patterns. The target item was always a global diamond constructed of small 'F' letters ('diamond-f'). Two conditions are plotted: global similarity (VS2) and global disparity (VS1). In the case of global similarity the distractor items are defined by global diamonds constructed of small 'E' letters ('diamond-e'). In the case of global disparity the distractors are given by global squares constructed of small letters, which were either 'E' or 'F' ('square-e' or 'square-f'). When there is global similarity the search appears to be "serial", whereas when there is global disparity the search appears to be "parallel".

subject taking part in all conditions during a single experimental session. There were three conditions: one corresponding to global similarity (VS2), and two corresponding to global disparity (VS1). In the case of global similarity the distractor items are defined by global diamonds constructed of small 'E' letters ('diamond-e'). In the case of global disparity the distractors are given by global squares constructed of small letters, which were either 'E' or 'F' ('square-e' or 'square-f'). After 15 practice trials the subjects had to complete 480 trials interrupted after every 120 trials by a break. No feedback was given. Figures 10.7b and 10.7c show typical displays for the global disparity and similarity conditions, respectively. Subjects were provided with a series of demonstration trials.

Figure 10.8 shows the mean reaction times over all trials and subjects as a function of the display size for the three conditions. The two conditions that pose a global disparity case (VS1) show a 'parallel' search. On the other hand, in the case of global similarity (VS2), a 'serial' search is observed. Only in the case of global disparity did the faster processing of the global structure (i.e. the global precedence effect) result in a significant facilitation of the search task.

In conclusion, these results confirm the theoretical prediction of our cortical model, and therefore the role of the resolution hypothesis in the attentional neurodynamics involved in the visual search of hierarchical patterns.

10.4 Visual conjunction search

We present, in this Section, an extension of the computational model presented in Chapter 9 that aims to explain the neurodynamics underlying visual search tasks that require the binding of different feature dimensions, i.e. conjunction search tasks.

10.4.1 The binding problem

In the analysis of the visual perception of objects, we can usually decompose perceptual objects into general feature components that are not unique for a particular object. Put differently, the individuality of objects results from the specific composition of elementary features and their spatial relations rather than from the specificity of the component features (see Section 8.2). Therefore, the brain should be able to represent objects by representing each elementary feature component and binding together the feature components constituting a particular object to distinguish them from features of other objects. The problem of grouping together the single feature components constituting an individual object is the so-called binding problem (Treisman and Gelade 1980, Treisman 1988, Gray and Singer 1987, Gray and Singer 1989, Eckhorn, Bauer, Jordan, Brosch, Munk and Reitboeck 1988). In brief, the binding problem in perception deals with the question of how we specify what goes with what and where, in order to achieve the perception of a coherent integration of shape, colour, movement, and size in individual objects. Three main theories have been formulated for achieving such linking of features corresponding to individual objects. The first theory (see Section 8.2.5 and Barlow (1972)) assumes the existence of a convergence of connections from neurons which respond to different elementary feature dimensions onto higher-order neurons that react to a specific combination of features. The highest level corresponds to a grandmother neuron, which responds only when a specific object is presented. Perrett, Rolls and Caan (1982) and Tanaka (1996) have shown that some single neurons in area IT respond specifically to relatively complex combination of features. These neurons are sensitive to the spatial arrangement of the features in the objects, and thus the features must be bound together with the correct spatial relations (Rolls, Tovee, Purcell, Stewart and Azzopardi 1994b, Tanaka 1996). A potential problem of this approach is the combinatorial explosion of the number of grandmother neurons required for the specific selection of all possible objects (i.e., of all possible combinations of single features) (Barlow 1972). Figure 10.9a illustrates this hypothesis schematically. Barlow's theory therefore achieves binding by convergence of fixed connections that link neurons coding component features with neurons coding the entire object. It is shown in Chapter 8 that a solution to this potential problem is to form neurons that encode only low-order combinations of features, to utilize a multi-layer hierarchical network architecture, and to take advantage of the statistical regularities of natural images.

A second alternative theory assumes a combination of the cell assembly theory of Hebb (1949) and the idea of timing coding (see Malsburg (1973), and Singer (1994), (1999) and (2000) for reviews). The basic idea is that neurons responding to single features and thereby

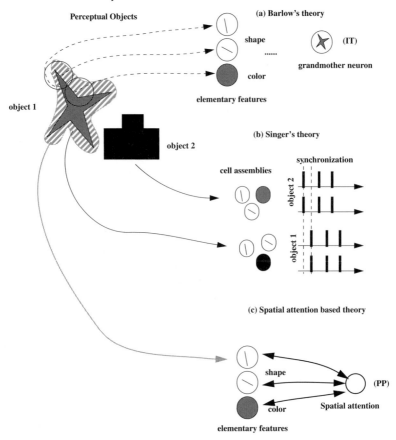

Fig. 10.9 Binding problem. (a) Barlow's theory of grandmother neurons (binding by convergence); (b) Singer's theory of cell assemblies (binding by synchronization); (c) Spatial attention-based integration of elementary features.

constituting an individual object are dynamically linked by the synchronization of their spikes. In other words, a cell assembly is a group of neurons firing coherently and coding the existence and binding of single features in a specific perceptual object (see Fig. 10.9b). In this case, there is no longer a combinatorial explosion of the number of neurons required for coding all possible objects. Experimental support for this theory is described in the above reviews and considered in Section 5.5.7. Nevertheless, this theory does not solve the binding problem as it was posed above. Synchrony is a possible way of holding on to a solution, of tagging the neurons that are responding to the same object once they have been identified, but we still need a way of finding which neurons those are, and of representing the spatial representations between the features (see Section 13.7).

The third theory suggests a possible mechanism for binding across dimensions through shared spatial locations (Treisman 1982, Treisman 1988). The idea is that one can code one object at a time, selected on the basis of its location at an early level where receptive fields are small. Spatial attentional mechanisms select that location by temporarily excluding stimuli from other spatial locations. All feature dimensions currently attended to are therefore bound. (A limitation of this approach is that it does not actually define the spatial relations between the parts, as the theory was developed to deal with for example binding colour and shape in a way which might be needed to distinguish a red triangle from a green square. In contrast, the theory developed in Chapter 8 does enable the spatial relations between features to be specified in the encoding scheme, as addressed specifically in Section 8.4.5). In this Section (10.4) though, we consider how local spatial attentional mechanisms can help to ameliorate the binding problem. The argument is as follows. The mutual biasing attentional interaction between the posterior parietal module, and the different early level modules (V1, V2, V4, ...) coding independent feature dimensions, put 'what' and 'where' together. The neural activity that dynamically emerges in the posterior parietal cortex selects a spatial location that is effectively object-based, and links together (or selectively enhances) the corresponding elementary features (shape, colour, motion, etc.). This is achieved by modulating the neural activity in the respective feature maps, so that only neurons associated with the spatially attended location remain active, and by temporarily suppressing the response of the other neurons corresponding to objects at other locations. Consequently, only the attended features corresponding to one object will be further analyzed at the object perception level implemented by the inferotemporal module. We suggest that illusory conjunctions[33] arise when this attentional process is given insufficient time to settle dynamically to allow high activity to occur only in corresponding locations in for example separate feature and colour maps.

Before presenting the details of the extension of our computational model for achieving binding (or more precisely selective enhancement) of multiple feature dimensions, we review briefly in the next subsection the main psychophysical findings from complex conjunction visual search experiments. Then, in the following subsections we show that our extended model can reproduce these experimental results, providing support therefore for a contribution to spatial binding of the spatial attention mechanisms that are part of the theory described in Chapters 9–11.

10.4.2 The time course of conjunction search: experimental evidence

Quinlan and Humphreys (1987) reported different time behaviours for visual search experiments in which the number of relevant shared features between targets and distractors was manipulated. They analyzed feature search and three different kinds of conjunction search, namely: standard conjunction, and two kinds of triple conjunction with the target differing from all distractors in one or two features, respectively (see Chapter 6). Let us define the different kinds of search tasks by using a pair of numbers m and n, where m is the number of distinguishing feature dimensions between target and distractors, and n is the number of features by which each distractor group differs from the target. Using this terminology, feature search corresponds to a 1,1-search; a standard conjunction search corresponds to a 2,1-search; a triple conjunction search can be a 3,1 or a 3,2-search depending of whether the target differs from all distractor groups by one or two features, respectively.

Quinlan and Humphreys (1987) showed that in feature search (1,1) the target is detected in parallel across the visual field. Furthermore, they show that the reaction time in both standard conjunction search and triple conjunction search conditions is a linear function of the display

---

[33] Illusory conjunctions are situations in which the binding of for example shape and colour occur incorrectly, e.g. when shape and colour may be misattributed between shapes with different colours.

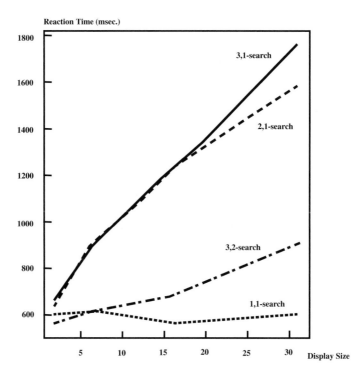

Fig. 10.10 Experimental search times for feature and conjunction searches measured by Quinlan and Humphreys (1987). The different kinds of search task are defined by using a pair of numbers m and n, where m is the number of features, and n is the number of features by which each distractor group differs from the target. Feature search corresponds to a 1,1-search; a standard conjunction search corresponds to a 2,1-search; a triple conjunction search can be a 3,1 or a 3,2-search depending on whether the target differs from all distractor groups by one or two features, respectively.

size. The slope of the function for the triple conjunction search task can be steeper or relatively flat, depending upon whether the target differs from the distractors in one (3,1) or two features (3,2), respectively. Figure 10.10 shows graphically the different slopes of reaction time as a function of the display size determined by Quinlan and Humphreys (1987). The reaction times for feature and standard conjunction search are calculated by averaging over experiments one, two and three of Quinlan and Humphreys (1987).

A realistic theoretical model of attentional dynamics in the context of multiple feature component binding should be able to distinguish between the different slopes associated with each type of visual search. In the next section, we extend the neurodynamical model described in Chapter 9 in order to achieve feature binding, and later we demonstrate that our computational model can account for the different slopes observed experimentally in complex conjunction visual search tasks.

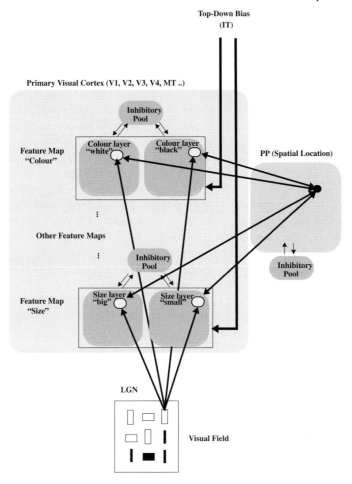

Fig. 10.11 Extended cortical architecture for visual attention and binding of multiple feature components (shape, colour, movement, ...). See text for details.

10.4.3 Extension of the computational cortical architecture

In this section, we extend the cortical architecture introduced in Chapter 9 in order to consider the binding of multiple feature components. We perform this extension in the framework of conjunction visual search. Figure 10.11 shows the overall hierarchical architecture of our extended model. The input retina is given as a matrix of visual items. The location of each item on the retina is specified by two indices ij, describing the position in row i and column j. The dimensions of this matrix are $S \times S$, i.e. the number of items in the display (display size) is S^2. Information is processed across the different spatial locations in parallel. Different feature maps in the primary visual cortex extract the local values of the features for an item at each position[34].

---

[34] As particular cases, some of these maps are the shape-maps given by pools extracting Gabor feature components such as that introduced in Section 9.2.3.

We hypothesize that selective attention results from independent competition mechanisms operating within each feature dimension. Let us assume that each visual item can be defined by M features. Each feature m can adopt $N(m)$ values, for example the feature 'colour' can have the values 'black' or 'white' (in this case $N(\text{colour}) = 2$). For each feature map m, there are $N(m)$ layers of neurons characterizing the presence of each feature value. A cell assembly consisting of a population of fully connected excitatory integrate-and-fire spiking neurons (pyramidal cells) is allocated to every location in each layer, in order to encode the presence of a specific feature value (e.g. colour 'white') at the corresponding position. This generates a sparsely distributed representation of the stimulus, in the sense that the activity of a population of neurons represents the presence of different features at a given position. The feature maps are topographically ordered, i.e. the receptive fields of the neurons belonging to cell assembly ij in one of these maps are limited to the location ij in the retinal input. We further assume that the cell assemblies in layers corresponding to one feature dimension are mutually inhibitory (e.g. at a given position the cell assembly coding the colour feature value 'white' inhibits the cell assembly coding 'black'). Let us denote by $A^{\mathrm{F}}_{pqmn}(t)$ the activity level of an excitatory pool at a location in the visual field, in the feature map m and layer (i.e. value) n.

The posterior parietal module is bidirectionally coupled with the different feature maps and serves to bind the different feature dimensions at each item location, in order to implement local conjunction detectors. The mutual coupling between a pool A^{F}_{pqmn} in the primary visual cortex and a posterior parietal pool A^{PP}_{ij} is defined, as before, by the Gaussian-like topographic connection given by equation 9.6.

We assume that the inferotemporal connections provide top-down information, consisting of the feature values for each feature dimension of the target item. We do not explicitly model the dynamics of the memory module that provides this information but only its output, i.e. the target definition. This information is fed into the system by including an extra excitatory input to the corresponding feature layers. For example, if the target is defined as small, vertical and black, then all the excitatory pools at each location in the layer coding 'small' in the feature map dimension 'size', in the layer coding 'vertical' in the feature map 'orientation', and in the layer coding 'black' in the feature map 'colour', receive an extra excitatory input from the inferotemporal (IT) module.

Let us now define the neurodynamical equations that regulate the evolution of the extended cortical system. The activity levels of the excitatory and the corresponding inhibitory pools in the feature maps are given by

$$
\tau \frac{\partial A^{\mathrm{F}}_{pqmn}(t)}{\partial t} = -A^{\mathrm{F}}_{pqmn} + \alpha \mathrm{F}(A^{\mathrm{F}}_{pqmn}(t)) - \beta \mathrm{F}(A^{\mathrm{I,F}}_{m}(t)) + I^{\mathrm{F}}_{pqmn}(t)
$$
$$
+ I^{\mathrm{F-PP}}_{pq}(t) + I^{\mathrm{F,A}}_{mn}(t) + I_0 + \nu \tag{10.3}
$$

$$
\tau_P \frac{\partial A^{\mathrm{I,F}}_{m}(t)}{\partial t} = -A^{\mathrm{F,V1}}_{m}(t) + \gamma \sum_{p}^{S^2} \sum_{q}^{S^2} \sum_{n}^{N(m)} \mathrm{F}(A^{\mathrm{F}}_{pqmn}(t)) - \delta \mathrm{F}(A^{\mathrm{I,F}}_{m}(t)) \tag{10.4}
$$

where the attentional biasing coupling $I^{\mathrm{F-PP}}_{pq}$ due to the intermodular 'where' connections with the pools in the parietal module PP is given by

$$
I^{\mathrm{F-PP}}_{pq} = \sum_{i,j} W_{pqij} \mathrm{F}(A^{\mathrm{PP}}_{ij}(t)) \tag{10.5}
$$

and the external attentional bias coming from the infero temporal areas $I_{mn}^{F,A}$ is equal to 0.005 for the layers which code the target properties and 0 otherwise. I_{pqmn}^{F} is the sensory input to the cells in feature map m sensitive to the value n and with receptive fields at the location pq on the retina. This sensory input I_{pqmn}^{F} characterizes the presence of the respective feature value at the corresponding position. A value of 0.05 corresponds to the presence of the respective feature value and a value of 0 to the absence of such value.

The posterior parietal integrating assemblies are also described by a system of differential equations. The activity of the excitatory pools in the posterior parietal module PP is given by

$$\tau \frac{\partial A_{ij}^{PP}(t)}{\partial t} = -A_{ij}^{PP} + \alpha F(A_{ij}^{PP}(t)) - \beta F(A^{I,PP}(t)) + I_{ij}^{PP-F}(t)$$
$$+I_0 + \nu \tag{10.6}$$

where intermodular attentional biasing I_{ij}^{PP-F} through the connections with the pools in the feature maps is

$$I_{ij}^{PP-F} = \sum_{p,q,m,n} W_{pqij} F(A_{pqmn}^{F}(t)) \tag{10.7}$$

and the activity of the common PP inhibitory pool evolves according to

$$\tau_P \frac{\partial A^{I,PP}(t)}{\partial t} = -A^{I,PP}(t) + \gamma \sum_i^{S^2} \sum_j^{S^2} F(A_{ij}^{PP}(t)) - \delta F(A_{ij}^{PP}(t)) \tag{10.8}$$

The additive Gaussian noise ν considered has a standard deviation of 0.002. The synaptic time constants were $\tau = 5$ ms for the excitatory populations and $\tau_P = 20$ ms for the inhibitory pools. The synaptic weights chosen were: $\alpha = 0.95, \beta = 0.8, \gamma = 2,$ and $\delta = 0.1. I_0 = 0.025$ is a diffuse spontaneous background input.

The system of differential equations (10.3)–(10.8) was integrated numerically until a convergence criterion was reached. This criterion was that the neurons in the PP module must be polarized, i.e.

$$F(A_{i_{max}j_{max}}^{PP}(t)) - \frac{1}{(S^2-1)} \sum_{i \neq i_{max}} \sum_{j \neq j_{max}} F(A_{i_{max}j_{max}}^{PP}(t)) > \theta \tag{10.9}$$

where the index $i_{max}j_{max}$ denotes the cell assembly in the PP module with maximal activity, and the threshold θ was chosen equal to 0.15. The second term in the l.h.s of the last inequality measures the mean distractor activity.

Let us now qualitatively analyze the dynamical behaviour of the system. The dynamics of the system, i.e. the temporal evolution of the activity level of each pool in the feature and posterior parietal maps, yields the formation of a focus of attention without explicitly assuming a spotlight, i.e. without the necessity of assuming a special serial scanning process. At each feature dimension the fixed point of the dynamics is given by a high activation of cell assemblies in the layers coding feature values which are shared by the target and sensitive to locations with items sharing this value. The remaining cell assemblies do not show any important activation. For example, if the target is 'black', in the colour map the activity in the 'white' layer will be suppressed, and the cell assemblies corresponding to 'black' items will be enhanced. This process implements a first competitive mechanism at the level of each feature dimension. The competitive mechanisms in each feature dimension are independent. In a given feature dimension, the pools coding different values (corresponding to different layers) compete at all locations with each other through the lateral inhibition given by the common inhibitory pool associated with the respective feature dimension. Only the pools receiving both excitatory inputs, i.e. the positive top-down input from the inferotemporal (IT) module, and the sensory input associated with the feature value at the corresponding location, will be able to win the competition. Therefore, if we are looking for 'black' objects, then the 'black' pools at locations corresponding to 'black' items will win the competition. In the posterior parietal map, the populations corresponding to locations which are simultaneously activated to a maximum in all feature dimensions will be enhanced, suppressing the others. In other words, the location that shows all feature dimensions corresponding to the top-down specification of the target is stimulated, and it will be further enhanced from top-down feedback when the target is at this location. This implements a second competitive mechanism at the level of conjunctive features. The latency of the dynamics, i.e. the speed of convergence, depends on all the competitive mechanisms. A way of analysing convergence is by monitoring the activity state just in the top posterior parietal map, because the state of this module reflects the total state of the dynamics of the system. Convergence depends therefore on the level of conflict within each feature dimension, i.e. on the properties of the distractors.

The whole system analyzes the information at all locations in parallel. Longer search times correspond to slower dynamical convergence at all levels[35]. In addition, as we will see in the next subsection, the model demonstrates that linear search-time functions with display size, usually associated with serial search, can be obtained as a result of parallel processing based on lateral inhibition, without the need to assume a special serial scanning process.

10.4.4 Computational results

In this section we present results simulating the visual search experiments of Quinlan and Humphreys (1987) involving feature and conjunction search. Let us assume that the items are defined by three feature dimensions ($M = 3$, e.g. size, orientation and colour), each having two values ($N(m) = 2$ for $m = 1, 2, 3$, e.g. size: big/small, orientation: horizontal/vertical, colour: white/black). Figure 10.12 shows examples for each kind of search. For each display size we repeat the experiment 100 times, each time with different randomly generated targets (i.e. different feature conjunctions), at random positions, and randomly generated distractors (depending on the target defined and the type of search task). To show the results, we plot the mean value T of the 100 simulated search times (in ms[36]) as a function of the display size. The search times are defined by the number of simulation steps required for the system to converge.

The results obtained for 1,1; 2,1; 3,1 and 3,2-searches are shown in Fig. 10.13. The slopes of the search time vs. display size curves for all simulations are consistent with the existing experimental results (Quinlan and Humphreys 1987). In feature search (1,1) the target is detected without any effects of display size on search time. Furthermore, the standard conjunction search and triple conjunction search tasks generate reaction times that are a linear function of the display size. The slope of the function for the triple conjunction search task can be steeper or relatively flat, depending upon whether the target differs from the distractors

---

[35] Even the effect of learning in visual search (Sireteanu and Rettenbach 1995) can be associated with faster dynamics at all levels.

[36] We assume that one time step update for the model is equivalent to one ms.

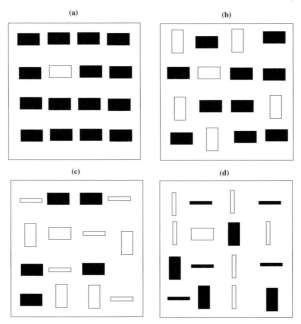

Fig. 10.12 Examples of visual search: (a) feature search (1,1-search). (b) standard conjunction search (2,1-search). (c) triple conjunction search with the target differing from all distractors in one feature (3,1-search). (d) triple conjunction search with the target differing from all distractors in two (3,2-search) features.

in one (3,1) or two features (3,2), respectively.

In order to illustrate why appropriate search functions emerge from the model, we demonstrate the dynamics of the search using a simple example. Let us assume that the display size is 4, i.e. 4 items positioned in a 2 x 2 matrix, and that the target is located at the top left position of the matrix in all cases. The saliency at each input location associated with each feature dimension can be illustrated by considering a coding matrix. The values of the matrix elements are either $(+1)$ for the items with the target feature value and (-1) for the items with feature values different from the target. Additionally, we define a 'Total Saliency' matrix, which represents the saliency after conjoining the different feature dimensions, and which is given by the sum of the saliency matrices of each feature dimension. We introduce these matrices only for illustration. (The model does not use this kind of saliency matrix.) Table 10.1 shows an example. If only one saliency matrix guides the competition between the different input locations, then the slopes of the search will be characterized by the level of conflict expressed by the polarization of the saliency matrix, i.e. how much more salient is one position (target) with respect to the others (distractors). If the search is guided by just one competition mechanism after the conjoining of feature dimensions at each location, then only the 'Total Saliency' matrix guides the competition. From Table 10.1, the 'Total Saliency' matrices show the same level of polarization to the location of the target for the different search types, and therefore such a mechanism cannot yield different slopes.

On the other hand, if we assume that the search is guided by several saliency matrices, then the slopes of the search will be characterized by the level of conflict expressed now by the

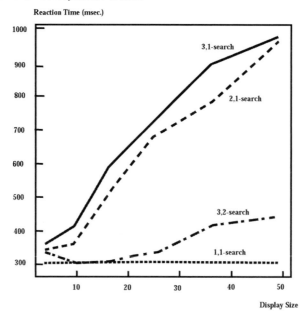

Fig. 10.13 Search times for feature and conjunction searches obtained utilizing the extended computational cortical model.

polarization of each saliency matrix to different conflicting positions. If the search is guided by a competition mechanism that takes place independently at the level of each feature map, i.e. prior to conjoining the feature dimensions, then the saliency matrices of features 1, 2 and 3 (see Table 10.1) guide the search. The polarization of the saliency matrix corresponding to a given feature dimension provides the input position relevant for the search with respect to that feature dimension. We have different levels of conflict between the saliency matrices corresponding to each feature dimension for different search types, and therefore a mechanism that can yield different slopes. In the case of search 1,1, only the matrix corresponding to feature 1 shows polarization at the position of the target, while the other two matrices do not show polarization at all (the same value of 1 is present at each position). Consequently, there is no conflict, and therefore search is very fast. Furthermore, increasing the number of inputs does not change the level of conflict, and therefore search will be parallel, i.e. at feature 1 we will have always a polarization to the position of the target, while at the other feature dimensions 2 and 3 there will be no polarization at all. Search 3,1 shows more conflicts between the different dimensions than search 1,1. In the case of 3,1-search, each feature dimension has a corresponding saliency matrix, all of which are polarized at different conflicting positions. For example if the target is 'black' and 'small' there will be several positions which are salient from the point of view of the colour, and there will also be other conflicting positions which are also salient from the point of view of the feature size. If the number of distractors increases, then the level of conflicting positions also increases, and therefore search will be serial. Search 3,1 shows more conflicts than 2,1, and 2,1 more than 3,2. As a consequence the slopes increase, as shown by the experiments (Fig. 10.10) and by the simulations (Fig. 10.13). In other words, what matters is the polarization of each saliency matrix at different conflicting positions, and not the sum of the polarizations (corresponding to

Table 10.1 Illustration of the different levels of competition

Search Type	Saliency of Feature 1	Saliency of Feature 2	Saliency of Feature 3	Total Saliency (conjunction)
1,1	$\begin{bmatrix} +1 & -1 \\ -1 & -1 \end{bmatrix}$	$\begin{bmatrix} +1 & +1 \\ +1 & +1 \end{bmatrix}$	$\begin{bmatrix} +1 & +1 \\ +1 & +1 \end{bmatrix}$	$\begin{bmatrix} +3 & +1 \\ +1 & +1 \end{bmatrix}$
2,1	$\begin{bmatrix} +1 & -1 \\ +1 & -1 \end{bmatrix}$	$\begin{bmatrix} +1 & +1 \\ -1 & +1 \end{bmatrix}$	$\begin{bmatrix} +1 & +1 \\ +1 & +1 \end{bmatrix}$	$\begin{bmatrix} +3 & +1 \\ +1 & +1 \end{bmatrix}$
3,2	$\begin{bmatrix} +1 & -1 \\ +1 & -1 \end{bmatrix}$	$\begin{bmatrix} +1 & -1 \\ -1 & +1 \end{bmatrix}$	$\begin{bmatrix} +1 & +1 \\ -1 & -1 \end{bmatrix}$	$\begin{bmatrix} +3 & -1 \\ -1 & -1 \end{bmatrix}$
2,1	$\begin{bmatrix} +1 & -1 \\ +1 & +1 \end{bmatrix}$	$\begin{bmatrix} +1 & +1 \\ -1 & +1 \end{bmatrix}$	$\begin{bmatrix} +1 & +1 \\ +1 & -1 \end{bmatrix}$	$\begin{bmatrix} +3 & +1 \\ +1 & +1 \end{bmatrix}$

the conflict level after conjunction). The level of conflict yields different dynamical latencies and therefore different search times.

In order to analyze in more detail the dynamical development of the system in Fig. 10.13 we plotted the temporal evolution of the rate activity corresponding to the target and to the distractors at the posterior parietal 'integration' map, and also separately at each feature dimension level, for each kind of visual task. We first consider the different types of search for a fixed display size in order to analyze the different levels of conflict and competition associated to the different definitions of the distractor. The display size used was 25. The level of conflict is reflected in the latency of the dynamics of the system. The convergence of the dynamics can be monitored by the state of polarization of the activity in the posterior parietal map, as shown in Fig. 10.14a. The left part of Fig. 10.14a shows the activity for the posterior parietal map corresponding to the pool at the target position, and the mean activity of the remaining pools coding the distractors. The right part of Fig. 10.14a shows the polarization of the posterior parietal activity, i.e. the difference between the target activity and the mean distractor activity. This difference provides the criterion for convergence. When the polarization state is large enough (i.e. larger than a given threshold) then we consider that the system has converged, and the corresponding time is taken as the search time. Particularly on the right part of Fig. 10.14a, the different latencies of convergence can be observed.

It is interesting to note that in the case of 1,1 and 3,2-search, the convergence times for all levels are very short, consistent with the standard interpretation of this kind of search as spatially parallel. In the case of 3,1-search, the latency of the dynamics is longer indicating apparently a "serial" component in the search, although the underlying mechanisms are working in parallel. In this case (see Fig. 10.14d) the strong competition present in each feature dimension delays the convergence of the dynamics, both within each feature dimension and also within the posterior parietal map. Note that in Fig. 10.14d the slow suppression of the distractor activity reflects the underlying competition. It is also interesting to note the dynamical development of the distractor-related activity at the feature map level for the 1,1-search in Fig. 10.14b. In this case only the distractor activity corresponding to the relevant feature dimension (colour) is suppressed. The other feature dimensions show always more or less the same activity, since there is no competition at all at the corresponding feature maps

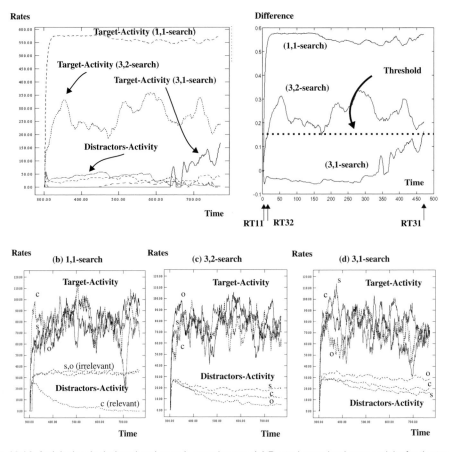

Fig. 10.14 Activity levels during visual search experiments. (a) Posterior-parietal-map activity for the target and for the mean value for the distractors (left), and their difference (right). The three search times RT11, RT32 and RT31 are indicated by arrows which denote the intersection with the threshold. (b) Feature-level map activity values for the target and one distractor for 1,1-search. There is one curve for each feature dimension (i.e. 3 for the target and 3 for the distractor). The corresponding feature dimensions are indicated by the letters 's' for size, 'o' for orientation, and 'c' for colour. (c) Same as (b) but for 3,2-search. (d) Same as (b) but for 3,1-search. The time scale is given in milliseconds.

(all items have the same values for the other two irrelevant dimensions). Furthermore, the convergence of the distractor and target activity at the level of the relevant feature map is very fast, because the conflict in this case is very easy to resolve (one 'black' item and the rest 'white'), yielding rapid convergence at the posterior parietal module (Fig. 10.14a).

In order to demonstrate the effect of display size (i.e. the number of distractors) on the latency of the system dynamics, Fig. 10.15 shows the dynamical development of the polarization of activity at the posterior parietal map for different set sizes in the case of a 3,1-search. When the number of distractors increases, the level of conflict, especially at the posterior parietal map, also increases. This results in a slower convergence, which yields longer search times for increasing numbers of distractors. The typical linear increase in

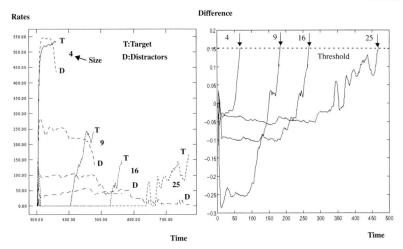

Fig. 10.15 Posterior parietal module activity rates for the target and for the mean of the distractors (left), and their difference (right), for different display sizes in the case of a 3,1-search. The search times are indicated by arrows which denote the intersection with the threshold. The numbers on the curves denote the corresponding display size.

the search time with the set size is clearly obtained as the result of a slower convergence (latency) of the dynamics. Linear increases in search times can be explained without special serial scanning mechanisms. In the case of negative trials, i.e. when no target is present, the difference between the rate of activity of the target and the mean rate of activity of the distractors does not reach the threshold and therefore the system does not stop. How subjects decide to generate a response in such a case is not included in our model.

10.5 Conclusion

In this Chapter, the neurodynamical model of attention described in Chapter 9 was extended to deal with the global precedence effect, in which low spatial frequencies in stimuli appear to be processed more rapidly than high spatial frequencies. The change to the model took the form of setting the membrane time constants of the neurons processing low spatial frequencies to have shorter values than those processing high spatial frequencies. With this change, the model was able to account for global precedence effects, and to account for how the high spatial resolution processing of fine detail in objects tends to settle where the low spatial frequency channels are already settling. The model led to a prediction which was confirmed in a new psychophysical experiment in humans in which both local and global features are present in stimuli.

The model was also extended to deal with conjunction feature search, that is the search for objects which are defined by a particular combination of features in an image. The extension involved adding several feature dimensions (e.g. orientation, colour, and size), providing for independent competition within each of the feature dimensions, and then providing a second level of competition, in the spatial, parietal, module in the network, whereby the neuronal populations that are activated by maximal inputs from all the feature dimension pools win the competition. The network is primed by a top-down feature conjunction bias to the object module, and correctly finds the location in the display where the conjunctive features are present.

In both types of attentional mode, the network produces response latencies that appear to be either serial or parallel, and which mimic human search in the same conditions. However, the network has no explicit serial search mechanism, but is instead a massively parallel network in which the settling time depends on how easily the constraints are satisfied. Indeed, neither explicit serial focal search nor saliency maps need to be assumed to account for performance in these search tasks. During the dynamical evolution of the network, information about the saliency at each input location emerges in the PP module. In object search the focus of attention is not included in the system, but only results after convergence of the dynamical behaviour of the neural networks. Attention and priority maps are emergent properties in our system, rather than separate mechanisms operating independently of other perceptual and cognitive processes. The neural population dynamics is described formally by a system of coupled differential equations derived in the framework of the mean-field approximation.

Of course, the fact that some apparently "serial" attentional processes in visual search can be explained without an explicitly serial mechanism but instead by a parallel dynamical system, does not imply that every attentional process is due to parallel dynamics. Indeed, our formulation applies to the covert situation where search using eye movements is excluded. In the case of overt attentional shifts eye movements (saccades) are made, and this is a much more serial process, though where the next eye movement is made to could be influenced by covert search.

The computational perspective that we have used provides not only a concrete mathematical description of all the mechanisms involved in the phenomenological and functional view of the problem, but also a model that allows a complete simulation of psychophysical experiments. Even more, the model can be used to help understand vision when different parts of the system are damaged in patients, and this is what we turn to in the next Chapter.

11 A computational approach to the neuropsychology of visual attention

11.1 Introduction

A computational perspective provides not only a concrete mathematical description of all the mechanisms involved in visual attention, but also a model that allows a complete simulation of psychophysical experiments. In addition, disruption of the operation of some of the computational modules corresponding to specific brain areas can be used in simulations to help understand and predict impairments in visual information processing in patients suffering from brain injury. In this Chapter, we study the computational neuropsychology of visual attention.

First, using as a computational basis the model described in Chapter 9, we investigate the specific visual cognitive impairment in brain-damaged patients known as visual spatial neglect. By damaging the model in different ways, a variety of dysfunctions associated with visual neglect can be simulated and explained by a disruption of specific subsystems. Essentially, the damage destabilizes the underlying intra- and inter-modular mutually biasing neurodynamical competition that macroscopically yields the functional deficits observed in visual neglect patients. In particular, we are able to explain the asymmetrical effect of spatial cueing on neglect, and the phenomenon of extinction in the framework of visual search. Also, we predict that patients with neglect can generate different saccade patterns depending on whether the object under observation is known or not. This finding explicitly demonstrates one of the main characteristics of 'active' vision, namely the influence of top-down processing, and its neuropsychological relevance.

Later on, in the context of the spatial resolution hypothesis described and analyzed in Chapter 10, we show that the model predicts a neuropsychologically relevant new impairment during the visual search for a target in conditions where the stimuli have either global (large stimulus features) differences or similarity. Finally, in the context of conjunction search, we show that the model predicts new behaviours for novel forms of conjunction search, and then describe tests of the predictions in normal human subjects. We thus show that neuropsychologically relevant modifications of visual search tests can be designed based on the present theoretical and computational framework.

11.2 The neglect syndrome

In neurological patients, attention deficiencies can result from damage to different brain regions. For example, unilateral parietal cortex damage can lead to symptoms of the neglect syndrome, in which patients fail to notice objects or events in the hemispace opposite to their lesion site (Humphreys and Riddoch 1994, Vallar and Pernai 1986, Samuelsson, Jensen, Ekholm, Naver and Blomstrand 1997). A significant characteristic of the neglect syndrome is extinction, the failure to perceive a stimulus contralateral to the lesion (contralesional) when a stimulus is simultaneously presented ipsilateral to the lesion (ipsilesional). In fact, extinction shows that neglect is not caused by a visual sensory defect, but rather by a disorder in visually attending. Despite the strong interest in the neglect syndrome, the underlying causes leading to this disorder are still obscure and an issue of controversy (Halligan and Marshall 1994, Rafal and Robertson 1997). A systems-level explanation of the neuropsychological findings on visual neglect would help us to better understand the mechanisms underlying the representation of visual space and the control of visual attention in space, and thus, the cognitive functioning of the brain (Kosslyn 1994). In this Section, we are interested in an attentional account of this neglect syndrome. We analyze the relation between visual neglect and the underlying neural basis of visual attention in the framework of the biased neurodynamical competition inherent in the computational cortical model introduced in Chapter 9. We try to understand the kinds of neuropsychological dysfunction that are produced after damage to specific subsystems.

Unilateral visual neglect is far from being a unitary phenomenon. Humphreys and Heinke (1998) analyzed different forms of unilateral visual neglect experimentally and theoretically (see also Heinke and Humphreys (1999)). Some patients have deficits in reacting to stimuli presented on one side of space relative to their body (Halligan and Marshall 1994). Let us call this kind of syndrome space-based neglect. Others show a form of neglect which appears to be independent of the spatial location of the object in the visual field relative to the viewer's body (Tipper and Behrmann 1996, Behrmann and Moscovitch 1994, Arguin and Bub 1993, Driver and Halligan 1991, Young, Newcombe and Ellis 1991). For instance, several studies have now found that patients may neglect the left side of an object (or perceptual group) whether or not it appears in the left or right hemifield with respect to the patient's body(Humphreys, Olson, Romani and Riddoch 1996, Walker 1995, Halligan and Marshall 1994, Arguin and Bub 1993, Driver, Baylis and Rafal 1992). This kind of visual disorder is called object-based neglect because only the left part of the object is neglected even when the object is rotated so that its left part fell in the right visual hemifield, or the object is translated to the right visual hemifield. In other words, in object-based neglect, parts (left or right) with respect to the object rather than with respect to egocentric space seem to be ignored.

11.2.1 A model of visual spatial neglect

Theoretical models of visual neglect can usually be divided into approaches based on a representational or on an attentional account of the syndrome (Pouget and Driver 1999, Humphreys and Heinke 1998, Bisiach 1996, Driver, Baylis, Goodrich and Rafal 1994). A representational account interprets neglect as the result of impairment of one side of a particular spatial representation, whereas an attentional account considers neglect as a deficit in orienting visual attention to the affected hemispace. The attentional account is strongly favoured by two types of evidence: first, the asymmetrical effect of spatial cueing on neglect (Posner, Walker, Friedrich and Rafal 1984, Rafal and Robertson 1997) (see Section 11.2.2); and second the phenomenon of extinction in the framework of visual search (Eglin, Robertson and Knight 1989) (see Section 11.2.3). In both cases, the degree of impairment increases when the stimulus on the affected side has to compete with a second stimulus at the unaffected side, relative to the condition when only one stimulus is on the affected side. This kind of asymmetry is consistent with the idea that the stimulus in the neglected field attracts attention only in a weak way. In the next subsections, we explain both experimental results in the theoretical framework of our cortical model based on the biased competition hypothesis. Our account of visual neglect is based on different kinds of damage to the posterior parietal module, which is the module representing and controlling the spatial location of visual attention.

Before modelling visual neglect, let us briefly mention the different syndrome of hemi-anopia. Hemianopic patients do not perceive information coming from one hemifield. This kind of disorder can be accounted for by damage to the early visual sensory system, at the level of the LGN cells in the thalamus or the cells in the primary visual cortex. In our model this can be simulated easily by cutting the input coming from the impaired hemifield, or by silencing in the primary module V1 the output of the neuronal pools whose receptive fields cover locations in the impaired hemifield. This type of disruption causes the system to behave in the same way as a patient with hemianopia, i.e. the subject will not perceive information coming from one hemifield. Objects or parts of objects falling in the impaired hemifield will be ignored as if they did not exist. This kind of visual disorder is not a true neglect because only the low-level system is disrupted. In fact, patients with damage to the visual system at this level do not show visual neglect, but do show impaired visually guided oculomotor scanning (Zihl 1995).

Let us now concentrate on visual neglect. We consider in this section two kinds of neglect: first, space-based; and second, object-based. (The object-based neglect considered in the simulations described next is for the case when there is one object in the visual field, the left half of which is neglected independently of where the object is in the visual field. We discuss the case of two objects in the visual field below.) We show that space-based visual neglect can be explained by unilateral damage to the PP module of our cortical architecture (Deco and Zihl 2001c). In outline, let us divide the neuronal pools in the PP module into two pools, for the left and right hemispheres (Fig. 11.1). The left (right) hemisphere is associated with the right (left) visual field. A group of neuronal pools in a given module can be impaired in an intrinsic way (Humphreys and Heinke 1998) by damaging only intrinsic inputs within the module. In mathematical terms, the intrinsic lesioning of a neuronal pool in the PP module is described by extending equation (9.11) to:

$$\tau\frac{\partial A_{ij}^{\mathrm{PP}}(t)}{\partial t} = -A_{ij}^{\mathrm{PP}} + L_{ij}\left\{\alpha\mathrm{F}(A_{ij}^{\mathrm{PP}}(t)) - \beta\mathrm{F}(A^{\mathrm{I,PP}}(t)) + I_0\right\} + I_{ij}^{\mathrm{PP-V1}}(t)$$
$$+I_{ij}^{\mathrm{PP,A}} + \nu \tag{11.1}$$

where L_{ij} is a lesioning factor. Values of L_{ij} equal to 1.0 leave the corresponding neuronal pools unaffected, whereas values of L_{ij} smaller than 1.0 damage the corresponding neuronal pools by reducing the influence of the intrinsic inputs A_{ij}^{PP}, $A^{\mathrm{I,PP}}$, and I_0.

Although visual neglect can be present in the left or right hemispace, we refer in the following to left-sided visual neglect (produced by damage to the right parietal cortex), because this type is more frequently observed (Vallar and Pernai 1986). However, the model can also be applied to right-sided visual neglect. We consider two different types of lesion pattern.

The first type of lesion pattern is associated with purely unilateral damage to the PP module in the right hemisphere. In this case, the damage factor is given by (see Fig. 11.1b)

$$\begin{cases} L_{ij} = 0.6 \text{ if } i < N/2 \\ L_{ij} = 1.0 \text{ if } i \geq N/2 \end{cases} \tag{11.2}$$

where the PP module consists of a $N \times N$ lattice of neurons.

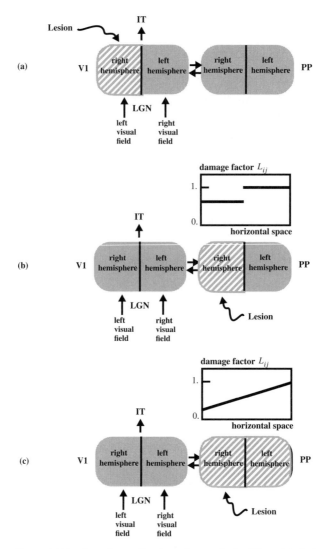

Fig. 11.1 Lesion patterns. (a) Hemianopia resulting from damage to the V1 module (primary visual cortex); (b) Left-sided space-based visual neglect resulting from unilateral damage to the PP (posterior parietal cortex) module in the right hemisphere; (c) Left-sided object-based visual neglect resulting from a gradient of damage to the PP module of the right hemisphere.

The second kind of lesion pattern consists of a gradient of impairment across the PP module. The corresponding damage factor is given by (see Fig. 11.1c)

$$L_{ij} = 0.2 + 0.8 \cdot i/N \tag{11.3}$$

This lesion pattern is designed to model the gradient of impairment in neglect following right parietal cortex damage (Pouget and Sejnowski 1997, Anderson 1996, Driver, Baylis, Goodrich and Rafal 1994, Mozer and Behrmann 1990, Kinsbourne 1993). The response to stimuli is

(a)   (b)   (c)

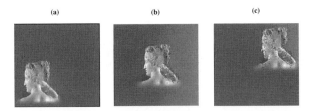

Fig. 11.2 Left-sided local space-based visual hemi-neglect for an object at different positions after unilateral damage to the PP module. The spots on the image show where activation in the parietal (PP) module is above a criterion after the network has settled. The results thus show that neurons in the PP module only reach the criterion when the object is in the right half of visual space. See text for details. This Figure appears in the colour plate Section.

increasingly impaired the further the stimuli are towards the patient's (egocentric) left. The damage factor equation 11.3 reflects the decreasing number of parietal cells in the lesioned system for positions toward the retinal left.

We will now show that the syndrome of space-based visual neglect can be explained by purely unilateral damage to the posterior parietal (PP) network of the cortical model. Further, object-based visual neglect can be explained by a gradient of impairment in the posterior parietal network. We damage the PP module of our model according to equations (11.1) and (11.2) to simulate space-based neglect, and according to equations (11.1) and (11.3) to simulate object-based neglect. In order to visualize the results of the different types of damage, we plot the final state of the PP module after dynamical convergence. This provides a spatial map of locations where attention has settled. In addition, we plot the topographically ordered final activation state of the PP module overlapped with the input image, so that the attentional spatial map can be easily interpreted. We symbolize with a gray point the pixels at locations that correspond to a PP neuronal pool that have an activity larger than a certain threshold $\vartheta = 0.08$. In this way, we can relate the attentional spatial map with the scan path of the covertly fixated locations during the perception of an object. Spatial locations in the PP module with high activity correspond to covertly spatially attended regions, and these potentially attract an overt visual fixation.

Deco and Zihl (2001c) used an input display given by a pixelized 66×66 image ($N = 66$) showing an object ('Paolina' from the Carnegie Mellon Database) translated to different locations. The V1 hypercolumns covered the entire image uniformly. They were distributed in 33×33 locations ($P = 33$), and each hypercolumn was sensitive to three spatial frequencies and to eight different orientations (i.e., $K = 3$ and $L = 8$). The PP module contained 4,356 pools corresponding to each possible spatial location, i.e. to each of the 66×66 pixels, and a common inhibitory pool. We first simulated the case of unilateral space-based neglect by disrupting the PP module with purely unilateral damage to the right hemisphere. Figure 11.2 shows the neglect of the entire left part of the visual field that resulted, and therefore the disrupted processing of objects (Fig. 11.2a) or parts of an object (Fig. 11.2b) in the impaired left hemifield.

If the object is completely in the intact right hemifield, the scan path is absolutely normal (Fig. 11.2c) and consequently spatial attention is deployed on both sides of the object. On the

(a)   (b)   (c)

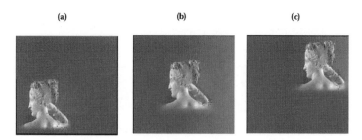

Fig. 11.3 Left-sided object-based visual hemi-neglect for an object at different translated positions after graded damage to the PP module. The damage increases gradually towards the right of the PP module. The spots on the image show where activation in the parietal (PP) module is above a criterion after the network has settled. The results thus show that neurons in the PP module only reach the criterion for the right half of the object, and do this independently of where the image is in visual space. See text for details. This Figure appears in the colour plate Section.

other hand, the neuronal pools corresponding to the right hemisphere are not able to achieve high activity due to the purely unilateral disruption affecting them. This occurs in spite of the positive bias coming from pools in the right hemisphere of V1 that excite the respective pools in the right hemisphere PP module. The low activity level of the PP neuronal pools after the right unilateral lesion can be interpreted as neglect for all sensory information coming from the left visual hemifield.

Figure 11.3 shows the attentional spatial map obtained after graded impairment of the PP which gradually increases to the right, produced as specified in equation 11.3. In this case a pure object-based neglect is obtained. The left side of the object is neglected. Translation of the object across the retina has no effect: the location of the neuronal pools in module PP with high activation is always restricted to the right part of the object with respect to the object frame. Although the absolute level of activity is lower when the object is in the left versus the right visual field, in both cases the left side of the object receives a weaker response than its right side, due to the horizontal activity gradient across the PP module. The intramodular inhibition between neuronal pools in PP drives the dynamical competition always in favour of the pools corresponding to the right side of the object, suppressing at the same time the activity of the pools associated with the left side of the object. Hence, in this case the underlying biased competition causes an attentional spatial map that explains the scan path of the fixation locations observed by object-based visual neglect.

11.2.2 Spatial cueing effect on neglect

A biased competition account of the underlying disturbances of attentional mechanisms that induce visual neglect is decisively supported by the asymmetrical repercussion of spatial cueing on neglect patients. In a typical spatial attention experiment using the precueing method, the subject is asked to respond, by pressing a key, to the appearance of a target at a peripheral location (Posner, Walker, Friedrich and Rafal 1984). The target is preceded by a precue that indicates to which visual hemifield the subject is to covertly attend. The cue is then followed by a target in either the correctly (valid cue) or the incorrectly (invalid cue)

(a) **Experiments**   (b) **Computer Simulations**

RT Difference (ms) (Contralesional-Ipsilesional)   **RT Difference (ms)** (Contralesional-Ipsilesional)

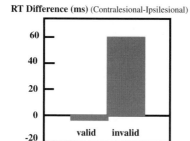
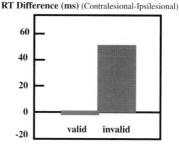

Fig. 11.4 Extinction-like reaction time patterns obtained in spatial precueing experiments when an invalid cue is given in the visual field contralateral to a parietal lesion (see text): (a) in patients with unilateral lesions of the parietal cortex (adapted from Rafal and Robertson, 1997); (b) in simulation results with the attentional model. The plot depicts the detection time difference between the contralesional and ipsilesional fields for the two cueing conditions.

cued location.

This kind of experiment for measuring covert shifts of attention has been applied using simple reaction time paradigms in patients with unilateral lesions of the parietal cortex. Although under normal conditions these patients may neglect contralesional stimuli, they can often be instructed to attend to the neglected field. In fact, in the precueing paradigm, reaction times to valid cued targets in the hemifield contralateral to the lesion are almost normal in the sense that they are not much slower than those for targets that occur in the ipsilesional hemifield when that field was precued. On the other hand, if the ipsilesional field was precued, the patients were unusually slow to respond to the target when it unexpectedly appeared in the contralesional hemifield. In other words, patients are able to orient contralesionally to benefit from a valid cue, but they are impaired when attention is summoned by an invalid cue in the ipsilesional field and the target occurs in the contralesional hemifield. Figure 11.4a shows an example of this measured experimental asymmetry by plotting the detection time difference between the contralesional and ipsilesional fields for the two cueing conditions (Rafal and Robertson 1997).

These asymmetric reaction time patterns were described as an extinction-like reaction time patterns to indicate their similarity to the clinical finding of extinction in these patients. The extinction-like asymmetric reaction time pattern has been found in several experimental studies (Posner, Walker, Friedrich and Rafal 1984, Posner, Walker, Friedrich and Rafal 1987, Baynes, Holtzman and Volpe 1986, Rafal and Robertson 1997). These studies demonstrate that the degree of asymmetry correlates with the severity of the clinical neglect (Morrow and Ratcliff 1988).

In order to understand the underlying attentional dynamical mechanisms involved in these extinction-like phenomena, Deco and Zihl (2001c) simulated the precueing experiments with the cortical architecture introduced in Chapter 9. Numerical experiments were performed by producing purely unilateral damage to the right hemisphere of the PP module (equations (11.1) and (11.2)) in order to simulate the behaviour of left neglect patients. The input display was given by a pixelized 66×66 image ($N = 66$). The target is a letter 'E' of 6×9 pixels that could appear centred in the left or right visual hemifield. The V1 hypercolumns covered the entire image uniformly. They were distributed in 33×33 locations ($P = 33$), and each hypercolumn was sensitive to three spatial frequencies and to eight different orientations (i.e., $K = 3$ and $L = 4$). The PP module contained 66×66 pools, and a common inhibitory pool. During the first precueing period of 250 ms no stimulus was presented in the visual field, and the left or right hemifields were cued by applying an external attentional location-specific bias $I_{ij}^{\text{PP},\text{A}} = 0.07$ to the neuronal pools in PP, corresponding to a 9×9 square centred in the right or left hemisphere, respectively. After 250 ms, the object to be recognized (the letter 'E') was presented as the stimulus in the left or right visual hemifield. Two pools in the IT module had been previously trained to learn the identity of the two stimuli 'E' and 'X', respectively. Four conditions were of interest: contralesional presentation of the stimulus preceded by a valid spatial cue (implemented by top-down bias to the part of the PP module for the right hemisphere) or by an invalid cue (top-down bias to the part of the PP module for the left hemisphere); and ipsilesional presentation of the stimulus preceded by a valid cue (implemented by top-down bias to the part of the PP module for the left hemisphere) or by an invalid cue (top-down bias to the part of the PP module for the right PP hemisphere). The reaction times were measured by detecting the time at which the IT pools reached a criterion of polarization in favour of the target, i.e. when the difference between the maximal activity of the IT pools corresponding to the target and non-targets $A_T^{\text{IT}} - A_D^{\text{IT}}$ (see Section 9.4) was larger than a certain threshold ζ. (We chose $\zeta = 0.05$). Figure 11.4b shows the numerical results. It is clear from the Figure, that the computational model reproduces the extinction-like asymmetric precueing effect very well.

Figure 11.5 shows the evolution of the underlying dynamics by plotting the maximum activities of the target assemblies and the distractor assemblies in the different modules IT, V1 and PP. The case of invalid precueing for the presentation of a contralesional and ipsilesional stimulus is shown. In the ipsilesional stimulus case, invalid precueing of the damaged right PP hemisphere polarized the PP activity in favour of the right hemisphere, but only very moderately, so that V1 was not even polarized at all (left versus right hemisphere). This was due to the fact that the pools in the PP module of the right hemisphere did not react to the cue as efficiently and quickly as usual because of the intrinsic damage. On the other hand, in the contralesional case, invalid precueing of the PP module in the left hemisphere polarized the PP activity strongly in favour of the left hemisphere. Consequently the activity in V1 was also strongly polarized in favour of the left hemisphere. Hence, after presentation of the stimulus in the contralesional visual field, the dynamics of the cortical system took much more time to invert the polarization in favour of the stimulus side in V1, and in the damaged and therefore slow right side of the PP module. Hence, in the contralesional stimulus case, the underlying biased intra- and inter-modular competition in V1 and PP explains the unusual delayed polarization of the IT pools that causes the asymmetric spatial precueing effect shown in Fig. 11.4.

11.2.3 Extinction and visual search

Another kind of extinction-like effect observed in patients with unilateral neglect is found in visual search experiments. Eglin, Robertson and Knight (1989) studied the behaviour of patients with unilateral parietal lesions during a visual search task. When visual search arrays were presented to only the contralesional or ipsilesional hemifields, the patients did not perform differently in the two hemifields. However, when search arrays were presented

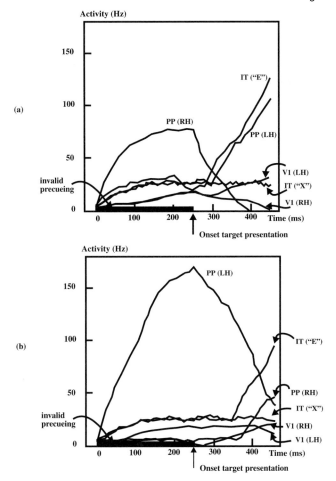

Fig. 11.5 Dynamical evolution of the computational model of attention with purely unilateral damage to the right PP module during the spatial precueing experiment. (a) ipsilesional stimulus presentation and invalid precueing; (b) contralesional stimulus presentation and invalid precueing. The letter 'E' was the target and 'X' was the distractor. The activities to these stimuli, which appeared at time = 250 ms, are shown for the parietal (PP), V1, and inferior temporal (IT) modules. LH and RH denote the left and right hemisphere respectively. The black horizontal bars at the bottom indicate the precuing period.

to both hemifields simultaneously, the patients' reaction time for target detection differed between the two hemifields. The reaction time for target detection in the intact hemifield was not affected by the presence of distractor stimuli in the contralesional hemifield. In contrast, the reaction time for target detection in the contralesional hemifield was much slower when distractor stimuli were presented in the intact field. Thus, in spatial cueing and visual search tasks, patients with unilateral lesions of the parietal lobe generally perform poorly at target detection on the contralesional side, and the performance declines even more when stimuli are presented to the ipsilesional side, or when they are precued to expect a target to be in the ipsilesional hemifield.

To examine the neurodynamical processes by which these effects could arise, Deco and

Zihl (2001c) simulated a simple version of visual search using the computational model described in Chapter 9. To simulate the behaviour of patients with left-sided neglect, purely unilateral damage of the PP module of the right hemisphere was produced as shown in equations (11.1) and (11.2). The same parameters and dimensions for the model were used as in the preceding subsection. The system functioned in an object attention mode. During the visual search task, the stimulus corresponding to the letter 'E' was defined as the target, and 'X' as a distractor. To implement object-biased attention to the target, the IT pool corresponding to the target letter 'E' received an external attentional object specific bias $I_i^{\mathrm{IT,A}} = 0.07$ (i corresponds to the 'E' pool). Four conditions were analyzed: 1) the target was in the ipsilesional hemifield (TI1); 2) the target was in the ipsilesional hemifield and distractor in the contralesional hemifield (TI2); 3) the target was in the contralesional hemifield (TC1); 4) the target was in the contralesional hemifield and the distractor was in the ipsilesional hemifield (TC2).

Figure 11.6a shows the input display for cases TI2 and TC2. The reaction time for target localization was measured as usual by the time for polarization in the PP module, $P = A_T^{\mathrm{PP}} - A_D^{\mathrm{PP}}$ (see Section 9.4) to reach a threshold $\varrho = 0.05$. The simulated reaction times for the four conditions are shown in Fig. 11.6b. The results show that target detection in the contralesional hemifield was strongly delayed by the presence of a distractor in the opposite hemifield. The results of the simulation are thus very similar to the performance observed by neglect patients in visual search tasks (Eglin, Robertson and Knight 1989).

Figure 11.7 shows the time course of the underlying dynamics by showing the maximum activities of the target neural assemblies A_T^{PP} and the distractor neural assemblies A_D^{PP} in module PP for the TC2 and TI2 conditions. When the target is in the neglected contralesional hemifield (condition TC2), the activity of the PP neuronal pools corresponding to the location of the distractor in the opposite undamaged hemifield was very high, and therefore the attentional spatial competition between the target and distractor pools was delayed. This delay was augmented by the slowness of the intrinsically impaired PP pools being activated by the target. In contrast, when the target was in the undamaged ipsilesional hemifield (TI2), the activity of the PP neuronal pools corresponding to the location of the distractor in the neglected hemifield was much lower than in the TC2 condition. This was because of the slowness of the impaired neural pools in the right PP hemisphere receiving the input from the distractor. This effect facilitated the attentional spatial competition between the target and distractor pools. The rapidly increasing target activity in this condition supported even more the polarization in the PP module in favour of the target location.

In conclusion, the model accounts for the effects observed in search tasks in patients with unilateral parietal lesions as follows. Purely unilateral damage in the parietal (PP) module unbalances the dynamical competition, so that it is more difficult to "disengage" spatial attention from stimuli in the undamaged ipsilesional hemifield.

11.2.4 Effect on neglect of top-down knowledge about objects

11.2.4.1 Simulations and resulting predictions

The influence of top-down knowledge about objects that have to be recognized can easily be studied in the framework of the computational model described in Chapter 9. We now describe the effect after unilateral parietal damage of object-specific external top-down bias coming from a brain area such as prefrontal cortex area v46 and impinging on IT neuronal

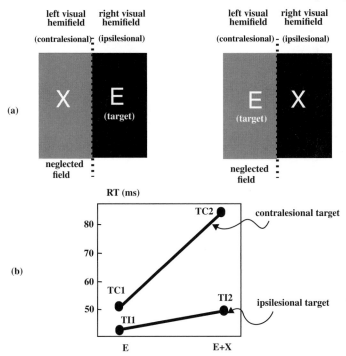

Fig. 11.6 Simulation of the extinction-like effect observed in visual search experiments in patients with unilateral neglect produced by unilateral parietal cortex damage. (a) The input display for cases TI2 and TC2, which were two of the four conditions analyzed as follows: 1) target in the ipsilesional hemifield (TI1); 2) target in the ipsilesional hemifield and distractor in the contralesional hemifield (TI2); 3) target in the contralesional hemifield (TC1); 4) target in the contralesional hemifield and distractor in the ipsilesional hemifield (TC2). (b) Simulated reaction times for the four conditions. 'E' was the target; and 'X' was the distractor used in conditions TC2 and II2.

pools during recognition. To be specific, we compare the dynamical evolution of the spatial attention-related neural activity in the PP module during the recognition of a previously learned object, when a top-down input from prefrontal area v46 to the IT module is present, and when it is not present.

To perform the simulation, we use an input display with a pixelized 203×265 image showing a particular woman's face from the Carnegie-Mellon Database. The V1 hypercolumns covered the entire image uniformly. They were distributed in 100×130 locations and each hypercolumn was sensitive to three spatial frequencies and to eight different orientations (i.e., $K = 3$ and $L = 8$). The PP module contained pools corresponding to each possible spatial location, i.e. to each of the 100×130 pixels, and a common inhibitory pool. An IT neuronal pool was trained for recognition of the particular face during a learning phase. Purely unilateral damage to the PP module of the right hemisphere was made as described previously in order to simulate the behaviour of left neglect patients. As in Section 11.1.2, we show with a gray point the pixels at locations that correspond to a PP neuronal pool which had a neural activity larger than a certain threshold $\vartheta = 0.08$. These points can be interpreted as the scan path of the covertly attended locations while the object was being looked at. During the recognition phase two conditions were analyzed: 1) top-down object-specific knowledge was

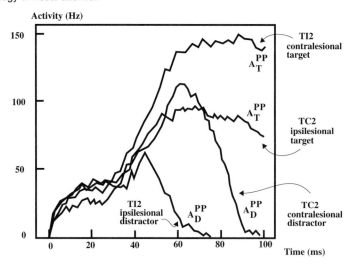

Fig. 11.7 Dynamical evolution during a visual search task of the PP module of the attentional model with purely unilateral damage to the PP module. The curves correspond to the maximum activities of the target neural assemblies A_T^{PP} and distractor neural assemblies A_D^{PP} in module PP for the TC2 (target in the contralesional hemifield and distractor in the ipsilesional hemifields) and TI2 (target in the ipsilesional hemifield and distractor in the contralesional hemifield) conditions.

used; 2) no top-down knowledge was used. The first condition was simulated by providing an external positive top-down bias $I_i^{\mathrm{IT,A}} = 0.07$ to the IT pool that was associated with the specific face. The second condition was simulated without the external top-down bias to the IT module. Figure 11.8 shows the simulated scan path of the attended locations. The effect of top-down knowledge about the object was to cause not only an increase in the activity in the PP module in the ipsilesional visual field, but also to a small extent in the contralesional (left) visual hemifield, which without the top-down object bias was almost completely neglected. The neural pool activity in the damaged (right) PP hemisphere was augmented by the stronger activity in the corresponding right hemisphere in the V1 module, which was strongly stimulated by the feedback originating from the top-down bias to the IT pool. The effects found in the simulation thus predict that, due to the mutually coupled intercortical dynamics, top-down knowledge of the object to be recognized should compensate a little for the lack of attention normally devoted to the left visual field in patients with neglect due to a unilateral right parietal lesion. This prediction was tested in a neglect patient as described in the next subsection.

11.2.4.2 Neuropsychological results: Top-down effect on saccadic patterns by a neglect patient

In this subsection the first neuropsychological results (Heinke, Deco, Humphreys and Zihl 2001b) that support the predictions made from the performance of the model described in the preceding subsection are described. The influence of object knowledge on the saccadic patterns was measured in a patient (MB) with mild left neglect. Although saccades, that is overt eye movements, were measured in the patient, and the location of covert attention was measured in the model, it is assumed, in order to make the prediction easily testable, that overt eye movements will be guided by the locations to which attention is paid covertly.

(a) neglect   (b) top-down effect on neglect

Fig. 11.8 Covert spatial attention after a right unilateral lesion while looking at (or 'scanning with an attentional spotlight' a face: (a) without a top-down object bias to the IT module; (b) with a top-down bias to the IT module (simulation results). The spots show where neural activity in the PP module exceeded a criterion. The effect of top-down knowledge about the object was to cause not only an increase in the activity in the PP module in the ipsilesional visual field, but also to a small extent in the contralesional (left) visual hemifield, which without the top-down object bias was almost completely neglected. This Figure appears in the colour plate Section.

MB suffered a stroke when aged 53 years. A CT scan revealed a right hemisphere lesion affecting the Sylvian fissure, the inferior frontal and superior temporal gyri, the inferior parietal lobe, the caudate and lenticular nuclei, and the insula. There was no field defect on confrontation testing. However, MB initially showed unilateral neglect on a variety of standard clinical tests including: drawing from memory, copying, line bisection, and star cancellation. On the star cancellation task from the Behavioural Inattention Test, MB omitted stars from the left-most third of the sheet. She also showed a consistent right-sided bias in line bisection (twelve 6 cm lines were all biased to the right of centre, with inaccuracy varying from 0.3 to 1.6 cm from the true centre). Apart from neglect there were few other signs of cognitive deficits. MB was well oriented with respect to time and space; she showed no problems with speaking or with speech comprehension, and no clinically-apparent memory deficit. The tests started 6 months after her lesion and continued over a 6-month period, during which MB's deficit was stable.

In this single-case study, four objects were used: 'car', 'lorry', 'ship' and 'cargo ship'. Each of the objects appeared at two different locations in the images. Figure 11.9 shows 4 examples of the images which were used in the experiment. The different locations were chosen so that the axis marked by a broken line in Fig. 11.9 was always aligned with the centre of the screen. Noise was applied to each image, so that the images imitate 'Mooney' pictures (an example is shown in Fig. 11.10). The noise was set to a level that made it difficult for MB to recognize the objects without any additional information.

The images were shown in two blocks to MB. In the first block MB had no knowledge about the possible objects hidden in the noise and she was asked to guess what objects were

Fig. 11.9 Illustration of the four objects shown to patient MB. For the analysis of the results the location of the fixations to the objects were categorized into two different areas (right and left).

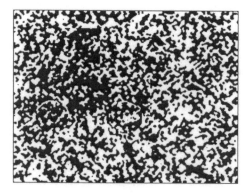

Fig. 11.10 Example of test image (a car) used in neuropsychological testing of patient MB.

hidden. Before the second block, the objects without noise were shown to MB and the names of the objects were pointed out to her. She was also informed that the object might appear at different locations. After that, the same images were shown to her in a different order. This time she was asked to name the object she could identify in the image. Eye movements were measured during the experiment.

Figure 11.11 shows the results. In order to analyze the eye movements, each of the four objects was separated into two right and left parts (see Fig. 11.9). The mean frequency of the fixations for the two regions of each image was determined. This was done separately for blocks one and two. In the first block, before learning, none of the objects was recognized.

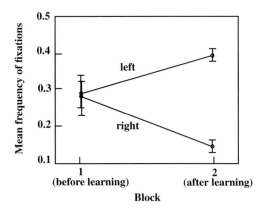

Fig. 11.11 Mean frequency of fixations on the left and right side of the different objects during the first block (before learning) and second block (after learning). The frequency of fixations on the side of the object that lies on the contralesional neglected hemifield increased after the object became known.

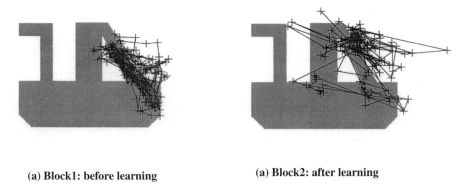

(a) Block1: before learning　　　　**(a) Block2: after learning**

Fig. 11.12 Example of alteration of the scan path due to the learning effect (left first block (before learning); right second block (after learning)). Some fixations appear in the contralesional (left) visual hemifield after learning.

After learning, 10 out of 16 pictures were correctly identified. The frequency of fixations in the left part of the object increased after learning according to the model predictions. An ANOVA with objects (car, lorry, ship, and cargo ship) as the random factor and the location of the object as a fixed factor was applied. The block factor (before and after learning) had a significant influence on the frequency of fixations on the left side ($F(1, 47) = 6.87, p < 0.05$) and on the right side ($F(1, 47) = 4.37, p < 0.05$). Figure 11.12 shows a typical example of the scan path of fixations before and after learning (for comparison with the predictions of the theoretical model see Fig. 11.8).

In conclusion, the neuropsychological data were as predicted by the neurodynamical model, namely that the frequency of fixations on the side of the object that lies in the contralesional neglected hemifield increases if the object is known. This after-learning effect is consistent with a top-down object-related bias influence. Interactions coming from the inferotemporal cortex through the 'what' path into the primary visual cortex are further transmitted through the 'where' path to the posterior parietal cortex. As a consequence, object-specific interactions enhance neural activity even in the damaged posterior parietal hemifield. This enhancement of neural activity can be interpreted as more deployment of spatial attention on the associated contralesional region that consequently attracts more saccades than before learning. The computational perspective developed here provides not only a concrete mathematical description of the mechanisms involved in attention, but also a model that allowed simulation and prediction of the results of new neuropsychological investigations. These investigation help to advance our understanding of the functional impairments resulting from brain damage in patients.

11.3 Search of hierarchical patterns: Neuropsychological predictions of a damaged model

The computational model described in Chapter 9 assumes different networks for the analysis of global and local information which can be independently 'damaged'. In this section we show that the model shows the equivalent of a neuropsychological double dissociation in visual search paradigms. For the present simulation the numerical experiment of visual search of hierarchical patterns described in Section 10.3.3 was repeated with selective damage to the network. In one case the neural pools in the V1 module associated with the processing of low spatial frequencies were damaged in order to simulate an impairment in the system when processing global information. This was achieved by selective lesioning of the appropriate pools, so that the normal advantage of global stimuli was disrupted.

In mathematical terms, the lesioning of the neuronal pools in the V1 module is described by writing equation (10.3) as follows:

$$\tau_k \frac{\partial A_{kpql}^{V1}(t)}{\partial t} = -A_{kpql}^{V1} + L_{ij}\left\{\alpha F(A_{kpql}^{V1}(t)) - \beta F(A_k^{I,V1}(t)) + I_0\right\}$$
$$+ I_{kpql}^{V1}(t) + I_{pq}^{V1-PP}(t) + I_{kpql}^{V1-IT}(t) + \nu \tag{11.4}$$

where the lesioning factor for damaging the lowest spatial frequency is given by

$$\begin{cases} L_{ij} = 0.6 \text{ if } k = K \\ L_{ij} = 1.0 \text{ if } k < K \ . \end{cases} \tag{11.5}$$

The results are plotted in Fig. 11.13a. We can observe that only parallel search is affected by this low spatial frequency impairment, i.e. the case where a target Hh has to be find in a distractor context of Fh items (global disparity). After damage to the low spatial resolution network in the V1 module, the global advantage for large stimuli is suppressed, and the neurodynamical evolution is marked by more competition between the input stimuli. In the case of searching for a target Hh among Hf distractors (global similarity), the potential advantage at the coarse resolution level does not yield any information, and therefore damage to the low spatial frequency system has no effect on the reaction times (see Fig. 11.13a).

In a second experiment, the neuronal pools in the V1 module corresponding to finer spatial resolution (high spatial frequency) were damaged. In this case the lesioning factor in equation (11.4) is given by

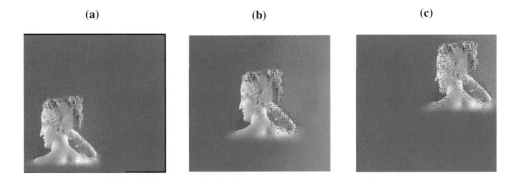

(a)   **(b)**   **(c)**

Fig. 11.2. Left-sided local space-based visual hemi-neglect for an object at different positions after unilateral damage to the PP module. The spots on the image show where activation in the parietal (PP) module is above a criterion after the network has settled. The results thus show that neurons in the PP module only reach the criterion when the object is in the right half of visual space. See text for details.

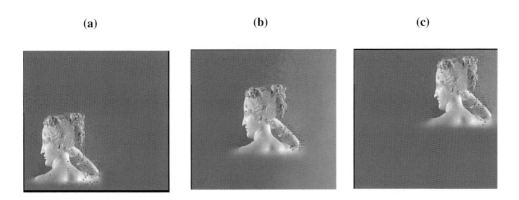

(a)   **(b)**   **(c)**

Fig. 11.3. Left-sided object-based visual hemi-neglect for an object at different translated positions after graded damage to the PP module. The damage increases gradually towards the right of the PP module. The spots on the image show where activation in the parietal (PP) module is above a criterion after the network has settled. The results thus show that neurons in the PP module only reach the criterion for the right half of the object, and do this independently of where the image is in visual space. See text for details.

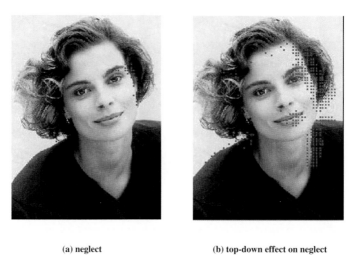

Fig. 11.8. Covert spatial attention after a right unilateral lesion while looking at (or "scanning with an attentional spotlight" a face: (a) without a top-down object bias to the IT module; (b) with a top-down bias to the IT module (simulation results). The red spots show where neural activity in the PP module exceeded a criterion. The effect of top-down knowledge about the object was to cause some activity in the PP module in the contralesional (left) visual hemifield.

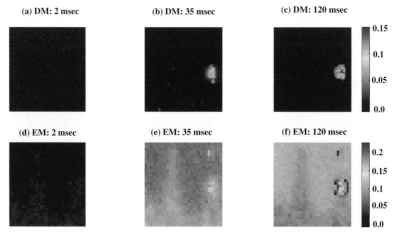

Fig. 9.12. Visual search in the object attention mode. Top row: (a-c) show the maximum activity pools in the spatial-position topological map in the Dorsal stream PP Module (DM) at the time 2 msec (a), 35 msec (b), and 120 msec (c) after stimulus onset. Bottom -row: (d-f) shows the maximum neural activity of each hypercolumn in the Early V1 visual Module (EM) at time 2 msec (d), 35 msec (e), 120 msec (f) after stimulus onset. Activities in both the early V1 module and the dorsal stream PP module exhibited a gradual contraction of responses to a localized area in the topological map due to lateral inhibition, corresponding to the localization of the searched for object.

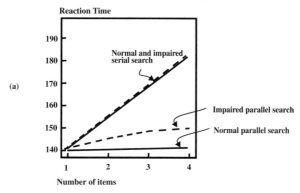

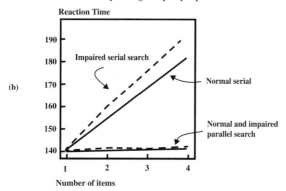

Fig. 11.13 Neuropsychological dissociation in visual search with hierarchical patterns. (a) After low spatial frequency impairment; (b) After high spatial frequency impairment. The continuous curves correspond to the normal case, and the dashed curves to the lesioned case. See text for details.

$$\begin{cases} L_{ij} = 0.6 \text{ if } k = 0 \\ L_{ij} = 1.0 \text{ if } k > K . \end{cases} \tag{11.6}$$

This damage simulates an impairment in the mechanism involved in the processing of local information. The results of the visual search experiments are plotted in Fig. 11.13b. In this case only serial search is affected (global similarity), such that the slope of the reaction time increases with the number of items. This damage has no effect on parallel search. The neurodynamical competition requires the analysis of higher spatial resolutions for unique and reliable recognition, and therefore more time is needed for the activation of the slower, lesioned, high spatial resolution pools. In the normal ('undamaged') case the activation of these extra pools, is required but they respond faster.

To summarize, the simulation predicts that impairments of low spatial frequency or high spatial frequency pools in V1 will produce a dissociation between serial and parallel search conditions (which can be shown in the detection time versus display size curves, as illustrated for the simulations in Fig. 11.13).

11.4   Conjunction search: Neuropsychological predictions from a damaged model

11.4.1   Simulations and predictions

The independent character of the search in each feature dimension postulated in Chapter 10 in our cortical model can be investigated in simulations in which parts of the model are damaged. The model suggests a neuropsychological test to assess the effect of injury to an attentional top-down connection in an individual feature dimension. In order to model damage to a feature dimension, i.e. m', we set the top-down attentional input corresponding to this feature dimension to 0, i.e. $I^A_{m'n} = 0$ for all n. Now, we can modify the 3,2-search (triple conjunction search with target differing from the distractors in two features) in such a way that the features of the distractors that can differ from the target are 'fixed' to the dimension which is not affected by the injury. Fixed 3,2-search can therefore be thought of as a 2,2-search on the feature dimensions not affected by the injury. Let us assume that we damage the third feature dimension (i.e. $m' = 3$), then the fixed 3,2-search consists of distractors which differ from the target only in the first and the second feature dimension, but never in the third dimension. For example, in a standard 3,2-search the target and the distractors can differ in size and colour, or orientation and colour, or size and orientation; while in a fixed 3,2 search the target and the distractors differ only in two a-priori fixed features, e.g. only in size and orientation. Therefore, in a fixed 3,2-search, independent competition within each feature map predicts that increasing the noise in the third feature dimension, i.e. damaging that channel, would not affect the slope of the search time versus size curves, due to the fact that the competition component associated with the third feature dimension contains no relevant information for the dynamical routing (all colours are equal). If the 3,2 search is not fixed, then damaging the third feature dimension should disrupt the search process, because in this case this information is relevant for the dynamical competition, and therefore the search time should be increased. These predictions can be numerically tested by simulations.

The results are shown in Fig. 11.14. In the not-fixed-case (i.e. standard 3,2 search), attentional perturbation in one feature dimension increases the slope of the search time as a function of the display size. In the fixed-case, no influence of the attentional perturbation is observed (in fact both curves overlap in Fig. 11.14).

It should be pointed out that the origin of this effect is based on the independence of the competition mechanisms at the level of the feature maps. If the competition mechanism acted after information from the feature dimensions was conjoined, then injury to one feature dimension would always disrupt the search process, therefore yielding steeper slopes of the search time functions for both cases (fixed and not-fixed). If independent competition mechanisms at the level of feature maps are considered in the model (i.e., prior to the conjunction of features), then injury to one dimension would only disrupt search in the cases where the damaged dimension was relevant. Accordingly, we observed an increase of the slope of the search time function with set size only in the not-fixed case. In general, other models

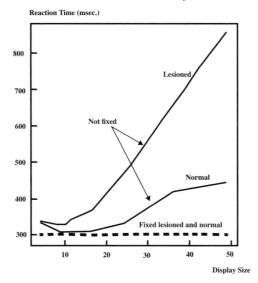

Fig. 11.14 Computational simulation results on the effects of damage to one feature dimension on the search times in both fixed and not–fixed 3,2-searches. The lesion consisted in neglecting the top-down attentional input for the colour dimension. The curves correspond to the lesioned and normal case for the not fixed (continuous) and fixed (dashed) case. In the not-fixed-case (i.e. standard 3,2 search) the attentional perturbation in one feature dimension increases the slope of the search time as a function of the display size. In the fixed-case, no influence of the attentional perturbation is observed (in fact both curves overlap in the Figure).

do not include independent competition at the level of the feature maps, and consequently other models cannot describe this effect.

This method of contrasting fixed and not-fixed search can be used to investigate whether a particular feature dimension shows impaired transmission to the attentional control mechanism. Even in a normal observer an effect can be measured by artificially introducing noise to a feature dimension, for example by making colour values very similar (red and pink) so that the information transmission is ambiguous. The outcome of such an experiment would support our theory.

11.4.2 Experimental test of the predictions in human subjects

An experiment to test the independent character of the search in each feature dimension postulated in our cortical model was performed by analysing the performance of normal subjects in visual search tasks where one feature dimension was artificially distorted (Pollatos 2000). Conjunction visual search of the 3,2 and 3,2-fixed types were tested on 30 normal subjects selected mainly from students of the Ludwig-Maximilian-University. Their ages ranged from 23 to 49 (mean value 35), and all reported having normal or corrected-to-normal vision.

The stimuli consisted of square figures which were defined by three dimensions: size, orientation, and colour. The size could adopt two values: small (0.5 cm x 0.5 cm) or big (1 cm x 1cm). The two possible orientations of the squares corresponded to angles of 0 or 45 degrees between one side of the square and the horizontal axis, respectively. Finally, the colours were

(a)   (b)

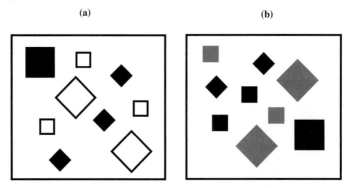

Fig. 11.15 (a) 'Not–lesioned' 3,2 visual search; (b) 'Lesioned' 3,2-visual search. Colours were manipulated in the 'not–lesioned' and 'lesioned' conditions. In the'not–lesioned' condition, the two possible colours were red and green (plotted here as black and white). In the 'lesioned' condition, the colour dimension was artificially damaged by using two very similar colours, namely red and pink (plotted here as black and grey).

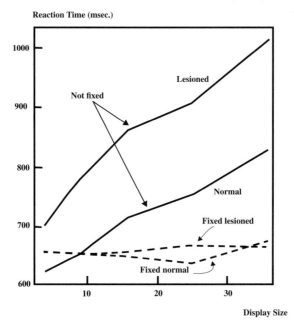

Fig. 11.16 Experimental results on the effects of damage to one feature dimension on the search times in both fixed and not–fixed 3,2-searches. The curves show the mean reaction time over all trials and subjects as a function of the display size. A clear and significant increase of the slope of the search curve for the lesioned case is only observed in the case of 'not–fixed' 3,2-search, as is predicted by our theory. The curves correspond to the 'lesioned' and 'normal' case for the 'not–fixed' (continuous) and 'fixed' (dashed) case.

manipulated in the 'not–lesioned' and 'lesioned' conditions. In the 'not–lesioned' condition, the two possible colours were red and green. Red corresponded to the value 8.5R and 5/16 of the Munsell Book of Color (1976) and green to the value 1.25G and 6/16. In the 'lesioned' condition, the colour dimension was artificially damaged by using two very similar colours, namely red and pink (with the value 5R and 5/14 of the Munsell Book of Color (1976)). The stimuli were presented on a 20" monitor at a distance of 40 cm. Display set sizes of 4, 9, 16, 25 and 36 items occurred equally often in all conditions. The items were presented at randomized locations inside a circle with a diameter of 28.5 cm. In all search conditions, subjects searched in displays in which the target could be present or absent. The sequence of events on each trial was as follows. The target was defined by displaying it at the centre of the search screen for 500 msec. The definition of the target was randomized so that in general the target was different for each trial. At the offset of target presentation, response timing was initiated and a search display was presented. This display remained visible until the computer picked up a key press. Timing stopped when the key press was detected. The subjects were instructed to respond as fast and accurately as possible if the target was present (left mouse key) or absent (right mouse key). Each experiment was run as a within-subjects design, with each subject taking part in all conditions during a single experimental session. There were four conditions: 3,2-search 'lesioned' and 'not lesioned', and 3,2-fixed search 'lesioned' and 'not lesioned'. Subjects were run through a series of demonstration trials. Typical examples of the search screen for the 3,2-search in the 'not–lesioned' and 'lesioned' case are shown in Fig. 11.15.

Figure 11.16 shows the mean reaction time over all trials and subjects as a function of the display size for the above four conditions. A detailed statistical analysis of the data was performed (Pollatos 2000). A clear and significant increase of the slope of the search curve for the lesioned case was only observed in the case of 'not–fixed' 3,2-search, as predicted by the theory. Therefore, these results confirm the independent character of the dynamical competition at each feature dimension.

11.5 Conclusion

We conclude that computational neuroscience provides a mathematical framework for studying the mechanisms involved in brain function, and allows complete simulation and prediction of neuropsychological syndromes. The disruption of submechanisms in the model can be used to simulate, understand, and predict impairments in visual information selection in patients suffering from brain injury. The simulations provide useful support for the explanation offered of the functional impairments resulting from brain damage in patients.

The simulation we describe of object-based neglect was for the case when there was one object in the visual field. The simulation was able to account for the finding in some patients that the left half of the object is neglected independently of where the object is in the visual field. When two or more objects are present in the visual field, the left half of each object may be neglected (see Farah (2000)). This seems surprising, given that the ignored left half of the right object is to the left of the seen right half of the left object. The model we describe accounts for this effect, we predict, if there is *local* spatial competition in an early topographically mapped module (e.g. V1 in the model). This prediction is being tested.

Further conclusions are provided in Chapter 13.

12 Outputs of visual processing

In this Chapter we consider primarily outputs of the ventral visual processing stream, from the inferior temporal visual cortical areas (IT) in particular. The neuronal outputs provide a distributed representation of 'what' object or face is being viewed. Specialized subregions, such as the cortex in the superior temporal sulcus, provide evidence about face expression, and about movements of objects and people which are likely to be useful in social interactions (as described in Chapter 5), and even in interpreting the intentions of others. In the following sections we consider many of the uses to which such information is put by the areas that receive from the inferior temporal visual cortex, and how the output from the temporal cortical visual areas is appropriate for the purposes to which it is put. We consider also the outputs of the parietal cortical visual areas to short term memory systems. We do not consider the outputs from the dorsal visual areas that are used to control the execution of visually guided actions such as arm reaching or eye movements towards a target, as this is a large further area in its own right, but point the reader to a number of books and reviews (Milner and Goodale 1995, Harris and Jenkin 1998, Matelli and Luppino 2000, Berthoz 2000, Andersen, Batista, Snyder, Buneo and Cohen 2000, Goldberg 2000, Krakauer and Ghez 2000).

We start this Chapter by considering the outputs of the temporal and parietal areas to short term memory systems, because this is an important part of the overall theory of attention described in this book. In Chapters 9–11 we assumed a top-down bias from prefrontal cortical areas to help provide the object or spatial bias needed to make the attentional system work. We next describe how that attentional bias could be implemented by the operation of short term memory systems, show why short term memory systems must in general be separate from the temporal lobe and parietal lobe perceptual systems, and provide a theory on how these short term memory systems work, that is how they are loaded with information, how they maintain the information, and how the information they hold influences other brain systems to implement an output of the short term memory systems. The short term memory systems described are not then in temporal and parietal areas, but instead in prefrontal cortical areas (see Fig. 12.1). The theory provides a fundamental reason for having a lateral prefrontal cortex, and suggests that short term memory functions are the main function performed by these lateral prefrontal cortical areas. It is suggested that the basis of the functions of these prefrontal brain areas in planning and 'executive function' is based on their short term memory attractor networks.

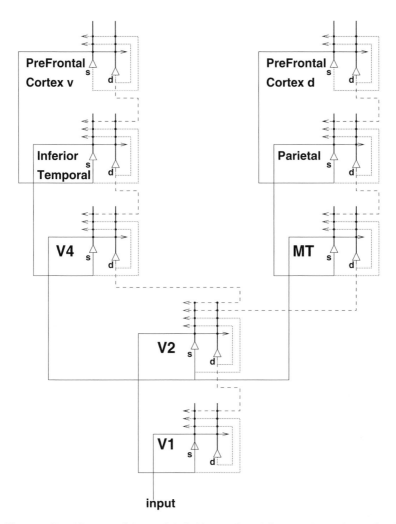

Fig. 12.1 The overall architecture of the model of object and spatial processing and attention, including the prefrontal cortical areas that provide the short term memory required to hold the object or spatial target of attention active. Forward projections between areas are shown as solid lines, and backprojections as dashed lines. The triangles represent pyramidal cell bodies, with the thick vertical line above them the dendritic trees. The cortical layers in which the cells are concentrated are indicated by s (superficial, layers 2 and 3) and d (deep, layers 5 and 6). The prefrontal cortical areas most strongly reciprocally connected to the inferior temporal cortex 'what' processing stream are labelled v to indicate that they are in the more ventral part of the lateral prefrontal cortex, area 46, close to the inferior convexity in macaques. The prefrontal cortical areas most strongly reciprocally connected to the parietal visual cortical 'where' processing stream are labelled d to indicate that they are in the more dorsal part of the lateral prefrontal cortex, area 46, in and close to the banks of the principal sulcus in macaques (see text).

12.1 Visual outputs to Short Term Memory systems

12.1.1 Prefrontal cortex short term memory networks, and their relation to temporal and parietal perceptual networks

A common way that the brain uses to implement a short term memory is to maintain the firing of neurons during a short memory period after the end of a stimulus (see Rolls and Treves (1998) and Fuster (2000)). In the inferior temporal cortex this firing may be maintained for a few hundred ms even when the monkey is not performing a memory task (Rolls and Tovee 1994, Rolls, Tovee, Purcell, Stewart and Azzopardi 1994b, Rolls, Tovee and Panzeri 1999b, Desimone 1996). In more ventral temporal cortical areas such as the entorhinal cortex the firing may be maintained for longer periods in delayed match to sample tasks (Suzuki, Miller and Desimone 1997), and in the prefrontal cortex for even tens of seconds (Fuster 1997, Fuster 2000). In the dorsolateral and inferior convexity prefrontal cortex the firing of the neurons may be related to the memory of spatial responses or objects (Goldman-Rakic 1996, Wilson, O'Sclaidhe and Goldman-Rakic 1993) or both (Rao, Rainer and Miller 1997), and in the principal sulcus / arcuate sulcus region to the memory of places for eye movements (Funahashi, Bruce and Goldman-Rakic 1989) (see Figs. 1.9 and 1.11). The firing may be maintained by the operation of associatively modified recurrent collateral connections between nearby pyramidal cells producing attractor states in autoassociative networks (see Section 7.3).

For the short term memory to be maintained during periods in which new stimuli are to be perceived, there must be separate networks for the perceptual and short term memory functions, and indeed two coupled networks, one in the inferior temporal visual cortex for perceptual functions, and another in the prefrontal cortex for maintaining the short term memory during intervening stimuli, provide a precise model of the interaction of perceptual and short term memory systems (Renart, Parga and Rolls 2000, Renart, Moreno, de al Rocha, Parga and Rolls 2001) (see Fig. 12.2). In particular, this model shows how a prefrontal cortex attractor (autoassociation) network could be triggered by a sample visual stimulus represented in the inferior temporal visual cortex in a delayed match to sample task, and could keep this attractor active during a memory interval in which intervening stimuli are shown. Then when the sample stimulus reappears in the task as a match stimulus, the inferior temporal cortex module showed a large response to the match stimulus, because it is activated both by the visual incoming match stimulus, and by the consistent backprojected memory of the sample stimulus still being represented in the prefrontal cortex memory module (see Fig. 12.2). This computational model makes it clear that in order for ongoing perception to occur unhindered implemented by posterior cortex (parietal and temporal lobe) networks, there must be a separate set of modules that is capable of maintaining a representation over intervening stimuli. This is the fundamental understanding offered for the evolution and functions of the dorsolateral prefrontal cortex, and it is this ability to provide multiple separate short term attractor memories that provides we suggest the basis for its functions in planning.

Renart, Parga and Rolls (2000) and Renart, Moreno, de al Rocha, Parga and Rolls (2001) performed analyses and simulations which showed that for working memory to be implemented in this way, the connections between the perceptual and the short term memory modules (see Fig. 12.2) must be relatively weak. As a starting point, they used the neurophysiological data showing that in delayed match to sample tasks with intervening stimuli, the neuronal activity in the inferior temporal visual cortex (IT) is driven by each new incoming

Inferior temporal cortex (IT)   Prefrontal cortex (PF)

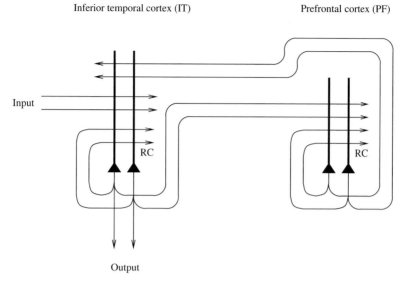

Fig. 12.2 A short term memory autoassociation network in the prefrontal cortex could hold active a working memory representation by maintaining its firing in an attractor state. The prefrontal module would be loaded with the to-be-remembered stimulus by the posterior module (in the temporal or parietal cortex) in which the incoming stimuli are represented. Backprojections from the prefrontal short term memory module to the posterior module would enable the working memory to be unloaded, to for example influence on-going perception (see text). RC - recurrent collateral connections.

visual stimulus (Miller, Li and Desimone 1993b, Miller and Desimone 1994), whereas in the prefrontal cortex, neurons start to fire when the sample stimulus is shown, and continue the firing that represents the sample stimulus even when the potential match stimuli are being shown (Miller, Erickson and Desimone 1996). The architecture studied by Renart, Parga and Rolls (2000) was as shown in Fig. 12.2, with both the intramodular (recurrent collateral) and the intermodular (forward IT to PF, and backward PF to IT) connections trained on the set of patterns with an associative synaptic modification rule. A crucial parameter is the strength of the intermodular connections, g, which indicates the relative strength of the intermodular to the intramodular connections. (This parameter measures effectively the relative strengths of the currents injected into the neurons by the inter-modular relative to the intra-modular connections, and the importance of setting this parameter to relatively weak values for useful interactions between coupled attractor networks was highlighted by Renart, Parga and Rolls (1999b) and Renart, Parga and Rolls (1999a), as shown in Sections 7.9 and 12.1.2.) The patterns themselves were sets of random numbers, and the simulation utilized a dynamical approach with neurons with continuous (hyperbolic tangent) activation functions (see Section 12.1.2 and (Shiino and Fukai 1990, Kuhn 1990, Kuhn, Bos and van Hemmen 1991, Amit and Tsodyks 1991)). The external current injected into IT by the incoming visual stimuli was sufficiently strong to trigger the IT module into a state representing the incoming stimulus. When the sample was shown, the initially silent PF module was triggered into activity by the weak ($g > 0.002$) intermodular connections. The PF module remained firing to the sample stimulus even when IT was responding to potential match stimuli later in the trial, provided that g was less than 0.024, because then the intramodular recurrent connections could dom-

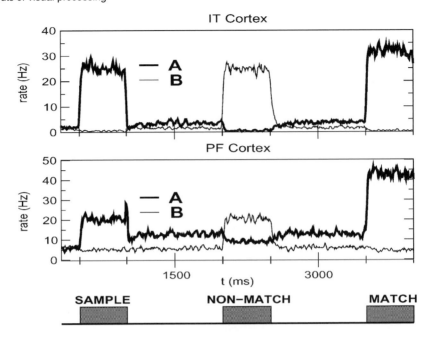

Fig. 12.3 Interaction between the prefrontal cortex (PF) and the inferior temporal cortex (IT) in a delayed match to sample task with intervening stimuli with the architecture illustrated in Fig. 12.2. Above: activity in the IT attractor module. Below: activity in the PF attractor module. The thick lines show the firing rates of the set of neurons with activity selective for the Sample stimulus (which is also shown as the Match stimulus, and is labelled **A**), and the thin lines the activity of the neurons with activity selective for the Non-Match stimulus, which is shown as an intervening stimulus between the Sample and Match stimulus and is labelled **B**. A trial is illustrated in which **A** is the Sample (and Match) stimulus. The prefrontal cortex module is pushed into an attractor state for the sample stimulus by the IT activity induced by the sample stimulus. Because of the weak coupling to the PF module from the IT module, the PF module remains in this Sample-related attractor state during the delay periods, and even while the IT module is responding to the non-match stimulus. The PF module remains in its Sample-related state even during the Non-Match stimulus because once a module is in an attractor state, it is relatively stable. When the Sample stimulus reappears as the Match stimulus, the PF module shows higher Sample stimulus-related firing, because the incoming input from IT is now adding to the activity in the PF attractor network. This in turn also produces a match enhancement effect in the IT neurons with Sample stimulus-related selectivity, because the backprojected activity from the PF module matches the incoming activity to the IT module. After Renart, Parga and Rolls, 2000 and Renart, Moreno, de la Rocha, Parga and Rolls, 2001.

inate the firing (see Fig. 12.3). If g was higher than this, then the PF module was pushed out of the attractor state produced by the sample stimulus. The IT module responded to each incoming potentially matching stimulus provided that g was not greater than approximately 0.024. Moreover, this value of g was sufficiently large that a larger response of the IT module was found when the stimulus matched the sample stimulus (the match enhancement effect found neurophysiologically, and a mechanism by which the matching stimulus can be identified). This simple model thus shows that the operation of the prefrontal cortex in short term memory tasks such as delayed match to sample with intervening stimuli, and its relation to posterior perceptual networks, can be understood by the interaction of two weakly coupled

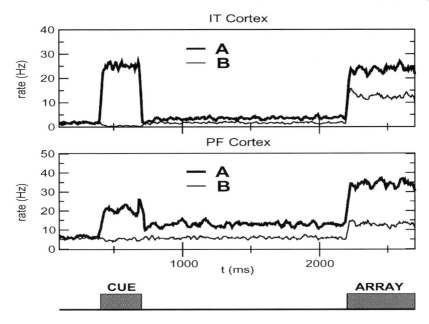

Fig. 12.4 Interaction between the prefrontal cortex (PF) and the inferior temporal cortex (IT) in a visual search task with the architecture illustrated in Fig. 12.2. Above: activity in the IT attractor module. Below: activity in the PF attractor module. The thick lines show the firing rates of the set of neurons with activity selective for search stimulus **A**, and the thin lines the activity of the neurons with activity selective for stimulus **B**. During the cue period either **A** or **B** is shown, to indicate to the monkey which stimulus to select when an array containing both **A** and **B** is shown after a delay period. The trial shown is for the case when **A** is the cue stimulus. When stimulus **A** is shown as a cue, then via the IT module, the PF module is pushed into an attractor state **A**, and the PF module remembers this state during the delay period. When the array **A** + **B** is shown later, there is more activity in the PF module for the neurons selective for **A**, because they have inputs both from the continuing attractor state held in the PF module and from the forward activity from the IT module which now contains both **A** and **B**. This PF firing to **A** in turn also produces greater firing of the population of IT neurons selective for **A** than in the IT neurons selective for **B**, because the IT neurons selective for **A** are receiving both **A**–related visual inputs, and **A**–related backprojected inputs from the PF module. After Renart, Parga and Rolls, 2000 and Renart, Moreno, de la Rocha, Parga and Rolls, 2001.

attractor networks, as shown in Figs. 12.2 and 12.3.

The same network can also be used to illustrate the interaction between the prefrontal cortex short term memory system and the posterior (IT or PP) perceptual regions in visual search tasks, as illustrated in Fig. 12.4.

12.1.2   Computational details of the model of short term memory

The model network of Renart, Parga and Rolls (2000) and Renart, Moreno, de al Rocha, Parga and Rolls (2001) consists of a large number of (excitatory) neurons arranged in two modules with the architecture shown in Fig. 12.2. Following (Kuhn 1990, Amit and Tsodyks 1991), each neuron is assumed to be a dynamical element which transforms an incoming afferent current into an output spike rate according to a given transduction function. A given afferent current I_{ai} to neuron i $(i = 1, \ldots, N)$ in module a $(a = \textbf{IT}, \textbf{PF})$ decays with a characteristic time constant τ but increases proportionally to the spike rates of the rest of the neurons in the network (both from inside and outside its module) connected to it, the contribution of each presynaptic neuron, e.g. neuron j from module b, and in proportion to the synaptic efficacy J_{ij}^{ab} between the two[37]. This can be expressed through the following equation

$$\frac{dI_{ai}(t)}{dt} = -\frac{I_{ai}(t)}{\tau} + \sum_{b,j} J_{ij}^{(a,b)} \nu_{bj} + h_{ai}^{(\text{ext})} \quad . \tag{12.1}$$

An external current $h_{ai}^{(\text{ext})}$ from outside the network, representing the stimuli, can also be imposed on every neuron. Selective stimuli are modelled as proportional to the stored patterns, i.e. $h_{ai}^{\mu(\text{ext})} = h_a \eta_{ai}^{\mu}$, where h_a is the intensity of the external current to module a.

The transduction function of the neurons transforming currents into rates was chosen as a threshold hyperbolic tangent of gain G and threshold θ. Thus, when the current is very large the firing rates saturate to an arbitrary value of 1.

The synaptic efficacies between the neurons of each module and between the neurons in different modules are respectively

$$J_{ij}^{(a,a)} = \frac{J_0}{f(1-f)N_t} \sum_{\mu=1}^{P} (\eta_{ai}^{\mu} - f)(\eta_{aj}^{\mu} - f) \quad i \neq j \; ; \; a = \textbf{IT}, \textbf{PF} \tag{12.2}$$

$$J_{ij}^{(a,b)} = \frac{g}{f(1-f)N_t} \sum_{\mu=1}^{P} (\eta_{ai}^{\mu} - f)(\eta_{bj}^{\mu} - f) \quad \forall \; i,j \; ; \; a \neq b \; . \tag{12.3}$$

The intra-modular connections are such that a number P of sparse independent configurations of neural activity are dynamically stable, constituting the possible sustained activity states in each module. This is expressed by saying that each module has learned P binary patterns $\{\eta_{ai}^{\mu} = 0, 1, \; \mu = 1, \ldots, P\}$, each of them signalling which neurons are active in each of the sustained activity configurations. Each variable η_{ai}^{μ} is allowed to take the values 1 and 0 with probabilities f and $(1 - f)$ respectively, independently across neurons and across patterns. The inter-modular connections reflect the temporal associations between the sustained activity states of each module. In this way, every stored pattern μ in the IT module has an associated pattern in the PF module which is labelled by the same index. The normalization constant $N_t = N(J_0 + g)$ was chosen so that the sum of the magnitudes of the inter- and the intra-modular connections remains constant and equal to 1 while their relative values are varied. When this constraint is imposed the strength of the connections can be expressed in terms of a single independent parameter g measuring the relative intensity of the inter- vs the intra-modular connections (J_0 can be set equal to 1 everywhere).

Both modules implicitly include an inhibitory population of neurons receiving and sending signals to the excitatory neurons through uniform synapses. In this case the inhibitory population can be treated as a single inhibitory neuron with an activity dependent only on the mean activity of the excitatory population. We chose the transduction function of the inhibitory neuron to be linear with slope γ.

---

[37] On this occasion we vert to the theoretical physicists' usual notation for synaptic weights or couplings, J_{ij}, from w_{ij}.

Since the number of neurons in a typical network one may be interested in is very large, e.g. $\sim 10^5 - 10^6$, the analytical treatment of the set of coupled differential equations (12.1) becomes untractable. On the other hand, when the number of neurons is large, a reliable description of the asymptotic solutions of these equations can be found using the techniques of statistical mechanics (Kuhn 1990). In this framework, instead of characterizing the states of the system by the state of every neuron, this characterization is performed in terms of *macroscopic* quantities called *order parameters* which measure and quantify some global properties of the network as a whole. The relevant order parameters appearing in the description of the system are the overlap of the state of each module with each of the stored patterns m_a^μ and the average activity of each module x_a, defined respectively as:

$$ m_a^\mu = \frac{1}{\chi N} \ll \sum_i (\eta_{ai}^\mu - f) \nu_{ai} \gg_\eta \; ; \; x_a = \frac{1}{N} \ll \sum_i \nu_{ai} \gg_\eta \; , \qquad (12.4) $$

where the symbol $\ll \ldots \gg_\eta$ stands for an average over the stored patterns.

Using the free energy per neuron of the system at zero temperature $\mathcal{F}$ (which is not written explicitly to reduce the technicalities to a minimum), Renart, Parga and Rolls (2000) and Renart, Moreno, de al Rocha, Parga and Rolls (2001) modelled the experiments by giving the order parameters the following dynamics:

$$ \tau \frac{\partial m_a^\mu}{\partial t} = -\frac{\partial \mathcal{F}}{\partial m_a^\mu} \; ; \; \tau \frac{\partial x_a}{\partial t} = -\frac{\partial \mathcal{F}}{\partial x_a} \; . \qquad (12.5) $$

These dynamics ensure that the stationary solutions, corresponding to the values of the order parameters at the attractors, correspond also to minima of the free energy, and that, as the system evolves, the free energy is always minimized through its gradient. The time constant of the macroscopical dynamics was chosen to be equal to the time constant of the individual neurons, which reflects the assumption that neurons operate in parallel. Equations (12.5) were solved by a simple discretizing procedure (first order Runge-Kutta method). An appropriate value for the time interval corresponding to one computer iteration was found to be $\tau/10$ and the time constant has been given the value $\tau = 10 \; ms$.

Since not all neurons in the network receive the same inputs, not all of them behave in the same way, i.e. have the same firing rates. In fact, the neurons in each of the modules can be split into different sub-populations according to their state of activity in each of the stored patterns. The mean firing rate of the neurons in each sub-population depends on the particular state realized by the network (characterized by the values of the order parameters). Associated with each pattern there are two large sub-populations denoted as foreground (all active neurons) and background (all inactive neurons) for that pattern. The overlap with a given pattern can be expressed as the difference between the mean firing rate of the neurons in its foreground and its background. The average was calculated over all other sub-populations to which each neuron in the foreground (background) belonged to, where the probability of a given sub-population is equal to the fraction of neurons in the module belonging to it (determined by the probability distribution of the stored patterns as given above). This partition of the neurons into sub-populations is appealing since, in neurophysiological experiments, cells are usually classified in terms of their response properties to a set of fixed stimuli, i.e. whether each stimulus is effective or ineffective in driving their response.

The modelling of the different experiments proceeded according to the macroscopic dynamics (12.5), where each stimulus was implemented as an extra current into free energy for a desired period of time.

Using this model, results of the type described in Section 12.1.1 were found (Renart, Parga and Rolls 2000, Renart, Moreno, de al Rocha, Parga and Rolls 2001). The paper by Renart, Moreno, de al Rocha, Parga and Rolls (2001) extended the earlier findings of Renart, Parga and Rolls (2000) to integrate-and-fire neurons, and it is results from the integrate-and-fire simulations that are shown in Figs. 12.3 and 12.4.

12.1.3 Computational necessity for a separate, prefrontal cortex, short term memory system

This approach emphasizes that in order to provide a good brain lesion test of prefrontal cortex short term memory functions, the task set should require a short term memory for stimuli over an interval in which other stimuli are being processed, because otherwise the posterior cortex perceptual modules could implement the short term memory function by their own recurrent collateral connections. This approach also emphasizes that there are many at least partially independent modules for short term memory functions in the prefrontal cortex (e.g. several modules for delayed saccades; one or more for delayed spatial (body) responses in the dorsolateral prefrontal cortex; one or more for remembering visual stimuli in the more ventral prefrontal cortex; and at least one in the left prefrontal cortex used for remembering the words produced in a verbal fluency task – see Section 10.3 of Rolls and Treves (1998)).

This computational approach thus provides a clear understanding of why a separate (prefrontal) mechanism is needed for working memory functions, as elaborated in Section 12.1.1. It may also be commented that if a prefrontal cortex module is to control behaviour in a working memory task, then it must be capable of assuming some type of executive control. There may be no need to have a single central executive additional to the control that must be capable of being exerted by every short term memory module. This is in contrast to what has traditionally been assumed for the prefrontal cortex (Shallice and Burgess 1996) (see also Section 9.6).

12.1.4 Role of prefrontal cortex short term memory systems in visual search and attention

The same model shown in Fig. 12.2 can also be used to help understand the implementation of visual search tasks in the brain (Renart, Parga and Rolls 2000). In such a visual search task, the target stimulus is made known beforehand, and inferior temporal cortex neurons then respond more when the search target (as compared to a different stimulus) appears in the receptive field of the IT neuron (Chelazzi, Miller, Duncan and Desimone 1993, Chelazzi, Duncan, Miller and Desimone 1998). The model shows that this could be implemented by the same system of weakly coupled attractor networks in PF and IT shown in Fig. 12.2 as follows. When the target stimulus is shown, it is loaded into the PF module from the IT module as described for the delayed match to sample task. Later, when the display appears with two or more stimuli present, there is an enhanced response to the target stimulus in the receptive field, because of the backprojected activity from PF to IT which adds to the firing being produced by the target stimulus itself (Renart, Parga and Rolls 2000, Renart, Moreno, de al Rocha, Parga and Rolls 2001) (see Fig. 12.4). The interacting spatial and object networks described in Chapters 9–11, and summarized in Fig. 12.1, take this analysis one stage further, and show that once the PF–IT interaction has set up a greater response to the search target in IT, this enhanced response can in turn by backprojections to topologically mapped earlier cortical visual areas move the "attentional spotlight" to the place where the search target is located. A further way in which attractor networks can help to account for the responses of IT neurons is described in Section 12.5.

12.1.5 Synaptic modification is needed to set up but not to reuse short term memory systems

To set up a new short term memory attractor, synaptic modification is needed to form the new stable attractor. Once the attractor is set up, it may be used repeatedly when triggered by an appropriate cue to hold the short term memory state active by continued neuronal firing even without any further synaptic modification (see section 7.3 and Kesner and Rolls (2001)). Thus manipulations that impair the long term potentiation of synapses (LTP) may impair the formation of new short term memory states, but not the use of previously learned short term memory states. Kesner and Rolls (2001) analyzed many studies of the effects of blockade of LTP in the hippocampus on spatial working memory tasks, and found evidence consistent with this prediction. Interestingly, it was found that if there was a large change in the delay interval over which the spatial information had to be remembered, then the task became susceptible, during the transition to the new delay interval, to the effects of blockade of LTP. The implication is that some new learning is required when the rat must learn the strategy of retaining information for longer periods when the retention interval is changed.

12.2 Visual outputs to Long Term Memory systems in the brain

The inferior temporal visual cortex projects via the perirhinal cortex and entorhinal cortex to the hippocampus (see Figs. 12.5 and 12.8), which is implicated in long term memory, of, for example, where objects are located in spatial scenes, which can be thought of as an example of episodic memory. The architecture shown in Fig. 12.5 indicates that the hippocampus provides a region where visual outputs from the inferior temporal visual cortex can, via the perirhinal cortex and entorhinal cortex, be brought together with outputs from the ends of other cortical processing streams. In this section, we consider how the visual input about objects is in the correct form for the types of memory implemented by the perirhinal and hippocampal systems, how the hippocampus of primates contains a representation of the visual space being viewed, how this may be similar computationally to the apparently very different representation of places that is present in the rat hippocampus, how these spatial representations are in a form that could be implemented by a continuous attractor which could be updated in the dark by idiothetic inputs in the way described in Section 7.5, and how a unified attractor theory of hippocampal function can be formulated using the concept of mixed attractors introduced in Section 7.5.8. The visual output from the inferior temporal visual cortex may be used to provide the perirhinal and hippocampal systems with information about objects that is useful in visual recognition memory, in episodic memory of where objects are seen, and for building spatial representations of visual scenes. Before summarizing the computational approaches to these issues, we first summarize some of the empirical evidence that needs to be accounted for in computational models.

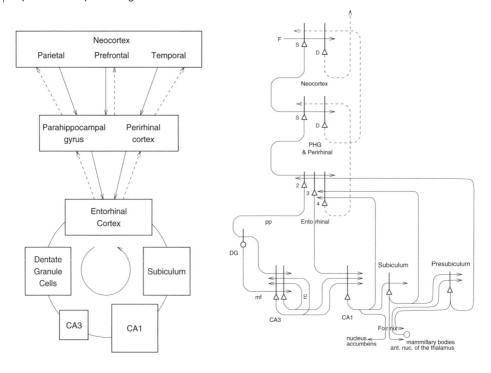

Fig. 12.5 Forward connections (solid lines) from areas of cerebral association neocortex via the parahippocampal gyrus and perirhinal cortex, and entorhinal cortex, to the hippocampus; and backprojections (dashed lines) via the hippocampal CA1 pyramidal cells, subiculum, and parahippocampal gyrus to the neocortex. There is great convergence in the forward connections down to the single network implemented in the CA3 pyramidal cells; and great divergence again in the backprojections. Left: block diagram. Right: more detailed representation of some of the principal excitatory neurons in the pathways. Abbreviations: D, Deep pyramidal cells; DG, dentate granule cells; F, forward inputs to areas of the association cortex from preceding cortical areas in the hierarchy. mf: mossy fibres; PHG, parahippocampal gyrus and perirhinal cortex; pp, perforant path; rc, recurrent collaterals of the CA3 hippocampal pyramidal cells; S, superficial pyramidal cells; 2, pyramidal cells in layer 2 of the entorhinal cortex; 3, pyramidal cells in layer 3 of the entorhinal cortex; 5, 6, pyramidal cells in the deep layers of the entorhinal cortex. The thick lines above the cell bodies represent the dendrites.

12.2.1 Effects of damage to the hippocampus and connected structures on object-place and episodic memory

Partly because of the evidence that in humans with bilateral damage to the hippocampus and nearby parts of the temporal lobe, anterograde amnesia is produced (Squire and Knowlton 2000), there is continuing great interest in how the hippocampus and connected structures operate in memory. The effects of damage to the hippocampus indicate that the very long-term storage of at least some types of information is not in the hippocampus, at least in humans. On the other hand, the hippocampus does appear to be necessary to learn certain types of information, that have been characterized as declarative, or knowing that, as contrasted with procedural, or knowing how, which is spared in amnesia. Declarative memory includes what can be declared or brought to mind as a proposition or an image. Declarative memory includes episodic memory (memory for particular episodes), and semantic memory (memory for facts)

(Squire and Knowlton 2000).

In monkeys, damage to the hippocampus or to some of its connections such as the fornix produces deficits in learning about where objects are and where responses must be made (see Rolls (1996b) and Buckley and Gaffan (2000)). For example, macaques and humans with damage to the hippocampus or fornix are impaired in object-place memory tasks in which not only the objects seen, but where they were seen, must be remembered (Gaffan and Saunders 1985, Parkinson, Murray and Mishkin 1988, Smith and Milner 1981). Such object-place tasks require a whole-scene or snapshot-like memory (Gaffan 1994). Also, fornix lesions impair conditional left-right discrimination learning, in which the visual appearance of an object specifies whether a response is to be made to the left or the right (Rupniak and Gaffan 1987). A comparable deficit is found in humans (Petrides 1985). Fornix sectioned monkeys are also impaired in learning on the basis of a spatial cue which object to choose (e.g. if two objects are on the left, choose object A, but if the two objects are on the right, choose object B) (Gaffan and Harrison 1989a). Further, monkeys with fornix damage are also impaired in using information about their place in an environment. For example, Gaffan and Harrison (1989b) found learning impairments when the position of the monkey in the room determined which of two or more objects the monkey had to choose. Rats with hippocampal lesions are impaired in using environmental spatial cues to remember particular places (Jarrard 1993, Martin, Grimwood and Morris 2000), and it has been argued that the necessity to utilize allocentric spatial cues (Cassaday and Rawlins 1997), to utilize spatial cues or bridge delays (Jackson, Kesner and Amann 1998, Kesner and Rolls 2001), or to perform relational operations on remembered material (Eichenbaum 1997), may be characteristic of the deficits.

One way of relating the impairment of spatial processing to other aspects of hippocampal function (including the memory of recent events or episodes in humans) is to note that this spatial processing involves a snapshot type of memory, in which one whole scene with its often unique set of parts or elements must be remembered. This memory may then be a special case of episodic memory, which involves an arbitrary association of a set of spatial and/or non-spatial events that describe a past episode. For example, the deficit in paired associate learning in humans (see Squire and Knowlton (2000)) may be especially evident when this involves arbitrary associations between words, for example window — lake.

It appears that the deficits in 'recognition' memory (tested for example for visual stimuli seen recently in a delayed match to sample task) produced by damage to this brain region are related to damage to the perirhinal cortex (Zola-Morgan, Squire, Amaral and Suzuki 1989, Zola-Morgan, Squire and Ramus 1994), which receives from high order association cortex and has connections to the hippocampus (see Figs. 12.5 and 12.8) (Suzuki and Amaral 1994a, Suzuki and Amaral 1994b). The functions of the perirhinal cortex in memory are discussed in Section 12.2.4.

12.2.2 Neurophysiology of the hippocampus and connected areas

In the rat, many hippocampal pyramidal cells fire when the rat is in a particular place, as defined for example by the visual spatial cues in an environment such as a room (O'Keefe 1990, O'Keefe 1991, Kubie and Muller 1991). There is information from the responses of many such cells about the place where the rat is in the environment. When a rat enters a new environment B connected to a known environment A, there is a period in the order of 10 minutes in which as the new environment is learned, some of the cells that formerly had place fields in A develop instead place fields in B. It is as if the hippocampus sets up a new spatial representation which can map both A and B, keeping the proportion of cells active at any one time approximately constant (Wilson and McNaughton 1993). Some rat hippocampal neurons are found to be more task-related, responding for example to olfactory stimuli to which particular behavioural responses must be made (Eichenbaum 1997), and some of these neurons may in different experiments show place-related responses.

It was recently discovered that in the primate hippocampus, many spatial cells have responses not related to the place where the monkey is, but instead related to the place where the monkey is looking (Rolls, Robertson and Georges-François 1997a, Rolls 1999d, Rolls 1999c). These are called 'spatial view cells', an example of which is shown in Fig. 12.6. These cells encode information in allocentric (world-based, as contrasted with egocentric, body-related) coordinates (Georges-Francois, Rolls and Robertson 1999, Rolls, Treves, Robertson, Georges-François and Panzeri 1998). They can in some cases respond to remembered spatial views in that they respond when the view details are obscured, and use idiothetic cues including eye position and head direction to trigger this memory recall operation (Robertson, Rolls and Georges-François 1998). Another idiothetic input that drives some primate hippocampal neurons is linear and axial whole body motion (O'Mara, Rolls, Berthoz and Kesner 1994), and in addition, the primate presubiculum has been shown to contain head direction cells (Robertson, Rolls, Georges-François and Panzeri 1999).

Part of the interest of spatial view cells is that they could provide the spatial representation required to enable primates to perform object-place memory, for example remembering where they saw a person or object, which is an example of an episodic memory, and indeed similar neurons in the hippocampus respond in object-place memory tasks (Rolls, Miyashita, Cahusac, Kesner, Niki, Feigenbaum and Bach 1989b). Associating together such a spatial representation with a representation of a person or object could be implemented by an autoassociation network implemented by the recurrent collateral connections of the CA3 hippocampal pyramidal cells (Rolls 1989b, Rolls 1996b, Rolls and Treves 1998). Some other primate hippocampal neurons respond in the object-place memory task to a combination of spatial information and information about the object seen (Rolls, Miyashita, Cahusac, Kesner, Niki, Feigenbaum and Bach 1989b). Further evidence for this convergence of spatial and object information in the hippocampus is that in another memory task for which the hippocampus is needed, learning where to make spatial responses conditional on which picture is shown, some primate hippocampal neurons respond to a combination of which picture is shown, and where the response must be made (Miyashita, Rolls, Cahusac, Niki and Feigenbaum 1989, Cahusac, Rolls, Miyashita and Niki 1993).

These primate spatial view cells are thus unlike place cells found in the rat (O'Keefe 1979, O'Keefe 1990, O'Keefe 1991, Kubie and Muller 1991, Wilson and McNaughton 1993). Primates, with their highly developed visual and eye movement control systems, can explore and remember information about what is present at places in the environment without having to visit those places. Such spatial view cells in primates would thus be useful as part of a memory system, in that they would provide a representation of a part of space that would not depend on exactly where the monkey or human was, and that could be associated with items that might be present in those spatial locations. An example of the utility of such a representation in humans would be remembering where a particular person had been seen. The primate spatial representations would also be useful in remembering trajectories through environments, of use for example in short-range spatial navigation (O'Mara, Rolls, Berthoz and Kesner 1994, Rolls 1999c).

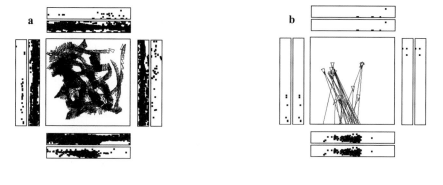

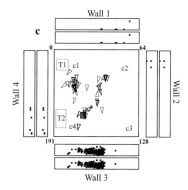

Fig. 12.6 Examples of the firing of a hippocampal spatial view cell when the monkey was walking around the laboratory. a. The firing of the cell is indicated by the spots in the outer set of 4 rectangles, each of which represents one of the walls of the room. There is one spot on the outer rectangle for each action potential. The base of the wall is towards the centre of each rectangle. The positions on the walls fixated during the recording sessions are indicated by points in the inner set of 4 rectangles, each of which also represents a wall of the room. The central square is a plan view of the room, with a triangle printed every 250 ms to indicate the position of the monkey, thus showing that many different places were visited during the recording sessions. b. A similar representation of the same 3 recording sessions as in (a), but modified to indicate some of the range of monkey positions and horizontal gaze directions when the cell fired at more than 12 spikes/s. c. A similar representation of the same 3 recording sessions as in (b), but modified to indicate more fully the range of places when the cell fired. The triangle indicates the current position of the monkey, and the line projected from it shows which part of the wall is being viewed at any one time while the monkey is walking. One spot is shown for each action potential. (After Georges-François, Rolls and Robertson, 1999)

The representation of space in the rat hippocampus, which is of the place where the rat is, may be related to the fact that with a much less developed visual system than the primate, the rat's representation of space may be defined more by the olfactory and tactile as well as distant visual cues present, and may thus tend to reflect the place where the rat is. An interesting hypothesis on how this difference could arise from essentially the same computational process in rats and monkeys is as follows (Rolls 1999c, de Araujo, Rolls and Stringer 2001). The starting assumption is that in both the rat and the primate, the dentate granule cells and the CA3 and CA1 pyramidal cells respond to combinations of the inputs received. In the case of the primate, a combination of visual features in the environment will over a typical viewing angle of perhaps 10–20 degrees result in the formation of a spatial view cell, the effective trigger for which will thus be a combination of visual features within a relatively small part of space. In contrast, in the rat, given the very extensive visual field which may extend over 180–270 degrees, a combination of visual features formed over such a wide visual angle would effectively define a position in space, that is a place. The actual processes by which the hippocampal formation cells would come to respond to feature combinations could be similar in rats and monkeys, involving for example competitive learning in the dentate granule cells, autoassociation learning in CA3 pyramidal cells, and competitive learning in CA1 pyramidal cells (Rolls 1989b, Rolls 1996b, Treves and Rolls 1994, Rolls and Treves 1998). Thus spatial view cells in primates and place cells in rats might arise by the same computational process but be different by virtue of the fact that primates are foveate and view a small part of the visual field at any one time, whereas the rat has a very wide visual field. Although the representation of space in rats therefore may be in some ways analogous to the representation of space in the primate hippocampus, the difference does have implications for theories, and modelling, of hippocampal function.

In rats, the presence of place cells has led to theories that the rat hippocampus is a spatial cognitive map, and can perform spatial computations to implement navigation through spatial environments (O'Keefe and Nadel 1978, O'Keefe 1991, Burgess, Recce and O'Keefe 1994, Burgess and O'Keefe 1996). The details of such navigational theories could not apply in any direct way to what is found in the primate hippocampus. Instead, what is applicable to both the primate and rat hippocampal recordings is that hippocampal neurons contain a representation of space (for the rat, primarily where the rat is, and for the primate primarily of positions 'out there' in space) which is a suitable representation for an episodic memory system. In primates, this would enable one to remember, for example, where an object was seen. In rats, it might enable memories to be formed of where particular objects (for example those defined by olfactory, tactile, and taste inputs) were found. Thus at least in primates, and possibly also in rats, the neuronal representation of space in the hippocampus may be appropriate for forming memories of events (which usually in these animals have a spatial component). Such memories would be useful for spatial navigation, for which according to the present hypothesis the hippocampus would implement the memory component but not the spatial computation component. Evidence that what neuronal recordings have shown is represented in the non-human primate hippocampal system may also be present in humans is that regions of the hippocampal formation can be activated when humans look at spatial views (Epstein and Kanwisher 1998, O'Keefe, Burgess, Donnett, Jeffery and Maguire 1998).

12.2.3 Hippocampal models

These neuropsychological and neurophysiological analyses are complemented by neuronal network models of how the hippocampus could operate to store and retrieve large numbers of memories (Rolls 1987, Rolls 1989b, Rolls 1996b, Treves and Rolls 1992, Treves and Rolls 1994, Rolls and Treves 1998)). One key hypothesis (adopted also by McClelland, McNaughton and O'Reilly (1995)) is that the hippocampal CA3 recurrent collateral connections which spread throughout the CA3 region provide a *single autoassociation network* that enables the firing of *any* set of CA3 neurons representing one part of a memory to be associated together with the firing of any other set of CA3 neurons representing another part of the same memory

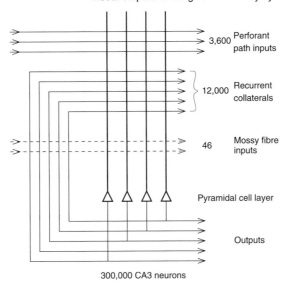

300,000 CA3 neurons

Fig. 12.7 The numbers of connections from three different sources onto each CA3 cell from three different sources in the rat. (After Treves and Rolls 1992, and Rolls and Treves 1998.)

(cf. Marr (1971)). The number of patterns p each representing a different memory that could be stored in the CA3 system operating as an autoassociation network would be as shown in equation 12.6 (see Section 7.3.3.7)

$$p \approx \frac{C^{RC}}{a \ln(\frac{1}{a})} k \qquad (12.6)$$

where C^{RC} is the number of synapses on the dendrites of each neuron devoted to the recurrent collaterals from other CA3 neurons in the network, a is the sparseness of the representation, and k is a factor that depends weakly on the detailed structure of the rate distribution, on the connectivity pattern, etc., but is roughly in the order of 0.2–0.3. Given that C^{RC} is approximately 12,000 in the rat, the resulting storage capacity would be greater than 12,000 memories, and perhaps up to 36,000 memories if the sparseness a of the representation was as low as 0.02 (Treves and Rolls 1992, Treves and Rolls 1994). Another part of the hypothesis is that the very sparse (see Fig. 12.7) but powerful connectivity of the mossy fibre inputs to the CA3 cells from the dentate granule cells is important during learning (but not recall) to force a new, arbitrary, set of firing onto the CA3 cells which dominates the activity of the recurrent collaterals, so enabling a new memory represented by the firing of the CA3 cells to be stored (Treves and Rolls 1992, Rolls 1989b, Rolls 1987). The perforant path input to the CA3 cells, which is numerically much larger but at the apical end of the dendrites, would be used to initiate recall from an incomplete pattern (Treves and Rolls 1992, Rolls and Treves 1998). The prediction of the theory about the necessity of the mossy fibre inputs to the CA3 cells during learning but not recall has now been confirmed (Lassalle, Bataille and Halley 2000). A way to enhance the efficacy of the mossy fibre system relative to the CA3 recurrent collateral connections during learning may be to increase the level of acetyl choline by increasing the firing of the septal cholinergic cells (Hasselmo, Schnell and Barkai 1995).

Another key part of the quantitative theory is that not only can retrieval of a memory to an incomplete cue be performed by the operation of the associatively modified CA3 recurrent collateral connections, but also that recall of that information to the neocortex can be performed via CA1 and the hippocampo-cortical and cortico-cortical backprojections (Treves and Rolls 1994, Rolls 1996b, Rolls and Treves 1998, Rolls 2000b) shown in Fig. 12.5. In this case, the number of memory patterns p^{BP} that can be retrieved by the backprojection system is

$$p^{BP} \approx \frac{C^{BP}}{a^{BP} \ln(\frac{1}{a^{BP}})} k^{BP} \tag{12.7}$$

where C^{BP} is the number of synapses on the dendrites of each neuron devoted to backprojections from the preceding stage (dashed lines in Fig. 12.5), a^{BP} is the sparseness of the representation in the backprojection pathways, and k^{BP} is a factor that depends weakly on the detailed structure of the rate distribution, on the connectivity pattern, etc., but is roughly in the order of 0.2–0.3. The insight into this quantitative analysis came from treating each layer of the backprojection hierarchy as being quantitatively equivalent to another iteration in a single recurrent attractor network (Treves and Rolls 1994, Treves and Rolls 1991). The need for this number of connections to implement recall, and more generally constraint satisfaction in connected networks (see Section 7.9), provides a fundamental and quantitative reason for why there are approximately as many backprojections as forward connections between the adjacent connected cortical areas in a cortical hierarchy (see Section 1.11). This, and other computational approaches to hippocampal function, are included in a special issue of the journal Hippocampus (1996), 6(6).

Another aspect of the theory is that the operation of the CA3 system to implement recall, and of the backprojections to retrieve the information, would be sufficiently fast, given the fast recall in associative networks built of neurons with continuous dynamics (see Section 7.6).

The fact that spatial patterns, which imply continuous representations of space, are represented in the hippocampus has led to the application of continuous attractor models to help understand hippocampal function. Such models have been developed by Samsonovich and McNaughton (1997), Battaglia and Treves (1998b), Stringer, Trappenberg, Rolls and de Araujo (2002c), Stringer, Rolls, Trappenberg and de Araujo (2002b), Stringer, Rolls and Trappenberg (2002a), and Stringer and Rolls (2002b), and are described in Section 7.5. Indeed, we have shown how a continuous attractor network could enable the head direction cell firing of presubicular cells to be maintained in the dark, and updated by idiothetic (self-motion) head rotation cell inputs (Stringer, Trappenberg, Rolls and de Araujo 2002c, Robertson, Rolls, Georges-François and Panzeri 1999). The continuous attractor model has been developed to understand how place cell firing in rats can be maintained and updated by idiothetic inputs in the dark (Stringer, Rolls, Trappenberg and de Araujo 2002b). The continuous attractor model has also been developed to understand how spatial view cell firing in primates can be maintained and updated by idiothetic eye movement and head direction inputs in the dark (Stringer, Rolls and Trappenberg 2002a, Robertson, Rolls and Georges-François 1998).

It has now been shown that attractor networks can store both continuous patterns and discrete patterns, and can thus be used to store for example the location in (continuous, physical) space where an object (a discrete item) is present (Rolls, Stringer and Trappenberg 2002c) (see Section 7.5.8). In this network, when events are stored that have both discrete (object) and continuous (spatial) aspects, then the whole place can be retrieved later by the object, and the object can be retrieved by using the place as a retrieval cue. Such networks are likely to be present in parts of the brain such as the hippocampus which receive and combine inputs both from systems that contain representations of continuous (physical) space, and from brain systems that contain representations of discrete objects, such as the inferior temporal visual cortex. The combined continuous and discrete attractor network described by Rolls, Stringer and Trappenberg (2002c) shows that in brain regions where the spatial and object processing streams are brought together, then a single network can represent and learn associations between both types of input (see Section 7.5.8). Indeed, in brain regions such as the hippocampal system, it is essential that the spatial and object processing streams are brought together in a single network, for it is only when both types of information are in the same network that spatial information can be retrieved from object information, and vice versa, which is a fundamental property of episodic memory.

12.2.4 The perirhinal cortex, recognition memory, and familiarity

As noted above, it appears that the deficits in 'recognition' memory (tested for example for visual stimuli seen recently in a delayed match to sample task) produced by damage to this brain region are related to damage to the perirhinal cortex (Zola-Morgan, Squire, Amaral and Suzuki 1989, Zola-Morgan, Squire and Ramus 1994), which receives from high order association cortex and has connections to the hippocampus (see Figs. 12.5 and 12.8) (Suzuki and Amaral 1994a, Suzuki and Amaral 1994b). Given that some topographic segregation is maintained in the afferents to the hippocampus through the perirhinal and parahippocampal cortices, it may be that these areas are able to subserve memory within one of these topographically separated areas; whereas the final convergence afforded by the hippocampus into a single network in CA3 which may operate by autoassociation (see Fig. 12.5 and above) allows arbitrary associations between any of the inputs to the hippocampus, e.g. spatial, visual object, and auditory, which may all be involved in typical episodic memories (see below and Rolls (1996b), Rolls and Treves (1998)).

Fascinating new insight into the contribution of the perirhinal cortex to recognition memory has recently been obtained by recordings from single perirhinal cortex neurons in macaques in which it was shown that the responses of the neurons are related to the long-term familiarity of visual stimuli (Hölscher and Rolls 2001, Hölscher, Rolls and Xiang 2002). It was found that the responses of many perirhinal cortex neurons were initially smaller to sets of 8 novel images than to very familiar images. The average response to the novel images was often only approximately 45% of that to highly familiar images. Then over several days, in which the initially familiar images were shown for approximately 500 1-sec presentations, the responses gradually increased to the initially novel stimuli to become as large as to the already very familiar stimuli (see Fig. 12.9). The effect was robust, repeated in seven different experiments in three macaques, and was found independently of whether the monkey was performing a delayed match to sample short term memory task, or was viewing the screen when not performing a memory task. The results suggest that in addition to a role in visual working memory shown in earlier investigations (Brown and Xiang 1998), some perirhinal cortex neurons reflect the long-term familiarity of visual stimuli. This computation may be performed by perirhinal cortex neurons, in that in control recordings made from inferior temporal cortex neurons, there was no overall tendency of the inferior temporal neurons to respond more to relatively novel than to very familiar images, with stimulus selectivity being the dominant response property of the neurons, irrespective of whether the images were relatively new in

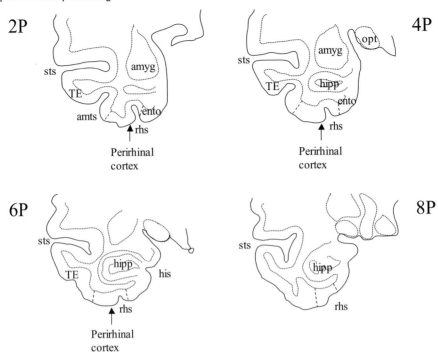

Fig. 12.8 The locations of the perirhinal cortex (areas 35 and 36 in and near the banks of the rhinal sulcus), the entorhinal cortex, and the hippocampus in macaques. The coronal (transverse) sections are at different distances behind (P, posterior to) an anatomical marker, the sphenoid bone, which is approximately at the anterior-posterior position of the optic chiasm and anterior commissure. amts - anterior middle temporal sulcus; amyg - amygdala; ento - entorhinal cortex; hipp - hippocampus; his - hippocampal sulcus; opt - optic tract; rhs - rhinal sulcus; sts - superior temporal sulcus; TE - inferior temporal visual cortex.

the task, or had been seen previously for many weeks.

Knowing whether a visual stimulus is very familiar could be a useful function performed by a memory system such as the perirhinal cortex which receives from the inferior temporal visual cortex. Indeed, the potential functions of computing the long-term familiarity of objects or visual images are many-fold, and include recognition of complex object-environment configurations such as members of one's own social and family group, recognition of one's own possessions, recognition of one's own territory, etc. It is notable that the loss of the feeling of familiarity with new people, objects and events that occur after medial temporal lobe damage is one of the important symptoms of medial temporal lobe (anterograde) amnesia, and this too may be related to the computation of long-term familiarity which may be being performed by the perirhinal cortex.

The perirhinal cortex neurons could develop their responses by increasing the synaptic strength from the IT inputs onto the perirhinal neurons by a small increment every time a particular object is being represented in the IT cortex. This, and the anatomical background that neuronal activity from any part of IT may converge into the visual part of the perirhinal cortex, would result in the perirhinal cortex neurons gradually becoming more responsive to any object that had produced object-related firing in IT neurons. A novel set of IT neurons

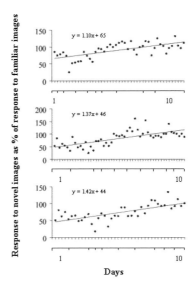

Fig. 12.9 Perirhinal cortex neurons reflect the long-term familiarity of visual stimuli. The average response of each neuron (shown by a full circle) to a set of 15 novel stimuli introduced on day 1 is shown as a proportion of the neuron's response to a very familiar set of stimuli. In each experiment on a neuron, each visual stimulus was shown for approximately 13 1-sec periods. Several neurons were analyzed on some days. Thus in the order of 500 exposures to each stimulus in the novel set were required before the response to the initially novel set of stimuli was as large as to the already very familiar set of stimuli. Each stimulus was an image of an object presented on a video monitor. The 3 separate graphs are for 3 separate replications of the whole experiment. (After Hölscher and Rolls 2001, and Hölscher, Rolls and Xiang 2002.)

responding to a relatively novel stimulus would not as a set have the same strong synapses onto perirhinal cortex neurons as a set of IT neurons representing a familiar stimulus, so that with global inhibition in the perirhinal cortex related to the number of IT neurons that are firing, the perirhinal cortex neurons would respond much better to any very familiar object or image than to a novel object or image.

12.2.4.1   The perirhinal cortex, short term memory, and the formation of invariant representations

As noted above, the perirhinal cortex has neurons that respond differently to the match and sample stimuli in short term memory tasks (Brown and Xiang 1998) over somewhat larger numbers of intervening stimuli than do inferior temporal cortex proper neurons in the paradigm experiment for this type of research (Baylis and Rolls 1987). Further, lesions of the perirhinal cortex can impair the performance of visual short term memory tasks. Hölscher and Rolls (2001) showed that some perirhinal cortex neurons have activity that is closely related to the performance of such short term memory tasks, in that their responses, typically greater to the sample than to the match stimuli, reset between trials of the task. The effect of this is that the response of the neurons is large to the stimulus when it is a sample stimulus, even if it has been seen recently as a match stimulus on the preceding trial.

In studies aimed to investigate the neural basis of semantic memory, Miyashita and colleagues (Miyashita 2000, Higuchi and Miyashita 1996) have shown that inferior temporal cortex neurons can have similar responses to pairs of visual stimuli which regularly occur separated from each other by 1–3 sec, and that the learning of these long-term associations may depend on the perirhinal cortex which has backprojections to the inferior temporal visual cortex. One way in which the perirhinal cortex could contribute to the formation of these long-term memory associations between visual stimuli is by producing maintained firing in response to one stimulus until the next stimulus appears a few seconds later, so that then associative learning between the neurons representing the two stimuli could take place.

A more natural use for any continuing firing in the perirhinal cortex after one stimulus is shown, and in which the firing would not need to be maintained for such long periods as in the experiments of Miyashita and colleagues, would be to bridge the gap until another view or arbitrary transform of a visual stimulus has occurred. Indeed, it is possible that the perirhinal cortex, or the entorhinal cortex in which firing after a match stimulus in a short term memory period has been observed (Suzuki, Miller and Desimone 1997), could contribute to the formation of invariant representations in the inferior temporal visual cortex by providing continuing firing between different transforms of stimuli, which is one way to implement the trace rule described in Chapter 8 as being useful in learning invariant representations of visual stimuli. In a test of this hypothesis, Buckley, Booth, Rolls and Gaffan (2002) (see Buckley, Booth, Rolls and Gaffan (1998) and for review Buckley and Gaffan (2000)) showed that perirhinal cortex lesions in macaques impair the learning of view-invariant representations of new objects, but not the use of view-invariant representations of objects seen and learned before damage to the perirhinal cortex.

12.3 Decoding of the reinforcement associations of visual stimuli for emotion and motivation: outputs of IT to the orbitofrontal cortex and amygdala

Learning about which visual and other stimuli in the environment are rewarding, punishing, or neutral is crucial for survival. For example, it takes just one trial to learn if a seen object is hot when we touch it, and associating that visual stimulus with the pain may help us to avoid serious injury in the future. Similarly, if we are given a new food which has an excellent taste, we can learn in one trial to associate the sight of it with its taste, so that we can select it in future. In these examples, the previously neutral visual stimuli become conditioned reinforcers by their association with a primary (unlearned) reinforcer such as taste or pain. Our examples show that learning about which stimuli are rewards and punishments is very important in the control of motivational behaviour such as feeding and drinking, and in emotional behaviour such as fear and pleasure. The type of learning involved is pattern association, between the conditioned and the unconditioned stimulus. This type of learning provides a major example of how the visual representations provided by the inferior temporal visual cortex are used by the other parts of the brain (Rolls 2000c). In this Section we consider where in sensory processing this stimulus-reinforcement association learning occurs, which brain structures are involved in this type of learning, how the neuronal networks for pattern association learning may actually be implemented in these regions, and how the distributed representation about objects provided by the inferior temporal cortex output is suitable for this pattern association learning.

The crux of the answer to the last question is that the inferior temporal cortex representation is ideal for this pattern association learning because it is a transform invariant representation of objects, and because the code can be read by a neuronal system which performs dot products using neuronal ensembles as inputs, which is precisely what pattern associators in the brain need, because they are implemented by neurons which perform as their generic computation a dot product of their inputs with their synaptic weight vectors (see Section 7.2).

A fascinating question is the stage in information processing at which such pattern association learning between visual stimuli and primary (unlearned) reinforcers occurs. Does it occur, for example, early on in the cortical processing of visual signals, or late in cortical visual processing, or even after the main stages of visual cortical processing? We shall see that at least in primates, the latter is the case, and that there are good reasons why this should be the case. Before we consider the neural processing itself, we introduce in more detail in Section 12.3.1 the brain systems involved in learning associations between visual stimuli and rewards and punishments, showing how they are very important in understanding the brain mechanisms involved in emotion and motivation. Section 12.3.1 is intended to be introductory, and may be omitted if the reader is familiar with this research area.

12.3.1 Emotion

Emotions can usefully be defined as states produced by instrumental reinforcing stimuli (see Rolls (1990) and Rolls (1999a), and earlier work by Millenson (1967), Weiskrantz (1968), Gray (1975) and Gray (1987)). (Instrumental reinforcers are stimuli which if their occurrence, termination, or omission is made contingent upon the making of a response, alter the probability of the future emission of that response.) Some stimuli are unlearned reinforcers (for example the taste of food if the animal is hungry, or pain); while others may become reinforcing by learning, because of their association with such primary reinforcers, thereby becoming 'secondary reinforcers'. This type of learning may thus be called 'stimulus-reinforcement association learning'. If a reinforcer increases the probability of emission of a response on which it is contingent, it is said to be a 'positive reinforcer' or 'reward'; if it decreases the probability of such a response it is a 'negative reinforcer' or 'punishment'[38]. For example, fear is an emotional state which might be produced by a sound (the conditioned stimulus) that has previously been associated with an electrical shock (the primary reinforcer). The converse reinforcement contingencies produce the opposite effects on behaviour. The omission or termination of a positive reinforcer ('extinction' and 'time out' respectively, sometimes described as 'punishing'), decrease the probability of responses. Responses followed by the omission or termination of a negative reinforcer increase in probability, this pair of negative reinforcement operations being termed 'active avoidance' and 'escape' respectively (see further Gray (1975), Mackintosh (1983), Rolls (1999a) and Rolls (2000f)).

The different emotions can be described and classified according to whether the reinforcer is positive or negative, and by the reinforcement contingency. An outline of such a classification scheme, elaborated more precisely by Rolls (1990), Rolls (1999a) and Rolls (2000f), is shown in Fig. 12.10.

---

[38]To make sure that our terms are clear, we can define a reward as anything for which an animal will work, and a punishment as anything an animal will work to avoid. We can also call the reward a positive reinforcer, and the punishment a negative reinforcer, where a reinforcer is defined as an event that alters the probability of a behavioural response on which it is made contingent. (Further treatment of these concepts is provided by Rolls (1999a) and Rolls (2000f)).

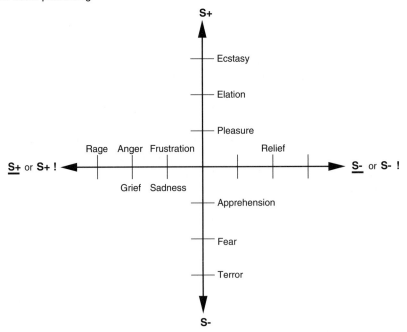

Fig. 12.10 Some of the emotions associated with different reinforcement contingencies are indicated. Intensity increases away from the centre of the diagram, on a continuous scale. The classification scheme created by the different reinforcement contingencies consists of: (1) the presentation of a positive reinforcer S+; (2) the presentation of a negative reinforcer S-; (3) the omission of a positive reinforcer S+ or the termination of a positive reinforcer S+!; (4) the omission of a negative reinforcer S- or the termination of a negative reinforcer S-!.

The processes described here would not be limited in the range of emotions for which they could account. Some of the factors which enable a very wide range of human emotions to be analyzed with this foundation are elaborated elsewhere (Rolls 1990, Rolls 1999a, Rolls 2000f), and include the following:

1. The reinforcement contingency (see Fig. 12.10).

2. The intensity of the reinforcer (see Fig. 12.10).

3. Any environmental stimulus might have a number of different reinforcement associations. For example, a stimulus might be associated both with the presentation of a reward and of a punishment, allowing states such as conflict and guilt to arise.

4. Emotions elicited by stimuli associated with different primary reinforcers will be different.

5. Emotions elicited by different secondary reinforcing stimuli will be different from each other (even if the primary reinforcer is similar).

6. The emotion elicited can depend on whether an active or passive behavioural response is possible. For example, if an active behavioural response can occur to the omission of an S+ (where S+ signifies a positively reinforcing stimulus), then anger might be produced, but if only passive behaviour is possible, then sadness, depression, or grief might occur.

By combining these six factors, it is possible to account for a very wide range of emotions (Rolls 1990, Rolls 1999a, Rolls 2000f). It is also worth noting that emotions can be produced just as much by the recall of reinforcing events as by external reinforcing stimuli; that cognitive processing (whether conscious or not) is important in many emotions, for very complex cognitive processing may be required to determine whether environmental events are reinforcing or not; that emotions normally consist of cognitive processing which determines the reinforcing valence of the stimulus, and then an elicited mood change if the valence is positive or negative; and that stability of mood implies that absolute levels of reinforcement must be represented over moderately long time spans by the firing of mood-related neurons, a difficult operation which may contribute to 'spontaneous' mood swings, depression which occurs without a clear external cause, and the multiplicity of hormonal and transmitter systems which seem to be involved in the control of mood (see further Rolls (1999a)).

In terms of the neural bases of emotion, the most important point from this introduction is that in order to understand the neural bases of emotion, we need to consider brain mechanisms involved in reward and punishment, and involved in learning about which environmental stimuli are associated, or are no longer associated, with rewards and punishments.

12.3.1.1 Functions of emotion

Before considering these pattern association learning mechanisms, it is also useful to summa-rize the functions of emotions, because these functions are important for understanding the output systems to which brain mechanisms involved in emotion must interface. Knowing the functions of emotion is also important when considering the brain mechanisms of emotion, for the functions of emotion provide an indication of which signals (for example the sight of a face, and touch) must be associated together. The functions, described more fully elsewhere (Rolls 1999a), can be summarized as follows:

1. The elicitation of autonomic responses (for example a change in heart rate) and en-docrine responses (for example the release of adrenaline (that is epinephrine)). These prepare the body for action. There are output pathways from the amygdala and orbitofrontal cortex directly, and via the hypothalamus, to the brainstem autonomic nuclei.

2. Flexibility of behavioral responses to reinforcing stimuli. Rolls' theory of emotion ((Rolls 1999a, Rolls 2000f) is that primary rewards and punishers are the goals for action specified by genes to promote their own survival, and that emotions are the states elicited by rewards and punishers to promote and maintain behaviour to those goals. Part of the argument is that it is more efficient for genes to specify goals than responses or actions. Emotional (and motivational) states allow genes to specify a simple interface between sensory inputs and motor outputs, because once the goal has been specified, the organism can learn any instrumental response to obtain the goal. In this sense, there is flexibility of behaviour. This is more flexible than simply learning a fixed behavioural response to a stimulus (Gray 1975, Rolls 1990, Rolls 1999a). Pathways from the amygdala and orbitofrontal cortex to the striatum are implicated in these functions.

Part of the importance then of the object-based representation provided by the inferior temporal cortex is that it provides an ideal representation for animals to learn which objects are associated with primary rewards and punishers, and are therefore goals for action. The actual pattern association learning is performed outside the inferior temporal cortex, where its outputs are brought together with representations of primary reinforcers, which are decoded in sensory modalities such as taste and touch (Rolls 1999a, Rolls 2000f). Most visual stimuli are not primary reinforcers (Rolls 1999a), and therefore stimulus-reinforcement association learning is the main way in which visual stimuli acquire reinforcing properties.

This function of emotions in supporting flexible behavioral responses is based on the crucial role which rewards and punishments have on behaviour. Animals are built with neural systems that enable them to evaluate which environmental stimuli, whether learned or not, are rewarding and punishing, that is are goals for actions, and will be worked for or avoided. A crucial part of this system is that with many competing rewards, goals, and priorities, there must be a selection system for enabling the most important of these goals to become the object of behaviour at any one time. This selection process must be capable of responding to many different types of reward decoded in different brain systems that have evolved at different times, even including the use in humans of a language system to enable long-term plans to be made (see Rolls (1997a) and Rolls (1999a)). These many different brain systems, some involving implicit evaluation of rewards, and others explicit, verbal, conscious, evaluation of rewards and planned long-term goals (see Rolls (1999a)), must all enter into the selector of behaviour. This selector, although itself poorly understood, might include a process of competition between all the competing calls on output, and might involve the basal ganglia (see Rolls (1999a)).

3. Emotion is motivating. For example, fear learned by stimulus-reinforcement association formation provides the motivation for actions performed to avoid noxious stimuli.

4. Communication. For example, monkeys may communicate their emotional state to others, by making an open-mouth threat to indicate the extent to which they are willing to compete for resources, and this may influence the behaviour of other animals. There are neural systems in the amygdala and overlying temporal cortical visual areas, and the orbitofrontal cortex, which are specialized for the face-related aspects of this processing (see Chapter 5).

5. Social bonding. Examples of this are the emotions associated with the attachment of the parents to their young, and the attachment of the young to their parents (see Dawkins (1989)).

6. The current mood state can affect the cognitive evaluation of events or memories (see Blaney (1986), Rolls and Stringer (2001b)), and this may have the function of facilitating continuity in the interpretation of the reinforcing value of events in the environment. A hypothesis on the neural pathways which implement this is presented in Section 12.3.6.

7. Emotion may facilitate the storage of memories. One way in which this occurs is that episodic memory (that is one's memory of particular episodes) is facilitated by emotional states. This may be advantageous in that storing many details of the prevailing situation when a strong reinforcer is delivered may be useful in generating appropriate behaviour in situations with some similarities in the future. This function may be implemented by the relatively non-specific projecting systems to the cerebral cortex and hippocampus, including the cholinergic pathways in the basal forebrain and medial septum, and the ascending noradrenergic pathways (see Section 7.1.5 of Rolls and Treves (1998)). A second way in which emotion may affect the storage of memories is that the current emotional state may be stored with episodic memories, providing a mechanism for the current emotional state to affect which memories are recalled. A third way in which emotion may affect the storage of memories is by guiding the cerebral cortex in the representations of the world which are set up. For example, in the visual system, it may be useful to build perceptual representations or analyzers which are different from each other if they are associated with different reinforcers, and to be less likely to build them if they have no association with reinforcement. Ways in which backprojections from parts of the brain important in emotion (such as the amygdala) to parts of the cerebral cortex could perform this function are discussed in Section 7.4.5 and by Rolls and Treves (1998).

8. Another function of emotion is that by enduring for minutes or longer after a reinforcing stimulus has occurred, it may help to produce persistent motivation and direction of behaviour.

9. Emotion may trigger the recall of memories stored in neocortical representations. Amygdala backprojections to the cortex could perform this for emotion in a way analogous to that in which the hippocampus could implement the retrieval in the neocortex of recent (episodic) memories (see Section 12.3.6).

For emotional behaviour, rapid learning is clearly important. For some types of emotional behaviour, rapid relearning of the reinforcement value of a stimulus may be useful. For example, in social behaviour, reinforcers are constantly being exchanged (for example a positive glance to indicate continuing cooperation), and the current reinforcement association therefore of a person must correspondingly continually be updated. It appears that, particularly in primates, brain systems have evolved to perform this very rapid stimulus-reinforcement relearning. The relearning is often tested by reversal, in which one (e.g. visual) stimulus is paired with reward (for example a sweet taste) and another with punishment (for example a salt taste), and then the reinforcement contingency is reversed.

Motivational behaviour, such as eating and drinking, also requires reinforcement-related learning, so that we can, for example, learn to associate the sight of a food with its taste (which is innately rewarding or aversive, and is described as a primary reinforcer).

In conclusion, a theme of this Section that is closely related to processing in the visual pathways is the importance of learning associations between previously neutral visual stimuli such as the sight of a particular object or face (an example of a secondary reinforcer) and innately rewarding stimuli (or primary reinforcers) such as the taste of food, or a pleasant or painful somatosensory input.

12.3.2 Reward is not processed in the temporal cortical visual areas

We now consider whether associations between visual stimuli and reinforcement are learned, and stored, in the visual cortical areas which proceed from the primary visual cortex, V1, through V2, V4, and the inferior temporal visual cortex (see Figs. 12.11, 12.12, 1.8, 1.9, and 1.10 for schematic diagrams of the organization in the brain of some of the systems being considered). Is the emotional or motivational valence of visual stimuli represented in these regions? A schematic diagram summarizing some of the conclusions that will be reached is shown in Fig. 12.13.

One way to answer this is to test monkeys in a learning paradigm in which one visual stimulus is associated with reward (for example glucose taste, or fruit juice taste), and another visual stimulus is associated with an aversive taste, such as saline. Rolls, Judge, and Sanghera did just such an experiment (Rolls, Judge and Sanghera 1977), and found that single neurons in the inferior temporal visual cortex did not respond differently to objects based on their reward association. To test whether a neuron might be influenced by the reward association, the monkey performed a visual discrimination task in which the reinforcement contingency could be reversed during the experiment. (That is, the visual stimulus, for example a triangle, to which the monkey had to lick to obtain a taste of fruit juice, was after the reversal associated with saline: if the monkey licked to the triangle after the reversal, he obtained mildly aversive salt solution.) An example of such an experiment is shown in Fig. 12.14. The neuron responded more to the triangle, both before reversal when it was associated with fruit juice, and after reversal, when the triangle was associated with saline. Thus the reinforcement association of the visual stimuli did not alter the response to the visual stimuli, which was based on the physical properties of the stimuli (for example their shape, colour, or texture). The same was true for the other neurons recorded in this study. This independence from reward association

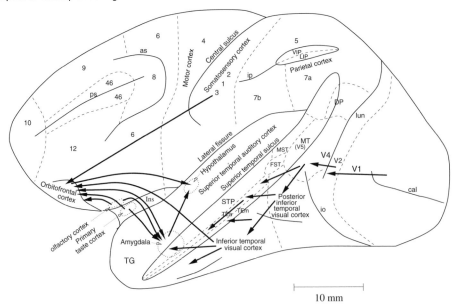

Fig. 12.11 Lateral view of the macaque brain showing the connections in the 'ventral visual system' from V1 to V2, V4, the inferior temporal visual cortex, etc., with some connections reaching the amygdala and orbitofrontal cortex. as, arcuate sulcus; cal, calcarine sulcus; cs, central sulcus; lf, lateral (or Sylvian) fissure; lun, lunate sulcus; ps, principal sulcus; io, inferior occipital sulcus; ip, intraparietal sulcus (which has been opened to reveal some of the areas it contains); sts, superior temporal sulcus (which has been opened to reveal some of the areas it contains). AIT, anterior inferior temporal cortex; FST, visual motion processing area; LIP, lateral intraparietal area; MST, visual motion processing area; MT, visual motion processing area (also called V5); PIT, posterior inferior temporal cortex; STP, superior temporal plane; TA, architectonic area including auditory association cortex; TE, architectonic area including high order visual association cortex, and some of its subareas TEa and TEm; TG, architectonic area in the temporal pole; V1 – V4, visual areas 1 – 4; VIP, ventral intraparietal area; TEO, architectonic area including posterior visual association cortex. The numerals refer to architectonic areas, and have the following approximate functional equivalence: 1,2,3, somatosensory cortex (posterior to the central sulcus); 4, motor cortex; 5, superior parietal lobule; 7a, inferior parietal lobule, visual part; 7b, inferior parietal lobule, somatosensory part; 6, lateral premotor cortex; 8, frontal eye field; 12, part of orbitofrontal cortex; 46, dorsolateral prefrontal cortex.

seems to be characteristic of neurons right through the temporal visual cortical areas, and must be true in earlier cortical areas too, in that they provide the inputs to the inferior temporal visual cortex.

12.3.3 Why the reward and punishment associations of stimuli are not represented early in information processing in the primate brain

The processing stream that has just been considered is that concerned with objects, that is with what is being looked at. Two fundamental points about pattern association networks for stimulus-reinforcement association learning can be made from what we have considered. The first point is that sensory processing in the primate brain proceeds as far as the invariant

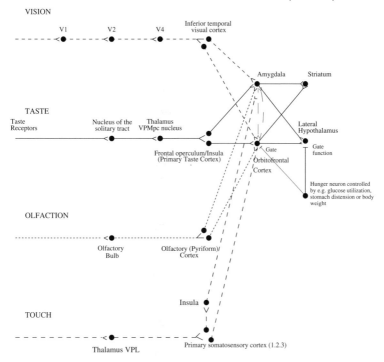

Fig. 12.12 Diagrammatic representation of some of the connections described in this chapter. V1, striate visual cortex. V2 and V4, cortical visual areas. In primates, sensory analysis proceeds in the visual system as far as the inferior temporal cortex and the primary gustatory cortex; beyond these areas, in for example the amygdala and orbitofrontal cortex, the hedonic value of the stimuli, and whether they are reinforcing or are associated with reinforcement, is represented (see text).

representation of objects (invariant with respect to, for example, size, position on the retina, and even view), independently of reward versus punishment association. Why should this be, in terms of systems level brain organization? The suggestion that is made is that the visual properties of the world about which reward associations must be learned are generally objects (for example the sight of a banana, or of an orange), and are not just raw pixels or edges, with no invariant properties, which is what is represented in the retina and V1. The implication is that the sensory processing must proceed to the stage of the invariant representation of objects before it is appropriate to learn reinforcement associations. The invariance aspect is important too, for if we had different representations for an object at different places in our visual field, then if we learned when an object was at one point on the retina that it was rewarding, we would not generalize correctly to it when presented at another position on the retina. If it had previously been punishing at that retinal position, we might find the same object rewarding when at one point on the retina, and punishing when at another. This is inappropriate given the world in which we live, and in which our brain evolved, in that the most appropriate assumption is that objects have the same reinforcement association wherever they are on the retina.

The same systems-level principle of brain organization is also likely to be true in other sensory systems, such as those for touch and hearing. For example, we do not generally want

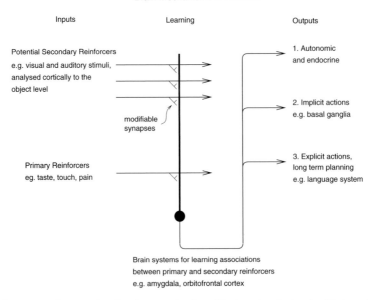

Fig. 12.13 Schematic diagram showing the organization of brain networks involved in learning reinforcement associations of visual and auditory stimuli. The learning is implemented by pattern association networks in the amygdala and orbitofrontal cortex. The visual representation provided by the inferior temporal cortex is in an appropriate form for this pattern association learning, in that information about objects can be read from a population of IT neurons by dot-product neuronal operations.

to learn that a particular pure tone is associated with reward or punishment. Instead, it might be a particular complex pattern of sounds such as a vocalization that carries a reinforcement signal, and this may be independent of the exact pitch at which it is uttered. Thus, cases in which some modulation of neuronal responses to pure tones in parts of the brain such as the medial geniculate nucleus (the thalamic relay for hearing) where tonotopic tuning is found (LeDoux 1994), may be rather special model systems (that is simplified systems on which to perform experiments), and not reflect the way in which auditory-to-reinforcement pattern associations are normally learned. The same may be true for touch in so far as one considers associations between objects identified by somatosensory input, and primary reinforcers. An example might be selecting a food object from a whole collection of objects in the dark.

So far we have been considering where the reward association of objects is represented in the systems that process information about what object is being looked at. It is also of importance in considering where pattern associations between representations of objects and their reward or punishment association are formed, to know where in the information processing of signals such as taste and touch the reward value of these primary reinforcers is decoded and represented. In the case of taste in primates, there is evidence that processing continues beyond the primary taste cortex (see Fig. 12.12) before the reward value of the taste is decoded. The evidence for this is that hunger does affect the responses of taste neurons in the secondary taste cortex, but not in the primary taste cortex (Rolls 1999a, Rolls, Sienkiewicz and Yaxley 1989c, Rolls, Scott, Sienkiewicz and Yaxley 1988, Yaxley, Rolls and Sienkiewicz 1988). (Hunger modulates the reward value of taste, in that if hunger is present,

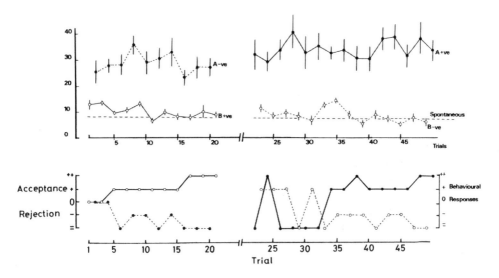

Fig. 12.14 Examples of the responses of a neuron in the inferior temporal visual cortex, showing that its responses (firing rate in spikes/s, upper panel) do not reverse when the reward association of the visual stimuli reverses. For the first 21 trials of the visual discrimination task, visual stimulus A was aversive (–ve, because if the monkey licked he obtained saline), and visual stimulus B was associated with reward (+ve, because if the monkey licked when he saw this stimulus, he obtained fruit juice). The neuron responded more to stimulus A than to stimulus B. After trial 21, the contingencies reversed (so that A was now +ve, and B –ve). The monkey learned the reversal correctly by about trial 35 (lower panel). However, the inferior temporal cortex neuron did not reverse when the reinforcement contingency reversed — it continued to respond to stimulus A after the reversal, even though the stimulus was now +ve. Thus this, and other inferior temporal cortex neurons, respond to the physical aspects of visual stimuli, and not to the stimuli based on their reinforcement contingency. (From Rolls, Judge and Sanghera, 1977.)

animals work to obtain the taste of a food, and if satiated, animals do not work to obtain the taste of a food.) The systems-level principle here is that identification of what the taste is should ideally be performed independently of how rewarding or pleasant the taste is. The adaptive value of this is that even when neurons that reflect whether the taste (or the sight or smell of the food) is still rewarding have ceased responding because of feeding to satiety, it may still be important to have a representation of the sight/smell/taste of food, for then we can still (with other systems) learn, for example, where food is in the environment, even when we do not want to eat it. It would not be adaptive, for example, to become blind to the sight of food after we have eaten it to satiety, and the same holds for taste. On the other hand, when associations must be made explicitly with primary reinforcers such as taste, the stage in processing where the taste representation is appropriate may be where its reward value is represented, and this appears to be in primates beyond the primary taste cortex, in regions such as the secondary taste cortex (Rolls 1989d, Rolls 1995a, Rolls 1997c, Rolls and Scott 2001). In the case of punishing somatosensory input, the situation may be different, for there pain may be decoded early in information processing, by being represented in a special set of sensory afferents, the C fibres. From a systems-level perspective, the primary reinforcer input to pattern associators for learning about which (for example visual or auditory) stimuli are associated with pain could in principle be derived early on in pain processing.

The second point, which complements the first, is that the visual system is not provided with the appropriate primary reinforcers for such pattern association learning, in that visual processing in the primate brain is mainly unimodal to and through the inferior temporal visual cortex (see Fig. 12.12). It is only after the inferior temporal visual cortex, when it projects to structures such as the amygdala and orbitofrontal cortex, that the appropriate convergence between visual processing pathways and pathways conveying information about primary reinforcers such as taste and touch/pain occurs (Fig. 12.12). We now, therefore, turn our attention to the amygdala and orbitofrontal cortex, to consider whether they might be the brain regions that contain the neuronal networks for pattern associations involving primary reinforcers. We note at this stage that in order to make the results as relevant as possible to brain function and its disorders in humans, the system being described is that present in primates such as monkeys. In rats, although the organization of the amygdala may be similar, the areas which may correspond to the primate inferior temporal visual cortex and orbitofrontal cortex are little developed.

12.3.4 Amygdala

Bilateral damage to the amygdala produces a deficit in learning to associate visual and other stimuli with a primary (that is unlearned) reward or punishment. For example, monkeys with damage to the amygdala when shown foods and non-foods pick up both and place them in their mouths. When such visual discrimination performance and learning is tested more formally, it is found that monkeys have difficulty in associating the sight of a stimulus with whether it produces a taste reward or is noxious and should be avoided (see Rolls (1999a), Rolls (2000f), Rolls (2000d), and Baxter and Murray (2000)). Similar changes in behaviour have been seen in humans with extensive damage to the temporal lobe.

12.3.4.1 Connections

The amygdala is a subcortical region in the anterior part of the temporal lobe. It receives massive projections in the primate from the overlying visual and auditory temporal lobe cortex (see Van Hoesen (1981) and Chapters in Aggleton (1992) and Aggleton (2000)) (see Fig. 12.15). These come in the monkey to overlapping but partly separate regions of the lateral and basal amygdala from the inferior temporal visual cortex, the superior temporal auditory cortex, the cortex of the temporal pole, and the cortex in the superior temporal sulcus. Thus the amygdala receives inputs from the inferior temporal visual cortex, but not from earlier stages of cortical visual information processing. Via these inputs, the amygdala receives inputs about objects that could become secondary reinforcers, as a result of pattern association in the amygdala with primary reinforcers. The amygdala also receives inputs that are potentially about primary reinforcers, for example taste inputs (from the secondary taste cortex, via connections from the orbitofrontal cortex to the amygdala), and somatosensory inputs, potentially about the rewarding or painful aspects of touch (from the somatosensory cortex via the insula) (Mesulam and Mufson 1982). The outputs of the amygdala include projections to the hypothalamus: from the lateral amygdala via the ventral amygdalofugal pathway to the lateral hypothalamus, potentially allowing autonomic outputs; and from the medial amygdala, which is relatively small in the primate, via the stria terminalis to the medial hypothalamus, potentially allowing influences on endocrine systems. The ventral amygdalofugal pathway is now known to contain some long descending fibres that project directly to the autonomic cen-

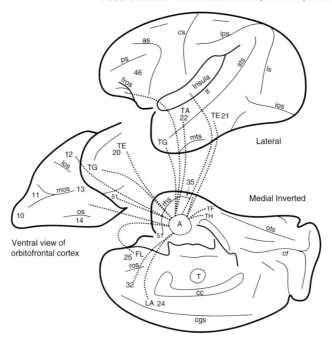

Fig. 12.15 Connections of the amygdala shown on lateral, ventral, and medial views of the monkey brain (after Van Hoesen, 1981). Abbreviations: as, arcuate sulcus; cc, corpus callosum; cf, calcarine fissure; cgs, cingulate sulcus; cs, central sulcus; ls, lunate sulcus; ios, inferior occipital sulcus; mos, medial orbital sulcus; os, orbital sulcus; ots, occipito-temporal sulcus; ps, principal sulcus; rhs, rhinal sulcus; sts, superior temporal sulcus; lf, Lateral (or Sylvian) fissure (which has been opened to reveal the insula); A, amygdala; INS, insula; T, thalamus; TE (21), inferior temporal visual cortex; TA (22), superior temporal auditory association cortex; TF and TH, parahippocampal cortex; TG, temporal pole cortex; 12, 13, 11, orbitofrontal cortex; 35, perirhinal cortex; 51, olfactory (prepyriform and periamygdaloid) cortex. The cortical connections shown provide afferents to the amygdala, but are reciprocated.

tres in the medulla oblongata (for example the dorsal motor nucleus of the vagus), and provide a route for cortically processed signals to influence autonomic responses. (Such autonomic responses include salivation and insulin release to visual stimuli associated with the taste of food, and heart rate changes to visual stimuli associated with anxiety.) A further interesting output of the amygdala is to the ventral striatum including the nucleus accumbens, for via this route information processed in the amygdala could gain access to the basal ganglia and thus influence behavioural output, such as whether we approach or avoid an object. In addition, the amygdala has direct projections back to many areas of the temporal, orbitofrontal, and insular cortices from which it receives inputs.

These anatomical connections of the amygdala indicate that it is placed to receive highly processed information from the cortex and to influence motor systems, autonomic systems, some of the cortical areas from which it receives inputs, and other limbic areas. The functions mediated through these connections will now be considered, using information available from the effects of damage to the amygdala and from the activity of neurons in the amygdala.

12.3.4.2   Effects of amygdala lesions

Bilateral removal of the amygdala in monkeys produces tameness, a lack of emotional responsiveness, excessive examination of objects, often with the mouth, and eating of previously rejected items such as meat (Weiskrantz 1956). These behavioural changes comprise much of the Kluver-Bucy syndrome which is produced in monkeys by bilateral anterior temporal lobectomy (Kluver and Bucy 1939). In analyses of the bases of these behavioural changes, it has been observed that there are deficits in learning to associate stimuli with primary reinforcement, including both punishments (Weiskrantz 1956) and rewards (Jones and Mishkin 1972, Aggleton 1993, Baxter and Murray 2000, Baylis and Gaffan 1991) (see Rolls (2000d)). The association learning deficit is present when the associations must be learned from a previously neutral stimulus (for example the sight of an object) to a primary reinforcing stimulus (such as the taste of food). Further evidence linking the amygdala to reinforcement mechanisms is that monkeys will work in order to obtain electrical stimulation of the amygdala, and that single neurons in the amygdala are activated by brain-stimulation reward of a number of different sites (Rolls 1974, Rolls 1975, Rolls 1976, Rolls, Burton and Mora 1980, Rolls 1999a).

The symptoms of the Kluver-Bucy syndrome, including the emotional changes, could be a result of this type of deficit in learning stimulus-reinforcement associations (Jones and Mishkin 1972, Rolls 1990, Rolls 1999a). For example, the tameness, the hypo-emotionality, the increased orality, and the altered responses to food would arise because of damage to the normal mechanism by which stimuli become associated with reward or punishment.

The amygdala is well placed anatomically for learning associations between objects and primary reinforcers, for it receives inputs from the higher parts of the visual system, and from systems processing primary reinforcers such as taste, smell, and touch (see Figs. 12.15 and 12.12). The association learning in the amygdala may be implemented by Hebb-modifiable synapses from visual and auditory neurons onto neurons receiving inputs from taste, olfactory or somatosensory primary reinforcers. Consistent with this, at least one type of associative learning in the amygdala (and not its retention) can be blocked by local application to the amygdala of an NMDA receptor blocker (Davis 2000). Once the association has been learned, the outputs from the amygdala could be driven by the conditioned as well as by the unconditioned stimuli. In line with this, LeDoux, Iwata, Cicchetti and Reis (1988) were able to show that lesions of the lateral hypothalamus (which receives from the central nucleus of the amygdala) blocked conditioned heart rate (autonomic) responses. Lesions of the central grey of the midbrain (which also receives from the central nucleus of the amygdala) blocked the conditioned freezing but not the conditioned autonomic response to the aversive conditioned stimulus. Further, Cador, Robbins and Everitt (1989) obtained evidence consistent with the hypothesis that the learned incentive (conditioned reinforcing) effects of previously neutral stimuli paired with rewards are mediated by the amygdala acting through the ventral striatum, in that amphetamine injections into the ventral striatum enhanced the effects of a conditioned reinforcing stimulus only if the amygdala was intact (see further Everitt and Robbins (1992)).

There is, thus, much evidence that the amygdala is involved in responses made to stimuli associated with primary reinforcement. There is evidence that it may also be involved in whether novel stimuli are approached, for monkeys with amygdala lesions place novel foods and non-food objects in their mouths, and rats with amygdala lesions have decreased neophobia, in that they more quickly accept new foods (Rolls and Rolls 1973, Dunn and Everitt 1988, Wilson and Rolls 1993).

12.3.4.3 Neuronal activity in the primate amygdala to reinforcing stimuli

Recordings from single neurons in the amygdala of the monkey have shown that some neurons do respond to visual stimuli, consistent with the inputs from the temporal lobe visual cortex (Sanghera, Rolls and Roper-Hall 1979). Other neurons responded to auditory, gustatory, olfactory, or somatosensory stimuli, or in relation to movements. In tests of whether the neurons responded on the basis of the association of stimuli with reinforcement, it was found that approximately 20% of the neurons with visual responses had responses which occurred primarily to stimuli associated with reinforcement, for example to food and to a range of stimuli which the monkey had learned signified food in a visual discrimination task (Sanghera, Rolls and Roper-Hall 1979, Wilson and Rolls 1993, Rolls 2000d). However, none of these neurons (in contrast to some neurons in the hypothalamus and orbitofrontal cortex described below) responded exclusively to rewarded stimuli, in that all responded at least partly to one or more neutral, novel, or aversive stimuli.

The degree to which the responses of these amygdala neurons are associated with reinforcement has also been assessed in learning tasks. When the association between a visual stimulus and reinforcement was altered by reversal (so that the visual stimulus formerly associated with juice reward became associated with aversive saline and vice versa), it was found that 10 of 11 neurons did not reverse their responses (and for the other neuron the evidence was not clear) (Sanghera, Rolls and Roper-Hall 1979, Wilson and Rolls 1993, Rolls 2000d). On the other hand, in a rather simpler relearning situation in which salt was added to a piece of food such as a water melon, the responses of 4 amygdala neurons to the sight of the water melon diminished (Nishijo, Ono and Nishino 1988). More investigations are needed to show the extent to which amygdala neurons do alter their activity flexibly and rapidly in relearning tests such as these (for discussion, see Rolls (2000d)). What has been found in contrast is that neurons in the orbitofrontal cortex do show very rapid reversal of their responses in visual discrimination reversal, and it therefore seems likely that the orbitofrontal cortex is especially involved when repeated relearning and re-assessment of stimulus-reinforcement associations is required, as described below, rather than initial learning, in which the amygdala may be involved.

12.3.4.4 Responses of these amygdala neurons to novel stimuli which are reinforcing

As described above, some of the amygdala neurons that responded to rewarding visual stimuli also responded to some other stimuli that were not associated with reward. Wilson and Rolls (see Rolls (2000d)) discovered a possible reason for this. They showed that these neurons with reward-related responses also responded to relatively novel visual stimuli in, for example, visual recognition memory tasks. When monkeys are given such relatively novel stimuli outside the task, they will reach out for and explore the objects, and in this respect the novel stimuli are reinforcing. Repeated presentation of the stimuli results in habituation of the responses of these amygdala neurons and of behavioural approach, if the stimuli are not associated with primary reinforcement. It is thus suggested that the amygdala neurons described provide an output if a stimulus is associated with a positive reinforcer, or is positively reinforcing because of relative novelty. The functions of this output may be to influence the interest shown in a stimulus; whether a stimulus is approached or avoided; whether an affective response occurs to a stimulus; and whether a representation of the stimulus is made or maintained via an action mediated through either the basal forebrain nucleus of Meynert (see Rolls and Treves (1998) Section 7.1.5) or the backprojections to the cerebral cortex (see Section 1.11).

It is an important adaptation to the environment to explore relatively novel objects or situations, for in this way potential advantage due to gene inheritance can become expressed and selected for. This function appears to be implemented in the amygdala in this way. Lesions of the amygdala impair the operation of this mechanism, in that objects are approached and explored indiscriminately, relatively independently of whether they are associated with positive or negative reinforcement, or are novel or familiar.

The details of the computational mechanisms that implement this in the amygdala are not currently known, but could be as follows. Cortical visual signals which do not show major habituation with repeated visual stimuli, as shown by recordings in the temporal cortical visual areas (see Chapter 5; Rolls, Judge and Sanghera (1977)), reach the amygdala. In the amygdala, neurons respond to these at first, and have the property that they gradually habituate unless the pattern association mechanism in the amygdala detects co-occurrence of these stimuli with a primary reinforcer, in which case the active synapses for that object are strengthened, so that the object continues to produce an output from amygdala neurons that respond to either rewarding or punishing visual stimuli. Neurophysiologically, the habituation condition would correspond in a pattern associator to long-term depression (LTD) of synapses with high presynaptic activity but low postsynaptic activity, that is to homosynaptic LTD (see Figs. 12.13 and 1.5). We thus know that some amygdala neurons can be driven by primary reinforcers such as taste or touch, that some can be activated by visual stimuli associated with primary reinforcers, and that in the rat associative conditioning may require NMDA receptor activation for the learning, though not necessarily for the later expression of the learning. We also know that autonomic responses learned to such stimuli can depend on outputs from the amygdala to the hypothalamus, and that the effects that learned incentives have on behaviour may involve outputs from the amygdala to the ventral striatum. We also know that there are similar neurons in the ventral striatum to some of those described in the amygdala (Williams, Rolls, Leonard and Stern 1993). All this is consistent with the hypothesis that there are neuronal networks in the amygdala that perform the required pattern association. Interestingly, there is somewhat of a gap in our knowledge here, for the microcircuitry of the amygdala has been remarkably little studied. It is known from Golgi studies in young rats in which sufficiently few amygdala cells are stained that it is possible to see them individually, that there are pyramidal cells in the amygdala with large dendrites and many synapses (Millhouse and DeOlmos 1983, McDonald and Aggleton 1992, Millhouse 1986). What these studies have not yet defined is whether visual and taste inputs converge anatomically onto some cells, and whether (as might be predicted) the taste inputs are likely to be strong (for example large synapses close to the cell body), whereas the visual inputs are more numerous, and on a part of the dendrite with NMDA receptors. Clearly, to bring our understanding fully to the network level, such evidence is required, together with evidence that the appropriate synapses are modifiable by a Hebb-like rule (such as might be implemented using the NMDA receptors), in a network of the type shown in Fig. 12.13.

There is a different population of neurons in the orbitofrontal cortex which respond to novel visual stimuli (Rolls, Browning, Inoue and Hernadi 2002a). These neurons respond to visual stimuli for only the first few times that they are shown, habituate over typically 5–10 presentations each 1 s long of visual stimuli, and retain the information about the stimuli for very long periods, of at least many days, in that they do not respond to these same stimuli again in that time. In contrast to the responses of the amygdala neurons just described, these neurons do not respond to reinforcing stimuli. This special population of neurons in a limited region of the macaque orbitofrontal cortex (Rolls, Browning, Inoue and Hernadi 2002a) thus respond in relation to a form of long term memory, and receive their inputs from temporal cortical visual areas in that their latencies are longer than those of inferior temporal cortex neurons, there are IT connections to the orbitofrontal cortex, and these orbitofrontal cortex neurons respond to new, often complex, images and objects.

12.3.5   Orbitofrontal cortex

The orbitofrontal cortex is strongly connected to the amygdala, and is involved in a related type of learning. Damage to the orbitofrontal cortex produces deficits in tasks in which associations learned to reinforcing stimuli must be extinguished or reversed. The orbitofrontal cortex receives inputs about primary reinforcers, such as the taste of food, and information about objects in the world, for example about what objects or faces are being seen. In that it receives information from the ventral visual system (see Chapters 1 and 5), and also from the taste, smell, and somatosensory systems (see Fig. 12.12), it is a major site in the brain of information about what stimuli are present in the world, and potentially for forming associations between these different '**what**' representations. This part of the cortex appears to be involved in the rapid learning and relearning of which objects in the world are associated with reinforcement, and probably does this using pattern association learning between a visual stimulus and a reward or punishment (unconditioned stimulus) as shown in Fig. 12.13. The orbitofrontal cortex develops very greatly in monkeys and humans compared to rats (in which the sulcal prefrontal cortex, which may be a homologue, is very little developed). The orbitofrontal cortex may be a pattern associator which is specialized to allow the very rapid re-evaluation of the current reinforcement value of many different objects or other animals or humans, which continually alter in, for example, social situations. Consistent with this, as described next, the impairments produced in humans by damage to the orbitofrontal cortex include emotional and social changes.

In that the orbitofrontal cortex incorporates a pattern association network for association between visual objects and faces, and receives its visual input from the temporal cortical visual areas, the invariant representation provided by the inferior temporal visual cortex, which is distributed and can be read by dot product decoding of a population of neurons carrying independent information (Chapter 5), is an excellent and appropriate input to the orbitofrontal cortex for this function.

12.3.5.1   Effects of orbitofrontal cortex lesions

The learning tasks impaired by orbitofrontal cortex damage include tasks in which associations learned to reinforcing stimuli must be extinguished or reversed. For example, in work with non-human primates, it has been found that if a reward (a positive reinforcer) is no longer delivered when a behavioural response is made, then that response normally gradually stops, and is said to show extinction. If the orbitofrontal cortex is damaged, then such behavioural responses continue for long periods even when no reward is given. In another example, in a visual discrimination task, a monkey learns that choice of one of two visual stimuli leads to a food reward and of the other to no reward. If the reinforcement contingency is then reversed, normal monkeys can learn to reverse their choices. However, if the orbitofrontal cortex is damaged, then the monkey keeps choosing the previously rewarded visual stimulus,

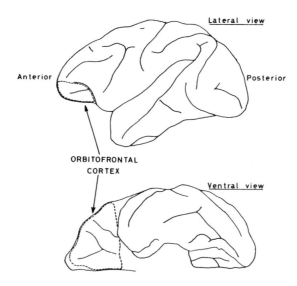

Fig. 12.16 The orbitofrontal cortex of the monkey. The effects of orbitofrontal lesions include: (a) changes in emotional behaviour; (b) changes in food-selection behaviour; (c) impaired extinction of previously learned stimulus-reinforcement associations; (d) difficulty in reversing previously learned stimulus-reinforcement associations during e.g. visual discrimination reversal.

even though it is no longer rewarded. This produces a visual discrimination reversal deficit (Butter 1969, Jones and Mishkin 1972, Rolls 1999a). The visual discrimination learning deficit shown by monkeys with orbitofrontal cortex damage (Jones and Mishkin 1972, Baylis and Gaffan 1991) may be due to the tendency of these monkeys not to withhold responses to non-rewarded stimuli (Rolls 1996a, Rolls 1999a, Rolls 2000e, Rolls 2002). Some of these effects of lesions of the orbitofrontal cortex are summarized in Fig. 12.16.

In humans, analogous deficits have been found after damage to the ventral part of the frontal lobe, in a region that includes the orbitofrontal cortex. It has long been known that in patients with frontal lobe damage there is an impairment in the Wisconsin Card Sorting Task, in which a special set of cards must be sorted into piles based on criteria such as the colour of the items on a card, their form, or the number of items on a card. After the subject has sorted correctly according to colour, the sorting rule is shifted to another dimension (form) without warning. Patients with frontal lobe damage often have problems shifting to the new sorting dimension, and continue sorting according to the previously reinforced sorting dimension. Because this task involves complex cognitive strategies including shifting attention from one stimulus dimension to another, Rolls, Hornak, Wade and McGrath (1994a) designed a much simpler visual discrimination reversal task, to test much more directly the hypothesis that damage to the orbitofrontal cortex impairs the rapid re-learning of which visual stimulus is associated with reward. Their patients touched one of two presented stimuli on a video monitor in order to obtain points, and had not to touch the other stimulus, otherwise they would lose points. After learning this, when the rewarded stimulus was reversed, the patients tended to keep choosing the previously rewarded stimulus, and were thus impaired at visual discrimination reversal (Rolls, Hornak, Wade and McGrath 1994a). This deficit is a perseveration to the previously rewarded stimulus, and not a response perseveration which may be produced by damage to other parts of the frontal lobe (see Rolls (1996a), Rolls (1999a), Rolls (2000e), and Rolls (2002)). Part of the evidence for this is that it is stimuli, such as visual and taste stimuli which are represented in the orbitofrontal cortex, and associated together by learning, and not motor responses, as shown by single neuron recording (see Section 12.3.5.3).

Damage to the caudal orbitofrontal cortex in the monkey produces emotional changes. These include decreased aggression to humans and to stimuli such as a snake and a doll, and a reduced tendency to reject foods such as meat (Butter, Snyder and McDonald 1970, Butter and Snyder 1972, Butter, McDonald and Snyder 1969). In the human, euphoria, irresponsibility, and lack of affect can follow frontal lobe damage (Damasio 1994, Kolb and Whishaw 1996, Rolls 1999a, Rolls 1999b). These changes in emotional and social behaviour may be accounted for, at least in part, by a failure to respond appropriately to stimuli on the basis of their previous and often rapidly changing association with reinforcement (Rolls 1996a, Rolls 1999a, Rolls, Hornak, Wade and McGrath 1994a). Interestingly, in a study prompted by our finding that there are some neurons in the orbitofrontal cortex that respond to faces, and by the inputs from face-selective areas in the inferior temporal visual cortex (see Chapter 5), we have found that some of these patients are impaired at discriminating facial expression correctly (though face recognition is not impaired) (Hornak, Rolls and Wade 1996). Face expression is used as a reinforcer in social situations (for example a smile to reinforce), and the hypothesis is that this is part of the input, equivalent to the unconditioned stimulus, to a pattern associator involved in learning associations between individual people or objects and their current reward status or value.

12.3.5.2 Connections

The inputs to the orbitofrontal cortex include many of those required to determine whether a visual or auditory stimulus is associated with a primary reinforcer such as taste or smell (see Fig. 12.17). The orbitofrontal cortex receives inputs about visual and auditory stimuli, both directly from the inferior temporal visual cortex, the cortex in the superior temporal sulcus (which contains visual, auditory, and somatosensory areas, see Baylis, Rolls and Leonard (1987), the temporal pole (another multimodal cortical area), and the amygdala (Jones and Powell 1970, Seltzer and Pandya 1989, Barbas 1988, Barbas 1995, Carmichael and Price 1995). The orbitofrontal cortex also receives inputs via the mediodorsal nucleus of the thalamus, pars magnocellularis, which itself receives afferents from temporal lobe structures such as the prepyriform (olfactory) cortex, amygdala and inferior temporal cortex (Ongur and Price 2000). It is now known that the orbitofrontal cortex contains the secondary taste cortex (in that it receives direct inputs from the primary taste cortex (Baylis, Rolls and Baylis 1994)), and that information about the reward value of taste, a primary or unlearned reinforcer, is represented here (in that neurons here only respond to, for example, the taste of glucose if the monkey is hungry (Rolls, Sienkiewicz and Yaxley 1989c, Rolls 1997c, Rolls 1999a, Rolls and Scott 2001)). The orbitofrontal cortex also receives somatosensory input (via the insula — see Mesulam and Mufson (1982)), and this could convey primary reinforcing information, for example about the reward value of touch (Francis, Rolls, Bowtell, McGlone, O'Doherty, Browning, Clare and Smith 1999), or of the texture of food in the mouth (Rolls, Critchley, Browning, Hernadi and Lenard 1999a), or about aversive, painful, stimulation (Rolls, Kringelbach, O'Doherty, Francis, Bowtell and McGlone 2001a). Consistent with the latter, patients with frontal damage may say that the pain is still present, but that it no longer

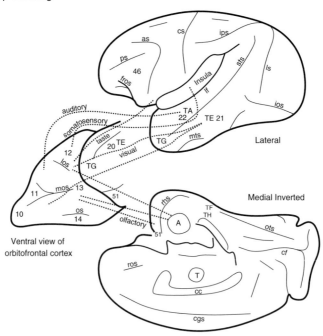

Fig. 12.17 Connections of the orbitofrontal cortex shown on lateral, ventral and medial views of the monkey brain. Abbreviations: as, arcuate sulcus; cc, corpus callosum; cf, calcarine fissure; cgs, cingulate sulcus; cs, central sulcus; ls, lunate sulcus; ios, inferior occipital sulcus; mos, medial orbital sulcus; os, orbital sulcus; ots, occipito-temporal sulcus; ps, principal sulcus; rhs, rhinal sulcus; sts, superior temporal sulcus; lf, Lateral (or Sylvian) fissure (which has been opened to reveal the insula); A, amygdala; INS, insula; T, thalamus; TE (21), inferior temporal visual cortex; TA (22), superior temporal auditory association cortex; TF and TH, parahippocampal cortex; TG, temporal pole cortex; 12, 13, 11, orbitofrontal cortex; 35, perirhinal cortex; 51, olfactory (prepyriform and periamygdaloid) cortex. The cortical connections shown provide afferents to the orbitofrontal cortex, but are reciprocated.

has its full emotional and motivational value. In a recent functional magnetic resonance imaging (fMRI) investigation Francis, Rolls, Bowtell, McGlone, O'Doherty, Browning, Clare and Smith (1999) have shown that the lateral orbitofrontal cortex of humans is activated much more by a pleasant than a neutral touch (relative to the primary somatosensory cortex), thus producing direct evidence that the orbitofrontal cortex is involved in the affective, reward-related, representation of touch. Olfactory inputs are received from the pyriform (primary olfactory) cortex, first to a caudal part of the orbitofrontal cortex designated area 13a, and then projecting from there more widely in the caudal orbitofrontal cortex (Carmichael, Clugnet and Price 1994, Ongur and Price 2000). There are also modulatory influences, which could modulate learning, in the form of a dopaminergic input. The orbitofrontal cortex has outputs, through which it can influence behaviour, to the striatum (Rolls 1999a) (including the ventral striatum), and also projections back to temporal lobe areas such as the inferior temporal cortex, and, in addition, to the entorhinal cortex and cingulate cortex (Van Hoesen, Pandya and Butters 1975, Ongur and Price 2000). The orbitofrontal cortex also projects to the preoptic region and lateral hypothalamus (through which learning could influence autonomic responses to conditioned stimuli) (Rolls 1999a).

The orbitofrontal cortex thus has inputs which provide it with information about primary reinforcing inputs such as taste and touch, inputs which provide it with information about objects in the world (for example visual inputs from the inferior temporal cortex), and outputs which could influence behaviour (via the striatum) and autonomic responses (via the preoptic area and lateral hypothalamus).

12.3.5.3 Neuronal activity in the primate orbitofrontal cortex to reinforcing stimuli

Orbitofrontal cortex neurons with taste responses were described by Thorpe, Rolls and Maddison (1983) and have now been analyzed further (Rolls 1997c, Rolls and Scott 2001). They are in a mainly caudal and lateral part of the orbitofrontal cortex, and can be tuned quite finely to gustatory stimuli such as a sweet taste (Rolls, Yaxley and Sienkiewicz 1990) or umami (protein) taste (Baylis and Rolls 1991, Rolls 2000g). Moreover, their activity is related to reward, in that those which respond to the taste of food do so only if the monkey is hungry (Rolls, Sienkiewicz and Yaxley 1989c). These orbitofrontal neurons receive their input from the primary gustatory cortex, in the frontal operculum (Scott, Yaxley, Sienkiewicz and Rolls 1986, Baylis, Rolls and Baylis 1994). There is also direct neurophysiological evidence that somatosensory inputs affect orbitofrontal cortex neurons, for the taste responses of some are affected by the texture of the food (Rolls 1997c, Rolls, Critchley, Browning, Hernadi and Lenard 1999a).

Visual inputs also influence some orbitofrontal cortex neurons, and in some cases convergence of visual and gustatory inputs onto the same neuron is found (Thorpe, Rolls and Maddison 1983, Rolls, Critchley, Mason and Wakeman 1996b). Moreover, in at least some cases the visual stimuli to which these neurons respond correspond to the taste to which they respond (Thorpe, Rolls and Maddison 1983, Rolls, Critchley, Mason and Wakeman 1996b). This suggests that learning of associations between visual stimuli and the taste with which they are associated influences the responses of orbitofrontal cortex neurons. This has been shown to be the case in formal investigations of the activity of orbitofrontal cortex visual neurons, which in many cases reverse their responses to visual stimuli when the taste with which the visual stimulus is associated is reversed by the experimenter (Thorpe, Rolls and Maddison 1983, Rolls, Critchley, Mason and Wakeman 1996b). An example of the responses of an orbitofrontal cortex cell that reverses the stimulus to which it responds during reward reversal is shown in Fig. 12.18.

This reversal by orbitofrontal visual neurons can be very fast, in as little as one trial, that is a few seconds (see for example Fig. 12.19).

Some of the visual cells in the orbitofrontal cortex are tuned to have responses selective for faces (see Rolls (1999a) and Rolls, Critchley, Browning and Inoue (2002b)). Some of these neurons potentially provide visual representations that are useful as the visual stimulus in stimulus–reinforcer association learning, and others are tuned to face expression and may provide a primary reinforcer useful in social and emotional interactions, acting as the reinforcer in stimulus-reinforcer association learning (Rolls, Critchley, Browning and Inoue 2002b, Kringelbach, Araujo and Rolls 2001).

Another class of neurons responds only in certain non-reward situations (Thorpe, Rolls and Maddison 1983). For example, some neurons responded in extinction, immediately after a lick had been made to a visual stimulus which had previously been associated with fruit juice reward, and other neurons responded in a reversal task, immediately after the monkey had responded to the previously rewarded visual stimulus, but had obtained punishment rather

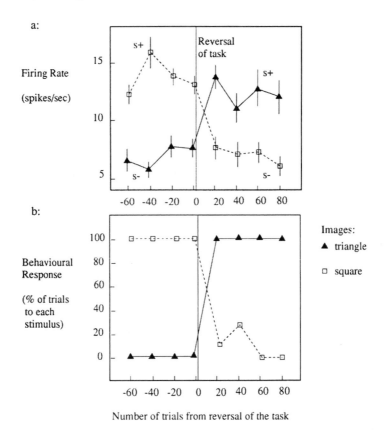

Fig. 12.18 Orbitofrontal cortex: visual discrimination reversal. The activity of an orbitofrontal visual neuron during performance of a visual discrimination task and its reversal. The stimuli were a triangle and a square presented on a video monitor. (a) Each point represents the mean poststimulus activity in a 500 ms period of the neuron to approximately 10 trials of the different visual stimuli. The standard errors of these responses are shown. After 60 trials of the task the reward associations of the visual stimuli were reversed. (+ indicates that a lick response to that visual stimulus produces fruit juice reward; − indicates that a lick response to that visual stimulus results in a small drop of aversive tasting saline). This neuron reversed its responses to the odorants following the task reversal. (b) The behavioural response of the monkey to the task. It is shown that the monkey performs well, in that he rapidly learns to lick only to the visual stimulus associated with fruit juice reward. (After Rolls, Critchley, Mason and Wakeman, 1996.)

than reward (Thorpe, Rolls and Maddison 1983). These neurons thus respond to a mismatch between the reward predicted from a visual stimulus by previous stimulus-reinforcement association learning, and the actual reinforcer that is obtained. This could not be computed in the temporal lobe visual cortical areas, as information about primary reinforcers (such as taste) is not represented there.

These cells thus reflect the information about which stimulus to make behavioural responses to during reversals of visual discrimination tasks. If a reversal occurs, then the taste cells provide the information that an unexpected taste reinforcer has been obtained, another group of cells shows a vigorous discharge which could signal that reversal is in progress,

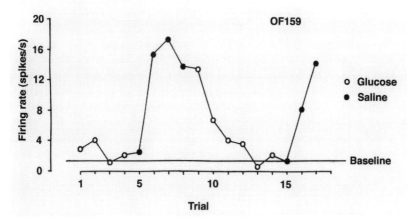

Fig. 12.19 Orbitofrontal cortex: one-trial visual discrimination reversal by a neuron. On trials 1–5, no response of the neuron occurred to the sight of a 2 ml syringe from which the monkey had been given orally glucose solution to drink on the previous trial. On trials 6-9, the neuron responded to the sight of the same syringe from which he had been given aversive hypertonic saline to drink on the previous trial. Two more reversal trials (10–15, and 16–17) were performed. The reversal of the neuron's response when the significance of the same visual stimulus was reversed shows that the responses of the neuron only occurred to the sight of the visual stimulus when it was associated with a negatively reinforcing and not with a positively reinforcing taste. Moreover, it is shown that the neuronal reversal took only one trial. (After Thorpe, Rolls and Maddison, 1983.)

and the visual cells with reinforcement association-related responses reverse the stimulus to which they are responsive. These neurophysiological changes take place rapidly, in as little as 5 s, and are presumed to be part of the neuronal learning mechanism that enables primates to alter their knowledge of the reinforcement association of visual stimuli so rapidly. This capacity is important whenever behaviour must be corrected when expected reinforcements are not obtained, in for example feeding, emotional, and social situations (see Rolls (1999a)).

The details of the neuronal network architecture which underlies this reversal learning have not been finalized for the orbitofrontal cortex. On the basis of the evidence described, it can be proposed that it is as shown in Fig. 12.13. It is likely that NMDA receptors help to implement the modifiability of the visual inputs onto taste neurons. This could be tested by using NMDA receptor blockers applied locally to individual orbitofrontal cortex visual neurons during visual discrimination reversal, or by perfusion of the orbitofrontal cortex with an NMDA receptor blocker to investigate whether this interferes with behavioural visual discrimination reversal. The difference predicted for this system from that in the amygdala is that the orbitofrontal associativity would be more rapidly modifiable (in one trial) than that in the amygdala (which might take very many trials). Indeed, the cortical reversal mechanism in the orbitofrontal cortex may be effectively a fast version of what is implemented in the amygdala, that has evolved particularly to enable rapid updating by received reinforcers in social and other situations in primates. This hypothesis, that the orbitofrontal cortex, as a rapid learning mechanism, effectively provides an additional route for some of the functions performed by the amygdala, and is very important when this stimulus-reinforcement learning must be rapidly readjusted, has been developed elsewhere (Rolls 1990, Rolls 1996a, Rolls

1999a, Rolls 1999b, Rolls 2000e, Rolls 2002).

Although the mechanism has been described so far for visual to taste association learning, this is because experiments on this are most direct. It is likely, given the evidence from the effects of lesions, that taste is only one type of primary reinforcer about which such learning occurs in the orbitofrontal cortex, and is likely to be an example of a much more general type of stimulus–reinforcement learning system. Some of the evidence for this is that humans with orbitofrontal cortex damage are impaired at visual discrimination reversal when working for a reward that consists of points (Rolls, Hornak, Wade and McGrath 1994a); and that the human orbitofrontal cortex has a medial area in which the activation reflects the amount of money won in a probabilistic reward task, and the lateral orbitofrontal cortex has an area the activation of which reflects the amount lost (O'Doherty, Kringelbach, Rolls, Hornak and Andrews 2001). Moreover, as described above, there is now evidence that the affective aspects of touch are represented in the human orbitofrontal cortex (Francis, Rolls, Bowtell, McGlone, O'Doherty, Browning, Clare and Smith 1999, Rolls, Kringelbach, O'Doherty, Francis, Bowtell and McGlone 2001a), and learning about what stimuli are associated with this class of primary reinforcer is also likely to be an important aspect of the stimulus–reinforcement association learning performed by the orbitofrontal cortex.

Further evidence on other types of stimulus–reinforcement association learning in which neurons in the orbitofrontal cortex are involved is starting to become available. One such example comes form learning associations between olfactory stimuli and tastes. It has been shown that some neurons in the orbitofrontal cortex respond to taste and to olfactory stimuli (Rolls and Baylis 1994). It is probably here in the orbitofrontal cortex that the representation of flavour in primates is built (Rolls 2001). To investigate whether these bimodal cells are built by olfactory to taste association learning, we measured the responses of these olfactory cells during olfactory to taste discrimination reversal (Rolls, Critchley, Mason and Wakeman 1996b). We found that some of these cells did reverse, others stopped discriminating between the odours during reversal, while others did not reverse. When reversal did occur, it was quite slow, often needing 20–50 trials. This evidence thus suggests that although there is some pattern association learning between olfactory stimuli and taste implemented in the orbitofrontal cortex, the mechanism involves quite slow learning, perhaps because in general it is important to maintain rather stable representations of flavours. Moreover, olfactory to taste association learning occurs for only some of the olfactory neurons. The olfactory neurons which do not reverse may be carrying information which is in some cases independent of reinforcement association (i.e. is about olfactory identity). In other cases, the olfactory representation in the orbitofrontal cortex may reflect associations of odours with other primary reinforcers (for example whether sickness has occurred in association with some smells), or may reflect primary reinforcement value provided by some olfactory stimuli. (For example, the smell of flowers may be innately pleasant and attractive, and some other odours may be innately unpleasant.) In this situation, the olfactory input to some orbitofrontal cortex neurons may represent an unconditioned stimulus input, with which other (for example visual) inputs may become associated.

Part of the importance of investigating what functions are performed by the orbitofrontal cortex, and how they are performed, is that these have implications for understanding the effects of damage to this part of the brain in patients. The patients we investigated (Rolls, Hornak, Wade and McGrath 1994a) with damage to the ventral part of the frontal lobe had high scores on a behaviour questionnaire which reflected the degree of disinhibited and socially inappropriate behaviour exhibited by the patients. They were also impaired at stimulus–reinforcement association reversal, and had impairments in identifying facial and voice expression (Hornak, Rolls and Wade 1996) which normally provide what are probably primary (unlearned) visual reinforcers (Rolls 1999a). Their altered social and emotional behaviour appear to be at least partly related to the impairment in learning correctly to alter their behaviour in response to changing reinforcing contingencies in the environment (see further (Rolls 1990, Rolls 1996a, Rolls 1999b, Rolls 1999a, Rolls 2000e, Rolls 2002). Consistent with this, it has recently been shown that there is an area of the human orbitofrontal cortex that is especially activated when humans must use face expression to alter their behaviour, compared to a condition in which just the face expression occurs (Kringelbach, Araujo and Rolls 2001).

In conclusion, some of the functions of the orbitofrontal cortex and amygdala are related to stimulus–reinforcement pattern association learning, involving systems-level connections of the form summarized in Fig. 12.12, and involving neuronal networks of the type shown in Fig. 12.13. For these functions, the visual inputs from the temporal cortical visual areas provide inputs that provide representations of faces and objects that are in an appropriate invariant and distributed form for the operation of such pattern association networks, and also provide representations of face expressions (Hasselmo, Rolls and Baylis 1989a) which may act in the orbitofrontal cortex as primary reinforcers (Rolls 1999a, Kringelbach, Araujo and Rolls 2001).

12.3.6 Effects of mood on memory and visual processing

As noted above, the current mood state can affect the cognitive evaluation of events or memories (see Blaney (1986), Rolls and Stringer (2001b)). An example is that when they are in a depressed mood, people tend to recall memories that were stored when they were depressed. The recall of depressing memories when depressed can have the effect of perpetuating the depression, and this may be a factor with relevance to the etiology and treatment of depression. A normal function of the effects of mood state on memory recall might be to facilitate continuity in the interpretation of the reinforcing value of events in the environment, or in the interpretation of an individual's behaviour by others, or simply to keep behaviour motivated to a particular goal. Another possibility is that the effects of mood on memory do not have adaptive value, but are a consequence of having a general cortical architecture with backprojections. According to the latter hypothesis, the selection pressure is great for leaving the general architecture operational, rather than trying to find a genetic way to switch off backprojections just for the projections of mood systems back to perceptual systems (cf. Rolls and Stringer (2000)).

Rolls and Stringer (2001b) (see also Rolls (1989b) and Rolls (1999a)) have developed a theory of how the effects of mood on memory and perception could be implemented in the brain. The architecture, shown in Fig. 12.20, uses the massive backprojections from parts of the brain where mood is represented, such as the orbitofrontal cortex and amygdala to the cortical areas such as the inferior temporal visual cortex and hippocampus-related areas (labelled IT in Fig. 12.20) that project into these mood-representing areas (Amaral, Price, Pitkanen and Carmichael 1992, Amaral and Price 1984). The model uses an attractor in the mood module (labelled amygdala in Fig. 12.20), which helps the mood to be an enduring state, and also an attractor in IT. The system is treated as a system of coupled attractors (see Sections 7.9 and 12.1.2), but with an odd twist: many different perceptual states are associated

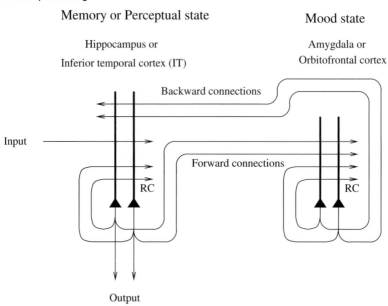

Memory or Perceptual state

Hippocampus or
Inferior temporal cortex (IT)

Mood state

Amygdala or
Orbitofrontal cortex

Backward connections

Input

Forward connections

RC

RC

Output

Fig. 12.20 Architecture used to investigate how mood can affect perception and memory. The IT module represents brain areas such as the inferior temporal cortex involved in perception and hippocampus-related cortical areas that have forward connections to regions such as the amygdala and orbitofrontal cortex involved in mood. (After Rolls and Stringer (2001b).)

with any one mood state. Overall, there is a large number of perceptual / memory states, and only a few mood states, so that there is a many-to-one relation between perceptual / memory states and the associated mood states. The network displays the properties that one would expect (provided that the coupling parameters g (see Section 7.9) are weak). These include the ability of a perceptual input to trigger a mood state in the 'amygdala' module if there is not an existing mood, but greater difficulty to induce a new mood if there is already a strong mood attractor present; and the ability of the mood to affect via the backprojections which memories are triggered.

An interesting property which was revealed by the model is that because of the many-to-few mapping of perceptual to mood states, an effect of a mood was that it tended to make all the perceptual or memory states associated with a particular mood more similar then they would otherwise have been. The implication is that the coupling parameter g for the backprojections must be quite weak, as otherwise interference increases in the perceptual / memory module (IT in Fig. 12.20).

12.4 Output to object selection and action systems

As shown in Chapter 5, translation invariance is a property of inferior temporal cortex neurons. This is a useful property of these neurons, because it enables the output of the ventral visual stream to provide an output about an object which, because it is invariant, enables the brain areas which receive from the inferior temporal cortex to perform tasks such as visual stimulus-reinforcement decoding (as described in Section 12.3), recognition using long term memory (see Section 12.2), episodic long-term memory including object-place memory (see Section 12.2), and short term memory for the object (see Section 12.1). Without the translation invariance, the output would alter whenever the object moved a little on the retina, and all these memory-related operations performed at the object level, which is the level that is ecologically useful, would fail.

The fact though that translation invariance is a property of inferior temporal cortex neurons raises problems. How are the coordinates of the object passed to the motor system for action when the location of the object with respect to the retina is not made explicit in the firing of inferior temporal cortex neurons? How also is the reinforcement association looked up correctly by the orbitofrontal cortex and amygdala, or the object recognized or remembered by memory systems (see Section 12.2), when there are several objects in the visual scene, and the objects are in a cluttered natural environment?

Part of the solution to these problems is that the receptive fields of inferior temporal cortex neurons become smaller in complex natural scenes. In particular, the results described in Section 5.6.3 and summarized in Fig. 5.20 show that the receptive fields of inferior temporal cortex neurons were large (65 deg) with a single stimulus in a blank background (top left), and were considerably reduced in size (to 36.6 deg) when the stimulus was presented in a complex natural scene (top right). As shown in Sections 5.6.3 and 8.4.9, part of the underlying mechanism here may be shunting effects produced by different objects on the potentially large receptive fields of IT neurons which are relatively more influenced by what is as the fovea as a result of the greater foveal cortical magnification factor. Consistent with this, almost all inferior temporal cortex neurons respond well when the effective stimulus is on the fovea, and most (but not all) inferior temporal cortex neurons have their greatest response when an effective stimulus is on the fovea (Tovee, Rolls and Azzopardi 1994). In addition, there is a further, smaller, effect of attention on the receptive field size of objects viewed in natural scenes, with the receptive field being a little larger when an object is selected as an object of attention in a visual search task (Fig. 5.20 bottom right), and a little smaller when the object is not being attended to in a visual search task (middle right). It is proposed that this reduced translation invariance in natural scenes, and as a function of whether attention is not being paid to an object, helps an unambiguous representation of an object (which may be used in memory operations or which may be the target for action) to be passed to the brain regions which receive from the primate inferior temporal visual cortex. It helps with the binding problem, by reducing in natural scenes the effective receptive field size of at least some inferior temporal cortex neurons to be approximately the size of an object in the scene.

The reduced receptive field size helps with the memory problem, but still leaves a major question with respect to how the spatial coordinates of the object being represented in the inferior temporal visual cortex are passed to the motor system, if the spatial coordinates of the object on the retina are not represented in the inferior temporal visual cortex. The solution proposed for this fundamental problem is that the spatial coordinates may not be explicitly passed as an output of the inferior temporal visual cortex. Instead, the coordinates may effectively be passed through the world, by taking the egocentric coordinates of the position being fixated as the coordinates of the target selected for action. This means that the action selection system, fed from the dorsal visual stream, can operate on the assumption that the target for action is on the fovea. Consistent with this, most inferior temporal cortex neurons in visual object search tasks in natural scenes respond in relation to whether the object for action is on or close to the fovea (see Section 5.6.3). Thus active vision, using explicit visual search by eye movements (cf. Ballard (1990) and Ballard (1993)), may be part of the coordinate passing mechanism with which the coordinates are passed effectively through the world. Once the fovea is at the target, the dorsal visual stream can provide the information needed to guide, for example, arm movements (cf. Andersen, Batista, Snyder, Buneo and Cohen (2000)), by making available local, egocentric, visual information available from for example stereo disparity and other cues that are made explicit in the dorsal visual system, in order to guide action.

Of course, we are not limited to performing actions that are exclusively at the fovea, although this may be the normal situation. When the object is not at the fovea in special situations, or just before an eye movement is made to acquire the object on the fovea, then the covert spatial attention mechanisms described in Chapters 9-10 may be used to make explicit the location of the target in egocentric space as represented in the parietal cortex. This it is suggested works as follows. The prefrontal cortex is loaded with the object that is the target for the visual search by for example visual inputs from the inferior temporal visual cortex (see Fig. 12.1). This short-term attractor memory for the object then biases the IT stream, and this in turn biases early visual areas (such as V2 and possibly V1) for features relevant to the object. For the part of the retina where those features are present, there is extra activity in V2 at that location (due to the summing of the top-down signals and the forward visual inputs from the retina), and this extra activity results in the bubble of activity in the parietal cortex continuous spatial attractor moving to the corresponding location, which is not necessarily at the fovea. The parietal mechanisms can then use the location of the covert attention bubble, and their knowledge of the current eye position, to select the appropriate arm movements to acquire the target.

This theory offers a fundamental understanding of why visual processing is divided into two main streams, a ventral 'what' stream and a dorsal 'where' stream (cf. Ungerleider and Mishkin (1982)). The ventral stream is needed so that the memory and related operations described in this Chapter including determining the reward value of objects, which are properties of objects in the world and not of their detailed position and size on the retina, can be computed about objects which are made explicit in the representation provided in the inferior temporal visual cortex. Having produced this ('what') representation of objects, emotional and motivational behaviour to the objects can be determined. It is important that the representation of objects in the inferior temporal cortex is though left untrammelled by reward associations, because the IT representations are needed for many different functions, as described in this Chapter, and these operations (such as remembering where fruit is located in the environment) must proceed independently of whether we currently are hungry for the fruit or not. Thus the 'what' representations need to be made explicit in a way that is independent of the location of the objects in egocentric space, and this is the major function performed by the ventral visual stream.

Correspondingly, the dorsal visual stream provides a representation of locations and motion in visual egocentric space, and this is needed in order to guide actions such as arm reaching or eye movements to the correct position in egocentric space. Because the same egocentric visual spatial analysis can be applied to most objects, there is no need to have a detailed representation of objects in the dorsal visual stream[39]. Indeed, it is more economical

---

[39] Of course it is useful to have some information about the 3D structure of the object being viewed represented in the parietal cortical visual areas, as this is needed to help define the details of the action. An example is shaping the hand in order to grasp a cup. Such representations have been described, by for example Murata, Gallese, Luppino,

in terms of neural hardware not to have a detailed visual representation of where all identified objects are in a 3D scene.

Not only is it economical not to have such a 3D visual scene representation with all objects in their correct positions, but it is also computationally extremely difficult to produce such a representation using natural visual scenes, and indeed there is no artificial intelligence or engineering-based solution to this problem currently. It may be partly because of the computationally enormous problem of trying to solve such a problem that our brains make do with a much simpler system, as described in this book, namely a ventral visual system that can represent the identity of objects and can be influenced by 3D cues but does not build a 3D scene-based representation, and a dorsal visual system which provides a 3D representation of space useful for guiding actions, but does not have object descriptions built-in.

It is in this situation that a solution is needed to how the object visual system interfaces to the spatial visual system, and the proposal just described is that it does so through the world.

According to the theory provided in this book, the ventral visual is needed for action as well as the dorsal visual system. (Milner and Goodale (1995) had argued that the ventral visual system was involved in perception, and the dorsal visual system in action.) The ventral visual system is involved in action by providing the basis for selecting the goal for the action. During action initiation to an object, it is proposed that the ventral visual system produces a representation of the object in the inferior temporal visual cortex. The object is then assessed as a possible goal or target for action by using stimulus-reinforcement association lookup in the orbitofrontal cortex and amygdala. If the object is reward-associated, and does not have too many punishment associations, then the positive net reward value identifies this as a target for action. The fact that it has been identified as an important goal (to approach or avoid, or to which to perform instrumental behaviour, see Rolls (1999a)) then influences the motor system (by structures such as the basal ganglia, including the ventral striatum) to perform actions towards the object. The coordinates of the object in egocentric space for the selected action are then available, as just described, to the action execution mechanism, which involves parietal cortical areas (see Rolls (1999a)).

One issue that this theory helps to understand is the fact that the representations in the parietal cortical areas are primarily egocentric (see Chapter 6). This is appropriate for a system which must guide actions towards locations already selected on the basis of the output of the ventral visual system. On the other hand, allocentric spatial representations, of places in the world independently of egocentric disposition, are represented in ventral brain regions such as the hippocampus (see Section 12.2), and the overlying parahippocampal and related cortical areas damage to which produces topographical agnosia (Habib and Sirigu 1987). Allocentric representations are formed to provide appropriate spatial links between items of object-based knowledge, such as the objects that might be represented in a city. As such, the close links from the object identification system in the inferior temporal visual cortex to the (temporal lobe) allocentric spatial representation systems are very appropriate. Of course, when the allocentric spatial representation needs to be linked to egocentric, idiothetic, information then special linking mechanisms are needed, and the possible linking mechanisms in the hippocampus are described in Sections 7.5.5 and 12.2.

One issue left to clarify is why the receptive fields of inferior temporal cortex neurons are so large, and show such translation invariance, when what is needed according to the

---

Kaseda and Sakata (2000). This does not mean that object identity needs to be represented in the parietal cortical visual areas.

above theory is primarily a representation of the object at the fovea. One reason for the large translation invariant receptive fields may be that the sizes of different objects vary enormously, that for identification of which object it may need all the properties of an object, and that therefore the size of the receptive fields of the object-identifying neurons must be capable of being very large. Of course, as described in Sections 5.6.3 and 8.4.9, the receptive fields may decrease in natural scenes to approximately the size of objects. However, flexibility in the size of the receptive fields may be needed for the reason just provided. Given the fact that the receptive fields of IT neurons may need, for at least some objects, to be large, the translation (and size etc) invariance is then needed both to ensure that the object can be recognized even when it is not aligned as a template with exactly the right position, size, and orientation (cf. Section 8.2.3), but also because the parts of the whole object may need to be identified independently of exactly how they appear on the retina.

The potential to have large receptive fields of IT neurons is also a useful property during visual search, for then the effective size of the receptive fields can be increased even in a complex natural scene by top-down bias from the object being searched for, as described in Section 8.4.9 and in Chapters 9 and 10.

12.5 Visual search

The outputs of the inferior temporal cortex are, according to the hypotheses presented, involved in visual search in a number of ways. First, the receptive fields of IT neurons will tend to be larger for the object that is the target of attention or search, as a result of the top-down bias effects that operate directly in IT as described in Section 8.4.9, and as a result of the constraint satisfaction implemented throughout the whole system shown in Fig. 12.1 described above and in Chapters 10 and 11. Second, the latter process, through the effects that occur in early visual modules such as V2, has the potential to adjust the shape of IT receptive fields to match the shape and size of the object that is the target object of attention and visual search. The second process implements an adjustable and potentially 'moving' region of enhanced responsiveness that is independent of or precedes eye movements, and is covert[40]. These covert attentional processes are particularly evident in reduced visual displays with only one or two objects in the scene (see Section 5.6.3), and are much less evident, at least in the responses of IT neurons, when objects are the targets for attention and visual search in a natural, complex, visual scene. Under such natural visual conditions, overt search may be relatively much more important than covert attention and search. However, it is possible that covert search may guide eye movements if a target object is starting to activate the spatial system as a result of the object bias, and the eyes are sufficiently close to features that are part of the selected target object. Under such conditions the covert location of attention might actually precede eye position changes. Although ETR has seen tantalizing evidence for IT neuronal activity to a target object starting at approximately the time that the eyes saccade to the target, rather than 100 ms after the eyes have acquired the target which would be the usual IT neuronal response latency after a visual stimulus is shown, the occasions on which this clearly happens are infrequent, and this is not the usual pattern of IT neuronal responses

---

[40]Covert attention refers to attention directed at a location that is away from the position being fixated. Overt attention refers to moving the eyes to fixate the location to which attention is being paid.

in visual search tasks. (The typical pattern is that the IT neurons start to respond 80–100 ms after an (overt) eye position change has acquired the target.)

In the context of covert attention, it may be noted that the covert attentional processes described in Chapters 9 and 10 are likely to be sufficiently rapid to allow the system to converge to a selected spatial location within the 300 ms that is typical of fixations between saccades (allowing 200 ms processing time given IT response latencies). This is shown not only by the simulations described in Chapters 9 and 10, but also by simulations revealing the dynamical properties of hierarchies of attractor networks implemented with integrate-and-fire neurons (see Section 7.26).

12.6   Visual outputs to behavioral response systems

Another class of visual output from the inferior temporal visual cortex reaches the striatum, with direct projections from the inferior temporal cortex to both the tail of the caudate nucleus, and the adjacent ventral putamen. It has been shown that neurons in the tail of the caudate nucleus of macaques respond to visual stimuli with latencies just longer than those of IT neurons, and can be orientation or shape selective, responding for example to objects depending on their orientation (Caan, Perrett and Rolls 1984). The tail of caudate neurons tend to habituate rapidly, within several trials, even though the monkey is attending to the stimuli and performing a visual discrimination task. It is thus suggested that these tail of caudate neurons may be involved in behavioural orienting to changing patterned visual stimuli, such as new shapes appearing in the visual scene (Caan, Perrett and Rolls 1984).

In the adjacent ventral putamen in the region which receives from the inferior temporal visual cortex, neurons may respond to the visual stimuli used in a delayed match to sample task (Johnstone and Rolls 1990). The neurons do not respond in an auditory delayed match to sample task, so their responses do not reflect movements made, but instead may represent a link between the type of short term memory implemented in the inferior temporal visual cortex and behavioural response systems.

In addition, a direct connection from the inferior temporal cortex to the inferolateral part of the prefrontal cortex may provide the visual input required for conditional visual to motor response learned mappings. In this type of task, one visual stimulus, e.g. a triangle, means 'go left', and another visual stimulus, e.g. a square, means 'go right'. The performance of this task is impaired by lesions of the inferior lateral prefrontal cortex (the inferior convexity cortex, area 12) (Passingham, Toni and Rushworth 2000).

12.7   Multimodal representations in different brain areas

The cortex in the anterior part of the superior temporal sulcus receives inputs from the inferior temporal visual cortex, and also inputs from auditory association cortical areas, and somatosensory association cortical areas (see Fig. 12.11). These areas are very probably involved in crossmodal associations, in that in this region neurons are found that respond to visual inputs (such as the mouth moving when vocalizing), and other neurons have auditory responses. The convergence between inputs corresponding in two or more modalities is a natural associative process, and may help to, for example, identify the sound in noisy conditions if there is a corresponding visual input. It is the operation of such processes that give rise to effects such as the McGurk effect, described in Section 7.9.

It should be noted that there are many areas of the cerebral cortex where multimodal representations are formed. Other such areas, which also receive from the inferior temporal visual cortex, are the orbitofrontal and amygdala, which build cross-modal associations of a special type, stimulus–reinforcement associations, by virtue of the fact that the orbitofrontal cortex and amygdala are brain regions where primary reinforcers such as taste, pleasant touch, and pain are represented (see Section 12.3).

Another brain region where multimodal representations are formed is the parietal cortex, where auditory and retinal inputs are mapped together, presumably for example to facilitate eye movements and attentional shift towards locations where sounds are heard (Andersen, Batista, Snyder, Buneo and Cohen 2000).

12.8   Visuo–spatial scratchpad, and change blindness

Given the fact that the responses of inferior temporal cortex neurons are quite locked to the stimulus being viewed, it is unlikely that IT provides the representation of the visual world which we think we see, with objects at their correct places in a visual scene. In fact, we do not really see the whole visual scene, as most of it is a memory reflecting what was seen when the eyes were last looking at a particular part of the scene. The evidence for this statement comes from change blindness experiments, which show that humans rather remarkably do not notice if, while they are moving their eyes and cannot respond to changes in the visual scene, a part of the scene changes (O'Regan, Rensink and Clark 1999, Rensink 2000). A famous example is that in which a baby was removed from the mother's arms during the subject's eye movement, and the subject failed to notice that the scene was any different. Similarly, unless we are fixating the location that is different in two alternated versions of a visual scene, we are remarkably insensitive to differences in the scene, such as a glass being present on a dining table in one but not another picture of a dining room. Given then that much of the apparent richness of our visual world is actually based on what was seen at previously fixated positions in the scene (with this being what IT represents), we may ask where this 'visuo-spatial' scratchpad (short-term memory) is located in the brain. One possibility is in the right parieto-occipital area, for patients with lesions in the parieto-occipital region have (dorsal) simultanagnosia, in which they can recognize objects, but cannot see more than one object at a time (Farah 1990).

The computational basis of this is not yet known, but might consist it is hypothesized of a number of separate, that is local, attractors each representing part of the space, and capable of being loaded by inputs from the inferior temporal visual cortex. According to this computational model, the particular attractor network in the visuo-spatial scratchpad memory would be addressed by information based on the position of the eyes, of covert attention, and probably of the head.

12.9   Conscious visual perception

Conscious visual perception may be more associated with the operation of the ventral than the dorsal processing stream (see Fig. 1.10) (Milner and Goodale 1995). Part of the evidence for this is that a patient with ventral visual system damage may be unable to report on the orientation of an object, but nevertheless be able to manipulate the object into the correct orientation for posting through a slot (Milner and Goodale 1995). The implication is that we are normally aware of processing in at least some part of our ventral visual system; and that many of the details of operation of the dorsal visual system can be performed without conscious processing, and may not normally reach consciousness.

Another type of evidence is that the depth of anticorrelated random dot stereograms is represented in the dorsal visual system but not the ventral visual system, and we are not conscious of the depth cues present in these anticorrelated visual displays (see Section 3.4.2). On the other hand, neurons that respond to correlated but not to anticorrelated random dot stereograms are present in the inferior temporal visual cortex, and we see depth cues consciously in the correlated but not the anticorrelated stereograms.

The possibility that consciousness is more closely associated with processing in the ventral visual pathways may be related to the fact that we can perform long-term planning and decision-making about objects, a process which when helped by higher-order thoughts used to correct the plans, may be closely related to consciousness, as argued by Rolls (1999a).

13 Principles and Conclusions

13.1 The inferior temporal visual cortex provides a transform invariant representation of objects and faces

The neurophysiological evidence described in Chapter 5 shows that the responses of many inferior temporal visual cortex neurons have transform invariant responses to objects and faces. The types of invariance include translation, size, spatial frequency, and even view for some neurons. Not all neurons have view invariance, and this is in accordance with the theory that view invariant representations are formed in the inferior temporal cortex by associating together different views of the same object. The transform invariance property reflects the point that information about objects and faces is made explicit in the neuronal output of the ventral visual processing stream.

13.2 The representation in the inferior temporal visual cortex (IT) is in a distributed form that provides high capacity and that can be read by neurons performing dot-product decoding

The neurophysiological evidence for this is described in Chapter 5. Part of the evidence is that IT neurons typically have an exponential distribution of firing rates to a set of stimuli.

Another part of the evidence is that the information provided by different IT neurons is almost independent, so that the number of stimuli rises exponentially with the number of neurons in the sample. This means that a receiving neuron needs to receive just a set of inputs from randomly selected IT neurons in order to obtain a great deal of information about which object is being viewed, and is a major factor in simplifying brain connectivity.

Another part of the evidence is that each neuron provides up to typically 0.3–0.5 bits of information about the stimulus set, so that to obtain information about which object in the high dimensional space of objects is being viewed, the responses of many neurons must be considered. A single neuron provides insufficient information to diagnose which of many objects is being seen.

Another part of the evidence is that much of the information available in the firing rates can be read by dot product decoding, which is very biologically plausible in that it is the simplest type of operation that neurons could perform.

Another part of the evidence is that the information can be read out from the firing rates of the neurons (in practice, the number of spikes they emit in a short time) without taking into account their relative time of firing.

This poses a major challenge to theories that synchronization between the spikes of different neurons is a part of the code in the ventral visual system, because quantitatively a very great deal of information is available in the rates, and it remains to be shown that synchronization adds quantitatively much or anything to the code, at least in the inferior temporal visual cortex, the end of the object processing stream.

13.3 Much of the information available from the responses of the neurons about shapes and objects is available in short time periods, of for example 20 ms.

This property enables processing to travel rapidly from stage to stage of the ventral visual system, as the next stage can read the information in short time periods from the previous stage. This property also shows that each cortical area can perform the computation necessary to support object recognition in a short period in the order of 20 ms. The evidence for this from information theoretic analyses of the responses of neurons, and from backward masking experiments when measuring neuronal responses and human psychophysical responses, is described in Chapter 5. The computation can be performed this rapidly because neurons have continuous dynamics, as shown in Chapter 7 (Section 7.6).

13.4 Implementing computations with neurons with continuous dynamics allows very rapid feedback processing within a cortical area, so that constraint satisfaction using recurrent collateral connections between neurons in a cortical area could be performed within approximately 15 ms per cortical area

The evidence for this is described in Section 7.6. The same type of rapid processing enables not only information to travel rapidly through a multiple layer hierarchy such as the visual cortical areas in the ventral stream with attractor-based feedback processing in each area, but also for top-down processing including attentional biasing effects to operate quite rapidly, as described in Chapters 9 and 10.

13.5 Of the approaches to the perception of objects described in Section 8.2, what is found in the ventral visual system appears to be closest to a hierarchical feature analysis system

Different abstract computational approaches to object recognition are described in Section 8.2. Of these, *feature lists, spaces or histograms* (Section 8.2.1) are inadequate to account for primate object perception and discrimination because they include no shape information, and cannot discriminate between objects with the same features in a different spatial arrangement. Nevertheless, the types of features, such as texture and colour, that are present in objects do provide valuable evidence in identification (and are incorporated very naturally into hierarchical feature analysis systems).

Template matching and the alignment approach (Section 8.2.3) have the major problem as potential solutions to primate vision that it is very difficult to conceive how the arbitrary shifts, rotations, and scale changes would be performed by a biological system that cannot perform very accurate matrix multiplication. It is also difficult to conceive how a biological system would, even if it had the capability to do such matrix multiplications, find the correct match in its database of the stored possible templates. Some form of canonical representation would be needed. Another difficulty is that reasonable segmentation of an object would be needed at an early stage of processing so that the transforms could be performed on the image of an object for later matching into the database, and segmentation early on in visual processing in complex natural scenes including occlusions is very difficult. Another difficulty with this proposal is that it is sufficiently powerful, if it could be implemented in the brain, to solve the problem of invariant object recognition in as little as one processing stage, and this does not match the gradual computation of invariant representations which seems to be being performed over many stages of the cortical visual hierarchy leading to IT as described in Chapters 5 and 8.

Syntactic 3D structural descriptions of objects (Section 8.2.2), which were chosen by Marr (1982) and by Biederman (1987), suffer from the problems that it is very difficult to conceive how the arbitrary 'on-the-fly' (dynamic, real-time) syntactic or relational linking between the limited set of parts could be performed by a biological system, as described in Section 8.2. The binding would need not only a link between part A and part B, but the link would have to provide some evidence about the spatial relation, e.g. 'is to the top left of', and this would mean that at least three items would be involved in every relational description (at least two parts and one relational descriptor). Synchronization does not seem to be a possible solution to this binding problem, because, to describe an object, not just knowing which features need to be bound, but the type of relational binding between every pair of features, would be needed (see further below). Another problem is that such a system only works well if the parts or features can be identified correctly (which includes early segmentation) before the structural description can be parsed, and this early segmentation is difficult in natural scenes.

Hierarchical feature analysis systems do have the capability of implementing the spatial relations between features into the analysis they perform, by incorporating fixed (non-dynamic) feature combination neurons, as described in Chapter 8. Such systems perform the computation gradually over several processing stages (matching what is found in the primate visual system) in order to limit the number of possible combinations that would potentially need to be tested if the analysis were to be globally over the whole image space at any one stage.

One potential problem with such systems is the difficulty of performing the analysis over the whole image because many different objects may be present in the scene, and the spatial relations between all the objects need to be encoded. It is proposed that the solution to this problem used by the brain is to limit the analysis so that it is not performed simultaneously over the whole visual field. This is achieved by using foveate vision (i.e. giving preference to a high resolution central region), by reducing receptive field size in complex natural scenes using shunting / competitive interactions (as described in Sections 5.6.3 and 8.4.9), and by using spatial or object attentional bias to limit what needs to be solved at any one time, as described in Chapters 9 and 10.

Another potential problem is the potential combinatorial explosion in the number of neurons required, but it is suggested in Chapter 8 that this is solved partly by taking low order combinations of the inputs from the preceding stage, partly by taking the inputs from only a small region of the previous stage, and partly by using the redundancy present in natural images, and matching by self-organization the feature analyzers to the features and feature combinations that are present in scenes.

Another potential problem is how the system might be trained to put together the different types of retinal feature that are transforms of each other in order to build transform invariant representations. The solution that is proposed to this in Chapter 8 is a short term memory trace-based local associative learning rule.

Another possible problem of such systems is that they most easily associate together different 2D views of objects, and may not include a full 3D description of the object. It is suggested that the work-around to this problem adopted by the visual system is to use a separate visual system, the dorsal visual system, to perform the 3D operations needed in space to perform actions on objects identified by the ventral visual system.

In summary, although there are problems and limitations of such systems, the processing being performed by the ventral visual system does appear to match a feature hierarchy analysis system, and solutions that the brain uses to many of the problems that arise are proposed and described in Chapter 8.

13.6   Invariant representations can be self-organized using a trace learning rule incorporated in a feature hierarchy network such as VisNet

Although the demonstration discussed in Section 9.6.2 shows that attention can contribute to help with providing useful translation invariant representations by limiting the area of the visual field being processed, and thus the number of objects that might otherwise be simultaneously represented in the inferior temporal visual cortex, attention-based models are not sufficient to solve the problem of forming invariant representations, as described in Section 13.7.

A sufficient and also biologically plausible way of achieving translation invariance using feedforward processing and a trace learning rule is described in Chapter 8. A feedforward system is more plausible in that with the masking experiments described in Section 5.5.6 and Appendix B.3.2, it is found that inferior temporal cortex neurons can still provide object-selective information even when each cortical area can fire for only 30 ms. This time is too short to enable activity once it has reached the inferior temporal cortex to be backprojected to V1, and forward projected again with dynamic settling. One property that may help the real visual cortical system with the feedforward aspect of its processing, especially when operating in cluttered natural scenes, may be the weighting given to whatever is at the fovea, given the larger magnification factor at the fovea (see Section 8.4.6).

Of course, in natural cluttered scenes when there is sufficient time for attentional processes to operate, good performance may be contributed both by the weighting given to whatever is at the fovea due to its greater magnification (which describes overt attention, where eye position is used to define what object in a scene to process), and to the types of covert attentional process described in this book, where attention can be paid to an object that is not at the fovea.

Another reason why the trace learning rule-based self-organizing processes described in Chapter 8 are more appropriate for forming invariant representations in general is that they provide an account of view-invariant object recognition, which a spatial attention–based approach to invariant object recognition cannot. Further, a spatial attention–based approach to invariant object recognition cannot, without special added mechanisms, account for the binding of features in which the relative spatial position of the features must be encoded, as discussed in Section 13.7.

13.7 Spatially selective shape feature binding, and invariant form-based object recognition, require neurons that respond to combinations of features in the correct spatial configuration, but not attention or synchronization

To obtain globally invariant representations of objects (i.e. representations that are invariant when the whole object is moved in the visual field), yet in which the local spatial arrangement of the features is critical, it is shown in Section 8.4.5 that a solution is to use neurons that become by a self-organizing process connected to respond to a low-order combination of features in the correct spatial arrangement from the previous layer. Such neurons might respond differently to an 'L' and a 'T', even though both are formed from a vertical and a horizontal line. Low-order combinations (combinations of a few inputs) are required in order to limit the combinatorial explosion, the solution to which is also helped by the layered architecture of the visual system (V1–V2–V4–Posterior IT–Anterior IT), by the redundancy in the statistics of the visual world, and by the large numbers of neurons in the visual system. It is also noted in Section 8.4.5 that the spatial binding of features must be implemented early on in the processing, so that the relative positions of the features become part of the object description, even though the object itself can be at any location in the space. Evidence consistent with this requirement and prediction of the theory of invariant object recognition described in Chapter 8 is now starting to appear. In particular, many neurons in V1 may respond better to combinations of spatial features with the correct spatial configuration and to single oriented lines or edges (see Section 2.5).

Attention per se (see Section 9.6.2) will not suffice to solve the computational problem, as shown by the following. If the mechanism of attention as described in Chapters 9 and 10 enables features in a small region of space to be highlighted (have increased activity) (produced by either spatial or object-based bias), then this still does not specify what the spatial relations are between the features in the highlighted zone. If we paid attention to an area where a 'T' or an 'L' was located and the description was in terms of only first order features such as a vertical line and a horizontal line, then we would still not know the spatial relations between the features, and just highlighting them would not solve the shape/object discrimination problem. Of course attention can be a major help in such a system, by helping with the selection of objects for action. Attention in this sense may enable us to select one object rather than another, which would otherwise be a major problem in a cluttered scene, as described in Sections 8.4.5, 8.4.6, 5.6.3 and 8.4.9. But attention would not itself enable us to tell one shape or object from another, where the shape is defined by the relative spatial locations of features within objects, and not just by the list of features contained within objects, as described in Section 8.2.

In a similar way, synchronization does not itself solve the problem of binding features in the correct spatial relation as required for shape and object recognition. Synchronization might enable the system to know that two features are for example part of the same object (see Sections 5.5.7 and 8.5), but would not by itself define the spatial relations between the features. Of course, there could be a special set of syntactic units, defining for example 'the first feature is to the left of and a bit above the second feature', which would be synchronously active with the first and second features, but this would raise all sorts of further problems, such as the system knowing the difference between the first and the second feature (itself needing a binding descriptor?), the enormous number of 'features + syntactic relation' pools that would need to be kept apart by the online synchronous binding process (cf. Malsburg (1973) and Malsburg (1999)) to describe a typical object, and the difficulty of specifying metric properties of the spatial relations (e.g. 'a little above and a lot to the left of') which would require an enormous number of spatial relation descriptors. All these are generic problems of syntactic structural description schemes, as described in Sections 8.2 and 13.5, which might not be a problem for computer implementations, but seem entirely implausible for the brain to implement for general-purpose object recognition. Of course, as noted elsewhere, the fact that the human brain can provide a structural description of objects is not the point at issue here, for this may be performed by a very different brain system present in humans which relies on the syntactic capabilities of our natural language processing.

We note that attention and/or synchronization could of course help when features such as shape, colour and motion are represented in different maps in intermediate-level visual analysis in the cortex. For example, as shown in Chapter 10, spatial attention could increase the activity of neurons in corresponding locations in shape and colour maps, so that the particular colour and form represented at that location will survive the competition in the object representation.

13.8 The representation at the end of the ventral, object processing, stream is in a form that is suitable for object-reward associations, recognition memory, short term memory, and episodic memory

The evidence for this is described in Chapter 5, and the implementations of these processes in areas that receive from the inferior temporal visual cortex are described in Chapter 12. The outputs of the ventral visual stream are appropriate for these functions because they provide evidence about the properties of objects in the world independently of how they appear on the retina, and it is objects that are associated with rewards and punishers, are recognized, or are found in particular places, not particular retinal images. This property enables the memory and other systems that receive from the inferior temporal visual cortex to generalize correctly, e.g. from one view to another view of the same object.

13.9 There are temporal cortical visual areas, found especially in the cortex in the anterior part of the superior temporal sulcus, with neurons specialized for face expression, for view-dependent representations, for face and body gestures, and for combining object and motion information

As described in Section 5.7, there are specialized populations of neurons that code for face expression and not face identity. These neurons are found primarily in the cortex in the superior temporal sulcus, while the neurons responsive to identity are found in the inferior temporal gyrus, and areas adjacent to this, such as TEa and TEm. Information about facial expression is of potential use in social interactions. A further way in which some of these neurons in the cortex in the superior temporal sulcus may be involved in social interactions is that some of them respond to gestures, e.g. to a face undergoing ventral flexion, or more generally turning towards or away from the observer. Many neurons in this region respond to objects or faces only when they are moving in particular ways, and this is a brain region in which evidence from the ventral and dorsal visual streams appears to be brought together for special functions. It is also important when decoding facial expression to retain some information about the direction of the head relative to the observer, for this is very important in determining whether a threat is being made in your direction. The presence of view-dependent, head and body gesture (Hasselmo, Rolls, Baylis and Nalwa 1989b), and eye gaze (Perrett, Smith, Potter, Mistlin, Head, Milner and Jeeves 1985b), representations in some of these cortical regions where face expression is represented is consistent with this requirement. These systems are likely to project to regions of the orbitofrontal cortex and amygdala, in which face expression and object–movement neurons are found.

13.10 Interactions between an object information processing stream and a spatial processing stream implemented by backprojections to an early topologically organized visual area can account for many properties of visual attention

The model of attention described in Chapters 9–11 represents an advance beyond the biased competition hypothesis, in that it shows how object and spatial attention can be produced by dynamic interactions between the 'what' and 'where' streams, and in that as a computational model that has been simulated, the details of the model have been made fully explicit and have been defined quantitatively. An interesting and important feature of the model is that the model does not use explicit multiplication as a computational method, but the modulation of attention (for example the effects of posterior parietal module (PP) activity on V1) appears to be like multiplication. This is an interesting contribution of the model, namely that multiplicative-like attentional gains are implemented without any explicit multiplicative operation (see Section 7.8.3).

In Chapters 9–11 we analyzed the neuronal ('microscopic-level') neurodynamical mechanisms that underlie visual attention. We formulated a computational model of cortical systems based on the 'biased competition' hypothesis. The model consists of interconnected populations of cortical neurons distributed in different brain modules, which are related to the different areas of the dorsal or 'where' and ventral or 'what' processing pathways of the primate visual cortex. The 'where' pathway incorporates mutual connections between a feature extracting module (V1–V4), and a parietal module (PP) that consists of pools coding the locations of the stimuli. The 'what' path incorporates mutual connections between the feature extracting module (V1–V4) and an inferotemporal module (IT) with pools of neurons coding for specific objects. External attentional top-down bias is defined as inputs coming from higher pre-frontal modules which are not explicitly modelled in Chapters 9–11 but are modelled in Section 12.1. Intermodular attentional biasing is modelled through the coupling between pools of different modules, which are explicitly modelled. Attention appears now as an emergent effect that supports the dynamical evolution to a state where the constraints given by the stimulus and the external bias are satisfied. Visual search and attention can be explained in this theoretical framework of a biased competitive neurodynamics. The top–down bias guides attention to concentrate at a given spatial location or on given features. The neural population dynamics are handled analytically in the framework of the mean-field approximation. Consequently, the whole process can be expressed as a system of coupled differential equations. The model was extended in order to include the resolution hypothesis, and a 'microscopic' physical (i.e. neuron-level) implementation of the global precedence effect. We analyzed the attentional neurodynamics involved in visual search of hierarchical patterns, and also modelled a mechanism for feature binding that can account for conjunction visual search tasks.

The essential contributions of this model of attention are:

1. Different functions involved in active visual perception have been integrated by a model based on the biased competition hypothesis. Attentional top-down bias guides the dynamics to concentrate at a given spatial location or on given (object) features. The model integrates, in a unifying form, the explanation of several existing types of experimental data obtained at different levels of investigation. At the microscopic neuronal level, we simulated single cell recordings, at the mesoscopic level of cortical areas we reproduced the results of fMRI (functional magnetic resonance imaging) studies, and at the macroscopic perceptual level we accounted for psychophysical performance. Specific predictions at different levels of investigation have also been made. These predictions inspired single cell, fMRI, and psychophysical experiments, that in part have been already performed and the results of which are consistent with our theory.

2. Attention is a dynamical emergent property in our system, rather than a separate mechanism operating independently of other perceptual and cognitive processes.

3. The computational perspective provides not only a concrete mathematical description of mechanisms involved in brain function, but also a model that allows complete simulation and prediction of neuropsychological experiments. Interference with the operation of some of the modules was used to predict impairment in visual information selection in patients suffering

from brain injury. The resulting experiments support our understanding of the functional impairments resulting from localized brain damage in patients.

13.11   Visual search

As discussed in Section 12.5, in complex natural scenes visual search may take place largely overtly, by eye movements, which are serial in nature. However, mechanisms for covert visual search are described in Chapters 9–11, and although perhaps contributing to performance more in simple visual displays with two or a few objects present, may contribute to performance in complex natural scenes by influencing the next eye movement that will be made, as suggested in Chapter 10.

We demonstrated that it is possible to build a neural system for visual search, which works across the visual field in parallel but, due to the different latencies of its dynamics, can show the two experimentally observed modes of visual attention, namely: serial focal attention, and the parallel spread of attention over space. Neither explicit serial focal search nor saliency maps need to be assumed.

The visual system works always in parallel, but the different latencies associated with different spatial resolutions allow the emergence of an early attentional focus based on coarse features. Spatial resolution then gradually increases at this focus as the high spatial frequency channels contribute to the processing being performed. The spatial form of the early focus is initially attached to the coarse (low spatial resolution) structure of an object. When finer details start to modulate the initially activated pools, a focus more attached to the details of the object emerges. Consequently, our neurodynamical cortical model explains the underlying massively parallel mechanisms that not only generate the emergence of a kind of attentional 'spotlight', but also its object-based character, and the associated spatially localized enhancement of spatial resolution suggested by the resolution hypothesis.

13.12   The parietal cortex contains egocentric spatial representations, and the medial temporal lobe system allocentric spatial representations

The representations in the parietal cortical areas are primarily egocentric (see Chapters 6 and 12). This is appropriate for a system which must guide actions towards locations already selected on the basis of the output of the ventral visual system. On the other hand, allocentric spatial representations, of places in the world independently of egocentric disposition, are represented in ventral brain regions such as the hippocampus (see Section 12.2), and the overlying parahippocampal and related cortical areas damage to which produces topographical agnosia (Habib and Sirigu 1987). Allocentric representations are formed to provide appropriate spatial links between items of object-based knowledge, such as the objects that might be represented in a city. As such, the close links from the object identification system in the inferior temporal visual cortex to the (temporal lobe) allocentric spatial representation systems are very appropriate. Of course, when the allocentric spatial representation needs to be linked to egocentric, idiothetic, information, then special linking mechanisms are needed, and the possible linking mechanisms in the hippocampus are described in Sections 7.5.5 and 12.2.

13.13 The controller of visual attention can now be understood in terms of the information represented in short term memory systems in the prefrontal cortex that biases earlier visual cortical spatial and object processing areas by backprojections

The model described in Chapter 9 shows that no mysterious controller of attention needs to be found, but that instead the control is performed by the information loaded into prefrontal cortex short term memories biasing earlier visual cortical spatial and object processing areas by backprojections. The short term memories are themselves loaded by presentation of the sample cue, in object or spatial working memory tasks such as delayed match to sample with intervening stimuli, as described in Section 12.1. The short term memories are loaded in visual search tasks with the object or location that is the subject of the search, as described in Section 12.1 and Chapter 9.

Other parts of the brain in addition to the prefrontal cortex might provide the top-down bias to the parietal spatial or the IT object modules. We do not wish to exclude these. One example is the auditory–verbal short term memory system in humans (which using rehearsal holds on-line a set of approximately 7 chunks of information), and which may be located in the cortex at the junction of the left parieto-occipito-temporal areas. The principle though is the same, that there is no mysterious controller of attention, and that what is needed is a short term memory system to hold the spatial or the object of attention active, and which provides top–down bias to the high–level spatial or object representation areas such as PP or IT.

This we believe is an important conceptual point, in that it removes the concern that there is some non-understood aspect of the control of attention, with a type of 'deus ex machina' or at least an unlocated (serial or parallel) 'spotlight controller' being needed. Indeed, the overall schematic architecture of the system described in this book is illustrated in Fig. 12.1. The architecture allows the target of attention to be analyzed in the spatial or object processing stream, then loaded into a prefrontal cortex (or other) short term memory system, from which it can exert its top-down biasing effect on the spatial or object stream, which in turn by interactive feedforward and feedback effects causes the whole system to settle to optimally satisfy the constraints. The constraint satisfaction we describe is not itself a mysterious process either, but can be understood as an energy minimization process now well understood in neural networks (Hopfield 1982, Amit 1989, Hertz, Krogh and Palmer 1991). This constraint satisfaction generally operates well in practice even when the conditions required for the formal analysis are not present, for example when the system does not have complete and reciprocal connectivity due to random asymmetric dilution of the connectivity, i.e. when synapses are missing at random (Treves 1991, Rolls and Treves 1998). For the memory retrieval properties of such systems to operate well, the number of synapses per neuron must be kept relatively high, above 1,000–2,000 (Rolls, Treves, Foster and Perez-Vicente 1997c), an important condition which significantly is well met by the actual numbers of synapses per neuron in the cortex. We note also that the large number of forward and backward connections between adjacent cortical areas in the architecture shown in Fig. 12.1 provides a suitable basis, given also some type of associative synaptic connection rule between the connected areas, for a system that can operate in the interactive constraint satisfaction way described in Sections 1.11, 7.9, and Chapters 9–11.

We believe that the conceptual framework for understanding attention described in this book may be useful in helping to understand the otherwise rather complicated pictures that are often produced in neuroimaging studies of attention in humans, in which large swathes of parietal and frontal cortical territory often show activation. We now have clear reasons for expecting frontal, parietal, temporal and even occipital lobe contributions to attention, given the architecture shown in Fig. 12.1 and the model described in Chapters 9–11. The model, and the specialization of function within the parietal cortex described in Chapter 4, lead us to understand that different parietal and even connected frontal areas may be activated during different types of spatial attention and memory, for example when attention must be paid to where a response is to be made with the arm or with the eyes, or when there are spatial cues in both visual fields, one of which is a target and others of which are distractors. We would also expect different temporal cortical areas to become activated while paying attention depending on whether the attention is to face identity, face expression, objects, objects undergoing motion, colour, etc (see Chapter 5). Given the tendency of neurons to cluster into small regions where similar neurons are found (due to the self-organizing map principles described in Section 7.4.6), we would even expect the exact loci of activation found in the temporal areas to be somewhat different for different classes of object, and to be in not necessarily the same relative positions in different humans. We would also expect some activation during attentional tasks to be found quite early on in cortical visual processing, perhaps as far back as V1, and have given reasons in Chapter 9 why this might though weak still be a useful feature of the attentional architecture.

Thus the fundamental understanding offered by our conceptualization of the operation of attentional processes in the brain may we hope help to provide a fundamental basis for understanding the phenomena that arise in imaging studies, but also of course in neurophysiological, psychophysical, and neuropsychological studies.

13.14　Output to object selection and action systems

How are the coordinates of a selected target object passed to the motor system for action, if there is little topology, and there is spatial invariance, in IT? It is suggested that the use of the position in visual space being fixated provides part of the interface between sensory representations of objects and their coordinates as targets for actions in the world (Section 12.4). The small receptive fields of IT neurons in natural scenes make this possible. After this, local, egocentric, processing implemented in the dorsal visual processing stream using e.g. stereodisparity may be used to guide action.

13.15　'What' versus 'where' processing streams

It is argued in Section 12.4 that the ventral stream is needed so that the memory and related operations described in Chapter 12 including determining the reward values of objects (which are properties of objects in the world and not of their detailed position and size on the retina), can be computed about objects that are made explicit in the representation provided in the inferior temporal visual cortex. Having produced this ('what') representation of objects, emotional and motivational behaviour to the objects can be determined. It is important though that the representation of objects in the inferior temporal cortex is left untrammelled by reward associations, because the IT representations are needed for many different functions, as described in Chapter 12, and these operations (such as remembering where fruit is located in the environment) must proceed independently of whether we currently are hungry for the fruit or not. Thus the 'what' representations need to be made explicit in a way that is independent of the location of the objects in egocentric space, and this is the major function performed by the ventral visual stream.

Correspondingly, the dorsal visual stream provides a representation of locations and motion in visual egocentric space, and this is needed in order to guide actions such as arm reaching or eye movements to the correct position in egocentric space.

Not only is it economical not to have such a 3D visual scene representation with all objects in their correct positions, but it is also computationally extremely difficult to produce such a representation using natural visual scenes. It may be partly because of the computationally enormous problem of trying to solve such a problem that our brains make do with a much simpler system, as described in this book, namely a ventral visual system that can represent the identity of objects and can be influenced by 3D cues but does not build a 3D scene-based representation, and a dorsal visual system that provides a 3D representation of space useful for guiding actions, but does not have object descriptions built-in.

13.16 Short Term Memory systems (in the frontal lobe) must be separate from perceptual mechanisms (in the temporal and parietal lobes)

A common method that the brain uses to implement a short term memory is to maintain the firing of neurons during a short memory period after the end of a stimulus (see Section 12.1). For the short term memory to be maintained during periods in which new stimuli are to be perceived, there must be separate networks for the perceptual and short term memory functions. Indeed two coupled networks, one in the inferior temporal visual cortex for perceptual functions, and another in the prefrontal cortex for maintaining the short term memory during intervening stimuli, provides a precise model of the interaction of perceptual and short term memory systems (Renart, Parga and Rolls 2000, Renart, Moreno, de al Rocha, Parga and Rolls 2001). In particular, this model shows how a prefrontal cortex attractor (autoassociation) network could be triggered by a sample visual stimulus represented in the inferior temporal visual cortex in a delayed match to sample task, and could keep this prefrontal attractor active during a memory interval in which intervening stimuli are shown. Then when the sample stimulus reappears in the task as a match stimulus, the inferior temporal cortex module shows a large response to the match stimulus, because it is activated both by the visual incoming match stimulus, and by the consistent backprojected memory of the sample stimulus still being represented in the prefrontal cortex memory module (see Figs. 12.2 and 12.3). This computational model makes it clear that in order for ongoing perception implemented by posterior cortex (parietal and temporal lobe) networks to occur unhindered, there must be a separate set of modules that is capable of maintaining a representation over intervening stimuli. This is the fundamental understanding offered for the evolution and functions of the dorsolateral prefrontal cortex, and it is this ability to provide multiple separate short term attractor memories that provides we suggest the basis for its functions in planning. One of the underlying computational constraints that drives these points is that a short term memory network implemented by continuing firing in an attractor state can usefully hold only one memory active at a time (see Section 7.3).

This approach emphasizes that in order to provide a good brain lesion test of prefrontal cortex short term memory functions, the task set should require a short term memory for stimuli over an interval in which other stimuli are being processed, because otherwise the posterior cortex perceptual modules could implement the short term memory function by their own recurrent collateral connections. This approach also emphasizes that there are many at least partially independent modules for short term memory functions in the prefrontal cortex (e.g. several modules for delayed saccades in the frontal eye fields; one or more for delayed spatial (body) responses in the dorsolateral prefrontal cortex; one or more for remembering visual stimuli in the more ventral prefrontal cortex; and at least one in the left prefrontal cortex used for remembering the words produced in a verbal fluency task – see Rolls and Treves (1998) Chapter 10).

This computational approach thus provides a clear understanding for why a separate (prefrontal) mechanism is needed for working memory functions. It may also be noted that if a prefrontal cortex module is to control behaviour in a working memory task, then it must be capable of assuming some type of executive control. There may be no need to have a single 'central executive' additional to the control that must be capable of being exerted by every short-term memory module. This is in contrast to what has traditionally been assumed for the prefrontal cortex (Shallice and Burgess 1996).

The same model shown in Fig. 12.2 can also be used to help understand the implementation of visual search tasks in the brain (Renart, Parga and Rolls 2000). In such a visual search task, the target stimulus is made known beforehand, and inferior temporal cortex neurons then respond more when the search target (as compared to a different stimulus) appears in the receptive field of the IT neuron (Chelazzi, Miller, Duncan and Desimone 1993, Chelazzi, Duncan, Miller and Desimone 1998). The model shows that this could be implemented by the same system of weakly coupled attractor networks in the prefrontal cortex (PF) and IT shown in Fig. 12.2 as follows. When the target stimulus is shown, it is loaded into the PF module from the IT module as described for the delayed match to sample task. Later, when the display appears with two or more stimuli present, there is an enhanced response to the target stimulus in the receptive field, because of the backprojected activity from PF to IT, which adds to the firing being produced by the target stimulus itself (Renart, Parga and Rolls 2000, Renart, Moreno, de al Rocha, Parga and Rolls 2001). The interacting spatial and object networks described in Chapters 9–11 (see Fig. 12.1) take this analysis one stage further, and show that once the PF–IT interaction has set up a greater response to the search target in IT, this enhanced response can in turn by backprojections to topologically mapped earlier cortical visual areas move the 'attentional spotlight' to the place where the search target is located. A further way in which attractor networks can help to account for the responses of IT neurons is described in Section 12.5.

13.17 Cortico-cortical backprojections must be weak relative to forward and intramodular recurrent connections

The evidence and reasons for this are described in Sections 1.11, 7.9 and 12.1.

13.18 Long-term potentiation is needed for the formation but not the reuse of short-term memories

To set up a new short term memory attractor, synaptic modification is needed to form the new stable attractor. Once the attractor connections are set up, the attractor may be used repeatedly when triggered by an appropriate cue to hold the short term memory state active by continued neuronal firing even without any further synaptic modification (see Sections 7.3, 12.1 and Kesner and Rolls (2001)). Thus manipulations that impair the long-term potentiation of synapses (LTP) may impair the formation of new short term memory states, but not the use of previously learned short term memory states.

13.19 "Executive control" functions of the prefrontal cortex may simply reflect the functions of the prefrontal cortex in providing short term memory systems, used for example for attentional targets to be maintained on-line.

The simple architecture described in Section 9.3 and more generally in Chapter 9 (see Fig. 12.1 and Deco and Lee (2001)) allows spatial attention (Helmholtz 1867) and object attention (James 1890) to be accounted for in a symmetric fashion. The two modes of attention emerge depending simply on whether a top-down bias is introduced to either the dorsal stream posterior parietal (PP) module or the ventral stream IT module. In this framework, attention is produced by a simple top-down bias communicated from the short term memory systems of the brain which hold the target object or location in memory (e.g. in the prefrontal cortex) to the dorsal stream or the ventral stream. Moreover, this conceptualization offers a way of understanding the "executive control" that is ascribed to the prefrontal cortex. It appears based on phenomenology to implement "executive control", but at least a major part of this function we suggest can be understood as providing the short term memory bias to posterior (parietal and temporal) perceptual systems to enable them to implement attentional effects as described. Of course, without a short term memory system in the prefrontal cortex to hold the target on-line in memory while the perceptual systems are processing sensory input (see Section 12.1), the whole organism would appear to an observer to be without "executive control", and indeed to be displaying a "dysexecutive syndrome" (Shallice and Burgess 1996).

Although we show in Fig. 12.1 separate working memory systems in the lateral prefrontal cortex, for spatial locations more dorsally (PFCd) and for objects more ventrally (PFCv) towards the inferior convexity, in line with where the inputs from the parietal cortical areas and the temporal lobe object areas may be focused, we do not require physically separated short term memory systems for the model to operate. The requirement of the model is to have a short term memory system, whether for spatial or object information, that is reciprocally connected back to the spatial (PP) and object (IT) processing systems (see Fig. 12.1). We note that there is evidence for at least some mixing of spatial and object short term memory systems (Rao, Rainer and Miller 1997), and indeed attractor networks can store both continuous representations (of for example physical space) and discrete representations (of for example objects) (see Section 7.5.8 and Rolls, Stringer and Trappenberg (2002c)). Our view is that there may in fact be partly separate and partly overlapping short term memory systems in the prefrontal cortex, with partial separation necessary in order to obtain a total memory capacity that is greater than that of just a single network covering the whole of the prefrontal cortex (see Section 7.9), which would be very limiting; but partly overlapping due to the short range spread of recurrent collateral connections in the cortex (see Chapter 1). Indeed, the short range connectivity within a cortical area of the recurrent collaterals with a relatively high density (up to perhaps 10%) within 1–2 mm (see Chapter 1) could be seen as a useful neocortical adaptation (and in contrast to the hippocampus) to enable partly separate operation of nearby cortical areas, in order to keep the total memory capacity high, in order to enable different computations to proceed simultaneously, and in order to enable several items to be kept in short term memory simultaneously by keeping the attractors separate (see Section 7.9).

13.20 Reward and punishment, and emotion and motivation, are not represented in the object processing stream

It is shown in Section 12.3 that visual sensory processing in the primate brain proceeds as far as the invariant representation of objects (invariant with respect to, for example, size, position on the retina, and even view), independently of reward versus punishment association. Why should this be, in terms of systems-level brain organization? The suggestion that is made is that the visual properties of the world about which reward associations must be learned are generally objects (for example the sight of a banana, or of an orange), and are not just raw pixels or edges, with no invariant properties, which are what is represented in the retina and V1. The implication is that the sensory processing must proceed to the stage of the invariant representation of objects before it is appropriate to learn reinforcement associations. The invariance aspect is important too, for if we had different representations for an object at different places in our visual field, then if we learned when an object was at one point on the retina that it was rewarding, we would not generalize correctly to it when presented at another position on the retina. If it had previously been punishing at that retinal position, we might find the same object rewarding when at one point on the retina, and punishing when at another. This is inappropriate given the world in which we live, and in which our brains evolved, in that the most appropriate assumption is that objects have the same reinforcement association wherever they are on the retina.

The same systems-level principle of brain organization is also likely to be true in other sensory systems, such as those for touch and hearing. For example, we do not generally want to learn that a particular pure tone is associated with reward or punishment. Instead, it might be a particular complex pattern of sounds such as a vocalization that carries a reinforcement signal, and this may be independent of the exact pitch at which it is uttered. The same may be true for touch in so far as one considers associations between objects identified by somatosensory input, and primary reinforcers. An example might be selecting a food object from a whole collection of objects in the dark.

The second point, which complements the first, is that the visual system is not provided with the appropriate primary reinforcers for such pattern association learning, in that visual processing in the primate brain is mainly unimodal to and through the inferior temporal visual cortex (see Fig. 12.12). It is only after the inferior temporal visual cortex, when it projects to structures such as the amygdala and orbitofrontal cortex, that the appropriate convergence between visual processing pathways and pathways conveying information about primary reinforcers such as taste and touch/pain occurs (Fig. 12.12).

Part of the functional significance of not representing the reward value of visual stimuli until after object representations have been formed is that the object representations may be needed for many different functions, including recognition, short term memory, the formation of long-term episodic memories, etc. The systems-level principle here is that identification of what the visual stimulus or taste is should ideally be performed independently of how rewarding or pleasant the visual stimulus or taste is. The adaptive value of this is that even when neurons that reflect whether the taste (or the sight or smell of the food) is still rewarding have ceased responding because of feeding to satiety, it may still be important to have a representation of the sight (and, for that matter, the smell and taste, see Rolls (1999a)) of food, for then we can still (with other systems) learn, for example, where food is in the environment, even when we do not want to eat it. It would not be adaptive, for example, to become blind to the sight of food after we have eaten it to satiety.

It is shown in Section 12.3 that the learning of associations of visual representations of objects with primary reinforcers occurs in brain regions that receive from IT, such as the orbitofrontal cortex and amygdala. These brain regions contain representations of primary reinforcers such as taste and touch, and receive distributed representations of objects directly from IT that are in a form suitable for pattern association networks to learn the associations between visual stimuli and primary reinforcers. In this way the orbitofrontal cortex and amygdala enable visual representations to become goals for actions.

Outputs from the amygdala and orbitofrontal cortex thus provide a pathway by which visual stimuli become the targets of actions. This function results in the orbitofrontal cortex and amygdala having important functions in emotion and in motivation, as described in Section 12.3.

13.21   Effects of mood on memory and visual processing

Backprojections from brain areas such as the orbitofrontal cortex and amygdala where mood is represented can influence the recall of memories and the visual images that are recalled, in ways analyzed by Rolls and Stringer (2001b) and described in Section 12.3.6.

13.22   Visual outputs to Long Term Memory systems

The representation of objects in the inferior temporal cortical areas is in a suitable form for an input to long term memory mechanisms, as described in Section 12.2. The representation is suitable in that objects are made explicit in the output, and transforms of the objects are not. The representation is also suitable in that it has high capacity with the neurons carrying almost independent information, and is suitable as an input to associative memories in that much of the information can be read by neurons that perform dot product operations. The outputs that reach the hippocampus may be used for episodic memory, especially when this has a spatial component. The outputs to the perirhinal cortex may be used for recognition memory, where interestingly it has been shown that the degree of long-term familiarity of stimuli may be represented. This has interesting implications for understanding amnesia. A part of the orbitofrontal cortex has neurons that respond only for the first few occasions on which novel stimuli are shown, and thus provides a long-term memory useful for detecting novel stimuli.

13.23 Episodic memory and the operation of mixed discrete and continuous attractor networks

As shown in Sections 12.2 and 7.5.8, the implementation of episodic memory which normally has continuous spatial and discrete (e.g. object) components, could be implemented by an attractor network that combines both continuous and discrete memory representations. This network may be located in medial temporal lobe regions, such as the hippocampus. The primate hippocampus provides with its spatial view cells an allocentric (world-based) representation of space 'out there', which is very appropriate for the spatial part of an episodic memory. The hippocampus can maintain this allocentric spatial representation in the dark without visual cues. Moreover, idiothetic, that is self-motion, cues are interfaced to visual representations in this brain region, in that eye movements update the allocentric representation even in the dark. A model for how this interfacing could operate using a continuous attractor network that performs 'path integration' on the idiothetic inputs is described in Sections 12.2 and 7.5.5.

13.24 Visual outputs to behavioural response systems

The inferior temporal cortex projects to brain regions such as the tail of the caudate nucleus that may be involved in orienting to changing objects (Caan, Perrett and Rolls 1984, Rolls 1999a), and to parts of the lateral prefrontal cortex which may be involved in conditional visual object to motor-response mapping (see Chapter 12).

13.25 Multimodal representations in different brain areas

Multimodal representations are found in many of the areas to which the inferior temporal visual cortex projects, for example in the orbitofrontal and amygdala (see Section 12.3) where they are involved in visual stimulus to reinforcement association learning, and in the cortex in the superior temporal sulcus, where visual and auditory processing streams converge (see Chapter 5 and Section 7.9).

13.26 Visuo–spatial scratchpad and change blindness

As argued in Section 12.8, the visual scene which we perceive may be implemented largely by at least partially separate and local short term memory attractor networks each representing different locations in the visual scene, and loaded with objects close to the fovea from the representation provided by the inferior temporal visual cortex when we are looking at the object. This system appears to be located in cortex in the parieto-occipital region.

13.27 A unified feature hierarchical model of invariant object recognition and dynamical attention

To help understand some of the computational issues involved in invariant object recognition using a feature hierarchy network, such as seems to be implemented in the brain, we described in Chapter 8 what for modelling tractability was kept as a feedforward model. The dynamical feedback model of attention described in Chapters 9–11 required dynamics, in order to account for many of the interacting processes involved in attention and its temporal phenomena, including "serial" vs "parallel" search. The approaches as described are complementary, but it is possible to combine both approaches into a single unified model. Indeed, the implementation of such a model is straightforward, in that it incorporates the overall network architecture and dynamical equations used in Chapter 9, with the local lateral inhibition within an area, the hierarchical multistage pyramidal architecture, and the trace learning rule, used in VisNet as described in Chapter 8. Indeed, we have now produced such a model, and have shown that it has the properties described for each model in Chapters 8–12. The computational principles described in this Chapter all arise in such an integrated overall approach to the operation of the ventral visual stream in object identification, the mechanisms of visual attention, and the role in these of interactions between the dorsal and ventral visual streams. Moreover, a single integrated model is allowing new aspects of the operation of the visual system to be explored, such as the fact that attentional processes produce larger effects in later than in earlier cortical stages of visual information processing (Kastner, Pinsk, De Weerd, Desimone and Ungerleider 1999).

13.28 Conscious visual perception

Conscious visual perception may be more associated with the operation of the ventral than the dorsal processing stream (see Fig. 1.10) (Milner and Goodale 1995) (Section 12.9). The possibility that consciousness is more closely associated with processing in the ventral visual pathways may be related to the fact that we can perform long-term planning about objects, a process that when helped by higher-order thoughts used to correct the plans, may be closely related to consciousness, as argued by Rolls (1999a).

13.29 Attention – future directions

We believe that the theoretical framework presented in this book not only offers a rigorous analysis of visual attention, but opens the possibility of posing new questions that can be addressed in future investigations. We see two main directions in which this kind of computational model could be further extended. The first is in the direction of the analysis of the functional role of rapid neural oscillations. Rapid oscillations of neural activity could have a relevant functional role, namely to facilitate dynamical cooperation between neuronal pools in the same or different brain areas. It is well known in the theory of dynamical systems that the synchronization of oscillators is a cooperative phenomenon. Cooperative mechanisms could complement the competitive mechanisms on which our computational cortical model is based. This would allow the activation of several competitive pools simultaneously, which is sometimes a convenient solution, for example for binding the features of two or more objects or parts of objects by simultaneous activity. As we have shown, attention helps to bind the different feature dimensions of each object. Therefore, we can imagine that attention can activate neuronal pools simultaneously and create dynamical states that are synchronized. Synchronization may then generate cooperative effects that allow the preferential activation of different clusters of pools simultaneously. Two rivalling clusters of synchronized pools could therefore coexist simultaneously if the clusters were dephased, allowing for example the perception of two or more objects simultaneously. Another example would be to help implement perceptual grouping effects. In this scenario, synchronized neuronal activity would be an emergent result of the attentional dynamics, and not the cause of attention, or even binding as it is usually understood. In order to provide this extension to the model, a new mean field theory for the description of the population activity of a group of identical neurons is required. The new theory should go beyond the adiabatic assumptions such as we have used in Chapters 9–11, and should allow rapid oscillations as possible dynamical solutions. Alternatively, the direct simulation of spiking neurons at the microscopic level (e.g. using integrate-and-fire models) would of course include the fine temporal resolution required for achieving rapid oscillations and synchronization effects. A definite advantage of this approach is that we would no longer need to assume the existence of pools of identical neurons as required by the mean field theory approach. Indeed, as shown in Chapter 5, neurons tend to have different response tuning profiles to sets of stimuli, and use this to encode information about the stimulus set. We would expect an integrate-and-fire implementation to have the same generic dynamical properties as those described in Chapters 9–11.

A second direction in which the kind of computational model described in Chapters 9–11 could be further extended is by the formulation of very large-scale models based on global brain connectivity. From an experimental point of view, this development of the theory would allow modelling of fMRI brain activity and MEG (Magnetic Electro Encephalogram) signals simultaneously, which is a trend observed in recent research, because this combination of techniques offers reasonable spatial resolution with fMRI and excellent temporal resolution with MEG. The formulation of very large-scale models of the brain is linked with the so-called inverse problem of extracting brain functional connectivity based on the measurement of fMRI or MEG signals. It is planned, in the near future, to organize standard global public databases of fMRI observations. Such data would allow, in principle, global modelling of the brain based on very large-scale simulations of massively connected neuronal pools. The problem is therefore the extraction of the correct functional connectivity that unifies all data, and consequently several brain functions, at the same time. An information-theoretic approach could be formulated such that it is possible to determine the set of observables that maximize the information about the functional connectivity, contributing to reducing the inverse problem. For example, fMRI activity is unlikely to be sufficient alone to extract the global functional connectivity, and measures describing the temporal or synchronization structure of this activity (by means of MEG for example) may help to complement the temporally averaged information yielded by fMRI signals. The potential clinical relevance of such an approach for neuropsychological assessment based on fMRI and MEG could provide an enormous breakthrough in computer- and model-based diagnosis.

13.30   Integrated approaches to understanding vision

One of the aims of this book has been to show how different approaches to analysing how vision is implemented in the brain can be combined in a complementary way to produce an integrated understanding that can be quantitatively implemented, tested, and explored in a quantitatively defined model. In doing this, bottom-up evidence from low levels of investigation (including detailed evidence of what is represented by the computing elements of the brain, single neurons) is taken to be very relevant to the computational theory. This is somewhat in contrast we note to the approach advocated by Marr (1982), who insisted that the first stage should be specification of the computational theory, the second specification of the algorithm, and the third specification of the implementation. In this book, we have shown that the experimental evidence and what is plausible in biological systems provide important constraints on the computational theory. Indeed, the result of these constraints is that in this book we have argued and produced evidence to support the view that the syntactic 3D structural description approach is not how the brain solves object recognition (see Section 8.2.2), but that instead the brain appears to implement what is close to a feature hierarchy system.

Moreover, again based on the experimental evidence, instead of adopting a primarily bottom-up or feed-forward approach (Marr 1982), we have incorporated top-down and interactive effects in the theory we describe of how attentional processes operate in vision.

However, we are very much in agreement with Marr in emphasizing the importance of the computational theory, and indeed it has been an aim of this book to develop computational theories of object recognition, of how other brain areas use these object representations, and of attention, but incorporating all the constraints from all levels of investigation (including disciplines from molecular biology to computational theory) that are available. Indeed, this interdisciplinary approach has led to quite remarkable advances in the last 35 years in understanding how some parts of the brain could function. Some of these advances, and how they are leading to an understanding of how the computations implemented in different parts of the brain link together to contribute to the overall systems-level operation of the brain to produce behaviour, are charted in this book.

13.31   Apostasis

[41] Extraordinary progress has been made in the last 35 years in understanding how the brain might actually work. Thirty-five years ago, observations on how neurons respond in different brain areas during behaviour were only just starting, and there was very limited understanding of how any part of the brain might work or compute. Many of the observations appeared as interesting phenomena, but there was no conceptual framework in place for understanding how the neuronal responses might be generated, nor of how they were part of an overall computational framework. Through the developments that have taken place since then, some of them charted in this book, we now have at least plausible and testable working models that are consistent with a great deal of experimental data about how invariant visual object representation and recognition may be implemented in the brain. We also have a model, also described in this book, and consistent with a great deal of experimental data, of how

---

[41] Apostasis – standing after.

short term memory is implemented in the brain, and how short term memory systems are related to perceptual systems. We also, and quite closely related to this, have detailed and testable models which make predictions about how attention works and is implemented in the brain. We also have testable models of long term memory, including episodic memory and spatial memory, which again are based on and are consistent with what is being found to be implemented in the primate brain. In addition, we have an understanding, at the evolutionary and adaptive level, of emotion, and at the computational level of how different brain areas implement the information processing underlying emotion (see also Rolls (1999a)). However, our understanding at the time of writing is developing to the stage where we can have not just a computational understanding of each of these systems separately, but also an understanding of how all these processes are linked and act together computationally. Even more than this, we have shown how many aspects of brain function, such as the operation of short term memory systems, of attention, and the effects of emotional state on memory and perception, can now start to be understood in terms of the reciprocal *interactions* of connected neuronal systems.

This leads us to emphasize that the understanding of how the brain actually operates is crucially dependent on a knowledge of the responses of single neurons, the computing elements of the brain, for it is through the connected operations in networks of single neurons, each with their own individual response properties, that the interesting collective computational properties of the brain arise in neuronal networks.

In describing in this book at least part of this great development in understanding in the last 35 years of how a significant number of parts of the brain actually operate, and how they operate together, we wish to make the point that we are at an exciting and conceptually fascinating time in the history of brain research. We are starting to see how a number of parts of the brain *could* work. There is much now that can be done and needs to be done to develop our understanding of how the brain actually works. But the work described in this book does give an indication of some of the types of information processing that take place in the brain, and an indication of the way in which we are now entering the age in which our conceptual understanding of brain function can be based on our developing understanding of how the brain computes.

Understanding how the brain works normally is of course an important foundation for understanding its dysfunctions and its functioning when damaged. Some examples have been given in this book. We regard this as a very important long term aim of the type of work described in this book.

Through neural computation, understanding

Appendix 1 Introduction to linear algebra for neural networks

In this Appendix we review some simple elements of linear algebra relevant to understanding neural networks. This will provide a useful basis for a quantitative understanding of how neural networks operate.

A.1 Vectors

A vector is an ordered set of numbers. An example of a vector is the set of numbers

$$\begin{bmatrix} 7 \\ 4 \end{bmatrix}$$

If we denote the jth element of this vector as w_j, then $w_1 = 7$, and $w_2 = 4$. We can denote the whole vector by $\mathbf{w}$. This notation is very economical. If the vector has 10,000 elements, then we can still refer to it in mathematical operations as $\mathbf{w}$. $\mathbf{w}$ might refer to the vector of 10,000 synaptic weights on the dendrites of a neuron. Another example of a vector is the set of firing rates of the axons that make synapses onto a dendrite, as shown in Fig. 1.2. The firing rate x of each axon forming the input vector can be indexed by j, and is denoted by x_j. The vector would be denoted by $\mathbf{x}$.

Certain mathematical operations can be performed with vectors. We start with the operation which is fundamental to simple models of neural networks, the inner product or dot product of two vectors.

A.1.1 The inner or dot product of two vectors

Recall the operation of computing the activation h of a neuron from the firing rate on its input axons multiplied by the corresponding synaptic weight:

$$h = \sum_j x_j w_j \tag{A.1}$$

where $\sum_j$ indicates that the sum is over the C input axons to each neuron, indexed by j. Denoting the firing rate vector as $\mathbf{x}$ and the synaptic weight vector as $\mathbf{w}$, we can write

$$h = \mathbf{x} \cdot \mathbf{w} \quad . \tag{A.2}$$

If the weight vector is

$$\mathbf{w} = \begin{bmatrix} 9 \\ 5 \\ 2 \end{bmatrix}$$

and the firing rate input vector is

$$\mathbf{x} = \begin{bmatrix} 3 \\ 6 \\ 7 \end{bmatrix}$$

then we can write

$$\mathbf{x} \cdot \mathbf{w} = (3 \cdot 9) + (6 \cdot 5) + (7 \cdot 2) = 71 \quad . \tag{A.3}$$

Thus in the inner or dot product, we multiply the corresponding terms, and then sum the result. As this is the simple mathematical operation that is used to compute the activation h in the most simplified abstraction of a neuron (see Chapter 1), we see that it is indeed the fundamental operation underlying many types of neural networks. We will shortly see that some of the properties of neuronal networks can be understood in terms of the properties of the dot product. We next review a number of basic aspects of vectors and inner products between vectors.

There is a simple geometrical interpretation of vectors, at least in low-dimensional spaces. If we define, for example, x and y axes at right angles to each other in a two-dimensional space, then any two-component vector can be thought of as having a direction and length in that space which can be defined by the values of the two elements of the vector. If the first element is taken to correspond to x and the second to y, then the x axis lies in the direction [1,0] in the space, and the y axis in the direction [0,1], as shown in Fig. A.1. The line to point [1,1] in the space then lies at 45° to both axes, as shown in Fig. A.1.

A.1.2   The length of a vector

Consider taking the inner product of a vector

$$\mathbf{w} = \begin{bmatrix} 4 \\ 3 \end{bmatrix}$$

with itself. Then

$$\|\mathbf{w}\| = \sqrt{\mathbf{w} \cdot \mathbf{w}} = \sqrt{4^2.3^2} = 5. \tag{A.4}$$

This is the length of the vector. We can represent this operation in the two-dimensional graph shown in Fig. A.1. In this case, the coordinates where vector $\mathbf{w}$ ends in the space are [1,1]. The length of the vector (from [0,0]) to [1,1] is obtained by Pythagoras' theorem. Pythagoras' theorem states that the length of the vector $\mathbf{w}$ is equal to the square root of the sum of the squares of the two sides. Thus we define the length of the vector $\mathbf{w}$ as

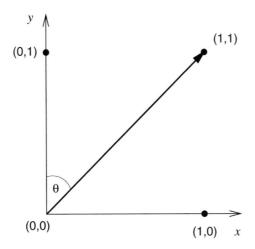

Fig. A.1 Illustration of a vector in a two-dimensional space. The basis for the space is made up of the x axis in the [1,0] direction, and the y axis in the [0,1] direction. (The first element of each vector is then the x value, and the second the y value. The values of x and y for different points, marked by a dot, in the space are shown. The origins of the axes are at point 0,0.) The [1,1] vector projects in the [1,1] (or $45°$) direction to the point 1,1, with length 1.414.

$$\|\mathbf{w}\| = \sqrt{\mathbf{w} \cdot \mathbf{w}} \tag{A.5}$$

In the [1,1] case, this value is $\sqrt{2} = 1.414$.

A.1.3 Normalizing the length of a vector

We can scale a vector in such a way that its length is equal to 1 by dividing it by its length. If we form the dot product of two normalized vectors, its maximum value will be 1, and its minimum value -1.

A.1.4 The angle between two vectors: the normalized dot product

The angle between two vectors $\mathbf{x}$ and $\mathbf{w}$ is defined in terms of the inner product as follows:

$$\cos \theta = \frac{\mathbf{x} \cdot \mathbf{w}}{\|\mathbf{x}\| \|\mathbf{w}\|} \tag{A.6}$$

For example, the angle between two vectors

$$\mathbf{x} = \begin{bmatrix} 0 \\ 1 \end{bmatrix} \quad \text{and} \quad \mathbf{w} = \begin{bmatrix} 1 \\ 1 \end{bmatrix}$$

where the length of vector $\mathbf{x}$ is $\sqrt{0.0 + 1.1} = 1$ and of vector $\mathbf{w}$ is $\sqrt{1.1 + 1.1} = \sqrt{2}$ is

$$\cos \theta = \frac{(0.1) + (1.1)}{1.\sqrt{2}} = 0.707. \tag{A.7}$$

Thus $\theta = \cos^{-1}(0.707) = 45°$.

We can give a simple geometrical interpretation of this as shown in Fig. A.1. However, equation A.6 is much easier to use in a high-dimensional space!

The dot product reflects the similarity between two vectors. Once the length of the vectors is fixed, the higher their dot product, the more similar are the two vectors. By normalizing the dot product, that is by dividing by the lengths of each vector as shown in equation A.6, we obtain a value that varies from -1 to $+1$. This normalized dot product is then just the cosine of the angle between the vectors, and is a very useful measure of the similarity between any two vectors, because it always lies in the range -1 to $+1$. It is closely related to the (Pearson product-moment) correlation coefficient between any two vectors, as we see if we write the equation in terms of its components

$$\cos \theta = \frac{\sum_j x_j w_j}{(\sum_j x_j^2)^{1/2}(\sum_j w_j^2)^{1/2}} \tag{A.8}$$

which is just the formula for the correlation coefficient between two sets of numbers with zero mean (or with the mean value removed by subtracting the mean of the components of each vector from each component of that vector).

Now consider two vectors which have a dot product of zero, that is where $\cos \theta = 0$ or the angle between the vectors is $90°$. Such vectors are described as orthogonal (literally at right angles) to each other. If our two orthogonal vectors were $\mathbf{x}$ and $\mathbf{w}$, then the activation of the neuron, measured by the dot product of these two vectors, would be zero. If our two orthogonal vectors each had a mean of zero, their correlation would also be zero: the two vectors can then be described as unrelated or independent.

If instead the two vectors had zero angle between them, that is if $\cos \theta = 1$, then the dot product would be maximal (given the vectors' lengths), the normalized dot product would be 1, and the two vectors would be described as identical to each other apart from their length. Note that in this case their correlation would also be 1, even if the two vectors did not have zero mean components.

For intermediate similarities of the two vectors, the degree of similarity would be expressed by the relative magnitude of the dot product, or by the normalized dot product of the two vectors, which is just the cosine of the angle between them. These measures are closely related to the correlation between two vectors.

Thus we can think of the simple operation performed by neurons as measuring the similarity between their current input vector and their synaptic weight vector. Their activation, h, is this dot product. It is because of this simple operation that neurons can generalize to similar inputs; can still produce useful outputs if some of their inputs or synaptic weights are damaged or missing, that is they can show graceful degradation or fault tolerance; and can be thought of as learning to point their weight vectors towards input patterns, which is very useful in enabling neurons to categorize their inputs in competitive networks (see Section 7.4).

A.1.5 The outer product of two vectors

Let us take a row vector having as components the firing rates of a set of output neurons in a pattern associator or competitive network, which we might denote as $\mathbf{y}$, with components y_i and the index i running from 1 to the number N of output neurons. $\mathbf{y}$ is then a shorthand for writing down each component, e.g. [7,2,5,2,...], to indicate that the firing rate of neuron 1 is 7, etc. To avoid confusion, we continue in the following to denote the firing rate of input neuron j as x_j. Now recall how the synaptic weights are formed in a pattern associator using a Hebb rule as follows:

$$\delta w_{ij} = \alpha y_i x_j \tag{A.9}$$

where δw_{ij} is the change of the synaptic weight w_{ij} which results from the simultaneous (or conjunctive) presence of presynaptic firing x_j and postsynaptic firing or activation y_i, and α is a learning rate constant which specifies how much the synapses alter on any one pairing. In a more compact vector notation, this expression would be

$$\delta \mathbf{w}_i = \alpha \mathbf{y}_i \mathbf{x}' \tag{A.10}$$

where the firing rates on the axons form a column vector with the values, for example, as follows[42]:

$$\mathbf{x}' = \begin{bmatrix} 2 \\ 0 \\ 3 \\ .... \end{bmatrix}$$

The weights are then updated by a change proportional (the α factor) to the following matrix (Table A.1):

Table A.1 Multiplication of a row vector [7 2 5 ....] by a column vector to form the external or tensor product, representing for example the changes to a matrix of synaptic weights **W**

	[7	2	5	.....]
[2]	14	4	10	......
[0]	0	0	0	......
[3]	21	6	15	......
.....	.....	.....	.....	.....

This multiplication of the two vectors is called the outer, or tensor, product, and forms a matrix, in this case of (alterations to) synaptic weights. Thus we see that the operation of altering synaptic weights in a network can be thought of as forming a matrix of weight changes, which can then be used to alter the existing matrix of synaptic weights.

A.1.6 Linear and non-linear systems

The operations with which we have been concerned in this Appendix so far are linear operations. We should note that if two matrices operate linearly, we can form their product by

---

[42]The prime after the **x** is used here to remind us that this vector is a column vector, which can be thought of as a transformed row vector, and the prime indicates the transformed vector. We do not use the prime for most of this book in order to keep the notation uncluttered.

matrix multiplication, and then replace the two matrices with the single matrix that is their product. We can thus effectively replace two synaptic matrices in a linear multilayer neural network with one synaptic matrix, the product of the two matrices. For this reason, multilayer neural networks if linear cannot achieve more than can be achieved in a single-layer linear network. It is only in non-linear networks that more can be achieved, in terms of mapping input vectors through the synaptic weight matrices, to produce particular mappings to output vectors. Much of the power of many networks in the brain comes from the fact that they are multilayer non-linear networks (in that the computing elements in each network, the neurons, have non-linear properties such as thresholds, and saturation at high levels of output). Because the matrix by matrix multiplication operations of linear algebra cannot be applied directly to the operation of neural networks in the brain, we turn instead back to other aspects of linear algebra, which can help us to understand which classes of pattern can be successfully learned by different types of neural network.

A.1.7 Linear combinations of vectors, linear independence, and linear separability

We can multiply a vector by a scalar (a single value, e.g. 2) thus:

$$2 \cdot \begin{bmatrix} 4 \\ 1 \\ 3 \end{bmatrix} = \begin{bmatrix} 8 \\ 2 \\ 6 \end{bmatrix}$$

We can add two vectors thus:

$$\begin{bmatrix} 4 \\ 1 \\ 3 \end{bmatrix} + \begin{bmatrix} 2 \\ 7 \\ 2 \end{bmatrix} = \begin{bmatrix} 6 \\ 8 \\ 5 \end{bmatrix}$$

The sum of the two vectors is an example of a linear combination of two vectors, which is in general a weighted sum of several vectors, component by component. Thus, the linear combination of vectors v_1, v_2, ... to form a vector v_s is expressed by the sum

$$v_s = c_1 v_1 + c_2 v_2 + \dots \tag{A.11}$$

where c_1 and c_2 are scalars.

By adding vectors in this way, we can produce any vector in the space spanned by a set of vectors as a linear combination of vectors in the set. If in a set of n vectors at least one can be written as a linear combination of the others, then the vectors are described as **linearly dependent**. If in a set of n vectors none can be written as a linear combination of the others, then the vectors are described as **linearly independent**. A linearly independent set of vectors has the properties that any vector in the space spanned by the set can be written in only one way as a linear combination of the set, and the space has dimension $d = n$. In contrast, a vector in a space spanned by a linearly dependent set can be written in an infinite number of equivalent ways, and the dimension d of the space is less than n.

Consider a set of linearly dependent vectors and the d-dimensional space they span. Two subsets of this set are described as **linearly separable** if the vectors of one subset (that is, their endpoints) can be separated from those of the other by a hyperplane, that is a subspace of dimension $d - 1$. *Subsets formed from a set of linearly independent vectors are always linearly separable.* For example, the four vectors:

$$\begin{bmatrix} 0 \\ 0 \end{bmatrix} \quad \begin{bmatrix} 0 \\ 1 \end{bmatrix} \quad \begin{bmatrix} 1 \\ 0 \end{bmatrix} \quad \begin{bmatrix} 1 \\ 1 \end{bmatrix}$$

are linearly dependent, because the fourth can be formed by a linear combination of the second and third (and also because the first, being the null vector, can be formed by multiplying any other vector by zero, a specific linear combination). In fact, $n = 4$ and $d = 2$. If we split this set into subset A including the first and fourth vector, and subset B including the second and third, the two subsets are not linearly separable, because there is no way to draw a line (which is the subspace of dimension $d - 1 = 1$) to separate the two subsets A and B. We have encountered this set of vectors in Chapter 7, and this is the geometrical interpretation of why a one-layer, one-output neuron network cannot separate these patterns. Such a network (a simple perceptron) is equivalent to its (single) weight vector, and in turn the weight vector defines a set of parallel $d - 1$ dimensional hyperplanes. (Here $d = 2$, so a hyperplane is simply a line, any line perpendicular to the weight vector.) No line can be found that separates the first and fourth vector from the second and third, whatever the weight vector the line is perpendicular to, and hence no perceptron exists that performs the required classification (see Section A.2.1). To separate such patterns, a multilayer network with non-linear neurons is needed (see Chapter 7).

Any set of linearly independent vectors comprise the basis of the space they span, and they are called basis vectors. All possible vectors in the space spanned by these vectors can be formed as linear combinations of these vectors. If the vectors of the basis are in addition mutually orthogonal, the basis is an orthogonal basis, and it is, further, an orthonormal basis if the vectors are chosen to be of unit length. Given any space of vectors with a preassigned meaning to each of their components (for example the space of patterns of activation, in which each component is the activation of a particular unit), the most natural, canonical choice for a basis is the set of vectors in which each vector has one component, in turn, with value 1, and all the others with value 0. For example, in the $d = 2$ space considered earlier, the natural choice is to take as basis vectors

$$\begin{bmatrix} 1 \\ 0 \end{bmatrix}$$

and

$$\begin{bmatrix} 0 \\ 1 \end{bmatrix}$$

from which all vectors in the space can be created. This can be seen from Fig. A.1. (A vector in the [-1,-1] direction would have the opposite direction of the vector shown in Fig. A.1.)

If we had three vectors that were all in the same plane in a three-dimensional (x, y, z) space, then the space they spanned would be less than three-dimensional. For example, the three vectors

$$\begin{bmatrix} 1 \\ 0 \\ 0 \end{bmatrix} \quad \begin{bmatrix} 0 \\ 1 \\ 0 \end{bmatrix} \quad \begin{bmatrix} -1 \\ -1 \\ 0 \end{bmatrix}$$

all lie in the same z plane, and span only a two-dimensional space. (All points in the space could be shown in the plane of the paper in Fig. A.1.)

A.2 Application to understanding simple neural networks

The operation of simple one-layer networks can be understood in terms of these concepts.

A.2.1 Capability and limitations of single-layer networks: linear separability and capacity

Single-layer perceptrons perform pattern classification, and can be trained by an associative (Hebb) learning rule or by an error-correction (delta) rule (see Chapter 7). That is, each neuron classifies the input patterns it receives into classes determined by the teacher. Single-layer perceptrons are thus supervised networks, with a separate teacher for each output neuron. The classification is most clearly understood if the output neurons are binary, or are strongly non-linear, but the network will still try to obtain an optimal mapping with linear or near-linear output neurons.

When each neuron operates as a binary classifier, we can consider how many input patterns p can be classified by each neuron, and the classes of pattern that can be correctly classified. The result is that the maximum number of patterns that can be correctly classified by a neuron with C inputs is

$$p_{\text{max}} = 2C \tag{A.12}$$

when the inputs have random continuous-valued inputs, but the patterns must be linearly separable (see Hertz et al., 1991). More generally, a network with a single binary unit can implement a classification between two subspaces of a space of possible input patterns provided that the p actual patterns given as examples of the correct classification are linearly separable.

The linear separability requirement can be made clear by considering a geometric interpretation of the logical AND problem, which is linearly separable, and the XOR (exclusive OR) problem, which is not linearly separable. The truth tables for the AND and XOR functions are shown first (there are two inputs, x_1 and x_2, and one output neuron):

Table A.2 Truth Table for AND and XOR functions performed by a single output neuron with two inputs. 1 = active or firing; 0 = inactive.

Inputs		Output	
x_1	x_2	AND	XOR
0	0	0	0
1	0	0	1
0	1	0	1
1	1	1	0

For the AND function, we can plot the mapping required in a 2D graph as shown in Fig. A.2.

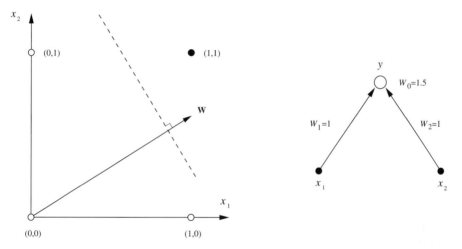

Fig. A.2 (Left) The AND function shown in a 2D space. Input values for the two neurons are shown along the two axes of the space. The outputs required are plotted at the coordinates where the inputs intersect, and the values of the output required are shown as an open circle for 0, and a filled circle for 1. The AND function is linearly separable, in that a line can be drawn in the space which separates the coordinates for which 0 output is required from those from which a 1 output is required. **w** shows the direction of the weight vector. (Right) A one-layer neural network can set its two weights w_1 and w_2 to values which allow the output neuron to be activated only if both inputs are present. In this diagram, w_0 is used to set a threshold for the neuron, and is connected to an input with value 1. The neuron thus fires only if the threshold of 1.5 is exceeded, which happens only if both inputs to the neuron are 1.

A line can be drawn to separate the input coordinates for which 0 is required as the output from those for which 1 is required as the output. The problem is thus linearly separable. A neuron with two inputs can set its weights to values which draw this line through this space, and such a one-layer network can thus solve the AND function.

For the XOR function, we can plot the mapping required in a 2D graph as shown in Fig. A.3. No straight line can be drawn to separate the input coordinates for which 0 is required as the output from those for which 1 is required as the output. The problem is thus not linearly separable. For a one-layer network, no set of weights can be found that will perform the XOR,

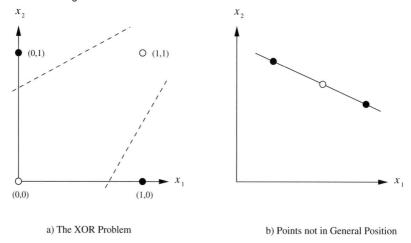

a) The XOR Problem   b) Points not in General Position

Fig. A.3 The XOR function shown in a 2D space. Input values for the two neurons are shown along the two axes of the space. The outputs required are plotted at the coordinates where the inputs intersect, and the values of the output required are shown as an open circle for 0, and a filled circle for 1. The XOR function is not linearly separable, in that a line cannot be drawn in the space to separate the coordinates for those from which a 0 output is required from those from which a 1 output is required. A one-layer neural network cannot set its two weights to values which allow the output neuron to be activated appropriately for the XOR function.

or any other non-linearly separable function.

Although the inability of one-layer networks with binary neurons to solve non-linearly separable problems is a limitation, it is not in practice a major limitation on the processing that can be performed in a neural network for a number of reasons. First, if the inputs can take continuous values, then if the patterns are drawn from a random distribution, the one-layer network can map up to $2C$ of them. Second, as described for pattern associators, and for one-layer error-correcting perceptrons (see Chapter 7), these networks could be preceded by an expansion recoding network such as a competitive network with more output than input neurons. This effectively provides a two-layer network for solving the problem, and multilayer networks are in general capable of solving arbitrary mapping problems. Ways in which such multilayer networks might be trained are discussed in Chapters 7 and 8.

More generally, a binary output unit provides by its operation a hyperplane (the hyperplane orthogonal to its synaptic weight vector as shown in Fig. A.2) that divides the input space in two. The input space is of dimension C, if C is the number of input axons or connections. A one-layer network with a number n of binary output units is equivalent to n hyperplanes, that could potentially divide the input space into as many as 2^n regions, each corresponding to input patterns leading to a different output. However the number p of *arbitrary* examples of the correct classification (each example consisting of an input pattern and its required correct output) that the network may be able to implement is well below 2^n, and in fact depends on C not on n. This is because for p too large it will be impossible to position the n weight vectors such that all examples of input vectors for which the first output unit is required to be 'on' fall on one side of the hyperplane associated with the first weight vector, all those for which it is required to be 'off' fall on the other side, and simultaneously the same holds with respect to the second output unit (a different dichotomy), the third, and so on. The limit on p, which can be thought of also as the number of independent associations implemented by the network, when this is viewed as a heteroassociator (i.e. pattern associator) with binary outputs, can be calculated with the Gardner method (Gardner 1987, Gardner 1988) and depends on the statistics of the patterns. For input patterns that are also binary, random and with equal probability for each of the two states on every unit, the limit is $p_c = 2C$ (see further Chapter 7 and Appendix A3 of Rolls and Treves (1998)).

A.2.2 Non-linear networks: neurons with non-linear activation functions

These concepts also help one to understand further the limitation of linear systems, and the power of non-linear systems. Consider the dot product operation by which the neuronal activation h is computed:

$$h = \sum_j x_j w_j. \tag{A.13}$$

If the output firing is just a linear function of the activation, any input pattern will produce a non-zero output unless it happens to be exactly orthogonal to the weight vector. For positive-only firing rates and synaptic weights, being orthogonal means taking non-zero values only on non-corresponding components. Since with distributed representations the non-zero components of different input firing vectors will in general be overlapping (i.e. some corresponding components in both firing rate vectors will be on, that is the vectors will overlap), this will result effectively in interference between any two different patterns that for example have to be associated to different outputs. Thus a basic limitation of linear networks is that they can perform pattern association perfectly only if the input patterns **x** are orthogonal; and for positive-only patterns that represent actual firing rates only if the different firing rate vectors are non-overlapping. Further, linear networks cannot of course perform any classification, just because they act linearly. (Classification implies producing output states that are clearly defined as being in one class, and not in other classes.) For example, in a linear network, if a pattern is presented which is intermediate between two patterns $\mathbf{v}_1$ and $\mathbf{v}_2$, such as $c_1\mathbf{v}_1 + c_2\mathbf{v}_2$, then the output pattern will be a linear combination of the outputs produced by $\mathbf{v}_1$ and $\mathbf{v}_1$ (e.g. $c_1\mathbf{o}_1 + c_2\mathbf{o}_2$), rather than being classified into $\mathbf{o}_1$ or $\mathbf{o}_2$. In contrast, with non-linear neurons, the patterns need not be orthogonal, only linearly separable, for a one-layer network to be able to correctly classify the patterns (provided that a sufficiently powerful learning rule is used – see Chapter 7).

The networks just described, and most of those described in this book, are trained with a local learning rule, in which the pre- and post-synaptic terms needed to alter the synaptic weights are available locally in the synapses, in terms for example of the release of transmitter from the presynaptic terminal, and the depolarization of the postsynaptic neuron. This type of network is considered because this is a biologically plausible constraint (see Section 7.12). It is much less biologically plausible to use an algorithm such as multilayer error backpropagation which calculates the correction to the value of a synapse that is needed taking into account the values of the errors of the neurons at alter stages of the system and the strengths of all the synapses to these neurons (see Section 7.11). This use of a local learning rule is a major difference of the networks described in this book, which is directed at neurally plausible computation, from connectionist networks, which typically assume non-local learning rules and which therefore operate very differently from real neural networks in the brain (see Section 7.11 and McLeod, Plunkett and Rolls (1998)).

A.2.3 Non-linear networks: neurons with non-linear activations

Most of the networks described in this book calculate the activation h of each neuron as the linear product of the input firing weighted by the synaptic weight vector (see equation A.13). This corresponds in a real neuron to receiving currents from each of its synapses which sum to produce depolarization of the neuronal cell body and the spike initiation region which is located very close to the cell body. This is a reasonable reflection of what does happen in many neurons, especially those with large dendrites such as pyramidal cells (Koch 1999). This calculation of the activation h by a linear summation not only approximates to what happens in many real neurons, but is also a useful simplification which makes tractable the analysis of many classes of network which utilize such neurons. These analyses provide insight into the operation of networks of neurons, even if the linear summation assumption is not perfectly realized. Having computed the activation linearly, the neurons do of course for essentially all the networks described in this book, then utilize a non-linear activation function, which, as described above, provides the networks with much of their interesting computational power. Given that the activation functions of the neurons are non-linear, some non-linearity in the summation expressed in equation A.13 may in practice be lumped into the non-linearity in the activation function.

However, another class of neuron that is implemented in some networks in the brain utilizes non-linearity in the calculation of the activation h of the neuron, which reflects a local product of two inputs to a neuron. This could arise for example if one synapse makes a presynaptic contact with another synapse which in turn connects to the dendrite, or if two synapses are close together on a thin dendrite. In such situations, the current injected into the neuron could reflect the conjoint firing of the two classes of input (Koch 1999). The dendrite as a whole could then sum all such products into the cell body, leading to the description **Sigma-Pi**. This could be expressed by equation A.14

$$h = \sum_j \sum_k w_{jk} x_j x^c_k \qquad \text{(A.14)}$$

where x_j is the firing rate of input cell j, x^c_k is the firing rate of input cell k of class c, and w_{jk} is the connection strength. Such Sigma-Pi neurons were utilized in the model described in Section 7.5.5 of how idiothetic inputs could update a continuous attractor network. Another possible application is to learning invariant representations in neural networks. For example, the r^c input in equation A.14 could be a signal that varies with the shift required to compute translation invariance, effectively mapping the appropriate set of x_j inputs through to the output neurons depending on the shift required (Mel, Ruderman and Archie 1998, Mel and Fiser 2000, Olshausen, Anderson and Van Essen 1993, Olshausen, Anderson and Van Essen 1995).

To train such a Sigma-Pi network requires that combinations of the two presynaptic inputs to a neuron be learned onto a neuron, using for example associativity with the post-synaptic term y, as exemplified in equation A.15

$$\delta w_{jk} = \alpha y x_j x^c_k. \qquad \text{(A.15)}$$

This learning principle is exemplified in the model described in Section 7.5.5 of how idiothetic inputs could update a continuous attractor network.

Sigma-Pi networks are clearly very powerful, but require rather specialized anatomical and biophysical arrangements (see Koch (1999)), and hence we do not use them unless they become very necessary in models of neural network operations in the brain.

Appendix 2 Information theory

Information theory provides the means for quantifying how much neurons communicate to other neurons, and thus provides a quantitative approach to fundamental questions about information processing in the brain. To investigate what in neural activity carries information, one must compare the amounts of information carried by different codes, that is different descriptions of the same activity, to provide the answer. To investigate the speed of information transmission, one must define and measure information rates from neuronal responses. To investigate to what extent the information provided by different cells is redundant or instead independent, again one must measure amounts of information in order to provide quantitative evidence. To compare the information carried by the number of spikes, by the timing of the spikes within the response of a single neuron, and by the relative time of firing of different neurons reflecting for example stimulus-dependent neuronal synchronization, information theory again provides a quantitative and well-founded basis for the necessary comparisons. To compare the information carried by a single neuron or a group of neurons with that reflected in the behaviour of the human or animal, one must again use information theory, that provides a single measure which can be applied to the measurement of the performance of all these different cases. In all these situations, there is no quantitative and well-founded alternative to information theory.

This Appendix briefly introduces the fundamental elements of information theory in the first Section. A more complete treatment can be found in many books on the subject (e.g. Abramson (1963), Hamming (1990), and Cover and Thomas (1991)), including also Rieke, Warland, de Ruyter van Steveninck and Bialek (1996) which is specifically about information transmitted by neuronal firing. The second Section discusses the extraction of information measures from neuronal activity, in particular in experiments with mammals, in which the central issue is how to obtain accurate measures in conditions of limited sampling, that is where the numbers of trials of neuronal data that can be obtained are usually limited by the available recording time. The third Section summarizes some of the results obtained so far. The essential terminology is summarized in a Glossary at the end of this Appendix in Section B.4. The approach taken in this Appendix is based on and updated from that provided by Rolls and Treves (1998).

B.1 Basic notions and their use in the analysis of formal models

Although information theory was a surprisingly late starter as a mathematical discipline, having being developed and formalized in 1948 by C. Shannon (Shannon (1948)), the intuitive notion of information is immediate to us. It is also very easy to understand why we use logarithms in order to quantify this intuitive notion, of how much we know about something, and why the resulting quantity is always defined in relative rather than absolute terms. An introduction to information theory is provided next, with a more formal summary given in the third subsection.

B.1.1 The information conveyed by definite statements

Suppose somebody, who did not know, is told that Reading is a town west of London. How much information is he given? Well, that depends. He may have known it was a town in England, but not whether it was east or west of London; in which case the new information amounts to the fact that of two *a priori* (i.e. initial) possibilities (E or W), one holds (W). It is also possible to interpret the statement in the more precise sense, that Reading is west of London, rather than east, north or south, i.e. one out of four possibilities; or else, west rather that north-west, north, etc. Clearly, the larger the number k of *a priori* possibilities, the more one is actually told, and a measure of information must take this into account. Moreover, we would like independent pieces of information to just add together. For example, our person may also be told that Cambridge is, out of l possible directions, north of London. Provided nothing was known on the mutual location of Reading and Cambridge, there are now overall $k \times l$ *a priori* (initial) possibilities, only one of which remains *a posteriori* (after receiving the information). Given that the number of possibilities for independent events are multiplicative, but that we would like the measure of information to be additive, we use logarithms when we measure information, as logarithms have this property. We thus define the amount I of information gained when we are informed in which of k possible locations Reading is located as

$$I(k) = \log_2 k. \tag{B.1}$$

Then when we combine independent information, for example producing $k \times l$ possibilities from independent events with k and l possibilities respectively, we obtain

$$I(k \times l) = \log_2(k \times l) = \log_2 k + \log_2 l = I(k) + I(l). \tag{B.2}$$

Thus in our example, the information about Cambridge adds up to that about Reading. We choose to take logarithms in base 2 as a mere convention, so that the answer to a yes/no question provides one unit, or bit, of information. Here it is just for the sake of clarity that we used different symbols for the number of possible directions with respect to which Reading and Cambridge are localized; if both locations are specified for example in terms of E, SE, S, SW, W, NW, N, NE, then obviously $k = l = 8$, $I(k) = I(l) = 3$ bits, and $I(k \times l) = 6$ bits. An important point to note is that the *resolution* with which the direction is specified determines the amount of information provided, and that in this example, as in many situations arising when analysing neuronal codings, the resolution could be made progressively finer, with a corresponding increase in information proportional to the log of the number of possibilities.

B.1.2 The information conveyed by probabilistic statements

The situation becomes slightly less trivial, and closer to what happens among neurons, if information is conveyed in less certain terms. Suppose for example that our friend is told, instead, that Reading has odds of 9 to 1 to be west, rather than east, of London (considering now just two *a priori* possibilities). He is certainly given some information, albeit less than in the previous case. We might put it this way: out of 18 equiprobable *a priori* possibilities (9 west + 9 east), 8 (east) are eliminated, and 10 remain, yielding

$$I = \log_2(18/10) = \log_2(9/5) \tag{B.3}$$

as the amount of information given. It is simpler to write this in terms of probabilities

$$I = \log_2 P^{\text{posterior}}(W)/P^{\text{prior}}(W) = \log_2(9/10)/(1/2) = \log_2(9/5). \tag{B.4}$$

This is of course equivalent to saying that the amount of information given by an uncertain statement is equal to the amount given by the absolute statement

$$I = -\log_2 P^{\text{prior}}(W) \tag{B.5}$$

minus the amount of uncertainty remaining after the statement, $I = -\log_2 P^{\text{posterior}}(W)$. A successive clarification that Reading is indeed west of London carries

$$I' = \log_2((1)/(9/10)) \tag{B.6}$$

bits of information, because 9 out of 10 are now the *a priori* odds, while *a posteriori* there is certainty, $P^{\text{posterior}}(W) = 1$. In total we would seem to have

$$I^{\text{TOTAL}} = I + I' = \log_2(9/5) + \log_2(10/9) = 1 \text{ bit} \tag{B.7}$$

as if the whole information had been provided at one time. This is strange, given that the two pieces of information are clearly not independent, and only independent information should be additive. In fact, we have cheated a little. Before the clarification, there was still one residual possibility (out of 10) that the answer was 'east', and this must be taken into account by writing

$$I = P^{\text{posterior}}(W) \log_2 \frac{P^{\text{posterior}}(W)}{P^{\text{prior}}(W)} + \tag{B.8}$$
$$P^{\text{posterior}}(E) \log_2 \frac{P^{\text{posterior}}(E}{P^{\text{prior}}(E)}$$

as the information contained in the first message. This little detour should serve to emphasize two aspects that are easy to forget when reasoning intuitively about information, and that in this example cancel each other. In general, when uncertainty remains, that is there is more than one possible *a posteriori* state, one has to average information values for each state with the corresponding *a posteriori* probability measure. In the specific example, the sum $I + I'$ totals slightly *more* than 1 bit, and the amount in excess is precisely the information 'wasted' by providing *correlated* messages.

B.1.3 Information sources, information channels, and information measures

In summary, the expression quantifying the information provided by a definite statement that event s, which had an *a priori* probability $P(s)$, has occurred is

$$I(s) = \log_2(1/P(s)) = -\log_2 P(s), \tag{B.9}$$

whereas if the statement is probabilistic, that is several *a posteriori* probabilities remain non-zero, the correct expression involves summing over all possibilities with the corresponding probabilities:

$$I = \sum_s \left[\mathrm{P^{posterior}}(s) \log_2 \frac{\mathrm{P^{posterior}}(s)}{\mathrm{P^{prior}}(s)} \right]. \tag{B.10}$$

When considering a discrete set of mutually exclusive events, it is convenient to use the metaphor of a set of *symbols* comprising an *alphabet* S. The occurrence of each event is then referred to as the emission of the corresponding symbol by an information *source*. The *entropy* of the source, H, is the average amount of information per source symbol, where the average is taken across the alphabet, with the corresponding probabilities

$$H(S) = -\sum_{s \in S} P(s) \log_2 P(s). \tag{B.11}$$

An information *channel* receives symbols s from an alphabet S and emits symbols s' from alphabet S'. If the *joint* probability of the channel receiving s and emitting s' is given by the product

$$P(s, s') = P(s)P(s') \tag{B.12}$$

for any pair s, s', then the input and output symbols are *independent* of each other, and the channel transmits zero information. Instead of joint probabilities, this can be expressed with conditional probabilities: the conditional probability of s' given s is written $P(s'|s)$, and if the two variables are independent, it is just equal to the unconditional probability $P(s')$. In general, and in particular if the channel does transmit information, the variables are not independent, and one can express their joint probability in two ways in terms of conditional probabilities

$$P(s, s') = P(s'|s)P(s) = P(s|s')P(s'), \tag{B.13}$$

from which it is clear that

$$P(s'|s) = P(s|s') \frac{P(s')}{P(s)}, \tag{B.14}$$

which is called Bayes' theorem (although when expressed as here in terms of probabilities it is strictly speaking an identity rather than a theorem). The information transmitted by the channel conditional to its having emitted symbol s' (or specific transinformation, $I(s')$) is given by equation B.10, once the unconditional probability $P(s)$ is inserted as the prior, and the conditional probability $P(s|s')$ as the posterior:

$$I(s') = \sum_s P(s|s') \log_2 \frac{P(s|s')}{P(s)}. \tag{B.15}$$

Symmetrically, one can define the transinformation conditional to the channel having received symbol s

$$I(s) = \sum_{s'} P(s'|s) \log_2 \frac{P(s'|s)}{P(s')}. \tag{B.16}$$

Finally, the average transinformation, or **mutual information**, can be expressed in fully symmetrical form

$$I = \sum_s P(s) \sum_{s'} P(s'|s) \log_2 \frac{P(s'|s)}{P(s')} \tag{B.17}$$

$$= \sum_{s,s'} P(s, s') \log_2 \frac{P(s, s')}{P(s)P(s')}.$$

The **mutual information** can also be expressed as the entropy of the source using alphabet S minus the *equivocation* of S with respect to the new alphabet S' used by the channel, written

$$I = H(S) - H(S|S') \equiv H(S) - \sum_{s'} P(s')H(S|s'). \qquad (B.18)$$

A channel is characterized, once the alphabets are given, by the set of conditional probabilities for the output symbols, $P(s'|s)$, whereas the unconditional probabilities of the input symbols $P(s)$ depend of course on the source from which the channel receives. Then, the *capacity* of the channel can be defined as the maximal mutual information across all possible sets of input probabilities $P(s)$. Thus, the information transmitted by a channel can range from zero to the lower of two independent upper bounds: the entropy of the source, and the capacity of the channel.

B.1.4 The information carried by a neuronal response and its averages

Considering the processing of information in the brain, we are often interested in the amount of information the response r of a neuron, or of a population of neurons, carries about an event happening in the outside world, for example a stimulus s shown to the animal. Once the inputs and outputs are conceived of as sets of symbols from two alphabets, the neuron(s) may be regarded an information channel. We may denote with $P(s)$ the *a priori* probability that the particular stimulus s out of a given set was shown, while the conditional probability $P(s|r)$ is the *a posteriori* probability, that is updated by the knowledge of the response r. The response-specific transinformation

$$I(r) = \sum_{s} P(s|r) \log_2 \frac{P(s|r)}{P(s)} \qquad (B.19)$$

takes the extreme values of $I(r) = -\log_2 P(s(r))$ if r unequivocally determines $s(r)$ (that is, $P(s|r)$ equals 1 for that one stimulus and 0 for all others); and $I(r) = \sum_{s} P(s) \log_2 (P(s)/P(s)) = 0$ if there is no relation between s and r, that is they are independent, so that the response tells us nothing new about the stimulus and thus $P(s|r) = P(s)$.

This is the information conveyed by each particular response. One is usually interested in further averaging this quantity over all possible responses r,

$$<I> = \sum_{r} P(r) \left[\sum_{s} P(s|r) \log_2 \frac{P(s|r)}{P(s)} \right]. \qquad (B.20)$$

The angular brackets $<>$ are used here to emphasize the averaging operation, in this case over responses. Denoting with $P(s,r)$ the *joint probability* of the pair of events s and r, and using Bayes' theorem, this reduces to the symmetric form (equation B.18) for the **mutual information**

$$<I> = \sum_{s,r} P(s,r) \log_2 \frac{P(s,r)}{P(s)P(r)} \qquad (B.21)$$

which emphasizes that responses tell us about stimuli just as much as stimuli tell us about responses. This is, of course, a general feature, independent of the two variables being in this instance stimuli and neuronal responses. In fact, what is of interest, besides the mutual information of equations B.20 and B.21, is often the information specifically conveyed about each stimulus,

$$I(s) = \sum_r \mathrm{P}(r|s) \log_2 \frac{\mathrm{P}(r|s)}{\mathrm{P}(r)} \qquad (B.22)$$

which is a direct quantification of the variability in the responses elicited by that stimulus, compared to the overall variability. Since $\mathrm{P}(r)$ is the probability distribution of responses averaged across stimuli, it is again evident that the stimulus-specific information measure of equation B.22 depends not only on the stimulus s, but also on all other stimuli used. Likewise, the mutual information measure, despite being of an average nature, is dependent on what set of stimuli has been used in the average. This emphasizes again the relative nature of all information measures. More specifically, it underscores the relevance of using, while measuring the information conveyed by a given neuronal population, stimuli that are either representative of real-life stimulus statistics, or of particular interest for the properties of the population being examined[43].

B.1.4.1   A numerical example

To make these notions clearer, we can consider a specific example in which the response of a cell to the presentation of, say, one of four visual stimuli (A, B, C, D) is recorded for 10 ms, during which the cell emits either 0, 1, or 2 spikes, but no more. Imagine that the cell tends to respond more vigorously to smell B, less to C, even less to A, and never to D, as described by the table of conditional probabilities $\mathrm{P}(r|s)$ shown in Table B.1. Then, if different visual

Table B.1 The conditional probabilities $\mathrm{P}(r|s)$ that different neuronal responses (r=0, 1 or 2 spikes) will be produced by each of four stimuli (A–D).

	r=0	r=1	r=2
s=A	0.6	0.4	0.0
s=B	0.0	0.2	0.8
s=C	0.4	0.5	0.1
s=D	1.0	0.0	0.0

stimuli are presented with equal probability, the table of joint probabilities $\mathrm{P}(s,r)$ will be as shown in Table B.2. From these two Tables one can compute various information measures

Table B.2 Joint probabilities $\mathrm{P}(s,r)$ that different neuronal responses (r=0, 1 or 2 spikes) will be produced by each of four equiprobable stimuli (A–D).

	r=0	r=1	r=2
s=A	0.15	0.1	0.0
s=B	0.0	0.05	0.2
s=C	0.1	0.125	0.025
s=D	0.25	0.0	0.0

[43]The quantity $I(s, R)$, which is what is shown in equation B.22 and where R draws attention to the fact that this quantity is calculated across the full set of responses R, has also been called the stimulus-specific surprise, see DeWeese and Meister (1999). Its average across stimuli is the mutual information $I(S, R)$.

by directly applying the definitions above. Since visual stimuli are presented with equal probability, $P(s) = 1/4$, the entropy of the stimulus set, which corresponds to the maximum amount of information any transmission channel, no matter how efficient, could convey on the identity of the stimuli, is $H_s = -\sum_s [P(s) \log_2 P(s)] = -4[(1/4) \log_2(1/4)] = \log_2 4 = 2$ bits. There is a more stringent upper bound on the mutual information that this cell's responses convey on the stimuli, however, and this second bound is the channel capacity T of the cell. Calculating this quantity involves maximizing the mutual information across prior visual stimulus probabilities, and it is a bit complicated to do, in general. In our particular case the maximum information is obtained when only stimuli B and D are presented, each with probability 0.5. The resulting capacity is $T = 1$ bit. We can easily calculate, in general, the entropy of the responses. This is not an upper bound characterizing the source, like the entropy of the stimuli, nor an upper bound characterizing the channel, like the capacity, but simply a bound on the mutual information for this specific combination of source (with its related visual stimulus probabilities) and channel (with its conditional probabilities). Since only three response levels are possible within the short recording window, and they occur with uneven probability, their entropy is considerably lower than H_s, at $H_r = -\sum_r P(r) \log_2 P(r) =$ $-P(0) \log_2 P(0) - P(1) \log_2 P(1) - P(2) \log_2 P(2) = -0.5 \log_2 0.5 - 0.275 \log_2 0.275 - 0.225 \log_2 0.225 = 1.496$ bits. The actual average information I that the responses transmit about the stimuli, which is a measure of the correlation in the variability of stimuli and responses, does not exceed the absolute variability of either stimuli (as quantified by the first bound) or responses (as quantified by the last bound), nor the capacity of the channel. An explicit calculation using the joint probabilities of the second table into expression B.21 yields $I = 0.733$ bits. This is of course only the average value, averaged both across stimuli and across responses.

The information conveyed by a particular response can be larger. For example, when the cell emits two spikes it indicates with a relatively large probability stimulus B, and this is reflected in the fact that it then transmits, according to expression B.19, $I(r = 2) = 1.497$ bits, more than double the average value.

Similarly, the amount of information conveyed about each individual visual stimulus varies with the stimulus, depending on the extent to which it tends to elicit a differential response. Thus, expression B.22 yields that only $I(s = C) = 0.185$ bits are conveyed on average about stimulus C, which tends to elicit responses with similar statistics to the average statistics across stimuli, and therefore not easily interpretable. On the other hand, exactly 1 bit of information is conveyed about stimulus D, since this stimulus never elicits any response, and when the cell emits no spike there is a probability of $1/2$ that the stimulus was stimulus D.

B.1.5   The information conveyed by continuous variables

A general feature, relevant also to the case of neuronal information, is that if, among a *continuum* of *a priori* possibilities, only one, or a discrete number, remains *a posteriori*, the information is strictly infinite. This would be the case if one were told, for example, that Reading is exactly $10'$ west, $1'$ north of London. The *a priori* probability of precisely this set of coordinates among the continuum of possible ones is zero, and then the information diverges to infinity. The problem is only theoretical, because in fact, with continuous distributions, there are always one or several factors that limit the resolution in the *a posteriori* knowledge, rendering the information finite. Moreover, when considering the mutual information in the conjoint probability of occurrence of two sets, e.g. stimuli and responses, it suffices that at least one of the sets is discrete to make matters easy, that is, finite. Nevertheless, the identification and appropriate consideration of these resolution-limiting factors in practical cases may require careful analysis.

B.1.5.1 Example: the information retrieved from an autoassociative memory

One example is the evaluation of the information that can be retrieved from an autoassociative memory. Such a memory stores a number of firing patterns, each one of which can be considered, as in Chapter 7, as a vector $\mathbf{r}^\mu$ with components the firing rates $\{r_i^\mu\}$, where the subscript i indexes the cell (and the superscript μ indexes the pattern). In retrieving pattern μ, the network in fact produces a distinct firing pattern, denoted for example simply as $\mathbf{r}$. The quality of retrieval, or the similarity between $\mathbf{r}^\mu$ and $\mathbf{r}$, can be measured by the average mutual information

$$< I(\mathbf{r}^\mu, \mathbf{r}) > = \sum_{\mathbf{r}^\mu, \mathbf{r}} P(\mathbf{r}^\mu, \mathbf{r}) \log_2 \frac{P(\mathbf{r}^\mu, \mathbf{r})}{P(\mathbf{r}^\mu)P(\mathbf{r})} \quad (\text{B.23})$$

$$\approx \sum_i \sum_{r_i^\mu, r_i} P(r_i^\mu, r_i) \log_2 \frac{P(r_i^\mu, r_i)}{P(r_i^\mu)P(r_i)}.$$

In this formula the 'approximately equal' sign $\approx$ marks a simplification that is not necessarily a reasonable approximation. If the simplification is valid, it means that in order to extract an information measure, one need not compare whole vectors (the entire firing patterns) with each other, and may instead compare the firing rates of individual cells at storage and retrieval, and sum the resulting single-cell information values. The validity of the simplification is a matter that will be discussed later and that has to be verified, in the end, experimentally, but for the purposes of the present discussion we can focus on the single-cell terms. If either r_i or r_i^μ has a continuous distribution of values, as it will if it represents not the number of spikes emitted in a fixed window, but more generally the firing rate of neuron i computed by convolving the firing train with a smoothing kernel, then one has to deal with probability densities, which we denote as $p(r)dr$, rather than the usual probabilities $P(r)$. Substituting $p(r)dr$ for $P(r)$ and $p(r^\mu, r)drdr^\mu$ for $P(r^\mu, r)$, one can write for each single-cell contribution (omitting the cell index i)

$$< I(r^\mu, r) >_i = \int dr^\mu dr \, p(r^\mu, r) \log_2 \frac{p(r^\mu, r)}{p(r^\mu)p(r)} \quad (\text{B.24})$$

and we see that the differentials $dr^\mu dr$ cancel out between numerator and denominator inside the logarithm, rendering the quantity well defined and finite. If, however, r^μ were to exactly determine r, one would have

$$p(r^\mu, r)dr^\mu dr = p(r^\mu)\delta(r - r(r^\mu))dr^\mu dr = p(r^\mu)dr^\mu \quad (\text{B.25})$$

and, by losing one differential on the way, the mutual information would become infinite. It is therefore important to consider what prevents r^μ from fully determining r in the case at hand — in other words, to consider the sources of noise in the system. In an autoassociative memory storing an extensive number of patterns (see Appendix A4 of Rolls and Treves (1998)), one source of noise always present is the interference effect due to the concurrent storage of all other patterns. Even neglecting other sources of noise, this produces a finite resolution width ρ, which allows one to write an expression of the type $p(r|r^{\mu})\mathrm{d}r = exp-(r-r(r^{\mu}))^2/2\rho^2\mathrm{d}r$ which ensures that the information is finite as long as the resolution ρ is larger than zero.

One further point that should be noted, in connection with estimating the information retrievable from an autoassociative memory, is that the mutual information between the current distribution of firing rates and that of the stored pattern does not coincide with the information *gain* provided by the memory device. Even when firing rates, or spike counts, are all that matter in terms of information carriers, as in the networks considered in this book, one more term should be taken into account in evaluating the information gain. This term, to be subtracted, is the information contained in the external input that elicits the retrieval. This may vary a lot from the retrieval of one particular memory to the next, but of course an efficient memory device is one that is able, when needed, to retrieve much more information than it requires to be present in the inputs, that is, a device that produces a large information gain.

Finally, one should appreciate the conceptual difference between the information a firing pattern carries about another one (that is, about the pattern stored), as considered above, and two different notions: (a) the information produced by the network in selecting the correct memory pattern and (b) the information a firing pattern carries about something in the outside world. Quantity (a), the information intrinsic to selecting the memory pattern, is ill defined when analysing a real system, but is a well-defined and particularly simple notion when considering a formal model. If p patterns are stored with equal strength, and the selection is errorless, this amounts to $\log_2 p$ bits of information, a quantity often, but not always, small compared with the information in the pattern itself. Quantity (b), the information conveyed about some outside correlate, is not defined when considering a formal model that does not include an explicit account of what the firing of each cell represents, but is well defined and measurable from the recorded activity of real cells. It is the quantity considered in the numerical example with the four visual stimuli, and it can be generalized to the information carried by the activity of several cells in a network, and specialized to the case that the network operates as an associative memory. One may note, in this case, that the capacity to retrieve memories with high fidelity, or high information content, is only useful to the extent that the representation to be retrieved carries that amount of information about something relevant — or, in other words, that it is pointless to store and retrieve with great care largely meaningless messages. This type of argument has been used to discuss the role of the mossy fibres in the operation of the CA3 network in the hippocampus (Treves and Rolls 1992, Rolls and Treves 1998).

B.2 Estimating the information carried by neuronal responses

B.2.1 The limited sampling problem

We now discuss in more detail the application of these general notions to the information transmitted by neurons. Suppose, to be concrete, that an animal has been presented with stimuli drawn from a discrete set, and that the responses of a set of C cells have been recorded following the presentation of each stimulus. We may choose any quantity or set of quantities to characterize the responses; for example let us assume that we consider the firing rate of each cell, r_i, calculated by convolving the spike response with an appropriate smoothing kernel. The response space is then C times the continuous set of all positive real numbers, $(\mathbf{R}/2)^C$. We want to evaluate the average information carried by such responses about which stimulus was shown. In principle, it is straightforward to apply the above formulas, e.g. in the form

$$< I(s, \mathbf{r}) > = \sum_s \mathrm{P}(s) \int \Pi_i dr_i \, p(\mathbf{r}|s) \log_2 \frac{p(\mathbf{r}|s)}{p(\mathbf{r})} \tag{B.26}$$

where it is important to note that $p(\mathbf{r})$ and $p(\mathbf{r}|s)$ are now probability densities defined over the high-dimensional vector space of multi-cell responses. The product sign Π signifies that this whole vector space has to be integrated over, along all its dimensions. $p(\mathbf{r})$ can be calculated as $\sum_s p(\mathbf{r}|s)\mathrm{P}(s)$, and therefore, in principle, all one has to do is to estimate, from the data, the conditional probability densities $p(\mathbf{r}|s)$ — the distributions of responses following each stimulus. In practice, however, in contrast to what happens with formal models, in which there is usually no problem in calculating the exact probability densities, real data come in limited amounts, and thus sample only sparsely the vast response space. This limits the accuracy with which, from the experimental *frequency* of each possible response, we can estimate its *probability*, in turn seriously impairing our ability to estimate $< I >$ correctly. We refer to this as the limited sampling problem. This is a purely technical problem that arises, typically when recording from mammals, because of external constraints on the duration or number of repetitions of a given set of stimulus conditions. With computer simulation experiments, and also with recordings from, for example, insects, sufficient data can usually be obtained that straightforward estimates of information are accurate enough (Strong, Koberle, de Ruyter van Steveninck and Bialek 1998, Golomb, Kleinfeld, Reid, Shapley and Shraiman 1994). The problem is, however, so serious in connection with recordings from monkeys and rats in which limited numbers of trials are usually available for neuronal data, that it is worthwhile to discuss it, in order to appreciate the scope and limits of applying information theory to neuronal processing.

In particular, if the responses are continuous quantities, the probability of observing exactly the same response twice is infinitesimal. In the absence of further manipulation, this would imply that each stimulus generates its own set of unique responses, therefore any response that has actually occurred could be associated unequivocally with one stimulus, and the mutual information would always equal the entropy of the stimulus set. This absurdity shows that in order to estimate probability densities from experimental frequencies, one has to resort to some *regularizing* manipulation, such as smoothing the point-like response values by convolution with suitable kernels, or binning them into a finite number of discrete bins.

B.2.1.1 Smoothing or binning neuronal response data

The issue is how to estimate the underlying probability distributions of neuronal responses to a set of stimuli from only a limited number of trials of data (e.g. 10–30) for each stimulus. Several strategies are possible. One is to discretize the response space into bins, and estimate the probability density as the histogram of the fraction of trials falling into each bin. If the bins are too narrow, almost every response is in a different bin, and the estimated information will be overestimated. Even if the bin width is increased to match the standard deviation of each underlying distribution, the information may still be overestimated. Alternatively, one may try to 'smooth' the data by convolving each response with a Gaussian with a width set to the standard deviation measured for each stimulus. Setting the standard deviation to this value may actually lead to an underestimation of the amount of information available, due to oversmoothing. Another possibility is to make a bold assumption as to what the general shape of the underlying densities should be, for example a Gaussian. This may produce closer estimates. Methods for regularizing the data are discussed further by Rolls and Treves (1998) in their Appendix A2, where a numerical example is given.

B.2.1.2   The effects of limited sampling

The crux of the problem is that, whatever procedure one adopts, limited sampling tends to produce distortions in the estimated probability densities. The resulting mutual information estimates are intrinsically biased. The bias, or average error of the estimate, is upward if the raw data have not been regularized much, and is downward if the regularization procedure chosen has been heavier. The bias can be, if the available trials are few, much larger than the true information values themselves. This is intuitive, as fluctuations due to the finite number of trials available would tend, on average, to either produce or emphasize differences among the distributions corresponding to different stimuli, differences that are preserved if the regularization is 'light', and that are interpreted in the calculation as carrying genuine information. This is illustrated with a quantitative example by Rolls and Treves (1998) in their Appendix A2.

Choosing the right amount of regularization, or the best regularizing procedure, is not possible *a priori*. Hertz, Kjaer, Eskander and Richmond (1992) have proposed the interesting procedure of using an artificial neural network to regularize the raw responses. The network can be trained on part of the data using backpropagation, and then used on the remaining part to produce what is in effect a clever data-driven regularization of the responses. This procedure is, however, rather computer intensive and not very safe, as shown by some self-evident inconsistency in the results (Heller, Hertz, Kjaer and Richmond 1995). Obviously, the best way to deal with the limited sampling problem is to try and use as many trials as possible. The improvement is slow, however, and generating as many trials as would be required for a reasonably unbiased estimate is often, in practice, impossible.

B.2.2   Correction procedures for limited sampling

The above point, that data drawn from a single distribution, when artificially paired, at random, to different stimulus labels, results in 'spurious' amounts of apparent information, suggests a simple way of checking the reliability of estimates produced from real data (Optican, Gawne, Richmond and Joseph 1991). One can disregard the true stimulus associated with each response, and generate a randomly reshuffled pairing of stimuli and responses, which should therefore, being not linked by any underlying relationship, carry no mutual information about each other. Calculating, with some procedure of choice, the spurious information obtained in this way, and comparing with the information value estimated with the same procedure for the real pairing, one can get a feeling for how far the procedure goes into eliminating the apparent information due to limited sampling. Although this spurious information, I_s, is only indicative of the amount of bias affecting the original estimate, a simple heuristic trick (called 'bootstrap'[44]) is to subtract the spurious from the original value, to obtain a somewhat

---

[44]In technical usage bootstrap procedures utilize random pairings of responses with stimuli with replacement, while shuffling procedures utilize random pairings of responses with stimuli without replacement.

'corrected' estimate. This procedure can result in quite accurate estimates (see Rolls and Treves (1998), Tovee, Rolls, Treves and Bellis (1993))[45].

A different correction procedure (called 'jack-knife') is based on the assumption that the bias is proportional to $1/N$, where N is the number of responses (data points) used in the estimation. One computes, beside the original estimate $< I_N >$, $/$;$/$; N auxiliary estimates $< I_{N-1} >_k$, by taking out from the data set response k, where k runs across the data set from 1 to N. The corrected estimate

$$< I > = N < I_N > -(1/N) \sum_k (N-1) < I_{N-1} >_k \qquad (B.27)$$

is free from bias (to leading order in $1/N$), if the proportionality factor is more or less the same in the original and auxiliary estimates. This procedure is very time-consuming, and it suffers from the same imprecision of any algorithm that tries to determine a quantity as the result of the subtraction of two large and nearly equal terms; in this case the terms have been made large on purpose, by multiplying them by N and $N - 1$.

A more fundamental approach (Miller 1955) is to derive an analytical expression for the bias (or, more precisely, for its leading terms in an expansion in $1/N$, the inverse of the sample size). This allows the estimation of the bias from the data itself, and its subsequent subtraction, as discussed in Treves and Panzeri (1995) and Panzeri and Treves (1996). Such a procedure produces satisfactory results, thereby lowering the size of the sample required for a given accuracy in the estimate by about an order of magnitude (Golomb, Hertz, Panzeri, Treves and Richmond 1997). However, it does not, in itself, make possible measures of the information contained in very complex responses with few trials. As a rule of thumb, the number of trials per stimulus required for a reasonable estimate of information, once the subtractive correction is applied, is of the order of the effectively independent (and utilized) bins in which the response space can be partitioned (Panzeri and Treves 1996).

B.2.3   The information from multiple cells: decoding procedures

The bias of information measures grows with the dimensionality of the response space, and for all practical purposes the limit on the number of dimensions that can lead to reasonably accurate direct measures, even when applying a correction procedure, is quite low, two to three. This implies, in particular, that it is not possible to apply equation A2.25 to extract the information content in the responses of several cells (more than two to three) recorded simultaneously. One way to address the problem is then to apply some strong form of regularization to the multiple cell responses. Smoothing has already been mentioned as a form of regularization that can be tuned from very soft to very strong, and that preserves the structure of the response space. Binning is another form, which changes the nature of the responses from continuous to discrete, but otherwise preserves their general structure, and which can also be tuned from soft to strong. Other forms of regularization involve much more radical transformations, or changes of variables.

Of particular interest for information estimates is a change of variables that transforms the response space into the stimulus set, by applying an algorithm that derives a predicted

---

[45] Subtracting the 'square' of the spurious fraction of information estimated by this bootstrap procedure as used by Optican, Gawne, Richmond and Joseph (1991) is unfounded and does not work correctly (see Rolls and Treves (1998) and Tovee, Rolls, Treves and Bellis (1993)).

stimulus from the response vector, i.e. the firing rates of all the cells, on each trial. Applying such an algorithm is called decoding. Of course, the predicted stimulus is not necessarily the same as the actual one. Therefore the term decoding should not be taken to imply that the algorithm works successfully, each time identifying the actual stimulus. The predicted stimulus is simply a function of the response, as determined by the algorithm considered. Just as with any regularizing transform, it is possible to compute the mutual information between actual stimuli s and predicted stimuli s', instead of the original one between stimuli s and responses r. Since information about (real) stimuli can only be lost and not be created by the transform, the information measured in this way is bound to be lower in value than the real information in the responses. If the decoding algorithm is efficient, it manages to preserve nearly all the information contained in the raw responses, while if it is poor, it loses a large portion of it. If the responses themselves provided all the information about stimuli, and the decoding is optimal, then predicted stimuli coincide with the actual stimuli, and the information extracted equals the entropy of the stimulus set.

The procedure of extracting information values after applying a decoding algorithm is indicated in Fig. 5.9 (in which s? is s'), and is schematized below

$$s \quad \Rightarrow \quad r \quad \rightarrow \quad s'$$
$$I(s, r)$$
$$I(s, s')$$

where the double arrow indicates the transformation from stimuli to responses operated by the nervous system, while the single arrow indicates the further transformation operated by the decoding procedure. $I(s, s')$ is the mutual information between the actual stimuli s and the stimuli s' that are predicted to have been shown based on the decoded responses.

A slightly more complex variant of this procedure is a decoding step that extracts from the response on each trial not a single predicted stimulus, but rather probabilities that each of the possible stimuli was the actual one. The joint probabilities of actual and posited stimuli can be averaged across trials, and information computed from the resulting probability matrix $(S \times S)$. Computing information in this way takes into account the relative uncertainty in assigning a predicted stimulus to each trial, an uncertainty that is instead not considered by the previous procedure based solely on the identification of the maximally likely stimulus (Treves 1997). *Maximum likelihood* information values I_{ml} based on a single stimulus tend therefore to be higher than *probability* information values I_p based on the whole set of stimuli, although in very specific situations the reverse could also be true.

The same correction procedures for limited sampling can be applied to information values computed after a decoding step. Values obtained from maximum likelihood decoding, I_{ml}, suffer from limited sampling more than those obtained from probability decoding, I_p, since each trial contributes a whole 'brick' of weight $1/N$ (N being the total number of trials), whereas with probabilities each brick is shared among several slots of the $(S \times S)$ probability matrix. The neural network procedure devised by Hertz, Kjaer, Eskander and Richmond (1992) can in fact be thought of as a decoding procedure based on probabilities, which deals with limited sampling not by applying a correction but rather by strongly regularizing the original responses.

When decoding is used, the rule of thumb becomes that the minimal number of trials per stimulus required for accurate information measures is roughly equal to the size of the stimulus set, if the subtractive correction is applied (Panzeri and Treves 1996).

B.2.3.1 Decoding algorithms

Any transformation from the response space to the stimulus set could be used in decoding, but of particular interest are the transformations that either approach optimality, so as to minimize information loss and hence the effect of decoding, or else are implementable by mechanisms that *could* conceivably be operating in the real system, so as to extract information values that could be extracted by the system itself.

The optimal transformation is in theory well-defined: one should estimate from the data the conditional probabilities $P(r|s)$, and use Bayes' rule to convert them into the conditional probabilities $P(s'|r)$. Having these for any value of r, one could use them to estimate I_p, and, after selecting for each particular real response the stimulus with the highest conditional probability, to estimate I_{ml}. To avoid biasing the estimation of conditional probabilities, the responses used in estimating $P(r|s)$ should not include the particular response for which $P(s'|r)$ is going to be derived (jack-knife cross validation). In practice, however, the estimation of $P(r|s)$ in usable form involves the fitting of some simple function to the responses. This need for fitting, together with the approximations implied in the estimation of the various quantities, prevents us from defining the really optimal decoding, and leaves us with various algorithms, depending essentially on the fitting function used, which are hopefully close to optimal in some conditions. We have experimented extensively with two such algorithms, that both approximate Bayesian decoding (Rolls, Treves and Tovee 1997b). Both these algorithms fit the response vectors produced over several trials by the cells being recorded to a product of conditional probabilities for the response of each cell given the stimulus. In one case, the single cell conditional probability is assumed to be Gaussian (truncated at zero); in the other it is assumed to be Poisson (with an additional weight at zero). Details of these algorithms are given by Rolls, Treves and Tovee (1997b).

Biologically plausible decoding algorithms are those that limit the algebraic operations used to types that could be easily implemented by neurons, e.g. dot product summations, thresholding and other single-cell non-linearities, and competition and contrast enhancement among the outputs of nearby cells. There is then no need for ever fitting functions or other sophisticated approximations, but of course the degree of arbitrariness in selecting a particular algorithm remains substantial, and a comparison among different choices based on which yields the higher information values may favour one choice in a given situation and another choice with a different data set.

To summarize, the key idea in decoding, in our context of estimating information values, is that it allows substitution of a possibly very high-dimensional response space (which is difficult to sample and regularize) with a reduced object much easier to handle, that is with a discrete set equivalent to the stimulus set. The mutual information between the new set and the stimulus set is then easier to estimate even with limited data, and if the assumptions about population coding, underlying the particular decoding algorithm used, are justified, the value obtained approximates the original target, the mutual information between stimuli and responses. For each response recorded, one can use all the responses except for that one to generate estimates of the average response vectors (the average response for each neuron in the population) to each stimulus. Then one considers how well the selected response vector matches the average response vectors, and uses the degree of matching to estimate, for all stimuli, the probability that they were the actual stimuli. The form of the matching embodies the general notions about population encoding, for example the 'degree of matching' might be simply the dot product between the current vector and the average vector ($\mathbf{r}^{\text{av}}$), suitably normalized over all average vectors to generate probabilities

$$P(s'|\mathbf{r}(s)) = \frac{\mathbf{r}(s) \cdot \mathbf{r}^{\text{av}}(s')}{\sum_{s''} \mathbf{r}(s) \cdot \mathbf{r}^{\text{av}}(s'')} \tag{B.28}$$

where s'' is a dummy variable. (This is called dot product decoding in Fig. 5.10.) One ends up, then, with a table of conjoint probabilities $P(s, s')$, and another table obtained by selecting for each trial the most likely (or predicted) single stimulus s^p, $P(s, s^p)$. Both s' and s^p stand for all possible stimuli, and hence belong to the same set S. These can be used to estimate mutual information values based on probability decoding (I_p) and on maximum likelihood decoding (I_{ml}):

$$<I_p> = \sum_{s \in S} \sum_{s' \in S} P(s, s') \log_2 \frac{P(s, s')}{P(s)P(s')} \tag{B.29}$$

and

$$<I_{ml}> = \sum_{s \in S} \sum_{s^p \in S} P(s, s^p) \log_2 \frac{P(s, s^p)}{P(s)P(s^p)} . \tag{B.30}$$

B.2.4 Information in the correlations between the spikes of different cells

Simultaneously recorded neurons sometimes shows cross-correlations in their firing, that is the firing of one is systematically related to the firing of the other cell. One example of this is neuronal response synchronization. The cross-correlation, to be defined below, shows the time difference between the cells at which the systematic relation appears. A significant peak or trough in the cross-correlation function could reveal a synaptic connection from one cell to the other, or a common input to each of the cells, or any of a considerable number of other possibilities. If the synchronization occurred for only some of the stimuli, then the presence of the significant cross-correlation for only those stimuli could provide additional evidence separate from any information in the firing rate of the neurons about which stimulus had been shown. Information theory in principle provides a way of quantitatively assessing the relative contributions from these two types of encoding, by expressing what can be learned from each type of encoding in the same units, bits of information. An information theory-based approach to this has now been developed as follows by Panzeri, Schultz, Treves and Rolls (1999a). A problem that must be overcome is the fact that with many simultaneously recorded neurons, each emitting perhaps many spikes at different times, the dimensionality of the response space becomes very large, the information tends to be overestimated, and even bias corrections cannot save the situation. The approach that has been adopted so far limits the problem by taking short time epochs for the information analysis, in which low numbers of spikes, in practice typically 0, 1 or 2, spikes are likely to occur from each neuron, as described next.

In a sufficiently short time window, at most two spikes are emitted from the population. Taking advantage of this, the response probabilities can be calculated in terms of pairwise correlations. These response probabilities are inserted into the Shannon information formula B.31

to obtain expressions quantifying the impact of the pairwise correlations on the information $I(t)$ transmitted in a short time t by groups of spiking neurons:

$$I(t) = \sum_{s \in S} \sum_{\mathbf{r}} P(s, \mathbf{r}) \log_2 \frac{P(s, \mathbf{r})}{P(s)P(\mathbf{r})} \tag{B.31}$$

where $\mathbf{r}$ is the firing rate response vector comprised by the number of spikes emitted by each of the cells in the population in the short time t, and $P(s, \mathbf{r})$ refers to the joint probability distribution of stimuli with their respective neuronal response vectors.

The information depends upon the following two types of correlation:

B.2.4.1 The correlations in the neuronal response variability from the average to each stimulus (sometimes called "noise" correlations) γ:

$\gamma_{ij}(s)$ (for $i \neq j$) is the fraction of coincidences above (or below) that expected from uncorrelated responses, relative to the number of coincidences in the uncorrelated case (which is $\overline{n}_i(s)\overline{n}_j(s)$, the bar denoting the average across trials belonging to stimulus s, where $n_i(s)$ is the number of spikes emitted by cell i to stimulus s on a given trial)

$$\gamma_{ij}(s) = \frac{\overline{n_i(s)n_j(s)}}{(\overline{n}_i(s)\overline{n}_j(s))} - 1, \tag{B.32}$$

and is named the 'scaled cross-correlation density'. It can vary from -1 to ∞; negative $\gamma_{ij}(s)$'s indicate anticorrelation, whereas positive $\gamma_{ij}(s)$'s indicate correlation[46]. $\gamma_{ij}(s)$ can be thought of as the amount of trial by trial concurrent firing of the cells i and j, compared to that expected in the uncorrelated case. $\gamma_{ij}(s)$ (for $i \neq j$) is the 'scaled cross-correlation density' (Aertsen, Gerstein, Habib and Palm 1989, Panzeri, Schultz, Treves and Rolls 1999a), and is sometimes called the "noise" correlation (Gawne and Richmond 1993, Shadlen and Newsome 1994, Shadlen and Newsome 1998).

B.2.4.2 The correlations in the mean responses of the neurons across the set of stimuli (sometimes called "signal" correlations) ν:

$$\nu_{ij} = \frac{< \overline{n}_i(s)\overline{n}_j(s) >_s}{< \overline{n}_i(s) >_s < \overline{n}_j(s) >_s} - 1 = \frac{< \overline{r}_i(s)\overline{r}_j(s) >_s}{< \overline{r}_i(s) >_s < \overline{r}_j(s) >_s} - 1 \tag{B.34}$$

where $\overline{r}_i(s)$ is the mean rate of response of cell i (among C cells in total) to stimulus s over all the trials in which that stimulus was present. ν_{ij} can be thought of as the degree of similarity

---

[46] $\gamma_{ij}(s)$ is an alternative, which produces a more compact information analysis, to the neuronal cross-correlation based on the Pearson correlation coefficient $\rho_{ij}(s)$ (equation B.33), which normalizes the number of coincidences above independence to the standard deviation of the number of coincidences expected if the cells were independent. The normalization used by the Pearson correlation coefficient has the advantage that it quantifies the strength of correlations between neurons in a rate-independent way. For the information analysis, it is more convenient to use the scaled correlation density $\gamma_{ij}(s)$ than the Pearson correlation coefficient, because of the compactness of the resulting formulation, and because of its scaling properties for small t. $\gamma_{ij}(s)$ remains finite as $t \to 0$, thus by using this measure we can keep the t expansion of the information explicit. Keeping the time-dependence of the resulting information components explicit greatly increases the amount of insight obtained from the series expansion. In contrast, the Pearson noise-correlation measure applied to short timescales approaches zero at short time windows:

$$\rho_{ij}(s) \equiv \frac{\overline{n_i(s)n_j(s)} - \overline{n}_i(s)\overline{n}_j(s)}{\sigma_{n_i(s)}\sigma_{n_j(s)}} \simeq t\,\gamma_{ij}(s)\,\sqrt{\overline{r}_i(s)\overline{r}_j(s)}, \tag{B.33}$$

where $\sigma_{n_i(s)}$ is the standard deviation of the count of spikes emitted by cell i in response to stimulus s.

in the mean response profiles (averaged across trials) of the cells i and j to different stimuli. ν_{ij} is sometimes called the "signal" correlation (Gawne and Richmond 1993, Shadlen and Newsome 1994, Shadlen and Newsome 1998).

B.2.4.3 Information in the cross-correlations in short time periods

In the short timescale limit, the first (I_t) and second (I_{tt}) information derivatives describe the information $I(t)$ available in the short time t

$$I(t) = t\, I_t + \frac{t^2}{2}\, I_{tt} \ . \tag{B.35}$$

(The zeroth order, time-independent term is zero, as no information can be transmitted by the neurons in a time window of zero length. Higher order terms are also excluded as they become negligible.)

The instantaneous information rate I_t is

$$I_t = \sum_{i=1}^{C} \left\langle \bar{r}_i(s) \log_2 \frac{\bar{r}_i(s)}{\langle \bar{r}_i(s') \rangle_{s'}} \right\rangle_s \ . \tag{B.36}$$

This formula shows that this information rate (the first time derivative) should not be linked to a high signal to noise ratio, but only reflects the extent to which the mean responses of each cell are distributed across stimuli. It does not reflect anything of the variability of those responses, that is of their noisiness, nor anything of the correlations among the mean responses of different cells.

The effect of (pairwise) correlations between the cells begins to be expressed in the second time derivative of the information. The expression for the instantaneous information 'acceleration' I_{tt} (the second time derivative of the information) breaks up into three terms[47]:

$$\begin{aligned} I_{tt} = {} & \frac{1}{\ln 2} \sum_{i=1}^{C} \sum_{j=1}^{C} \langle \bar{r}_i(s) \rangle_s \langle \bar{r}_j(s) \rangle_s \left[\nu_{ij} + (1 + \nu_{ij}) \ln\left(\frac{1}{1 + \nu_{ij}}\right) \right] \\ & + \sum_{i=1}^{C} \sum_{j=1}^{C} \left[\langle \bar{r}_i(s)\bar{r}_j(s)\gamma_{ij}(s) \rangle_s \right] \log_2\left(\frac{1}{1 + \nu_{ij}}\right) \\ & + \sum_{i=1}^{C} \sum_{j=1}^{C} \left\langle \bar{r}_i(s)\bar{r}_j(s)(1 + \gamma_{ij}(s)) \log_2\left[\frac{(1 + \gamma_{ij}(s))\, \langle \bar{r}_i(s')\bar{r}_j(s') \rangle_{s'}}{\langle \bar{r}_i(s')\bar{r}_j(s')(1 + \gamma_{ij}(s')) \rangle_{s'}}\right] \right\rangle_s \ . \end{aligned} \tag{B.37}$$

The first of these terms is all that survives if there is no noise correlation at all. Thus the *rate component* of the information is given by the sum of I_t (which is always greater than or equal to zero) and of the first term of I_{tt} (which is instead always less than or equal to zero).

The second term is non-zero if there is some correlation in the variance to a given stimulus, even if it is independent of which stimulus is present; this term thus represents the contribution of *stimulus-independent noise correlation* to the information.

---

[47]Note that s' is used in equation B.37 just as a dummy variable to stand for s, as there are two summations performed over s.

The third component of I_{tt} represents the contribution of *stimulus-modulated noise correlation*, as it becomes non-zero only for stimulus-dependent noise correlations. These last two terms of I_{tt} together are referred to as the correlational components of the information.

The application of this approach to measuring the information in the relative time of firing of simultaneously recorded cells, together with the bias corrections for limited sampling, is described by Panzeri, Treves, Schultz and Rolls (1999b) and in Section B.3.4. They showed that with the responses of simultaneously recorded inferior temporal cortex neurons tested with static stimuli, the information increases approximately linearly with the number of neurons, and that there is little extra information available in the relative time of firing of neurons, that is in the correlational components of the information. A further approach is described by Brenner, Strong, Koberle and Bialek (2000), in which they show that when moving stimuli are used, information may be evident in the relative time of firing of (single) neurons in the fly visual system. It may be that with multiple cell simultaneously recorded data in primates, information from the relative time of firing of the different neurons may be especially evident primarily when moving stimuli are used.

B.3 Main results obtained from applying information-theoretic analyses

Although information theory provides the natural mathematical framework for analysing the performance of neuronal systems, its applications in neuroscience have been for many years rather sparse and episodic (e.g. MacKay and McCulloch (1952); Eckhorn and Popel (1974); Eckhorn and Popel (1975); Eckhorn, Grusser, Kroller, Pellnitz and Popel (1976)). One reason for this lukewarm interest in information theory applications has certainly been the great effort that was apparently required, due essentially to the limited sampling problem, in order to obtain reliable results. Another reason has been the hesitation towards analysing as a single complex 'black-box' large neuronal systems all the way from some external, easily controllable inputs, down to neuronal outputs in some central cortical area of interest, for example including all visual stations from the periphery to the end of the ventral visual stream in the temporal lobe. In fact, two important bodies of work, that have greatly helped revive interest in applications of the theory in recent years, both sidestep these two problems. The problem with analyzing a huge black-box is avoided by considering systems at the sensory periphery; the limited sampling problem is avoided either by working with insects, in which sampling can be extensive (Bialek, Rieke, de Ruyter van Steveninck and Warland 1991, de Ruyter van Steveninck and Laughlin 1996, Rieke, Warland, de Ruyter van Steveninck and Bialek 1996) or by utilizing a formal model instead of real data (Atick and Redlich 1990, Atick 1992). Both approaches have provided insightful quantitative analyses that are in the process of being extended to more central mammalian systems (see e.g. Atick, Griffin and Relich (1996)).

B.3.1 Temporal codes versus rate codes within the spike train of a single neuron

In the third of a series of papers that analyze the response of single units in the primate inferior temporal cortex to a set of static visual stimuli, Optican and Richmond (1987) applied information theory in a particularly direct and useful way. To ascertain the relevance of stimulus-locked temporal modulations in the firing of those units, they compared the amount of information about the stimuli that could be extracted from just the firing rate, computed over a relatively long interval of 384 ms, with the amount of information that could be extracted from a more complete description of the firing, that included temporal modulation. To derive this latter description (the temporal code within the spike train of a single neuron) they applied principal component analysis (PCA) to the temporal response vectors recorded for each neuron on each trial. The PCA helped to reduce the dimensionality of the neuronal response measurements. A temporal response vector was defined as a vector with as components the firing rates in each of 64 successive 6 ms time bins. The (64 × 64) covariance matrix was calculated across all trials of a particular neuron, and diagonalized. The first few eigenvectors of the matrix, those with the largest eigenvalues, are the principal components of the response, and the weights of each response vector on these four to five components can be used as a reduced description of the response, which still preserves, unlike the single value giving the mean firing rate along the entire interval, the main features of the temporal modulation within the interval. Thus a four to five-dimensional temporal code could be contrasted with a one-dimensional rate code, and the comparison made quantitative by measuring the respective values for the mutual information with the stimuli. Although the initial claim (Optican, Gawne, Richmond and Joseph 1991, Eskandar, Richmond and Optican 1992), that the temporal code carried nearly three times as much information as the rate code, was later found to be an artefact of limited sampling, and more recent analyses tend to minimize the additional information in the temporal description (Tovee, Rolls, Treves and Bellis 1993, Heller, Hertz, Kjaer and Richmond 1995), this type of application has immediately appeared straightforward and important, and it has led to many developments. By concentrating on the code expressed in the output rather than on the characterization of the neuronal channel itself, this approach is not affected much by the potential complexities of the preceding black box. Limited sampling, on the other hand, is a problem, particularly because it affects much more codes with a larger number of components, for example the four to five components of the PCA temporal description, than the one-dimensional firing rate code. This is made evident in the paper by Heller, Hertz, Kjaer and Richmond (1995), in which the comparison is extended to several more detailed temporal descriptions, including a binary vector description in which the presence or not of a spike in each 1 ms bin of the response constitutes a component of a 320-dimensional vector. Obviously, this binary vector must contain at least all the information present in the reduced descriptions, whereas in the results of Heller, Hertz, Kjaer and Richmond (1995), despite the use of a sophisticated neural network procedure to control limited sampling biases, the binary vector appears to be the code that carries the least information of all. In practice, with the data samples available in the experiments that have been done, and even when using the most recent procedures to control limited sampling (Panzeri and Treves 1996), reliable comparison can be made only with up to two- to three-dimensional codes.

Tovee, Rolls, Treves and Bellis (1993) and Tovee and Rolls (1995) obtained further evidence that little information was encoded in the temporal aspects of firing within the spike train of a single neuron in the inferior temporal cortex by taking short epochs of the firing of neurons, lasting 20 ms or 50 ms, in which the opportunity for temporal encoding would be limited (because there were few spikes in these short time intervals). They found that a considerable proportion (30%) of the information available in a long time period of 400 ms utilizing temporal encoding within the spike train was available in time periods as short as 20 ms when only the number of spikes was taken into account.

Overall, the main result of these analyses applied to the responses to static stimuli in the temporal visual cortex of primates is that not much more information (perhaps only up to 10% more) can be extracted from temporal codes than from the firing rate measured over a judiciously chosen interval (Tovee, Rolls, Treves and Bellis 1993, Heller, Hertz, Kjaer and Richmond 1995). Indeed, it turns out that even this small amount of 'temporal information' is related primarily to the onset latency of the neuronal responses to different stimuli, rather than to anything more subtle (Tovee, Rolls, Treves and Bellis 1993). Consistent with this point, in earlier visual areas the additional 'temporally encoded' fraction of information can be larger, due especially to the increased relevance, earlier on, of precisely locked transient responses (Kjaer, Hertz and Richmond 1994, Golomb, Kleinfeld, Reid, Shapley and Shraiman 1994, Heller, Hertz, Kjaer and Richmond 1995). This is because if the responses to some stimuli are more transient and to others more sustained, this will result in more information if the temporal modulation of the response of the neuron is taken into account. However, the relevance of more substantial temporal codes for static visual stimuli remains to be demonstrated. For non-static visual stimuli and for other cortical systems, similar analyses have largely yet to be carried out, although clearly one expects to find much more prominent temporal effects e.g. in the auditory system (Nelken, Prut, Vaadia and Abeles 1994, deCharms and Merzenich 1996), for reasons similar to those just annunciated.

B.3.2 The speed of information transfer from single neurons

It is intuitive that if short periods of firing of single cells are considered, there is less time for temporal modulation effects. The information conveyed about stimuli by the firing rate and that conveyed by more detailed temporal codes become similar in value. When the firing periods analyzed become shorter than roughly the mean interspike interval, even the statistics of firing rate values on individual trials cease to be relevant, and the information content of the firing depends solely on the mean firing rates across all trials with each stimulus. This is expressed mathematically by considering the amount of information provided as a function of the length t of the time window over which firing is analyzed, and taking the limit for $t \to 0$ (Skaggs, McNaughton, Gothard and Markus 1993, Panzeri, Biella, Rolls, Skaggs and Treves 1996). To first order in t, only two responses can occur in a short window of length t: either the emission of an action potential, with probability tr_s, where r_s is the mean firing rate calculated over many trials using the same window and stimulus; or no action potential, with probability $1 - tr_s$. Inserting these conditional probabilities into equation B.22, taking the limit and dividing by t, one obtains for the derivative of the stimulus-specific transinformation

$$\mathrm{d}I(s)/\mathrm{d}t = r_s \log_2(r_s/<r>) + (<r> - r_s)/\ln 2, \qquad (B.38)$$

where $<r>$ is the grand mean rate across stimuli. This formula thus gives the rate, in bits/s, at which information about a stimulus begins to accumulate when the firing of a cell is recorded. Such an information rate depends only on the mean firing rate to that stimulus and on the grand mean rate across stimuli. As a function of r_s, it follows the U-shaped curve in Fig. B.1. The curve is universal, in the sense that it applies irrespective of the detailed firing statistics of the cell, and it expresses the fact that the emission or not of a spike in a short window conveys information in as much as the mean response to a given stimulus is above or below the overall mean rate. No information is conveyed about those stimuli the mean response to which is the same as the overall mean. In practice, although the curve describes only the universal behaviour of the initial slope of the specific information as a function of time, it approximates

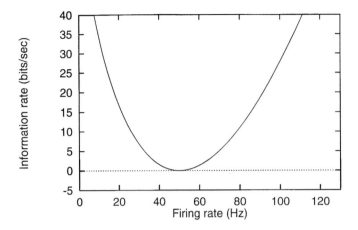

Fig. B.1 Time derivative of the stimulus specific information as a function of firing rate, for a cell firing at a grand mean rate of 50 Hz. For different grand mean rates, the graph would simply be rescaled.

well the full specific information computed even over rather long periods (Rolls, Critchley and Treves 1996a, Rolls, Treves, Tovee and Panzeri 1997d).

Averaging equation B.38 across stimuli one obtains the time derivative of the mutual information. Further dividing by the overall mean rate yields the adimensional quantity

$$\chi = \sum_s P(s)(r_s / <r>) \log_2(r_s / <r>)$$ (B.39)

which measures, in bits, the mutual information per spike provided by the cell (Bialek, 1991; Skaggs et al., 1993). One can prove that this quantity can range from 0 to $\log_2(1/a)$

$$0 < \chi < \log_2(1/a),$$ (B.40)

where a is the sparseness defined in Chapter 1. For mean rates r_s distributed in a nearly binary fashion, χ is close to its upper limit $\log_2(1/a)$, whereas for mean rates that are nearly uniform, or at least unimodally distributed, χ is relatively close to zero (Panzeri, Biella, Rolls, Skaggs and Treves 1996). In practice, whenever a large number of more or less 'ecological' stimuli are considered, mean rates are not distributed in arbitrary ways, but rather tend to follow stereotyped distributions (which approximate an exponential distribution of firing rates — see Treves, Panzeri, Rolls, Booth and Wakeman (1999), Baddeley, Abbott, Booth, Sengpiel, Freeman, Wakeman and Rolls (1997), and Rolls and Treves (1998)), and as a consequence χ and a (or, equivalently, its logarithm) tend to covary (rather than to be independent variables) (Skaggs and McNaughton 1992)). Therefore, measuring sparseness is in practice nearly equivalent to measuring information per spike, and the rate of rise in mutual information, $\chi <r>$, is largely determined by the sparseness a and the overall mean firing rate $<r>$.

The important point to note about the single-cell information rate $\chi <r>$ is that, to the extent that different cells express non-redundant codes, as discussed below, the instantaneous information flow across a population of C cells can be taken to be simply $C\chi <r>$, and this quantity can easily be measured directly without major limited sampling biases, or else inferred indirectly through measurements of the sparseness a. Values for the information rate $\chi < r >$ that have been published range from 2–3 bits/s for rat hippocampal cells (Skaggs, McNaughton, Gothard and Markus 1993), to 10–30 bits/s for primate temporal cortex visual cells (Rolls, Treves and Tovee 1997b), and could be compared with analogous measurements in the sensory systems of frogs and crickets, in the 100–300 bits/s range (Rieke and Bialek 1993).

If the first time-derivative of the mutual information measures information flow, successive derivatives characterize, at the single-cell level, different firing modes. This is because whereas the first derivative is universal and depends only on the mean firing rates to each stimulus, the next derivatives depend also on the variability of the firing rate around its mean value, across trials, and take different forms in different firing regimes. Thus they can serve as a measure of discrimination among firing regimes with limited variability, for which, for example, the second derivative is large and positive, and firing regimes with large variability, for which the second derivative is large and negative. Poisson firing, in which in every short period of time there is a fixed probability of emitting a spike irrespective of previous firing, is an example of large variability, and the second derivative of the mutual information can be calculated to be

$$\mathrm{d}^2 I/\mathrm{d}t^2 = [\ln a + (1-a)] < r >^2 /(a \ln 2), \qquad (B.41)$$

where a is the sparseness defined in Chapter 1. This quantity is always negative. Strictly periodic firing is an example of zero variability, and in fact the second time-derivative of the mutual information becomes infinitely large in this case (although actual information values measured in a short time interval remain of course finite even for exactly periodic firing, because there is still some variability, ± 1, in the number of spikes recorded in the interval). Measures of mutual information from short intervals of firing of temporal cortex visual cells have revealed a degree of variability intermediate between that of periodic and of Poisson regimes (Rolls, Treves, Tovee and Panzeri 1997d). Similar measures can also be used to contrast the effect of the graded nature of neuronal responses, once they are analyzed over a finite period of time, with the information content that would characterize neuronal activity if it reduced to a binary variable (Panzeri, Biella, Rolls, Skaggs and Treves 1996). A binary variable with the same degree of variability would convey information at the same instantaneous rate (the first derivative being universal), but in for example 20–30% reduced amounts when analyzed over times of the order of the interspike interval or longer.

Utilizing these approaches, Tovee, Rolls, Treves and Bellis (1993) and Tovee and Rolls (1995) measured the information available in short epochs of the firing of single neurons, and found that a considerable proportion of the information available in a long time period of 400 ms was available in time periods as short as 20 ms and 50 ms (see Section 5.5.6). Further, Rolls, Tovee and Panzeri (1999b) showed in a backward masking paradigm in which a test stimulus shown for 16 ms was followed after different delays by a masking stimulus (Rolls, Tovee, Purcell, Stewart and Azzopardi 1994b, Rolls and Tovee 1994), that reasonable amounts of information were available even from the 16 ms presentation, and that this was gradually reduced as the mask was brought closer to the stimulus (see Section 5.5.6). The rate of information transfer from single neurons is sufficiently high that the ceiling of the amount of information that can be obtained from a single neuron is reached (for typical stimulus set sizes) in less than 100 ms (Tovee, Rolls, Treves and Bellis 1993).

The conclusions from the single cell information analyses are thus that most of the information is encoded in the spike count; that large parts of this information are available in short temporal epochs of e.g. 20 ms or 50 ms; and that any additional information which appears to be temporally encoded is related to the latency of the neuronal response, and reflect sudden changes in the visual stimuli.

B.3.3 The information from multiple cells: independent information versus redundancy across cells

The rate at which a single cell provides information translates into an instantaneous information flow across a population (with a simple multiplication by the number of cells) only to the extent that different cells provide different (independent) information. To verify whether this condition holds, one cannot extend to multiple cells the simplified formula for the first time-derivative, because it is made simple precisely by the assumption of independence between spikes, and one cannot even measure directly the full information provided by multiple (more than two to three) cells, because of the limited sampling problem discussed above. Therefore one has to analyze the degree of independence (or conversely of redundancy) either directly among pairs — at most triplets — of cells, or indirectly by using decoding procedures to transform population responses. Obviously, the results of the analysis will vary a great deal with the particular neural system considered and the particular set of stimuli, or in general of neuronal correlates, used. For many systems, before undertaking to quantify the analysis in terms of information measures, it takes only a simple qualitative description of the responses to realize that there is a lot of redundancy and very little diversity in the responses. For example, if one selects pain-responsive cells in the somatosensory system and uses painful electrical stimulation of different intensities, most of the recorded cells are likely to convey pretty much the same information, signalling the intensity of the stimulation with the intensity of their single-cell response. Therefore, an analysis of redundancy makes sense only for a neuronal system that functions to represent, and enable discriminations between, a large variety of stimuli, and only when using a set of stimuli representative, in some sense, of that large variety.

Redundancy can be defined with reference to a multiple channel of capacity $T(C)$ which can be decomposed into C separate channels of capacities $T_i, i = 1, ..., C$:

$$R = 1 - T(C) / \sum_i T_i \qquad (B.42)$$

so that when the C channels are multiplexed with maximal efficiency, $T(C) = \sum_i T_i$ and $R = 0$. What is measured more easily, in practice, is the redundancy defined with reference to a specific source (the set of stimuli with their probabilities). Then in terms of mutual information

$$R' = 1 - I(C) / \sum_i I_i. \qquad (B.43)$$

Gawne and Richmond (1993) measured the redundancy R' among pairs of nearby primate inferior temporal cortex visual neurons, in their response to a set of 32 Walsh patterns. They found values with a mean $< R' > = 0.1$ (and a mean single-cell transinformation of 0.23 bits). Since to discriminate 32 different patterns takes 5 bits of information, in principle one would need at least 22 cells providing each 0.23 bits of strictly orthogonal information to represent the full entropy of the stimulus set. Gawne and Richmond reasoned, however, that, because of the overlap, y, in the information they provided, more cells would be needed than if the redundancy had been zero. They constructed a simple model based on the notion that the overlap, y, in the information provided by any two cells in the population always corresponds to the average redundancy measured for nearby pairs. A redundancy $R' = 0.1$ corresponds to an overlap $y = 0.2$ in the information provided by the two neurons, since, counting the overlapping information only once, two cells would yield 1.8 times the amount transmitted by one cell alone. If a fraction of $1 - y = 0.8$ of the information provided by a cell is novel with respect to that provided by another cell, a fraction $(1 - y)^2$ of the information provided by a third cell will be novel with respect to what was known from the first pair, and so on, yielding an estimate of $I(C) = I(1) \sum_{i=0}^{C-1} (1 - y)^i$ for the total information conveyed by C cells. However such a sum saturates, in the limit of an infinite number of cells, at the level $I(\infty) = I(1)/y$, implying in their case that even with very many cells, no more than $0.23/0.2 = 1.15$ bits could be read off their activity, or less than a quarter of what was available as entropy in the stimulus set! Gawne and Richmond (1993) concluded, therefore, that the average overlap among non-nearby cells must be considerably lower than that measured for cells close to each other.

The model above is simple and attractive, but experimental verification of the actual scaling of redundancy with the number of cells entails collecting the responses of several cells interspersed in a population of interest. Gochin, Colombo, Dorfman, Gerstein and Gross (1994) recorded from up to 58 cells in the primate temporal visual cortex, using sets of two to five visual stimuli, and applied decoding procedures to measure the information content in the population response. The recordings were not simultaneous, but comparison with simultaneous recordings from a smaller number of cells indicated that the effect of recording the individual responses on separate trials was minor. The results were expressed in terms of the novelty N in the information provided by C cells, which being defined as the ratio of such information to C times the average single-cell information, can be expressed as

$$N = 1 - R' \tag{B.44}$$

and is thus the complement of the redundancy. An analysis of two different data sets, which included three information measures per data set, indicated a behaviour $N(C) \approx 1/\sqrt{C}$, reminiscent of the improvement in the overall noise-to-signal ratio characterizing C independent processes contributing to the same signal. The analysis neglected however to consider limited sampling effects, and more seriously it neglected to consider saturation effects due to the information content approaching its ceiling, given by the entropy of the stimulus set. Since this ceiling was quite low, for 5 stimuli at $\log_2 5 = 2.32$ bits, relative to the mutual information values measured from the population (an average of 0.26 bits, or 1/9 of the ceiling, was provided by single cells), it is conceivable that the novelty would have taken much larger values if larger stimulus sets had been used.

A simple formula describing the approach to the ceiling, and thus the saturation of information values as they come close to the entropy of the stimulus set, can be derived from a natural extension of the Gawne and Richmond (1993) model. In this extension, the information provided by single cells, measured as a fraction of the ceiling, is taken to coincide with the average overlap among pairs of randomly selected, not necessarily nearby, cells from the population. The actual value measured by Gawne and Richmond would have been, again, $1/22 = 0.045$, below the overlap among nearby cells, $y = 0.2$. The assumption that y, measured across any pair of cells, would have been as low as the fraction of information provided by single cells is equivalent to conceiving of single cells as 'covering' a random portion y of information space, and thus of randomly selected pairs of cells as overlapping in a fraction $(y)^2$ of that space, and so on, as postulated by the Gawne and Richmond (1993) model, for higher numbers of cells. The approach to the ceiling is then described by the formula

$$I(C) \approx H\{1 - exp[C\ln(1 - y)]\} \tag{B.45}$$

that is, a simple exponential saturation to the ceiling. This simple law indeed describes remarkably well the trend in the data analyzed by Rolls, Treves and Tovee (1997b). Although the model has no reason to be exact, and therefore its agreement with the data should not be expected to be accurate, the crucial point it embodies is that deviations from a purely linear increase in information with the number of cells analyzed are due solely to the ceiling effect. Aside from the ceiling, due to the sampling of an information space of finite entropy, the information contents of different cells' responses are independent of each other. Thus, in the model, the observed redundancy (or indeed the overlap) is purely a consequence of the finite size of the stimulus set. If the population were probed with larger and larger sets of stimuli, or more precisely with sets of increasing entropy, and the amount of information conveyed by single cells were to remain approximately the same, then the fraction of space 'covered' by each cell, again y, would get smaller and smaller, tending to eliminate redundancy for very large stimulus entropies (and a fixed number of cells). The actual data were obtained with limited numbers of stimuli, and therefore cannot probe directly the conditions in which redundancy might reduce to zero. The data are consistent, however, with the hypothesis embodied in the simple model, as shown also by the near exponential approach to lower ceilings found for information values calculated with reduced subsets of the original set of stimuli Rolls, Treves and Tovee (1997b), as described in Section 5.5.3.

B.3.4 The information from multiple cells: the effects of cross-correlations between cells

Rather than only demonstrating that cross-correlations can be found between simultaneously recorded neurons, and that the cross-correlations may depend on the experimental parameters, it is very useful to be able to apply information theory to provide an objective measure of how much information may be carried by the number of spikes, and how much may be gained or lost by the correlations that can occur between cells. Using the methods described in Section B.2.4, the information available from the number of spikes vs that from the cross-correlations between simultaneously recorded cells has been analyzed for a population of neurons in the inferior temporal visual cortex (experiments of Rolls, Aggelopoulos, Booth, Zheng, Panzeri and Treves). The stimuli were a set of 20 objects, faces, and scenes presented while the monkey performed a visual discrimination task. If synchronization was being used to bind the parts of each object into the correct spatial relationship to other parts, this might be expected to be revealed by stimulus-dependent cross-correlations in the firing of simultaneously recorded groups of 2–4 cells using multiple single-neuron microelectrodes. A typical cross-correlogram between the responses of simultaneously recorded IT neurons is shown in Fig. B.2. In this typical case, there were no significant cross-correlations between the responses of the neurons.

A typical result from the information analysis described in Section B.2.4 on a set of three simultaneously recorded cells from the experiment shown in Fig. B.2 is shown in Fig. B.3. This shows that most of the information available in a 100 ms time period was available in the rates, and that there was little contribution to the information from stimulus-dependent

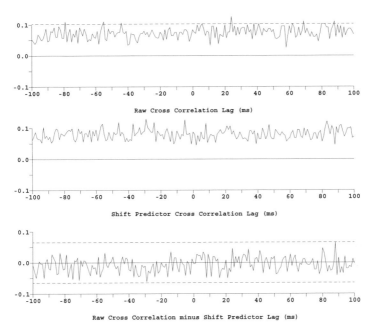

Fig. B.2 A typical cross-correlogram between the responses of simultaneously recorded inferior temporal cortex neurons from an experiment in which 20 stimuli of the type that activate inferior temporal cortex neurons were shown in random sequence. Top: the raw cross-correlogram. Middle: the shift predictor data, showing the values obtained for the cross-correlation coefficients when data for each cell were shuffled between trials. Bottom: the cross-correlogram corrected by subtracting the shift predictor, with the $p < 0.01$ confidence limits indicated.

('noise') correlations (which would have shown as positive values if for example there was stimulus-dependent synchronization of the neuronal responses); or from stimulus-independent 'noise' correlation effects, which might if present have reflected common input to the different neurons so that their responses tended to be correlated independently of which stimulus was shown.

In some other sets of simultaneously recorded inferior temporal cortex neurons some statistically significant cross-correlations were found between particular pairs of IT neurons. In some cases these cross-correlations showed a broad symmetrical peak extending up to 30 ms, which can reflect common input to the neurons. In other cases, the cross-correlations were evident as short asymmetrical peaks in the cross-correlogram extending over a few ms. In both types of case, the predominant result of the information theoretical analysis with the methods described in Section B.2.4 was that the stimulus-independent cross-term (for $i \neq j$) made a small negative contribution to the total information, that is the cross-correlation reflected a small amount of redundancy. No large positive stimulus-dependent contributions were found that would arise from stimulus-dependent synchronization in any of these simultaneously recorded sets of inferior temporal cortex neurons when shown natural stimuli.

The results for the 13 experiments completed at the time of writing with groups of 2–4

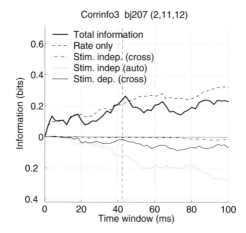

Fig. B.3 A typical result from the information analysis described in Section B.2.4 on a set of 3 simultaneously recorded inferior temporal cortex neurons in the experiment in which 20 complex stimuli effective for IT neurons (objects, faces and scenes) were shown. The graphs show the contributions to the information from the different terms in equations B.36 and B.37, as a function of the length of the time window, which started 100 ms after stimulus onset, which is when IT neurons start to respond. The rate information is the sum of the term in equation B.36 and the first term of equation B.37. The contribution of the stimulus-independent noise correlation to the information is the second term of equation B.37, and is separated into components arising from the correlations between cells (the cross component, for $i \neq j$) and from the autocorrelation within a cell (the auto component, for $i = j$). This term is non-zero if there is some correlation in the variance to a given stimulus, even if it is independent of which stimulus is present. The contribution of the stimulus-dependent noise correlation to the information is the third term of equation B.37, and only the cross term is shown (for $i \neq j$), as this is the term of interest.

simultaneously recorded inferior temporal cortex neurons are shown in Table B.3. The total information is the total from equations B.36 and B.37 in a 100 ms time window, and is not expected to be the sum of the contributions shown in Table B.3 because only the information from the cross terms (for $i \neq j$) is shown in the Table for the contributions related to the stimulus-dependent contributions and the stimulus-independent contributions arising from the 'noise' correlations. The results show that the greatest contribution to the information is that from the rates, that is from the numbers of spikes from each neuron in the time window of 100 ms. The average value of -0.03 for the cross term of the stimulus independent 'noise' correlation–related contribution is consistent with on average a small amount of common input to neurons in the inferior temporal visual cortex. A positive value for the cross term of the stimulus independent 'noise' correlation related contribution would be consistent with on average a small amount of stimulus-dependent synchronization, but the actual value found, 0.01, is so small that it is less than that which can arise by chance statistical fluctuations of the time of arrival of the spikes, as shown by MonteCarlo control rearrangements of the same data. Thus on average there was no significant contribution to the information from stimulus-dependent synchronization effects.

Thus, this data set provides evidence for considerable information available from the number of spikes that each cell produces to different stimuli, and evidence for little impact of common input, or of synchronization, on the amount of information provided by sets of simultaneously recorded inferior temporal cortex neurons. So far, we know of no analyses

Table B.3 The average contributions (in bits) of different components of equations B.36 and B.37 to the information available in a 100 ms time window from 13 sets of simultaneously recorded inferior temporal cortex neurons when shown 20 stimuli effective for the cells.

rate	0.17
stimulus–dependent "noise" correlation–related, cross term	0.01
stimulus–independent "noise" correlation–related, cross term	-0.03
total information	0.18

which have shown with information theoretic methods that considerable amounts of information are available about the stimulus shown from the correlations between the responses of neurons in the ventral visual system. The use of such methods is needed to test quantitatively the hypothesis that synchronization contributes to the encoding of information by neurons.

B.3.5 Conclusions

The conclusions emerging from this set of information theoretic analyses, all performed in cortical areas towards the end of the ventral visual stream of the monkey, are as follows. The representation of at least some classes of objects in those areas is achieved with minimal redundancy by cells that are allocated each to analyze a different aspect of the visual stimulus (Abbott, Rolls and Tovee 1996, Rolls, Treves and Tovee 1997b) (as shown in Sections 5.5.3 and B.3.3). This minimal redundancy is what would be expected of a self-organizing system in which different cells acquired their response selectivities through processes that include some randomness in the initial connectivity , and local competition among nearby cells (see Chapter 7). Towards the end of the ventral visual stream redundancy may thus be effectively minimized, a finding consistent with the general idea that one of the functions of the early visual system is indeed that of progressively minimizing redundancy in the representation of visual stimuli (Attneave 1954, Barlow 1961). Indeed, the evidence described in Sections 5.5.3 and B.3.3 shows that the exponential rise in the number of stimuli that can be decoded when the firing rates of different numbers of neurons are analyzed indicates that the encoding of information using firing rates (in practice the number of spikes emitted by each of a large population of neurons in a short time period) is a very powerful coding scheme used by the cerebral cortex, and that the information carried by different neurons is close to independent provided that the number of stimuli being considered is sufficiently large.

Quantitatively, the encoding of information using firing rates (in practice the number of spikes emitted by each of a large population of neurons in a short time period) is likely to be far more important than temporal encoding, in terms of the number of stimuli that can be encoded. Moreover, the information available from an ensemble of cortical neurons when only the firing rates are read, that is with no temporal encoding within or between neurons, is made available very rapidly (see Fig. 5.14). Further, the neuronal responses in most ventral or 'what' processing streams of behaving monkeys show sustained firing rate differences to different stimuli (see for example Fig. 5.5, and Rolls and Treves (1998) for examples for the olfactory pathways (Rolls, Critchley and Treves 1996a), for spatial view cells in the hippocampus (Rolls, Treves, Robertson, Georges-François and Panzeri 1998), and for head direction cells in the presubiculum (Robertson, Rolls, Georges-François and Panzeri 1999)), so that it may not usually be necessary to invoke temporal encoding for the information about the stimulus to be present. Further, as indicated in Section B.2.4, information theoretic approaches have enabled the information that is available from the firing rate and from the relative time of firing (synchronization) of inferior temporal cortex neurons to be directly compared with the same metric, and most of the information appears to be encoded in the numbers of spikes emitted by a population of cells in a short time period, rather than by the temporal synchronization of the responses of different neurons when certain stimuli appear (see Panzeri, Schultz, Treves and Rolls (1999a)).

Information theoretic approaches also have enabled different types of readout or decoding that could be performed by the brain of the information available in the responses of cell populations to be compared (Rolls, Treves and Tovee 1997b, Robertson, Rolls, Georges-François and Panzeri 1999). It has been shown for example that the multiple cell representation of information used by the brain in the inferior temporal visual cortex (Rolls, Treves and Tovee 1997b), olfactory cortex (Rolls, Critchley and Treves 1996a), hippocampus (Rolls, Treves, Robertson, Georges-François and Panzeri 1998), and presubiculum (Robertson, Rolls, Georges-François and Panzeri 1999) can be read fairly efficiently by the neuronally plausible dot product decoding, and that the representation has all the desirable properties of generalization and graceful degradation, as well as exponential coding capacity (see Sections 5.5.3 and B.3.3). Information theoretic approaches have also enabled the information available about different aspects of stimuli to be directly compared. For example, it has been shown that inferior temporal cortex neurons make explicit much more information about what stimulus has been shown rather than where the stimulus is in the visual field (Tovee, Rolls and Azzopardi 1994), and this is part of the evidence that inferior temporal cortex neurons provide translation invariant representations. In a similar way, information theoretic analysis has provided clear evidence that view invariant representations of objects and faces are present in the inferior temporal visual cortex, in that for example much information is available about what object has been shown from any single trial on which any view of any object is presented (Booth and Rolls 1998). Information theory has also enabled the information available in neuronal representations to be compared with that available to the whole animal in its behaviour (Zohary, Shadlen and Newsome 1994) (but see Section 5.5.5). Finally, information theory also provides a metric for directly comparing the information available from neurons in the brain (see Chapter 5 and this Appendix) with that available from single neurons and populations of neurons in simulations of visual information processing (see Chapter 8).

B.4 Information theory terms — a short glossary

1. The **amount of information**, or **surprise**, in the occurrence of an event (or symbol) s_i of probability $P(s_i)$ is

$$I(s_i) = \log_2 1/P(s_i) = -\log_2 P(s_i). \tag{B.46}$$

(The measure is in bits if logs to the base 2 are used.). This is also the amount of **uncertainty** removed by the occurrence of the event.

2. The average amount of information per source symbol over the whole alphabet (S) of symbols s_i is the **entropy**,

$$H(S) = -\sum_i P(s_i) \log_2 P(s_i) \tag{B.47}$$

(or *a priori* entropy).

3. The probability of the pair of symbols s and s' is denoted $P(s, s')$, and is $P(s)P(s')$ only when the two symbols are **independent**.

4. Bayes theorem (given the output s', what was the input s ?) states that

$$P(s'|s) = \frac{P(s|s')P(s)}{P(s')} \tag{B.48}$$

where $P(s'|s)$ are the **forward** conditional probabilities (given the input s, what will be the output s'?), and $P(s|s')$ are the **backward** conditional probabilities (given the output s', what was the input s ?).

5. **Mutual information**. Prior to reception of s', the probability of the input symbol s was $P(s)$. This is the *a priori* probability of s. After reception of s', the probability that the input symbol was s becomes $P(s|s')$, the conditional probability that s was sent given that s' was received. This is the *a posteriori* probability of s. The difference between the *a priori* and *a posteriori* uncertainties measures the gain of information due to the reception of s'. Once averaged across the values of both symbols s and s', this is the **mutual information**, or **transinformation**

$$I(S, S') = \sum_{s,s'} P(s, s')\{\log_2[1/P(s)] - \log_2[1/P(s|s')]\} \tag{B.49}$$

$$= \sum_{s,s'} P(s, s') \log_2[P(s|s')/P(s)].$$

Alternatively,

$$I(S, S') = H(S) - H(S|S'). \tag{B.50}$$

$H(S|S')$ is sometimes called the **equivocation** (of S with respect to S').

References

Abbott, L. F. (1991). Realistic synaptic inputs for model neural networks, *Network* **2**: 245–258.

Abbott, L. F. and Blum, K. I. (1996). Functional significance of long-term potentiation for sequence learning and prediction, *Cerebral Cortex* **6**: 406–416.

Abbott, L. F. and Nelson, S. B. (2000). Synaptic plasticity: taming the beast, *Nature Neuroscience* **3**: 1178–1183.

Abbott, L. F., Rolls, E. T. and Tovee, M. J. (1996). Representational capacity of face coding in monkeys, *Cerebral Cortex* **6**: 498–505.

Abeles, M. (1991). *Corticonics: Neural Circuits of the Cerebral Cortex*, Cambridge University Press, Cambridge.

Abeles, M., Vaadia, E. and Bergman, H. (1990). Firing patterns of single units in the prefrontal cortex and neural network models, *Network* **1**: 13–25.

Abeles, M., Bergman, H., Margalit, E. and Vaadia, E. (1993). Spatiotemporal firing patterns in the frontal cortex of behaving monkeys, *Journal of Neurophysiology* **70**: 1629–1638.

Abramson, N. (1963). *Information Theory and Coding*, McGraw-Hill, New York.

Ackley, D. H., Hinton, G. E. and Sejnowski, T. J. (1985). A learning algorithm for Boltzmann machines, *Cognitive Science* **9**: 147–169.

Aertsen, A. M. H. J., Gerstein, G. L., Habib, M. K. and Palm, G. (1989). Dynamics of neuronal firing correlation: modulation of 'effective connectivity', *Journal of Neurophysiology* **61**: 900–917.

Aggleton, J. P. (1993). The contribution of the amygdala to normal and abnormal emotional states, *Trends in Neurosciences* **16**: 328–333.

Aggleton, J. P. (ed.) (1992). *The Amygdala*, Wiley-Liss, New York.

Aggleton, J. P. (ed.) (2000). *The Amygdala, A Functional Analysis. Second Edition*, Oxford University Press, Oxford.

Ahmad, S. (1992). Visit: A neural model of covert visual attention, *in* J. Moody, S. Hanson and R. Lippman (eds), *Advances in Neural Information Processing Systems 4*, Morgan Kaufmann Publishers, pp. 420–427.

Albus, J. S. (1971). A theory of cerebellar function, *Mathematical Biosciences* **10**: 25–61.

Amaral, D. G. (1986). Amygdalohippocampal and amygdalocortical projections in the primate brain, *in* R. Schwarcz and Y. Ben-Ari (eds), *Excitatory Amino Acids and Epilepsy*, Plenum Press, New York, pp. 3–18.

Amaral, D. G. (1987). Memory: Anatomical organization of candidate brain regions, *in* F. Plum and V. Mountcastle (eds), *Higher Functions of the Brain. Handbook of Physiology, Part I*, American Physiological Society, Washington, DC, pp. 211–294.

Amaral, D. G. and Price, J. L. (1984). Amygdalo-cortical projections in the monkey (Macaca fascicularis), *Journal of Comparative Neurology* **230**: 465–496.

Amaral, D. G., Price, J. L., Pitkanen, A. and Carmichael, S. T. (1992). Anatomical organization of the primate amygdaloid complex, *in* J. P. Aggleton (ed.), *The Amygdala*, Wiley-Liss, New York, chapter 1, pp. 1–66.

Amari, S. (1977). Dynamics of pattern formation in lateral-inhibition type neural fields,

Biological Cybernetics **27**: 77–87.

Amari, S. (1982). Competitive and cooperative aspects in dynamics of neural excitation and self-organization, *in* S. Amari and M. A. Arbib (eds), *Competition and Cooperation in Neural Nets*, Springer, Berlin, chapter 1, pp. 1–28.

Amari, S., Yoshida, K. and Kanatani, K.-I. (1977). A mathematical foundation for statistical neurodynamics, *SIAM Journal of Applied Mathematics* **33**: 95–126.

Amit, D. J. (1989). *Modelling Brain Function*, Cambridge University Press, New York.

Amit, D. J. (1995). The Hebbian paradigm reintegrated: local reverberations as internal representations, *Behavioral and Brain Sciences* **18**: 617–657.

Amit, D. J. and Tsodyks, M. V. (1991). Quantitative study of attractor neural network retrieving at low spike rates. I. substrate – spikes, rates and neuronal gain, *Network* **2**: 259–273.

Amit, D. J., Gutfreund, H. and Sompolinsky, H. (1987). Statistical mechanics of neural networks near saturation, *Annals of Physics (New York)* **173**: 30–67.

Andersen, P., Dingledine, R., Gjerstad, L., Langmoen, I. A. and Laursen, A. M. (1980). Two different responses of hippocampal pyramidal cells to application of gamma-aminobutyric acid, *Journal of Physiology* **307**: 279–296.

Andersen, R. A., Asanuma, C., Essick, G. K. and Siegel, R. M. (1990a). Cortico-cortical connections of anatomically and physiologically defined subdivisions within inferior parietal lobule, *Journal of Computational Neuroscience* **296**: 65–113.

Andersen, R. A., Bracewell, R. M., Barash, S., Gnadt, J. W. and Fogassi, L. (1990b). Eye position effects on visual, memory, and saccade-related activity in areas lip and 7a of a macaque, *Journal of Neuroscience* **10**: 1176–1196.

Andersen, R. A., Snyder, L. H., Bradley, D. C. and Xing, J. (1997). Multimodal representation of space in the posterior parietal cortex and its use in planning movements, *Annual Review of Neuroscience* **20**: 303–330.

Andersen, R. A., Batista, A. P., Snyder, L. H., Buneo, C. A. and Cohen, Y. E. (2000). Programming to look and reach in the posterior parietal cortex, *in* M. Gazzaniga (ed.), *The New Cognitive Neurosciences*, 2 edn, MIT Press, Cambridge, MA, chapter 36, pp. 515–524.

Anderson, B. (1996). A mathematical model of line bisection behaviour in neglect, *Brain* **119**: 841–850.

Arbib, M. A. (1964). *Brains, Machines, and Mathematics*, McGraw-Hill, New York (2nd Edn 1987 Springer).

Arguin, M. and Bub, D. N. (1993). Evidence for an independent stimulus-centered reference frame from a case of visual hemineglect, *Cortex* **29**: 349–357.

Artola, A. and Singer, W. (1993). Long-term depression: related mechanisms in cerebellum, neocortex and hippocampus, *Synaptic Plasticity: Molecular, Cellular and Functional Aspects*, MIT Press, Cambridge, Mass., chapter 7, pp. 129–146.

Atick, J. J. (1992). Could information theory provide an ecological theory of sensory processing?, *Network* **3**: 213–251.

Atick, J. J. and Redlich, A. N. (1990). Towards a theory of early visual processing, *Neural Computation* **2**: 308–320.

Atick, J. J., Griffin, P. A. and Relich, A. N. (1996). The vocabulary of shape: principal shapes for probing perception and neural response, *Network* **7**: 1–5.

Attneave, F. (1954). Some informational aspects of visual perception, *Psychological Review* **61**: 183–193.

Badcock, J., Whitworth, F., Badcock, D. and Lovegrove, W. (1990). Low-frequency filtering and the processing of local-global stimuli, *Perception* **19**: 617–629.

Baddeley, R. J., Abbott, L. F., Booth, M. J. A., Sengpiel, F., Freeman, T., Wakeman, E. A. and Rolls, E. T. (1997). Responses of neurons in primary and inferior temporal visual cortices to natural scenes, *Proceedings of the Royal Society B* **264**: 1775–1783.

Baizer, J. S., Ungerleider, L. G. and Desimone, R. (1991). Organization of visual inputs to the inferior temporal and posterior parietal cortex in macaques, *Journal of Neuroscience* **11**: 168–190.

Balint, R. (1909). Die Seelenlahmung des Schauens, optische Ataxie, räumliche Störung der Aufmerksamkeit, *Monatsschrift für Psychiatrie und Neurologie* **25**: 51–81.

Ballard, D. H. (1990). Animate vision uses object-centred reference frames, *in* R. Eckmiller (ed.), *Advanced Neural Computers*, North-Holland, Amsterdam, pp. 229–236.

Ballard, D. H. (1993). Subsymbolic modelling of hand-eye co-ordination, *in* D. E. Broadbent (ed.), *The Simulation of Human Intelligence*, Blackwell, Oxford, chapter 3, pp. 71–102.

Barbas, H. (1988). Anatomic organization of basoventral and mediodorsal visual recipient prefrontal regions in the rhesus monkey, *Journal of Comparative Neurology* **276**: 313–342.

Barbas, H. (1995). Anatomic basis of cognitive-emotional interactions in the primate prefrontal cortex, *Neuroscience and Biobehavioral Reviews* **19**: 499–510.

Barlow, H. (1995). The neuron doctrine in perception, *in* M. Gazzaniga (ed.), *The Cognitive Neurosciences*, MIT Press, Cambridge, Mass., chapter 26, pp. 415–435.

Barlow, H. B. (1953). Summation and inhibition in the frog's retina, *Journal of Physiology* **119**: 69–88.

Barlow, H. B. (1961). Possible principles underlying the transformation of sensory messages, *in* W. Rosenblith (ed.), *Sensory Communication*, MIT Press, Cambridge, Mass.

Barlow, H. B. (1972). Single units and sensation: A neuron doctrine for perceptual psychology, *Perception* **1**: 371–394.

Barlow, H. B. (1985). Cerebral cortex as model builder, *in* D. Rose and V. G. Dobson (eds), *Models of the Visual Cortex*, Wiley, Chichester, pp. 37–46.

Barlow, H. B. (1989). Unsupervised learning, *Neural Computation* **1**: 295–311.

Barlow, H. B., Blakemore, C. and Pettigrew, J. D. (1967). The neural mechanisms of binocular depth discrimination, *Journal of Physiology (London)* **193**: 327–342.

Barlow, H. B., Kaushal, T. P. and Mitchison, G. J. (1989). Finding minimum entropy codes, *Neural Computation* **1**: 412–423.

Bartlett, M. S. and Sejnowski, T. J. (1997). Viewpoint invariant face recognition using independent component analysis and attractor networks, *in* M. Mozer, M. Jordan and T. Petsche (eds), *Advances in Neural Information Processing Systems 9*, MIT Press, Cambridge, MA.

Barto, A. G. (1985). Learning by statistical cooperation of self-interested neuron-like computing elements, *Technical Report COINS Tech. Rep. 85 11*, University of Massachusetts, Department of Computer and Information Science: Amherst.

Battaglia, F. and Treves, A. (1998a). Stable and rapid recurrent processing in realistic autoassociative memories, *Neural Computation* **10**: 431–450.

Battaglia, F. P. and Treves, A. (1998b). Attractor neural networks storing multiple space representations: A model for hippocampal place fields, *Physical Review E* **58**: 7738–7753.

Baxter, M. G. and Murray, E. A. (2000). Reinterpreting the behavioural effects of amygdala lesions in non-human primates, *in* J. Aggleton (ed.), *The Amygdala: Second Edition. A Functional Analysis*, Oxford University Press, Oxford, chapter 16, pp. 546–568.

Baylis, G. C. and Rolls, E. T. (1987). Responses of neurons in the inferior temporal cortex in short term and serial recognition memory tasks, *Experimental Brain Research* **65**: 614–622.

Baylis, G. C., Rolls, E. T. and Leonard, C. M. (1985). Selectivity between faces in the responses of a population of neurons in the cortex in the superior temporal sulcus of the monkey, *Brain Research* **342**: 91–102.

Baylis, G. C., Rolls, E. T. and Leonard, C. M. (1987). Functional subdivisions of temporal lobe neocortex, *Journal of Neuroscience* **7**: 330–342.

Baylis, L. L. and Gaffan, D. (1991). Amygdalectomy and ventromedial prefrontal ablation produce similar deficits in food choice and in simple object discrimination learning for an unseen reward, *Experimental Brain Research* **86**: 617–622.

Baylis, L. L. and Rolls, E. T. (1991). Responses of neurons in the primate taste cortex to glutamate, *Physiology and Behavior* **49**: 973–979.

Baylis, L. L., Rolls, E. T. and Baylis, G. C. (1994). Afferent connections of the orbitofrontal cortex taste area of the primate, *Neuroscience* **64**: 801–812.

Baynes, K., Holtzman, H. and Volpe, V. (1986). Components of visual attention: Alterations in response pattern to visual stimuli following parietal lobe infarction, *Brain* **109**: 99–144.

Becker, S. and Hinton, G. E. (1992). Self-organizing neural network that discovers surfaces in random-dot stereograms, *Nature* **355**: 161–163.

Behrman, M. and Tipper, S. P. (1994). Object-based attentional mechanisms: Evidence from patients with unilateral neglect, *in* C. Umilta and M. Moscovitch (eds), *Attention and Performance XV*, MIT Press, Cambridge MA, pp. 351–375.

Behrmann, M. and Moscovitch, M. (1994). Object-centered neglect in patients with unilateral neglect: Effects of left-right coordinates of objects, *Journal of Cognitive Neuroscience* **6**: 151–155.

Behrmann, M., Zemel, R. and Mozer, M. (1998). Object-based attention and occlusion: Evidence from normal participants and a computational model, *Journal of Experimental Psychology: Human Perception and Performance* **24**: 1011–1036.

Bell, A. J. and Sejnowski, T. J. (1995). An information-maximation approach to blind separation and blind deconvolution, *Neural Computation* **7**: 1129–1159.

Bell, A. J. and Sejnowski, T. J. (1997). The independent components of natural scenes are edge filters, *Vision Research* **37**: 3327–3338.

Bennett, A. (1990). Large competitive networks, *Network* **1**: 449–462.

Berthoz, A. (2000). *The Brain's Sense of Movement*, Harvard University Press, Cambridge, Mass.

Bi, G.-Q. and Poo, M.-M. (1998). Activity-induced synaptic modifications in hippocampal culture, dependence on spike timing, synaptic strength and cell type, *Journal of Neuroscience* **18**: 10464–10472.

Bialek, W., Rieke, F., de Ruyter van Steveninck, R. R. and Warland, D. (1991). Reading a neural code, *Science* **252**: 1854–1857.

Biederman, I. (1987). Recognition-by-components: A theory of human image understanding, *Psychological Review* **94**(2): 115–147.

Bienenstock, E. L. C. L. N. and Munro, P. W. (1982). Theory for the development of neuron selectivity: orientation specificity and binocular interaction in visual cortex, *Journal of Neuroscience* **2**: 32–48.

Binford, T. O. (1981). Inferring surfaces from images, *Artificial Intelligence* **17**: 205–244.

Bishop, C. M. (1995). *Neural Networks for Pattern Recognition*, Clarendon Press, Oxford.

Bisiach, E. (1996). Unilateral neglect and the structure of space representation, *Current Directions in Psychological Science* **5**: 62–65.

Bisiach, E. and Luzzatti, C. (1978). Unilateral neglect of representational space, *Cortex* **14**: 129–133.

Blair, R. J., Morris, J. S., Frith, C. D., Perrett, D. I. and Dolan, R. J. (1999). Dissociable neural responses to facial expressions of sadness and anger, *Brain* **122**: 883–893.

Blaney, P. H. (1986). Affect and memory: a review, *Psychological Bulletin* **99**: 229–246.

Bliss, T. V. P. and Collingridge, G. L. (1993). A synaptic model of memory: long-term potentiation in the hippocampus, *Nature* **361**: 31–39.

Block, H. D. (1962). The perceptron: a model for brain functioning, *Reviews of Modern Physics* **34**: 123–135.

Bloomfield, S. (1974). Arithmetical operations performed by nerve cells, *Brain Research* **69**: 115–124.

Bolles, R. C. and Cain, R. A. (1982). Recognizing and locating partially visible objects: The local-feature-focus method, *International Journal of Robotics Research* **1**: 57–82.

Booth, M. C. A. and Rolls, E. T. (1998). View-invariant representations of familiar objects by neurons in the inferior temporal visual cortex, *Cerebral Cortex* **8**: 510–523.

Boussaoud, D. D. R. and Ungerleider, L. G. (1991). Visual topography of area TEO in the macaque, *Journal of Comparative Neurology* **306**: 554–575.

Brady, M., Ponce, J., Yuille, A. and Asada, H. (1985). Describing surfaces, *A. I. Memo 882, The Artificial Intelligence* **17**: 285–349.

Braitenberg, V. and Schuz, A. (1991). *Anatomy of the Cortex*, Springer-Verlag, Berlin.

Breitmeyer, B. and Breier, J. (1994). Effects of background color on reaction time to stimuli varying in size and contrast: Inferences about human M channels, *Vision Research* **34**: 1039–1045.

Brenner, N., Strong, S. P., Koberle, R. and Bialek, W. (2000). Synergy in a neural code, *Neural Computation* **12**: 1531–1532.

Bridle, J. S. (1990). Probabilistic interpretation of feedforward classification network outputs, with relationships to statistical pattern recognition, *in* F. Fogelman-Soulie and J. Herault (eds), *Neurocomputing: Algorithms, Architectures and Applications*, Springer-Verlag, New York, pp. 227–236.

Broadbent, D. E. (1958). *Perception and Communication*, Pergamon Press, London.

Brooks, L. R. (1978). Nonanalytic concept formation and memory for instances, *in* E. Rosch and B. B. Lloyd (eds), *Cognition and Categorization*, Erlbaum, Hillsdale, NJ.

Brotchie, P., Andersen, R., Snyder, L. and Goodman, S. (1995). Head position signals used by parietal neurons to encode locations of visual stimuli, *Nature, London* **375**: 232–235.

Brothers, L., Ring, B. and Kling, A. S. (1990). Response of neurons in the macaque amygdala to complex social stimuli, *Behavioural Brain Research* **41**: 199–213.

Brown, D. A., Gähwiler, B. H., Griffith, W. H. and Halliwell, J. V. (1990a). Membrane currents in hippocampal neurons, *Progress in Brain Research* **83**: 141–160.

Brown, M. and Xiang, J. (1998). Recognition memory: neuronal substrates of the judgement of prior occurrence, *Progress in Neurobiology* **55**: 149–189.

Brown, T. H., Kairiss, E. W. and Keenan, C. L. (1990b). Hebbian synapses: biophysical mechanisms and algorithms, *Annual Review of Neuroscience* **13**: 475–511.

Bruce, C., Desimone, R. and Gross, C. G. (1981). Visual properties of neurons in a polysensory area in superior temporal sulcus of the macaque, *Journal of Neurophysiology* **46**: 369–384.

Bruce, V. (1988). *Recognising Faces*, Erlbaum, Hillsdale, NJ.

Buckley, M. J. and Gaffan, D. (2000). The hippocampus, perirhinal cortex, and memory in the monkey, *in* J. J. Bolhuis (ed.), *Brain, Perception, and Memory: Advances in Cognitive Neuroscience*, Oxford University Press, Oxford, pp. 279–298.

Buckley, M. J., Booth, M. C. A., Rolls, E. T. and Gaffan, D. (1998). Selective perceptual impairments following perirhinal cortex ablation, *Society for Neuroscience Abstracts* **24**: 18.

Buckley, M. J., Booth, M. C. A., Rolls, E. T. and Gaffan, D. (2002). Selective perceptual impairments following perirhinal cortex ablation, *Journal of Neuroscience, in press*.

Buhl, E. H., Halasy, K. and Somogyi, P. (1994). Diverse sources of hippocampal unitary inhibitory postsynaptic potentials and the number of synaptic release sites, *Nature* **368**: 823–828.

Buhmann, J., Lades, M. and von der Malsburg, C. (1990). Size and distortion invariant object recognition by hierarchical graph matching, *Proceedings of the IJCNN International Joint Conference on Neural Networks*, pp. 411–416.

Buhmann, J., Lange, J., von der Malsburg, C., Vorbrüggen, J. C. and Würtz, R. P. (1991). Object recognition in the dynamic link architecture: Parallel implementation of a transputer network, *in* B. Kosko (ed.), *Neural Networks for Signal Processing*, Prentice Hall, Englewood Cliffs, New Jersey, pp. 121–159.

Bullier, J. and Nowak, L. (1995). Parallel versus serial processing: new vistas on the distributed organization of the visual system, *Current Opinion in Neurobiology* **5**: 497–503.

Burgess, N. and O'Keefe, J. (1996). Neuronal computations underlying the firing of place cells and their role in navigation, *Hippocampus* **6**: 749–762.

Burgess, N., Recce, M. and O'Keefe, J. (1994). A model of hippocampal function, *Neural Networks* **7**: 1065–1081.

Bushnell, C., Goldberg, M. E. and Robinson, D. L. (1981). Behavioral enhancement of visual responses in monkey cerebral cortex. I. Modulation in posterior parietal cortex related to selective visual attention, *Journal of Neurophysiology* **46**: 755–772.

Butter, C. M. (1969). Perseveration in extinction and in discrimination reversal tasks following selective prefrontal ablations in Macaca mulatta, *Physiology and Behavior* **4**: 163–171.

Butter, C. M. and Snyder, D. R. (1972). Alterations in aversive and aggressive behaviors following orbitofrontal lesions in rhesus monkeys, *Acta Neurobiologica Experimentalis* **32**: 525–565.

Butter, C. M., McDonald, J. A. and Snyder, D. R. (1969). Orality, preference behavior, and reinforcement value of non-food objects in monkeys with orbital frontal lesions, *Science* **164**: 1306–1307.

Butter, C. M., Snyder, D. R. and McDonald, J. A. (1970). Effects of orbitofrontal lesions on aversive and aggressive behaviors in rhesus monkeys, *Journal of Comparative and Physiological Psychology* **72**: 132–144.

Caan, W., Perrett, D. I. and Rolls, E. T. (1984). Responses of striatal neurons in the behaving monkey. 2. Visual processing in the caudal neostriatum, *Brain Research* **290**: 53–65.

Cador, M., Robbins, T. W. and Everitt, B. J. (1989). Involvement of the amygdala in stimulus-reward associations: interaction with the ventral striatum, *Neuroscience* **30**: 77–86.

Cahusac, P. M. B., Rolls, E. T., Miyashita, Y. and Niki, H. (1993). Modification of the responses of hippocampal neurons in the monkey during the learning of a conditional spatial response task, *Hippocampus* **3**: 29–42.

Calvert, G. A., Bullmore, E. T., Brammer, M. J., Campbell, R., Williams, S. C. R., McGuire, P. K., Woodruff, P. W. R., Iversen, S. D. and David, A. S. (1997). Activation of auditory cortex during silent lip-reading, *Science* **276**: 593–596.

Carmichael, S. T. and Price, J. L. (1995). Sensory and premotor connections of the orbital and medial prefrontal cortex of macaque monkeys, *Journal of Comparative Neurology* **363**: 642–664.

Carmichael, S. T., Clugnet, M. C. and Price, J. L. (1994). Central olfactory connections in the macaque monkey, *Journal of Comparative Neurology* **346**: 403–434.

Carrasco, M. and Frieder, K. (1997). Cortical magnification neutralizes the eccentricity effect in visual search, *Visual Research* **37**: 63–82.

Carrasco, M. and Yeshurun, Y. (1998). The contribution of covert attention to the set-size and eccentricity effects in visual search, *Journal of Experimental Psychology: Human Perception and Performance* **24**: 673–692.

Cassaday, H. J. and Rawlins, J. N. (1997). The hippocampus, objects, and their contexts, *Behavioural Neuroscience* **111**: 1228–1244.

Celebrini, S., Thorpe, S., Trotter, Y. and Imbert, M. (1993). Dynamics of orientation coding in area V1 of the awake primate, *Visual Neuroscience* **10**: 811–825.

Cerella, J. (1986). Pigeons and perceptrons, *Pattern Recognition* **19**: 431–438.

Chakravarty, I. (1979). A generalized line and junction labeling scheme with applications to scene analysis, *IEEE Transactions PAMI* pp. 202–205.

Chelazzi, L. (1998). Serial attention mechanisms in visual search: A critical look at the evidence, *Psychological Research* **62**: 195–219.

Chelazzi, L., Miller, E. K., Duncan, J. and Desimone, R. (1993). A neural basis for visual search in inferior temporal cortex, *Nature (London)* **363**: 345–347.

Chelazzi, L., Duncan, J., Miller, E. and Desimone, R. (1998). Responses of neurons in inferior temporal cortex during memory-guuided visual search, *Journal of Neurophysiology* **80**: 2918–2940.

Christie, B. R. (1996). Long-term depression in the hippocampus, *Hippocampus* **6**: 1–2.

Colby, C. L. (1991). The neuroanatomy and neurophysiology of attention, *Journal of Child Neurology* **6**: 90–118.

Colby, C. L. and Goldberg, M. E. (1999). Space and attention in parietal cortex, *Annual Review of Neuroscience* **22**: 319–349.

Colby, C. L., Duhamel, J. R. and Goldberg, M. E. (1993). Ventral intraparietal area of the macaque – anatomic location and visual response properties, *Journal of Neurophysiology* **69**: 902–914.

Connor, C. E., Gallant, J. L. and Van Essen, D. C. (1993). Effects of focal attention on receptive field profiles in area V4, *Society for Neuroscience Abstracts* **19**: 15–16.

Connor, C. E., Gallant, J. L., Preddie, D. and Van Essen, D. (1996). Responses in area V4 depend on the spatial relationship between stimulus and attention, *Journal of Neurophysiology* **75**: 1306–1308.

Corbetta, M. and Shulman, G. (1998). Human cortical mechanisms of visual attention during orienting and search, *Philosophical Transactions of the Royal Society of London* **353**: 1353–1362.

Corchs, S. and Deco, G. (2001). Large-scale neural model for visual attention: Integration of experimental single cell and fMRI data, *Cerebral Cortex, in press*.

Cortes, C., Jaeckel, L. D., Solla, S. A., Vapnik, V. and Denker, J. S. (1996). Learning curves: asymptotic values and rates of convergence, *Neural Information Processing Systems* **6**: 327–334.

Cover, T. M. (1965). Geometrical and statistical properties of systems of linear inequalities with applications in pattern recognition, *IEEE Transactions on Electronic Computers* **14**: 326–334.

Cover, T. M. and Thomas, J. A. (1991). *Elements of Information Theory*, Wiley, New York.

Cowey, A. (1979). Cortical maps and visual perception, *Quarterly Journal of Experimental Psychology* **31**: 1–17.

Cowey, A. and Rolls, E. T. (1975). Human cortical magnification factor and its relation to visual acuity, *Experimental Brain Research* **21**: 447–454.

Crick, F. (1984). Function of the thalamic reticular complex: The searchlight hypothesis, *Proceedings of the National Academy of Science, USA* **81**: 4586–4590.

Crick, F. and Koch, C. (1990). Towards a neurobiological theory of consciousness, *Seminars in Neuroscience* **2**: 263–275.

Critchley, H. D. and Rolls, E. T. (1996). Olfactory neuronal responses in the primate orbitofrontal cortex: analysis in an olfactory discrimination task, *Journal of Neurophysiology* **75**: 1659–1672.

Damasio, A. R. (1994). *Descartes' Error*, Putnam, New York.

Dane, C. and Bajcsy, R. (1982). An object-centred three-dimensional model builder, *Proceedings of the 6th International Conference on Pattern Recognition*, pp. 348–350.

Daugman, J. (1988). Complete discrete 2D-Gabor transforms by neural networks for image analysis and compression, *IEEE Transactions on Acoustic, Speech, and Signal Processing* **36**: 1169–1179.

Daugman, J. (1997). Neural image processing strategies applied in real-time pattern recognition, *Real-Time Imaging* **3**: 157–171.

Davis, M. (2000). The role of the amygdala in conditioned and unconditioned fear and anxiety, *in* J. P. Aggleton (ed.), *The Amygdala, Second Edition*, Oxford University Press, Oxford, pp. 213–287.

Dawkins, R. (1989). *The Selfish Gene*, 2nd edn, Oxford University Press, Oxford.

de Araujo, I. E. T., Rolls, E. T. and Stringer, S. M. (2001). A view model which accounts for the response properties of hippocampal primate spatial view cells and rat place cells, *Hippocampus* **11**: in press.

de Ruyter van Steveninck, R. R. and Laughlin, S. B. (1996). The rates of information transfer at graded-potential synapses, *Nature* **379**: 642–645.

De Sieno, D. (1988). Adding a conscience to competitive learning, *IEEE International Conference on Neural Networks (San Diego 1988)*, Vol. 1, IEEE, New York, pp. 117–124.

De Valois, R. L. and De Valois, K. K. (1988). *Spatial Vision*, Oxford University Press, New York.

DeAngelis, G. C., Cumming, B. G. and Newsome, W. T. (2000). A new role for cortical area MT: the perception of stereoscopic depth, *in* M. Gazzaniga (ed.), *The New Cognitive*

Neurosciences, Second Edition, MIT Press, Cambridge, MA, chapter 21, pp. 305–314.

deCharms, R. C. and Merzenich, M. M. (1996). Primary cortical representation of sounds by the coordination of action-potential timing, *Nature* **381**: 610–613.

Deco, G. and Lee, T. S. (2001). An interactive neurodynamical model of biased competition for attentive object recognition and visual search.

Deco, G. and Obradovic, D. (1996). *An Information-Theoretic Approach to Neural Computing*, Springer, New York.

Deco, G. and Schürmann, B. (2001). *Information Dynamics: Foundations and Applications*, Springer, New York.

Deco, G. and Zihl, J. (2001a). A neurodynamical model of visual attention: Feedback enhancement of spatial resolution in a hierarchical system, *Journal of Computational Neuroscience* **10**: 231–253.

Deco, G. and Zihl, J. (2001b). Top-down selective visual attention: A neurodynamical approach, *Visual Cognition* **8**: 119–140.

Deco, G. and Zihl, J. (2001c). A visual account of visual neglect, *Journal of Computational Neuroscience*.

Desimone, R. (1991). Face-selective cells in the temporal cortex of monkeys, *Journal of Cognitive Neuroscience* **3**: 1–8.

Desimone, R. (1996). Neural mechanisms for visual memory and their role in attention, *Proceedings of the National Academy of Sciences USA* **93**: 13494–13499.

Desimone, R. and Duncan, J. (1995). Neural mechanisms of selective visual attention, *Annual Review of Neuroscience* **18**: 193–222.

Desimone, R. and Gross, C. G. (1979). Visual areas in the temporal lobe of the macaque, *Brain Res.* **178**: 363–380.

Desimone, R., Albright, T. D., Gross, C. G. and Bruce, C. (1984). Stimulus-selective properties of inferior temporal neurons in the macaque, *Journal of Neuroscience* **4**: 2051–2062.

DeWeese, M. R. and Meister, M. (1999). How to measure the information gained from one symbol, *Network* **10**: 325–340.

Diamond, M. E., Huang, W. and Ebner, F. F. (1994). Laminar comparison of somatosensory cortical plasticity, *Science* **265**: 1885–1888.

Dolan, R. J., Fink, G. R., Rolls, E. T., Booth, M., Holmes, A., Frackowiak, R. S. J. and Friston, K. J. (1997). How the brain learns to see objects and faces in an impoverished context, *Nature* **389**: 596–599.

Douglas, R. J. and Martin, K. A. C. (1990). Neocortex, *in* G. M. Shepherd (ed.), *The Synaptic Organization of the Brain*, 3rd edn, Oxford University Press, Oxford, chapter 12, pp. 389–438.

Douglas, R. J., Mahowald, M. A. and Martin, K. A. C. (1996). Microarchitecture of cortical columns, *in* A. Aertsen and V. Braitenberg (eds), *Brain Theory: Biological Basis and Computational Theory of Vision*, Elsevier, Amsterdam.

Dow, B. W., Snyder, A. Z., Vautin, R. G. and Bauer, R. (1981). Magnification factor and receptive field size in foveal striate cortex of the monkey, *Experimental Brain Research* **44**: 213–218.

Driver, J. and Baylis, G. (1989). Movement and visual attention: The spotlight metaphor breaks down, *Journal of Experimental Psychology: Human Perception and Performance* **17**: 561–570.

Driver, J. and Halligan, P. (1991). Can visual neglect operate in object-centred co-ordinates?

An affirmative single-case study, *Cognitive Neuropsychology* **8**: 475–496.

Driver, J., Baylis, G. and Rafal, R. (1992). Preserved figure-ground segregation and symmetry perception in visual neglect, *Nature* **360**: 73–75.

Driver, J., Baylis, G. C., Goodrich, S. and Rafal, R. (1994). Axis-based neglect of visual shapes, *Neuropsychologia* **32**: 1353–1365.

Duhamel, J. R., Colby, C. L. and Goldberg, M. E. (1991). Congruent representation of visual and somatosensory space in single neurons of monkey ventral intra-parietal cortex (area VIP), *in* J. Paillard (ed.), *Brain and Space*, Oxford University Press, Oxford, pp. 223–236.

Duhamel, J. R., Colby, C. L. and Goldberg, M. E. (1992a). The updating of the representation of visual space in parietal cortex by intended eye movements, *Science* **255**: 90–92.

Duhamel, J. R., Goldberg, M. E., Fitzgibbon, E. J., Sirigu, A. and Grafman, J. (1992b). Saccadic dysmetria in a patient with a right frontoparietal lesion: The importance of corollary discharge for accurate spatial orientation, *Brain* **115**: 1387–1402.

Duhamel, J. R., Bremmer, F., BenHamed, S. and Werner, G. (1997). Spatial invariance of visual receptive fields in parietal cortex neurons, *Nature* **389**: 845–848.

Duncan, J. (1980). The locus of interference in the perception of simultaneous stimuli, *Psychological Review* **87**: 272–300.

Duncan, J. (1984). Selective attention and the organization of visual information, *Journal of Experimental Psychology: General* **113**: 501–517.

Duncan, J. (1996). Cooperating brain systems in selective perception and action, *in* T. Inui and J. L. McClelland (eds), *Attention and Performance XVI*, MIT Press, Cambridge, pp. 549–578.

Duncan, J. and Humphreys, G. (1989). Visual search and stimulus similarity, *Psychological Review* **96**: 433–458.

Duncan, J., Humphreys, G. and Ward, R. (1997). Competitive brain activity in visual attention, *Current Opinion in Neurobiology* **7**: 255–261.

Dunn, L. T. and Everitt, B. J. (1988). Double dissociations of the effects of amygdala and insular cortex lesions on conditioned taste aversion, passive avoidance, and neophobia in the rat using the excitotoxin ibotenic acid, *Behavioral Neuroscience* **102**: 3–23.

Durbin, R. and Mitchison, G. (1990). A dimension reduction framework for understanding cortical maps, *Nature* **343**: 644–647.

Eccles, J. C. (1984). The cerebral neocortex: a theory of its operation, *in* E. G. Jones and A. Peters (eds), *Cerebral Cortex: Functional Properties of Cortical Cells*, Vol. 2, Plenum, New York, chapter 1, pp. 1–36.

Eckhorn, R. and Popel, B. (1974). Rigorous and extended application of information theory to the afferent visual system of the cat. I. Basic concepts, *Kybernetik* **16**: 191–200.

Eckhorn, R. and Popel, B. (1975). Rigorous and extended application of information theory to the afferent visual system of the cat. II. Experimental results, *Kybernetik* **17**: 7–17.

Eckhorn, R., Grusser, O. J., Kroller, J., Pellnitz, K. and Popel, B. (1976). Efficiency of different neural codes: information transfer calculations for three different neuronal systems, *Biological Cybernetics* **22**: 49–60.

Eckhorn, R., Bauer, R., Jordan, W., Brosch, M., Munk, M. and Reitboeck, H. (1988). Coherent oscillations: A mechanism of feature linking in the visual cortex? Multiple electrode and correlation analysis in the cat, *Biological Cybernetics* **60**: 121–128.

Edelman, S. (1999). *Representation and Recognition in Vision*, MIT Press, Cambridge, Massachusetts.

Eglin, M., Robertson, L. and Knight, R. (1989). Visual search performance in the neglect syndrome, *Journal of Cognitive Neuroscience* **1**: 372–385.

Eichenbaum, H. (1997). Declarative memory: insights from cognitive neurobiology, *Annual Review of Psychology* **48**: 547–572.

Elliffe, M. C. M., Rolls, E. T., Parga, N. and Renart, A. (2000). A recurrent model of transformation invariance by association, *Neural Networks* **13**: 225–237.

Elliffe, M. C. M., Rolls, E. T. and Stringer, S. M. (2001). Invariant recognition of feature combinations in the visual system, *Biological Cybernetics* **85**(5): in press.

Engel, A. K., Konig, P. and Singer, W. (1991). Direct physiological evidence for scene segmentation by temporal encoding, *Proceedings of the National Academy of Sciences of the USA* **88**: 9136–9140.

Engel, A. K., Konig, P., Kreiter, A. K., Schillen, T. B. and Singer, W. (1992). Temporal coding in the visual system: new vistas on integration in the nervous system, *Trends in Neurosciences* **15**: 218–226.

Epstein, R. and Kanwisher, N. (1998). A cortical representation of the local visual environment, *Nature* **392**: 598–601.

Eriksen, C. W. and Hoffmann, J. (1973). The extent of processing of noise elements during selective encoding from visual displays, *Perception and Psychophysics* **14**: 155–160.

Erwin, E. and Miller, K. (1998). Correlation based development of ocularly matched orientation and ocular dominance maps: determination of required input activities, *Journal of Neuroscience* **18**: 9870–9895.

Erwin, E., Obermayer, K. and Schulten, K. (1995). Models of orientation and ocular dominance coumns in the visual cortex: a critical comparison, *Neural Computation* **7**: 425–468.

Eskandar, E. N., Richmond, B. J. and Optican, L. M. (1992). Role of inferior temporal neurons in visual memory. I. Temporal encoding of information about visual images, recalled images, and behavioural context, *Journal of Neurophysiology* **68**: 1277–1295.

Everitt, B. J. and Robbins, T. W. (1992). Amygdala-ventral striatal interactions and reward-related processes, *in* J. P. Aggleton (ed.), *The Amygdala*, Wiley, Chichester, pp. 401–430.

Farah, M. J. (1990). *Visual Agnosia*, MIT Press, Cambridge, Mass.

Farah, M. J. (2000). *The Cognitive Neuroscience of Vision*, Blackwell, Oxford.

Farah, M. J., Meyer, M. M. and McMullen, P. A. (1996). The living/nonliving dissociation is not an artifact: giving an *a priori* implausible hypothesis a strong test, *Cognitive Neuropsychology* **13**: 137–154.

Faugeras (1993). *The Representation, Recognition and Location of 3-D Objects*, MIT Press.

Faugeras, O. D. and Hebert, M. (1986). The representation, recognition and location of 3-D objects, *International Journal of Robotics Research* **5**: 27–52.

Fazeli, M. S. and Collingridge, G. L. (eds) (1996). *Cortical Plasticity: LTP and LTD*, Bios, Oxford.

Feldman, J. A. (1985). Four frames suffice: a provisional model of vision and space, *Behavioural Brain Sciences* **8**: 265–289.

Felleman, D. J. and Van Essen, D. C. (1991). Distributed hierarchical processing in the primate cerebral cortex, *Cerebral Cortex* **1**: 1–47.

Ferster, D. and Miller, K. D. (2000). Neural mechanisms of orientation selectivity in the visual cortex, *Annual Review of Neuroscience* **23**: 441–471.

Ferster, D. and Spruston, N. (1995). Cracking the neuronal code, *Science* **270**: 756–757.

Field, D. J. (1987). Relations between the statistics of natural images and the response properties of cortical cells, *Journal of the Optical Society of America, A* **4**: 2379–2394.

Field, D. J. (1994). What is the goal of sensory coding?, *Neural Computation* **6**: 559–601.

Finkel, L. H. and Edelman, G. M. (1987). Population rules for synapses in networks, *in* G. M. Edelman, W. E. Gall and W. M. Cowan (eds), *Synaptic Function*, John Wiley & Sons, New York, pp. 711–757.

Földiák, P. (1989). Adaptive network for optimal linear feature extraction, *Proceedings of the IEEE/INNS International Joint Conference on Neural Networks*, pp. 401–405.

Földiák, P. (1991). Learning invariance from transformation sequences, *Neural Computation* **3**: 193–199.

Földiák, P. (1992). Models of sensory coding, *Technical Report CUED/F–INFENG/TR 91*, University of Cambridge, Department of Engineering.

Francis, S., Rolls, E. T., Bowtell, R., McGlone, F., O'Doherty, J., Browning, A., Clare, S. and Smith, E. (1999). The representation of pleasant touch in the brain and its relationship with taste and olfactory areas, *NeuroReport* **10**: 453–459.

Frégnac, Y. (1996). Dynamics of cortical connectivity in visual cortical networks: an overview, *Journal of Physiology, Paris* **90**: 113–139.

Frolov, A. A. and Medvedev, A. V. (1986). Substantiation of the "point approximation" for describing the total electrical activity of the brain with use of a simulation model, *Biophysics* **31**: 332–337.

Fukushima, K. (1975). Cognitron: a self-organizing neural network, *Biological Cybernetics* **20**: 121–136.

Fukushima, K. (1980). Neocognitron: a self-organizing neural network model for a mechanism of pattern recognition unaffected by shift in position, *Biological Cybernetics* **36**: 193–202.

Fukushima, K. (1988). Neocognitron: A hierarchical neural network model capable of visual pattern recognition unaffected by shift in position, *Neural Networks* **1**: 119–130.

Fukushima, K. (1989). Analysis of the process of visual pattern recognition by the neocognitron, *Neural Networks* **2**: 413–420.

Fukushima, K. (1991). Neural networks for visual pattern recognition, *IEEE Transactions E* **74**: 179–190.

Fukushima, K. and Miyake, S. (1982). Neocognitron: A new algorithm for pattern recognition tolerant of deformations and shifts in position, *Pattern Recognition* **15**(6): 455–469.

Funahashi, S., Bruce, C. and Goldman-Rakic, P. (1989). Mnemonic coding of visual space in monkey dorsolateral prefrontal cortex, *Journal of Neurophysiology* **61**: 331–349.

Fuster, J. (1997). *The Prefrontal Cortex*, 3rd edn, Raven Press, New York.

Fuster, J. (2000). *Memory Systems in the Brain*, Raven Press, New York.

Gaffan, D. (1994). Scene-specific memory for objects: a model of episodic memory impairment in monkeys with fornix transection, *Journal of Cognitive Neuroscience* **6**: 305–320.

Gaffan, D. and Harrison, S. (1989a). A comparison of the effects of fornix section and sulcus principalis ablation upon spatial learning by monkeys, *Behavioural Brain Research* **31**: 207–220.

Gaffan, D. and Harrison, S. (1989b). Place memory and scene memory: effects of fornix transection in the monkey, *Experimental Brain Research* **74**: 202–212.

Gaffan, D. and Saunders, R. C. (1985). Running recognition of configural stimuli by fornix transected monkeys, *Quarterly Journal of Experimental Psychology* **37B**: 61–71.

Gallant, J. L., Connor, C. E. and Van-Essen, D. C. (1998). Neural activity in areas V1, V2 and V4 during free viewing of natural scenes compared to controlled viewing, *NeuroReport* **9**: 85–90.

Galletti, C., Battaglini, P. P. and Fattori, P. (1991). Functional properties of neurons in the anterior bank of the parieto-occipital sulcus of the macaque monkey, *European Journal of Neuroscience* **3**: 452–461.

Gardner, E. (1987). Maximum storage capacity in neural networks, *Europhysics Letters* **4**: 481–485.

Gardner, E. (1988). The space of interactions in neural network models, *Journal of Physics A* **21**: 257–270.

Gattass, R., Sousa, A. P. B. and Covey, E. (1985). Cortical visual areas of the macaque: possible substrates for pattern recognition mechanisms, *Experimental Brain Research, Supplement 11*.

Gawne, T. J. and Richmond, B. J. (1993). How independent are the messages carried by adjacent inferior temporal cortical neurons?, *Journal of Neuroscience* **13**: 2758–2771.

Georges-Francois, P., Rolls, E. T. and Robertson, R. G. (1999). Spatial view cells in the primate hippocampus: allocentric view not head direction or eye position or place, *Cerebral Cortex* **9**: 197–212.

Gerstner, W. (1995). Time structure of the activity in neural network models, *Physical Review E* **51**: 738–758.

Gerstner, W. (2000). Population dynamics of spiking neurons: Fast transients, asynchronous states, and locking, *Neural Computation* **12**: 43–89.

Gerstner, W., Ritz, R. and Van Hemmen, L. (1993). A biologically motivated and analytically solvable model of collective oscillations in the cortex, *Biological Cybernetics* **68**: 363–374.

Gerstner, W., Kreiter, A. K., Markram, H. and Herz, A. V. (1997). Neural codes: firing rates and beyond, *Proceedings of the National Academy of Sciences USA* **94**: 12740–12741.

Gibson, J. J. (1950). *The Perception of the Visual World*, Houghton Mifflin, Boston.

Gibson, J. J. (1979). *The Ecological Approach to Visual Perception*, Houghton Mifflin, Boston.

Ginsburg, A. (1986). Spatial filtering and visual form perception, *in* K. Boff, L. Kaufman and K. Thomas (eds), *Handbook of Perception and Human Performance. Cognitive Processes and Performance*, John Wiley, New York.

Gnadt, J. W. and Andersen, R. A. (1988). Monkey related motor planning activity in posterior parietal cortex of macaque, *Experimental Brain Research* **70**: 216–220.

Gochin, P. M., Colombo, M., Dorfman, G. A., Gerstein, G. L. and Gross, C. G. (1994). Neural ensemble encoding in inferior temporal cortex, *Journal of Neurophysiology* **71**: 2325–2337.

Goldberg, M. E. (2000). The control of gaze, *in* E. R. Kandel, J. H. Schwartz and T. M. Jessell (eds), *Principles of Neural Science*, 4th edn, McGraw-Hill, New York, chapter 39, pp. 782–800.

Goldberg, M. E., Colby, C. L. and Duhamel, J. R. (1990). Representation of visuomotor space in the parietal lobe of the monkey, *Cold Spring Harbor Symposium on Quantitative Biology* **55**: 729–739.

Goldman-Rakic, P. S. (1996). The prefrontal landscape: implications of functional architecture for understanding human mentation and the central executive, *Philosophical Transactions of the Royal Society of London, Series B* **351**: 1445–1453.

Golomb, D., Kleinfeld, D., Reid, R. C., Shapley, R. M. and Shraiman, B. (1994). On temporal codes and the spatiotemporal response of neurons in the lateral geniculate nucleus, *Journal of Neurophysiology* **72**: 2990–3003.

Golomb, D., Hertz, J. A., Panzeri, S., Treves, A. and Richmond, B. J. (1997). How well can we estimate the information carried in neuronal responses from limited samples?, *Neural Computation* **9**: 649–665.

Goodale, M. A. and Milner, A. D. (1992). Separate visual pathways for perception and action, *Trends in Neurosciences* **15**: 20–25.

Gottlieb, J. P., Kusunoki, M. and Goldberg, M. E. (1998). The representation of visual salience in monkey parietal cortex, *Nature* **391**: 481–484.

Gray, C. M. and Singer, W. (1987). Stimulus-specific neuronal oscillations in the cat visual cortex: A cortical functional unit, *Society for Neuroscience Abstracts* **13**: 403.

Gray, C. M. and Singer, W. (1989). Stimulus-specific neuronal oscillations in orientation columns of cat visual cortex, *Proceedings of the National Academy of Sciences USA* **86**: 1698–1702.

Gray, C. M., Konig, P., Engel, A. K. and Singer, W. (1989). Oscillatory responses in cat visual cortex exhibit inter-columnar synchronization which reflects global stimulus properties, *Nature* **338**: 334–337.

Gray, C. M., Engel, A. K. and Singer, W. (1992). Synchronization of oscillatory neuronal responses in cat striate cortex: temporal properties, *Visual Neuroscience* **8**: 337–347.

Gray, J. A. (1975). *Elements of a Two-Process Theory of Learning*, Academic Press, London.

Gray, J. A. (1987). *The Psychology of Fear and Stress*, 2nd edn, Cambridge University Press, Cambridge.

Graziano, M. S. and Gross, C. G. (1993). A bimodal map of space: somatosensory receptive fields in the macaque putamen with corresponding visual receptive fields, *Experimental Brain Research* **97**: 96–109.

Gregory, R. L. (1970). *The Intelligent Eye*, McGraw-Hill, New York.

Grenander, U. (1976). *Lectures in Pattern Theory I, II and III: Pattern Analysis, Pattern Synthesis and Regular Structures*, Springer, Berlin.

Grimson, W. E. L. (1990). *Object Recognition by Computer*, MIT Press, Cambridge, MA.

Griniasty, M., Tsodyks, M. V. and Amit, D. J. (1993). Conversion of temporal correlations between stimuli to spatial correlations between attractors, *Neural Computation* **35**: 1–17.

Gross, C. G. (1973). Visual function of inferotemporal cortex, *in* R. Jung (ed.), *Handbook of Sensory Physiology*, Springer, Berlin, pp. 451–482.

Gross, C. G., Desimone, R., Albright, T. D. and Schwartz, E. L. (1985). Inferior temporal cortex and pattern recognition, *Experimental Brain Research, Supplement* **11**: 179–201.

Grossberg, S. (1976a). Adaptive pattern classification and universal recoding: I. Parallel development and coding of neural feature detectors, *Biological Cybernetics* **23**: 121–134.

Grossberg, S. (1976b). Adaptive pattern classification and universal recoding: II. Feedback, expectation, olfaction, illusions, *Biological Cybernetics* **23**: 187–202.

Grossberg, S. (1987). Competitive learning: from interactive activation to adaptive resonance, *Cognitive Science* **11**: 23–63.

Grossberg, S. (1988). Nonlinear neural networks: principles, mechanisms, and architectures, *Neural Networks* **1**: 17–61.

Grossberg, S. and Mingolla, E. (1985a). Neural dynamics of form perception: Boundary completion, illusory figures, and neon colour spreading, *Psychological Review* **92**: 173–211.

Grossberg, S. and Mingolla, E. (1985b). Neural dynamics of perceptual grouping: Textures, boundaries, and form completion, *Perception and Psychophysics* **38**: 141–171.

Habib, M. and Sirigu, A. (1987). Pure topographical disorientation: a definition and anatomical basis, *Cortex* **23**: 73–85.

Haenny, P. E. and Schiller, P. H. (1988). State dependent activity in monkey visual cortex. I. Single cell activity in V1 and V4 on visual tasks, *Experimental Brain Research* **69**: 225–244.

Haenny, P. E., Maunsell, J. H. and Schiller, P. H. (1988). State dependent activity in monkey visual cortex. II. Retinal and extraretinal factors in V4, *Experimental Brain Research* **69**: 245–259.

Halligan, P. and Marshall, J. (1994). *Spatial Neglect: Position Papers on the Theory and Practice*, Lawrence Erlbaum, Hillsdale.

Hamker, F. (1999). The role of feedback connections in task-driven visual search, *in* D. Heinke, G. Humphreys and A. Olson (eds), *Connectionist Models in Cognitive Neuroscience – The 5th Neural Computation and Psychology Workshop*, Springer, Berlin, pp. 252–261.

Hamming, R. W. (1990). *Coding and Information Theory*, 2nd edn, Prentice-Hall, Englewood Cliffs, New Jersey.

Harris, A. E., Ermentrout, G. B. and Small, S. L. (1997). A model of ocular dominance column development by competition for trophic factor, *Proceedings of the National Academy of Sciences USA* **94**: 9944–9949.

Harris, L. R. and Jenkin, M. (eds) (1998). *Vision and Action*, Cambridge University Press, New York.

Hasselmo, M. E., Rolls, E. T. and Baylis, G. C. (1989a). The role of expression and identity in the face-selective responses of neurons in the temporal visual cortex of the monkey, *Behavioural Brain Research* **32**: 203–218.

Hasselmo, M. E., Rolls, E. T., Baylis, G. C. and Nalwa, V. (1989b). Object-centered encoding by face-selective neurons in the cortex in the superior temporal sulcus of the monkey, *Experimental Brain Research* **75**: 417–429.

Hasselmo, M. E., Schnell, E. and Barkai, E. (1995). Learning and recall at excitatory recurrent synapses and cholinergic modulation in hippocampal region CA3, *Journal of Neuroscience* **15**: 5249–5262.

Hawken, M. J. and Parker, A. J. (1987). Spatial properties of the monkey striate cortex, *Proceedings of the Royal Society, London [B]* **231**: 251–288.

Haxby, J. V., Horwitz, B., Ungerleider, L. G., Maisog, J. M., Pietrini, P. and Grady, C. L. (1994). The functional organization of human extrastriate cortex: A PET-rCBF study of selective attention to faces and locations, *Journal of Neuroscienc* **14**: 6336–6353.

Hebb, D. O. (1949). *The Organization of Behavior*, Wiley, New York.

Heinke, D. and Humphreys, G. (1999). Modelling emergent attentional properties, *in* D. Heinke, G. Humphreys and A. Olson (eds), *Connectionist Models in Cognitive Neuroscience – The 5th Neural Computation and Psychology Workshop*, Springer, Berlin, pp. 240–251.

Heinke, D., Deco, G., Humphreys, G. and Zihl, J. (2001a). Visual search of hierarchical patterns.

Heinke, D., Deco, G., Humphreys, G. W. and Zihl, J. (2001b). Top-down effect of object knowledge on the scan path of fixations by visual neglect.

Heller, J., Hertz, J. A., Kjaer, T. W. and Richmond, B. J. (1995). Information flow and temporal coding in primate pattern vision, *Journal of Computational Neuroscience* **2**: 175–193.

Helmholtz, H. v. (1867). *Handbuch der physiologischen Optik*, Voss, Leipzig.

Hering, E. (1878 (republished 1964)). *Outlines of a Theory of the Light Sense*, Harvard University Press, Cambridge, MA.

Herrnstein, R. J. (1984). Objects, categories, and discriminative stimuli, *in* H. L. Roitblat, T. G. Bever and H. S. Terrace (eds), *Animal Cognition*, Lawrence Erlbaum and Associates, Hillsdale, NJ.

Hertz, J. A., Krogh, A. and Palmer, R. G. (1991). *Introduction to the Theory of Neural Computation*, Addison-Wesley, Wokingham, UK.

Hertz, J. A., Kjaer, T. W., Eskander, E. N. and Richmond, B. J. (1992). Measuring natural neural processing with artificial neural networks, *International Journal of Neural Systems* **3 (Suppl.)**: 91–103.

Higuchi, S. I. and Miyashita, Y. (1996). Formation of mnemonic neuronal responses to visual paired associates in inferotemporal cortex is impaired by perirhinal and entorhinal lesions, *Proceedings of the National Academy of Sciences of the USA* **93**: 739–743.

Hinton, G. E. (1989). Deterministic Boltzmann learning performs steepest descent in weight-space, *Neural Computation* **1**: 143–150.

Hinton, G. E. and Anderson, J. A. (1981). *Parallel Models of Associative Memory*, Erlbaum, Hillsdale, NJ.

Hinton, G. E. and Ghahramani, Z. (1997). Generative models for discovering sparse distributed representations, *Philosophical Transactions of the Royal Society of London, B* **352**: 1177–1190.

Hinton, G. E. and Sejnowski, T. J. (1986). Learning and relearning in Boltzmann machines, *in* D. Rumelhart and J. L. McClelland (eds), *Parallel Distributed Processing*, Vol. 1, MIT Press, Cambridge, Mass., chapter 7, pp. 282–317.

Hinton, G. E., Dayan, P., Frey, B. J. and Neal, R. M. (1995). The "wake-sleep" algorithm for unsupervised neural networks, *Science* **268**: 1158–1161.

Hodgkin, A. L. and Huxley, A. F. (1952). A quantitative description of membrane current and its application to conduction and excitation in nerve, *Journal of Physiology* **117**: 500–544.

Hoffman, E. A. and Haxby, J. V. (2000). Distinct representations of eye gaze and identity in the distributed neural system for face perception, *Nature Neuroscience* **3**: 80–84.

Hölscher, C. and Rolls, E. T. (2001). Perirhinal cortex neurons in macaques with activity related to the long-term familiarity of visual stimuli, and to working memory, *Society for Neuroscience Abstracts* **27**: in press.

Hölscher, C., Rolls, E. T. and Xiang, J. (2002). Perirhinal cortex neurons in macaques with activity related to the long-term familiarity of visual stimuli.

Hopfield, J. J. (1982). Neural networks and physical systems with emergent collective computational abilities, *Proceedings of the National Academy of Sciences of the U.S.A.* **79**: 2554–2558.

Hopfield, J. J. (1984). Neurons with graded response have collective computational properties like those of two-state neurons, *Proceedings of the National Academy of Sciences of the USA* **81**: 3088–3092.

Hornak, J., Rolls, E. T. and Wade, D. (1996). Face and voice expression identification in patients with emotional and behavioural changes following ventral frontal lobe damage, *Neuropsychologia* **34**: 247–261.

Hubel, D. H. and Wiesel, T. N. (1959). Receptive fields of single neurons in the cat's visual cortex, *Journal of Physiology, London* **148**: 574–591.

Hubel, D. H. and Wiesel, T. N. (1962). Receptive fields, binocular interaction, and functional architecture in the cat's visual cortex, *Journal of Physiology* **160**: 106–154.

Hubel, D. H. and Wiesel, T. N. (1968). Receptive fields and functional architecture of monkey striate cortex, *Journal of Physiology, London* **195**: 215–243.

Hubel, D. H. and Wiesel, T. N. (1972). Laminar and columnar distribution of geniculo-cortical fibers in the macaque monkey, *Journal of Computational Neurology* **146**: 421–450.

Hubel, D. H. and Wiesel, T. N. (1977). Functional architecture of the macaque monkey visual cortex, *Proceedings of the Royal Society, London [B]* **198**: 1–59.

Hummel, J. E. and Biederman, I. (1992). Dynamic binding in a neural network for shape recognition, *Psychological Review* **99**: 480–517.

Humphreys, G. W. and Heinke, D. (1998). Spatial representation and selection in the brain: Neuropsychological and computational constraints, *Visual Cognition* **5**: 9–47.

Humphreys, G. W. and Müller, H. J. (1993). Search via recursive rejection (SERR): A connectionist model of visual search, *Cognitive Psychology* **25**: 43–110.

Humphreys, G. W. and Riddoch, J. (1994). Attention to within-object and between-object spatial representations: Multiple sites for visual selection, *Cognitive Neuropsychology* **11**: 207–241.

Humphreys, G. W. and Riddoch, M. J. (1992). Interactions between objects and space-vision revealed through neuropsychology, *in* D. E. Meyers and S. Kornblum (eds), *Attention and Performance XIV*, Lawrence Erlbaum Associates, Hillsdale, pp. 143–162.

Humphreys, G. W., Olson, A., Romani, C. and Riddoch, J. (1996). Competitive mechanisms of selection by space and object: A neuropsychological approach, *in* A. Kramer, M. Coles and G. Logan (eds), *Converging Operations in the Study of Visual Selective Attention*, American Psychological Association, Washington, DC, pp. 365–393.

Hupe, J. M., James, A. C., Payne, B. R., Lomber, S. G., Girard, P. and Bullier, J. (1998). Cortical feedback improves discrimination between figure and background by V1, V2 and V3 neurons, *Nature* **394**: 784–787.

Huttenlocher, D. P. and Ullman, S. (1990). Recognizing solid objects by alignment with an image, *International Journal of Computer Vision* **5**: 195–212.

Ishai, A., Ungerleider, L. G., Martin, A., Schouten, J. L. and Haxby, J. V. (1999). Distributed representation of objects in the human ventral visual pathway, *Proceedings of the National Academy of Sciences USA* **96**: 9379–9384.

Ito, M. (1984). Neuronal network model, *The Cerebellum and Neural Control*, Raven Press, New York, chapter 10, pp. 115–130.

Ito, M. (1989). Long-term depression, *Annual Review of Neuroscience* **12**: 85–102.

Ito, M. (1993a). Cerebellar mechanisms of long-term depression, *in* M. Baudry, R. F. Thompson and J. Davis (eds), *Synaptic Plasticity: Molecular, Cellular and Functional Aspects*, MIT Press, Cambridge, Mass., chapter 6, pp. 117–128.

Ito, M. (1993b). Synaptic plasticity in the cerebellar cortex and its role in motor learning, *Canadian Journal of Neurological Science, Suppl.* **3**: S70–74.

Ito, M. and Gilbert, C. D. (1999). Attention modulates contextual influences in the primary visual cortex of alert monkeys, *Neuron* **22**: 593–604.

Itti, L. and Koch, C. (2001). Computational modelling of visual attention, *Nature Reviews Neuroscience* **2**: 194–203.

Jackson, P. A., Kesner, R. P. and Amann, K. (1998). Memory for duration: role of hippocampus and medial prefrontal cortex, *Neurobiology of Learning and Memory* **70**: 328–348.

Jacoby, L. L. (1983a). Perceptual enhancement: persistent effects of an experience, *Journal of Experimental Psychology: Learning, Memory, and Cognition* **9**: 21–38.

Jacoby, L. L. (1983b). Remembering the data: analyzing interaction processes in reading, *Journal of Verbal Learning and Verbal Behavior* **22**: 485–508.

James, W. (1890). *The Principles of Psychology*, Henry Holt, New York.

Janssen, P., Vogels, R. and Orban, G. A. (1999). Macaque inferior temporal neurons are selective for disparity-defined three-dimensional shapes, *Proceedings of the National Academy of Sciences* **96**: 8217–8222.

Janssen, P., Vogels, R. and Orban, G. A. (2000). Selectivity for 3D shape that reveals distinct areas within macaque inferior temporal cortex, *Science* **288**: 2054–2056.

Jarrard, E. L. (1993). On the role of the hippocampus in learning and memory in the rat, *Behavioral and Neural Biology* **60**: 9–26.

Johnstone, S. and Rolls, E. T. (1990). Delay, discriminatory, and modality specific neurons in striatum and pallidum during short-term memory tasks, *Brain Research* **522**: 147–151.

Jonas, E. A. and Kaczmarek, L. K. (1999). The inside story: subcellular mechanisms of neuromodulation, *in* P. S. Katz (ed.), *Beyond Neurotransmission*, Oxford University Press, New York, chapter 3, pp. 83–120.

Jones, B. and Mishkin, M. (1972). Limbic lesions and the problem of stimulus-reinforcement associations, *Experimental Neurology* **36**: 362–377.

Jones, E. G. and Peters, A. (eds) (1984). *Cerebral Cortex, Functional Properties of Cortical Cells*, Vol. 2, Plenum, New York.

Jones, E. G. and Powell, T. P. S. (1970). An anatomical study of converging sensory pathways within the cerebral cortex of the monkey, *Brain* **93**: 793–820.

Kahneman, D. (1973). *Attention and Effort*, Prentice-Hall, Englewood Cliffs, NJ.

Kammen, D. M. and Yuille, A. L. (1988). Spontaneous symmetry-breaking energy functions and the emergence of orientation selective cortical cells, *Biological Cybernetics* **59**: 23–31.

Kandel, E. R., Schwartz, J. H. and Jessel, T. H. (2000). *Principles of Neural Science*, 4 edn, McGraw-Hill, New York.

Kanter, I. and Sompolinsky, H. (1987). Associative recall of memories without errors, *Physical Review A* **35**: 380–392.

Kanwisher, N., McDermott, J. and Chun, M. M. (1997). The fusiform face area: a module in human extrastriate cortex specialized for face perception, *Journal of Neuroscience* **17**: 4301–4311.

Kastner, S., De Weerd, P., Desimone, R. and Ungerleider, L. G. (1998). Mechanisms of directed attention in the human extrastriate cortex as revealed by functional MRI, *Science* **282**: 108–111.

Kastner, S., Pinsk, M., De Weerd, P., Desimone, R. and Ungerleider, L. G. (1999). Increased activity in human visual cortex during directed attention in the absence of visual stimulation, *Neuron* **22**: 751–761.

Kesner, R. and Rolls, E. T. (2001). Role of long term synaptic modification in short term memory, *Hippocampus* **11**: 240–250.

Keysers, C., Xiao, D., Foldiak, P. and Perrett, D. (2001). The speed of sight, *Journal of Cognitive Neuroscience* **13**: 90–101.

Kinchla, R. (1974). Detecting target elements in multi-element arrays: A confusability model, *Perception and Psychophysics* **15**: 149–158.

Kinsbourne, M. (1993). Orientational bias model of unilateral neglect: Evidence from attentional gradients within hemispace, *in* I. Robertson and J. Marshall (eds), *Unilateral Neglect: Clinical and Experimental Studies*, Erlbaum, Hove, pp. 63–86.

Kjaer, T. W., Hertz, J. A. and Richmond, B. J. (1994). Decoding cortical neuronal signals: networks models, information estimation and spatial tuning, *Journal of Computational Neuroscience* **1**: 109–139.

Kleinfeld, D. (1986). Sequential state generation by model neural networks, *Proceedings of the National Academy of Sciences of the USA* **83**: 9469–9473.

Kluver, H. and Bucy, P. C. (1939). Preliminary analysis of functions of the temporal lobes in monkeys, *Archives of Neurology and Psychiatry* **42**: 979–1000.

Koch, C. (1999). *Biophysics of Computation*, Oxford University Press, Oxford.

Koch, C. and Ullman, S. (1985). Shifts in selective visual attention: Towards the underlying neural circuitry, *Human Neurobiology* **4**: 219–227.

Koenderink, J. J. (1990). *Solid Shape*, MIT Press, Cambridge, Mass.

Koenderink, J. J. and Van Doorn, A. J. (1979). The internal representation of solid shape with respect to vision, *Biological Cybernetics* **32**: 211–217.

Koenderink, J. J. and van Doorn, A. J. (1991). Affine structure from motion, *Journal of the Optical Society of America, A* **8**: 377–385.

Kohonen, T. (1977). *Associative Memory: A System Theoretical Approach*, Springer, New York.

Kohonen, T. (1982). Clustering, taxonomy, and topological maps of patterns, *in* M. Lang (ed.), *Proceedings of the Sixth International Conference on Pattern Recognition*, IEEE Computer Society Press, Silver Spring, MD, pp. 114–125.

Kohonen, T. (1988). *Self-Organization and Associative Memory*, 2nd edn, Springer-Verlag, New York.

Kohonen, T. (1989). *Self-Organization and Associative Memory*, 3rd edn, (1984, 1st edn; 1988, 2nd edn, Springer-Verlag, Berlin.

Kohonen, T. (1995). *Self-Organizing Maps*, Springer-Verlag, Berlin.

Kolb, B. and Whishaw, I. Q. (1996). *Fundamentals of Human Neuropsychology*, 4th edn, Freeman, New York.

Kosslyn, S. M. (1994). *Image and Brain: The Resolution of the Imagery Debate*, MIT Press, Cambridge, Mass.

Kovacs, G., Vogels, R. and Orban, G. (1995). Cortical correlate of pattern backward masking, *Proceedings of the National Academy of Sciences of the USA* **92**: 5587–5591.

Krakauer, J. and Ghez, C. (2000). Voluntary movement, *in* E. R. Kandel, J. H. Schwartz and T. M. Jessell (eds), *Principles of Neural Science*, 4th edn, McGraw-Hill, New York, chapter 38, pp. 756–781.

Kramer, A. and Jacobson, A. (1991). Perceptual organization and focused attention: The role of objects and proximity in visual processing, *Perception and Psychophysics* **50**: 267–284.

Kramer, A. and Watson, S. (1995). Object-based visual selection and the principle of uniform connectedness, *in* A. Kramer, M. Coles and G. Logan (eds), *Converging Operations in the Study of Visual Attention*, American Psychological Association, Washington, DC, pp. 395–414.

Krieman, G., Koch, C. and Fried, I. (2000). Category-specific visual responses of single neurons in the human medial temporal lobe, *Nature Neuroscience* **3**: 946–953.

Kringelbach, M. L., Araujo, I. and Rolls, E. T. (2001). Face expression as a reinforcer activates the orbitofrontal cortex in an emotion-related reversal task, *NeuroImage* **13**: S433.

Kubie, J. L. and Muller, R. U. (1991). Multiple representations in the hippocampus, *Hippocampus* **1**: 240–242.

Kuffler, S. (1953). Discharge patterns and functional organization of the mammalian retina, *Journal of Neurophysiology* **16**: 37–68.

Kuhn, R. (1990). Statistical mechanics of neural networks near saturation, *in* L. Garrido (ed.), *Statistical Mechanics of Neural Networks*, Springer-Verlag, Berlin.

Kuhn, R., Bos, S. and van Hemmen, J. L. (1991). Statistical mechanics for networks of graded response neurons, *Physical Review A* **243**: 2084–2087.

Kulikowski, J. and Bishop, P. O. (1981). Fourier analysis and spatial representation in the visual cortex, *Experientia* **37**: 160–163.

Lamme, V. A. F. (1995). The neurophysiology of figure-ground segregation in primary visual cortex, *Journal of Neuroscience* **15**: 1605–1615.

Land, M. F. (1999). Motion and vision: why animals move their eyes, *Journal of Comparative Physiology A* **185**: 341–352.

Land, M. F. and Collett, T. S. (1997). A survey of active vision in invertebrates, *in* M. V. Srinivasan and S. Venkatesh (eds), *From Living Eyes to Seeing Machines*, Oxford University Press, Oxford.

Lanthorn, T., Storn, J. and Andersen, P. (1984). Current-to-frequency transduction in CA1 hippocampal pyramidal cells: slow prepotentials dominate the primary range firing, *Experimental Brain Research* **53**: 431–443.

Lassalle, J. M., Bataille, T. and Halley, H. (2000). Reversible inactivation of the hippocampal mossy fiber synapses in mice impairs spatial learning, but neither consolidation nor memory retrieval, in the Morris navigation task, *Neurobiology of Learning and Memory* **73**: 243–257.

Lavie, N. and Driver, J. (1996). On the spatial extent of attention in object-based visual selection, *Perception and Psychophysics* **58**: 1238–1251.

LeDoux, J. E. (1994). Emotion, memory and the brain, *Scientific American* **220 (June)**: 50–57.

LeDoux, J. E., Iwata, J., Cicchetti, J. P. and Reis, D. J. (1988). Different projections of the central amygdaloid nucleus mediate autonomic and behavioral correlates of conditioned fear, *Journal of Neuroscience* **8**: 2517–2529.

Lee, T. S. (1996). Image representation using 2D Gabor wavelets, *IEEE Transactions on Pattern Analysis and Machine Intelligence* **18**: 959–971.

Lee, T. S. and Nguyen, M. (2001). Dynamics of subjective contour formation in the early visual cortex, *Proceedings of the National Academy of Science* **98**: 1907–1911.

Lee, T. S., Mumford, D., Romero, R. and Lamme, V. A. F. (1998). The role of primary visual cortex in higher level vision, *Vision Research* **38**: 2429–2454.

Lee, T. S., Romero, R. D. and Mumford, D. (2000). Neural correlate of asymmetric visual search behaviors in the primary visual cortex.

Leonard, C. M., Rolls, E. T., Wilson, F. A. W. and Baylis, G. C. (1985). Neurons in the amygdala of the monkey with responses selective for faces, *Behavioural Brain Research*

15: 159–176.

LeVay, S., Hubel, D. H. and Wiesel, T. N. (1975). The pattern of ocular dominance columns in macaque visual cortex revealed by a reduced silver stain, *Journal of Comparative Neurology* **159**: 559–575.

Levitt, J. B., Lund, J. S. and Yoshioka, T. (1996). Anatomical substrates for early stages in cortical processing of visual information in the macaque monkey, *Behavioural Brain Research* **76**: 5–19.

Levy, W. B. (1985). Associative changes in the synapse: LTP in the hippocampus, *in* W. B. Levy, J. A. Anderson and S. Lehmkuhle (eds), *Synaptic Modification, Neuron Selectivity, and Nervous System Organization*, Erlbaum, Hillsdale, NJ, chapter 1, pp. 5–33.

Levy, W. B. and Baxter, R. A. (1996). Energy efficient neural codes, *Neural Computation* **8**: 531–543.

Levy, W. B. and Desmond, N. L. (1985). The rules of elemental synaptic plasticity, *in* W. B. Levy, J. A. Anderson and S. Lehmkuhle (eds), *Synaptic Modification, Neuron Selectivity, and Nervous System Organization*, Erlbaum, Hillsdale, NJ, chapter 6, pp. 105–121.

Levy, W. B., Wu, X. and Baxter, R. A. (1995). Unification of hippocampal function via computational/encoding considerations, *International Journal of Neural Systems* **6, Suppl.**: 71–80.

Lewicki, M. and Sejnowski, T. J. (2000). Learning overcomplete representations, *Neural Computation* **12**: 337–365.

Li, Z. (2000). Pre-attentive segmentation in the primary visual cortex, *Spatial Vision* **13**: 25–50.

Linsker, E. (1986). From basic network principles to neural architecture, *Proceedings of the National Academy of Science of the USA* **83**: 7508–7512, 8390–8394, 8779–8783.

Linsker, E. (1988). Self-organization in a perceptual network, *Computer* **March**: 105–117.

Linsker, R. (1992). Local synaptic learning rule suffice to maximize mutual information in a linear network, *Neural Computation* **4**: 691–702.

Lisman, J. E., Fellous, J. M. and Wang, X. J. (1998). A role for NMDA-receptor channels in working memory, *Nature Neuroscience* **1**: 273–275.

Lissauer, H. (1890). Ein Fall von Seelenblindt nebst einem Beitrage zue Theorie derselben, *Archiv für Psychiatrie und Nervenkrankheiten* **22**: 222–270.

Little, W. A. (1974). The existence of persistent states in the brain, *Mathematical Bioscience* **19**: 101–120.

Livingstone, M. and Hubel, D. (1988). Segregation of form, colour, movement, and depth: Anatomy, physiology, and perception, *Science* **240**: 740–749.

Livingstone, M. S. and Hubel, D. H. (1984). Specificity of intrinsic connections in primate primary visual cortex, *Journal of Neuroscience* **4**: 2830.

Logothetis, N. K. and Sheinberg, D. L. (1996). Visual object recognition, *Annual Review of Neuroscience* **19**: 577–621.

Logothetis, N. K., Pauls, J., Bulthoff, H. H. and Poggio, T. (1994). View-dependent object recognition by monkeys, *Current Biology* **4**: 401–414.

Logothetis, N. K., Pauls, J. and Poggio, T. (1995). Shape representation in the inferior temporal cortex of monkeys, *Current Biology* **5**: 552–563.

Lovegrove, W., Lehmkuhle, S., Baro, J. and Garzia, R. (1991). The effect of uniform field flicker and blurring on the global precedence effect, *Bulletin of the Psychonomic Society* **29**: 289–291.

Lowe, D. (1985). *Perceptual Organization and Visual Recognition*, Kluwer, Boston.

Luck, S. J., Chelazzi, L., Hillyard, S. A. and Desimone, R. (1997). Neural mechanisms of spatial selective attention in areas V1, V2, and V4 of macaque visual cortex, *Journal of Neurophysiology* **77**: 24–42.

Lund, J. S. (1984). Spiny stellate neurons, *in* A. Peters and E. Jones (eds), *Cerebral Cortex, Vol. 1, Cellular Components of the Cerebral Cortex*, Plenum, New York, chapter 7, pp. 255–308.

MacGregor, R. J. (1987). *Neural and Brain Modelling*, Academic Press, San Diego.

MacKay, D. J. C. and Miller, K. D. (1990). Analysis of Linsker's simulation of Hebbian rules, *Neural Computation* **2**: 173–187.

MacKay, D. M. and McCulloch, W. S. (1952). The limiting information capacity of a neuronal link, *Bulletin of Mathematical Biophysics* **14**: 127–135.

Mackintosh, N. J. (1983). *Conditioning and Associative Learning*, Oxford University Press, Oxford.

Mallot, H. A. (2000). *Computational Vision*, MIT Press, Cambridge, Mass.

Malsburg, C. v. d. (1973). Self-organization of orientation-sensitive columns in the striate cortex, *Kybernetik* **14**: 85–100.

Malsburg, C. v. d. (1990). A neural architecture for the representation of scenes, *in* J. L. McGaugh, N. M. Weinberger and G. Lynch (eds), *Brain Organization and Memory: Cells, Systems and Circuits*, Oxford University Press, New York, chapter 19, pp. 356–372.

Malsburg, C. v. d. (1999). The what and why of binding: the modeler's perspective, *Neuron* **24**: 95–104.

Malsburg, C. v. d. and Bienenstock, E. (1986). Statistical coding and short-term synaptic plasticity: a scheme for knowledge representation in the brain, *in* E. Bienenstock, F. Fogelman-Soulie and G. Weisbuch (eds), *Disordered Systems and Biological Organization, NATO ASI series Vol. F20*, Springer, Berlin, pp. 247–272.

Malsburg, C. v. d. and Schneider, W. (1986). A neural cocktail-party processor, *Biological Cybernetics* **54**: 29–40.

Marcelja, S. (1980). Mathematical description of the responses of simple cortical cells, *Journal of the Optical Society of America* **70**: 1297–1300.

Markram, H. and Tsodyks, M. (1996). Redistribution of synaptic efficacy between neocortical pyramidal neurons, *Nature* **382**: 807–810.

Markram, H., Lübke, J., Frotscher, M. and Sakmann, B. (1997). Regulation of synaptic efficacy by coincidence of postsynaptic APs and EPSPs, *Science* **275**: 213–215.

Markram, H., Pikus, D., Gupta, A. and Tsodyks, M. (1998). Information processing with frequency-dependent synaptic connections, *Neuropharmacology* **37**: 489–500.

Markus, E. J., Qin, Y. L., Leonard, B., Skaggs, W., McNaughton, B. L. and Barnes, C. A. (1995). Interactions between location and task affect the spatial and directional firing of hippocampal neurons, *Journal of Neuroscience* **15**: 7079–7094.

Marr, D. (1969). A theory of cerebellar cortex, *Journal of Physiology* **202**: 437–470.

Marr, D. (1970). A theory for cerebral cortex, *Proceedings of The Royal Society of London, Series B* **176**: 161–234.

Marr, D. (1971). Simple memory: a theory for archicortex, *Philosophical Transactions of The Royal Society of London, Series B* **262**: 23–81.

Marr, D. (1982). *Vision*, Freeman, San Francisco.

Marr, D. and Nishihara, H. K. (1978). Representation and recognition of the spatial organization of three dimensional structure, *The Proceedings of the Royal Society, London [B]* **200**: 269–294.

Martin, K. A. C. (1984). Neuronal circuits in cat striate cortex, *in* E. Jones and A. Peters (eds), *Cerebral Cortex, Vol. 2, Functional Properties of Cortical Cells*, Plenum, New York, chapter 9, pp. 241–284.

Martin, S. J., Grimwood, P. D. and Morris, R. G. (2000). Synaptic plasticity and memory: an evaluation of the hypothesis, *Annual Review of Neuroscience* **23**: 649–711.

Mason, A. and Larkman, A. (1990). Correlations between morphology and electrophysiology of pyramidal neurones in slices of rat visual cortex. I. Electrophysiology, *Journal of Neuroscience* **10**: 1415–1428.

Matelli, M. and Luppino, G. (2000). Parietofrontal circuits for action and space perception in the macaque monkey, *NeuroImage* **14**: S27–S32.

Maunsell, J. H. R. (1995). The brain's visual world: representation of visual targets in cerebral cortex, *Science* **270**: 764–769.

Maunsell, J. H. R. and Newsome, W. T. (1987). Visual processing in monkey extrastriate cortex, *Annual Review of Neuroscience* **10**: 363–401.

McAdams, C. and Maunsell, J. (1999). Effects of attention on orientation-tuning functions of single neurons in macaque cortical area V4, *The Journal of Neuroscience* **19**: 431–441.

McClelland, J. L. and Rumelhart, D. E. (1981). An interactive activation model of context effects in letter perception. Part I: an account of basic findings, *Psychological Review* **88**: 375–407.

McClelland, J. L. and Rumelhart, D. E. (1986). A distributed model of human learning and memory, *in* J. L. McClelland and D. E. Rumelhart (eds), *Parallel Distributed Processing*, Vol. 2, MIT Press, Cambridge, Mass., chapter 17, pp. 170–215.

McClelland, J. L. and Rumelhart, D. E. (1988). *Explorations in Parallel Distributed Processing*, MIT Press, Cambridge, Mass.

McClelland, J. L., McNaughton, B. L. and O'Reilly, R. C. (1995). Why there are complementary learning systems in the hippocampus and neocortex: insights from the successes and failures of connectionist models of learning and memory, *Psychological Review* **102**: 419–457.

McDonald, A. J. and Aggleton, J. P. (1992). Cell types and intrinsic connections of the amygdala, *in* J. P. Aggleton (ed.), *The Amygdala*, Wiley-Liss, New York, chapter 2, pp. 67–96.

McGurk, H. and MacDonald, J. (1976). Hearing lips and seeing voices, *Nature* **264**: 746–748.

McKeefry, D. J. and Zeki, S. M. (1997). The position and topography of the human color center as revealed by functional magnetic resonance imaging, *Brain* **120**: 2229–2242.

McLeod, P., Driver, J. and Crisp, J. (1988). Visual search for a conjunction of movement and form is parallel, *Nature* **332**: 154–155.

McLeod, P., Plunkett, K. and Rolls, E. T. (1998). *Introduction to Connectionist Modelling of Cognitive Processes*, Oxford University Press, Oxford.

McNaughton, B. L., Barnes, C. A. and O'Keefe, J. (1983). The contributions of position, direction, and velocity to single unit activity in the hippocampus of freely-moving rats., *Experimental Brain Research* **52**: 41–49.

Medin, D. L. and Schaffer, M. M. (1978). Context theory of classification learning, *Psychological Review* **85**: 207–238.

Mel, B. W. (1997). Seemore: Combining color, shape, and texture histogramming in a neurally-inspired approach to visual object recognition, *Neural Computation* **9**: 777–804.

Mel, B. W. and Fiser, J. (2000). Minimizing binding errors using learned conjunctive features, *Neural Computation* **12**: 731–762.

Mel, B. W., Ruderman, D. L. and Archie, K. A. (1998). Translation-invariant orientation tuning in visual "complex" cells could derive from intradendritic computations, *Journal of Neuroscience* **18**(11): 4325–4334.

Mesulam, M. M. and Mufson, E. J. (1982). Insula of the old world monkey. III. Efferent cortical output and comments on function, *Journal of Comparative Neurology* **212**: 38–52.

Michimata, C., Okubo, M. and Mugishima, Y. (1999). Effects of background color on the global and local processing of hierarchically organized stimuli, *Journal of Cognitive Neuroscience* **11**: 1–8.

Mikami, A., Nakamura, K. and Kubota, K. (1994). Neuronal responses to photographs in the superior temporal sulcus of the rhesus monkey, *Behavioural Brain Research* **60**: 1–13.

Millenson, J. R. (1967). *Principles of Behavioral Analysis*, MacMillan, New York.

Miller, E. K. and Desimone, R. (1994). Parallel neuronal mechanisms for short-term memory, *Science* **263**: 520–522.

Miller, E. K., Gochin, P. and Gross, C. (1993a). Suppression of visual responses of neurons in inferior temporal cortex of the awake macaque by addition of a second stimulus, *Brain Research* **616**: 25–29.

Miller, E. K., Li, L. and Desimone, R. (1993b). Activity of neurons in anterior inferior temporal cortex during a short-term memory task, *Journal of Neuroscience* **13**: 1460–1478.

Miller, E. K., Erickson, C. and Desimone, R. (1996). Neural mechanism of visual working memory in prefrontal cortex of the macaque, *Journal of Neuroscience* **16**: 5154–5167.

Miller, G. A. (1955). Note on the bias of information estimates, *Information Theory in Psychology; Problems and Methods II-B* pp. 95–100.

Miller, K. D. (1994). Models of activity-dependent neural development, *Progress in Brain Research* **102**: 303–308.

Millhouse, O. E. (1986). The intercalated cells of the amygdala, *Journal of Comparative Neurology* **247**: 246–271.

Millhouse, O. E. and DeOlmos, J. (1983). Neuronal configuration in lateral and basolateral amygdala, *Neuroscience* **10**: 1269–1300.

Milner, A. D. and Goodale, M. A. (1995). *The Visual Brain in Action*, Oxford University Press, Oxford.

Milner, P. (1974). A model for visual shape recognition, *Psychological Review* **81**: 521–535.

Minai, A. A. and Levy, W. B. (1993). Sequence learning in a single trial, *International Neural Network Society World Congress of Neural Networks* **2**: 505–508.

Minsky, M. L. and Papert, S. A. (1969). *Perceptrons*, expanded 1988 edn, MIT Press, Cambridge, MA.

Miyashita, A. (2000). Visual associative long-term memory: encoding and retrieval in inferotemporal cortex of the primate, *in* M. Gazzaniga (ed.), *The New Cognitive Neurosciences*, 2nd edn, MIT Press, Cambridge, MA, chapter 27, pp. 379–392.

Miyashita, Y. (1988). Neuronal correlate of visual associative long–term memory in the primate temporal cortex, *Nature* **335**: 817–820.

Miyashita, Y. (1993). Inferior temporal cortex: where visual perception meets memory, *Annual Review of Neuroscience* **16**: 245–263.

Miyashita, Y. and Chang, H. S. (1988). Neuronal correlate of pictorial short-term memory in the primate temporal cortex, *Nature* **331**: 68–70.

Miyashita, Y., Rolls, E. T., Cahusac, P. M. B., Niki, H. and Feigenbaum, J. D. (1989). Activity of hippocampal neurons in the monkey related to a conditional spatial response task, *Journal of Neurophysiology* **61**: 669–678.

Montague, R., Gally, J. and Edelman, G. (1991). Spatial signalling in the development and function of neural connections, *Cerebral Cortex* **1**: 199–220.

Moran, J. and Desimone, R. (1985). Selective attention gates visual processing in the extrastriate cortex, *Science* **229**: 782–784.

Morris, J. S., Fritch, C. D., Perrett, D. I., Rowland, D., Young, A. W., Calder, A. J. and Dolan, R. J. (1996). A differential neural response in the human amygdala to fearful and happy face expressions, *Nature* **383**: 812–815.

Morris, R. G. M. (1989). Does synaptic plasticity play a role in information storage in the vertebrate brain?, *in* R. G. M. Morris (ed.), *Parallel Distributed Processing: Implications for Psychology and Neurobiology*, Oxford University Press, Oxford, chapter 11, pp. 248–285.

Morrow, L. and Ratcliff, G. (1988). The disengagement of covert attention and the neglect syndrome, *Psychobiology* **16**: 261–269.

Motter, B. C. (1993). Focal attention produces spatially selective processing in visual cortical areas V1, V2, and V4 in the presence of competing stimuli, *Journal of Neurophysiology* **70**: 909–919.

Motter, B. C. (1994). Neural correlates of attentive selection for colours or luminance in extrastriate area V4, *Journal of Neuroscience* **14**: 2178–2189.

Motter, B. C. and Mountcastle, V. B. (1981). The functional properties of the light-sensitive neurons of the posterior parietal cortex studied in waking monkeys: foveal sparing and opponent vector organization, *Journal of Neuroscience* **1**: 3–26.

Mountcastle, V. B. (1957). Modality and topographic properties of single neurons of cat's somatosensory cortex, *Journal of Neurophysiology* **20**: 408–434.

Mountcastle, V. B. (1984). Central nervous mechanisms in mechanoreceptive sensibility, *in* I. Darian-Smith (ed.), *Handbook of Physiology, Section 1: The Nervous System, Vol III, Sensory Processes, Part 2*, American Physiological Society, Bethesda, MD, pp. 789–878.

Movshon, J. A., Adelson, E. H., Gizzi, M. S. and Newsome, W. T. (1985). The analysis of moving visual patterns, *in* C. Chagas, R. Gattas and C. Gross (eds), *Pattern recognition mechanisms*, Springer, New York, pp. 117–151.

Mozer, M. (1991). *The Perception of Multiple Objects: a Connectionist Approach*, MIT Press, Cambridge, MA.

Mozer, M. and Behrmann, M. (1990). On the interaction of selective attention and lexical knowledge: A connectionist account of neglect dyslexia, *Journal of Cognitive Neuroscience* **2**: 96–123.

Mozer, M. and Sitton, M. (1998). Computational modeling of spatial attention, *in* H. Pashler (ed.), *Attention*, Psychology Press, pp. 341–393.

Muller, R. U., Kubie, J. L., Bostock, E. M., Taube, J. S. and Quirk, G. J. (1991). Spatial firing correlates of neurons in the hippocampal formation of freely moving rats, *in* J. Paillard (ed.), *Brain and Space*, Oxford University Press, Oxford, pp. 296–333.

Muller, R. U., Ranck, J. B. and Taube, J. S. (1996). Head direction cells: properties and functional significance, *Current Opinion in Neurobiology* **6**: 196–206.

Mumford, D. (1991). On the computational architecture of the neocortex. I. The role of the thalamo-cortical loop, *Biological Cybernetics* **65**: 135–145.

Mumford, D. (1992). On the computational architecture of the neocortex. II. The role of the cortico-cortical loop, *Biological Cybernetics* **66**: 241–251.

Mundy, J. and Zisserman, A. (1992). Introduction–towards a new framework for vision, *in* J. Mundy and A. Zisserman (eds), *Geometric Invariance in Computer Vision*, MIT Press, Cambridge, MA, pp. 1–39.

Munsell (1976). *Book of Color*, Gregtag Macbeth, Baltimore.

Murata, A., Gallese, V., Luppino, G., Kaseda, M. and Sakata, H. (2000). Selectivity for the shape, size, and orientation of objects for grasping in neurons of monkey parietal area AIP, *Journal of Neurophysiology* **83**: 2580–2601.

Nakayama, K. and Silverman, G. H. (1986). Serial and parallel processing of visual feature conjunctions, *Nature* **320**: 264–265.

Navon, D. (1977). Forest before trees: the precedence of global features in visual perception, *Cognitive Psychology* **9**: 353–383.

Neisser, U. (1967). *Cognitive Psychology*, Appleton-Century-Crofts, New York.

Nelken, I., Prut, Y., Vaadia, E. and Abeles, M. (1994). Population responses to multifrequency sounds in the cat auditory cortex: one- and two-parameter families of sounds, *Hearing Research* **72**: 206–222.

Newsome, W. T., Britten, K. H. and Movshon, J. A. (1989). Neuronal correlates of a perceptual decision, *Nature* **341**: 52–54.

Nicoll, R. A. and Malenka, R. C. (1995). Contrasting properties of two forms of long-term potentiation in the hippocampus, *Nature* **377**: 115–118.

Niebur, E. and Koch, C. (1994). A model for the neural implementation of selective visual attention based on temporal correlation among neurons, *Journal of Computational Neuroscience* **1**: 141–158.

Nishijo, H., Ono, T. and Nishino, H. (1988). Single neuron responses in amygdala of alert monkey during complex sensory stimulation with affective significance, *Journal of Neuroscience* **8**: 3570–3583.

Nowak, L. and Bullier, J. (1997). The timing of information transfer in the visual system, *in* K. Rockland, J. Kaas and A. Peters (eds), *Cerebral Cortex: Extrastriate Cortex in Primate*, Plenum, New York, p. 870.

O'Doherty, J., Kringelbach, M. L., Rolls, E. T., Hornak, J. and Andrews, C. (2001). Abstract reward and punishment representations in the human orbitofrontal cortex, *Nature Neuroscience* **4**: 95–102.

Oja, E. (1982). A simplified neuron model as a principal component analyzer, *Journal of Mathematical Biology* **15**: 267–273.

O'Kane, D. and Treves, A. (1992). Why the simplest notion of neocortex as an autoassociative memory would not work, *Network* **3**: 379–384.

O'Keefe, J. (1979). A review of the hippocampal place cells, *Progress in Neurobiology* **13**: 419–439.

O'Keefe, J. (1984). Spatial memory within and without the hippocampal system, *in* W. Seifert (ed.), *Neurobiology of the Hippocampus*, Academic Press, London, pp. 375–403.

O'Keefe, J. (1990). A computational theory of the cognitive map, *Progress in Brain Research* **83**: 301–312.

O'Keefe, J. (1991). The hippocampal cognitive map and navigational strategies, *in* J. Paillard (ed.), *Brain and Space*, Oxford University Press, Oxford, chapter 16, pp. 273–295.

O'Keefe, J. and Dostrovsky, J. (1971). The hippocampus as a spatial map: preliminary evidence from unit activity in the freely moving rat, *Brain Research* **34**: 171–175.

O'Keefe, J. and Nadel, L. (1978). *The Hippocampus as a Cognitive Map*, Clarendon Press, Oxford.

O'Keefe, J., Burgess, N., Donnett, J. G., Jeffery, K. J. and Maguire, E. A. (1998). Place cells, navigational accuracy, and the human hippocampus, *Philosophical Transactions of the Royal Society, London [B]* **353**: 1333–1340.

Olshausen, B. A. and Field, D. J. (1996). Emergence of simple-cell receptive field properties by learning a sparse code for natural images, *Nature* **381**: 607–609.

Olshausen, B. A., Anderson, C. H. and Van Essen, D. C. (1993). A neurobiological model of visual attention and invariant pattern recognition based on dynamic routing of information, *Journal of Neuroscience* **13**: 4700–4719.

Olshausen, B. A., Anderson, C. H. and Van Essen, D. C. (1995). A multiscale dynamic routing circuit for forming size- and position-invariant object representations, *Journal of Computational Neuroscience* **2**: 45–62.

Olson, C. and Gettner, S. (1995). Object-centred direction selectively in the macaque supplementary eye field, *Science* **269**: 985–988.

O'Mara, S. M., Rolls, E. T., Berthoz, A. and Kesner, R. P. (1994). Neurons responding to whole-body motion in the primate hippocampus, *Journal of Neuroscience* **14**: 6511–6523.

Ongur, D. and Price, J. L. (2000). The organization of networks within the orbital and medial prefrontal cortex of rats, monkeys and humans, *Cerebral Cortex* **10**: 206–219.

Optican, L. M. and Richmond, B. J. (1987). Temporal encoding of two-dimensional patterns by single units in primate inferior temporal cortex: III. Information theoretic analysis, *Journal of Neurophysiology* **57**: 162–178.

Optican, L. M., Gawne, T. J., Richmond, B. J. and Joseph, P. J. (1991). Unbiased measures of transmitted information and channel capacity from multivariate neuronal data, *Biological Cybernetics* **65**: 305–310.

Oram, M. W. and Perrett, D. I. (1994). Modeling visual recognition from neurophysiological constraints, *Neural Networks* **7**: 945–972.

O'Regan, J. K., Rensink, R. A. and Clark, J. J. (1999). Change-blindness as a result of 'mudsplashes', *Nature* **398**: 1736–1753.

O'Reilly, J. and Munakata, Y. (2000). *Computational Explorations in Cognitive Neuroscience*, MIT Press, Cambridge, Mass.

O'Reilly, R. and Johnson, M. (1994). Object recognition and sensitive periods: A computational analysis of visual imprinting, *Neural Computation* **6**: 357–389.

O'Sclaidhe, S. P., Wilson, F. A. W. and Goldman-Rakic, P. S. (1999). Face-selective neurons during passive viewing and working memory performance of rhesus monkeys: evidence for intrinsic specialization of neuronal coding, *Cerebral Cortex* **9**: 459–475.

Palmer, S. E. (1999). *Vision Science. Photons to Phenomenology*, MIT Press, Cambridge, Mass.

Panzeri, S. and Treves, A. (1996). Analytical estimates of limited sampling biases in different information measures, *Network* **7**: 87–107.

Panzeri, S., Biella, G., Rolls, E. T., Skaggs, W. E. and Treves, A. (1996). Speed, noise, information and the graded nature of neuronal responses, *Network* **7**: 365–370.

Panzeri, S., Schultz, S. R., Treves, A. and Rolls, E. T. (1999a). Correlations and the encoding of information in the nervous system, *Proceedings of the Royal Society B* **266**: 1001–1012.

Panzeri, S., Treves, A., Schultz, S. and Rolls, E. T. (1999b). On decoding the responses of a population of neurons from short time epochs, *Neural Computation* **11**: 1553–1577.

Panzeri, S., Rolls, E. T., Battaglia, F. and Lavis, R. (2001). Speed of information retrieval in multilayer networks of integrate-and-fire neurons, *Network: Computation in Neural Systems* **12**: 423–440.

Parga, N. and Rolls, E. T. (1998). Transform invariant recognition by association in a recurrent network, *Neural Computation* **10**: 1507–1525.

Parker, A. J., Cumming, B. G. and Dodd, J. V. (2000). Binocular neurons and the perception of depth, *in* M. Gazzaniga (ed.), *The New Cognitive Neurosciences, Second Edition*, MIT Press, Cambridge, MA, chapter 18, pp. 263–277.

Parker, J. R. (1997). *Algorithms for Image Processing and Computer Vision*, John Wiley & Sons, New York.

Parkinson, J. K., Murray, E. A. and Mishkin, M. (1988). A selective mnemonic role for the hippocampus in monkeys: memory for the location of objects, *Journal of Neuroscience* **8**: 4059–4167.

Pashler, H. (1996). *The Psychology of Attention*, MIT Press, Cambridge, MA.

Passingham, R. E. P., Toni, I. and Rushworth, M. F. S. (2000). Specialisation within the prefrontal cortex: the ventral prefrontal cortex and associative learning, *Experimental Brain Research* **133**: 103–113.

Pasupathy, A. and Connor, C. E. (1999). Responses to contour features in macaque area V4, *Journal of Neurophysiology* **82**: 2490–2502.

Peng, H. C., Sha, L. F., Gan, Q. and Wei, Y. (1998). Energy function for learning invariance in multilayer perceptron, *Electronics Letters* **34**(3): 292–294.

Perrett, D. and Oram, M. (1993). Neurophysiology of shape processing, *Image and Vision Computing* **11**(6): 317–333.

Perrett, D. I., Rolls, E. T. and Caan, W. (1982). Visual neurons responsive to faces in the monkey temporal cortex, *Experimental Brain Research* **47**: 329–342.

Perrett, D. I., Smith, P. A. J., Mistlin, A. J., Chitty, A. J., Head, A. S., Potter, D. D., Broennimann, R., Milner, A. D. and Jeeves, M. A. (1985a). Visual analysis of body movements by neurons in the temporal cortex of the macaque monkey: a preliminary report, *Behavioural Brain Research* **16**: 153–170.

Perrett, D. I., Smith, P. A. J., Potter, D. D., Mistlin, A. J., Head, A. S., Milner, D. and Jeeves, M. A. (1985b). Visual cells in temporal cortex sensitive to face view and gaze direction, *Proceedings of the Royal Society of London, Series B* **223**: 293–317.

Personnaz, L., Guyon, I. and Dreyfus, G. (1985). Information storage and retrieval in spin-glass-like neural networks, *Journal de Physique Lettres (Paris)* **46**: 359–365.

Peters, A. (1984a). Bipolar cells, *in* A. Peters and E. G. Jones (eds), *Cerebral Cortex, Vol. 1, Cellular Components of the Cerebral Cortex*, Plenum, New York, chapter 11, pp. 381–407.

Peters, A. (1984b). Chandelier cells, *in* A. Peters and E. G. Jones (eds), *Cerebral Cortex, Vol. 1, Cellular Components of the Cerebral Cortex*, Plenum, New York, chapter 10, pp. 361–380.

Peters, A. and Jones, E. G. (eds) (1984). *Cerebral Cortex, Vol. 1, Cellular Components of the Cerebral Cortex*, Plenum, New York.

Peters, A. and Regidor, J. (1981). A reassessment of the forms of nonpyramidal neurons in area 17 of the cat visual cortex, *Journal of Comparative Neurology* **203**: 685–716.

Peters, A. and Saint Marie, R. L. (1984). Smooth and sparsely spinous nonpyramidal cells forming local axonal plexuses, *in* A. Peters and E. G. Jones (eds), *Cerebral Cortex, Vol. 1, Cellular Components of the Cerebral Cortex*, New York, Plenum, chapter 13, pp. 419–445.

Peterson, C. and Anderson, J. R. (1987). A mean field theory learning algorithm for neural networks, *Complex Systems* **1**: 995–1015.

Petrides, M. (1985). Deficits on conditional associative-learning tasks after frontal- and temporal-lobe lesions in man, *Neuropsychologia* **23**: 601–614.

Phaf, H., Van der Heijden, A. and Hudson, P. (1990). A connectionist model for attention in visual selection tasks, *Cognitive Psychology* **22**: 273–341.

Phillips, W. A., Kay, J. and Smyth, D. (1995). The discovery of structure by multi-stream networks of local processors with contextual guidance, *Network* **6**: 225–246.

Poggio, G. F. and Fischer, B. (1977). Binocular interaction and depth sensitivity in striate and prestriate cortex of behaving rhesus monkey, *Journal of Neurophysiology* **40**: 1392–1405.

Poggio, T. and Edelman, S. (1990). A network that learns to recognize three-dimensional objects, *Nature* **343**: 263–266.

Poggio, T. and Girosi, F. (1990a). Networks for approximation and learning, *Proceedings of the IEEE* **78**: 1481–1497.

Poggio, T. and Girosi, F. (1990b). Regularization algorithms for learning that are equivalent to multilayer networks, *Science* **247**: 978–982.

Pollatos, O. (2000). Kontextabhängige visuelle Suchprozesse, *Diplomarbeit*, Ludwig-Maximilian-University, Munich.

Pollen, D. and Ronner, S. (1981). Phase relationship between adjacent simple cells in the visual cortex, *Science* **212**: 1409–1411.

Posner, M. and Dehaene, S. (1994). Attentional networks, *Trends in Neurosciences* **17**: 75–79.

Posner, M. and Snyder, C. (1975). Attention and cognitive controls, *in* R. Solso (ed.), *Information Processing and Cognition: The Loyola Symposium*, Lawrence Erlbaum Associates, Hillsdale, NJ, pp. 55–85.

Posner, M., Walker, J., Friedrich, F. and Rafal, B. (1984). Effects of parietal injury on covert orienting of attention, *Journal of Neuroscience* **4**: 1863–1874.

Posner, M., Walker, J., Friedrich, F. and Rafal, R. (1987). How do the parietal lobes direct covert attention?, *Neuropsychologia* **25**: 135–146.

Posner, M. I. and Keele, S. W. (1968). On the genesis of abstract ideas, *Journal of Experimental Psychology* **77**: 353–363.

Pouget, A. and Driver, J. (1999). Visual neglect, *in* R. Wilson and F. Keil (eds), *MIT Encyclopedia of Cognitive Sciences*, MIT Press, Cambridge.

Pouget, A. and Sejnowski, T. (1997). New view of hemineglect based on the response properties of parietal neurons, *Philosophical Transactions of the Royal Society: Series B* **352**: 1449–1459.

Powell, T. P. S. (1981). Certain aspects of the intrinsic organisation of the cerebral cortex, *in* O. Pompeiano and C. Ajmone Marsan (eds), *Brain Mechanisms and Perceptual*

Awareness, Raven Press, New York, pp. 1–19.

Prinzmetal, W. (1981). Principle of feature integration in visual perception, *Perception and Psychophysics* **30**: 330–340.

Pylyshyn, Z. W. and Storm, R. W. (1988). Tracking multiple independent targets: Evidence for a parallel tracking mechanism, *Spatial Vision* **3**: 1–19.

Quinlan, P. T. and Humphreys, G. W. (1987). Visual search for targets defined by combination of color, shape, and size: An examination of the task constraints on feature and conjunction searches, *Perception and Psychophysics* **41**: 455–472.

Rafal, R. and Robertson, L. (1997). The neurology of visual attention, *in* M. Gazzaniga (ed.), *The Cognitive Neuroscience*, MIT Press, Cambridge.

Rall, W. and Segev, I. (1987). Functional possibilities for synapses on dendrites and dendritic spines, *in* G. M. Edelman, E. E. Gall and W. M. Cowan (eds), *Synaptic Function*, Wiley, New York, pp. 605–636.

Ranck, Jr., J. B. (1985). Head direction cells in the deep cell layer of dorsolateral presubiculum in freely moving rats, *in* G. Buzsáki and C. H. Vanderwolf (eds), *Electrical Activity of the Archicortex*, Akadémiai Kiadó, Budapest.

Rao, R. and Ballard, D. H. (1999). Predictive coding in the visual cortex: A functional interpretation of some extra-classical receptive-field effects, *Nature Neuroscience* **2**: 79–87.

Rao, S., Rainer, G. and Miller, E. (1997). Integration of what and where in the primate prefrontal cortex, *Science* **276**: 821 824.

Redlich, A. (1993). Redundancy reduction as a strategy for unsupervised learning, *Neural Computation* **5**: 289–304.

Renart, A., Parga, N. and Rolls, E. T. (1999a). Associative memory properties of multiple cortical modules, *Network* **10**: 237–255.

Renart, A., Parga, N. and Rolls, E. T. (1999b). Backprojections in the cerebral cortex: implications for memory storage, *Neural Computation* **11**: 1349–1388.

Renart, A., Parga, N. and Rolls, E. T. (2000). A recurrent model of the interaction between the prefrontal cortex and inferior temporal cortex in delay memory tasks, *in* S. Solla, T. Leen and K.-R. Mueller (eds), *Advances in Neural Information Processing Systems*, Vol. 12, MIT Press, Cambridge Mass, pp. 171–177.

Renart, A., Moreno, R., de al Rocha, J., Parga, N. and Rolls, E. T. (2001). A model of the IT–PF network in object working memory which includes balanced persistent activity and tuned inhibition, *Neurocomputing* **38–40**: 1525–1531.

Rensink, R. A. (2000). Seeing, sensing, and scrutinizing, *Vision Research* **40**: 1469–1487.

Reynolds, J. and Desimone, R. (1999). The role of neural mechanisms of attention in solving the binding problem, *Neuron* **24**: 19–29.

Reynolds, J. H., Chelazzi, L. and Desimone, R. (1999). Competitive mechanisms subserve attention in macaque areas V2 and V4, *Journal of Neuroscience* **19**: 1736–1753.

Rhodes, P. (1992). The open time of the NMDA channel facilitates the self-organisation of invariant object responses in cortex, *Society for Neuroscience Abstracts* **18**: 740.

Richmond, B. J. and Optican, L. (1987). Temporal encoding of two-dimensional patterns by single units in primate inferior temporal cortex. II. Quantification of response waveform, *Journal of Neurophysiology* **57**: 147–161.

Richmond, B. J. and Optican, L. (1990). Temporal encoding of two dimensional patterns by single units in primate primary visual cortex. II. Information transmission, *Journal of*

Neurophysiology **64**: 351–369.

Rieke, F., Warland, D., de Ruyter van Steveninck, R. R. and Bialek, W. (1996). *Spikes: Exploring the Neural Code*, MIT Press, Cambridge, Mass.

Rieke, F. W. D. and Bialek, W. (1993). Coding efficiency and information rates in sensory neurons, *Europhysics Letters* **22**: 151–156.

Riesenhuber, M. and Poggio, T. (1998). Just one view: Invariances in inferotemporal cell tuning, *in* M. I. Jordan, M. J. Kearns and S. A. Solla (eds), *Advances in Neural Information Processing Systems*, Vol. 10, MIT Press, Cambridge, Massachusetts, pp. 215–221.

Riesenhuber, M. and Poggio, T. (1999a). Are cortical models really bound by the "binding problem"?, *Neuron* **24**: 87–93.

Riesenhuber, M. and Poggio, T. (1999b). Hierarchical models of object recognition in cortex, *Nature Neuroscience* **2**: 1019–1025.

Riesenhuber, M. and Poggio, T. (2000). Models of object recognition, *Nature Neuroscience Supplement* **3**: 1199–1204.

Robertson, R. G., Rolls, E. T. and Georges-François, P. (1998). Spatial view cells in the primate hippocampus: Effects of removal of view details, *Journal of Neurophysiology* **79**: 1145–1156.

Robertson, R. G., Rolls, E. T., Georges-François, P. and Panzeri, S. (1999). Head direction cells in the primate pre-subiculum, *Hippocampus* **9**: 206–219.

Robinson, D. L., Goldberg, M. E. and Stanton, G. B. (1978). Parietal association cortex in the primate: sensory mechanisms and behavioral modulations, *Journal of Neurophysiology* **41**: 910–932.

Robinson, D. L., Bowman, E. M. and Kertzman, C. (1991). Covert orienting of attention in macaque. II. A signal in parietal cortex to disengage attention, *Society for Neuroscience Abstracts* **17**: 442.

Roelfsema, P. R., Lamme, V. A. and Spekreijse, H. (1998). Object-based attention in the primary visual cortex of the macaque monkey, *Nature* **395**: 376–381.

Roland, P. E. and Friberg, L. (1985). Localization of cortical areas activated by thinking, *Journal of Neurophysiology* **53**: 1219–1243.

Rolls, E. T. (1974). The neural basis of brain-stimulation reward, *Progress in Neurobiology* **3**: 71–160.

Rolls, E. T. (1975). *The Brain and Reward*, Pergamon Press, Oxford.

Rolls, E. T. (1976). The neurophysiological basis of brain-stimulation reward, *in* A. Wauquier and E. Rolls (eds), *Brain-Stimulation Reward*, North Holland, Amsterdam, pp. 65–87.

Rolls, E. T. (1984). Neurons in the cortex of the temporal lobe and in the amygdala of the monkey with responses selective for faces, *Human Neurobiology* **3**: 209–222.

Rolls, E. T. (1987). Information representation, processing and storage in the brain: analysis at the single neuron level, *in* J.-P. Changeux and M. Konishi (eds), *The Neural and Molecular Bases of Learning*, Wiley, Chichester, pp. 503–540.

Rolls, E. T. (1989a). Functions of neuronal networks in the hippocampus and cerebral cortex in memory, *in* R. Cotterill (ed.), *Models of Brain Function*, Cambridge University Press, Cambridge, pp. 15–33.

Rolls, E. T. (1989b). Functions of neuronal networks in the hippocampus and neocortex in memory, *in* J. Byrne and W. Berry (eds), *Neural Models of Plasticity: Experimental and Theoretical Approaches*, Academic Press, San Diego, chapter 13, pp. 240–265.

Rolls, E. T. (1989c). Information processing and basal ganglia function, *in* C. Kennard and M. Swash (eds), *Hierarchies in Neurology*, Springer-Verlag, London, chapter 15, pp. 123–142.

Rolls, E. T. (1989d). Information processing in the taste system of primates, *Journal of Experimental Biology* **146**: 141–164.

Rolls, E. T. (1989e). Parallel distributed processing in the brain: implications of the functional architecture of neuronal networks in the hippocampus, *in* R. Morris (ed.), *Parallel Distributed Processing: Implications for Psychology and Neurobiology*, Oxford University Press, Oxford, chapter 12, pp. 286–308.

Rolls, E. T. (1989f). The representation and storage of information in neuronal networks in the primate cerebral cortex and hippocampus, *in* R. Durbin, C. Miall and G. Mitchison (eds), *The Computing Neuron*, Addison-Wesley, Wokingham, England, chapter 8, pp. 125–159.

Rolls, E. T. (1990). A theory of emotion, and its application to understanding the neural basis of emotion, *Cognition and Emotion* **4**: 161–190.

Rolls, E. T. (1992a). Neurophysiological mechanisms underlying face processing within and beyond the temporal cortical visual areas, *Philosophical Transactions of the Royal Society* **335**: 11–21.

Rolls, E. T. (1992b). Neurophysiology and functions of the primate amygdala, *in* J. Aggleton (ed.), *The Amygdala*, Wiley-Liss, New York, chapter 5, pp. 143–165.

Rolls, E. T. (1992c). The processing of face information in the primate temporal lobe, *in* V. Bruce and M. Burton (eds), *Processing Images of Faces*, Ablex, Norwood, New Jersey, chapter 3. 41–68.

Rolls, E. T. (1994). Brain mechanisms for invariant visual recognition and learning, *Behavioural Processes* **33**: 113–138.

Rolls, E. T. (1995a). Central taste anatomy and neurophysiology, *in* R. Doty (ed.), *Handbook of Olfaction and Gustation*, Dekker, New York, chapter 24, pp. 549–573.

Rolls, E. T. (1995b). Learning mechanisms in the temporal lobe visual cortex, *Behavioural Brain Research* **66**: 177–185.

Rolls, E. T. (1996a). The orbitofrontal cortex, *Philosophical Transactions of the Royal Society B* **351**: 1433–1444.

Rolls, E. T. (1996b). A theory of hippocampal function in memory, *Hippocampus* **6**: 601–620.

Rolls, E. T. (1997a). Consciousness in neural networks?, *Neural Networks* **10**: 1227–1240.

Rolls, E. T. (1997b). A neurophysiological and computational approach to the functions of the temporal lobe cortical visual areas in invariant object recognition, *in* M. Jenkin and L. Harris (eds), *Computational and Psychophysical Mechanisms of Visual Coding*, Cambridge University Press, Cambridge, chapter 9, pp. 184–220.

Rolls, E. T. (1997c). Taste and olfactory processing in the brain and its relation to the control of eating, *Critical Reviews in Neurobiology* **11**: 263–287.

Rolls, E. T. (1999a). *The Brain and Emotion*, Oxford University Press, Oxford.

Rolls, E. T. (1999b). The functions of the orbitofrontal cortex, *Neurocase* **5**: 301–312.

Rolls, E. T. (1999c). The representation of space in the primate hippocampus, and its role in memory, *in* N. Burgess, K. Jeffrey and J. O'Keefe (eds), *The Hippocampal and Parietal Foundations of Spatial Cognition*, Oxford University Press, Oxford, chapter 17, pp. 320–344.

Rolls, E. T. (1999d). Spatial view cells and the representation of place in the primate hippocampus, *Hippocampus* **9**: 467–480.

Rolls, E. T. (2000a). Functions of the primate temporal lobe cortical visual areas in invariant visual object and face recognition, *Neuron* **27**: 205–218.

Rolls, E. T. (2000b). Hippocampo-cortical and cortico-cortical backprojections, *Hippocampus* **10**: 380–388.

Rolls, E. T. (2000c). Memory systems in the brain, *Annual Review of Psychology* **51**: 599–630.

Rolls, E. T. (2000d). Neurophysiology and functions of the primate amygdala, and the neural basis of emotion, *in* J. Aggleton (ed.), *The Amygdala: Second Edition. A Functional Analysis*, Oxford University Press, Oxford, chapter 13, pp. 447–478.

Rolls, E. T. (2000e). The orbitofrontal cortex and reward, *Cerebral Cortex* **10**: 284–294.

Rolls, E. T. (2000f). Précis of The Brain and Emotion, *Behavioral and Brain Sciences* **23**: 177–233.

Rolls, E. T. (2000g). The representation of umami taste in the taste cortex, *Journal of Nutrition* **130**: S960–S965.

Rolls, E. T. (2001). The rules of formation of the olfactory representations found in the orbitofrontal cortex olfactory areas in primates, *Chemical Senses* **26**: 595–604.

Rolls, E. T. (2002). The functions of the orbitofrontal cortex, *in* D. T. Stuss and R. T. Knight (eds), *The Frontal Lobes*, Oxford University Press, Oxford, chapter 23.

Rolls, E. T. and Baylis, G. C. (1986). Size and contrast have only small effects on the responses to faces of neurons in the cortex of the superior temporal sulcus of the monkey, *Experimental Brain Research* **65**: 38–48.

Rolls, E. T. and Baylis, L. L. (1994). Gustatory, olfactory and visual convergence within the primate orbitofrontal cortex, *Journal of Neuroscience* **14**: 5437–5452.

Rolls, E. T. and Cowey, A. (1970). Topography of the retina and striate cortex and its relationship to visual acuity in rhesus monkeys and squirrel monkeys, *Experimental Brain Research* **10**: 298–310.

Rolls, E. T. and Milward, T. (2000). A model of invariant object recognition in the visual system: learning rules, activation functions, lateral inhibition, and information-based performance measures, *Neural Computation* **12**: 2547–2572.

Rolls, E. T. and Rolls, B. J. (1973). Altered food preferences after lesions in the basolateral region of the amygdala in the rat, *Journal of Comparative and Physiological Psychology* **83**: 248–259.

Rolls, E. T. and Scott, T. R. (2001). Central taste anatomy and neurophysiology, *in* R. Doty (ed.), *Handbook of Olfaction and Gustation, Second Edition*, Dekker, New York, chapter 32, p. in press.

Rolls, E. T. and Stringer, S. M. (2000). On the design of neural networks in the brain by genetic evolution, *Progress in Neurobiology* **61**: 557–579.

Rolls, E. T. and Stringer, S. M. (2001a). Invariant object recognition in the visual system with error correction and temporal difference learning, *Network: Computation in Neural Systems* **12**: 111–129.

Rolls, E. T. and Stringer, S. M. (2001b). A model of the interaction between mood and memory, *Network: Computation in Neural Systems* **12**: 89–109.

Rolls, E. T. and Tovee, M. J. (1994). Processing speed in the cerebral cortex and the neurophysiology of visual masking, *Proceedings of the Royal Society, B* **257**: 9–15.

Rolls, E. T. and Tovee, M. J. (1995a). The responses of single neurons in the temporal visual cortical areas of the macaque when more than one stimulus is present in the visual field, *Experimental Brain Research* **103**: 409–420.

Rolls, E. T. and Tovee, M. J. (1995b). Sparseness of the neuronal representation of stimuli in the primate temporal visual cortex, *Journal of Neurophysiology* **73**: 713–726.

Rolls, E. T. and Treves, A. (1990). The relative advantages of sparse versus distributed encoding for associative neuronal networks in the brain, *Network* **1**: 407–421.

Rolls, E. T. and Treves, A. (1998). *Neural Networks and Brain Function*, Oxford University Press, Oxford.

Rolls, E. T., Judge, S. J. and Sanghera, M. (1977). Activity of neurones in the inferotemporal cortex of the alert monkey, *Brain Research* **130**: 229–238.

Rolls, E. T., Burton, M. J. and Mora, F. (1980). Neurophysiological analysis of brain-stimulation reward in the monkey, *Brain Research* **194**: 339–357.

Rolls, E. T., Baylis, G. C. and Leonard, C. M. (1985). Role of low and high spatial frequencies in the face-selective responses of neurons in the cortex in the superior temporal sulcus, *Vision Research* **25**: 1021–1035.

Rolls, E. T., Baylis, G. C. and Hasselmo, M. E. (1987). The responses of neurons in the cortex in the superior temporal sulcus of the monkey to band-pass spatial frequency filtered faces, *Vision Research* **27**: 311–326.

Rolls, E. T., Scott, T. R., Sienkiewicz, Z. J. and Yaxley, S. (1988). The responsiveness of neurones in the frontal opercular gustatory cortex of the macaque monkey is independent of hunger, *Journal of Physiology* **397**: 1–12.

Rolls, E. T., Baylis, G. C., Hasselmo, M. and Nalwa, V. (1989a). The representation of information in the temporal lobe visual cortical areas of macaque monkeys, *in* J. Kulikowski, C. Dickinson and I. Murray (eds), *Seeing Contour and Colour*, Pergamon, Oxford.

Rolls, E. T., Miyashita, Y., Cahusac, P. M. B., Kesner, R. P., Niki, H., Feigenbaum, J. and Bach, L. (1989b). Hippocampal neurons in the monkey with activity related to the place in which a stimulus is shown, *Journal of Neuroscience* **9**: 1835–1845.

Rolls, E. T., Sienkiewicz, Z. J. and Yaxley, S. (1989c). Hunger modulates the responses to gustatory stimuli of single neurons in the caudolateral orbitofrontal cortex of the macaque monkey, *European Journal of Neuroscience* **1**: 53–60.

Rolls, E. T., Yaxley, S. and Sienkiewicz, Z. J. (1990). Gustatory responses of single neurons in the orbitofrontal cortex of the macaque monkey, *Journal of Neurophysiology* **64**: 1055–1066.

Rolls, E. T., Hornak, J., Wade, D. and McGrath, J. (1994a). Emotion-related learning in patients with social and emotional changes associated with frontal lobe damage, *Journal of Neurology, Neurosurgery and Psychiatry* **57**: 1518–1524.

Rolls, E. T., Tovee, M. J., Purcell, D. G., Stewart, A. L. and Azzopardi, P. (1994b). The responses of neurons in the temporal cortex of primates, and face identification and detection, *Experimental Brain Research* **101**: 474–484.

Rolls, E. T., Critchley, H. D. and Treves, A. (1996a). The representation of olfactory information in the primate orbitofrontal cortex, *Journal of Neurophysiology* **75**: 1982–1996.

Rolls, E. T., Critchley, H., Mason, R. and Wakeman, E. A. (1996b). Orbitofrontal cortex neurons: role in olfactory and visual association learning, *Journal of Neurophysiology* **75**: 1970–1981.

Rolls, E. T., Robertson, R. G. and Georges-François, P. (1997a). Spatial view cells in the primate hippocampus, *European Journal of Neuroscience* **9**: 1789–1794.

Rolls, E. T., Treves, A. and Tovee, M. J. (1997b). The representational capacity of the distributed encoding of information provided by populations of neurons in the primate temporal visual cortex, *Experimental Brain Research* **114**: 149–162.

Rolls, E. T., Treves, A., Foster, D. and Perez-Vicente, C. (1997c). Simulation studies of the CA3 hippocampal subfield modelled as an attractor neural network, *Neural Networks* **10**: 1559–1569.

Rolls, E. T., Treves, A., Tovee, M. and Panzeri, S. (1997d). Information in the neuronal representation of individual stimuli in the primate temporal visual cortex, *Journal of Computational Neuroscience* **4**: 309–333.

Rolls, E. T., Treves, A., Robertson, R. G., Georges-François, P. and Panzeri, S. (1998). Information about spatial view in an ensemble of primate hippocampal cells, *Journal of Neurophysiology* **79**: 1797–1813.

Rolls, E. T., Critchley, H. D., Browning, A. S., Hernadi, A. and Lenard, L. (1999a). Responses to the sensory properties of fat of neurons in the primate orbitofrontal cortex, *Journal of Neuroscience* **19**: 1532–1540.

Rolls, E. T., Tovee, M. J. and Panzeri, S. (1999b). The neurophysiology of backward visual masking: information analysis, *Journal of Cognitive Neuroscience* **11**: 335–346.

Rolls, E. T., Webb, B. and Booth, M. C. A. (2000). Responses of inferior temporal cortex neurons to objects in natural scenes, *Society for Neuroscience Abstracts* **26**: 1331.

Rolls, E. T., Kringelbach, M. L., O'Doherty, J., Francis, S., Bowtell, R. and McGlone, F. (2001a). Pleasant and painful touch are represented in the human orbitofrontal cortex, *NeuroImage* **13**: S468.

Rolls, E. T., Zheng, F. and Aggelopoulos, N. (2001b). Responses of inferior temporal cortex neurons to objects in natural scenes, *Society for Neuroscience Abstracts* **27**: in press.

Rolls, E. T., Browning, A. S., Inoue, K. and Hernadi, S. (2002a). Novel visual stimuli activate a population of neurons in the primate orbitofrontal cortex.

Rolls, E. T., Critchley, H. D., Browning, A. S. and Inoue, K. (2002b). Face-selective and auditory neurons in the primate orbitofrontal cortex.

Rolls, E. T., Stringer, S. M. and Trappenberg, T. P. (2002c). A unified model of spatial and episodic memory.

Rosch, E. (1975). Cognitive representations of semantic categories, *Journal of Experimental Psychology: General* **104**: 192–233.

Rose, D. and Dobson, V. G. (1985). *Models of the Visual Cortex*, Wiley, Chichester.

Rosenblatt, F. (1961). *Principles of Neurodynamics: Perceptrons and the Theory of Brain Mechanisms*, Spartan, Washington, DC.

Rossi, A. F., Desimone, R. and Ungerleider, L. G. (2001). Contextual modulation in primary visual cortex of macaques, *Journal of Neuroscience* **21**: 1698–1709.

Rubner, J. and Tavan, P. (1989). A self-organization network for principal-component analysis, *Europhysics Letters* **10**: 693–698.

Rumelhart, D. E. and McClelland, J. L. (1986). *Parallel Distributed Processing*, Vol. 1: Foundations, MIT Press, Cambridge, Massachusetts.

Rumelhart, D. E., Hinton, G. E. and Williams, R. J. (1986a). Learning internal representations by error propagation, *in* D. E. Rumelhart, J. L. McClelland and the PDP Research Group (eds), *Parallel Distributed Processing: Explorations in the Microstructure of Cognition*, Vol. 1, MIT Press, Cambridge, Mass., chapter 8.

Rumelhart, D. E., Hinton, G. E. and Williams, R. J. (1986b). Learning representations by back-propagating errors, *Nature* **323**: 533–536.

Rupniak, N. M. J. and Gaffan, D. (1987). Monkey hippocampus and learning about spatially directed movements, *Journal of Neuroscience* **7**: 2331–2337.

Sagi, D. and Julesz, B. (1986). Enhanced detection in the aperture of focal attention during simple shape discrimination tasks, *Nature* **321**: 693–695.

Sakai, K. and Miyashita, Y. (1991). Neural organisation for the long-term memory of paired associates, *Nature* **354**(6349): 152–155.

Salinas, E. and Abbott, L. (1996). A model of multiplicative neural responses in parietal cortex, *Proceedings of the National Academy of Science, USA* **93**: 11956–11961.

Salinas, E. and Abbott, L. F. (1997). Invariant visual responses from attentional gain fields, *Journal of Neurophysiology* **77**: 3267–3272.

Samsonovich, A. and McNaughton, B. (1997). Path integration and cognitive mapping in a continuous attractor neural network model, *Journal of Neuroscience* **17**: 5900–5920.

Samuelsson, H., Jensen, C., Ekholm, S., Naver, H. and Blomstrand, C. (1997). Anatomical and neurobiological correlates of acute and chronic visuo-spatial neglect following right hemispheres stroke., *Cortex* **33**: 271–285.

Sanghera, M. K., Rolls, E. T. and Roper-Hall, A. (1979). Visual responses of neurons in the dorsolateral amygdala of the alert monkey, *Experimental Neurology* **63**: 610–626.

Sato, T. (1989). Interactions of visual stimuli in the receptive fields of inferior temporal neurons in macaque, *Experimental Brain Research* **77**: 23–30.

Schmolesky, M., Wang, Y., Hanes, D., Thompson, K., Leutgeb, S., Schall, J. and Leventhal, A. (1998). Signal timing across the macaque visual system, *Journal of Neurophysiology* **79**: 3272–3277.

Scott, T. R., Yaxley, S., Sienkiewicz, Z. J. and Rolls, E. T. (1986). Taste responses in the nucleus tractus solitarius of the behaving monkey, *Journal of Neurophysiology* **55**: 182–200.

Selfridge, O. G. (1959). Pandemonium: A paradigm for learning, *The Mechanization of Thought Processes*, H. M. Stationery Office, London.

Seltzer, B. and Pandya, D. N. (1978). Afferent cortical connections and architectonics of the superior temporal sulcus and surrounding cortex in the rhesus monkey, *Brain Research* **149**: 1–24.

Seltzer, B. and Pandya, D. N. (1989). Frontal lobe connections of the superior temporal sulcus in the rhesus monkey, *Journal of Comparative Neurology* **281**: 97–113.

Sereno, A. B. and Maunsell, J. H. (1998). Shape selectivity in primate lateral intraparietal cortex, *Nature* **395**: 500–503.

Shadlen, M. and Movshon, J. (1999). Synchrony unbound: A critical evaluation of the temporal binding hypothesis, *Neuron* **24**: 67–77.

Shadlen, M. and Newsome, W. (1994). Is there a signal in the noise?, *Current Opinion in Neurobiology* **5**: 248–250.

Shadlen, M. and Newsome, W. (1998). The variable discharge of cortical neurons: implications for connectivity, computation and coding, *Journal of Neuroscience* **18**: 3870–3896.

Shallice, T. and Burgess, P. (1996). The domain of supervisory processes and temporal organization of behaviour, *Philosophical Transactions of the Royal Society of London. Series B Biological Sciences* **351**: 1405–1411.

Shannon, C. E. (1948). A mathematical theory of communication, *AT&T Bell Laboratories Technical Journal* **27**: 379–423.

Shapley, R. (1995). Parallel neural pathways and visual function, *in* M. S. Gazzaniga (ed.), *The Cognitive Neurosciences*, MIT Press, Cambridge, Massachusetts, pp. 315–324.

Shapley, R. and Perry, V. H. (1986). Cat and monkey retinal ganglion cells and their visual functional roles, *Trends in Neurosciences* **9**: 229–235.

Shashua, A. (1995). Algebraic functions for recognition, *IEEE Transactions on Pattern Analysis and Machine Intelligence* **17**: 779–789.

Shaw, M. (1978). A capacity allocation of cognitive resources to spatial locations, *Journal of Experimental Psychology: Human Perception and Performance* **4**: 586–598.

Shaw, M. and Shaw, P. (1977). Optimal allocation of cognitive resources to spatial locations, *Journal of Experimental Psychology: Human Perception and Performance* **3**: 201–211.

Sheinberg, D. L. and Logothetis, N. K. (2001). Noticing familiar objects in real world scenes: The role of temporal cortical neurons in natural vision, *Journal of Neuroscience* **21**: 1340–1350.

Shepherd, G. M. (1998). *The Synaptic Organisation of the Brain*, 4th edn, Oxford University Press, Oxford.

Shevelev, I. A., Novikova, R. V., Lazareva, N. A., Tikhomirov, A. S. and Sharaev, G. A. (1995). Sensitivity to cross-like figures in cat striate neurons, *Neuroscience* **69**: 51–57.

Shiino, M. and Fukai, T. (1990). Replica-symmetric theory of the nonlinear analogue neural networks, *Journal of Physics A: Math. Gen.* **23**: L1009–L1017.

Shulman, G. and Wilson, J. (1987). Spatial frequency and selective attention to local and global information, *Perception* **16**: 89–101.

Shulman, G., Sullivan, M., Gish, K. and Sakoda, W. (1986). The role of spatial-frequency channels in the perception of local and global structure, *Perception* **15**: 259–273.

Siegel, R. M. and Read, H. L. (1997). Analysis of optic flow in the monkey parietal area 7a, *Cerebral Cortex* **7**: 1–20.

Sillito, A. M. (1984). Functional considerations of the operation of GABAergic inhibitory processes in the visual cortex, *in* E. G. Jones and A. Peters (eds), *Cerebral Cortex, Vol. 2, Functional Properties of Cortical Cells*, Plenum, New York, chapter 4, pp. 91–117.

Sillito, A. M., Grieve, K. L., Jones, H. E., Cudeiro, J. and Davis, J. (1995). Visual cortical mechanisms detecting focal orientation discontinuities, *Nature* **378**: 492–496.

Simmen, M. W., Rolls, E. T. and Treves, A. (1996). On the dynamics of a network of spiking neurons, *in* F. Eekman and J. Bower (eds), *Computations and Neuronal Systems: Proceedings of CNS95*, Kluwer, Boston.

Singer, W. (1987). Activity-dependent self-organization of synaptic connections as a substrate for learning, *in* J. P. Changeux and M. Konishi (eds), *The Neural and Molecular Bases of Learning*, Chichester, Wiley, pp. 301–335.

Singer, W. (1994). The role of synchrony in neocortical processing and synaptic plasticity, *in* E. Domany, L. Van Hemmen and K. Schulten (eds), *Model of Neural Networks II*, Springer, Berlin.

Singer, W. (1995). Development and plasticity of cortical processing architectures, *Science* **270**: 758–764.

Singer, W. (1999). Neuronal synchrony: A versatile code for the definition of relations?, *Neuron* **24**: 49–65.

Singer, W. (2000). Response synchronisation: A universal coding strategy for the definition of relations, *in* M. Gazzaniga (ed.), *The New Cognitive Neurosciences*, 2nd edn, MIT Press, Cambridge, MA, chapter 23, pp. 325–338.

Singer, W. and Gray, C. M. (1995). Visual feature integration and the temporal correlation hypothesis, *Annual Review of Neuroscience* **18**: 555–586.

Singer, W., Gray, C., Engel, A., Konig, P., Artola, A. and Brocher, S. (1990). Formation of cortical cell assemblies, *Cold Spring Harbor Symposium on Quantitative Biology* **55**: 939–952.

Sireteanu, R. and Rettenbach, R. (1995). Perceptual learning in visual search: Fast enduring but non-specific, *Vision Research* **35**: 2037–2043.

Skaggs, W. E. and McNaughton, B. L. (1992). Quantification of what it is that hippocampal cell firing encodes, *Society for Neuroscience Abstracts* **18**: 1216.

Skaggs, W. E., McNaughton, B. L., Gothard, K. and Markus, E. (1993). An information theoretic approach to deciphering the hippocampal code, *in* S. Hanson, J. D. Cowan and C. L. Giles (eds), *Advances in Neural Information Processing Systems*, Vol. 5, Morgan Kaufmann, San Mateo, CA, pp. 1030–1037.

Skaggs, W. E., Knierim, J. J., Kudrimoti, H. S. and McNaughton, B. L. (1995). A model of the neural basis of the rat's sense of direction, *in* G. Tesauro, D. S. Touretzky and T. K. Leen (eds), *Advances in Neural Information Processing Systems*, Vol. 7, MIT Press, Cambridge, Massachusetts, pp. 173–180.

Sloper, J. J. and Powell, T. P. S. (1979a). An experimental electron microscopic study of afferent connections to the primate motor and somatic sensory cortices, *Philosophical Transactions of the Royal Society of London, Series B* **285**: 199–226.

Sloper, J. J. and Powell, T. P. S. (1979b). A study of the axon initial segment and proximal axon of neurons in the primate motor and somatic sensory cortices, *Philosophical Transactions of the Royal Society of London, Series B* **285**: 173–197.

Smith, M. L. and Milner, B. (1981). The role of the right hippocampus in the recall of spatial location, *Neuropsychologia* **19**: 781–793.

Snyder, L. H., Grieve, K. L., Brotchie, P. R. and Andersen, R. A. (1998). Separate body- and world-referenced representations in posterior parietal cortex, *Nature* **394**: 887–890.

Somogyi, P. and Cowey, A. C. (1984). Double bouquet cells, *in* A. Peters and E. G. Jones (eds), *Cerebral Cortex, Vol. 1, Cellular Components of the Cerebral Cortex*, Plenum, New York, chapter 9, pp. 337–360.

Somogyi, P., Kisvarday, Z. F., Martin, K. A. C. and Whitteridge, D. (1983). Synaptic connections of morphologically identified and physiologically characterized large basket cells in the striate cortex of the cat, *Neuroscience* **10**: 261–294.

Sompolinsky, H. and Kanter, I. (1986). Temporal association in asymmetric neural networks, *Physical Review Letters* **57**: 2861–2864.

Sperling, G. and Weichselgartner, J. (1995). Episodic theory of the dynamics of spatial attention, *Psychological Review* **102**: 503–532.

Spitzer, H., Desimone, R. and Moran, J. (1988). Increased attention enhances both behavioral and neuronal performance, *Science* **240**: 338–340.

Squire, L. R. and Knowlton, B. J. (2000). The medial temporal lobe, the hippocampus, and the memory systems of the brain, *in* M. Gazzaniga (ed.), *The New Cognitive Neurosciences*, 2nd edn, MIT Press, Cambridge, MA, chapter 53, pp. 765–779.

Stankiewicz, B. and Hummel, J. (1994). Metricat: A representation for basic and subordinate-level classification, *in* G. W. Cottrell (ed.), *Proceedings of the 18th Annual Conference of the Cognitive Science Society*, Erlbaum, San Diego, pp. 254–259.

Steinmetz, M. A., Connor, C. E. and MacLeod, K. M. (1992). Focal spatial attention suppresses responses of visual neurons in monkey posterior parietal cortex, *Society for Neuroscience Abstracts* **18**: 148.

Stent, G. S. (1973). A psychological mechanism for Hebb's postulate of learning, *Proceedings of the National Academy of Sciences of the USA* **70**: 997–1001.

Stringer, S. M. and Rolls, E. T. (2000). Position invariant recognition in the visual system with cluttered environments, *Neural Networks* **13**: 305–315.

Stringer, S. M. and Rolls, E. T. (2002a). Invariant object recognition in the visual system with novel views of 3D objects., *Neural Computation.*

Stringer, S. M. and Rolls, E. T. (2002b). Self-organizing continuous attractor network models of spatial view cells for an agent that is able to move freely through different locations.

Stringer, S. M., Rolls, E. T. and Trappenberg, T. P. (2002a). Self-organizing continuous attractor network models of hippocampal spatial view cells.

Stringer, S. M., Rolls, E. T., Trappenberg, T. P. and de Araujo, I. E. T. (2002b). Self-organizing continuous attractor networks and path integration II: Two-dimensional models of place cells, *Network, in press.*

Stringer, S. M., Trappenberg, T. P., Rolls, E. T. and de Araujo, I. E. T. (2002c). Self-organizing continuous attractor networks and path integration I: One-dimensional models of head direction cells, *Network, in press.*

Strong, S. P., Koberle, R., de Ruyter van Steveninck, R. R. and Bialek, W. (1998). Entropy and information in neural spike trains, *Physical Review Letters* **80**: 197–200.

Sugase,.Y., Yamane, S., Ueno, S. and Kawano, K. (1999). Global and fine information coded by single neurons in the temporal visual cortex, *Nature* **400**: 869–873.

Sutherland, N. S. (1968). Outline of a theory of visual pattern recognition in animal and man, *Proceedings of the Royal Society, B* **171**: 297–317.

Sutherland, R. J. and Rudy, J. W. (1991). Exceptions to the rule of space, *Hippocampus* **1**: 250–252.

Sutton, R. S. (1988). Learning to predict by the methods of temporal differences, *Machine Learning* **3**: 9–44.

Sutton, R. S. and Barto, A. G. (1981). Towards a modern theory of adaptive networks: expectation and prediction, *Psychological Review* **88**: 135–170.

Sutton, R. S. and Barto, A. G. (1998). *Reinforcement Learning*, MIT Press, Cambridge, Mass.

Suzuki, W. A. and Amaral, D. G. (1994a). Perirhinal and parahippocampal cortices of the macaque monkey: cortical afferents, *Journal of Comparative Neurology* **350**: 497–533.

Suzuki, W. A. and Amaral, D. G. (1994b). Topographic organization of the reciprocal connections between the monkey entorhinal cortex and the perirhinal and parahippocampal cortices, *Journal of Neuroscience* **14**: 1856–1877.

Suzuki, W. A., Miller, E. K. and Desimone, R. (1997). Object and place memory in the macaque entorhinal cortex, *Journal of Neurophysiology* **78**: 1062–1081.

Szentagothai, J. (1978). The neuron network model of the cerebral cortex: a functional interpretation, *Proceedings of the Royal Society of London, Series B* **201**: 219–248.

Tagametz, M. and Horwitz, B. (1998). Integrating electrophysiological and anatomical experimental data to create a large-scale model that simulates a delayed match-to-sample human brain study, *Cerebral Cortex* **8**: 310–320.

Tanaka, K. (1993). Neuronal mechanisms of object recognition, *Science* **262**: 685–688.

Tanaka, K. (1996). Inferotemporal cortex and object vision, *Annual Review of Neuroscience* **19**: 109–139.

Tanaka, K., Saito, C., Fukada, Y. and Moriya, M. (1990). Integration of form, texture, and color information in the inferotemporal cortex of the macaque, *in* E. Iwai and M. Mishkin (eds), *Vision, Memory and the Temporal Lobe*, Elsevier, New York, chapter 10, pp. 101–109.

Tanaka, K., Saito, H., Fukada, Y. and Moriya, M. (1991). Coding visual images of objects in the inferotemporal cortex of the macaque monkey, *Journal of Neurophysiology* **66**: 170–189.

Taube, J. S., Muller, R. U. and Ranck, Jr., J. B. (1990). Head-direction cells recorded from the postsubiculum in freely moving rats. I. Description and quantitative analysis, *Journal of Neuroscience* **10**: 420–435.

Taube, J. S., Goodridge, J. P., Golob, E. G., Dudchenko, P. A. and Stackman, R. W. (1996). Processing the head direction signal: a review and commentary, *Brain Research Bulletin* **40**: 477–486.

Taylor, J. G. (1999). Neural 'bubble' dynamics in two dimensions: foundations, *Biological Cybernetics* **80**: 393–409.

Thomson, A. M. and Deuchars, J. (1994). Temporal and spatial properties of local circuits in neocortex, *Trends in Neurosciences* **17**: 119–126.

Thorpe, S. J. and Imbert, M. (1989). Biological constraints on connectionist models, *in* R. Pfeifer, Z. Schreter and F. Fogelman-Soulie (eds), *Connectionism in Perspective*, Elsevier, Amsterdam, pp. 63–92.

Thorpe, S. J., Rolls, E. T. and Maddison, S. (1983). Neuronal activity in the orbitofrontal cortex of the behaving monkey, *Experimental Brain Research* **49**: 93–115.

Thorpe, S. J., O'Regan, J. K. and Pouget, A. (1989). Humans fail on XOR pattern classification problems, *in* L. Personnaz and G. Dreyfus (eds), *Neural Networks: From Models to Applications*, I.D.S.E.T., Paris, pp. 12–25.

Thorpe, S. J., Fize, D. and Marlot, C. (1996). Speed of processing in the human visual system, *Nature* **381**: 520–522.

Tipper, S. P. and Behrmann, M. (1996). Object-centered not scene-based visual neglect, *Journal of Experimental Psychology, Human Perception and Performance* **22**: 1261–1278.

Tootell, R. B., Silverman, M. S., De Valois, R. L. and Jacobs, G. H. (1983). Functional organization of the second functional area in primates, *Science* **220**: 737–739.

Tootell, R. B. H., Silverman, M. S., Switkes, E. and De Valois, R. L. (1982). Deoxyglucose analysis of retinotopic organization in primate striate cortex, *Science* **218**: 902–904.

Tou, J. T. and Gonzalez, A. G. (1974). *Pattern Recognition Principles*, Addison-Wesley, Reading, MA.

Tovee, M. J. and Rolls, E. T. (1992). Oscillatory activity is not evident in the primate temporal visual cortex with static stimuli, *Neuroreport* **3**: 369–372.

Tovee, M. J. and Rolls, E. T. (1995). Information encoding in short firing rate epochs by single neurons in the primate temporal visual cortex, *Visual Cognition* **2**: 35–58.

Tovee, M. J., Rolls, E. T., Treves, A. and Bellis, R. P. (1993). Information encoding and the responses of single neurons in the primate temporal visual cortex, *Journal of Neurophysiology* **70**: 640–654.

Tovee, M. J., Rolls, E. T. and Azzopardi, P. (1994). Translation invariance and the responses of neurons in the temporal visual cortical areas of primates, *Journal of Neurophysiology* **72**: 1049–1060.

Tovee, M. J., Rolls, E. T. and Ramachandran, V. S. (1996). Rapid visual learning in neurones of the primate temporal visual cortex, *Neuroreport* **7**: 2757–2760.

Trappenberg, T. P., Rolls, E. T. and Stringer, S. M. (2002). Effective size of receptive fields of inferior temporal visual cortex neurons in natural scenes, *Advances in Neural Information*

Processing Systems, in press, MIT Press, Cambridge, MA.

Treisman, A. (1982). Perceptual grouping and attention in visual search for features and for objects, *Journal of Experimental Psychology: Human Perception and Performance* **8**: 194–214.

Treisman, A. (1988). Features and objects: The fourteenth Bartlett memorial lecture, *The Quarterly Journal of Experimental Psychology* **40A**: 201–237.

Treisman, A. and Gelade, G. (1980). A feature-integration theory of attention, *Cognitive Psychology* **12**: 97–136.

Treisman, A. and Sato, S. (1990). Conjunction search revisited, *Journal of Experimental Psychology: Human Perception and Performance,* **16**: 459–478.

Treves, A. (1991). Dilution and sparse encoding in threshold-linear nets, *Journal of Physics A* **23**: 1–9.

Treves, A. (1993). Mean-field analysis of neuronal spike dynamics quantitative estimate of the information relayed by the Schaffer collaterals, *Network* **4**: 259–284.

Treves, A. (1995). Quantitative estimate of the information relayed by the Schaffer collaterals, *Journal of Computational Neuroscience* **2**: 259–272.

Treves, A. (1997). On the perceptual structure of face space, *Biosystems* **40**: 189–196.

Treves, A. and Panzeri, S. (1995). The upward bias in measures of information derived from limited data samples, *Neural Computation* **7**: 399–407.

Treves, A. and Rolls, E. T. (1991). What determines the capacity of autoassociative memories in the brain?, *Network* **2**: 371–397.

Treves, A. and Rolls, E. T. (1992). Computational constraints suggest the need for two distinct input systems to the hippocampal CA3 network, *Hippocampus* **2**: 189–199.

Treves, A. and Rolls, E. T. (1994). A computational analysis of the role of the hippocampus in memory, *Hippocampus* **4**: 374–391.

Treves, A., Rolls, E. T. and Simmen, M. (1997). Time for retrieval in recurrent associative memories, *Physica D* **107**: 392–400.

Treves, A., Panzeri, S., Rolls, E. T., Booth, M. and Wakeman, E. A. (1999). Firing rate distributions and efficiency of information transmission of inferior temporal cortex neurons to natural visual stimuli, *Neural Computation* **11**: 611–641.

Tsodyks, M. V. and Feigel'man, M. V. (1988). The enhanced storage capacity in neural networks with low activity level, *Europhysics Letters* **6**: 101–105.

Tsotsos, J. (1990). Analyzing vision at the complexity level, *Behavioral and Brain Sciences* **13**: 423–469.

Tsotsos, J. K. (1991). Localizing stimuli in a sensory field using an inhibitory attention beam, *Technical Report, RBCV-TR-91-37, Department of Computer Science*, University of Toronto.

Tuckwell, H. (1988). *Introduction to Theoretical Neurobiology*, Cambridge University Press, Cambridge.

Turner, B. H. (1981). The cortical sequence and terminal distribution of sensory related afferents to the amygdaloid complex of the rat and monkey, *in* Y. Ben-Ari (ed.), *The Amygdaloid Complex*, Elsevier, Amsterdam, pp. 51–62.

Ullman, S. (1996). *High-Level Vision. Object Recognition and Visual Cognition*, Bradford / MIT Press, Cambridge, Mass.

Ungerleider, L. G. (1995). Functional brain imaging studies of cortical mechanisms for memory, *Science* **270**: 769–775.

Ungerleider, L. G. and Haxby, J. V. (1994). 'What' and 'Where' in the human brain, *Current Opinion in Neurobiology* **4**: 157–165.

Ungerleider, L. G. and Mishkin, M. (1982). Two cortical visual systems, *in* D. Ingle, M. A. Goodale and R. Mansfield (eds), *Analysis of Visual Behaviour*, MIT Press, Cambridge, Mass.

Usher, M. and Niebur, E. (1996). Modelling the temporal dynamics of IT neurons in visual search: A mechanism for top-down selective attention, *Journal of Cognitive Neuroscience* **8**: 311–327.

Vallar, G. and Pernai, D. (1986). The anatomy of unilateral neglect after right-hemisphere stroke lesions. A clinical/CT-scan correlation study in man, *Neuropsychologia* **24**: 609–622.

Van de Laar, P., Heskes, T. and Gielen, S. (1997). Task-dependent learning of attention, *Neural Networks* **10**: 981–992.

Van Essen, D. C. (1985). Functional organization of primate visual cortex, *in* A. Peters and E. G. Jones (eds), *Cerebral Cortex, vol. 3*, Plenum, New York, pp. 259–329.

Van Essen, D. C. and DeYoe, E. A. (1995). Concurrent processing in the primate visual cortex, *in* M. S. Gazzaniga (ed.), *The Cognitive Neurosciences*, MIT Press, Cambridge, Massachusetts, pp. 383–400.

Van Essen, D. C., Felleman, D. J., DeYoe, E. A., Olavarria, J. and Knierim, J. (1990). Modular and hierarchical organization of extrastriate visual cortex in the macaque monkey, *Cold Spring Harbor Symposia on Quantitative Biology* **55**: 679–696.

Van Essen, D., Anderson, C. H. and Felleman, D. J. (1992). Information processing in the primate visual system: an integrated systems perspective, *Science* **255**: 419–423.

Van Hoesen, G. W. (1981). The differential distribution, diversity and sprouting of cortical projections to the amygdala in the rhesus monkey, *in* Y. Ben-Ari (ed.), *The Amygdaloid Complex*, Elsevier, Amsterdam, pp. 77–90.

Van Hoesen, G. W., Pandya, N. D. and Butters, N. (1975). Some connections of the entorhinal (area 28) and perirhinal (area 35) cortices of the rhesus monkey. II. Frontal lobe afferents, *Brain Research* **95**: 25–38.

Vandenberghe, R., Dupont, P., Debruyn, B., Bormans, G., Michiels, J., Mortelmans, L. and Orban, G. (1996). The influence of stimulus location on the brain activation pattern in detection and orientation discrimination – a PET study of visual attention, *Brain* **119**: 1263–1276.

Vandenberghe, R., Duncan, J., Dupont, P., Ward, R., Poline, J., Bormans, G., Michiels, J., Mortelmans, L. and Orban, G. (1997). Attention to one or two features in left and right visual field: a positron emission tomography study, *Journal of Neuroscience* **17**: 3739–3750.

Vecera, S. and Farah, M. (1994). Does visual attention select objects or location?, *Journal of Experimental Psychology: General* **123**: 146–160.

Walker, R. (1995). Spatial and object-based neglect, *Neurocase* **1**: 371–383.

Wallis, G. and Baddeley, R. (1997). Optimal unsupervised learning in invariant object recognition, *Neural Computation* **9**: 883–894.

Wallis, G. and Bulthoff, H. (1999). Learning to recognize objects, *Trends in Cognitive Sciences* **3**: 22–31.

Wallis, G. and Rolls, E. T. (1997). Invariant face and object recognition in the visual system, *Progress in Neurobiology* **51**: 167–194.

Wallis, G., Rolls, E. T. and Foldiak, P. (1993). Learning invariant responses to the natural transformations of objects, *International Joint Conference on Neural Networks* **2**: 1087–1090.

Wandell, B. A. (2000). Computational neuroimaging: color representations and processing, *in* M. Gazzaniga (ed.), *The New Cognitive Neurosciences, Second Edition*, MIT Press, Cambridge, MA, chapter 20, pp. 291–303.

Wang, X. J. (1999). Synaptic basis of cortical persistent activity: the importance of NMDA receptors to working memory, *Journal of Neuroscience* **19**: 9587–9603.

Wasserman, E., Kirkpatrick-Steger, A. and Biederman, I. (1998). Effects of geon deletion, scrambling, and movement on picture identification in pigeons, *Journal of Experimental Psychology – Animal Behavior Processes* **24**: 34–46.

Watanabe, S., Lea, S. E. G. and Dittrich, W. H. (1993). What can we learn from experiments on pigeon discrimination?, *in* H. P. Zeigler and H.-J. Bischof (eds), *Vision, Brain, and Behavior in Birds*, MIT Press, Cambridge, MA, pp. 351–376.

Watson, A. and Robson, J. (1981). Discrimination at threshold: labelled detectors in human vision, *Vision Research* **18**: 107–110.

Webster, M. A. and De Valois, R. L. (1985). Relationships between spatial frequency and orientation tuning of striate cortex cells, *Journal of the Optical Society America* **2**: 1124–1132.

Weiskrantz, L. (1956). Behavioral changes associated with ablation of the amygdaloid complex in monkeys, *Journal of Comparative and Physiological Psychology* **49**: 381–391.

Weiskrantz, L. (1968). Emotion, *in* L. Weiskrantz (ed.), *Analysis of Behavioral Change*, Harper and Row, New York, pp. 50–90.

Weiskrantz, L. (1998). *Blindsight. A Case Study and Implications*, 2nd edn, Oxford University Press, Oxford.

Werblin, F. S. and Dowling, J. E. (1969). Organization of the retina in the mudpuppy, *Necturus maculosus*: II Intracellular recording, *Journal of Neurophysiology* **32**: 339–355.

Whittlesea, B. W. A. (1983). *Representation and Generalization of Concepts: the Abstractive and Episodic Perspectives Evaluated*, Unpublished doctoral dissertation, MacMaster University.

Widrow, B. and Hoff, M. E. (1960). Adaptive switching circuits, *1960 IRE WESCON Convention Record, Part 4 (Reprinted in Anderson and Rosenfeld, 1988)*, IRE, New York, pp. 96–104.

Widrow, B. and Stearns, S. D. (1985). *Adaptive Signal Processing*, Prentice-Hall.

Williams, G. V., Rolls, E. T., Leonard, C. M. and Stern, C. (1993). Neuronal responses in the ventral striatum of the behaving macaque, *Behavioural Brain Research* **55**: 243–252.

Willshaw, D. J. (1981). Holography, associative memory, and inductive generalization, *in* G. E. Hinton and J. A. Anderson (eds), *Parallel Models of Associative Memory*, Erlbaum, Hillsdale, NJ, chapter 3, pp. 83–104.

Willshaw, D. J. and Longuet-Higgins, H. C. (1969). The holophone – recent developments, *in* D. Mitchie (ed.), *Machine Intelligence*, Vol. 4, Edinburgh University Press, Edinburgh.

Willshaw, D. J. and von der Malsburg, C. (1976). How patterned neural connections can be set up by self-organization, *Proceedings of The Royal Society of London, Series B* **194**: 431–445.

Willshaw, D. J., Buneman, O. P. and Longuet-Higgins, H. C. (1969). Non-holographic associative memory, *Nature* **222**: 960–962.

Wilson, F. A. W. and Rolls, E. T. (1993). The effects of stimulus novelty and familiarity on neuronal activity in the amygdala of monkeys performing recognition memory tasks, *Experimental Brain Research* **93**: 367–382.

Wilson, F. A. W., O'Sclaidhe, S. P. and Goldman-Rakic, P. S. (1993). Dissociation of object and spatial processing domains in primate prefrontal cortex, *Science* **260**: 1955–1958.

Wilson, H. (1978). Quantitative characterization of two types of line-spread function near the fovea, *Vision Research* **18**: 971–981.

Wilson, H. and Cowan, J. (1972). Excitatory and inhibitory interactions in localised populations of model neurons, *Biophysics Journal* **12**: 1–24.

Wilson, M. A. and McNaughton, B. L. (1993). Dynamics of the hippocampal ensemble code for space, *Science* **261**: 1055–1058.

Winston, P. H. (1975). Learning structural descriptions from examples, *in* P. H. Winston (ed.), *The Psychology of Computer Vision*, McGraw-Hill, New York.

Wolfe, J. M. (1994). Guided search 2.0: A revised model of visual search, *Psychonomic Bulletin and Review* **1**: 202–238.

Wolfe, J. M., Cave, K. R. and Franzel, S. L. (1989). Guided search: An alternative to the feature integration model for visual search, *Journal of Experimental Psychology: Human Perception and Performance* **15**: 419–433.

Wong, E. and Weisstein, N. (1998). Sharp targets are detected better against a figure, and blurred targets are detected better against a background, *Journal of Experimental Psychology: Human Perception and Performance* **9**: 194–202.

Wörgöter, F., Suder, K., Zhao, Y., Kerscher, N., Eysel, U. and Funke, K. (1998). State-dependent receptive field restructuring in the visual cortex, *Nature* **396**: 165–168.

Wu, X., Baxter, R. A. and Levy, W. B. (1996). Context codes and the effect of noisy learning on a simplified hippocampal CA3 model, *Biological Cybernetics* **74**: 159–165.

Xiang, J. Z. and Brown, M. W. (1998). Differential neuronal encoding of novelty, familiarity and recency in regions of the anterior temporal lobe, *Neuropharmacology* **37**: 657–676.

Yamane, S., Kaji, S. and Kawano, K. (1988). What facial features activate face neurons in the inferotemporal cortex of the monkey?, *Experimental Brain Research* **73**: 209–214.

Yaxley, S., Rolls, E. T. and Sienkiewicz, Z. J. (1988). The responsiveness of neurones in the insular gustatory cortex of the macaque monkey is independent of hunger, *Physiology and Behavior* **42**: 223–229.

Yeshurun, Y. and Carrasco, M. (1998). Attention improves or impairs visual performance by enhancing spatial resolution, *Nature* **395**: 72–75.

Yeshurun, Y. and Carrasco, M. (1999). Spatial attention improves performance in spatial resolution tasks, *Vision Research* **39**: 293–305.

Young, A., Newcombe, F. and Ellis, A. (1991). Different impairments contribute to neglect dyslexia, *Cognitive Neuropsychology* **8**: 177–191.

Yuille, A. L., Kammen, D. M. and Cohen, D. S. (1989). Quadrature and the development of orientation selective cortical cells by Hebb rules, *Biological Cybernetics* **61**: 183–194.

Zeki, S. (1993). *A Vision of the Brain*, Blackwell Scientific Publications, Osney Mead, Oxford.

Zeki, S., Watson, J. D. G., Lueck, C. J., Friston, K. J., Kennard, C. and Frackowiak, R. S. J. (1991). A direct demonstration of functional specialization in human visual cortex, *Journal of Neuroscience* **11**: 641–649.

Zeki, S. M. (1976). The functional organization of projections from striate to prestriate visual cortex in rhesus monkey, *Cold Spring Harbor Symposium on Quantitative Biology*

40: 591–600.

Zhang, K. (1996). Representation of spatial orientation by the intrinsic dynamics of the head-direction cell ensemble: A theory, *Journal of Neuroscience* **16**: 2112–2126.

Zihl, J. (1995). Visual scanning behavior in patients with homonymous hemianopia, *Neuropsychologia* **33**: 287–303.

Zihl, J. (2000). *Rehabilitation of Visual Disorders After Brain Injury*, Psychology Press, Brighton.

Zihl, J. and von Cramon, D. (1979). The contribution of the second visual system to directed visual attention in man, *Brain* **102**: 853–856.

Zihl, J., Von Cramon, D. and Mai, N. (1983). Selective disturbance of movement vision after bilateral brain damage, *Brain* **106**: 313–340.

Zipf, G. (1949). *Human Behavior and the Principle of Least Effort*, Addison-Wesley, Cambridge.

Zipser, D. and Andersen, R. (1988). A backprojection programmed network that simulates response properties of a subset of posterior parietal neurons, *Nature* **331**: 679–684.

Zohary, E., Shadlen, M. N. and Newsome, W. T. (1994). Correlated neuronal discharge rate and its implications for psychophysical performance, *Nature* **370**: 140–143.

Zola-Morgan, S., Squire, L. R., Amaral, D. G. and Suzuki, W. A. (1989). Lesions of perirhinal and parahippocampal cortex that spare the amygdala and hippocampal formation produce severe memory impairment, *Journal of Neuroscience* **9**: 4355–4370.

Zola-Morgan, S., Squire, L. R. and Ramus, S. J. (1994). Severity of memory impairment in monkeys as a function of locus and extent of damage within the medial temporal lobe memory system, *Hippocampus* **4**: 483–494.

Zucker, S. W., Dobbins, A. and Iverson, L. (1989). Two stages of curve detection suggest two styles of visual computation, *Neural Computation* **1**: 68–81.

Index

3D model, 245–247
3D object recognition, 301–307
7a, 74

action, 451, 471
  object selection for, 448–452
activation, 3
activation function, 3, 161
active vision, 248
agnosia, 60, 79
alignment approach to object recognition, 247–248
allocentric space, 413–421, 451, 464
amygdala, 118, 434–439
AND, 484
area 7a, 74
associative learning, 145–171
ataxia, 79
attention, 126–144, 313–319, 322–403
  a neurodynamical model, 322–403
  and executive control, 469
  and short term memory, 404, 412, 469
  biased competition, 126, 132–135, 140–142
  binding, 127–132, 460
  computational models, 129–133
  controller, 464
  covert, 127
  disengagement, 77, 392
  feature integration theory, 127
  future directions, 473–474
  high-resolution buffer hypothesis, 139–140
  inferior temporal cortex, 136
  model of, 462–464
  neuroimaging, 140–142, 465
  neurophysiology, 133–135
  neuropsychology, 382–403
  non-spatial, 136–139
  overt, 127
  saliency or priority map, 129–132
  serial vs parallel, 126–144
  V4, V2 and V1, 135–136
attentional bias, 136
attentional spotlight, 126–132, 464
attractor network, 159–171, 224–228, 307–319
attractor network dynamics, 209–215
attractor networks
  discrete and continuous, 203
autoassociation, 34
autoassociation network, 26, 159–171, 307–313
autocorrelation memory, 162

backprojections, 30–35, 55, 170, 180–182, 257, 413–421, 447–448
backpropagation of error network, 13, 236–240
Balint's syndrome, 77
Bayes' theorem, 519
biased competition, 126, 132–135, 140–142, 220–221, 313–319, 326–352, 462–464
binding, 84, 107–111, 114, 127–132, 246–247, 256, 284–295, 369–382, 458, 460–461
biologically plausible networks, 13, 239–240
blindsight, 124
Boltzmann machine, 241–242

capacity, 484
  autoassociator, 166–168
  competitive network, 175
  pattern associator, 153–156, 158–159
catastrophic changes in the qualitative shape descriptors, 302
categorization, 171, 174
central tendency, 151, 164
change blindness, 454
clamped inputs, 163
classification, 171
cluster analysis, 178
cluttered environment, 295–301
coarse coding, 12
coding, 12, 84–111, 158
Cognitron, 253
colour, 61–65
colour vision, 37
columnar organization, 45, 119
combinatorial explosion, 257, 295
competitive network, 171–192, 254
completion, 102, 162, 163
complex cells, 44
conditional behavioral response learning, 453
conditioning, 145–159
conductances, 205–209
connectionism, 13
conscious visual perception, 454
constraint satisfaction, 34
content addressability, 159
context, 168
continuous attractor networks, 192–204
contrastive Hebbian learning, 241–242
convergence, 253
correlation, 479
cortical structure, 21–35
coupled attractor networks, 224–228
covert attention, 127

covert visual search, 352–382, 452–453
cross-correlation, 514–517
cross-correlations
  between neuronal spike trains, 504–507

decoding, 425
delta rule, 230
depth, 67–69
difference of Gaussian (DOG) filtering, 44, 263–264
discrete attractor network, 159–171
distributed encoding, 84–111, 257
distributed representation, 11–13, 84–111, 150, 158
  advantages, 12–13, 98–103
divisive inhibition, 24
dorsal visual pathway, 16–20, 57–80, 322–413, 448–
          452, 466
dot product, 243, 477
  neuronal operation, 149
dot product decoding, 93–111
dynamics, 204–228
dyscalculia, 79
dysgraphia, 79

egocentric space, 451, 464
emotion, 425–448, 470
  definition and classification, 425–427
  functions, 427–429
encoding, 84–111, 425
energy
  in an autoassociator, 162
episodic memory, 170, 203, 413–421
error correction, 228–240
error correction learning, 275–284
excitatory cortical neurons, 21
executive control, 468, 469
expansion recoding, 157, 179
expression, 118–120
extinction, 382–403
extrastriate cortex, 57–69

face expression, 118–120, 441, 443, 446
face identity, 118–120
face processing, 81–125
  neuroimaging, 119
false binding errors, 294–295
familiarity memory, 421–424
fault tolerance, 150, 164
feature
  binding, 256
  combinations, 256
feature analysis, 173
feature analyzer, 178
feature binding, 107–111, 284–295
feature combinations, 84, 284–295
feature hierarchies, 249–322
  introduction, 249–253
feature hierarchy, 458
feature integration theory, 127
feature spaces, 244–245

feature subsets and supersets, 284–288
feedback inhibition, 24
feedback processing, 209–215
feedforward processing, 212–215
firing rate distribution, 85–90, 168

GABA, gamma-amino-butyric acid, 23–25, 161
Gabor wavelets, 43–44, 327, 332–333
generalization, 102, 150, 164, 291–295
generic view, 302
genetic algorithm, 186
Gerstmann's syndrome, 79
global precedence principle, 353–369
graceful degradation, 102, 150, 164
grandmother cell, 11, 84–111, 254

habituation, 453
Hebb rule, 147, 160, 166
hidden units, 239
hierarchical feature analysis networks for object recog-
          nition, 249–322
hierarchical processing, 182, 212–215
hierarchy, 253
hippocampal models, 418–421
hippocampus, 179, 413–421
Hodgkin-Huxley equations, 205–206
Hopfield, 162
hypotheses, invariant object recognition, 253–257

idiothetic (self-motion) inputs, 198–202
imaging, 119
inferior temporal cortex
  attention, 136
  feature combinations, 84
  invariance, 111–118
  models, 242–322
  reward is not represented, 429–434
  sparse distributed representation, 84–111
  subdivisions, 81–83, 118–120
  topology, 184
inferior temporal visual cortex, 81–125
inferotemporal cortex, 81–125
information
  continuous variables, 496–498
  in the cross-correlation, 514–517
  in the rate, 514–517
  limited sampling problem, 498–504
  multiple cell, 501–507, 512–518
  mutual, 493, 519
  neuronal, 498–518
  single neuron, 494–496
  speed of processing, 106–107
  speed of transfer, 122–125, 509–512
  temporal encoding within a spike train, 105–107,
          507–509
information encoding, 84–111, 489–518
information theoretic analysis of neuronal responses,
          90–111, 507–518
information theory, 489–519

inhibitory cortical neurons, 23–25
inhibitory neurons, 161
inner product, 149, 477
integrate-and-fire models, 474
integrate-and-fire neuronal networks, 204–215
interacting attractor networks, 224–228
interference, 156
invariance
    dendritic computation, 321
    rotation, 255
    scale, 255
    size, 112, 255
    spatial frequency, 112, 255
    translation, 113–114, 255, 291–295, 448–452
    view, 115, 256, 301–307
invariant object recognition
    approaches, 244–253
    hypotheses, 253–257
    models, 242–322
invariant representations, 111–118
invertible networks, 248–249
IT, 322–403
IT Cortex, 81–125

Kluver-Bucy syndrome, 436
Kohonen map, 182–186

latency, 122
lateral geniculate nucleus, 37–43, 55
lateral inhibition, 24
learning, 31
    associative, 145–171
learning in IT, 120
learning rule
    local, 152, 166
least mean squares rule, 230
linear separability, 176, 482–488
linearity, 481
linearly independent vectors, 482–488
Linsker's network, 50–53
LIP, 71–73, 130
local learning rule, 7, 13, 152, 166
local representation, 11, 84–111, 158
long term memory, 413–424
long-term depression, 7–11
long-term potentiation, 7–11, 147, 469

Magnocellular (M) pathway, 36, 38–49, 59–69
map, 119, 182–186
Marr, 475
masking, 107, 123
McGurk effect, 182, 226–228
mean field neurodynamics, 216–224, 331–336, 373–376
memory, 437, 438
memory, effects of mood, 447–448
mixture states, 169
modular organization, 185
modularity, 26–30

mood, 447–448
motion, 65–67
MST, 57–69, 74
MT, 57–69
multilayer perceptron, 236–240
multimodal representations, 181, 453–454, 470, 472
multimodular dynamics, 221–228
multiple cell information, 93–111, 512–518
multiple objects, 294, 313–319
multiplicative interactions, 223
mutual information, 519

natural scene, 295–301
natural scenes, 452–453
neglect, 75–80, 383–403
Neocognitron, 253
neuroimaging, 119, 120, 140–142, 465
neurons, 2–13, 21–35
neuropsychology of attention, 382–403
NMDA receptors, 7–11, 189, 202
noise reduction, 151, 164
non-linear networks, 487–489
non-linearity, 481
    in the learning rule, 189
normalization
    of neuronal activity, 24
    of synaptic weights, 188
novel stimuli, 437, 438, 453

object identification, 119, 321
occlusion, 300–301
ocular ataxia, 78
Oja learning rule, 189
orbitofrontal cortex, 118, 439–447
orientation, behavioral, 453
orthogonalization, 174, 179
overt attention, 127
overt search, 452–453

parietal cortex, 70–80
Parvocellular (P) pathway, 36, 38–49, 59–69
path integration, 198–202
pattern association memory, 145–159, 436–437, 443–446
pattern separation, 176
perceptron, 228–240
perceptual learning, 120
perirhinal cortex, 421–424
planning, 467
pop out, 129
PP, 322–403
prefrontal cortex, 404–413
    46 d and 46 v, 322–403
    computational necessity for, 406–413, 467
primary visual cortex, 36–56
priming, 34
principal component analysis (PCA), 178
prototype extraction, 151, 164

radial basis function (RBF) networks, 187–188
recall, 33, 148
  in autoassociation memories, 162
receptive field size, 253
  of IT neurons, 451
recognition memory, 421–424
redundancy, 173, 178
reinforcement learning, 240
resistance to noise, 102
retina, 37–43
reversal, 443
reward, 424–448, 470
Riesenhuber and Poggio model of invariant recognition, 320–321
Rolls model of invariant recognition, 249–322

search, 352–382, 452–453, 464
  and short term memory, 412
  conjunction, 128, 369–382
  feature, 128
  hierarchical patterns, 358–369
  neuropsychology, 382–403
  serial vs parallel, 352–382, 464
segregation of visual pathways, 36–49, 57–69
  principles, 186
  principles(, 448, 466
  principles), 452, 467
self-organization, 173
sequence memory, 169
shift invariance, 255
short term memory, 121, 159, 170, 224, 404–413, 453, 464–470
  for visual stimuli, 121
shunting inhibition, 24
Sigma-Pi neurons, 6, 321, 488–489
simple cells, 43–54
simultanagnosia, 77–79, 454
single cell information, 90–93
size invariance, 112
soft competition, 189–190
soft max, 190
sparseness, 12, 175
sparseness maximization principle, 53
sparseness of IT representations, 84–111
sparsification, 179
spatial configuration of features, 84
spatial frequency, 112
spatial memory, 413–421
spatial neglect, 75–80
spatial processing, 70–80
spatial resolution hypothesis, 353–369
spatial view cells, 415–418
speed of processing, 26, 102, 106–107, 122–125, 151, 165, 204–215
spike response model, 215–216
spin glass, 162
spontaneous firing rate, 26
statistical mechanics, 162
stereoscopic vision, 67–69

stimulus-reinforcement association, 424–448
striate cortex, 36–56
structural shape descriptions, 245–247
symmetric synaptic weights, 160
synaptic modification, 4–11
synaptic weight vector, 149
synchronization, 107–111, 293, 319, 458, 514–517
  binding, 461
syntactic binding, 319
syntactic pattern recognition, 245–247

TE, 81–125
template matching, 247–248, 458
temporal difference (TD) learning, 275–284
temporal encoding within a spike train, 105–107, 507–509
temporal synchronization, 107–111
top-down processing, 30–35, 142–144
topographic map, 182–186
trace learning rule, 255, 258–260, 275–284
trace rule value η, 270, 274–275
translation invariance, 113–114, 255, 291–295, 448–452

unclamped inputs, 163
unsupervised learning, 257
unsupervised networks, 171–192

V1, 36–56, 134, 135, 322–403
V2, 57–69, 134, 135
V4, 57–69, 133–136
V5, 57–69
vector, 477
  angle, 479
  correlation, 479
  dot product, 477
  length, 478
  linear combination, 482–488
  linear independence, 482–488
  linear separability, 482–488
  normalization, 479
  normalized dot product, 479
  outer product, 480
  weight, 149
vector quantization, 174
ventral visual pathway, 16–20, 57–69, 81–125, 242–403, 448–452, 466
  outputs, 404–454
  principles, 186
ventral visual stream, 454
view invariance, 256, 272–274, 301–307
view invariant representations, 115
view-based object recognition, 249–322
VIP, 73
VisNet, 257–322
  3D transforms, 301–307
  architecture, 258–264
  attention, 313–319
  attractor version, 307–313

capacity, 307–313
cluttered environment, 295–301
feature binding, 284–295
generalization, 291–295
multiple objects, 313–319
natural scenes, 313–319
occlusion, 300–301
performance measures, 264–266
receptive field size, 313–319
trace learning rule, 258–260
trace rule, 275–284
trace rule value η, 270, 274–275
translation invariance, 270–272
view invariance, 272–274
visual attention, 382–403
visual search, 352–382, 452–453, 468
and short term memory, 412
visuo-spatial scratchpad, 454

weight normalization, 188
weight vector, 149
what visual stream, 16–20, 36, 57–69, 81–125, 242–413, 448–452, 462, 466
where visual stream, 16–20, 36, 57–69, 322–413, 448–452, 462, 466
Widrow-Hoff rule, 230
winner-take-all, 172, 254
wiring length, 185

XOR, 157, 484